AF182252

# POISONING BY PLANTS, MYCOTOXINS, AND RELATED TOXINS

# Poisoning by Plants, Mycotoxins, and Related Toxins

*Edited by*


**Franklin Riet-Correa**
*Hospital Veterinário, Universidade Federal de Campina Grande,*
*Patos, Paraíba, 58700-000, Brazil*

**Jim Pfister**
*USDA-ARS Poisonous Plant Research Laboratory*
*Logan, Utah 84341, USA*

**Ana Lucia Schild**
*Laboratória Regional de Diagnóstico, Faculdade de Medicina Veterinária, Universidade Federal de Pelotas, Pelotas-RS, Brazil*

**Terrie Wierenga**
*USDA-ARS Poisonous Plant Research Laboratory*
*Logan, Utah 84341, USA*



www.cabi.org

**CABI is a trading name of CAB International**

CABI Head Office
Nosworthy Way
Wallingford
Oxfordshire OX10 8DE
UK

Tel: +44 (0)1491 832111
Fax: +44 (0)1491 833508
E-mail: cabi@cabi.org
Website: www.cabi.org

CABI North American Office
875 Massachusetts Avenue
7th Floor
Cambridge, MA 02139
USA

Tel: +1 617 395 4056
Fax: +1 617 354 6875
E-mail: cabi-nao@cabi.org

© CAB International 2011. All rights reserved. No part of this publication may be reproduced in any form or by any means, electronically, mechanically, by photocopying, recording or otherwise, without the prior permission of the copyright owners.

A catalogue record for this book is available from the British Library, London, UK.

**Library of Congress Cataloging-in-Publication Data**

International Symposium on Poisonous Plants (8th : 2009 : Paraíba, Brazil)
Poisoning by plants, mycotoxins, and related toxins / edited by Franklin Riet-Correa ... [et al.].
p. cm.
Includes bibliographical references and index.
ISBN 978-1-84593-833-8 (alk. paper)
1. Livestock poisoning plants--Toxicology--Congresses. 2. Poisonous plants--Toxicology--Congresses. 3.
Plant toxins--Physiological effect--Congresses. 4. Mycotoxins--Physiological effect--Congresses. 5.
Livestock poisoning plants--Congresses. 6. Poisonous plants--Congresses. I. Riet-Correa, Franklin. II. Title.
SF757.5.I56 2009
636.089'5952--dc22

2010053920

ISBN-13: 978 1 84593 833 8

Commissioning editor: Rachel Cutts
Production editor: Fiona Chippendale

Printed and bound in the UK from copy supplied by the authors by MPG Books Group.

# Contents

**Toxic Plants Affecting Other Systems**

**Mycotoxins and Other Toxins**



**Toxic Compounds and Chemical Methods**



**Control Measures**

# Preface

The chapters published in this book were presented at the 8th International Symposium on Poisonous Plants (ISOPP8) held in Joâo Pessoa, Brazil, May 2009.

The idea of the poisonous plant symposia began with Dr Lynn F. James, Research Leader of the USDA-ARS Poisonous Plant Research Laboratory in Logan, Utah, USA. In 1973, Dr James presented an invited paper at the IV International Association of Rumen Physiologists in Sydney, Australia. Dr James arranged to visit many laboratories where research on poisonous plants was being done and presented seminars in Sydney, Melbourne, Adelaide, and Perth highlighting the poisonous plant research in the USA with the purpose of proposing a joint US Australian symposium on poisonous plants. After presenting a lecture at the University of Queensland to the Queensland Poisonous Plants Committee, the committee agreed to assist Dr James in this endeavor and the concept of the first joint US-Australian Symposium on Poisonous Plants was created. Dr J.H. Whitten (scientific attache, Australian Embassy, Washington, DC) acted as the coordinator between the two countries. Dr James was the US coordinator and program chairman, Dr Selwyn Everist was the Australian Coordinator, and Dr Alan Seawright from the Queensland Poisonous Plants Committee was the program co-chair.

The first joint US-Australian Symposium on Poisonous Plants was held in Logan, Utah, June 19–24, 1977 and the proceedings *Effects of Poisonous Plants on Livestock* was published in 1978. As agreed in the early plans, the second symposium was held in Brisbane, Australia under the direction of the Queensland Poisonous Plants Committee in 1984. The proceedings *Plant Toxicology* was published by the Queensland Poisonous Plants Committee in 1985. This joint poisonous plant symposium had an international interest from the beginning and the third symposium was returned to Logan, Utah, USA in 1989, again under the chairmanship of Dr Lynn F. James. This symposium was called the 3rd International Symposium On Poisonous Plants. The proceedings *Poisonous Plants* was published by Iowa State Press in 1992. In 1993, the 4th International Symposium On Poisonous Plants was held on September 26-October 1 in Fremantle, Western Australia, under the chairmanship of Peter Dorling and the acronym ISOPP® was coined (ISOPP4). The proceedings *Plant-Associated Toxins, Agricultural, Phytochemical and Ecological Aspects* was published by CABI in 1994. ISOPP5 was held in San Angelo, Texas, USA on May 18-23, 1997, under the co-chairmanship of Murl Bailey and Tam Garland and the proceedings *Toxic Plants and Other Natural Toxicants* was published by CABI in 1998. ISOPP6 was held on August 6-10, 2001 in Glasgow, Scotland under the chairmanship of Tom Acamovic and the proceedings *Poisonous Plants and Related Toxins* was published by CABI in 2004. ISOPP7 was held again in Logan, Utah, USA, June 6-10, 2005. *Poisonous Plants: Global Research and Solutions* was published by CABI in 2007. ISOPP8, held in João Pessoa, Brazil on May 4-8, 2009, was the first held in a non-English-speaking country. ISOPP9 will be held in Inner Mongolia, China in 2013.

The ISOPP series evolved from joint meetings between the USA and Australia into international conferences. Exchange of information between disciplines including chemistry, veterinary medicine, toxicology, plant physiology, rangeland management, biomedical research, etc. is encouraged at this meeting. This multi-disciplinary approach is what makes this meeting the great success it has been and will continue to be. Interest in the international scope of the symposium continues and we anticipate a great meeting in 2013.

The Editors

# Acknowledgements

The 8th International Symposium on Poisonous Plants (ISOPP8) was sponsored by the Federal University of Campina Grande and Federal University of Paraíba, both in the state of Paraíba, Brazil, by the USDA-ARS Poisonous Plant Research Laboratory, Logan, Utah, and by the Brazilian College of Pathology. The meeting was financially supported by Brazilian Council of Science and Technology (CNPq-grant 454084/2008-0), Coordination for the Improvement of Higher Education Personnel (CAPES-grant 0017/09-4), Research Foundation of the Sate of Paraíba (FAPESQ, grant 502239/2004-2, FAPESQ MCT), and by the National Institute for Science and Technology for the Control of Plant Poisonings (CNPq and MCT-grant 573534/2008-0). The organizers kindly acknowledge all these Institutions.

The editors thank the researchers at the USDA-ARS Poisonous Plant Research Laboratory in Logan, Utah for their assistance in reviewing the chapters.

# Carlos Tokarnia

Prof. Carlos Maria Antonio Hubinger Tokarnia was born in the city of Rio de Janeiro on the 24th of March, 1929. Dr Tokarnia has devoted his life's work to research, diagnostic work and teaching in the field of Veterinary Science. He graduated in 1952 from the Escola Nacional de Veterinária (National College of Veterinary Medicine), which is today called the Universidade Federal Rural do Rio de Janeiro (Federal Rural University of Rio of Janeiro -*UFRRJ).*

During his university studies, he was especially interested in Veterinary Pathology, under the influential guidance of Prof. Paulo Dacorso Filho, of whom he always considered himself a disciple. Once Dr Tokarnia graduated, he got a contract as a pathologist at the former Instituto de Biologia Animal (Institute of Animal Biology – IBA) of the Ministry of Agriculture, situated in the area known as Km 47, in the state of Rio de Janeiro. In 1953, he made his first research trip to the northeast of Brazil to study a disease of unknown etiology. His initial suspicion was a mineral deficiency, which he later confirmed. In 1955, his career was definitively influenced by his decision to get advanced training, sponsored by a fellowship from FAO, at the Ondestepoort Veterinary Research Institute, South Africa, where he stayed for one year. With this decision he lost his position at the Instituto de Biologia Animal, Km 47, but his study abroad expanded his vision for field research, especially for the diagnosis of diseases of unknown etiology that were economic burdens for the livestock industry. The application of methods acquired in South Africa was fundamental for the success of his investigations.

After his return to Brazil in 1956, he accepted a research grant from the Conselho Nacional de Pequisas (National Research Council–CNPq) and moved to the northeast, at that time a relatively inhospitable region, in order to investigate diseases of cattle. Initially, because of the harsh landscape and limited transportation, he went from farm to farm on horseback. Later, driving a jeep, Dr Tokarnia teamed with Dr Jürgen Döbereiner and veterinary surgeon Camillo F.C. Canella as they continued to investigate the main discases in livestock. His partnership with these research workers has continued up to the present day.

In 1959, he began his teaching activities, when he decided to return to Rio de Janeiro to become an Assistant of Prof. Jefferson Andrade dos Santos, in the Chair of Animal Pathology at Universidade Federal Fluminense (UFF), Niteroi. In 1965, he defended his thesis for *Docência Livre* (Doctorate) at the Universidade Federal do Rio Grande do Sul (Federal University of Rio Grande do Sul), Porto Alegre. He has maintained his research activities at the IBA, which later changed to the Federal agricultural research agency Empresa Brasileira de Pesquisa Agropecuaria (Embrapa).

Since 1960, he has given lectures at veterinary graduate courses, and from 1974 on, also for post-graduate courses at the UFRRJ, where he created the disciplines of Poisonous Plants and Mineral Deficiencies and Metabolic Diseases and continued with his lectures in Animal Pathology at *UFF*. In 1978, he officially transferred his professorship from UFF to UFRRJ at Km 47. In collaboration, he has been giving lectures in post-graduate courses at other Brazilian universities, in the same disciplines. Although forced to retire ten years ago, he did not change his activities of lecturing, extension activities and research, and continues to be a holder of a fellowship of CNPq.

The research group to which he belongs described plant poisonings in ruminants due to the following plant species: *Cestrum laevigatum*, *Tetrapterys multiglandulosa*, *T. acutifolia*, *Mascagnia rigida*, *M. pubiflora*, *M.* aff. *rigida*, *M. elegans*, *Thiloa glaucocarpa*, *Polygala klotzschii*, *Arrabidaea japurensis*, *Piptadenia macrocarpa*, *P. viridiflora*, *Manihot glaziovii*, *M. piauyensis*, *Ditaxis desertorum*, *Palicourea juruana*, *P. grandiflora*, *P. aeneofusca*, *Lantana tiliaefolia*, *Baccharis megapotamica* var. *weirii*, *Ipomoea carnea* var. *fistulosa*, and *I. asarifolia.* The first association with the ingestion of *Pteridium aquilinum* (in Brazil, today classified as *P. arachnoideum*) and carcinomas of the upper digestive tract was suggested by the same group. Several other current diseases due to plant poisoning, already studied in other countries, were characterized by them in Brazil, among these, poisoning by *Solanum malacoxylon*, *Lantana camara*, *Baccharis coridifolia*, and *Ricinus communis.* Regarding mineral deficiencies in livestock, the group established the etiology of various conditions related to cobalt, copper, phosphorus, and sodium deficiencies. They described, for the first time in Brazil, epizootic botulism secondary to phosphorus deficiency. It was Prof. Tokarnia who established, in 1978, the diagnosis of Africana Swine Fever in Brazil. In his research travels Dr Tokarnia has visited all the Brazilian states.

Dr Tokarnia was senior author of the influential book *Plantas Tóxicas do Brasil* (Poisonous Plants of Brazil), published in 2000, with a second edition coming next year. In this work, Prof. Tokarnia has compiled the results of his research and other dispersed information on the subject of toxic plants in Brazil. In 2007 the second edition of the book *Plantas Tóxicas da Amazonia* (1976) was published, and this book is based on research studies done under his leadership. The first edition of the book *Deficêencias Minerais em Animais de Produção* (Mineral Deficiencies of Livestock) is currently being published.

A life's work with the depth and thoroughness of Dr Tokarnia demands, of course, a lot of dedication. It is said that behind every great man, there is always a great woman behind the scenes. Those who know Prof. Tokarnia and his wife, Maria Luiza, certainly agree with that axiom.

Beside the great knowledge, persistence, rigor with scientific information, and innate facility in the identification of plants, Prof. Tokarnia also developed a rare capacity of organization, that allows him, consulting his notebooks, maintained from the 1950s to today, to recall the farms he visited on each specific day as well as each individual consultation during the investigation of each disease.

It is very difficult to describe in words the enormous contribution that Prof. Tokarnia has made, and continues to make, to Brazilian veterinary science, and the positive impact of his research on animal husbandry. Poisonous plants are among the main causes of death of adult cattle in Brazil. Estimates based on sampling of necropsies indicate that at least 1,000,000 (0.5%) cattle die annually from poisonous plants in Brazil, while the losses caused by mineral deficiencies are incalculable. A significant part of what is known today about the diseases caused by these two conditions in Brazil is due to his efforts. His pioneering research work and achievements in the two scientific areas are outstanding. Working under harsh and precarious conditions, he investigated diseases of unknown etiology in the Amazon, the Pantanal, Sertão, Cerrado, Agreste, Caatinga, and Serra and in the coastal areas of Brazil.

The magnitude and exactitude of information which he produced is impressive. He wrote more than 200 scientific papers published in national and international journals. In conclusion, those who know Dr Carlos Tokarnia agree that with all his successes, he exemplifies two personal traits that have characterized his interaction with other people: simplicity and humility. For his lifelong work on toxic plants and animal diseases, we pay tribute to Dr Tokarnia.

**Dr Paulo Vargas Peixoto**

# Jürgen Döbereiner

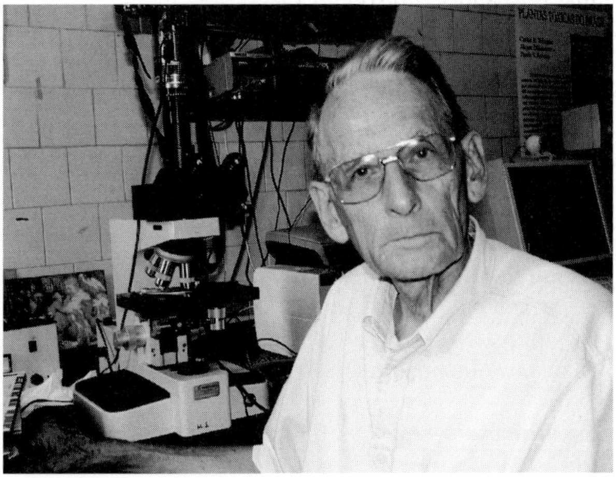

To begin this tribute to Dr Jürgen Döbereiner, I would like to make a brief account of his life: He was born in Königsberg, the former capital of East Prussia, Germany, on November 1, 1923, and while still a young man participated in the Second World War. He studied Veterinary Medicine at the University of Munich from 1947 to 1950, and immigrated to Brazil in 1950. He received a degree in Veterinary Medicine from the National Veterinary School of the Rural University of Brazil in Rio de Janeiro (today the Federal Rural University of Rio de Janeiro – UFRRJ) in 1954. He began working as a researcher for the Ministry of Agriculture at the Pathology Section of the Institute of Animal Biology (IBA), which later was changed to the Animal Health Project of Embrapa/UFRRJ. In 1963, he completed a Master's degree at the University of Wisconsin in Madison, USA, as a Rockefeller Foundation fellow. In 1970-71, he studied at the Royal Veterinary College in London, England, sponsored by the Queen's Scholarship Programme of the British Council. In 1977, he was awarded the title of Dr *Honoris Causa* in Veterinary Medicine of the Justus-Liebig-University, Giessen, Germany, for his research work carried out in Brazil. From the beginning of his professional career, he has dedicated himself to the research of cattle diseases caused by toxic plants and mineral deficiencies, and more recently to the elucidation of the etiology of a multifactorial periodontitis ('swollen face') of cattle in Brazil. He was a research fellow for The National Council for Scientific and Technological Development (CNPq) most of his professional life. Under the sponsorship of CNPq and DAAD – a German academic exchange program – he did 'swollen face' studies at the Universities of Giessen and Berlin. He has published over 170 papers and has supervised several graduate dissertations. Dr Jürgen has always been concerned about the publication of scientific research done in Brazil and has dedicated much of his time to the publishing of scientific journals. From 1959 to 1961, he was responsible for the edition of *Arquivos do Instituto de Biologia Animal*, and from 1966 to 1976 of *Pesquisa Agropecuária Brasileira*. Since 1981, he has edited, through the Brazilian College of Animal Pathology, the journal *Pesquisa Veterinária Brasileira*, undoubtedly the best scientific journal in veterinary medicine in Brazil. Furthermore, he is the co-author of the books *Plantas Tóxicas da Amazônia* (1979, 2007), *Plantas Tóxicas do Brasil* (2000), and *Deficiências Minerais em Animais de Produção* (2010).

Throughout his research career he had his wife, Johanna Döbereiner D.Sc. 1924-2000), an agronomist, and like him an internationally recognized researcher, as his partner. She is famous for her work in discovering the role of soil bacteria in nitrogen fixation.

For Dr Jürgen's lifetime of work in animal diseases and toxic plants, he is considered a pathfinder, a pioneer who initiated, together with Prof. Dr Carlos Tokarnia, the study of toxic plants in Brazil. As we have paid tribute to Dr Tokarnia today, we must also include Dr Jürgen Döbereiner because in many ways they were a dedicated team. Everything that has been said about Dr Tokarnia also applies to Dr Jürgen. Therefore, it is a great privilege to pay homage to both of these dedicated scientists at this ISOPP meeting.

I first met Dr Jürgen in 1984 at a Congress in Fortaleza, Ceará, and since then he has become an example for me and many of my generation, for his inexhaustible capacity for hard work and dedication to professional activities for over 50 years. Without question he is an example for the next generation, and for the young professionals and students who are participating in this symposium. From all of us convened here, and from all researchers worldwide in toxic plants, we thank you Dr Jürgen Döbereiner.

**With Sincerity and Admiration,**
**Dr Ana Lucia Schild**

# Severo Sales de Barros

In this event when we pay homage to Severo Sales de Barros, it is fair to say that he laid the foundation for veterinary pathology in the Brazilian state of Rio Grande do Sul (RS), and has shaped the careers of several veterinary pathologists that were directly or indirectly influenced by him.

Severo was born on March 18, 1932 in Júlio de Castilhos, RS, and received a degree in Veterinary Medicine, finishing first in his class in 1954 at the Universidade Federal Rural do Rio de Janeiro. At the start of a brilliant career he worked from May to October on two sheep farms located in the Argentinean Tierra del Fuego and in the Patagonian Province of Chubut. Back in Brazil, he worked from February 1957 to March 1958 as the veterinarian responsible for livestock inspection and sanitation in the municipality of Tupanciretã, RS, a position known as Veterinary Inspector, under the State Secretary of Agriculture of RS. Shortly thereafter he was the first to hold a similar position in the neighboring municipality of Júlio de Castilhos, his hometown. In December 1958, he was transferred to the Veterinary Research Institute 'Desidério Finamor' (IPVDF), another institution under the State Secretary of Agriculture of RS. At IPVDF he developed and implemented the laboratory of veterinary pathology. Unfortunately at that time in RS, microbiological methods were regarded as the most important, if not the sole methods for the diagnosis of livestock diseases, and veterinary anatomical pathology had not yet reached the position it deserved in this process. Discontented with this approach to the diagnosis of veterinary diseases at IPVDF, he resigned. With an invitation from Dr Edgardo Trein, Severo then assumed a position as resident at the Veterinary School of the Federal University of Rio Grande do Sul (UFRGS), working under Professors Wilhelm Brass and Hans Merkt, from April 1959 to March 1961. In March 1964, amidst uncertain political developments that shook the country at that time, he got a position in the newly founded School of Veterinary Medicine of the Federal University of Santa Maria (UFSM). There, at the same time, he alone developed the course of veterinary pathology and was the first professor to teach this course at the UFSM. Severo remained there until 1996, with only a sabbatical leave from January 1969 to April 1970, when he was awarded a fellowship from the Alexander von Humboldt Foundation to study Veterinary Pathology in the famous Veterinary School of Hannover, Germany.

After his retirement from UFSM in 1991, Severo worked in the same institution as a Guest Professor until 1996; during this time he developed several research projects and was the head of the Electron Microscopy Laboratory of the Department of Pathology of the UFSM, a section for which he had been the founder and organizer back in the late 1970s. From 1996 to 2007 Severo worked at the Federal University of Pelotas (UFPel), RS, where he again created and organized the Electron Microscopy Laboratory, and gave the ultrastructural support to several experiments that were ongoing not only at the UFPel, but also at the UFSM and UFRGS. During this period (1996-2007) his work was supported by research fellowships from the Brazilian governmental agencies CNPq, CAPES and FAPERGS; during the last quarter of this period he was hired as a faculty member at UFPel.

The above is a brief summary of Severo's career trajectory, but several achievements and the human factor are not revealed within these accomplishments, and it is important that these be recognized.

Most importantly, Severo Barros established the basis for diagnostic pathology in Rio Grande do Sul, back in 1964 when he founded the Veterinary Diagnostic Laboratory at the Department of Pathology of UFSM, where he introduced the notion of field research and necropsies to diagnose livestock diseases. The cause of several diseases was elucidated following this approach, and several students, many of whom are distinguished pathologists today in their own right, were trained in this manner. Before that, pathology laboratories and research institutes alike in RS approached diagnosis as restricted to the boundaries of the lab, examining mailed-in tissue specimens.

Another legacy of Severo Barros to his students is the notion that one's professional competence is only achieved through hard work and constantly keeping abreast with the literature in one's field of specialty; it is as simple as that, there are no shortcuts.

Severo Barros was involved in several important historical events related to veterinary medicine – not only veterinary pathology – research and teaching. He was critical in the introduction of electron microscopy to improve research in veterinary medicine in RS. He was also a key participant in the successful efforts to introduce embryo transfer techniques in the Laboratory of Reproductive Physiopathology at the UFSM.

One of the many research interests of Severo involved the effects of poisonous plants on livestock. He diagnosed for the first time in 1968 a form of calcinosis that affected sheep in RS. He called the disease 'enzootic calcinosis of sheep' and dedicated a great part of his prolific career as a veterinary pathologist and electron microscopist studying aspects of this condition. This evolved and he continued to study the intricate mechanisms of soft tissue mineralization, and made important original contributions to the subject, many of which are published in such journals as *Veterinary Pathology*, *Journal of Comparative Pathology*, *Cell*, and *Pesquisa Veterinária Brasileira*.

Many generations to come will be indebted to the contributions of Prof. Severo Sales de Barros, and we pay tribute to his invaluable lifelong contributions to veterinary science.

**Claudio S.L. Barros**

# OVERVIEW

# Chapter 1

# Caatinga of Northeastern Brazil: Vegetation and Floristic Aspects

O.F. de Oliveira

*Former Botany Professor, Department of Plant Sciences, Universidade Federal Rural do Semi-Árido, Mossoró-RN-Brazil – Present address: Caixa Postal 117, 59600-970 Mossoró-RN-Brazil; e-mail: odaci@uol.com.br*

The biome known as caatinga (from the Tupi word meaning 'white forest') or caatingas in northeastern Brazil has its origin possibly long after the splitting of the South American and African continents as a result of geological, edaphic, and climatic interactions, with its floristic composition and physiognomy attained through periods of decreasing rainfall and prevailing irregular pluviometric regime, and its xerophytic identity derived along the Tertiary-Quaternary. This biome, characteristically unique in the world, occupies an area of 844,453 km$^2$, which corresponds to roughly 10% of the Brazilian territory (IBGE 2004), extending along undulated pediplanes of erosive origin that exposed the Brazilian Precambrian crystalline bedrock (Cole 1960; Andrade and Lins 1965) and formed numerous exorheic ephemeral water courses (Ab'Sáber 1974), which drain in a radial pattern to the north, east, and south, due to the presence of a mountain range in the center of the biome (Sampaio 1995).

The caatinga vegetation is identified by its xerophytic character together with the presence of a considerable number of spiny plant species. It constitutes a well-defined phytogeographic unity and is the dominant vegetation form that occurs from the state of Piauí (except in the center and southwest portions) to the northernmost portion of the state of Minas Gerais (*c.* 17°S latitude), occupying almost the entire area comprised by the states of Ceará, Rio Grande do Norte, Paraíba, Pernambuco, Alagoas, Sergipe, and Bahia, reaching the littoral in the northern portion of the Brazilian northeast in the state of Rio Grande do Norte, where it is found to occur near shore sands. Its domain is surrounded by two characteristically different biomes, e.g. cerrado(s) and Atlantic forest, and restricted to, depending on the opinion of the author, the inside of the portion bounded by either the 800 mm/year isohyet (Figueiredo 1992; Mello-Netto *et al.* 1992; Souza *et al.* 1994; Velloso *et al.* 2002) – which roughly coincides with the boundaries of what is called the Drought Polygon of northeastern Brazil – or the 1000 mm/year isohyet (Nimer 1972; Reis 1976; Andrade-Lima 1981).

The origin of the flora of caatinga is still a matter of debate. The number of endemic taxa suggests that it may have had, at least in part, an autochthonous origin. Other evidence suggests that the Amazon forest, the Atlantic forest, and the cerrado contributed with genetic stocks in different times.

©CAB International 2011. *Poisoning by Plants, Mycotoxins, and Related Toxins*
(eds F. Riet-Correa, J. Pfister, A.L. Schild, and T.L. Wierenga)

Despite its apparent unique physiognomy due to the presence of widely distributed species and deciduous nature of most of the species, in some areas throughout the caatinga-dominated area, although maintaining most of its common phenological characteristics, the vegetation shows particular physiognomies, which have been interpreted as geographically and ecologically related. In each of these areas, now identified as ecoregions (Velloso *et al.* 2002), there occur a number of species that are exclusive, some being so restrictedly localized that the hazard of extinction is undeniable.

Over the years the caatinga vegetation has undergone accelerated processes of degradation as a consequence of the growing pressure of human activities as to land use for agriculture, extensive cattle raising, and intense extractivistic wood exploitation. Although some policies and strategies have been devised, the level of conservation of its biodiversity is still insignificant.

## Geologic, Edaphic, and Climatic Aspects

The caatinga occupies basically the areas of the interplanaltic depressions (Ab'Sáber 1974), but also extends to areas of low tablelands, uplands, and plateaus (Andrade-Lima 1981; Queiroz 2006). In general the vegetation follows the undulated pediplanes (Precambrian basement) that were exposed as results of erosive processes of the Cretacean or Tertiary sediments (Cole 1960; Andrade and Lins 1965). The calcareous outcrops very common in the area are also Cretacean formations (Oliveira and Leonardos 1978). Intense pediplanation processes during the Cenozoic (Late Tertiary to Early Quaternary) resulted in the Precambrian rock (gneisses, granites, and schists) outcrops leaving only isolated vestiges (inselbergs, mountains, and tablelands) of the younger surfaces (Ab'Sáber 1974). The tablelands still present the complete characteristics of the original sand sediments of the Tertiary, whereas the mountains are undergoing advanced pediplanation processes.

The geological formation of the area resulted in a complex mosaic of soil types with extremely different characteristics. Soils on the sedimentary areas are mostly deep and sandy, usually classified as latosol, podzolic, and quartz sand soils, but those on the crystalline basement are predominantly shallow, clayey and rocky, and usually classified as lithosols, regosols, and non-calcic brown soils (Sampaio 1995).

In comparison with the other Brazilian continental biomes, the caatinga presents many extreme characteristics with regard to meteorological parameters, e.g. high annual total solar radiation (from 3000 h in the northernmost portion to 2400 h in the southernmost portion), high annual mean temperature (23-28°C), high annual evapotranspiration potential (1500-2000 mm), and low annual pluviometric precipitation (250-1000 mm), which is irregularly distributed and concentrated in a usually very short period of the year (3-5 months), according to a combination of data from Hueck (1972), Reis (1976), Sampaio (1995), and Prado (2003). However, over most of the biome area the average annual rainfall is between 500-750 mm and, as a general rule, 20% of the annual rainfall occurs on a single day and 60% in a single month (Sampaio 1995). Temperatures rise and rainfall decreases from the biome boundaries toward the center and north (Sampaio 1995).

The semiarid nature of most of the northeastern Brazil region is due chiefly to the predominant stable air masses that are pushed southeastwards by the trade winds that blow from the South Atlantic. The east coast of Brazil consists of a narrow strip of lowlands backed by a strip of mountains that extends from the state of Rio Grande do Norte to the state of Rio Grande do Sul. When the trade winds carry the Atlantic-Equatorial water-vapor-loaded air masses against the Brazilian northeastern east coast, they humidify and precipitate over the Atlantic forest. So while the Atlantic-Equatorial system loses most of its humidity, the caatinga is submitted to the effect of dry, stable air masses (Andrade and Lins 1965).

A low-pressure zone (Intertropical Front) is formed where the trade winds from both hemispheres meet. This zone is positioned almost parallel to the Equator at *c.*10°N and when it moves southwards from the Equator in the summer it causes the climate of the northern half of the northeastern region to be highly unstable during February to April, which is the rainier period in the major part of the caatinga (Reis 1976). Additionally, the humid equatorial-continental air mass, which originates along the Amazon and causes convectively strong precipitation, may reach the western portion of the caatinga during November to January, particularly when it meets the southward moving Intertropical Front, thus increasing the possibilities of longer rainy periods (Reis 1976).

Floods usually occur as a result of the confluence of these systems. If these systems are prevented from reaching the region by the influence of the trades, catastrophic droughts commonly occur (Andrade and Lins 1965; Reis 1976) and may last for a several years or longer. Although concrete evidence is missing, it is suspected that the El Niño South Oscillation phenomenon also plays a role in the caatinga climate.

The caatinga acquired its characteristic physiognomy of the vegetation while evolving under pressure from climatic changes along with drastic erosive processes that altered the soil composition, as the older soils were being washed away and replaced continuously by newly formed soils (Ratter *et al.* 1988). These pedogenic processes reconfigured soil composition and nutrient balance in such a manner that the old vegetation (savanna) elements were forced to either adapt to the newly changing conditions or gradually disappear from the area with time. It is possible that the chemical composition of the soils of the old savanna areas was not much different from those of the present day cerrado areas, since higher aluminum concentrations are found in areas paved with remnants of older sandy sediments, for instance those of the Barreiras group formation, in which some flowering plant species common to cerrado vegetation are also found.

Also it is not unreasonable to think that before the caatinga emerged as a phytogeographic unit as seen today, the Brazilian diagonal dry area (which could have been the center of an older vegetation composed of a mixture of savanna and dry forest) that is covered by the present-day seasonally dry vegetation, was occupied by an Amazonian-like forest that extended to the Brazilian eastern coast which the Atlantic forest occupies nowadays. This spreading forest would be the result of a very humid climate and high temperatures that lasted for a long period of time. Then when climate became dryer again after the last glacial maximum, this forest retreated gradually, allowing not only the old savanna-like vegetation to re-cover the northward areas, but also new vegetation types to be formed in some areas it had occupied. This sequence of events may be abstracted by combining evidences of species shared occurrence and the results obtained in several studies (Pennington *et al.* 2006), although some of these may lead to different conclusions, as is the case when the long-distance dispersal theory is considered.

## Vegetation Physiognomy and Classification

The caatinga vegetation has a characteristic seasonally dry physiognomy with its floristic elements presenting variable habits and distribution densities. This vegetation is predominantly composed of deciduous shrubs and trees with heights usually not reaching over 8 m, and these elements being mostly spiny. In some areas the plants are sparsely distributed; in some they compose denser formations. So the caatinga may show, depending on the area, any of the following aspects: arboreal, shrubby-arboreal, or shrubby. In the shrubby formations plants may be densely or sparsely distributed. However, the vegetation in some areas is predominantly composed of an herbaceous component with scattered shrubs, an aspect acquired as a result of intense human activities, although the vegetation in a number of these areas may have been formed through natural processes.

The caatinga has long been recognized as a vegetation unit due to its overall similar physiognomic and phenological aspects. Nonetheless, in spite of the apparent physiognomic singleness of the caatinga vegetation, there has been much debate about the classification of the different vegetation physiognomies that can be recognized in the caatinga biome.

In a broad sense the caatinga vegetation has been classified into two types: hyperxerophilous, occupying the dryer area within the caatinga biome, and hypoxerophilous, showing a less 'aggressive' aspect and occupying the surroundings of the hyperxerophilous type, where the climate is less dry due to the influence of the other biomes. According to Sá *et al.* (2004), these two types cover respectively 34.3% and 43.2% of the caatinga-dominated areas, the rest of the area being represented by humid vegetation 'islands' (9.0%), which occur spottily in places of higher altitudes, and patches of *agreste* and transition vegetations (13.5%).

Other classifications (e.g. Luetzelburg 1922; Duque 1973) were developed taking into account some ecological aspects and utilized popular terms like sertão, seridó, agreste, carrasco, and cariri for defining vegetation units that differed from their concept of typical caatinga.

Andrade-Lima (1981) proposed a classification in which he recognized six types of caatinga on the basis of physiognomy, ecological aspects, and genera associations. Prado (2003) followed this classification and rearranged it into six units and 13 subunits or communities (Table 1). However, these units cannot be precisely mapped since they gradually intergrade (Sampaio and Rodal 2000) (Figure 1). Perhaps soil type variations in the caatinga biome also account for the varying physiognomies and distribution of plant species throughout the biome, but, besides the great exceptions in some soil characteristics, there are not enough data for evidencing correlations as such (Sampaio 1995). Also it is likely that altitude affects plant species distribution patterns and vegetation physiognomy, as appears to be the case of some species or places (Alcoforado-Filho 1993; Oliveira *et al.* 1997; Araújo *et al.* 1998b), but studies have not been extensively carried out in this regard.

Rodal (1983) and Oliveira *et al.* (1997) recognized that there is a particular type of caatinga with characteristic physiognomy and flora that occurs in areas of sedimentary basins with sandy and deep soils, although this type of caatinga (caatinga of sand) also occurs in areas where the crystalline basement is covered with pediment. Lemos and Rodal (2002), through comparisons of several phytosociological surveys, concluded that the results suggested that the deciduous vegetation found on sedimentary plateaus shows a physiognomic pattern distinct from that of the spiny vegetation (caatinga) observed in some crystalline basement areas. Recently Queiroz (2006) recognized two major floristic units as inferred by the distribution of the family Leguminosae: one that remained characteristically on the sedimentary areas and the other that occupies the exposed crystalline bedrock zone. This new approach is more realistic, according to Queiroz (2006), since it is based on a larger volume of data and more accurate methods of analysis than those based on the surveys carried out during the 1950s through 1970s, e.g. Andrade-Lima (1954, 1971, 1977).

The carrasco – xerophytic shrubby non-spiny vegetation that was recognized as a different vegetation unit by Andrade-Lima (1978) – which occurs on sedimentary plateaus inside the caatinga biome has been a subject of much debate. According to Fernandes (1996), carrasco and caatinga are different vegetation types characteristic to the semiarid northeastern Brazil. However, floristic studies (Araújo *et al.* 1998a,b; Araújo and Martins 1999) have shown that a considerable number of species are common to both types of vegetation, making it difficult to infer whether caatinga and carrasco are different phytogeographic units. Also, due to the large number of plant species common to both carrasco and cerrado, there is a possibility that carrasco is a degraded form of cerradão (a denser type of cerrado with more woody elements and less herbaceous components) (Araújo *et al.* 1998a). However, it is possible that carrasco and caatinga represent distinct phytogeographic units that were formed through different historical processes (Queiroz 2006; Cardoso and Queiroz 2007).

**Table 1.** Classification of caatinga vegetation according to typical genera associations, general aspects, and typical basement type (C – crystalline; S – sedimentary).

| Units/ Subunits[1] | Aspect[2] | Genera associations[3] | Basement type |
|---|---|---|---|
| I.1 | H | *Tabebuia-Aspidosperma-Astronium-Cavanillesia* | Calcareous/C |
| II.2 | M | *Astronium-Schinopsis-Caesalpinia* | C |
| II.3 | M | *Caesalpinia-Spondias-Bursera-Aspidosperma* | C |
| II.4 | M/L | *Mimosa-Syagrus-Spondias-Cereus* | C |
| II.6 | M/L | *Cnidoscolus-Bursera-Caesalpinia* | C |
| II.13 | M | *Auxemma-Mimosa-Luetzelburgia-Thiloa* | S/C |
| III.5 | M/L | *Pilosocereus-Poeppigia-Dalbergia-Piptadenia* | S |
| IV.7 | M/L | *Caesalpinia-Aspidosperma-Jatropha* | C |
| IV.8 | M/L | *Caesalpinia-Aspidosperma* | C |
| IV.9 | M/L | *Mimosa-Caesalpinia-Aristida* | C |
| IV.10 | L | *Aspidosperma-Pilosocereus* | C |
| V.11 | L | *Calliandra-Pilosocereus* | C |
| VI.12 | H/M(G) | *Copernicia-Geoffroea-Licania* | Alluvial/C |

[1] Subunit 13 may be considered as a unit (Prado 2003).
[2] H = high; M = median; L = low; G = gallery forest.
[3] *Astronium* is now partly in *Myracrodruon* and the *Bursera* of caatinga is in *Commiphora*.

Additionally, inside the caatinga biome there occur some patches of other vegetation types – brejos, cerrados, and campos rupestres. The brejos (upland forests) are enclaves (relicts) of Atlantic forest with elements of both the Atlantic forest and caatinga (Vasconcelos Sobrinho 1971; Porto *et al.* 2004; Silva *et al.* 2007) that occur in places of altitude usually over 500 m (in the states of Paraíba, Pernambuco, and Bahia), where the climate is more humid and the soils are more profound; similar vegetation also occurs in the state of Ceará (Uruburetama and Baturité mountains), but it is possibly more related to the Amazon forest biome than to the Atlantic forest.

Enclaves of cerrado (or at least cerrado-like vegetation) occur in the states of Ceará – municipalities of Iguatu and Salgado, Araripe plateau, and Caririaçu and Ibiapaba mountains (Figueiredo 1989, 1997; Fernandes 1990); Rio Grande do Norte – municipality of São Miguel (Figueiredo *et al.* 1991) and Portalegre mountain; and Bahia – middle portion of the Diamantina Plateau (Stannard 1995). Disjunctions of cerrado also occur in areas of the eastern portion of the Brazilian northeast (Rio Grande do Norte, Paraíba, Pernambuco Alagoas, Sergipe, and northern Bahia) stretched between the caatinga ecoregion and the littoral vegetation (Veloso 1964; Sarmento and Soares 1971; Tavares 1988a,b; Oliveira-Filho and Carvalho 1993). The existence of these cerrado patches suggests that the cerrado is a form of vegetation older than the Amazonian forest, but there are pros and cons to this opinion (Ratter *et al.* 2006).

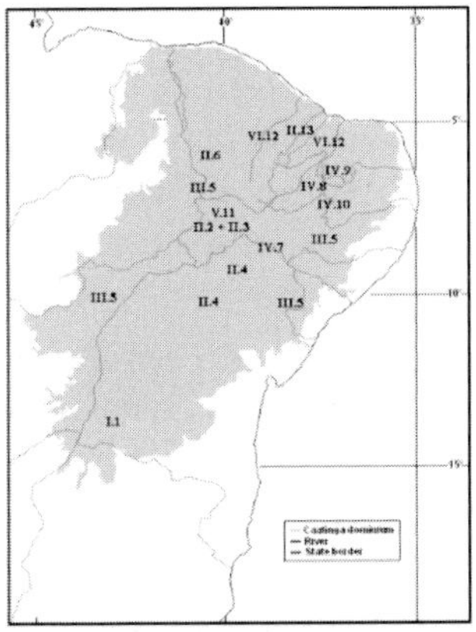


**Figure 1.** The caatinga ecoregion with units/subunits reflecting different types of vegetation that occur throughout the ecoregion.

In the Diamantina plateau there are also the campos rupestres, a form of vegetation composed basically of herbs and shrubs, with trees usually restricted to places where the soil is deeper and less subjected to desiccation (Conceição 2006), probably derived from cerrado-type vegetation (Stannard 1995).

The present day gallery forests (Andrade-Lima's unit 6) that line rivers and large streams in the caatinga ecoregion, where carnauba (*Copernicia prunifera*), oiticica (*Licania rigida*), and marizeiro (*Geoffroea spinosa*) are predominant elements, seem to be also relicts (or refugia) of the older vegetation that remained in the biota after replacement of the rain forest during the last glacial maximum, as a result of drier climate in combination with lower water table associated with lowered sea levels (Pennington *et al.* 2000).

Recently most of the floristic surveys and attempts to classify the caatinga vegetation have taken into account the concept of ecoregions proposed by Velloso *et al.* (2002). According to these authors, the caatinga biome comprises eight ecoregions: (i) Campo Maior complex, an area of low altitude located in northern Piauí, where floods periodically occur and the vegetation is a transition between caatinga and cerrado; (ii) Ibiapaba-Araripe Plateau, located in the areas near the borders of the states of Piauí, Ceará, and Pernambuco, and characterized by the presence of a spineless vegetation (carrasco) that is distributed between cerrado and typical caatinga vegetations; (iii) Northern Sertaneja Depression, which comprises almost entirely the areas of the states of Ceará and Rio Grande do Norte, as well as the central western portion of the state of Paraíba, where the vegetation cover is the typical caatinga of the crystalline; (iv) Borborema Plateau, an area with varying types of vegetation (typical caatinga and brejos) and characterized by irregularly undulated terrain that extends across the eastern portion of the states of Rio Grande do Norte, Paraíba, and Pernambuco, between the Northern Sertaneja Depression and the Atlantic forest zone; (v) Raso da Catarina, a sedimentary basin with sandy soils covered by a type of vegetation called caatinga of sand as an opposition to that of the crystalline; (vi) Continental Dunes or

São Francisco Dunes, where the vegetation is bushy and not so dense; (vii) Diamantina Complex, which includes the main chain of the mountains that divide the Bahian semiarid and extends to the northernmost portion of the state of Minas Gerais – in this complex there occurs a mosaic of vegetation, which includes caatinga, cerrado, campos rupestres, and humid forest-like vegetation patches; and (viii) Southern Sertaneja Depression, which includes the rest of the Bahian semiarid, the center-western portion of the state of Pernambuco, and the western portions of the states of Alagoas and Sergipe, and reaches the cerrado of central Brazil and the transition zone toward the Atlantic forest.

## Origin of the Flora

The origin of the flora of caatinga is an issue that has been considerably debated. First it was thought that the caatinga flora had been derived through an African connection (Thorne 1973; Smith 1973), but this idea was soon abandoned for the lack of a reasonable representative number of angiosperm genera/species sharing occurrence in both African and South American continents. According to Gillett (1980), the only American species of *Commiphora*, genus of Burseraceae composed of about 185 species, almost all African, is *C. leptophloeos*, a species previously placed in the genus *Bursera*, which seems to have originated in the New World, despite some taxonomic problems. It is uncertain when *C. leptophloeos* genetic stock dispersed from Africa to Brazil. According to Becerra (2003) this dispersion occurred before the major continental fragmentations of Gondwana and the complete separation of Africa from South America, which occurred between 95 and 100 million years ago, but according to Weeks and Simpson (2007) it is a recent event. Another example of disjunct occurrence between the Americas and Africa is the genus *Cochlospermum* (Cochlospermaceae), but it seems it migrated from South America to Africa. The genus *Ziziphus* (Rhamnaceae), with two of its species occurring in the caatinga biome (Lima 1995), is regarded as having had its center of both distribution and differentiation in South and Southeast Asia (Liu and Cheng 1995). As matter of fact, it cannot be denied that a great number of ancestors of the present-day South American plant species might have evolved from the old stock of the Gondwanan flora. However, such an event is too remote to be considered for explaining the evolution of the South American angiosperm species.

Some species that occur in the caatinga seem to have originated from sibling stocks of the Caribbean dry coast (north of Colombia and Venezuela). This view is supported by Sarmiento (1975), who considers the following pairs, for instance, as possible vicariants: *Copernicia prunifera*/*Copernicia tectorum* (Arecaceae), *Licania rigida*/*Licania arborea* (Chrysobalanaceae), *Pereskia aureiflora*/*Pereskia guamacho* (Cactaceae), and *Spondias tuberosa*/*Spondias mombin* (Anacardiaceae); the first of each pair occurs only in Brazil and almost exclusively in the caatinga biome. Besides those examples, the distribution of *Cochlospermum vitifolium*, as certified by herbarium vouchers, suggests the existence of such a floristic connection. A strong support to this view is the disjunct distribution of *Chloroleucon mangense* and *Mimosa tenuiflora* (Fabaceae *s.l.*), which occurs in the caatinga and from Venezuela to Mexico, but not in the intermediate areas.

There are two other dispersion routes that a number of plant species may have followed in either different time periods or concomitantly to reach the caatinga: (i) the Andean – from Colombia and Peru through the Chaco (Bolivia and Paraguay) to northeastern Brazil; and (ii) the Transamazonian – from Central America through the dry

Amazonian plains that appear to have existed in a remote past. Nonetheless it is possible that some plant species have migrated inversely on the same routes.

On the other hand there is strong evidence that the seasonally dry forests of South America are relicts of a biota that reached its maximum expansion during the driest periods of the Pleistocene (Prado and Gibbs 1993). The present-day flora distribution describes an arc-like strip (the Pleistocene arc) from caatinga southwards through southeastern Brazil, to the confluence of rivers in northern Argentina, then curving northwards to northwestern Argentina and southeastern Bolivia, and extending sporadically through dry valleys of the Peruvian Andes and west coast of Ecuador. These areas have been considered (Pennington *et al.* 2000; Prado 2000) as part of a new phytogeographic unit of South America (Neotropical Seasonally Dry Forests), the caatinga being the largest and most isolated of its nuclei. The flora of the arc includes a considerable number of endemic plant genera (for instance, in Fabaceae *s.l.* – *Amburana* and *Pterogyne*, in Boraginaceae – *Patagonula*, in Sapindaceae – *Diatenopteryx*, in Anacardiaceae – *Myracrodruon*, and in Bignoniaceae – *Perianthomega*) and species (Prado 2003).

Rizzini (1963, 1979) and Andrade-Lima (1982) interpreted the caatinga as a poor biota under the assumption that in this biome there were very few endemic taxa. These authors also considered its flora as representing an impoverished composition as compared to those of the Chaco, cerrado, and Atlantic forest. However in more recent studies (Giulietti *et al.* 2002; Prado 2003; Queiroz 2006) the number of taxa reported to be endemic suggests that the flora of the caatinga may have had, in some part, an autochthonous origin.

Queiroz's (2006) analyses led to the conclusion that there are 17 species of Leguminosae pantropically distributed and 39 widely distributed in the Neotropics. Twenty-one species of the caatinga have extended distribution to eastern Brazil (including Atlantic forest areas, dunes, and restingas) and 23 to central Brazil, with 27 widely distributed in the caatinga. These data reflect the recent dynamics of the flora and imply that the caatinga vegetation elements have been widening their distribution areas toward the nearby biomes as a result of interactions of climatic, edaphic, and anthropogenic factors.

It seems that none of the theories regarding the origin of the flora of the caatinga, except those based on the African and Chacoan connections, can be discarded, because a number of species from the Amazon forest, cerrado, and Atlantic forest might have dispersed into the caatinga biome in different times, therefore evolving into new species, and thence dispersing to areas outside the caatinga biome, as well as taking routes back to their ancestors' place of origin. Since dispersion is a very dynamic and random process it is extremely difficult to trace back the origin of species populations.

## Floristics

There are 385 endemic (or possibly endemic) species (including subspecies and varieties) distributed in 151 genera (22 endemic) of 40 angiosperm families. Table 2 is a combination of lists (Giulietti *et al.* 2002; Barbosa *et al.* 2006) with the taxa screened through virtual NYBG, MBG, MICH, BGBM (Röpert 2000) and WU databases, as well as Lorenzi *et al.* (2004) for Arecaceae; Smith and Downs (1979) for Bromeliaceae; Flora Brasiliensis Revisitada (2009), Taylor (1991), Zappi (1994), and Taylor and Zappi (2004) for Cactaceae; and Rogers and Appan (1973), Govaerts *et al.* (2000), and Melo (2000) for Euphorbiaceae. If no vouchers or type locality citations were available for any taxon listed by Giulietti *et al.* (2002), these authors' statements were maintained.

**Table 2.** Flowering plants endemic (and possibly endemic) to the caatinga biome.

| Families | Gen/Sp | Species[1] |
|---|---|---|
| Anacardiaceae | 2/2 | ***Apterokarpos*** *gardneri* (Engl.) Rizzini |
| | | *Spondias tuberosa* Arruda Cam. |
| Annonaceae | 1/1 | *Annona vepretorum* Mart. |
| Arecaceae | 3/5 | *Attalea seabrensis* Glassman |
| | | *Copernicia prunifera* (Mill.) H.E.Moore |
| | | *Syagrus microphylla* Burnet |
| | | *Syagrus vagans* (Bondar) Hawkes |
| | | *Syagrus x matafome* (Bondar) Glassman |
| Asclepiadaceae | 5/10 | *Ditassa dolichoglossa* Schlecht. |
| | | **Gonolobus cordatus* Malme |
| | | **Marsdenia queirozii* Fontella |
| | | *Marsdenia ulei* Rothe |
| | | *Marsdenia zehntneri* Fontella |
| | | **Matelea harleyi* Fontella & Morillo |
| | | **Matelea morilloana* Fontella |
| | | **Matelea nigra* (Decne.) Morillo & Fontella |
| | | *Matelea roulinioides* Agra & Stevens |
| | | **Metastelma giuliettianum* Fontella |
| Asteraceae | 3/3 | *Argyrovernonia harleyi* (H.Rob.) MacLeish |
| | | ***Blanchetia*** *heterotricha* DC. |
| | | ***Telmatophila*** *scolymastrum* Mart. |
| Bignoniaceae | 8/12 | **Adenocalymma apparicianum* J.C.Gomes |
| | | **Adenocalymma dichilum* A.H.Gentry |
| | | **Adenocalymma reticulatum* Bureau ex K.Schum. |
| | | **Amphilophium blanchetii* (DC.) Bureau & K.Schum. |
| | | *Anemopaegma laeve* DC. |
| | | *Arrabidaea harleyi* A.Gentry ex ex M.M.Silva & L.P.Queiroz |
| | | *Godmania dardanoi* (J.C.Gomes) A.H.Gentry |
| | | **Jacaranda microcalyx* A.H.Gentry |
| | | **Jacaranda rugosa* A.H.Gentry |
| | | *Sparattosperma catingae* A.H.Gentry |
| | | **Tabebuia selachidentata* A.H.Gentry |
| | | *Tabebuia spongiosa* Rizzini |
| Bombacaceae | 2/2 | *Ceiba glaziovii* K.Schum. ex Chod. & Hassl. |
| | | *Pseudobombax simplicifolium* A.Robyns |
| Boraginaceae[2] | 3/5 | ***Auxemma*** *glazioviana* Taub. |
| | | ***Auxemma*** *oncocalyx* (Allemão) |
| | | *Cordia dardani* Taroda |
| | | *Cordia leucocephala* Moric. |
| | | *Patagonula bahiensis* Moric. |
| Bromeliaceae | 7/14 | *Aechmea leucolepis* L.B.Sm. |
| | | *Billbergia euphemiae* E.Morren |
| | | *Billbergia fosteriana* L.B.Sm. |
| | | *Billbergia elongata* Mex |
| | | *Dyckia limae* L.B.Sm. |
| | | *Dyckia maracasensis* Ule |
| | | *Dyckia pernambucana* L.B.Sm. |
| | | *Encholirium spectabile* Mart. ex. Schultes & Schultes f. |
| | | *Hohenbergia catingae* Ule |
| | | *Hohenbergia utriculosa* Ule |
| | | *Neoglaziovia variegata* (Arruda) Mez. |
| | | *Orthophytum maracasense* L.B.Sm. |

**Table 2.** (Continued)

| Families | Gen/Sp | Species[1] |
|---|---|---|
| | | *Orthophytum rubrum* L.B.Sm. |
| | | *Orthophytum saxicola* (Ule) L.B.Sm. |
| Cactaceae | 14/49 | **Arrojadoa marylanae* Soares-Filho & M.Machado |
| | | *Arrojadoa bahiensis* (P.J. Braun & E. Esteves Pereira) N.P. Taylor & Eggli |
| | | *Arrojadoa dinae* Buining & Brederoo [2 subsp.] |
| | | *Arrojadoa penicillata* (Gürke) Britton & Rose |
| | | *Arrojadoa rhodantha* (Gürke) Britton & Rose |
| | | ***Brasilicereus*** *phaeacanthus* (Gürke) Backeberg |
| | | ***Brasilicereus*** *markgrafii* Backeb. & Voll |
| | | *Coleocephalocerus goebelianus* (Vaupel) Buining. |
| | | *Discocactus bahiensis* Britton & Rose |
| | | *Discocactus zehntneri* Britton & Rose [2 subsp.] |
| | | ***Espostoopsis*** *dybowskii* (Roland-Goss.) Backbg. |
| | | ***Facheiroa*** *cephaliomelana* Buining & Brederoo [2 subsp.] |
| | | ***Facheiroa*** *squamosa* (Gürke) P.J.Braun & E.Esteves Pereira |
| | | ***Facheiroa*** *ulei* (Gürke) Werderm. |
| | | *Harrisia adscendens* Britton & Rose |
| | | ***Leocereus*** *bahiensis* Britton & Rose |
| | | *Melocactus azureus* Buining & Brederoo |
| | | *Melocactus bahiensis* (Britton & Rose) Luetzelb. subsp. *bahiensis* |
| | | *Melocactus concinus* Buining & Brederoo |
| | | *Melocactus conoideus* Buining & Brederoo |
| | | *Melocactus deinacanthus* Buining & Brederoo |
| | | *Melocactus ernestii* Vaupel |
| | | *Melocactus ferreophilus* Buining & Brederoo |
| | | *Melocactus lanssersianus* P.J.Braun |
| | | *Melocactus levitestatus* Buining & Brederoo |
| | | *Melocactus oreas* Miq. [2 subsp.] |
| | | *Melocactus pachyacanthus* Buining & Brederoo [2 subsp.] |
| | | *Melocactus paucispinus* Heimen & R.J.Paul |
| | | *Melocactus zehntneti* (Britton & Rose) Luetzelb. |
| | | *Pilosocereus catingicola* (Gürke) Byles & G.D.Rowley subsp. *catingicola* |
| | | *Pilosocereus chrysostele* (Voupel) Byles & G.D.Rowley |
| | | *Pilosocereus glaucochrous* (Werderm.) Byles & G.D.Rowley |
| | | *Pilosocereus gounellei* subsp. *zehntneri* (Britton & Rose) Zappi |
| | | *Pilosocereus pachycladus* Ritter [2 subsp.] |
| | | *Pilosocereus pentaedrophorus* (Cels) Byles & G.D.Rowley [2 subsp.] |
| | | *Pilosocereus piauhyensis* (Gürke) Byles & G.D.Rowley |
| | | *Pilosocereus tuberculatus* (Werderm.) Byles & G.D.Rowley |
| | | *Pereskia aureiflora* Ritter |
| | | *Pereskia bahiensis* Gürke |
| | | *Pereskia stenantha* Ritter |
| | | *Pseudoacanthocereus brasiliensis* (Britton & Rose) Ritter |
| | | ***Stephanocereus*** *leucostele* (Gürke) A.Berger |
| | | ***Stephanocereus*** *luetzelburgii* (Vaupel) N.P.Taylor & Eggli |
| | | *Tacinga braunii* E.Esteves Pereira |
| | | *Tacinga funalis* Britton & Rose |
| | | *Tacinga inamoena* (K.Schum.) N.P.Taylor & Stuppy |
| | | *Tacinga palmadora* (Britton & Rose) N.P.Taylor & Stuppy |

**Table 2.** (Continued)

| Families | Gen/Sp | Species[1] |
|---|---|---|
| | | *Tacinga saxatilis* (Ritter) N.P.Taylor & Stuppy [2 subsp.]<br>*Tacinga werneri* (Eggli) N.P.Taylor & Stuppy |
| Capparaceae | 2/3 | *Capparis jacobinae* Moric.<br>*Capparis yco* Mart.<br>***Haptocarpum** bahiense* Ule |
| Caricaceae | 1/1 | *Jacaratia heptaphylla* (Vell.) A.DC. |
| Celastraceae | 2/3 | *Fraunhofera multiflora* Mart.<br>*Maytenus rigida* Mart.<br>*Maytenus catingarum* Reissek |
| Chrysobalanaceae | 1/1 | *Licania rigida* Benth. |
| Combretaceae | 1/2 | *Combretum monetaria* Mart.<br>*Combretum rupicola* Ridley |
| Commelinacee | 1/1 | *Dichorisandra glaziovii* Taub. |
| Convolvulaceae | 2/9 | *Evolvulus chamaepitys* Mart. var. *desertorum* (Mart. ex Choisy) Ooststr.<br>*Evolvulus gnaphaloides* Moric.<br>*Evolvulus flexuosus* Helwig.<br>*Evolvulus speciosus* Moric.<br>*Ipomoea decipiens* Dammer<br>*Ipomaea franciscana* Choisy<br>*Ipomaea longistaminea* O'Donnell<br>*Ipomoea marsellia* Meisn.<br>*Ipomoea pintoi* O'Donnell |
| Cucurbitaceae | 1/7 | *Apodanthera congestiflora* Cogn.<br>*Apodanthera fasciculata* Cogn.<br>*Apodanthera glaziovii* Cogn.<br>*Apodanthera hatschbachii* C.Jeffrey<br>*Apodanthera succulenta* C.Jeffrey<br>*Apodanthera trifoliata* Cogn.<br>*Apodanthera villosa* C.Jeffrey |
| Cyperaceae | 1/1 | *Rhynchospora calderana* D.A.Simpson |
| Euphorbiaceae | 8/49 | *Cnidoscolus bahianus* (Ule) Pax. & K.Hoffm.<br>**Cnidoscolus pubescens* Pohl<br>**Cnidoscolus urnigerus* (Pax) Pax<br>**Croton acradenius* Pax & K.Hoffm.<br>**Croton anisodontus* Müll.Arg.<br>*Croton araripensis* Croizat (= *Croton luetzelburgii* Pax & K. Hoffm.)<br>**Croton betulaster* Müll.Arg.<br>**Croton catinganus* Müll.Arg.<br>**Croton cordiifolius* Baill.<br>**Croton echioides* Baill.<br>**Croton eichleri* Müll.Arg.<br>**Croton eremophilus* Müll.Arg.<br>**Croton gardnerianus* Baill.<br>**Croton jacobinensis* Baill.<br>*Croton japirensis* Müll.Arg.<br>**Croton lachnocladus* Mart. ex Müll.Arg.<br>**Croton linearifolius* Müll.Arg.<br>**Croton mucronifolius* Müll.Arg.<br>*Croton muscicarpa* Müll.Arg.<br>**Croton mysinites* Baill. |

**Table 2.** (Continued)

| Families | Gen/Sp | Species[1] |
|---|---|---|
| | | **Croton nummularius* Baill. |
| | | **Croton pulegioides* Müll.Arg. |
| | | **Croton regelianus* Müll.Arg. [2 varieties] |
| | | **Croton salzmannii* (Baill.) G.L.Webster |
| | | **Croton schultesii* Müll.Arg. |
| | | **Croton sonderianus* Müll.Arg. |
| | | **Croton triangularis* Müll.Arg. |
| | | **Croton tridentatus* Mart. ex Müll.Arg. |
| | | **Croton velutinus* Baill. |
| | | *Croton virgultosus* Müll.Arg. |
| | | *Croton zehntneri* Pax & K.Hoffm. |
| | | *Ditaxis desertorum* (Müll.Arg.) Pax. & K.Hoffm. |
| | | *Ditaxis malpighiacea* (Ule) Pax. & K.Hoffm. |
| | | *Jatropha mollissima* Baill. var. *mollissima* |
| | | *Jatropha mutabilis* (Pohl) Baill. |
| | | *Jatropha ribifolia* Baill. var. *ribifolia* |
| | | *Manihot brachyandra* Pax. & K.Hoffm. [sect. Glaziovianae] |
| | | *Manihot catingae* Ule [sect. Glaziovianae] |
| | | *Manihot dichotoma* Ule [sect. Glaziovianae] |
| | | *Manihot epruinosa* Pax. & K.Hoffm. [sect. Glaziovianae] |
| | | *Manihot glaziovii* Müll.Arg. [sect. Glaziovianae] |
| | | *Manihot heptaphylla* Ule [sect. Caerulescentes] |
| | | *Manihot maracasensis* Ule [sect. Glaziovianae] |
| | | *Manihot pseudoglaziovii* Pax. & K.Hoffm. [sect. Glaziovianae] |
| | | **Microstachys revoluta* (Ule) Esser |
| | | **Sebastiania uleana* (Pax & K.Hoffm.) Esser |
| | | **Sebastiania brevifolia* (Müll.Arg.) Müll.Arg. |
| | | **Sebastiania echinocarpa* Müll.Arg. |
| | | **Stillingia uleana* Pax & K.Hoffm. |
| Fabaceae (*s.l.*) | 29/117 | **Acacia bahiensis* Benth. |
| | | *Acacia piauhiensis* Benth. |
| | | **Acacia santosii* G.P.Lewis |
| | | *Aeschynomene carvalhoi* G.P.Lewis |
| | | *Aeschynomene monteiroi* Afr.Fern. & J.L.Bezerra |
| | | *Aeschynomene martii* Benth. |
| | | *Aeschynomene soniae* G.P.Lewis |
| | | *Aeschynomene venulosa* Afr. Fern. |
| | | **Apuleia grazielana* Afr. Fern. |
| | | *Arachis dardani* Krapov. & W.C.Greg. |
| | | **Arachis sylvestris* (A.Chev.) A.Chev. |
| | | *Arachis triseminata* Krapov. & W.C.Gregory |
| | | *Bauhinia flexuosa* Moric. |
| | | ***Blanchetiodendron*** *blanchetii* (Benth.) Barneby & J.W. Grimes |
| | | *Caesalpinia calycina* Benth. |
| | | *Caesalpinia laxiflora* Tul. |
| | | *Caesalpinia microphylla* Mart. ex G.Don |
| | | *Caesalpinia pyramidalis* Tul. var. *Pyramidalis* |
| | | *Calliandra aeschynomenoides* Benth. |
| | | **Calliandra blanchetii* Benth |
| | | **Calliandra calycina* Benth. |
| | | **Calliandra coccinea* Renvoize [2 varieties] |

**Table 2.** (Continued)

| Families | Gen/Sp | Species[1] |
|---|---|---|
| | | *Calliandra depauperata* Benth. |
| | | **Calliandra debilis* Renvoize |
| | | **Calliandra elegans* Renvoize |
| | | *Calliandra duckei* Barneby |
| | | **Calliandra erubescens* Renvoize |
| | | **Calliandra fernandesii* Barneby |
| | | **Calliandra fuscipila* Harms |
| | | **Calliandra ganevii* Barneby |
| | | **Calliandra hirsuticaulis* Harms |
| | | *Calliandra imperialis* Barneby |
| | | **Calliandra involuta* Mackinder & G.P.Lewis |
| | | *Calliandra leptopoda* Benth. |
| | | *Calliandra lintea* Barneby |
| | | *Calliandra longipinna* Benth. |
| | | *Calliandra macrocalyx* Benth. [2 varieties] |
| | | *Calliandra mucugeana* Renvoize |
| | | *Calliandra pilgeriana* Harms |
| | | *Mimosa setuligera* Harms |
| | | *Mimosa subenervis* Benth. |
| | | *Mimosa ulbrichiana* Harms |
| | | *Mimosa xiquexiquensis* Barneby |
| | | ***Mysanthus*** *uleanus* (Harms) G.P.Lewis & A.Delgado |
| | | **Ormosia bahiensis* Monach. |
| | | *Parapiptadenia zehntneri* (Harms) M.P.Lima & H.C.de Lima |
| | | *Piptadenia viridiflora* (Kunth) Benth. |
| | | **Pithecellobium diversifolium* Benth. |
| | | *Pterocarpus monophyllus* Klitgaard, L.P.Queiroz & G.P.Lewis |
| | | *Senna acuruensis* (Benth.) H.S.Irwin & Barneby [3 varieties] |
| | | *Senna aversiflora* (Herb.) H.S.Irwin & Barneby |
| | | *Senna gardneri* (Benth.) H.S.Irwin & Barneby |
| | | *Senna harleyi* H.S.Irwin & Barneby |
| | | *Senna martiana* (Benth.) H.S.Irwin & Barneby |
| | | *Senna rizzinii* H.S.Irwin & Barneby |
| | | *Stylosanthes pilosa* M.B.Ferreira & Sousa Costa |
| | | *Trischidium molle* (Benth.) H.E.Ireland |
| | | *Zapoteca filipes* (Benth.) H.M.Hern. |
| | | *Zornia afranioi* R.Vanni |
| | | *Zornia cearensis* Huber |
| | | *Zornia echinocarpa* (Moric.) Benth. |
| | | *Zornia harmsiana* Standl. |
| | | *Zornia ulei* Harms |
| Gentianaceae | 1/1 | **Schultesia crenuliflora* Mart. |
| Lamiaceae | 2/9 | *Hyptidendron amethystoides* (Benth.) Harley |
| | | *Hyptis calida* Mart. ex Benth. |
| | | *Hyptis leptostachys* Epling subsp. *caatingae* Harley |
| | | *Hyptis leucocephala* Mart. ex Benth. |
| | | *Hyptis martiusii* Benth. |
| | | *Hyptis pinheiroi* Harley |
| | | *Hyptis platanifolia* Mart. ex Benth. |
| | | *Hyptis simulans* Epling |
| | | *Hyptis viatica* Harley |
| Malpighiaceae | 9/13 | *Barnebya harleyi* W.R.Anderson & B.Gates |
| | | **Byrsonima morii* W.R.Anderson |

**Table 2.** (Continued)

| Families | Gen/Sp | Species[1] |
|---|---|---|
| | | *Byrsonima pedunculata* W.R.Anderson |
| | | **Byrsonima triopterifolia* A.Juss. |
| | | **Camarea elongata* Mamede |
| | | **Heteropterys arenaria* Markgr. |
| | | **Heteropterys catingarum* A.Juss. |
| | | **Heteropterys perplexa* W.R.Anderson |
| | | ***Mcvaughia*** *bahiana* W.R.Anderson |
| | | **Peixotoa spinensis* C.E.Anderson |
| | | *Stigmaphyllon harleyi* W.R.Anderson |
| | | **Tetrapterys cardiophylla* Nied. |
| | | **Verrucularina glaucophylla* (A.Juss.) Rauschert |
| Malvaceae | 4/9 | *Gossypium mustelinum* Miers ex Watt |
| | | *Herissantia tiubae* (K.Schum.) Brizicky |
| | | *Pavonia erythrolema* Gürke |
| | | *Pavonia glazioviana* Gürke |
| | | *Pavonia repens* Fryxell |
| | | *Pavonia spinistipula* Gürke |
| | | *Pavonia varians* Moric |
| | | *Pavonia zehntneri* Ulbr. |
| | | *Sida galheirensis* Ulbr. |
| Molluginaceae | 1/1 | ***Glischrothamnus*** *ulei* Pilg. |
| Myrtaceae | 1/1 | *Campomanesia eugenioides* (Cambess.) D.Legrand var. *desertorum* (DC.) Landrum |
| Poaceae | 2/2 | ***Neesiochloa*** *barbata* (Nees) Pilger |
| | | *Panicum caatingense* Renvoize |
| Polygonaceae | 1/1 | *Ruprechtia glauca* Meisn. |
| Pontederiaceae | 2/2 | *Heteranthera seubertiana* Solms |
| | | *Hydrothrix gardneri* Hook. |
| Rhamnaceae | 4/4 | ***Alvimiantha*** *tricamerata* C.Grey-Wilson |
| | | *Crumenaria decumbens* Mart. [but Gardner 2314 is from Rio de Janeiro] |
| | | *Rhamnidium molle* Reissek |
| | | *Ziziphus joazeiro* Mart. |
| Rubiaceae | 3/4 | *Alseis involuta* K. Schum. [but the type is from Rio de Janeiro] |
| | | *Guettarda angelica* Mart. ex. Müll.Arg. |
| | | *Guettarda sericea* Mull.Arg |
| | | *Simira gardneriana* M.R.Barbosa & A.L.Peixoto |
| Rutaceae | 4/6 | *Balfourodendron molle* (Miq) Pirani |
| | | *Esenbeckia decidua* Pirani |
| | | *Pilocarpus sulcatus* Skorupa |
| | | *Pilocarpus trachylophus* Holmes |
| | | *Zanthoxylum hamadryadicum* Pirani |
| | | *Zanthoxylum stelligerum* Turcz. |
| Sapindaceae | 3/4 | *Averrhoidium gardnerianum* Baill. |
| | | *Cardiospermum oliveirae* Ferrucci |
| | | *Serjania coradinii* Ferrucci |
| | | **Serjania bahiana* Ferrucci |
| Scrophulariaceae | 6/9 | ***Ameroglossum*** *pernambucense* Eb.Fisch., S.Vogel & A.V. Lopes |
| | | ***Anamaria*** *heterophylla* (Giul. & V.C.Souza) V.C.Souza |
| | | *Angelonia campestris* Nees & Mart. |

**Table 2.** (Continued)

| Families | Gen/Sp | Species[1] |
|---|---|---|
| | | *Angelonia cornigera* Hook f. |
| | | *Bacopa angulata* (Benth.) Edwall |
| | | *Bacopa depressa* (Benth.) Edwall |
| | | ***Dizygostemon*** *angustifolium* Giulietti |
| | | ***Dizygostemon*** *floribundum* Benth. ex Radlk. |
| | | *Monopera micrantha* (Benth.) Barringer |
| Solanaceae | 2/2 | ***Heteranthia*** *decipiens* Needs & Mart. |
| | | *Solanum jabrense* M.F.Agra |
| Sterculiaceae | 4/7 | *Ayenia blanchetiana* K.Schum. |
| | | *Ayenia erecta* Mart. ex K.Schum. |
| | | *Ayenia hirta* St.-Hil. ex Naud. |
| | | **Ayenia noblickii* Cristóbal |
| | | *Melochia betonicifolia* St.-Hil. |
| | | ***Rayleya*** *bahiensis* Cristobal |
| | | *Waltheria brachypetala* Turcz. |
| Turneraceae | 2/9 | *Piriqueta asperifolia* Arbo. |
| | | *Piriqueta assuruensis* Urb. |
| | | *Piriqueta densiflora* Urb. var. *densiflora* |
| | | *Piriqueta dentata* Arbo |
| | | *Piriqueta duarteana* (St.-Hil.) Urb. var. *ulei* Urb. |
| | | *Piriqueta scabrida* Urb. |
| | | **Turnera caatingana* Arbo |
| | | **Turnera cearensis* Urb. |
| | | **Turnera hebepetala* Urb. |
| Velloziaceae | 1/1 | *Vellozia cinerascens* (Mart. ex Schult. f.) Mart. ex Schult. f. = *Xerophyta cinerascens* Roem. & Schult. |
| Verbenaceae | 2/3 | *Lantana caatingensis* Moldenke |
| | | *Lippia bahiensis* Moldenke |
| | | *Lippia gracilis* Schauer |

[1] Asterisks refer to possible endemics; boldfaced genera are endemic.
[2] *Auxemma* is now placed under *Cordia* (Gottschling & Miller 2006).

A considerable number of the species listed in Table 2, especially Cactaceae, are endemic to the state of Bahia, and mostly to the vegetation of the Diamantina plateau and near surroundings. As shown in Table 2, Fabaceae (*s.l.*), Cactaceae, and Euphorbiaceae are the richest endemically represented families with regards to species, but Cactaceae is the richest in terms of endemic genera, all listed as endangered. The number of species in Fabaceae is 29 lower than that reported by Queiroz (2006), but this author included infraspecific taxa as units and inedited species, as well as some not endemic.

Some species, despite widely distributed in the caatinga, are not endemic as previously thought, for instance: *Aspidosperma pyrifolium* (Apocynaceae) – from the caatinga to Argentina, Paraguay, and Bolivia, through the Pleistocenic arc, although not continuously (MBG and NYBG databases); *Commiphora leptophloeos* (Burseraceae) – also recorded for the states of Minas Gerais, Goiás, and Mato Grosso, and for Bolivia and Venezuela; *Cereus jamacaru* (Cactaceae) – subspecies *jamacaru* distribution expands to the Atlantic forest and to areas of the state of Maranhão, while subspecies *calcurupicola* distribution extends to areas of cerrado and cerrado variants (Flora... 2009); *Pilosocereus gounellei* subsp. *gounellei* (Cactaceae) populations extend to areas outside the bordering limits of the caatinga biome, reaching as far as the eastern portion of the state of Maranhão

(Zappi 1994); *Combretum leprosum* (Combretacese) – distributed from the caatinga to the state of Maranhão, Mato Grosso do Sul, Argentina, Paraguay, and Bolivia (MBG and NYBG databases); *Bauhinia cheilantha* (Fabaceae Caesalpinioideae) – distribution recorded also for the states of Maranhão and Mato Grosso, Bolivia and Paraguay (MBG and NYBG databases); *Mimosa caesalpiniifolia* (Fabaceae Mimosoideae) – distribution slightly extended westwards to the state of Maranhão (Queiroz 2006); and *Erythrina velutina* (Fabaceae Papilionoideae) – widely distributed toward southern Brazil and recorded for western and northern South America (MBG and NYBG databases).

The macambira (*Bromelia laciniosa*), a widely distributed element in the caatinga, has been usually excluded from the list of endemics, perhaps because of a couple of specimens collected from outside the caatinga biome in the state of Espírito Santo (see list in Smith and Downs 1979). However, these collections may represent populations derived from cultivation escapes.

Among the woody species commonly occurring in the caatinga biome, besides those mentioned here, are *Myracrodruon urundeuva* and *Schinopsis brasiliensis* (Anacardiaceae), *Tabebuia aurea* and *T. impetiginosa* (Bignoniaceae), *Cordia trichotoma* (Boraginaceae), *Combretum glaucocarpum* (=*Thiloa glaucocarpa*) (Combretaceae), *Amburana cearensis* (Fabaceae Papilionoideae), and *Sideroxylum obtusifolium* subsp. *obtusifolium* (Sapotaceae).

A number of plant species toxic to farm animals are also represented in the flora of the caatinga (Riet-Correa *et al.* 2009). In Amaranthaceae, *Amaranthus spinosus* and *A. viridis*, invaders of degraded areas and crops, are nephrotoxic to sheep and swine, respectively, and *Froelichia humboldtiana* causes photosensitization in horses. In Apocynaceae, *Aspidosperma pyrifolium* causes abortion at least in goats. In Asclepiadaceae, at least three species of *Marsdenia* affect the nervous system in cattle and sheep. In Asteraceae, the widespread *Centratherum punctatum* (=*C. brachylepis*) affects the digestive system of cattle and goats. In Convolvulaceae, *Ipomoea carnea* subsp. *fistulosa*, *I. batatoides* (=*I. riedelii*), and *Turbina cordata* affect the nervous system, chiefly in goats. In Euphorbiaceae, *Cnidoscolus quercifolius* (=*C. phyllacanthus*) and *Manihot* spp. are cyanogenic, and *Ditaxis desertorum* causes hemolytic anemia in cattle. In Fabaceae Caesalpinioideae, *Senna occidentalis*, invader of crops, pastures, and disturbed areas, causes malformations in farm animals. Some Fabaceae Mimosoideae affect farm animals in one of several ways: *Anadenanthera colubrina* var. *cebil* (=*A. macrocarpa*) and *Piptadenia viridiflora* are cyanogenic; *Enterolobium contortisiliquum* and *E. gummiferum* (=*E. timbouva*) cause digestive disorders, abortion, and photosensitization in cattle; the pods of *Stryphnodendron coriaceum*, which occurs in areas of the cerrado, cause death to livestock by affecting the digestive system and photosensitization in the surviving animals; the very common *Mimosa tenuiflora* causes malformations and abortion in livestock. In Fabaceae Papilionoideae, some species are the cause of several intoxication problems in farm animals: *Crotalaria retusa* (hepatic necrosis and fibrosis), *Indigofera suffruticosa* (hemolytic anemia), *Pterodon emarginatus* (liver necrosis), *Riedeliella graciliflora* (necrosis of lymphatic tissues), and *Tephrosia cinerea* (liver fibrosis). In Malpighiaceae, *Amorimia rigida* (=*Mascagnia rigida*) is a frequent problem for cattle since it causes death by cardiac failure. Cases of intoxication by *Plumbago scandens* (Plumbaginaceae) are rare, since the animals do not usually graze on it. Intoxication by *Dodonaea viscosa* (Sapindaceae) or *Trema micrantha* (Ulmaceae) causes hepatic necrosis in cattle. The widespread *Solanum paniculatum* (Solanaceae) may affect the nervous system. The Verbenaceae *Lantana camara* and *L. tiliifolia* affect animals by causing hepatogenous photosensitization.

At present the exact number of species occurring in the caatinga biome is not known, since many taxa are still being reviewed and the floras of the northeastern states are not yet completely surveyed. Among these species, many are sources of wood for rural construction and charcoal, many have medicinal properties, many serve as forage for livestock, a reasonable number are important forage sources for honey bees, but just a few bear fleshy fruits for animal and human consumption.

## Degradation and Conservation

The spatial heterogeneity of the caatinga contributes to its great diversity, but it also makes it difficult to evaluate whether the alterations of the biome reflect the action of natural factors or effects of anthropic pressure (Sampaio *et al.* 1994; Barbosa *et al.* 2005).

The characteristics of the caatinga vegetation from the pre-colonization period in Brazil are not known. However, if historical facts of the last 150-200 years are taken into account, the vegetation of the caatinga biome was very different from the present day, at least in those areas with deeper soils and higher altitudes. Doors and windows of churches, chapels, and homes built in that period were made of solid wood boards as wide as 60 cm, all obtained from local plants. About 40 years ago, large areas of arboreal caatinga were still reasonably common throughout the biome. Also the shrubby-arboreal caatingas were floristically more diversified, denser, and perhaps taller. Some plant species collected about 35 years ago are no longer represented in some areas. What is seen today is vegetation that has been impoverished through time as a consequence of the intense pressure imposed by human activities. Although not so evident, the impact of these human factors is greater than the resilience of the vegetation.

The following activities have caused heavy impacts on the caatinga biome. Deforestation is a common practice utilized for opening new areas for agriculture and cattle raising. This practice causes a process of fragmentation in the remaining vegetation that creates isolated, smaller plant populations. Extensive cattle raising is the factor of alteration that encompasses the largest area in the biome, thus altering directly all native species populations, either decreasing population sizes or influencing their nature as a result of the introduction of alien species. The increasing number of farm animals raised extensively is certainly affecting the availability of natural forages, thus possibly contributing to toxic plants becoming more abundant, as reflected by the increased number of reported intoxication cases (Riet-Correa *et al.* 2009). Cutting of woody plants for firewood and production of charcoal, whether the areas have been utilized as natural pasture or not, is the second major form of exploitation of the vegetation in the biome (Barbosa *et al.* 2005), and because of that, in some states the remaining vegetation is critically endangered.

Estimations of the level of impact on the caatinga vegetation vary. According to Mendes (1997), approximately 80% of the original ecosystems of the caatinga biome are already altered by human activities such as deforestation and burning. Castelletti *et al.* (2003), analyzing the effect of roads, estimated that the level of human impaction on the caatinga is about 50%. Another diagnosis (Sá *et al.* 2004), based on edaphic, management level, and intensity of exploitation criteria, led to the conclusion that about 66% of the driest area of semiarid northeastern Brazil is degraded, including levels ranging from low (7.07%) to severe (38.42%). On the other side, according to the Brazilian Ministry of the Environment (*apud* Queiroz 2006), only 3.2% of this biome can be considered as unaltered. Thus, all endemic centers of the biome are altered at some level, which puts all endemic species in danger of extinction.

In the caatinga biome there are seven centers of endemisms (Queiroz 2006): (i) The Northern Sertaneja Depression; (ii) the Southern Sertaneja Depression; (iii) the sedimentary tablelands of the Tucano-Jatobá basin (Raso da Catarina); (iv) the dunes of mid São Francisco valley; (v) the Ibiapaba-Araripe Plateaus; (vi) Borborema Plateau; and (vii) the Diamantina Plateau. The first two centers are related to the crystalline basement surfaces and the others to sandy sedimentary surfaces, except the Diamantina Plateau, which is of mixed origin.

According to Velloso *et al.* (2002), the ecoregions Ibiapaba-Araripe Plateau and Dunes of São Francisco have about 30% and 45% of the area conserved; the area conserved in the Borborema Plateau is almost 0% and in the Northern Sertaneja Depression is less than 2%; in the other ecoregions the area conserved varies from 3% to 6%. So there is an urgent necessity of activating preservation programs in each ecoregion of the caatinga biome, prioritizing the areas where the centers of endemisms are located, along with reinforcing sustainable development policies in the region. Nonetheless, this is not an easy task, since the greater the human population grows the greater the pressure on the natural resources.

## Acknowledgements

The participation of Dr Odací F. de Oliveira to the 8th International Symposium on Poisonous Plants was financially supported by Conselho Nacional de Desenvolvimento Científico e Tecnológico (CNPq), grant 454084/2008-0, and by Coordenação de Aprefeiçoamento de Pessoal de Nível Superior (CAPES), grant 0017/09-4.

## References

Ab'Sáber AN (1974). O domínio morfoclimático semi-árido das Caatingas brasileiras. *Geomorfologia* 43:1-39.

Alcoforado-Filho FG (1993). *Composição florística e fitossociologia de uma área de caatinga arbórea no município de Caruaru-PE*, 220 pp. M.Sc. Dissertation. Universidade Federal Rural de Pernambuco, Recife-PE.

Andrade GO and Lins RC (1965). Introdução ao estudo dos 'Brejos' pernambucanos. *Arquivos do Instituto de Ciências da Terra* 3-4:17-28.

Andrade-Lima D (1954). *Contribution to the study of the flora of Pernambuco. Brazil,* 154 pp. Universidade Rural de Pernambuco, Recife-PE. (Monografias, 1).

Andrade-Lima D (1971). Vegetação da área Jaguaquara-Maracás, Bahia. *Ciência e Cultura* 23(3):317-319.

Andrade-Lima D (1977). Flora de áreas erodidas de calcário Bambuí, em Bom Jesus da Lapa, Bahia. *Revista Brasileira de Biologia* 37:179-194

Andrade-Lima D (1978). Vegetação. In *Bacia do Parnaíba: aspectos fisiográficos* (RC Lins, ed), pp. 131-135. Instituto Joaquim Nabuco de Pesquisas Sociais, Recife-PE (Série Estudos e Pesquisas, 9).

Andrade-Lima D (1981). The Caatingas dominium. *Revista Brasileira de Botânica* 4:149-163.

Andrade-Lima D (1982). Present day forest refuges in northeastern Brazil. In *Biological Diversification in the Tropics* (GT Prance, ed.), p. 245. Columbia University Press, New York.

Araújo FS and Martins FR (1999). Fisionomia e organização da vegetação do carrasco no planalto da Ibiapaba, Estado do Ceará. *Acta Botanica Brasilica* 13(1):1-14.

Araújo FS, Sampaio EVSB, Figueiredo MA, Rodal MJN, and Fernandes AG (1998a). Composição florística da vegetação de carrasco, Novo Oriente, CE. *Revista Brasileira de Botânica* 21(2):105-116.

Araújo FS, Sampaio EVSB, Rodal MJN, and Figueiredo MA (1998b). Organização comunitária do componente lenhoso de três áreas de carrasco em Novo Oriente - CE. *Revista Brasileira de Biologia* 58(1):85-95.

Barbosa MR de V, Castro R, Araújo FS de, and Rodal MJN (2005). Estratégias para conservação da biodiversidade e prioridades para a pesquisa científica no bioma Caatinga. In *Análise das variações da biodiversidade do bioma Caatinga* (FS de Araújo, MJN Rodal, and MR de V Barbosa, orgs), pp. 416-429. Ministério do Meio Ambiente/Secretaria de Biodiversidade e Florestas, Brasília-DF,

Barbosa MR de V, Sothers C, Mayo S, Gamarra-Rojas CFL, and Mesquita AC de (2006). *Checklist das Plantas do Nordeste Brasileiro: Angiospermas e Gymnospermas,* 156 pp. Ministério da Ciência e Tecnologia, Brasília-DF.

Becerra JX (2003). Synchronous coadaptation in an ancient case of herbivory. In *Proceedings of the National Academy of Sciences* 100(22):12804-12807.

Cardoso DBOS and Queiroz LP (2007). Diversidade de leguminosae nas caatingas de Tucano, Bahia: implicações para a fitogeografia do semi-árido do Nordeste do Brasil. *Rodriguésia* 58(2):379-391.

Castelletti CHM, Santos AMM, Tabarelli M, and Silva JMC da (2003). Quanto ainda resta da caatinga? Uma estimativa preliminar. In *Ecologia e Conservação da Caatinga* (IR Leal, M Tabarelli, and JMC da Silva, orgs), ch. 18, pp. 719-734. Ed. Universitária da UFPE, Recife-PE.

Cole MM (1960). Cerrado, Caatinga and Pantanal: The distribution and origin of the savanna vegetation of Brazil. *The Geographical Journal* 126(2):168-179.

Conceição AA (2006). Ecologia Vegetal em Campos Rupestres da Chapada Diamantina. In *Rumo ao Amplo Conhecimento da Biodiversidade do Semi-árido Brasileiro* [*Towards Greater Knowledge of the Brazilian Semi-Arid Biodiversity*] (LP Queiroz, A Rapini, and AM Giulietti, eds), ch. 9, pp. 61-66. Ministério da Ciência e Tecnologia, Brasília-DF. (Published on the Internet http://www.uefs.br/ppbio/cd/portugues/capitulo9.htm and http://www.mct.gov.br/upd_blob/ 0010/10823.pdf).

Duque JG (1973). *Solo e Água no Polígono das Secas,* 4 edn, 223 pp. DNOCS, Fortaleza-CE (DNOCS. Publicação n. 154 – Série I,A).

Fernandes A (1990). *Temas fitogeográficos*, 116 pp. Stylus Comunicações, Fortaleza-CE.

Fernandes A (1996). Fitogeografia do semi-árido. In *Reunião Especial da Sociedade Brasileira para o Progresso da Ciência*, 4. *Anais*, pp. 215-219. Sociedade Brasileira para o Progresso da Ciência, Feira de Santana-BA.

Figueiredo MA (1989). *Nordeste do Brasil – Relíquias vegetacionais no semi-árido cearense (Cerrados),* 13 pp. ESAM, Mossoró-RN (Coleção Mossoroense, B, 646).

Figueiredo MA (1992). The Caatinga Ecosystem. *Proceedings of the Third International Botanic Gardens Conservation Congress*, Rio de Janeiro, 19-25 October 1992. Published on the Internet http://www.bgci.org/congress/congress_rio_1992/ figueiredo. html (accessed July 31, 2007).

Figueiredo MA (1997). Unidades Fitoecológicas. In *Atlas do Ceará*. Editora IPLANCE, Fortaleza-CE.

Figueiredo MA, Fernandes A, Oliveira OF de, and Araújo FS de (1991). Expedição botânica ao Rio Grande do Norte. *Reunião Nordestina de Botânica*, 15, Maceió-AL, September 25-29, 1991. *Resumos*, p. 77.

Flora Brasiliensis Revisitada (= Flora Brasiliensis Revisited) (2009). Published on the Internet http://flora.cria.org.br/taxonCard?id=FBR4248 (accessed April 11, 2009).

Gillett JB (1980). *Commiphora* (Burseraceae) in South America and its relationships to *Bursera*. *Kew Bulletin* 34:569–587.

Giulietti AM, Harley RM, Queiroz LP, Barbosa MRV, Bocage AL, and Figueiredo MA (2002). *Plantas endêmicas da caatinga.* In *Vegetação e flora das caatingas* (EVSB Sampaio, AM Giulietti, J Virgínio and CFL Gamarra-Rojas, eds), pp. 103-115. APNE/CNiP, Recife-PE.

Gottschling M and Miller JS (2006). Clarification of the taxonomic position of *Auxemma*, *Patagonula* and *Saccelium* (Cordiaceae, Boraginales). *Systematic Botany* 31:361-367.

Govaerts R, Frodin DG, and Radcliffe-Smith A (2000). *World checklist and bibliography of Euphorbiaceae (with Pandaceae)*, 4 vols. The Royal Botanic Gardens, Kew.

Hueck K (1972). *As Florestas da América do Sul* [Hans Reichardt, transl.], 466 pp. Polígono, São Paulo.

IBGE (2004). Instituto Brasileiro de Geografia e Estatística. http://www.ibge.gov.br (accessed July 2007).

IBGE (2007). [*Biomas Continentais Brasileiros*]. Instituto Brasileiro de Geografia e Estatística. Comunicação Social, May 21, 2004. Published on the Internet http://www1.ibge.gov.br/home/presidencia/noticias/noticia_visualiza.php?id_noticia=169&id_pagina=1 (accessed July 15, 2007).

Lemos JR and Rodal MHN (2002). Fitossociologia do componente lenhoso de um trecho da vegetação de caatinga no Parque Nacional Serra da Capivara, Piauí, Brasil. *Acta Botanica Brasilica* 16(1):23-42.

Lima RB (1995). Rhamnaceae de Pernambuco: aspectos taxonômicos, 220 pp. M.Sc. Dissertation. Universidade Federal Rural de Pernambuco, Recife-PE.

Liu MJ and Cheng CY (1995). A taxonomic study on the genus *Ziziphus*. *Acta Horticulturae* 390:181-165.

Lorenzi H, Souza M de S, Costa JT de M, Cerqueira LSC de V, and Ferreira E (2004). *Palmeiras Brasileiras e Exóticas Cultivadas,* 416 pp. Instituto Plantarum, Nova Odessa-SP.

Luetzelburg P von (1922). *Estudo Botânico do Nordeste*. Inspetoria Federal de Obras Contra as Secas, Rio de Janeiro, v. 2 and 3 (IFOCS. Publicação n. 57 – Série I, A).

MBG. Missouri Botanical Garden. $W^3$ Tropicos. Published on the Internet http://www.tropicos.org/ (several accessions).

Mello-Netto AV, Lins RC, and Coutinho SF (1992). Áreas de exceção úmidas e subúmidas do semi-árido do Nordeste do Brasil: estudo especial. In *Impactos de variações climáticas e desenvolvimento sustentável em regiões semi-áridas*. pp. 1-12. Fundação Joaquim Nabuco/ICID, Recife-PE.

Melo AL de (2000). Estudos taxonômicos sobre o gênero Cnidoscolus Pohl (Crotonoideae-Euphorbiaceae) no Estado de Pernambuco – Brasil, 152 pp. M.Sc. Dissertation. Universidade Federal Rural de Pernambuco, Recife-PE.

Mendes BV (1997). *Biodiversidade e Desenvolvimento Sustentável do Semi-Árido,* 108 pp. SEMACE, Fortaleza-CE.

MICH. University of Michigan Herbarium. Published on the Internet http://herbarium.lsa.umich.edu/databases.html (several accessions).

NYBG. The New York Botanical Garden. Virtual Herbarium database. Published on the Internet http://sciweb.nybg.org/Science2/vii2.asp [several accessions].

Nimer E (1972). Climatologia da região nordeste do Brasil. Introdução a climatologia dinâmica (subsídios a geografia regional do Brasil). *Revista Brasileira de Geografia* 34:3-51.

Oliveira AI de and Leonardos OH (1978). *Geologia do Brasil.* 3edn., 813 pp. ESAM/Coord. de Ensino de Problemas Brasileiros, Mossoró-RN (Coleção Mossoroense, C, 72).

Oliveira MEA, Sampaio EVSB, Castro AAJF, and Rodal MJN (1997). Flora e fitossociologia de uma área de transição caatinga de areia-carrasco em Padre Marcos-PI. *Naturalia* 22:131-150.

Oliveira-Filho AT and Carvalho DA (1993). Florística e fisionomia da vegetação no extremo norte do litoral da Paraíba. *Revista Brasileira de Botânica* 16(1):115-130.

Pennington RT, Prado DE, and Pendry CA (2000). Neotropical seasonally dry forests and Quaternary vegetation changes. *Journal of Biogeography* 27(2):261-273.

Pennington RT, Lewis GT, and Ratter JA, eds (2006). *Neotropical Savannas and Seasonally Dry Forests,* 484 pp. CRC Press/Taylor & Francis Group, Boca Raton.

Porto KC, Cabral JJP, and Tabarelli M, eds (2004). *Brejos de altitude em Pernambuco e Paraíba: história natural, ecologia e conservação,* 324 pp. Ministério do Meio Ambiente, Brasília-DF (Série Biodiversidade, 9).

Prado DE (2000). Seasonally dry forests of tropical South America: from forgotten ecosystems to a new phytogeographical unit. *Edinburgh Journal of Botany* 57(3):437-461.

Prado DE (2003). As Caatingas da América do Sul. In *Ecologia e Conservação da Caatinga* (IR Leal, M Tabarelli, and JMC da Silva, orgs), ch. 1, pp. 3-74. Ed. Universitária da UFPE, Recife-PE.

Prado DE and Gibbs PE (1993). Patterns of species distributions in the dry seasonal forests of South America. *Annals of the Missouri Botanical Garden* 80:902-927.

Queiroz LP (2006). The Brazilian Caatinga: phytogeographical patterns inferred from distribution data of the Leguminosae. In *Neotropical Savannas and Seasonally Dry Forests* (RT Pennington, GP Lewis, and JA Ratter, eds), pp. 121-157. CRC Press/Taylor & Francis Group, Boca Raton.

Ratter JA, Pott A, Pott VJ, Cunha CN, and Haridasan M (1988). Observations on wood vegetation types in the Pantanal and at Corumbá, Brazil. *Notes of the Royal Botanic Garden of Edinburgh* 45:503-525.

Ratter JA, Bridgewater S, and Ribeiro JF (2006). Biodiversity Patterns of the Woody Vegetation of the Brazilian Cerrado. In *Neotropical Savannas and Seasonally Dry Forests* (RT Pennington, GP Lewis, and JA Ratter, eds), pp. 31-66. CRC Press/Taylor & Francis Group, Boca Raton.

Reis AC (1976). Clima da Caatinga. *Anais da Academia Brasileira de Ciências*. 48:325-335.

Riet-Correa F, Medeiros RMT, Pfister J, Schild AL, and Dantas AFM (2009). *Poisonings by plants, mycotoxins and related substances in Brazilian livestock,* 246 pp. Editora da Universidade Federal de Campina Grande, Campina Grande-PB.

Rizzini CT (1963). Nota prévia sobre a divisão fitogeográfica do Brasil. *Revista Brasileira de Geografia* 25(1):3-64.

Rizzini CT (1979). *Tratado de Fitogeografia do Brasil.* 747 pp. Hucitec/Universidade de São Paulo, São Paulo.

Rodal MJN. 1983. Fitoecologia de uma área do médio vale do Moxotó, Pernambuco. 132 pp. M.Sc. Dissertation. Universidade Federal Rural de Pernambuco. Recife-PE.

Rogers DJ and Appan SG (1973). Manihot Maniothoides (Euphorbiaceae): a computer-assisted study. *Flora Neotropica Monograph* 13:1-272.

Röpert D (2000-continuously updated). Digital specimen images at the Herbarium Berolinense. Published on the Internet http://ww2.bgbm.org/herbarium/default.cfm (accessed April 12, 2009).

Sá IB, Riché GR, and Fotius GA (2004). As paisagens e o processo de degradação do semi-árido nordestino. In *Biodiversidade da Caatinga: áreas e ações prioritárias para a conservação* (JMC da Silva, M Tabarelli, MT da Fonseca, and LV Lins, orgs), pp. 17-36. Ministério do Meio Ambiente/Universidade Federal de Pernambuco, Brasília-DF.

Sampaio E and Rodal M de J (2000). Fitofisionomias da caatinga. In *Avaliação e identificação de ações prioritárias para a conservação, utilização sustentável e repartição de benefícios da biodiversidade do bioma caatinga.* Petrolina, PE (Documento para o GT Botânica). Published on the Internet at www.biodiversitas.org.br/caatinga/relatorios/fitofisionomias.pdf (accessed on January 12, 2009).

Sampaio EVSB (1995). Overview of the Brazilian caatinga. In *Seasonally Dry Tropical forests* (SH Bullock, HA Mooney, and E Medina, eds), pp. 35-63. Cambridge University Press, Cambridge.

Sampaio EVSB, Souto A, Rodal MJN, Castro AAJF, and Hazin C (1994). Caatingas e cerrados do NE – biodiversidade e ação antrópica. In *Conferência Nacional e Seminário Latino-americano da desertificação. Anais*, pp. 1-15, Fortaleza-CE.

Sarmento AC and Soares CMC (1971). Nova área de cerrado em Pernambuco. *Anais do ICB - Universidade Federal Rural de Pernambuco*, Recife-PE 1(1):75-82.

Sarmiento G (1975). The dry plant formations of South America and their floristic connections. *Journal of Biogeography* 2(4):233-251.

Silva EAES, Guedes RSA, Santos AMM, and Tabarelli M (2007). Distribuição de plantas da caatinga nos brejos de altitude em um gradiente de continentalidade. In *Congresso de Ecologia do Brasil*, 7, Caxambu-MG, September 23-28, 2007; *Anais*, Sociedade de Ecologia do Brasil. Published on the Internet at http://www.seb-ecologia.org.br/viiiceb/pdf/1794.pdf (accessed March 30, 2009).

Smith AC (1973). Angiosperms evolution and the relationship of the floras of Africa and America. In *Tropical forest ecosystems in Africa and South America: a comparative review* (BJ Meggers, ES Ayensu and WD Duckworth, eds), pp. 49-61. Smithsonian Institution, Washington-DC.

Smith LB and Downs RJ (1979). Bromelioideae (Bromeliaceae). *Flora Neotropica Monograph* 14(3):1493-2142.

Souza MJN de, Martins MLR, Soares ZML, Freitas Filho MR, Almeida MAG, Sampaio MAB, Carvalho GBS, Soares AMR, Gomes SCB, and Silva EA (1994). Redimensionamento da região semi-árida do Nordeste do Brasil. In *Conferência Nacional e Seminário Latino-Americano da Desertificação*, pp. 1-25. Fundação Esquel do Brasil, Fortaleza-CE.

Stannard BL (1995). *Flora of the Pico das Almas – Chapada Diamantina, Bahia, Brazil*, p. 15. Royal Botanic Gardens, Kew.

Tavares S (1988a). *Inventário da vegetação dos tabuleiros do Nordeste*. 2 pp. ESAM/FGD, Mossoró-RN, (Coleção Mossoroense, B, 493).

Tavares S (1988b). *Contribuição para o estudo da cobertura vegetal dos tabuleiros do nordeste*, 25 pp. ESAM/FGD, Mossoró-RN (Coleção Mossoroense, B, 494).

Taylor NP (1991). *The genus* Melocactus *in Central and South America*. Royal Botanic Gardens, Kew (Reprint from *Bradleya* 9:1-80).

Taylor N and Zappi D (2004). *Cacti of Eastern Brazil*. 499 pp. Royal Botanic Gardens, Kew.

Thorne RF (1973). Floristic relationship between Tropical Africa and Tropical America. In *Tropical forest ecosystems in Africa and South America: a comparative review* (BJ Meggers, ES Ayensu, and WD Duckworth, eds), pp. 27-47. Smithsonian Institution, Washington DC.

Vasconcelos Sobrinho J (1971). *As Regiões Naturais do Nordeste, o Meio e a Civilização,* 441 pp. Conselho de Desenvolvimento de Pernambuco, Recife-PE.

Velloso AL, Sampaio EVSB, and Pereyn FGC (2002). *Ecorregiões Propostas para o Bioma Caatinga*. Associação Plantas do Nordeste, Recife-PE.

Veloso HP (1964). Os Grandes Climaces do Brasil. IV- Considerações Gerais sobre a Vegetação da Região Nordeste. *Memórias do Instituto Oswaldo Cruz* 62:203-223.

Weeks A and Simpson BB (2007). Molecular phylogenetic analysis of *Commiphora* (Burseraceae) yields insight on the evolution and historical biogeography of an 'impossible' genus. *Molecular Phylogenetics and Evolution* 42(1):62-79.

WU. Institute of Botany, University of Vienna. Published on the Internet http://herbarium.univie.ac.at/ (several accessions).

Zappi D (1994). *Pilosocereus* (Cactaceae). The genus in Brazil. In *Succulent Plant Research* (D Hunt and NP Taylor, eds), vol. 3, pp. 1-160. Royal Botanic Gardens, Kew.

# Chapter 2

# Toxic Plants and Mycotoxins Affecting Cattle and Sheep in Uruguay

R. Rivero[1], F. Riet-Correa[2], F. Dutra[3], and C. Matto[1]

*[1]DILAVE 'Miguel C. Rubino', Laboratorio Regional Noroeste, Casilla de Correo 57037, CP 60.000, Paysandú, Uruguay; [2]Hospital Veterinário, CSTR, UFCG, Patos PB, Brazil; [3]DILAVE 'Miguel C. Rubino', Laboratorio Regional Este, Treinta y Tres, Uruguay*

## Introduction

Toxic plants affecting livestock in Uruguay have been reviewed (Rivero *et al.* 1989, 2000; Riet-Correa *et al.* 1993; Rivero and Riet-Correa 2004). In Uruguay data from the last 10 years of the Regional Diagnostic Laboratories East at Treinta y Tres and Northwest at Paysandú showed that plant intoxications in cattle represent 16% and 10%, respectively, of the field diagnostic cases of both diagnostic centers. For sheep, plant poisonings represent 11% and 15% of the cases diagnosed in Treinta y Tres and Paysandú, respectively (Matto 2008).

In Uruguay toxic plants affecting cattle and sheep include 31 species and 26 genera (Table 1). Bloat caused by *Trifolium* spp., nitrite intoxication caused by different grasses, and cyanide poisoning caused by *Sorghum* spp. are also frequent. Chronic phytogen intoxication by copper caused by *Trifolium repens* and *Trifolium pratense* in sheep is often seen. Mycotoxicosis caused by *Ramaria flavo-brunnescens*, *Fusarium solani* (*Ipomoea batatas*), and *Pithomyces chartarum* are reported. Despite this large number of toxic plants, few have been identified as very important. In the area served by the Northwest Regional Laboratory at Paysandú, the five principal plant poisonings affecting cattle during the last 10 years were intoxications by *Cestrum parqui*, *Senecio* spp., and *Baccharis coridifolia*, bloat by *Trifolium* spp., and nitrate intoxication by grasses. Bovine mortality by plant poisoning during that period was 6.5% in cattle and 4.7% in sheep. The main sheep intoxications were chronic phytogen copper poisoning caused by *Trifolium repens* and *T. pratense*, and poisonings by *Anagallis arvensis*, *Nierembergia repens*, and *Sessea vestiode.* For the East Regional Laboratory the mortality registered for the same period of time was 6.75% in bovine and 13.5% in sheep, and the most important intoxications in cattle were *Senecio* spp. poisoning and bloat by *Trifolium* spp. Copper intoxication caused by *Trifolium* spp. was also the main poisoning in sheep.

In Uruguay bloat caused mainly by *Trifolium repens* and *T. pratense* is considered the most important cause of death in adult cattle. *Baccharis coridifolia* is also a very important cause of death in animals transferred from areas free of the plant to areas in which it exists. Weeds such as *Nierembergia hippomanica* and *Anagallis arvensis* caused many outbreaks

©CAB International 2011. *Poisoning by Plants, Mycotoxins, and Related Toxins* (eds F. Riet-Correa, J. Pfister, A.L. Schild, and T.L. Wierenga)

of intoxication in sheep and cattle in cultivated pastures (Odini *et al.* 1995; Rivero *et al.* 1998). In areas of non-cultivated pastures the intoxication by *Senecio* spp. is the most important intoxication in cattle. Recently an acute liver necrosis caused by *Sessea vestiode* in cattle and sheep in northern Uruguay was investigated and experimentally reproduced. This chapter will report some of these plant intoxications in cattle and sheep.

**Table 1.** Plant intoxications and mycotoxicoses in ruminants in Uruguay.

| | |
|---|---|
| Hepatotoxic plants and mycotoxins | |
| Plants causing hepatic necrosis | *Cestrum parqui*, *Xanthium cavanillesii*, *Wedelia glauca*, *Cycas revoluta*, and *Sessea vestiode* |
| Plants causing hepatic fibrosis | *Senecio* spp., *Echium plantagineum*, and *Erichtites hieracifolia* |
| Plants and mycotoxins causing hematogenous photosensitization | *Myoporum laetum*, *Lantana camara*, and *Pithomyces chartarum* |
| Plants causing primary photosensitization | *Ammi magus* |
| Plants affecting the heart | *Nerium oleander* |
| Plants and mycotoxins causing neurological disorders | *Solanum bonariense*, *Paspalum notatum*, *Paspalum dilatatum*, *Phalaris* spp., *Halimium brasiliense*, *Cynodon dactylon*, and *Ramaria flavo-brunnescens* (in sheep) |
| Plants causing nephrosis | *Amaranthus* spp., *Anagallis arvensis*, *Quercus* spp. |
| Plants affecting the digestive tract | *Baccharis coridifolia*, *Nierembergia hippomanica*, *Chicorium intybus*, *Trifolium repens*, *Trifolium pretense*, and *Medicago sativa* |
| Cyanogenic plants | *Sorghum* spp. |
| Plants causing systemic calcification | *Solanum malacoxylon* and *Nierembergia repens* |
| Plants with estrogenic activity | *Trifolium pratense* |
| Mycotoxins affecting the respiratory system | *Fusarium solani* toxins (*Ipomoea batata*) |
| Plants causing nitrate/nitrite intoxication | *Lolium multiflorum*, *Triticum aestivum*, *Avena sativa*, *Trifolium repens*, *Trifolium pratense*, and *Lotus corniculatus* |
| Plants causing chronic phytogenic copper intoxication | *Trifolium repens* and *Trifolium pratense* |
| Mycotoxicosis causing ergotism | *Festuca arundinacea* and *Claviceps purpurea* |
| Mycotoxicosis affecting different systems | *Ramaria flavo-brunnescens* |

## Intoxication by *Xanthium cavanillesii* in Cattle

Outbreaks of this plant-caused intoxication have been observed in Rio Grande do Sul and Uruguay during spring (September and October). The poisoning occurs on the banks of rivers or creeks, in sandy soils, and after floods. One or two weeks after the water recedes there is a massive germination of the plant, and the animals can eat sufficient amounts of the newly germinated seedlings in their cotyledonary stage to became intoxicated. Mortality varies between 3% and 82% (Mendez *et al.* 1997). Clinical signs, observed a few hours after the ingestion of the plant, are characterized by depression, muscle fasciculation, increased respiratory and cardiac frequencies, opisthotonos, sternal or lateral recumbence, and terminal paddling movements. The animals die after clinical manifestation periods of 12-24 h. At necropsy the liver is swollen and dark reddish, and the wall of the gall bladder is edematous. The cavities have yellowish fluid. Petechiae and ecchymosis are seen on serous membranes. Dry feces with blood or mucus are frequently observed in the rectum. Microscopically, the liver has hemorrhagic centrilobular necrosis, frequently extending to the periportal hepatocytes (Mendez *et al.* 1997). The intoxication was produced experimentally in calves dosed with 7.5-10 g/kg of body weight (BW) of cotyledons (Mendez *et al.* 1997).

## Intoxication by *Sessea vestioides* in Cattle

*Sessea vestioides*, known as Linillo Paraguayo, belongs to the Solanaceae family. The intoxication by this plant was studied in ten farms in the county of Salto, northern Uruguay. The main clinical signs, characteristic of acute hepatic encephalopathy, are aggressiveness and diarrhea. Gross and microscopic lesions are periacinar hepatic necrosis. The intoxication was reproduced experimentally in four bovines that received doses of 40 and 14 g/kg of fresh green plant, and 40 and 60 g/kg of dry plant, respectively. In three animals these doses were lethal. The dose of 14 g/kg of fresh plant caused clinical signs, but the animal recovered (Alonso *et al.* 2006).

## Intoxication by *Cycas revoluta* in Cattle

An outbreak of acute intoxication by *Cycas revoluta* was observed in Uruguay in September 1995 (Riet-Correa *et al.* 1996). Two bulls had signs of aggressiveness, incoordination and diarrhea, 7-10 days after being introduced into an area where *C. revoluta* had been cultivated as an ornamental plant. At necropsy the liver was swollen, dark reddish, and mottled. The gall bladder wall, the mesentery, and the abomasum wall were edematous. Hemorrhages were observed in the digestive tract. Microscopically there was a centrilobular liver necrosis. Hepatocytes of the midzonal and periportal regions were vacuolated. The disease was produced in a calf given 20 g/kg BW of green leaves of *C. revoluta* collected in the area where the outbreak was observed. Clinical signs and lesions were similar to those observed in natural cases.

## Intoxication by *Lantana camara* in Cattle and Sheep

Two outbreaks of intoxication by *Lantana camara* in sheep and one in cattle were observed in northwestern Uruguay. Mortalities of 87% and 33% were observed in two flocks of 200 and 600 sheep, respectively. The outbreaks occurred after the transportation of the flocks to parks where *L. camara* had been cultivated as an ornamental plant. Sheep stayed in the paddocks for 24 h. Many animals showed clinical signs after being removed from the area. Another outbreak affected two cows introduced into a park where the plant was also present. Clinical signs in sheep were characterized by severe photodermatitis affecting mainly the face and ears, anorexia, restlessness, jaundice, brown urine, weight loss, ruminal stasis, drooling of saliva, lacrimation, and occasionally keratitis. Serum GGT and AST were increased. Some sheep died 24-48 h after the onset of signs, but in most animals the clinical manifestation period varied from 5 to 20 days. Jaundice, subcutaneous yellow edema and swollen ochre-coloured liver with distended and edematous gall bladder were observed at necropsies. Microscopically, the liver had severe vacuolation of periportal hepatocytes and mild proliferation of bile duct cells. A mild tubular nephrosis was also observed. The outbreak observed in cattle affected two cows that died after being sick for 24-48 h. Clinical signs and lesions were similar to those observed in sheep. The green plant was administered experimentally in cattle and sheep, in unique doses of 25-40 g/kg BW. Clinical signs were similar to those observed in field cases. The cattle that received 25 g/kg BW died 7 days after the administration of the green plant. The two sheep died 24-36 h after the administration of a single dose of 40 g/kg BW of leaves and flowers (Riet-Correa *et al.* 1996).

## Intoxication by *Myoporum laetum* in Cattle

The intoxications took place in the southeast and southern regions of Uruguay, in the counties of Canelones, Lavalleja, Rocha and San José, during the winter of 2005 (García y Santos *et al.* 2008). The disease affected Holstein, Hereford, Aberdeen Angus, and cross breed cows and young steers, which had access to fallen branches of trees after a big storm. Clinical signs were observed 4 to 6 days after the storm, and were characterized by colic, edema of the mammary gland, serous ocular discharge, generalized jaundice, severe dermatitis in white areas of the skin exposed to the sun, abortion in heifers, and death 24 to 48 h after the beginning of clinical signs. Gross lesions included subcutaneous edema, generalized jaundice, large amount of liquid in serous cavities, hemorrhages in the epicardium and endocardium, and yellowish liver with petechial hemorrhages. A large quantity of *Myoporum laetum* leaves were observed in the ruminal contents by microhistological analysis. The main histopathology lesions were diffuse periportal and midzonal necrosis, with canalicular proliferation and hepatocytic hypertrophy and vacuolization.

## Intoxication by *Senecio* spp.

*Senecio* spp. is the main poisonous plant in cattle in eastern Uruguay, and the second most important plant in the West Littoral of the country, with a 19% prevalence among all plant intoxications, for both East and Northwest Regional Laboratories (Matto 2008). There are 25 species of *Senecio* identified in Uruguay, but *S. brasiliensis*, *S. grisebachii*, *S. selloi*, and *S. madagascariensis* are the most important. *Senecio grisebachii* is a weedy member of the Compositae family, commonly known as 'spring weed' or 'Maria Mole'. It is generally associated with death in bovines in the regions where it is abundant, especially in the west of Uruguay and the north of Argentina. A research study was conducted to study the intoxication by *S. grisebachii* in cattle (Preliasco and Monroy 2008). The plant was administered experimentally to three calves at doses of 45, 24 and 15 g/kg BW. All experimental calves showed loss of weight and body mass with pronounced depression, anorexia, abdominal pain, tenesmus, dry grey feces, drooling, recumbency, dehydration, and death of the three animals. The necropsy findings revealed a general edema pattern, ascites, gray diminished liver with increased consistency, and an increased gall bladder.

The histopathology showed hepatic fibrosis with hepatic degeneration and necrosis, megalocytosis, bile duct cells proliferation, and fibroblastic proliferation with abundant collagen tissue. The clinical pattern and postmortem and histopathological findings confirmed the hepatotoxic nature of this weed. In the last few years, *S. madagascariensis* has invaded the counties of Colonia and Soriano in the Southwest Littoral of Uruguay, but no cases of poisoning by this plant were reported. A study conducted by Ferreira and Fumerol (2008) was not able to successfully reproduce the intoxication after the administration of dry and milled *S. madagascariensis* at doses of 49.4, 65, and 80 g/kg BW.

## Intoxication by *Erechtites hieracifolia* in Cattle

The intoxication by *Erechtites hieracifolia* was observed in eastern Uruguay in March 1993, in a herd of 120 one-year-old Aberdeen Angus cattle (Riet-Correa *et al.* 1996). Eight animals were affected and died. Clinical signs were characterized by progressive weight loss, wasting, abdominal straining, protracted scouring, and prolapsed rectum. At necropsies there was excessive abdominal fluid, edema of the mesentery and wall of the abomasum, and pale hard liver with enlarged and edematous gall bladder. Microscopic lesions of the liver were characterized by diffuse fibrosis, megalocytosis, and proliferation of bile duct cells. The plant contained 0.2% pyrrolizidine alkaloids.

## Intoxication by *Nerium oleander* in Cattle

*Nerium oleander* is an ornamental plant found commonly in Uruguay. The toxicity of *N. oleander* results from several cardiac glycosides, mainly oleandrin. The intoxication was observed in northwestern Uruguay in a paddock where a eucalyptus forest had been trimmed and an oleander plant was also cut. Eighty, 2-year-old heifers were introduced into the area and five of them died 24-48 h after being in the paddock. Some animals were found dead. Others had clinical signs characterized by depression, weakness, anorexia, ataxia, and diarrhea. No significant lesions were observed at necropsies. Oleander leaves were found in the rumen. The disease was produced in three calves given singles doses of 1, 0.5, and 0.25 g/kg BW of leaves of *N. oleander* collected at the farm. The animals that received a single dose of 1 and 0.5 g/kg BW died between 6 to 36 h after the administration, with clinical signs of weakness, ataxia, anorexia, tachypnea, and severe tachycardia with arrhythmia. Postmortem findings were of little significance. The principal histological lesions were in the heart and consisted of multifocal myocardial edema, degeneration, and necrosis. The calf intoxicated with 0.25 g/kg BW showed anorexia, weakness, and bradycardia in the first 24 h, but returned to normal after 72 h (Riet-Correa *et al.* 1996).

## Intoxication by *Halimium brasiliense* in Sheep

Poisoning by *Halimium brasiliense* in sheep is characterized by transient seizures with muscular tremors, ventroflexion of the neck, opisthotonous, nystagmus, tetanic spasms and limb paddling movements. The intoxication has been observed on two farms in the municipality of Rio Grande in Rio Grande do Sul, Brazil, and on at least 36 farms in the departments of Lavalleja, Maldonado, Cerro Largo, Durazno and Treinta y Tres in

Uruguay. The illness is seasonal with most cases occurring from August to November, but a few cases are also observed from May to July. Most sheep recovered when moved to other pastures. The frequency varies between farms and between years. There are also variations between different paddocks within farms. Morbidity varies between 1% and 15%, but some farmers reported a frequency of up to 50% in years when drought conditions prevailed. On farms where affected sheep are removed from the paddocks after the observation of the first clinical signs, mortality is between 1% and 5%. Nevertheless, in drought conditions, on some farms where this measure is not practised, mortality may be as high as 35% (Riet-Correa *et al.* 2009). Macroscopic lesions are not significant. The main histological lesion was the presence of vacuoles, sometimes containing macrophages or axonal residues, in the white matter of the brain and spinal cord. Under electron microscopy the lesions were characterized by axonal degeneration followed by ballooned myelin sheaths with disappearance of the axoplasm. Convulsions are probably secondarily inducing the death of neurons which results in Wallerian degeneration, or perhaps the plant causes axonal degeneration. A pigment identified as ceroid-lipofuscin is also present in neurons, astrocytes, Kupffer cells, and macrophages of the spleen and lymphonodes. This pigmentation was apparently not related to the clinical signs. Feeding trials in sheep demonstrated that the disease is caused by the ingestion of *Halimium brasiliense* in amounts ranging from 2100 to 3000 g/kg BW total plant material given over many days (Riet-Correa *et al.* 2009).

## Intoxication by *Cynodon dactylon* in Cattle

Two outbreaks of intoxication by *Cynodon dactylon* (Bermuda grass) in cattle were reported in northwestern Uruguay in July 1996 and August 1998. Both outbreaks occurred during winter time after heavy frosts. One outbreak affected Hereford heifers in a paddock covered by a dense and dry pasture composed mainly by *C. dactylon*. The other affected 2- and 3-year-old Hereford steers grazing in a eucalyptus forest with an abundant presence of the plant. Morbidity was 23.6% and 8.7%, and mortality was 1% and 1.4%, respectively. Clinical signs were muscle tremors and twitching, marked incoordination, weaving and bobbing of the head, and inability to rise. Some animals appeared to be stiff legged, and others showed marked weakness of the hind limbs. Most animals died accidentally as a result of the nervous disorder, as they drowned in streams or ditches.

## Intoxication by *Anagallis arvensis* in Cattle and Sheep

Ten outbreaks of intoxication by *A. arvensis* were diagnosed in the Department of Paysandú, Uruguay, during December 1994, January 1995, and December 1996 and 1997 (Rivero *et al.* 1998). Cattle morbidity varied between 3.2% and 53.2% and lethality between 42.6% and 100%. Sheep morbidity was between 2.8% and 42.9% and case fatality rates between 81.3% and 100%. In nine outbreaks the animals were grazing on wheat or barley stubble. On eight occasions the animals were introduced in the stubble 2-10 days before first clinical signs, and in one outbreak, 25 days before. In all outbreaks *A. arvensis* was in the vegetative state and in bloom, covering the soil and dominating the other species. Flowers of red and blue color were observed. Other nephrotoxic plants, such as *Amaranthus* spp. or *Quercus* spp., were not observed.

The remaining outbreak occurred on a paddock plowed in winter but not cultivated, that remained without animals until December when it was covered by dense green pasture basically composed of *A. arvensis*. At the beginning of December, 1200 yearling sheep were introduced into this paddock and started to die 36 h later. In four outbreaks where cattle and sheep had been grazing together, both species were equally affected. In one outbreak cattle were less affected because they remained in the paddock for no more than 36 h while sheep were kept in the pasture until the first clinical signs occurred. No differences in frequency were observed in animals of different age or sex. Two outbreaks occurred in the same paddock of the same farm in different years (November 1994 and December 1996). The farms were located on basaltic or cretaceous soils.

Clinical signs for both species were weakness, loss of body condition, ruminal atony, diarrhea (occasionally stained with blood), slow gait, staggers, convulsions in some animals, coma, and death within 12-48 h. Blood serum values of creatinine, urea, and magnesium were increased, and levels of calcium were decreased. Gross lesions were characterized by a ventral subcutaneous edema and petechiae, submandibular edema, presence of fluids in cavities, and edema and petechial hemorrhages of the mesenterium. Kidneys were edematous, pale or yellowish in colour with petechiae on the cortex. Erosive and ulcerative lesions were observed in the esophagus. Edema of the abomasal submucosa, and hemorrhagic abomasitis and enteritis were also observed. Cardiac and skeletal muscles were pale and flaccid. Some animals had congestion and edema of the lungs. The most significant histological lesion was a severe nephrosis with tubular degeneration and necrosis, hyaline cylinders, moderate intratubular hemorrhages, and interstitial edema and congestion. Catarrhal enteritis with hemorrhages, focal necrosis, and mononuclear infiltration of the mucosa and lamina propria with gland hyperplasia and increased secretion were observed in the gut. Liver congestion, and in some cases moderate hepatocytic granular degeneration and mild proliferation of Kupffer cells, were also seen.

*A. arvensis* in the vegetative stage was collected on a farm where an outbreak was occurring in December 1996. The plant was immediately carried to the laboratory and maintained at 4-5°C until administration. Leaves, fruits, and fine stems were milled and administered to two sheep through a stomach tube. The administration started 24 h after the plant collection. One ewe received a daily dose of 40 g/kg BW for 4 consecutive days. Another was dosed daily with 32 g/kg BW for 7 days. The two sheep died as a consequence of the experimental intoxication and were necropsied and examined histologically. One sheep showed depression, anorexia, and weakness, and died on day 5, 12 h after the onset of signs. The second sheep had weakness, anorexia, ruminal atony, diarrhea, and staggers; it died on day 9, 36 h after the onset of signs. Macroscopic and histological lesions were similar to those observed in field cases. Prevention of intoxication should be based on avoiding grazing in areas severely infested by the weed during the season and under the conditions of this report. One reason for the presence of large amounts of the plant in certain areas could be the use of commercial seeds contaminated by seeds of *A. arvensis*.

## Intoxication by *Quercus* spp. in Cattle

An outbreak of intoxication by *Quercus* spp. was observed in eastern Uruguay in May 1997 in a herd of 90, one-year-old Hereford, Aberdeen Angus, and crossbred cattle. They were grazing in a forest of *Quercus* spp. with accumulation of acorns from the trees and dead forage. Morbidity was 16.6% and mortality 4.4%. Clinical signs were characterized by weakness, loss of body condition, and dark diarrhea. Gross lesions were characterized by erosive and ulcerative lesions in the esophagus, edema and petechial hemorrhages of the mesenterium, necrosis of buccal and rumen papilla. Kidneys were edematous, pale or yellowish in color with petechiae on the cortex. The most significant histological lesion was a severe nephrosis with tubular necrosis, diffuse epithelial regeneration, hyaline cylinders, moderate intratubular hemorrhages, discrete fibrosis, and mononuclear infiltration.

## Intoxication by *Nierembergia hippomanica* in Cattle

Outbreaks of intoxication by *Nierembergia hippomanica* have been frequently diagnosed in cattle in northwestern Uruguay. Morbidity is 10-80% and deaths do not occur. Most outbreaks are observed in milking cows or in 3- to 4-year-old steers. Younger cattle appear to be more resistant. The intoxication occurs at any time of the year from January to November. All outbreaks occurred in cultivated pastures or in wheat or barley stubble fields. Invasion of pastures by the plant is apparently due to the use of seeds contaminated by *N. hippomanica* seeds. Clinical signs are characterized by salivation, diarrhea, restlessness, abdominal pain, and periodic motion of the head and limbs. Milking cows have decreased milk production. Affected animals recovered within 1 week after removal from the pastures (Odini *et al.* 1995). The green plant was administered experimentally to cattle and sheep at 10-50 g/kg BW. The lowest toxic dose was 10-15 g/kg BW. No differences were observed in the toxicity of plant samples collected in winter or spring. Clinical signs were similar to those observed in field cases. All animals recovered in 1-8 days, except one calf that died after the ingestion of 50 g/kg BW. The main lesions were focal hemorrhages in the large intestine and enteritis in the small intestine. The dried plant was not toxic to cattle and sheep. One steer that received 10 daily doses of 5 g/kg BW showed clinical signs after the last dose, demonstrating a cumulative effect of the plant (Odini *et al.* 1995). Two sheep that received 20 g/kg BW of the plant presented anorexia, diarrhea, abdominal pain, restlessness, and excessive salivation (Odini *et al.* 1995). A previous description of the spontaneous intoxication in sheep reported nervous signs and deaths of some animals (Riet-Alvariza 1979). A pyrrole-3-carbamidine has been identified as the toxic principle of *N. hippomanica* (Buschi and Pomilio 1987).

## Chronic Phytogen Copper Intoxication in Sheep

This intoxication is associated with pastures containing normal levels of copper but very low levels of molybdenum. In Uruguay, the condition occurs in sheep grazing pastures of *Trifolium repens* and *T. pratense.* From 1980 to 1985, 12 outbreaks were diagnosed in different regions of the country. From 1983 to 1988, 25 outbreaks of the intoxication were diagnosed in northwestern Uruguay. Twelve of these outbreaks occurred during 1988. The increase in the frequency of the intoxication was due to an incremental increase in sheep production in Uruguay due to a good international wool price. Areas previously used for agriculture or cattle production, like the northwestern region, were partially used for sheep breeding in *T. repens* and/or *T. pratense* pastures. After 1988 the outbreaks of chronic phytogen copper toxicosis decreased because it was not profitable to graze sheep due to a drop in the wool price. From 1997 to 1999, the frequency of the disease in the northwestern region increased again due to the use of mainly *T. pratense* pastures for the production of fattening lambs for exportation. First cases are often seen after 3 months grazing in pastures of *T. repens* and/or *T. pratense*, mainly in animals in good nutritional state. The intoxication occurs during the entire year but is more frequent in spring. The onset of the disease is commonly associated with stress factors like vaccination, insemination, dipping, transportation, and reduction in forage availability. Morbidity varies between 1% and 12%, and lethality is nearly 100%. There was no variation in susceptibility between breeds (Corriedale, Ideal, Romney Marsh, Merilin, and Merino), but most outbreaks occurred in Corriedale because this breed represents 70% of the sheep population in Uruguay. The animals showed depression, anorexia, jaundice, hemoglobinuria, anemia and liquid, fetid and dark feces. In most sheep the death occurred in 24-96 h. Few sheep survived to the hemolytic crisis.

At necropsies, the main gross lesions were jaundice; subcutaneous yellow edema; serous liquid in cavities; swollen friable ochre-coloured liver with distended and edematous gall bladder; dark kidneys with edema and diminished consistency; and dark urine. Microscopically, the liver had enlarged pleomorphic and vacuolated hepatocytes, and, occasionally, centrilobular necrosis, biliary stasis, mild proliferation of bile duct cells with fibrosis in the portal space, and proliferation of Kupffer cells with abundant cytoplasm and granules containing copper. The most significant histological lesion in kidneys was a severe hemoglobinuric nephrosis, with tubular degeneration and necrosis, with the presence of hemoglobin and copper in the epithelial cells. Chronic phytogen copper intoxication occurs in pastures with low molybdenum, less than 0.36 ppm (Pereira and Rivero 1993), and normal copper concentrations in pastures of *T. repens* and/or *T. pratense*. The diagnosis is based on the epidemiological data, clinical signs, macroscopic and histologic lesions, and the determination of Cu levels in the liver (over 500 ppm) and kidneys (over 80 ppm). For the prevention of the intoxication in Uruguay, grazing periods of no more than 3 months in pastures with predominance of *T. repens* or *T. pratense* are recommended.

## Intoxication by *Ramaria flavo-brunnescens* in Sheep

Several outbreaks of poisoning by *Ramaria flavo-brunnescens*, a well known disease in cattle in Uruguay and Rio Grande do Sur, Brazil, have been recently reported in sheep in Uruguay. The disease occurred mainly in the northwestern region with a morbidity of 7% to 35% and a mortality of 7% to 26% (Riet-Correa *et al.* 1996). The intoxication occurs between March and July in eucalyptus forests where the fungus is found. Clinical signs in sheep are characterized by nervous disorders with convulsions, muscle tremors, ataxia, hypermetria, nystagmus and opisthotonous. Some animals remain recumbent and die. Hypertemia, polyuria, ulcers in the tongue, and necrotic lesions in the extremities characterized by a hyperemic line with crusts at the coronary band were also observed in experimental intoxications.

The increase in the frequency of this intoxication in the last few years in cattle and sheep in Uruguay is due to an increase of the forested area with eucalyptus trees. These new forests are also used by livestock during the breeding season.

## Acknowledgements

The participation of Dr Rodolfo Rivero to the 8th International Symposium on Poisonous Plants was financially supported by Conselho Nacional de Desenvolvimento Científico e Tecnológico (CNPq), grant 454084/2008-0, and by Coordenação de Aprefeiçoamento de Pessoal de Nível Superior (CAPES), grant 0017/09-4.

## References

Alonso M, Bianchi JA, and Nuñes Jr (2006). Intoxicacion por *Sessea vestioides* en bovinos del Uruguay, 26 pp. Tesis de Grado, Facultad de Veterinaria, Uruguay.

Buschi CA and Pomilio AB (1987). Pyrrole-3-Carbamidine: a lethal principle from *Nierembergia hippomanica*. *Phytochemistry* 26:863-865.

Ferreira S and Fumero L (2008). Investigación sobre la toxicidad de *Senecio madagascariensis* en bovinos, 32 pp. Tesis de Grado, Facultad de Veterinaria, Uruguay.

García y Santos C, Pérez W, Capella A, and Rivero R (2008). Intoxicación espontánea por *Myoporum laetum* en bovinos en Uruguay. *Veterinaria*, Uruguay, 43:25-29.

Matto C (2008). Caracterización de los Laboratorios Regionales de diagnóstico veterinario Este y Noroeste de la DILAVE 'Miguel C. Rubino' y principales enfermedades diagnosticadas utilizando una base de datos relacional, 90 pp. Tesis de Grado, Facultad de Veterinaria, Uruguay.

Méndez MC, Santos RC, and Riet-Correa F (1997). Intoxication by *Xanthium cavanillesii* in cattle and sheep in southern Brazil. *Veterinary and Human Toxicology* 40:144-147.

Odini A, Rivero R, Riet-Correa F, Mendez MC, and Giannechinni E (1995). Intoxicación por *Nierembergia hippomanica* en bovinos y ovinos. *Veterinaria*, Uruguay 30:3-12.

Pereira D and Rivero R (1993). Intoxicação crónica fitógena por Cobre. In *Intoxicações por Plantas e Micotoxicoses en animais domésticos* (F Riet-Correa, MC Mendez, AL Schild, eds) pp. 299-307. Editora Hemisfério Sul do Brasil, Brasil.

Preliasco M and Monroy, N (2008). Investigación sobre la toxicidad de *Senecio grisebachii* en bovinos del Uruguay, 66 pp. Tesis de Grado, Facultad de Veterinaria, Uruguay.

Riet-Alvariza F (1979). Comunicación de un caso de intoxicación por *Nierembergia hippomanica.* Apuntes de toxicología veterinaria. Dirección General de Extensión Universitaria, Montevideo, pp. 165-166.

Riet-Correa F, Méndez MC, and Schild AL (1993). *Intoxicações por plantas e micotoxicoses em animais domésticos*, 340 pp. Editorial Hemisferio Sur, Montevideo.

Riet-Correa F, Rivero R, Dutra F, and Mendez MC (1996). Intoxicaciones en Rumiantes en Río Grande del Sur y Uruguay. Publicación en CD. *VI Congreso Nacional de Veterinaria.* Montevideo, Uruguay.

Riet-Correa F, Barros SS, Mendez MC, Fernandes CG, Pereira Neto O, and McGavin D (2009). Axonal degeneration in sheep caused by the ingestion of *Halimium brasiliense*. *Journal of Veterinary Diagnostic Investigation* 21:478-486.

Rivero R and Riet-Correa F (2004). Toxic plants affecting cattle and sheep in Uruguay, pp. 54-55. *1° Simposio Latinoamericano de Plantas Tóxicas.* Bahia, Brasil.

Rivero R, Quintana S, Feola R, and Haedo F (1989). Principales enfermedades diagnosticadas en el área de influencia del Laboratorio de Diagnóstico Regional Noroeste del C.I Vet. Miguel C. Rubino, I1-I73. *Publicación XVII Jornadas Uruguayas de Buiatría*, Paysandú, Uruguay.

Rivero R, Zabala A, Gil J, Gianneechini RE, and Moraes J (1998). Intoxicación por *Anagallis arvensis* en Bovinos y Ovinos del Uruguay, pp. 26-29. *Publicación de las XXVI Jornadas Uruguays de Buiatria*, Paysandú, Uruguay.

Rivero R, Riet-Correa F, and Dutra F (2000). Toxic plants affecting cattle and sheep in Uruguay, Libro de abstracts, N°748, P 10. *XXI World Buiatric Congress*, Diciembre, Uruguay.

# Chapter 3

# Poisoning by Plants, Mycotoxins, and Algae in Argentinian Livestock

E. Odriozola

*Department of Animal Production, EEA INTA Balcarce, Buenos Aires, Argentina*

## Hepatotoxic Plants

### Plants causing hepatic necrosis

*Wedelia glauca* (yuyo sapo, espanta colono) is a hepatotoxic plant that is widespread in Argentina and causes significant animal losses. The toxic principle is a carboxyatractyloside, similar to that of *Cestrum parqui* and *Xanthium* spp. The risk period includes spring and summer. After the first frosts in autumn, the aerial part of the plant disappears until the next spring. It expands by seeds and stolons, and grows in clumps. Within the period of risk there are two peaks of mortality: one when the plant sprouts in September/October; and another one during the flowering period, February/March, when the plant is eaten voluntarily. Poisonings due to the consumption of hay occur throughout the year because the plant is still toxic during the haymaking season. There are records of poisoning in ovines, bovines, and swine (Lopez *et al.* 1991).

*Cestrum* species are found throughout Argentina. *Cestrum parqui* (duraznillo negro, palqui, hediondilla) is widespread. Cases of poisoning throughout the year appear in our records, reaching the highest number of deaths during spring and autumn. It is ingested when forage or water are scarce. Poisioning by *C. estrigillatum* (dama de la noche) is confined to the province of Santa Fe (Lopez *et al.* 1991).

*Xanthium cavanillessi* (abrojo grande) is a widespread weed in Argentina. Burr poisoning is markedly seasonal, from September to November, depending on rainfall, as dry weather can delay the sprouting of cotyledons, which are the only source of poisoning (Campero *et al.* 1993).

All of these plants display similarities in their clinical manifestation and necropsy findings; therefore, the differential diagnosis is based on two aspects: the presence of the plant with evidence of consumption, and the time of the year in which the poisoning occurs. In general, most of the animals are found dead or display signs of aggressiveness, lateral recumbency, paddling, and death.

Edema of the wall of the gall bladder and duodenum are typical findings in these poisonings, accompanied by the presence of free blood commonly in intestine and sometimes in abomasum, and petechiae and suffusions in endocardium and epicardium. Red spots caused by centrilobular necrosis are found in the liver.

©CAB International 2011. *Poisoning by Plants, Mycotoxins, and Related Toxins* (eds F. Riet-Correa, J. Pfister, A.L. Schild, and T.L. Wierenga)

## Plants causing hepatic fibrosis

Pyrrolizidine alkaloidosis in Argentina is caused by two genera, *Senecio* and *Echium*. The *Senecio* species that cause toxicity in bovines are *S. tweediei*, *S. pampeanus*, *S. grisebachii*, *S. selloi*, and *S. madagascariensis*. The outbreaks are related to the use of high stocking rates or to winters with low grass availability. Both situations force animals to ingest these weeds during the sprouting period. Poisoning by *E. plantagineum* (flor morada) took place in the first year after pasture plantings when the weed was more frequent. Unlike plants that produce acute hepatic necrosis, these plants must be ingested in large amounts over a long period of time.

The affected animals display submandibular, pectoral, and abdominal edema; they usually show a certain degree of aggressiveness. Edemas of the lower parts of the body and ascites are observed at necropsy. The liver is hard, whitish, or with nodular aspect. Histologically there is proliferation of fibrous tissue mainly in the portal area, bile duct cell proliferation, and megalocytosis (Peterson 1984).

## Plants, fungi, and algae causing hepatogenic photosensitivity

*Lantana camara* (bandera española) is an ornamental plant that has spread to pastures, and livestock may ingest it when forage is lacking. Natural cases have been recorded in the province of Jujuy (Marin *et al.* 2005).

*Myoporum laetum* are shrubs commonly used in fences, and when pruned, animals consume them (Odriozola *et al.* 1987).

All outbreaks of *Microsystis aeruginosa* (alga verde azulada) poisoning in Argentina have involved a very high number of dead bovines. Animal mortalities occur when certain conditions occur, namely: (i) temperatures oscillating between 25-20°C; (ii) permanent winds coming from a specific direction which allows the accumulation of algae in the lagoon coast (flowering); and (iii) thirsty animals (previously penned). The main toxins in water contaminated by algae are neurotoxins (anatoxin) and hepatotoxins (microcystin). Quantification of the latter indicates the risk from ingesting the water (Odriozola *et al.* 1984).

*Panicum milliaceum* (mijo) is cultivated around the province of Buenos Aires. In certain situations it becomes a toxic plant by accumulating lithogenic steroidal saponins, causing photosensitivity and death due to bile duct obstruction. Younger animals are more susceptible than adult ones. There are records of cases of photosensitivity and death in sheep caused by the consumption of *Tribulus terrestris* (roseta), a very common weed in some regions of the southwest of the province of Buenos Aires. The toxic principle is similar to that of mijo (Tokarnia *et al.* 2000).

*Pithomyces chartarum* is a saprophytic fungus widely disseminated in Argentina and is considered the main cause of hepatogenic photosensitivity. This fungus, present in dead leaves of Gramineae, produces a toxin known as sporidesmin which causes obstructive canalicular injuries. Not all strains of *P. chartarum* produce sporidesmin. For the diagnosis of poisoning, it is necessary to determine the number of spores by gram of grass and determine if the strains are sporidesmin producers by ELISA (Odeon *et al.* 1983; Collin *et al.* 1998).

*Kochia scoparia* (morenita) is a weed that can be found in the northern province of Buenos Aires, and in the provinces of Córdoba, Santa Fe, La Pampa, Río Negro, Chubut, and Santa Cruz. The toxic principles responsible for the hepatic injury and consequent photosensitivity are saponins. However, other toxins are recognized in the plant, such as thiaminase, which causes polioencephalomalacia. The poisoning occurs in previously flooded fields with abundant weeds which are voluntarily consumed by livestock (Keeler *et al.* 1978).

## Plants Causing Primary Photosensitivity

The only plant proven to cause primary photosensitization is *Ammi majus* (falsa biznaga, apio cimarrón), which is very common in wheat stubble. Although *A. viznaga* is widely spread, there are no records of natural poisoning by this species (Odriozola 1984; Lopez and Odriozola 1987).

## Plants Affecting the Central Nervous Systems

### Tremorgenic plants and fungus

Gramineae of the genus *Paspalum* (*P. dilatatum*, *P. distichum*, and *P. notatum*) are susceptible to infection by the *Claviceps paspali* fungus. The florets of the flowering grasses are infected by fungal spores. Fungal mycelium then destroy the plant ovary and use plant nutrients to grow into sclerotia. Consumption of these alkaloid-containing kernels causes nervous signs, predominantly tremorgenic. As it is related to the flowering period of these Gramineae, the poisoning is seasonal, restricted to summer and autumn (February-April). With the first frosts sclerotia fall to the ground where they will remain until the next summer. In general, they do not cause death and the clinical manifestation period is nearly 20 days (Lopez *et al.* 1985).

Poisoning by consumption of *Cynodon dactylon* (gramilla helada) appears every year after the first frosts, mainly in natural fields and maize stubble. The toxic principle is unknown, and the clinical signs remain for around 20 days after animals cease consumption (Odriozola *et al.* 2001).

*Lolium perenne* (raigrás perenne) is a high quality forage. Most of the varieties planted in Argentina come from seed plots from New Zealand, and are prized for their properties such as good tilling, good resistance to insects, and drought tolerance. All these advantages are attributed to the presence of the endophytic fungus *Neothipodium lolii*. This fungus produces tremorgenic substances called lolitrems. Consumption produces clinical signs similar to those reported in *C. paspali* and *C. dactylon*, with high morbidity and low or no lethality (Odriozola *et al.* 1993).

*Phalaris angusta* is a Gramineae typically found in wetlands and riparian areas prone to floods. The toxic principles responsible for the nervous signs and death are tryptamine alkaloids. Toxin concentration in the plant depends on solar light: less light, greater concentration. The toxins also vary with humidity: the level of alkaloids increases after rains preceded by periods of drought. Clinical signs are tremors and noticeable incoordination (Del Potro *et al.* 1984; Odriozola *et al.* 1991a).

Survivors display sequelae such as a noticeable loss of condition and cachexia that concludes with the animal's death after a course of 3 to 4 months. This is caused by lesions in the CNS that affect prehension.

### Plants and fungi producing ataxia

*Condalia microphylla* (piquillín) is a perennial shrub of the Rhamnaceae family found in Patagonia and elsewhere. The toxic principle is unknown (Blanco Viera *et al.* 2000). In the provinces of Buenos Aires, Río Negro, and La Pampa, clearing of the native vegetation results in replacement by winter annual grasses, and populations of *Condalia microphylla* remain in the paddock. During grazing of these pastures bovines of up to 2 years of age are affected by nervous signs characterized by progressive hind limb paresis and death. Axonal degeneration is observed on histologic examination. *Stenocarpella maydis* is a fungus that produces a maize disease called stalk rot and white ear rot. This fungus also produces a toxin currently not well characterized, called diplodia toxin, whose consumption by ruminants produces nervous disruptions initially characterized by hindlimb ataxia, abortion, and death. There are no necropsy findings and the histopathology study reveals, although not consistently, a diffuse cerebellar myelomalacia. This poisoning occurs annually in Argentina (Riet-Correa 1993; Odriozola *et al.* 2005).

### Other plants affecting the nervous system

*Centaurea solstitialis* (abrepuño amarillo) is a weed found in several provinces of Argentina. The toxic principle is unknown. It exclusively affects equines, which consume it voluntarily. The occurrence of the poisoning is considered to have two phases: a peak in June-July and a second peak in November-December. Affected equines are not older than 18 months. Clinical signs appear abruptly after prolonged plant consumption. Animals lose their capacity to feed and to drink water. They die due to starvation and aspiration pneumonia. Lesions are mainly bilateral affecting the substantia nigra and globus pallidus. Cavitations ranging from 0.5 to 1 cm are observed. Coagulation necrosis is observed exclusively in the grey matter (Martin *et al.* 1971).

*Prosopis caldenia* is a tree that belongs to the Leguminosae family. It is widespread in Argentina. Clinical cases have been reported in the provinces of Mendoza and La Pampa due to exclusive consumption of the beans of this tree. Clinical signs include unilateral or bilateral paralysis of the pinna (Lopez *et al.* 1991).

## Nephrotoxic Plants

*Amaranthus quitensis* (yuyo colorado) is a weed found throughout Argentina. It can be found from the province of Rio Negro to the provinces of Catamarca and Formosa. *Chenopodium albun* (Quinoa) is found in the provinces of La Rioja, Entre Rios, Santa Fé, La Pampa, Mendoza, Chubut, Santa Cruz, and Buenos Aires. *Rumex crispus* (lengua de vaca) is present throughout the country.

These three species accumulate oxalates and produce intoxication when they are ingested in large amounts as the only dietary element. These weeds are present in maize and wheat stubble and are eagerly eaten by bovines. When the amount of oxalates ingested exceeds the degradation by the ruminal flora, they form calcium oxalate crystals which produce mechanical injuries in the rumen and kidneys. In the blood, oxalates fix calcium with formation of calcium oxalate crystals and consequently cause hypocalcemia. Animals die of renal failure due to deposition of oxalate crystals in the kidneys (Lopez *et al.* 1991).

*Quercus* spp. (roble) are trees that have been planted in farms in the humid pampas and the consumption of new shoots produces mortality in bovines. The toxic principle is tannin which causes nephrosis (Odriozola *et al.* 1990).

## Plants Affecting the Digestive System

*Baccharis coridifolia* (romerillo, mio mio) was the first toxic plant reported in Argentina, Uruguay, and southern Brazil. Poisonings occur when animals raised in areas without the plant are transported to pastures infested by *B. coridifolia.*

It is spread all over the country and poisonings occur throughout the year. The toxic principles are macrocyclic trichothecenes produced by the fungi *Mirothecium roridum* and *M. verrucaria*, which live in the soil. The mycotoxins roridin A, D, and E, verrucarin A, and miotoxins A and E produce injuries in the gastrointestinal tract causing the death of the animals (Tokarnia *et al.* 2000).

*Asclepia mellodora* (yerba de la víbora) is a weed found in several areas of Argentina. The toxic principles are cardenolides and resinoids (galitoxin). Animals only consume this plant when forage is limited. The plant is believed to cause mortalities and the poisoning has been experimentally reproduced; however, there are no records of natural cases (Tokarnia *et al.* 2000).

Bloat is one of the main causes of cattle deaths during winter. It occurs mainly in *Medicago sativa* (alfalfa) pastures, but cases produced by the ingestion of *Trifolium repens* (trébol blanco) and *T. pratense* (trébol rojo) have been also recorded. Although several control measures have been implemented, none are 100% effective.

## Plants with Mutagenic and Anti-Hematopoietic Effects

Poisoning by *Pteridium aquilinum* causes enzootic hematuria and squamous cell carcinomas of the upper digestive tract and is reported in the province of Jujuy (Marin *et al.* 2004).

## Cyanogenic Plants

*Sorghum* spp. is currently being used in pastures for the grazing of beef cattle. Its toxicity is caused by the ingestion of sprouting plants, rich in hydrocyanic acid. Sorghum has been recognized for many years as a cause of death in Argentina. Recently the poisoning appeared under different clinical characteristics and, apparently, the toxic principles were not the same. Clinical signs, observed mainly in horses but also in bovines, were characterized by locomotive dysfunction and irreversible urinary incontinence.

## Plants Storing Nitrates

For the last 2 years (2008-2009), Argentina has been suffering from widespread drought. Consequently, crops like sorghum and maize, used to feed animals, accumulate nitrates in their tissues, causing nitrate poisoning in ruminants. Cases have been identified from animals grazing and from the consumption of entire cobs of maize stored in silage. It is believed that silage processing lowers nitrate concentration in the plant, however, there are cases of animal deaths due to nitrate intoxication in animals ingesting silage. Many poisoning cases respond well to methylene blue treatment. Most of the pregnant animals that overcame acute poisoning had abortions.

## Systemic Toxics

*Festuca arundinacea* is a Graminae valuable for forage which is irreplaceable in several areas of Argentina. However, when parasitized by the endophyte fungus *Neothipodium cohenophialum* it causes different toxic manifestations depending on the time of year in which it is ingested; for example, hyperthermic syndrome is noted in summer and gangrene in winter (Odriozola *et al.* 2000).

*Vicia villosa* and *V. sativa* are Leguminosae generally found in pastures mixed with oats. These plants are widely used in the province of Buenos Aires where several mortalities have been registered. The poisoning occurs when periods of drought are followed by rain, which causes the oats to disappear and promotes the complete dominance of *Vicia* spp. The toxic principle remains unknown. The first clinical sign characteristic of this kind of poisoning is alopecic dermatitis (Odriozola *et al.* 1991b).

*Claviceps purpurea* is a fungus widely spread in our country. The toxins are ergot alkaloids and vary in their concentration depending on the host. It parasitizes cereal grains and cultivated and native Graminae pastures. Sources of poisoning can be either by direct consumption of contaminated pastures, or by consumption of contaminated grains or their by-products. Clinical signs are similar to those caused by *Festuca* (Khallou *et al.* 2009).

## Calcinogenic Plants

*Solanum glaucophyllum* (duraznillo blanco) is the toxic plant which causes the greatest economic loss in Argentina. A large area of the Salado River basin cannot be used for grazing livestock until after the plant loses its leaves (6 months each year). Cattle and sheep are affected producing signs known as 'enteque seco' (enzootic calcinosis). Affected animals suffer emaciation without diarrhea and have a tucked-up abdomen. Walking with short steps and kyphosis is observed. The pathology consists of calcium accumulation in soft tissues, mainly arteries, tendons, and lungs (Campero and Odriozola 1990).

## Plants Causing Sudden Death

*Taxus baccata* (tejo) is an ornamental plant commonly found in farms. The toxic principles are alkaloids called taxines A and B. All species, even man, are susceptible to the poisoning. If a large dose is ingested, the animal may die suddenly, next to the plant (Keeler *et al.* 1978).

## Estrogenic Plants

Lack of water and fungal infections increase the concentration of isoflavones, coumestans, formononetin–biochanin A, genistein in *Trifolium subterraneum* (subterranean clover) and alfalfa. The consumption of forages with these toxins is responsible for the clinical signs of hyperestrogenism, which are marked edema of the vulva and development of mammary glands. The former can quickly be resolved by the removal of the animals from that pasture (Lopez and Odriozola 1988).

## Teratogenic Plants

*Conium maculatum* is a weed belonging to the Umbelliferae family and is widespread. It is biannual and voluntarily ingested by animals. The toxic principles are piperidine alkaloids: coniine, N-methyl coniine, conhydrine, gamma-coniceine, and pseudo-conhydrine. In our country, it is common to use bovines to control weeds in hills and near mills where this weed frequently grows. In bovines ingestion during days 45 to 75 of gestation causes arthrogryposis, cleft palate, and kyphosis in calves (Lopez and Odriozola 1988).

## References

Blanco Viera FJ, Salvat A, Godoy H, Antonacci L, Rivera G, Jagle L, and Carrillo B (2000). Intoxicación por *Condalia megacarpa*, (piquillin). Descripción de un caso natural y reproducción experimental. *Comunicación preliminar, Reunión de la Asociación de Veterinarios de Diagnóstico*. Merlo, Sal Luis, pp. 18-20

Campero C and Odriozola E (1990). A case of *Solanum malacoxylon* toxicity in pigs. *Veterinary and Human Toxicology* 32:238-239.

Campero C, Odriozola E, and Casaro AP (1993). Mortandad en bovinos de cría por ingestión de abrojo grande (*Xanthium cavanillesii*). *Veterinaria Argentina* 99:591-596.

Collin RG, Odriozola E, and Towers NR (1998). Sporidesmin production by *Pithomyces chartarum* isolates from New Zealand, Australia and South America. *Mycology Research* 102:163-166.

Del Potro D, Odriozola E, Odeon A, and Larralde S (1984). Intoxicación de ovinos con Falaris. *Veterinaria Argentina* 1:763-766.

Keeler, Richard F., Kent R. Van Kampen, and Lynn F. James, eds (1978). *Effects of Poisonous Plants on Livestock*. Academic Press, New York.

Khallou P, Diab S, Licoff N, Bengolea A, Lazaro L, Canton G, and Odriozola E (2009). Efecto del consumo de *Claviceps purpurea* em novillos em engorde. *Revista de Medicina Veterinaria* 88:78-82.

Lopez TA and Odriozola E (1987). Grado de riesgo de fotosensibilización en pastoreo de rastrojos de trigo con falsa viznaga (*Ammi majus*). *Revista de Medicina Veterinaria* 68:98-101.

Lopez T.A. and Odriozola E (1988). Riesgos de toxicidad asociados con la utilización de tréboles en las pasturas. *Boletín Informativo y de Extensión. E.E.A. Balcarce* 88:1-6.

Lopez TA, Odriozola E, and Mutti G (1985). Intoxicación de bovinos con *Paspalum Dilatatum* poir (pasto miel) contaminado con *Claviceps paspali* Stivens et Hall. *Veterinaria Argentina* 3:863-870.

Lopez T, Odriozola E, and Eyherabide J (1991). *Toxicidad vegetal para el ganado. Patología, prevención y control*, 58 pp. De Defalco impresores S.A, Mar del Plata.

Marin R, Lloberas M, Vignale D, and Odriozola E (2004). Toxicidad natural del *Pteridium aquilinum* (helecho) en bovinos y su importancia en humanos. *Veterinaria Argentina* 21:413-420.

Marin R, Lloberas M, Vignale D, and Odriozola E (2005). Intoxicación natural y experimental de bovinos por consumo de *Lantana camara. Veterinaria Argentina* 22:215-218.

Martìn A, Yamarela F, Maurel R, and Roager J (1971). Intoxicación en equinos con abrepuño. *Proyección Rural, Buenos Aires* 40:1.

Odeon A, Steffan P, Salamanco A, and Odriozola E (1983). Fotosensibilización hepatógena del ganado bovino. *Revista de Medicina Veterinaria* 64:110.

Odriozola E (1984). Fotosensibilización y queratoconjuntivitis en rumiantes por consumo de semillas de falsa viznaga (*Ammi majus*). *Veterinaria Argentina* 1:684-688.

Odriozola E, Ballabene N, and Salamanco A (1984). Intoxicación en ganado bovino por algas verdes-azuladas. *Revista Argentina de Microbiología* 16:219-224.

Odriozola E, Tapia O, Lopez TA, Casaro A, and Calandra W (1987). Intoxicación natural de bovinos con transparente (*Myoporum laetum*). *Revista de Medicina Veterinaria* 68:230-232.

Odriozola E, Lopez TA, Daguere S, Cacace P, and Viejo R (1990). Nefropatía tóxica natural en bovinos por consumo de roble (*Quercus* spp.). *Veterinaria Argentina* 6:608-611.

Odriozola E, Lopez T, Campero CM, and Gimenez Placeres C (1991a). Neuropathological effects and deaths of cattle and sheep in Argentina from *Phalaris angusta*. *Veterinary and Human Toxicology* 33:465-467.

Odriozola E, Campero C, Lopez TA, Andrada M, and Casaro G. (1991b). An outbreak of *Vicia villosa* (hairyvetch) poisoning in grazing Aberdeen Angus bulls in Argentina. *Veterinary and Human Toxicology* 33:278-280.

Odriozola E, Lopez T, Campero, CM, and Gimenez Placeres C (1993). Ryegrass staggers in heifers: a new mycotoxicosis in Argentina. *Veterinary and Human Toxicology* 35:144-146.

Odriozola E, Iraguen Pagate I, Lloberas M, Cosentino I, and Porley J. (2000). Festuca tóxica su efecto en diferentes razas bovinas. *Veterinaria Argentina* 18:12-21.

Odriozola E, Bretschneider G, Pagalday M, Odriozola H, Quiroz J, and Ferreria J (2001). Intoxicación natural con *Cynodon dactylon* (pata de perdiz) en un rodeo de cría. *Veterinaria Argentina* 19:579-583.

Odriozola E, Odeón A, Canton G, Clemente G, and Escande A (2005). *Diplodia Maydis:* a cause of death of cattle in Argentina. *New Zealand Veterinary Journal* 53:161-163.

Peterson JE (1984). The toxicity of *Echium plantagineum* (Paterson's Curse), pp.191-199. *Plant Toxicology* (AA Seawright, MP Hegarty, LF James, and RF Keeler, eds), Brisbane, Australia.

Riet-Correa F (1993). Intoxicação por *Diplodia maydis* (Diplodiose). In *Intoxicações por plantas e micotoxicoses em animais domésticos* (F Riet-Correa, MC Méndez, and AL Schild, eds), pp. 142-145. Editorial Hemisfério Sul do Brasil, Pelotas, RS.

Tokarnia C, Dobereiner J, and Peixoto P (2000). *Plantas Tóxicas do Brasil*, 310 pp. Editora Helianthus, Rio de Janeiro.

# Chapter 4

# Toxic Plants of Cuba

E. Marrero, C.B. Goicochea, L.M.S. Perera, and I.P. Páez

*Centro Nacional de Sanidad Agropecuaria, Apdo.# 10, San José de Las Lajas, La Habana, Cuba*

## Introduction

A great spectrum of plants, particularly in the tropics, offer great chemical diversity, which not only represents a source of new therapeutic molecules but also compounds that can produce severe intoxication in animals with potential hazard to humans as well. In the series *Flora de Cuba*, 6500 plant species were reported as endemic with more than 50% of those as vascular plants. During the 1970s and 1980s we observed an increase in clinical plant toxicoses in farm animals which occurred in parallel with the intensification of commercial cattle production. This hazardous situation has been appreciably ameliorated in the last two decades due to better knowledge of the factors contributing to the toxic accidents, proper diagnosis, and control of the clinical process by the farmers and technicians from improved animal management. Multidisciplinary toxicological research studies combined with scientifically documented information were important tools that contributed to train cattle producers in preventing intoxications. In Cuba, 388 plant species were previously reported as toxic, 28 of these endemic, and grouped into 260 genera from 98 families (Roig Mesa 1974; Alfonso *et al.* 1998; Marrero *et al.* 2006). In this context, experimental and clinical multidisciplinary research has been conducted at the Centro Nacional de Sanidad Agropecuaria (CENSA) in the last three decades to help farmers with toxic plant problems. Relevant cases of intoxication which affect animal production, with either acute or chronic diseases, are caused by plants producing the following primary health effects: heart and circulatory damage: *Urechites lutea* (L.) Britton (Apocynaceae), *Nerium oleander* L. (Apocynaceae), *Melanthera deltoidea* L. C. Rich *ex* Michx. (Asteraceae); hepatotoxicity/photodermatitis: *Crotalaria* spp. such as *Crotalaria retusa* L. (Fabaceae), *Lantana camara* L. (Verbenaceae); *Ageratum houstonianum* Mill. (Asteraceae); cell respiratory uncoupling: *Cynodon nlemfuensis* Vanderyst (Poaceae); *Manihot esculenta* Crantz. (Euphorbiaceae), *Achyranthes aspera* L. var. *indica* (Amaranthaceae); *Amaranthus viridis* L. (Amaranthaceae), among other plants compromising the health of animals. The third edition of the book *Toxic Plants in the Tropics* (Marrero *et al.* 2008) summarizes most of the studies of plant toxicosis observed in animals, and some in humans, under these climatic and geographic conditions. The complex process of the animal intoxication from feeding on undesirable plants depends on many factors. They involve the animal species, the kind of husbandry, the type of soil, the original flora, environmental factors, and others contributing to condition the growth of the

©CAB International 2011. *Poisoning by Plants, Mycotoxins, and Related Toxins*
(eds F. Riet-Correa, J. Pfister, A.L. Schild, and T.L. Wierenga)

undesirable plants and influence the course of the clinical process. By recognizing the risk factors involved, it may be possible to reduce or avoid the occurrence of plant-caused intoxications.

## Plants Affecting the Cardio-Circulatory System

*Urechites lutea* (Apocynaceae) contains steroidal digitalis-like glycosides. Cattle are affected by spontaneous intoxication, whereas guinea pigs are very susceptible to experimental intoxication (Marrero *et al.* 2004). The myocardium is affected at an early stage of the clinical course of the disease. *U. lutea* was responsible for two different clinical forms in cattle, acute or chronic, depending on the type of animal management. An acute toxicity appeared when cattle had consumed *U. lutea* mixed with milled forages (total mixed rations, TMR). It occurred under intensive production systems, where animals were fed TMR and not allowed to select their diets. The acute poisoning was clinically characterized by diarrhea, initially catarrhal with quick evolution to bloody, anorexia, general depression, dyspnea, heart arrhythmia, and irregular arterial pulses with bradycardia. Most animals died after a few days eating the toxic feed, but some recovered from the intoxication by eliminating the toxic forage and being provided palliative care such as rehydration and feeding hay. Chronic intoxication in cattle occurs under extensive pasture grazing. The animals are apparently healthy and do not present any prodromal signs, but when they are stressed, for example treated with tick baths or another physical intervention, they may fall down drastically and then die some minutes later (Marrero 1996).

The effects of the glycoside on heart activity were investigated by electrocardiography monitoring of six crossbreed Holstein calves of 6 months of age that received 0.50-0.30 mg of total glycosides per kg body weight intravenously. All the animals died following severe heart electrical conduction disturbances that ended with ventricular fibrillation (Marrero *et al.* 1984). Necropsies of experimental and clinical cases showed degeneration and necrosis of myocardium cells, enlarged lymph nodes, severe nephritis, some degree of liver degeneration, and hemorrhagic enteritis with ulcers in the duodenum; all more evident in the acute intoxication. When 200 ml *U. lutea* total aqueous extracts were administered orally to eight crossbreed Holstein calves and repeated for 7 days, necropsy showed: lung congestion and edema; hemorrhages in the epicardium, myocardium, and endocardium; myocardial necrosis; congestion, and dissociation of the hepatic cords; lymphoid follicular hyperplasia; glomerular nephritis; intestinal congestion and hemorrhage; encephalic congestion and edema with perivasculitis and glial reaction (Joa *et al.* 1985). In general, the findings were similar to those observed in natural intoxications; duodenal ulcerations were also interesting lesions observed both in natural or experimental cattle poisoning by *U. lutea* (Marrero *et al.* 2008). The presence of cardiac glycosides in the meat of intoxicated animals was determined by TLC (thin-layer chromatography) (Sánchez *et al.* 1990).

*Nerium oleander* (Apocynaceae), originally from the Mediterranean region, is a very common ornamental plant in Cuba, with a variety of pink and white flowers. It also contains a complex mix of cardiac glycosides chemically related to those of digitalis, with oleandrine as one of their powerful toxins. However, the plant does not represent a relevant toxic hazard for animals. Cattle and horses have become intoxicated when this ornamental plant has been cut by gardeners and the leaves accidentally consumed by animals. Severe intoxication by *N. oleander* was observed in geese (Alfonso *et al.* 1994).

*Melanthera deltoidea* (Asteraceae), popularly known as silver button in Cuba, is widely distributed over the entire island. It is also present in Florida, Bahamas, Jamaica, Yucatan, and in other countries of the American continent. The *Melanthera* genus has two other species in Cuba, *M. hastata* and *M. angustifolia*. A preliminary phytochemical study of the leaves showed alkaloids and another toxic substance identified as methyl-para-coumarate. In general all animal species consuming high doses of the *M. deltoidea* alkaloid were susceptible to being intoxicated. Naturally intoxicated cattle developed weakness, muscular tremors, mydriasis, discrete tympanic abdomen, dyspnea, convulsions, and eventually died. Necropsies showed generalized vascular congestion and urinary bladder repletion. Biochemical studies of blood showed abnormal disturbances of transaminases, bilirubin, cholesterol, and total lipids. Hemoconcentration, leucocytosis with granulocytic left deviation, and yellow colored plasma were also observed. In general, the anamnesis, clinical signs, and characteristic lesions in addition to the presence of the toxic plants in the area were considered for the diagnosis of plants affecting the cardiovascular system. In some cases, the presence of the toxic compound was confirmed in animal tissues.

There were no specific treatments for these cardiovascular toxicities; some of the intoxicated animals recovered with rest and symptomatic treatment.

## Plants Causing Hepatotoxicity/Photodermatitis

*Crotalaria* spp. such as *C. retusa* and *C. incana* (Fabaceae) are well distributed on the island and cause cattle intoxication. Others that could be of toxicological interest are *C. mucronata*, *C. sagittalis*, *C. incana*, *C. spectabilis*, *C. verrucosa*, *C. pilosa*, *C. pumila*, C. *vitellina, and C. tuerkheimil*. Most of these have been used by farmers in association with other plants as natural soil fertilizers and as protein supplements for animal feeding. Nevertheless, *Crotalaria* spp. represent a hazard to cattle production. It is well known that they contain pyrrolizidine alkaloids (PA), with monocrotaline as the major alkaloid that is bio-transformed into very reactive pyrrolic compounds, which are very hepato- and pneumo-toxic, and/or nephrotoxic as well. Inhibition of mitosis is a process found in the intoxication (megalocytosis). The quantity and rate of pyrrolizidine alkaloids converted into pyrrolic compounds were among the factors that influenced clinical progress (Seawright 2000). The presence of saponins was found in a primary phytochemical study; the plant stems also contained alkaloids, steroids, nitrates, tannins, and triterpenes. Cattle with acute intoxication showed anorexia, weakness, ocular and nasal discharges, bloody feces and jaundice, with death occurring approximately on the seventh day of the process. In cattle with chronic toxicity, death occurred some months after *Crotalaria* spp. consumption. Clinical signs are apparent a few days before death. They are characterized by bloody feces, dry hair, sunken eyes, diarrhea, jaundice, and progressive weakness followed by prostration. However, photosensitization was the relevant sign observed by farmers in many cases of suspected cattle intoxication. Necropsies of acutely intoxicated animals showed hepatic necrosis and hemorrhages, whereas the chronic events showed fibrosis, degeneration, and hepatic megalocytosis. The lungs showed edema, fibrosis, proliferation of type II pneumocytes, and emphysema. Nephritis with megalocytosis in the kidneys and ulcerations in intestines were also observed (Figueredo *et al.* 2001).

*Ageratum houstonianum* is a weed that grows in some humid locations and on river banks. A primary phytochemical study of an alcoholic leaf extract showed the presence of coumarin and triterpene compounds. Triterpenes are hepatotoxic and often produce photodermatitis lesions in intoxicated cattle (Sánchez *et al.* 1993). Coumarin is transformed into dicoumarol in the plant by different environmental conditions (weather or fungal contaminations). Dicoumarol is known to produce blood anticoagulant activity interfering with liver prothrombin production. Intoxication leads to hemorrhagic lesions throughout the animal, but may be mainly localized in the subcutaneous tissue and organ cavities at the beginning of the process. The phytochemical study of an organic extract of *A. houstonianum* leaves showed $C_{27}H_{56}$, $C_{29}H_{60}$, $C_{31}H_{64}$, phytoesterols (β-sitosterol, stigmasterol, and campesterol), 16C and 18C fatty acids, and fatty acid ethylic esters of 20C and 34C (Aparicio 2000). Essential oils such as precocene I, precocene II, and β-caryophyllene were also isolated from this species. *A. houstonianum* induced clinical signs such as severe gait deformities caused by hemorrhages on the bony prominences of the extremities. Bloody diarrhea was also observed. Some animals died without showing clinical signs. Cattle intoxication by *A. houstonianum* also starts with general depression, decreased milk production, enteritis, increased cardiac and respiratory frequency with a fatal course to hypothermia, prostration, and hemorrhagic flow from all natural orifices. Photodynamic dermatitis appeared in a second phase of the intoxication with a more chronic nature (Alfonso *et al.* 1989).

*Lantana camara* (Verbenaceae) is a common bush in hills, woods, gardens, and, in general, on calcareous soils. More frequent in northern areas of Cuba it is, however, well distributed throughout the country. There are different *L. camara* varieties: L. *camara.* var, *aculeata*; *L. camara* var. *mutabilis*; *L. camara* var. *involucrata*; *L, camara* var. *trifolia*; *L. camara* var. *crocea*; and *L. camara* var. *reticulata. L. camara* is known to contain polycyclic hepatotoxic triterpenes, lantadene A and B, which are responsible for liver injury. Cattle intoxicated by *L. camara* showed enough phylloerythrin concentration in the skin (from the breakdown of chlorophyll in the rumen) 5 to 10 days later to cause photosensitivity after exposure to intense sunlight. The cattle intoxication is similar to that reported in other countries. Approximately 40 g of fresh leaves or 10 g/kg daily for 5 days were sufficient to elicit photosensitization lesions. Other relevant clinical signs observed in cattle were anorexia and weight loss, ruminal stasis, restlessness, salivation, eye discharge, jaundice, and brown urine. Corneal inflammation with opacity was occasionally observed. Necropsy showed ochre colored livers, distended and edematous gall bladder, jaundice, and subcutaneous yellow edema (Alfonso *et al.* 1998).

## Toxicoses Caused by Cell Respiratory Disruption

*Cynodon nlemfuensis* (Poaceae) pastures are well adapted to tropical regions; the plant was introduced into Cuba in the 1980s while *Manihot esculenta* (Euphorbiaceae) was introduced from Africa during the Spanish colonization period. *Cynodon nlenfuensis* has been responsible for episodes of acute intoxication in cattle during bad weather periods, for example tropical hurricanes (Aguilera 1986). Moreover, *C. nlenfuensis* also caused acute intoxication in a herd of 105 buffaloes in an extensive grazing area of the western region of the country with a prevalence of 26.6% morbidity and 22.8% mortality. Anamnesis indicated that the animals were confined during a hurricane period of about 96 h, with restricted access to feed and water. The toxicological study revealed cyanide concentrations up to 56.16 mg/kg in the green pasture. The environment and herd management were predisposing factors for high intake of the toxic pasture in a very short time after the animals were released from confinement. The animals that survived showed dyspnea with respiratory frequency over 40/min, decreased response to general stimulus, weakness, disorientation, prostration, rectal tenesmus, and hyperemia in mucosa. Blood samples showed that leukocytes increased moderately with predominance of neutrophils. Average body temperature was between 38.3°C and 39.2°C. The necropsy results were consistent with those observed in bovine cyanogenic intoxication (Marrero *et al.* 1996).

*Manihot esculenta* (Euphorbiaceae) is a cause of death in different animal species with relative frequency, but due to the mainly individual occurrence of clinical cases, its economic impact is less important when compared *C. nlemfuensis*. Some farmers have reported intoxications in pigs from ingestion of fresh plants.

*Achyranthes aspera* var. *indica* (Amaranthaceae), *A. fructicosa pubescens*, *Centrostachys aspera*, and *Stachyarpagophora aspera* are perennial herbaceous plants that are considered weeds in Cuba. Clinical intoxication in cattle and other species of economic interest have been reported. Clinical signs were those corresponding to intoxication by nitrates/nitrites. Cattle showed respiratory disorders, nervous disturbances, incoordination, muscular tremors, and cyanotic mucosa. Profuse salivation and nasal secretions were also seen with muscular spasms mainly of the intercostal muscles and death. A primary phytochemical study showed differences in the chemical composition of secondary metabolites, such as nitrates, tannins, steroids, triterpenes, and alkaloids (Sánchez *et al.* 1995). Hydrocarbons of high molecular weight and aliphatic chain carboxylic acid were also identified. Further research is needed to study the doses and dynamics of the intoxication by *A. aspera* in different species. Necropsies showed dark chocolate-like color in blood and tissues, and internal organs presented congestion and hemorrhages, with petechiae on mucosal surfaces.

*Amaranthus viridis* (Amaranthaceae) compromises animals' health. *A. viridis*, *A. emarginatus*, *A. gracilis*, and *Euxolus viridus* are very common weeds on the island. Other species of the *Amaranthus* genus have also caused intoxication in animals: *A. hybridus*; *A. hypochondriacus*; *A. palmeri*; *A. paniculatum*; *A. retroflexus*; and *A. spinosus*. The *Amaranthus* genus accumulates nitrates that are converted into nitrites in ruminants by the microbial flora, thus, nitrites are quickly absorbed from the gastrointestinal system to the blood stream. It is also possible that a small amount of nitrate passes to the blood and then is reduced to nitrite by tissues. Clinical signs are the same as those above described for *A. aspera.* Nevertheless, the *Amaranthus* genus has been widely used as forage in Cuba and other countries because of its high level of protein. *A. graecizans* has 20% crude protein in dry material. It is not frequently reported to cause toxic problems.

The third edition of the book *Toxic Plants in the Tropics* (Marrero *et al.* 2008) summarizes most of the studies on plant toxicity observed in animals and some in humans in these wet and humid weather conditions. The complex process of animal intoxication by ingesting toxic plants depends on many factors involving the animal species, husbandry, characteristics of the soil, the environment, the original flora, and others that could contribute to the establishment and growth of weeds and undesirable plants (Marrero 2007). By recognizing the risk factors involved, it will be possible to reduce or avoid the occurrence of plant poisoning episodes.

## Acknowledgements

The participation of Dr Evangelina Marrero to the 8th International Symposium on Poisonous Plants was financially supported by Conselho Nacional de Desenvolvimento Científico e Tecnológico (CNPq), grant 454084/2008-0, and by Coordenação de Aprefeiçoamento de Pessoal de Nível Superior (CAPES), grant 0017/09-4.

Thanks are also due to the Ministry of Higher Education and the Institute of Veterinary Medicine of the Ministry of Agriculture, Republic of Cuba, for their contribution to the educational programs and the financial support for research on toxic plants. The authors would like to thank to Dr Eduardo Sistach for the valuable English language corrections to the manuscript.

## References

Aguilera JM (1986). Comportamiento del potencial cianogénico en pasto estrella (*Cynodon nlenfuensis*). VII (Final): Intoxicación experimental aguda por extractos acuosos y forraje con alto contenido de cianuro. *Revista de Salud Animal* 8:251-254.

Alfonso HA, Rivera M, Aparicio M, Ancisar J, Marrero E, and Cabrera JM (1989). Intoxicación natural con *Ageratum houstonianum* Mill (celestina azul). *Revista Cubana de Ciencias Veterinarias* 20:113-120.

Alfonso HA, Luz Ma, Sánchez M, Merino BC, and Gómez (1994). Intoxication due to *Nerium oleander* in geese. *Veterinary and Human Toxicology* 36(1):47.

Alfonso HA, Marrero E, Fuentes V, and Sánchez LM (1998). *Toxic Plants of the Tropics*, 140 pp. Ed. CENSA, Havana, Cuba.

Aparicio JM (2000). *Ageratum houstonianum* Mill. (*celestina azul*) toxicosis in ruminants. PhD Thesis, Havana Agriculture University, Havana, Cuba.

Figueredo MA, Sánchez LM, Marrero E, and Rodríguez JG (2001). Residualidad de monocrotalina en músculo de ternero tratados con dosis única *Crotalaria retusa* (L.) desecada. *Revista de Salud Animal* 23:23-26.

Joa R, Merino N, Marrero E, Bulnes C, and González A (1985). Estudio anatomopatológico en bovinos intoxicados experimentalmente con glicósidos aislados de *Urechites lutea* (L) Britton. *Revista Cubana de Ciencias Veterinarias* 16:41-52.

Marrero E (1996). Poisoning by *Urechites lutea* (L) Britton in cattle. *Veterinary and Human Toxicology* 38:313-314.

Marrero E, Fernández O, Pompa A, Hernández L, and Fajardo H (1984) Electrocardiographic variations in experimental poisoning with *Urechites lutea* L. Britton glycosides in calves. *Revista Cubana de Ciencias Veterinarias* 15:179-189.

Marrero E, Aparicio M, Figueredo MA, Bulnes C, Sánchez LM, Palenzuela I, and Durand R (2004). More frequent natural and experimental plant intoxication in animals reported in Cuba. In *Toxic plants and other Natural toxicants* (T Garland and C Barr, eds), pp. 335-340. CABI Publishing UK.

Marrero E, Alfonso HA, Fuentes V, Sánchez LM, and Palenzuela I (2006). *Plantas tóxicas en el Trópico,* 226 pp., 2nd edn. Edi CENSA/Univ. Sta Catarina, Brasil.

Marrero E, Alfonso HA, Fuentes V, Sánchez LM, Palenzuela I, and Tablada R (2008). *Plantas tóxicas en el Trópico*, 229 pp., 3rd edn. Edi CENSA, Havana.

Roig y Mesa JT (1974). *Plantas Medicinales, Aromáticas o Tóxicas de Cuba.* 2nd edn. Ciencia y Técnica, Instituto del Libro, Cuba.

Sánchez Luz M, Noa M, and Alfonso HA (1990) Determinación de glicósidos cardiotónicos de *U. lutea* (L) en productos cárnicos por cromatografía de capa delgada. *Revista de Salud Animal* 12(1-3):85-87.

Sánchez LM, Alfonso HA, and Palenzuela Iris (1993) Principales compuestos tóxicos presentes en plantas contaminantes de áreas forrajeras y de pastos. *Revista de Salud Animal* 13(3):256-261.

Sánchez Perere LM, Alfonso HA, Noa M, M. Figueredo MA, and Gómez BC (1995). Intoxication due to *Achyrantes aspera* L. *Veterinary and Human Toxicology* 37(6):582.
Seawright A (2000). Pathogenesis of hepatogenous photosensitization. Conference. In *Proceedings of the 1st International Workshop on Toxic Plants for Human and Animals*. (NRCT Australia, ed.). National Centre for Animal and Plant Health, Cuba.

# Chapter 5

# Toxic Plants Affecting Grazing Cattle in Colombia

G.J. Diaz[1] and H.J. Boermans[2]

*[1]Laboratorio de Toxicología, Facultad de Medicina Veterinaria y de Zootecnia, Universidad Nacional de Colombia, Bogotá, Colombia; [2]Department of Biomedical Sciences, Ontario Veterinary College, University of Guelph, Guelph, Ontario, Canada*

## Introduction

Plant biodiversity in Colombia is very high, comprising about 25,000 species of vascular plants, both native and naturalized. This biodiversity corresponds to about 8% of the total vascular plants on earth, which makes the country the second largest in plant biodiversity in the world. The impact of toxic plants on Colombian livestock has not been fully evaluated, although it is estimated that more than 40 million ha of the country are used for livestock production of which 26 million ha are for bovines. Colombia is counted amongst the ten largest countries in the world in terms of cattle production. However, livestock production is mostly extensive with a very low population of animals kept under intensive production systems. Conservative estimates indicate that in Colombia, toxic plants cause an annual mortality rate of about 0.5% (Peña *et al.* 1980) which is equivalent to about 130,000 cattle/year. The present review describes a selection of the most important native and introduced toxic plants that affect grazing cattle in Colombia.

The selected plants are grouped based on the major organ system affected by the consumption of the plant. Common names given to these plants in Colombia are provided in brackets after the Latin name or indicated in one of the tables. It is important to note that Colombia is a tropical country located on the equator, and that there are no seasons such as winter, spring, summer, or fall. The average environmental temperature is mostly determined by the altitude, with lowlands being hotter and highlands colder. There are 'dry' and 'rainy' seasons, when low or high precipitation is expected every year (the highest precipitation occurs during March-April and October-November). Some plants tend to accumulate more toxins in one of the seasons compared to the other or may accumulate one type of compound in the dry season and another type during the rainy season.

## Plants That Affect the Digestive System

*Ricinus communis* (castor, higuerilla, palmacristi, ricino) is a naturalized plant common in Colombia that grows from sea level to 2600 m. *R. communis* seeds contain

©CAB International 2011. *Poisoning by Plants, Mycotoxins, and Related Toxins* (eds F. Riet-Correa, J. Pfister, A.L. Schild, and T.L. Wierenga)

ricin, one of the most potent lectins known. All animal species are sensitive to the effects of ricin. The toxicosis, however, is uncommon and is usually associated with feeding garden clippings or with contamination of forage grasses with *R. communis* trimmings. Clinical signs include weakness, salivation, profuse aqueous diarrhea, dehydration, mydriasis, teeth grinding, hypothermia, and recumbence; severe gastroenteritis is the major postmortem finding (Aslani *et al.* 2007). Other plants that accumulate lectins and may eventually affect cattle are *Jatropha curcas* (piñón de Fraile, purga de Fraile), *Abrus precatorius* (chochos de pinta negra, jetiriquí), and *Canavalia ensiformis* (canavalia, fríjol blanco). The lectins contained in these plants are known as curcin, abrin, and concanavalin A, respectively.

## Plants That Affect the Blood

### Plants causing hemolytic anemia

Feeding culled onions has been associated with hemolytic anemia in cattle and other animal species. *Allium cepa*, which includes all types of onions, is capable of causing toxicosis in both large and small animals due to its content of organic sulfoxides, especially alkyl or alkenyl cysteinyl sulfoxides (Parton 2000). Onion toxicosis, which occurs sporadically in cattle, has been extensively documented in the literature with the first case reported in 1909. The toxicosis occurs because cattle readily consume culled onions and usually prefer them to high quality forages or grains.

### Plants causing methemoglobinemia

The nitrite ion, which is formed by bacteria in the rumen from plant nitrate, is the major cause of methemoglobinemia in ruminants. Methemoglobin is an abnormal form of hemoglobin in which its normal ferrous moiety ($Fe^{2+}$) is oxidized to the abnormal ferric form ($Fe^{3+}$). The oxidized form is not capable of transporting oxygen and therefore a reduction in the oxygenating capacity of the blood occurs. The severity of the clinical signs and effects depends on the amount of methemoglobin formed. Signs of hypoxia develop when 20-30% of the hemoglobin is converted to methemogloblin and death usually occurs at 70-80% methemoglobin levels (Vermunt and Visser 1987). Many plants have been identified as accumulators of toxic nitrate levels in Colombia as this is a common plant-caused toxicosis in cattle. As shown in Table 1, the most important group of plants capable of accumulating high nitrate levels are forage grasses, with at least nine species associated with nitrate poisoning. An example of these is *Panicum maximum*. Samples of this grass collected from the north part of the country (Departments of Córdoba and Sucre) had nitrate levels of 1209 and 5260 ppm for fresh plants collected during the dry season and after the onset of the rainy season, respectively (Trheebilcock *et al.* 1978). Also in Colombia, Guzmán *et al.* (1978) reported a case in the Valle del Cauca Department that caused acute mortality in 19 out of 64 steers that were fed cut *Pennisetum purpureum*. The grass was found to contain 445 ppm nitrate and 971 ppm nitrite; the unusually high nitrite content was attributed to oxidative microbial processes.

The Amaranthaceae family also contains plants associated with nitrate poisoning in cattle. *Amaranthus dubius* and *Amaranthus spinosus* are two species of *Amaranthus* common in Colombia, which have been associated with nitrate intoxication, especially during the transitions between the dry and the rainy seasons (Torres 1984). Another Amaranthaceae is *Chenopodium album*, a plant recently reported in Colombia (Fernández-

Alonso and Hernández-Schmidt 2007), which can cause lethal intoxication in ruminants because of its high nitrate levels (although it also accumulates soluble oxalates). Levels of 2500 ppm nitrate-nitrogen were reported in *Chenopodium album* hay associated with mortality in cattle (Ozmen *et al.* 2003). *Bonafousia sananho*, also known as *Tabernaemontana sananho*, is a plant of the Apocynaceae family common in Colombia that has been shown to accumulate up to 2630 ppm nitrate. This plant also accumulates cyanogenic glycosides and is considered to be one of the most important poisonous plants for cattle in the Arauca River valley (Vargas *et al.* 1998). Another plant associated with high nitrate content is *Mascagnia concinna*, a vine of the Malpighiaceae family native to the Magdalena valley of Colombia. Nitrate concentrations ranging from 5300 to 29,200 ppm dry matter were reported by Torres (1984) and from 1555 to 10,763 in fresh material by Trheebilcock *et al.* (1978). The Phytolaccaceae *Petiveria alliacea* can also accumulate toxic levels of nitrate. Studies conducted in Colombia with fresh plants showed that during the dry season the plant accumulated an average of 1155 ppm nitrate, but during the rainy season the average level was 7867 ppm (Trheebilcock *et al.* 1978). *Heliotropium indicum* is a Boraginaceae also known to accumulate toxic concentrations of nitrates. In samples collected in the north part of the country, Trheebilcock *et al.* (1978) found average nitrate levels of 178 and 7195 ppm in fresh material collected before the rainy season and immediately after the start of the rainy season, respectively.

**Table 1.** Major nitrate-accumulating plants affecting livestock in Colombia.

| Family | Latin name | Common name |
|---|---|---|
| Amaranthaceae | *Amaranthus dubius* | Adormidera, bledo liso |
| | *Amaranthus hybridus* | Amaranto, bledo chico |
| | *Chenopodium album* | Quenopodio |
| Apocynaceae | *Bonafousia sananho* | Guachamacá, lirio blanco |
| Boraginaceae | *Heliotropium indicum* | Verbena, rabo de alacrán |
| Malpighiaceae | *Mascagnia concinna* | Mindaca, mataganado |
| Poaceae | *Andropogon bicornis* | Barba de indio, cola de zorro |
| | *Brachiaria mutica* | Pasto pará |
| | *Lolium perenne* | Balico, raigrás inglés, raigrás perenne |
| | *Panicum maximum* | Pasto guinea, siempreverde |
| | *Paspalum paniculatum* | Paja brava, paja del camino |
| | *Paspalum virgatum* | Gramalote, yerba peluda |
| | *Penisetum purpureum* | Pasto elefante |
| | *Sorghum bicolor* | Sorgo, sorgo forrajero |
| | *Sorghum halepense* | Pasto Johnson, capim argentino |
| Phytolaccaceae | *Petiveria alliacea* | Anamú |
| Solanaceae | *Solanum nigrum* | Campano, yerbamora |

## Cardiotoxic Plants

Cardiac glycosides are a special type of toxic glycosides that affect the cardiac muscle, sometimes causing fatal acute or subacute toxicosis. Cardiac glycosides increase the contraction force of the heart by inhibiting the myocardial Na-K ATP-ase, which can lead to cardiac arrest. At least four plants containing toxic levels of cardiac glycosides are present in Colombia: the Plantaginacea known as *Digitalis purpurea* (dedalera, digital, guargüeron), and the plants of the Apocynaceae family *Nerium oleander* (oleander, delfa, adelfa, azuceno de La Habana), *Thevetia peruviana* (catapis, oleander amarillo), and

*Asclepias curassavica* (bencenuco, mataganado). All these plants have sporadically caused toxicosis in herbivores.

## Hepatotoxic Plants

Hepatotoxic plants may affect the liver of animals either by causing hepatocellular necrosis or by inducing intrahepatic cholestasis. Pyrrolizidine alkaloids (PAs) are a large group of hepatotoxins characterized by the presence of a pyrrolizidine nucleus in their structure capable of causing hepatocellular necrosis. Compounds in plants known to cause cholestasis are the lantadenes from *Lantana* spp., sporidesmin (a mycotoxin formed by the fungus *Pithomyces chartarum* in some grasses), and the steroidal saponins present in several grasses. Hepatotoxins may cause secondary photosensitization in ruminants due to an alteration in the metabolism of chlorophyll which results in the abnormal accumulation of phylloerythrin, the pigment responsible for the photosensitive skin damage.

### Plants containing substances that cause hepatocellular necrosis

The PAs are a large group of hepatotoxins and PA toxicosis has been reported in livestock, poultry, pigs, and humans (Diaz 2001). More than 6000 plants are believed to have PAs, many of which are present in Colombia, in all kinds of ecosystems. The most important PA-producing plants from the toxicological standpoint belong to one of the families Asteraceae, Boraginaceae, or Fabaceae. Table 2 summarizes the major PA-containing plants present in Colombia.

**Table 2.** Major pyrrolizidine alkaloid-producing plants reported in Colombia.

| Family | Latin name | Common name |
|---|---|---|
| Asteraceae | *Eupatorium* spp. | Amarguero, chilico, hierba de chivo |
| | *Senecio formosus* | Árnica, árnica de páramo / de Bogotá |
| | *Senecio madagascariensis* | Manzanilla del llano |
| Boraginaceae | *Borago officinalis* | Borraja |
| | *Cynoglossum* spp. | Cinoglosa, lengua de perro |
| | *Heliotropium europeum, H. indicum* | Verbena, rabo de alacrán |
| | *Symphytum officinale* | Consuelda, consuelda mayor |
| Fabaceae | *Crotalaria* spp. | Crotalaria, cascabel, cascabelito |

Among the Asteraceae family the most important hepatotoxic genera are *Senecio* and *Eupatorium*. The toxic species *S. formosus* and *S. madagascariensis* are common in Colombia. The former is a plant native to the highlands that grows between 3000 and 4000 m above sea level, commonly found in the Colombian Andean regions of Cundinamarca, Cauca, and Nariño. There are no reports of toxicosis in animals caused by this plant; however, *S. formosus* has caused irreversible hepatic damage in human patients that ingested infusions made with its dry leaves. The clinical history, signs, lesions, and postmortem findings of almost 20 fatal cases reported in Bogotá were documented by Toro *et al.* (1997). *S. madagascariensis* is an annual or perennial herb native to South Africa reported for the first time in Colombia in the 1980s. It is an aggressive weed that propagates rapidly and has already colonized all the high plateau of the regions of Cundinamarca and Boyacá (Fernández-Alonso and Hernández-Schmidt 2007). In Colombia, *S. madagascariensis* has been associated with sudden death in cows immediately after parturition. The cause of this sudden death syndrome is unknown but it is possible that the metabolic changes associated with parturition and the onset of lactation pose an extra load to a liver that has been severely affected by the chronic ingestion of the plant. The other genus of the Asteraceae family reported to accumulate PA is *Eupatorium*. Several species of this genus have been reported in Colombia (Powell and King 1969) including *E. inulaefolium*, which has been reported as hepatotoxic for cattle in other countries. Another toxic *Eupatorium* species in Colombia is *E. stochaedifolium*, whose leaves and flowers were reported as toxic by Pérez-Arbeláez (1931).

Within the Fabaceae family, the genus *Crotalaria* is notorious for the high PA content of some of its plants. In Colombia, *Crotalaria* spp. grow from sea level to about 3000 m above sea level, especially in areas with clearly defined dry periods such as the inter-Andean valleys, the northern part of the country, and the eastern savannas known as the 'llanos'. These plants grow as weeds in well fertilized soils used to grow maize, sorghum, or soybeans and their seeds may contaminate these agricultural crops. At least 19 species of *Crotalaria* are present in Colombia, some of them recognized as toxic, including *C. spectabilis*, *C. retusa*, *C. sagittalis*, and *C. pallida* (Diaz *et al.* 2003). *Crotalaria* poisoning in Colombia has been reported in pigs, goats, laying hens, and broiler chickens. In 2001 large losses were caused to the poultry and pig industry when sorghum grain contaminated with *C. retusa* seeds was used to prepare mixed feeds for monogastric animals. The level of contamination in sorghum lots with *C. retusa* seeds ranged between 2 to 5% (GJ Diaz, unpublished data). The shrub *C. pallida* was reported to cause a natural outbreak of poisoning in goats in the Department of Santander (Canchila 2001) and experimentally *C. pallida* seeds were found to be highly toxic for broiler chickens (Diaz *et al.* 2003).

A third family of plants known to accumulate high levels of PA is Boraginaceae. Thirteen genera of the Boraginaceae family have been reported in Colombia, including the toxic genera *Heliotropium*, *Symphytum*, and *Cynoglossum* (Barajas-Meneses *et al.* 2005).

**Plants that cause intrahepatic cholestasis**

*Lantana camara* (venturosa, sanguinaria, lantana) is a tree of the Verbenaceae family native to tropical America. In Colombia it is common in all ecosystems, from 0 to 2500 m above sea level. The hepatotoxic action of *L. camara* has been attributed to two pentacyclic triterpenes known as lantadenes A and B. The primary toxic action of the lantadenes may result in secondary photosensitization due to the reduced excretion of phylloerythrin, a natural metabolite product of the anaerobic fermentation of chlorophyll, which is normally excreted in bile. *L. camara* toxicosis can affect cattle, sheep, goats, horses, and buffaloes. Apart from *L. camara* there are at least 14 species of *Lantana* present in Colombia, whose toxicology and potential adverse effects in animals have not been investigated.

Plants that contain steroidal saponins may also cause intrahepatic cholestasis in cattle. The toxic effect of the steroidal saponins is related to their normal metabolism in the ruminant, which leads to the accumulation of insoluble calcium salts of sapogenin glucuronate that precipitate inside and around the biliary ducts. These glucuronate crystals block the normal secretion of bile which in turn disrupts the normal secretion of phylloerythrin. Most of the plants that contain toxic levels of steroidal saponins in Colombia belong to the Poaceae family (grasses) and include *Brachiaria brizantha* (pasto alambre), *B. decumbens* (braquiaria), *Panicum coloratum* (pasto Klein), *P. maximum* (pasto guinea), and *Penisetum clandestinum* (kikuyo). Toxic effects have been reported but not confirmed. Alternatively, *B. brizantha* and *B. decumbens* can also induce secondary photosensitization in cattle, sheep, and goats due to hepatic damage from the hepatotoxic compound sporidesmin, a product of the fungus *Pithomyces chartarum*. This toxicosis has been observed sporadically in Colombia.

## Plants That Affect the Urinary System

Urinary bladder tumors in cattle have been associated with the intake of *Pteridium aquilinum* (helecho macho, helecho liso). This weed is distributed worldwide and grows in well-drained acid soils and open lands and is common in the eastern savannas of the country. In Colombia the toxicosis by *P. aquilinum* has been mainly associated with a disease in cattle known as 'bovine enzootic hematuria', which causes economic losses in some Departments where dairy cattle are raised (Pedraza *et al.* 1983).

High levels of soluble oxalates, which chemically correspond to sodium or potassium salts of oxalic acid (Diaz 2001), are a common cause of plant-induced nephrotoxicity. Soluble oxalates are readily absorbed in the systemic circulation where they can react with the blood calcium causing hypocalcemia and tetania. Oxalates may eventually form insoluble calcium oxalate crystals that block the renal tubules. Most of the soluble oxalate-accumulating plants of toxicological interest in Colombia belong to the Poaceae, Amaranthaceae, and Polygonaceae families.

Native or naturalized grasses known to accumulate potentially toxic levels of soluble oxalates include *Brachiaria humidicola* (braquiaria alambre), *Cenchrus ciliaris* (pasto buffel), *Digitaria decumbens* (pasto pangola), *Panicum maximum* (pasto guinea, india, siempreverde), *Penisetum clandestinum* (kikuyo), *Penisetum purpureum* (pasto elefante), and *Setaria sphacelata* (setaria, pasto miel). In horses, prolonged intake of tropical grasses containing soluble oxalates can lead to secondary hyperparathyroidism or osteodystrophia fibrosa (Cheeke 1995). From the Amaranthaceae family, the highly toxic plant *Halogeton glomeratus* has not been reported in Colombia, but there are about 20 *Amaranthus* species, including *A. retroflexus* and *A. hybridus*, two introduced invasive and toxic weeds. These two weeds contain both soluble oxalates and nitrates, although the toxicosis is generally associated with their oxalate contents. Acute renal failure and perirenal edema have been reported worldwide in cattle, sheep, pigs, and horses that ate these plants (Last *et al.* 2007). Another plant common in Colombia that accumulates potentially toxic levels of soluble oxalates is the Polygonaceae *Rumex crispus* (lengua de vaca, romaza).

## Plants That Affect the Nervous System

*Conium maculatum* (Umbelliferae) is native to Europe, naturalized in Colombia, and is commonly found along roadsides and close to irrigation ditches, usually between 1200 and 2800 m above sea level. *Conium maculatum* contains at least five main piperidine alkaloids, with the most important being coniine (mainly in the seeds) and γ-coniceine (in vegetative tissue). These compounds cause paralysis of the musculature due to the blockade of the neuromuscular junctions. The initial signs of the acute toxicosis include muscle weakness, tremors, incoordination, and mydriasis, followed by bradycardia, depression, coma, and death from respiratory failure. The closely related toxic plant of the same family known as 'water hemlock' (*Cicuta* spp.) has not been reported in Colombia.

*Ipomoea carnea* (Convolvulaceae), a native to tropical and subtropical America, grows spontaneously in the east part of Colombia and other warm parts of the country. It is used as an ornamental and can become a weed in pastures. This plant has been shown to affect the central nervous system of cattle (Tokarnia *et al.* 2002; Antoniassi *et al.* 2007), sheep, and goats in Brazil. The toxic compound of this plant was found to be the indolizidine alkaloid swainsonine, which inhibits lysosomal hydroxylases, causing a cellular alteration known as 'lysosomal storage disease'. Cattle exposed to the toxin fail to gain weight and exhibit neurological alterations including failure to prehend and swallow feed, hypermetria, and ataxia. *Ipomoea carnea* is considered to be one of the most important toxic plants for cattle in the Arauca River valley (Vargas *et al.* 1998). *Ipomoea* spp. can also accumulate ergot alkaloids.

## Plants That Affect the Musculoskeletal System and Connective Tissue

The genus *Senna* (Fabaceae) includes several species of plants known to induce myopathy in cattle, horses, and pigs. *Senna* toxicosis causes myocardial degeneration, congestive heart failure, and generalized degeneration of skeletal muscles. Among the toxic species in Colombia are *S. occidentalis* (café de brusca, cafelillo), *S. obtusifolia* (bicho, chilinchil), *S. reticulata* (bajagua, dorancé), *S. tora*, and *S. roemeriana* (Torres *et al.* 2003).

*Petiveria alliacea* (anamú, Phytolaccaceae) is an herb native to tropical America known in some places as 'garlic weed' because of its strong garlic odor. Reported only in Colombia, *P. alliacea* produces a unique subchronic toxicosis in young cattle known as 'dystrophic muscular emaciation'. The disease is observed mainly in calves 2-12 months old and has been reproduced by feeding 3 g of the plant daily during 30 days (Torres 1984). Experimental intoxication of cattle and sheep shows decreased serum cholinesterases, incoordination, severe flexion of the fetlock, and severe muscle atrophy. Also, the meat from these animals develops a strong garlic odor and is usually rejected by the consumer. The compound responsible for the toxic affects of *P. alliacea* has not been identified but could potentially be dibenzyltrisulfide. Two plants of the Phytolaccaceae family reported as toxic in Colombia are *Phytolacca icosandra* (altasara, yerba de culebra) and *P. bogotensis* (cargamanta, guaba, yerba de culebra). The roots, leaves, and fruits are toxic.

Two plants of the Malpighiaceae family native to Colombia have been associated with a disease of cattle and sheep characterized by the deposition of an abnormal pink or violet pigment in connective tissues (including bones and teeth): *Bunchosia pseudonitida* (mamey, tomatillo, pateperro, cuatrecasas) and *B. armeniaca* (mamey de tierra fría, manzano de monte). Mortality is usually low (<5%) but morbidity, represented by animals abnormally pigmented, can be higher than 90% (Peña 1982). This disease occurs in the Departments of Tolima and Huila, particularly during periods of drought, and it is known as 'bovine chromatosis'. Besides producing a discoloration of the connective tissues, *B. pseudonitida* has been associated with ataxia and incoordination, excretion of discolored urine, and hepatic effects characterized by increased activity of serum gamma glutamyl transpeptidase (GGT) and mild degenerative histologic lesions (Mejía 1984).

## Plants Containing Systemic Poisons

Systemic poisons interfere with biochemical processes common to all cells and usually do not have a particular target organ. Examples of these poisons are cyanide, an inhibitor of the respiratory chain in the mitochondria, and monofluoroacetic acid, a compound capable of blocking the Krebs cycle. The Rubiaceae *Palicourea margravii*, also known as *P. crocea* (café de monte, cafecillo, café bravo, flor de muerto), contains monofluoroacetic acid and is considered to be one of the main toxic plants for cattle in the Arauca River valley (Vargas *et al.* 1998). This plant typically causes sudden death, especially if the animals are forced to walk or run.

Plants containing cyanogenic glycosides that release cyanide upon ingestion are a common cause of acute toxicosis in cattle in Colombia (Table 3). Levels of 20 mg HCN/100 g of fresh plant or higher are considered to be potentially toxic. The plant of the Apocynaceae family *Bonafousia sananho*, mentioned earlier as a nitrate accumulator, is also a cyanogenic plant that has been found to contain up to 260 mg HCN/100 g (dry matter) during the dry season (Vargas *et al.* 1998). *Tanaecium exitiosum* (bejuco blanco, mataganado) and *T. jaroba* (calabacillo, bejuco blanco) are two plants of the Bignonaceae family native to Colombia, which are toxic for livestock. *T. exitiosum* is a vine native to the Magdalena River valley that was reported for the first time by the Colombian botanist Armando Dugand in 1942 (Dugand 1942). Even though the plant is known to be toxic and highly palatable to cattle (hence the common name of 'mataganado', which translates into 'cattle killer'), no studies have been conducted to determine its phytochemistry. Mora (1943) found that 70 g of the plant was lethal for cows and goats within 24 h and postulated that the plant accumulates cyanogenic glycosides, but no attempt was made to isolate these compounds. *Mascagnia concinna*, another Colombian native plant, accumulates toxic levels of both cyanogenic glycosides and nitrate. Torres (1984) reported that this plant may contain concentrations of HCN greater than 40 mg HCN/100 g, and Gómez (1975) found that less than 2 g of fresh leaves per kg body weight was lethal for cattle during the dry season. He also found that the plant accumulates more cyanogenic glycosides during the dry season, as compared to the rainy season.

## Conclusion

The present review describes some of the plants potentially toxic for cattle in Colombia. Even though many of the toxic plants present in Colombia have been described and studied in other countries, it is important to investigate if the same toxic compounds reported elsewhere are present in the plants that actually grow in Colombia. Some of the plants, however, are only known in Colombia and their toxicology needs to be further investigated. These include, for example, three *Tanaecium* spp. reported as toxic (*T. exitiosum*, *T. jaroba*, and *T. nocturnum*), two *Bunchosia* spp. (*Bunchosia pseudonitida* and *B. armeniaca*), two *Phytolacca* spp. (*P. bogotensis* and *P. icosandra*), and *Mascagnia concinna*.

The information on toxic plant chemistry in Colombia is mostly limited to their nitrate or cyanide content. Research is needed in order to determine not only which plants represent a potential risk for animal health and production but also their phytochemistry and toxicology. It would be very useful if veterinarians were able to document plant poisoning cases through government reporting services. Furthermore, university and government researchers could fully investigate toxicoses and publish in specialized journals. This would help identify toxic species for further phytochemical and toxicological studies and possibly pharmacological activity.

**Table 3.** Major plants known to accumulate cyanogenic glycosides in Colombia.

| Family | Latin name | Common name |
|---|---|---|
| Apocynaceae | *Bonafousia sananho* | Guachamacá, lirio blanco |
| Bignoniaceae | *Tanaecium exitiosum, T. jaroba** | Mataganado, bejuco blanco |
| | *Tanaecium nocturnum** | Unknown |
| Euphorbiaceae | *Manihot esculenta* | Yuca agria, yuca blanca |
| Fabaceae | *Lotus* spp. | Trébol pata de pájaro, trébol de cuernos |
| | *Phaseolus lunatus* | Fríjol lima |
| | *Trifolium repens* | Trébol blanco |
| Linaceae | *Linum usitatissimum* | Lino, linaza |
| Malpighiaceae | *Mascagnia concinna* | Mindaca, mataganado |
| Poaceae | *Cynodon dactylon* | Pasto argentina, pasto bermuda |
| | *Digitaria sanguinalis* | Guardarocío, pata de gallina |
| | *Sorghum bicolor* | Sorgo, sorgo forrajero |
| | *Sorghum halepense* | Pasto Johnson, capim argentino |
| | *Zea mays* | Maíz |
| Rosaceae | *Prunus* spp. | Cerezo, duraznillo, manzano criollo |

*These plants are considered to be cyanogenic but the toxic compound has not been isolated.

## Acknowledgements

The participation of Dr Gonzalo Diaz to the 8th International Symposium on Poisonous Plants was financially supported by Conselho Nacional de Desenvolvimento Científico e Tecnológico (CNPq), grant 454084/2008-0, and by Coordenação de Aprefeiçoamento de Pessoal de Nível Superior (CAPES), grant 0017/09-4.

## References

Antoniassi NAB, Ferreira EV, dos Santos CP, de Arruda LP, Campos JLE, Nakazato L, and Colodel ED (2007). Intoxicação espontânea por *Ipomoea carnea* subsp. *fistulosa* (Convolvulaceae) em bovinos no Pantanal Matogrossense. *Pesquisa Veterinaria Brasileira* 27:415-418.

Aslani MR, Maleki M, Mohri M, Sharifi K, Najjar-Nezhad V, and Afshari E (2007). Castor bean (*Ricinus communis*) toxicosis in a sheep flock. *Toxicon* 49:400-406.

Barajas-Meneses F, Fernández-Alonso JL, and Galindo-Tarazona R (2005). Diversidad y composición de la familia Boraginaceae en el Departamento de Santander (Colombia). *Caldasia* 27:151-172.

Canchila A (2001). Intoxicación en cabras por consumo de *Crotalaria pallida* en Santander. *Revista Colombiana de Ciencias Pecuarias* 14(Supl. 2001):65.

Cheeke PR (1995). Endogenous toxins and mycotoxins in forage grasses and their effects on livestock. *Journal of Animal Science* 73:909-918.

Diaz GJ (2001). Naturally occurring toxins relevant to poultry nutrition. In *Scott's Nutrition of the Chicken* (S Leeson, and JD Summers, eds), 4th edn, pp. 544-591. University Books, Guelph, Canada.

Diaz GJ, Roldán LP, and Cortés A (2003). Intoxication of *Crotalaria pallida* seeds to growing broiler chickens. *Veterinary and Human Toxicology* 45:187-189.

Dugand A (1942). Dos nuevas Bignonaceas del Valle del Magdalena. *Caldasia* 1:29-35.
Fernández-Alonso JL and Hernández-Schmidt M (2007). Catálogo de la flora vascular de la cuenca alta del río Subachoque (Cundinamarca, Colombia). *Caldasia* 29:73-104.
Gomez WB (1975). *Mascagnia concinna* (Morton), a plant poisonous to cattle. *Revista Instituto Colombiano Agropecuario* 10(4):513-514.
Guzmán VH, Morales GA, and Ochoa R (1978). Intoxicación en bovinos por nitratos acumulados en pasto elefante (*Pennisetum purpureum*, Shum). *Revista ICA* 13:113-118.
Last RD, Hill JH, and Theron G (2007). An outbreak of perirenal oedema syndrome in cattle associated with ingestion of pigweed (*Amaranthus hybridus* L.). *Journal of the South African Veterinary Association* 78(3):171-174.
Mejía B (1984). Toxicología de la Bunchosia pseudonitida, estudio clinico-patológico de la cromatosis bovina. Tesis de maestría, Programa Escuela de Graduados Universidad Nacional de Colombia, Instituto Colombiano Agropecuario, Bogotá, Colombia.
Mora CR (1943). Contribución al estudio de las plantas tóxicas en medicina. *Revista de Medicina Veterinaria* 12:5-38.
Ozmen O, Mor F, and Ayhan U (2003). Nitrate poisoning in cattle fed *Chenopodium album* hay. *Veterinary and Human Toxicology* 45:83-84.
Parton K (2000). Onion toxicity in farmed animals. *New Zealand Veterinary Journal* 48(3): 89-89
Pedraza C, Villafañe F, and Torrenegra RD (1983). Hematuria vesical bovina y su relación con algunas especies vegetales. *Revista ACOVEZ* 7:11-19.
Peña NE (1982). Contribución a la epidemiología de la cromatosis bovina en algunos municipios del Huila y Tolima (Colombia). *Revista Colombiana de Ciencias Pecuarias* 4:51-63.
Peña NE, Villamil LC, Parra D, and Lobo CA (1980). *Las enfermedades de los animales en Colombia. Situación por regiones naturales*. Ministerio de Agricultura, Instituto Colombiano Agropecuario - ICA. Bogotá, Colombia.
Pérez-Arbeláez E (1931). Plantas venenosas de Colombia. *Revista de Medicina Veterinaria* 3:189-198.
Powell AM and King RM (1969). Chromosome numbers in the Compositae, Colombian species. *American Journal of Botany* 56:116-121.
Tokarnia CH, Döbereiner J, and Peixoto PV (2002). Poisonous plants affecting livestock in Brazil. *Toxicon* 40:1635-1660.
Toro G, Rojas E, and Arango G (1997). Seneciosis. Enfermedad veno-ocusiva del hígado (EVOH) en Colombia. 1964-1996. Identificación, manejo y solución de un problema. *Revista de la Academia Colombiana de Ciencias* 21:35-56.
Torres JE (1984). Plantas tóxicas para el ganado. I Parte. *Carta Ganadera* 21(4):14-19.
Torres LM, Diaz P, and Diaz GJ (2003). Efectos de la adición de semillas de *Senna obtusifolia* (cafelillo, bajagua) en la dieta para aves de postura y pollos de engorde. *El Cerealista* 67:35-38.
Trheebilcock E, Villafañe F, and Gil A (1978). Síndrome caída del ganado – Contribución a su estudio. *Revista ICA* 13:119-125.
Vargas OM, Quiñonez LM, and Parra JL (1998). *Plantas tóxicas para los bovinos en la vega del río Arauca*, 31 pp. Manual de Asistencia Técnica No. 03, CORPOICA Regional 8, Villavicencio, Meta, Colombia.
Vermunt J and Visser R (1987). Nitrate toxicity in cattle. *New Zealand Veterinary Journal* 35(8):136-137.

# Chapter 6

# Poisonous Plants Affecting Livestock in Central America, with Emphasis on Panama

A.G. Armién[1], P.V. Peixoto[2], and C.H. Tokarnia[2]

[1]*Minnesota Veterinary Diagnostic Laboratory, Department Veterinary Population Medicine, University of Minnesota, USA;* [2]*Departamento de Nutricao Animal e Pastagem, Instituto de Zootecnia, Universidade Federal Rural do Rio de Janeiro, Brazil*

## Introduction

Plant poisoning is an important cause of economic losses worldwide. In the USA, plant poisoning is responsible for losses estimated to surpass US$250 million per year. These costs are attributable not only to livestock deaths and diminished productivity related to plant poisoning, but also the cost of managing forage in areas invaded by poisonous plants (Nielsen 1988). In Brazil, a rough estimate indicates that poisonous plants cause the loss of a million adult cattle per year due to mortality alone (Riet-Correa and Medeiros 2001; Tokarnia *et al.* 2002). In contrast to North and South America, literature related to the most important poisonous plants affecting the livestock industry is lacking in Central America. In Panama, field surveys and necropsies during 1994-1995 (A. Armién, unpublished data) indicate that, similar to Brazil, plant poisoning, rabies, and enzootic botulism are the most important cause of death for adult grazing cattle (Tokarnia *et al.* 2002).

The existing literature about the poisonous plants of Panama is limited and mostly focused on human public health (Allen 1943; Escobar 1972). The data here presented about poisonous plants that affect livestock in Panama were obtained from field surveys and necropsies in 1994-1995, 2000-2001, and 2009 (A. Armién, unpublished data). Only plants that induce spontaneous intoxication on livestock raised under range conditions are considered here. Parameters of inclusion in this group were mainly plants that: (i) invade dense pastures; (ii) have a regional or global geographic distribution; (iii) are accessible for grazing cattle; (iv) had a history of poisoning cattle when ingested spontaneously; (v) cause signs of disease after ingestion; and (vi) cause death after ingestion. Because of large plant biodiversity (Correa *et al.* 2004), it is expected that indigenous poisonous plant species are numerous and have a remarkable impact on the Central American livestock as well. Nevertheless, so far in Panama the most common toxicological syndromes seem to be associated with plants that are also a worldwide problem or that were introduced for cultivation. Ornamental plants, which are known to have toxic properties and occasionally cause the death of ruminants when ingested are, with few exceptions, native to Central America. Yet, it has been shown elsewhere that, in general, ornamental plants such as

©CAB International 2011. *Poisoning by Plants, Mycotoxins, and Related Toxins* (eds F. Riet-Correa, J. Pfister, A.L. Schild, and T.L. Wierenga)

*Nerium oleander, Rhododendrum* spp., *Allamanda cathartica, Thevetia peruviana,* and *Ricinus communis*, are of little natural risk for ranching livestock, and therefore were considered beyond discussion in this paper (Armién *et al.* 1994, 1995; Armién and Tokarnia 1994; Brito *et al.* 1996; Tokarnia *et al.* 1996). In this paper we present the poisonous plants affecting livestock in Central America with emphasis on Panama.

## Poisonous Plants Affecting Livestock in Central America

Poisonous plants affecting livestock in Panama are few. In Table 1 we present the main data about plants that cause or are suspected to induce poisoning in Panama and Central America. In Figure 1 we present a geographic location of poisoning outbreaks studied in Panama.

In Costa Rica, Nicaragua, Honduras, El Salvador, Guatemala and Belize, the scientific references are narrowed to *Pteridium caudatum, Melochia pyramidata* and *Brachiaria radicans*.

**Table 1.** Poisonous plants affecting livestock in Central America with emphasis in Panama.

| Plants categorized by the clinical-pathological signs they cause | Country | Plant ID | History and/or clinical signs | Lesions found | Experimental reproduction |
|---|---|---|---|---|---|
| Radiomimetic plants | | | | | |
| *Pteridium caudatum, P. arachnoideum* | P, CR | + | + | + | - |
| Photosensitivity | | | | | |
| *Bracharia decumbens, B. humidicola* | P | + | + | + | |
| *Lantana camara* | P | + | + | + | - |
| *Enterolobium cyclocarpum* | G, P | + | - | - | + |
| Hepatotoxic plants | | | | | |
| *Cestrum* spp. | P, CR | + | - | + | - |
| Nephrotoxic plants | | | | | |
| *Amaranthus* spp. | P | + | + | + | - |
| Cyanogenic plants | | | | | |
| *Sorgum* spp. | P, CR | + | + | - | - |
| *Manihot esculenta* | P, CR | + | + | - | - |
| Neurotoxic plants | | | | | |
| *Melochia pyramidata* | ES, N | + | + | + | + |
| Plants that affect the gastrointestinal tract | | | | | |
| *Thevetia ahouai* | P | + | + | - | - |
| Plants that cause nitrate/nitrite poisoning | | | | | |
| *Brachiaria radicans* | P, CR | + | + | + | + |

P= Panama; CR= Costa Rica; N= Nicaragua; ES= El Salvador; G= Guatemala

### Radiomimetic plants

*Pteridium caudatum* and *P. arachnoideum* are a common cause of poisoning at middle (1000 m) to high (3000 m) altitude regions of Panama and Costa Rica (Villalobos-Salazar 1985; Amelot 1999). These (Correa *et al.* 2004) invade grazing areas affecting juvenile to adult cattle. Depending on the period during which the plant is eaten and on the amount ingested, the radiomimetic principle (ptaquiloside) is responsible for three different clinic-pathological features, observed mainly in cattle: hemorrhagic diathesis, bovine enzootic hematuria, and carcinomas of the upper digestive tract (Tokarnia *et al.* 2000; França *et al.* 2002). All three forms of intoxication by *P. caudatum* and *P. arachnoideum* are common in Panama (Figure 1). Urinary bladder neoplasias leading to hematuria and upper digestive tract tumors that induce impaired chewing and swallowing, reflux, and ruminal bloating, occur in the Cordillera Central. Heavily invaded pastures or lack of pasture and hunger are the most important predisposing factors. Outbreaks with animals presenting bone marrow depletion follow by pancytopenia, and hemorrhagic diastase has been diagnosed in juvenile cattle in Azuero. Occurrences of gastric carcinomas and urinary bladder tumors in humans have been associated with *Pteridium* spp. or raw milk consumption from cattle that are ingesting this plant (França *et al.* 2002). Circumstantial evidence indicates that similar situations may be occurring in Panama as well (B. Armién, personal communication).

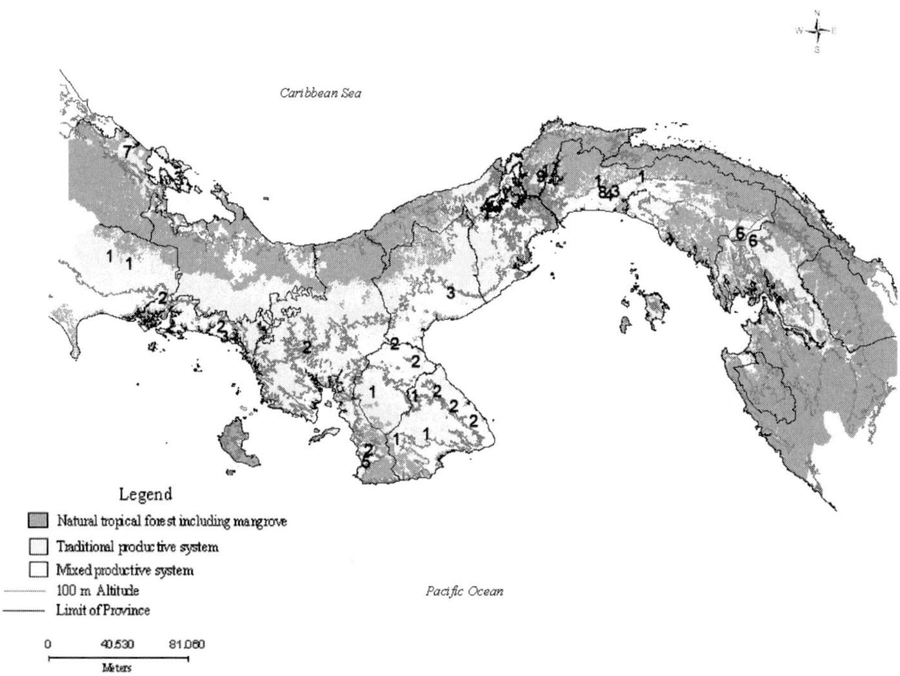


**Figure 1.** Map of ecosystems and location of plant poisoning cases in livestock diagnosed in 1994-1995 and 2000-2001. (1) *Pteridium caudatum*; (2) *Brachiaria humidicola* and *B. decumbens*; (3) *Lantana camara*; (4) *Cestrum* spp.; (5) *Amaranthus* spp.; (6) *Thevetia ahouai*; (7) *Brachiaria radicans*; (8) Nephrotoxicosis; (9) Sudden death. Source: Map of vegetation cover of the Republic of Panama. Satellite images LandSat TM5-TM7, Panamanian National Environment Authority (ANAM), 2000.

## Plants that cause photosensitivity

Photosensitivity in Panama has been associated with both *Brachiaria decumbens* and *B. humidicola* grasses and pastures invaded with *Lantana camara*. Photosensitivity appears to be observed in most regions of Panama and in most instances the cause of some outbreaks has not been established. These grasses, *B. decumbens* and *B. humidicola*, have recently been introduced and have extensively replaced the traditional pastures of *Hyparrhemia rufa* in the majority of the land allocated for grazing livestock. Cases of photosensitivity have been largely observed in cattle (European and Zebu) that were grazing in *B. decumbens* and *B. humidicola* (Figure 1). We have recorded two outbreaks of photosensitivity in sheep. Cases of photosensitivity seem to occur most commonly in juvenile cattle throughout the year and mortality has not been recorded. However, microscopic changes consistent with secondary hepatogenous photosensitivity associated with *Brachiaria* spp. has been identified as an incidental finding in cattle. As described in Brazil, these animals histologically show the presence of foam cells in the liver and lymph nodes, intraluminal crystals in biliary ducts, biliary duct proliferation, and pericolangitis (Tokarnia *et al.* 2000). A relation of the disease and the presence of the fungus *Pithomyces chartarum* have not yet been investigated. However, most of the seeds of *B. decumbens* and *B. humidicola* used to establish pastures in Panama have been imported from Brazil.

There are seven species within the genus *Lantana* in Panama (Correa *et al.* 2004). *L. camara* encompasses two varieties, var. *aculata* and var. *mista*. The shrub is found in the entire country from 0 to 1000 m above sea level. Secondary photosensitivity induced by *L. camara* has been observed in cattle in Panama. In two outbreaks, the affected cattle were transfered to poor pastures invaded by this weed in traditional livestock productions on sloped topography located at 300 m above sea level (Figure 1). These outbreaks occurred in the wet season. Animals showed subacute skin lesions of photosensitivity and light icterus. Lesion characteristics of *Lantana* hepato- and nephro-toxicity were confirmed by histopathology. In general, the predisposing conditions necessary to the poisoning, disease, and lesions are similar as those described in Brazil (Tokarnia *et al.* 1999).

*Enterolobium cyclocarpum* is a deciduous tree of approximately 30 m in height, which grows in lowlands in dry to humid climates. The fruits have the shape of a human ear and are palatable to cattle. The tree is extensively distributed in Central America. *E. cyclocarpum* is popularly known as 'Corotú' tree in Panama, and as 'Conacaste' tree in other countries of Central America. Across the Pacific lowland of Panama, its fruit is available at the end of the dry season from March to April. Evidence of cattle poisoning has been reported in Mexico and Guatemala (Sovalbarro 1990; Avedano-Reyes and Flores-Gudino 1999). Cattle present colic, diarrhea, and skin lesions consistent with secondary photosensitivity (Sovalbarro 1990). In Brazil, hepatogenous photosensitivity has been experimentally induced with *E. gummiferum* (Tokarnia *et al.* 2000). In many areas of the Azuero peninsula, outbreaks of photosensitivity can overlap with photosensitivity induced by others plants such as *Bracharia* spp. Seasonal photosensitivity associated with *E. cyclocarpum* in Panama remains unknown. Experimental studies to characterize the clinical-pathological features of the fruits of the *E. cyclocarpum* are necessary. The toxic principle appears to be a saponin (Sovalbarro 1990).

## Hepatotoxic plants

*Cestrum* sp. is a shrub that is suspected to cause hepatotoxicity in cattle and horses in Panama. There are around 21 species of the genus *Cestrum* in Panama alone (Correa *et al.*

2004). The plant is used as an ornamental plant as well as for fencing. *Cestrum* species are found from 0 to 3000 m above sea level. Most people are unaware that *Cestrum* may cause toxicity in livestock. *C. nocturnum* has nevertheless been suspected to be the etiology in two poisonings of cattle and of one in horses in the Panama province. The diagnosis of poisoning by this plant was based on the presence of typical acute coagulative hepatocellular necrosis affecting central and midzonal areas of the hepatic lobule and the presence of the plant in this area. *C. laevigatum*, the most important hepatotoxic plant in Brazil, is also found in Panama. The toxic principles are saponins (ditogenin and digitogenin) (Tokarnia *et al.* 2000). An important potential differential diagnosis includes poisoning by *Trema micratha*, a deciduous tree or shrub largely distributed across the Pacific and Atlantic lowlands, for which association with spontaneous intoxication in ruminants has not been noted in Panama.

**Nephrotoxic plants**

The only plant of this group known to have caused mortality of cattle in Panama has been *Amaranthus* spp. (Amaranthaceae). This outbreak of *Amaranthus* spp. intoxication was diagnosed by histopathology and the abundance of the plants in a harvested rice plantation in the Azuero peninsula. Several animals became sick and died acutely after being transferred to this field. In addition, a case of phytotoxic nephropathy was suspected by histopathology in the Panama province in cattle that were in a grass field heavily invaded by weeds and some Fabaceae-Papilionoideae trees. No known nephrotoxic plants were identified after field inspection.

**Cyanogenic plants**

The diagnosis of intoxication by cyanogenic plants is difficult and has been based on clinical history in Panama. Cattle poisoned with *Sorghum* spp. has been reported in the plantation in the Azuero peninsula. Poisoning with *Manihot esculenta* (Euphorbiaceae) has been reported in swine which were fed the root of this plant. In the provinces of Panama, Colon, and Darien poisoning by cyanogenic plants is suspected in areas where cattle died suddenly; these sudden deaths generally occurred in newly established pastures in fragmented forest. These areas are normally heavily invaded by regrowth of native vegetation. Differential diagnoses with plants that cause sudden cardiac death, such as plants that contain monofluoroacetic acid, has been investigated, but so far no toxic *Palicourea* has been identified. Plants that cause sudden cardiac death are responsible for 50% of cattle mortality in Brazil (Tokarnia *et al.* 2000).

**Neurotoxic plants**

In El Salvador (Palmer and Woodham 1975) and Nicaragua *Melochia pyramidata* L (Sterculiaceae) is responsible for a disease named ‘derrengue’. In these countries, the losses of cattle due to the ingestion of *M. pyramidata* are from tens to hundreds of animal each year. Hungry animals ingest the plant during severe droughts when the shrub is the only available fodder. Large amounts of the plant are necessary to induce disease. Clinical signs are evident a few weeks after ingestion of the plant and are characterized by hind leg paresis progressing to tetraparalysis. Animals die by starvation 1 to 2 weeks later. Postmortem changes are unspecific and consistent with serous fat atrophy and edema of the subcutaneous tissue and cavities. The most important lesion is found in the peripheral nerve and skeletal muscle. Nerve lesions are characterized by axonal and myelin degeneration with infiltration of macrophages and later proliferation of Schwann cells. There is also neurogenic muscle atrophy. In the central nervous system, demyelination of the inferior cerebellar peduncle and the spinocerebellar tract is found (Palmer and Woodham 1975). The toxic principle is the alkaloid melochinine (Breuer *et al.* 1982).

**Plants that affect the gastrointestinal tract**

Plants that specifically affect the gastrointestinal tract in cattle are unknown in Panama. However, field observation and clinical history indicate that the Apocynaceae *Thevetia ahouai* has been frequently associated with cattle mortality in the provinces of Panama and Darien. *T. ahouai* is a tree or shrub approximately 8 m high with yellow flowers that produces a milky sap. Drupe fruit are red when ripe. It grows in lowlands in dry to very humid climates. It is found from Mexico to Venezuela (Carrasquilla 2008). In Panama this plant is largely distributed from the Chiriquí to the Darien province. Field surveys and clinical history indicate that animals presented severe diarrhea and died after grazing in fields densely invaded with plant regrowth. No pathological examination or experimental studies have been performed to confirm the toxic properties of this plant. *Thevetia peruviana* is a native plant of the same genus found in Panama as an ornamental plant and has been experimentally toxic for cattle and sheep (Armién *et al.* 1995; Tokarnia *et al.* 1996). *Thevetia peruviana* has induced tachycardia, ruminal atony, and diarrhea in cattle and sheep. Experimentally, a single dose larger than 15g/kg liveweight has been lethal to cattle. Postmortem changes are characterized by a severe necrotizing ruminitis, reticulitis and enteritis (Armién and Tokarnia 1994; Tokarnia *et al.* 1996).

**Plants that cause nitrate/nitrite poisoning**

Cattle poisoning with *Brachiaria radicans* (Tanner grass) has been described in Costa Rica (Villalobos *et al.* 1981). In Panama hemoglobinuria has been suspected in cattle that graze in Tanner grass in Bocas del Toro province. Symptoms of intoxication are noted a few days after animals are introduced to Tanner grass pasture. Animals show severe anemia, weakness, diarrhea, hemoglobinuria, oliguria, and dehydration. Animals with pronounced clinical signs present significantly increased urea and creatinine values in urine and high methemoglobin values. Data suggest that intoxication by *B. radicans* results from the nitrate-nitrite factor causing production of methemoglobin in addition to other factors causing degenerative and necrotic hepatocellular and renal lesions and intravascular hemolysis (Villalobos *et al.* 1981).

## Conclusion and Perspectives

The diagnosis of plant poisoning basically depends on the correct evaluation of the historical and epidemiological information, identification of the clinical signs, gross and microscopical lesions, and unequivocal evidence that the suspected toxic plants have been ingested by the affected animals. In cases of unknown poisonous plants, experimental poisoning may be necessary to confirm not only the toxic property of the plants in question and the sensitivity of the affected animal species but to characterize the clinical and pathological features. This information is vital for prompt identification of a problem by local field veterinarians and diagnosticians (Tokarnia *et al.* 2000).

Central America has one of the largest plant biodiversities in the world, and it is expected that indigenous poisonous plant species are numerous and have a remarkable impact on the livestock in newly established pastures in recently fragmented pristine forests. However, the native vegetation has been extensively replaced by foreign vegetation as a consequence of the introduction of modern agriculture. With this, poisonous plants of global significance have been favored and, in consequence, are rising as an important issue in this country.

Currently, the literature regarding poisonous plants in Central America is lacking. To foster the understanding of poisonous plants in Central America, a program must first be implemented that reinforces field veterinarians and veterinary diagnostic laboratories and that focuses on the identification, characterization, and documentation of disease induced by poisonous plants of importance for livestock production in Central America. Secondly, collaborative research that helps to identify the toxic principles and studies the ecology of new poisonous plants must also be considered.

## Acknowledgements

We are grateful for the support of the Laboratory of Diagnosis and Livestock Research of the Ministry of Agricultural Development of Panama (MIDA), and Panamanian Institute of Livestock and Agricultural Research (IDIAP). We thank Dr A. Espinosa of the University of Panama for the plants' identification and Drs Oneida Calderon, Carmen Sousa, R. Villareal, and Venancio Gonzalez for their support in the field survey.

The financial assistance for Dr Anibal Armién to attend the 8th International Symposium on Toxic Plants was provided by the Conselho Nacional de Desenvolvimento Científico e Tecnológico (CNPq), grant 454084/2008-0, and by Coordenação de Aprefeiçoamento de Pessoal de Nível Superior (CAPES), grant 0017/09-4.

## References

Allen PH (1943). Poisonous and injurious plants of Panama. *American Journal Tropical Medicine,* Supplement 23(1):1-76.

Amelot A (1999). Bracken fern, animal and human health. *Revista de la Universidad de Agronomia*, Universidad del Zulia 16(5):528-547.

Armién AG and Tokarnia CH (1994). Experiments on the toxicity of some ornamentals plants toxicity in sheep. *Pesquisa Veterinária Brasileira* 14(2/3):69-73.

Armién AG, Peixoto PV, Barbosa JD, and Tokarnia CH (1994). Experimental poisoning of sheep by *Nerium oleander* (Apocinaceae). *Pesquisa Veterinária Brasileira* 14(2/3):85-93.

Armién AG, Peixoto PV, Barbosa JD, and Tokarnia CH (1995). Experimental poisoning by *Rhododendron ledifolium* (Ericaceae) in sheep. *Pesquisa Veterinária Brasileira* 15(1):1-9.

Avedano-Reyes S and Flores-Gudino JS (1999). Registro de plantas tóxicas en el estado de Veracruz, Mexico. *Veterinaria Mexico* 30(1):79-94.

Breuer H, Rangel M, and Medina E (1982). Pharmacological properties of melochinine, an alkaloid producing Central American cattle paralysis. *Toxicology* 25(2-3):223-42.

Brito MF, Armién AG, and Tokarnia CH (1996). Experimental poisoning by *Abrus precatorius* seeds (Leg. Papilionoidea) in sheep. *Pesquisa Veterinária Brasileira* 16(2/3):59-66.

Carrasquilla LG (2008). *Trees and shrubs of Panama.* 478 pp. Editora Novo S.A., Panama.

Correa AMD, Gal Dames C, and Stapf de M (2004). *Catalogo de las plantas vasculares de Panama*, 599 pp. Quebecor World Bogota, S.A., Colombia.

Escobar NA (1972). *Introducción al estudio de la flora tóxica de Panamá.* 13 pp. Ministerio de Agricultura y Ganadería, Dirección General de Investigación, Extensión y Educación Agropecuaria. Panamá.

França TN, Tokarnia CH, and Vargas Peixoto P (2002). Disease caused by the radiomimetic principle of *Pteridium aquilinum* (Polypodiaceae). *Pesquisa Veterinária Brasileira* 22(3):85-96.

Nielsen DB (1988). Economical impact of poisoning plants on rangeland livestock industry. *Journal Animal Science* 66:2330-2333.

Palmer AC and Woodham (1975). Derrengue, a paralysis of cattle in El Salvador ascribed to ingestion of *Melochia pyramidata. Veterinary Record* 21(25):547-8.

Riet-Correa F and Medeiros RMT (2001). Intoxicações por plantas em ruminantes no Brasil e no Uruguai: importância econômica, controle e riscos para a saúde pública. *Pesquisa Veterinária Brasileira* 21:38-42.

Sovalbarro AA (1990). Identification and typing of potential poisonous plants of southern Pacific coast of Guatemala. Toxicity of fruits of the tree *Enterolobium cyclocarpum* (Mimoseae) for ruminants and laboratory animals. 149 pp. Doctoral degree Dissertation, Tierärztliche Hochschule Hannover, Germany.

Tokarnia CH, Armién AG, Peixoto PV, Barbosa JD, Brito MF, and Döbereiner J (1996). Experiments on the toxicity of some ornamental plants in cattle. *Pesquisa Veterinária Brasileira* 16(1):5-20.

Tokarnia CH, Armién AG, Barros SS, Peixoto PV, and Döbereiner J (1999). Complementary studies on the toxicity of *Lantana camara* (Verbenaceae) in cattle. *Pesquisa Veterinária Brasileira* 19(3/4):128-132.

Tokarnia CH, Döbereiner J, and Peixoto PV (2000). *Plantas tóxicas do Brazil.* pp.120-122. Editora Helianthus, Rio de Janeiro.

Tokarnia CH, Dobereiner J, and Peixoto PV (2002). Poisonous plants affecting livestock in Brazil. *Toxicon* 40:1635-1660.

Villalobos-Salazar J (1985(. Carcinogenicidad del *Pteridium aquilinum* y alta incidentia del cancer gástrico en Costa Rica. *Revista Costa Rica de Ciencias Medicas* 6:131-139.

Villalobos SJ, Meneses GA, Leon CS, and Carballo CG (1981). Symptoms and pathology of poisoning (of cattle) with tanner grass, *Brachiaria radicans* Napper. *Ciencias Veterinarias*, Costa Rica. 3(2/3):163-169.

# Chapter 7

# Plant Poisonings in Mato Grosso do Sul

R.A.A. de Lemos[1], E.B. Guimarães[1], N.M. Carvalho[2], A.P.A. Nogueira[3], B.S. Santos[3], R.I.C. Souza[3], S.G. Cardinal[3], and H.O. Kassab[4]

[1]*Departamento de Medicina Veterinária (DMV/FAMEZ);* [2]*Divisão Clínica, Médico Veterinário (FAMEZ);* [3]*Programa de Mestrado em Ciência Animal, FAMEZ, UFMS, Campo Grande, MS 79070-900, Brazil;* [4]*Aluno de graduação do curso de Medicina veterinária. FAMEZ, UFMS, Campo Grande, MS 79070-900, Brazil*

## Introduction

Despite the economic losses caused by poisonous plants in Brazil, few systematic surveys about epidemiology and economic impact of these intoxications have been conducted in some Brazilian regions (Silva *et al.* 2006; Santos *et al.* 2008).

Poisonous plants can be classified in many ways. Categorizing plants according to the clinical and pathological signs they cause is the most common and useful, and has been the primary classification used in Brazilian poisonous plant books (Tokarnia *et al.* 2000; Riet-Correa and Méndez 2007; Riet-Correa *et al.* 2007).

The present study reports the epidemiology, clinical signs, gross and histological findings, and control measures of the poisoning by plants in herbivores in the state of Mato Grosso do Sul, central-western Brazil during the period of 1994 to 2008.

## Plants Causing Hepatogenous Photosensitization

During 1994 to 2008 plants causing hepatogenous photosensitization were responsible for serious economic looses. *Brachiaria* spp. was the main cause of outbreaks in bovines and ovines. Thirty-six outbreaks were diagnosed in cattle, 32 in sheep, and two in goats. Outbreaks occurred throughout the year. Clinical and pathological findings characteristic of hepatogenous photosensitization were similar in all species. The main histological alteration in the liver was the presence of optically active retractile crystals in the lumen of canaliculi and bile ducts, and infiltration of foamy macrophages with peripheral nucleus. No significant amounts of *Pithomyces chartarum* spores were found in the pastures. In one outbreak in sheep, concentrations of non-steroidal saponins (protodioscin) in the pastures were 2.36%. Clinical and pathological signs observed, as well as the absence of *P. chartarum*, were similar to other reports of natural or experimental cases of intoxication by *Brachiaria* spp. Despite its toxicity, *Brachiaria* spp. is very important forage in central-western Brazil; however, since the factors affecting saponin concentration in the plant are unknown, there are no effective means to prevent the intoxication.

©CAB International 2011. *Poisoning by Plants, Mycotoxins, and Related Toxins* (eds F. Riet-Correa, J. Pfister, A.L. Schild, and T.L. Wierenga)

*Enterolobium contortisiliquum* is also an important cause of hepatogenous photosensitization in Mato Grosso do Sul. It was responsible for eight outbreaks of poisoning during the period. Clinical signs and gross findings were similar to those observed in the intoxication by *Brachiaria* spp. In two outbreaks in pregnant cows the poisoning was associated with abortion. Histologic lesions were swelling and individual necrosis of hepatocytes, and in some cases, discrete proliferation of epithelial bile duct cells. Optically active crystals were also found occasionally. Besides the small number of cases sent to the diagnostic laboratory, many farmers and field veterinarians mention the occurrence of outbreaks of photosensitization and abortion in cows associated with the ingestion of *E. contortisiliquum*. The pods of this tree were experimentally toxic to bovines, causing intestinal disorders and mild photosensitization, however, abortion was not reproduced (Lemos and Purisco 2002). As the intoxication occurs from August to November, a period in which the fruits of *E. contortisiliquum* fall on the ground, the main control measure is to avoid animal exposure during this time of the year.

Sporadic outbreaks of hepatogenous photosensitization were also associated with ingestion of *Pterodon emarginatus* and *Stryphnodendron fissuratun.* Outbreaks of poisoning by *P. emarginatus* occurred after rainfalls that caused branches to fall and leaves to be ingested by bovines. Outbreaks of poisoning by *S. fissuratun* occurred between August and September, when the fruits of this tree fall to the ground. Spontaneous poisoning by *P. emarginatus* and *S. fissuratun* are also reported in the state of Mato Grosso (Arruda *et al.* 2008; Ferreira *et al.* 2008). No photosensitization was observed in the experimental reproduction of these intoxications, but, 48 h after the ingestion of *S. fissuratun*, the animals showed dry feces with mucus. Five days after dosing, mild hyperthermia, ruminal hypotonia, with accentuation of depression was observed. Death occurred 11 days after the beginning of the experiment.

## Plants Causing Sudden Death

*Mascagnia pubiflora* was responsible for ten outbreaks of sudden death during the period. Outbreaks occurred during January, March, May, November, and December (rainy season) and morbidity varied from 1% to 3.5%. Case fatality rate was 100%. In eight of these outbreaks, handling of the animals was the triggering factor for the observation of clinical signs. No gross lesions were observed at necropsy. Vacuolar hydropic degeneration of the epithelium of the contorted distal duct, a typical lesion of poisoning by plants causing sudden deaths, was not observed on histologic examination. The prevention of the intoxication consists of eradicating the plant by using herbicides or keeping animals from grazing in areas where the plant is present.

## Plants That Affect the Heart

Two outbreaks by *Tetrapterys multiglandulosa* were diagnosed in cattle, both on the same farm. The plant was present in an ecological reserve that was grazed. The first outbreak affected a herd of 290 pregnant cows causing death by cardiac insufficiency in seven (2.4%) cows and abortion or delivery of weak calves that died after parturition in 230 (79.3%) cows. In a second outbreak only non-pregnant heifers were affected by neurological disturbances and cardiac insufficiency. Nine of 285 open heifers showed clinical signs and died. Morbidity and case fatality rates were 3% and 100%, respectively.

Clinical and pathological signs were characteristic of chronic heart failure. Status spongiosus of the brain was also observed histologically. These lesions were similar in cows and calves. Because the plant seems to be palatable, the main control measure is to limit animal access to the plant (Carvalho *et al.* 2006).

## Hepatotoxic Plants

*Vernonia rubricaulis* was the main cause of liver necrosis in bovines. During 1994 to 2008, 21 outbreaks were diagnosed, from September to February in the wet season when the plant was sprouting. The main condition associated with outbreaks was the regrowth of the plant after burns, mowing, and rotational grazing. In two outbreaks, none of these conditions were present and the cases occurred only in bovines that had been moved from areas where the plant did not exist to areas where it was present. Morbidity varied from 2% to 21% and case fatality was virtually 100%. Clinical signs were characterized by apathy, tremors, dry muzzle (dehydration), dry feces with blood, and aggressive behavior. Histological findings were severe coagulation necrosis, mainly in the centrilobular region of the liver, and hemorrhages, occasionally affecting the whole lobule. Cases with massive necrosis of the liver and bleeding were frequent. The epidemiology, clinical signs, and pathology observed in these outbreaks are similar to previous reports (Brum *et al.* 2002). To prevent the intoxication it is necessary to avoid grazing during sprouting of the plant after burns, mowing, and rotational grazing. Caution is necessary if animals are moved from areas were *V. rubricaulis* does not exist and are then introduced into paddocks invaded by this plant.

Two outbreaks of *Crotalaria* spp. poisoning were diagnosed as a cause of liver fibrosis being associated with interstitial pneumonia. In the first outbreak, 14 of 1300 bovines were affected, and 12 died. In the other, 3 out of 2800 bovines were affected and died. Liver fibrosis and interstitial pneumonia were found in two cows. The herds were raised extensively in the Pantanal region. Clinical signs were progressive weight loss, respiratory distress, and soft feces. Gross and histological findings were hepatic cirrhosis and chronic interstitial pneumonia. In addition to this, another outbreak was diagnosed in calves on a dairy farm; they exhibited clinical and pathological signs of respiratory failure. In this outbreak, a *Brachiaria* sp. pasture was invaded by *Crotalaria* sp.

## Calcinogenic Plants

Three outbreaks of *Solanum malacoxylon* poisoning were diagnosed in the months of June, August, and October. Morbidity varied from 1.7% to 3.75% and lethality from 2% to 100%. In one outbreak, the area was deforested to plant a pasture, but *S. malacoxylon* invaded the area in large amounts. In the other two outbreaks, special conditions that led to the growth of the plant were not identified. Clinical signs were progressive weight loss, rough hair, stiff gait with lameness, tucked up abdomen, kyphosis, and heart murmurs. The clinical course is chronic, and death can occur up to 4 months from extreme malnutrition and cachexia if animals are not removed from pastures where the plant occurs. At necropsy the arteries are thickened and rigid. Rough plates are observed in the internal surface of the arteries and endocardium. The main histological finding is the presence of calcium deposits in the arteries and other soft tissues and pulmonary emphysema. As the plant is disseminated in the Pantanal region, the intoxication is known by farmers and practitioners, and cases of intoxication are probably under reported. In these outbreaks epidemiology, clinical signs, and lesions were similar to those previously reported (Tokarnia *et al.* 2000). Poisoning can be prevented by restricting animal access to areas invaded by the plant.

## Plants Causing Segmental Muscular Necrosis

Three outbreaks of *Senna occidentalis* poisoning were diagnosed in bovines grazing in pastures invaded by the plant. Morbidity varied from 6.4% to 17% and lethality was 100% in all outbreaks. In another outbreak that affected bovines under rotational grazing in a pasture severely invaded by *S. occidentalis*, morbidity was 62%. This high morbidity was probably due to the high stocking rate in a paddock severely invaded by *S. occidentalis*, and low forage availability due to previous grazing. All outbreaks occurred when the plant was in seed. Clinical signs were diarrhea, muscle weakness, ataxia of the hind limbs, restlessness, and recumbence followed by death. At necropsy pale areas of some muscles and bleeding and congestion of the fascia were observed. The bladder contained dark urine and *S. occidentalis* seeds were observed in the reticulum. Main histological findings were segmental necrosis of the muscles. Clinical signs and lesions were similar to previous reports (Barros *et al.* 1999), but this is the first time that the disease has been reported in cattle grazing pastures invaded by the plant. In previous reports in Brazil the disease occurred in confined bovines due to contamination of grains by seeds of the plant during harvesting (Barros *et al.* 1999). To prevent intoxication, avoid grazing animals in paddocks severely invaded by *S. occidentalis* and avoid feeding grain contaminated with *S. occidentalis* seeds.

## Other Poisonous Plants in Mato Grosso do Sul

Other poisonous plants in the state of Mato Grosso do Sul are *Stryphnodendron abovatum* causing abortion and photosensitization, *Brachiaria radicans* as a cause of hemolytic anemia, and *Manihot* spp. causing cyanide poisoning, but outbreaks of these intoxications were not reported to our laboratory during 1994 to 2008. However, farmers and practitioners mentioned outbreaks of poisoning by these plants. Sporadic outbreaks of nephrosis probably caused by an unknown toxic plant were also noted.

## Acknowledgements

This work has been financially supported by the National Institute for Science and Technology for the Control of Toxic Plants (CNPq), grant number 573534/2008-0.

## References

Arruda LP, Ferreira EV, Boabaid FM, Rocha PRD, Cruz RAS, Ubiali DG, Mendonça FS, Gasparetto ND, Nespoli JE, Souza F, and Colodel EM (2008). Intoxicação por *Pterodon emarginatus* (Fabaceae) em bovinos. In *Anais Encontro Nacional de Diagnóstico Veterinário*. Campo Grande, Mato Grosso do Sul.

Barros CSL, Ilha MRS, Bezerra Jr PS, Langhor IM, and Kommers GD (1999). Intoxicação por *Senna occidentalis* (Leg. Caesalpinoideae) em bovinos em pastoreio. *Pesquisa Veterinária Brasileira* 19:68-70.

Brum KB, Purisco E, Lemos RAA, and Riet-Correa F (2002). Intoxicação por *Vernonia rubricaulis* em bovinos no Mato Grosso do Sul. *Pesquisa Veterinária Brasileira* 22:119-128.

Carvalho NM, Alonso LA, Cunha TG, Ravedutti J, Barros CSL, and Lemos RAA (2006). Intoxicação de bovinos por *Tetrapterys multiglandulosa* (Malpighiaceae) em Mato Grosso do Sul. *Pesquisa Veterinária Brasileira* 26:139-146.

Ferreira EV, Boabaid FM, Arruda LP, Gasparetto ND, Rocha PRD, Cruz RAS, Souza MA, Nakazato L, and Colodel EM (2008). Intoxicação espontânea e experimental por *Stryphnodendron fissuratum* Mart. em bovinos na região Centro-Oeste. In *Anais Encontro Nacional de Diagnóstico Veterinário*. Campo Grande, Mato Grosso do Sul.

Lemos RAA and Purisco E (2002). Plantas que causam fotossensibilização hepatógena. In *Enfermidades de interesse econômico em bovinos de corte: perguntas e respostas* (RAA Lemos, N Barros, e KB Brum, eds), 292 pp. UFMS, Campo Grande.

Riet-Correa F and Méndez MC (2007). Intoxicações por plantas e micotoxicoses. In *Doenças de Ruminantes e Eqüídeos* (F Riet-Correa, AL Schild, RAA Lemos, and JRJ Borges, eds), vol.2, p.99-221. Palloti, Santa Maria.

Riet-Correa F, Medeiros RMT, Tokarnia CH, and Döbereiner J (2007). Toxic plants for livestock in Brazil: Economic impact, toxic species, control measures and public health implications. In *Poisonous Plants: Global research and solutions* (KE Panter, TL Wierenga, and JA Pfister, eds), pp. 2-14. CAB International, Wallingford.

Santos JCA, Riet-Correa F, Simões SVD, and Barros CLS (2008). Patogênese, sinais clínicos e patologia das doenças causadas por plantas hepatotóxicas em ruminantes e eqüinos no Brasil. *Pesquisa Veterinária Brasileira* 28(1):1-14.

Silva DM, Riet-Correa F, Medeiros RT, and Oliveira OD (2006). Plantas tóxicas para ruminantes e eqüídeos no Seridó Ocidental e Oriental do Rio Grande do Norte. *Pesquisa Veterinária Brasileira* 26(1):223-236.

Tokarnia CH, Döbereiner J, and Peixoto PV (2000). *Plantas Tóxicas do Brasil*. Editora Helianthus, Rio de Janeiro, 310 pp.

# Chapter 8

# Poisonous Plants Affecting Sheep in Southern Brazil

D.R. Rissi, F.J.F. Sant'Ana, F. Pierezan, B.L. Anjos, and C.S.L. Barros

*Laboratório de Patologia Veterinária, Universidade Federal de Santa Maria, Santa Maria, RS, 97105-900, Brazil*

## Introduction

The ingestion of toxic plants is an important cause of direct and indirect losses of livestock worldwide. Associated problems include death of animals, reproductive failure such as abortion and congenital defects, and chronic disease leading to poor development and weight loss. A review of the most important poisonous plants that affect cattle in southern Brazil was conducted by the Laboratory of Veterinary Pathology (LVP) of the Universidade Federal de Santa Maria (UFSM). The results of that study demonstrated that the main cause of death in cattle was liver failure induced by the ingestion of *Senecio* spp. In an effort to compile a similar data set for sheep affected by plant toxicosis in southern Brazil, an 18-year (1990-2007) database search was done in the files of the LVP-UFSM and the results are reported here. Necropsy reports of sheep that died due to ingestion of toxic plants were evaluated.

## Epidemiology

Five hundred and seventeen sheep were examined in the LVP-UFSM over the past 18 years. Of these, 361 had a conclusive diagnosis, 67 (18.5%) of which were cases of toxic plant poisoning. Plants involved in the death of sheep in this period included *Nierembergia veitchii* (35 cases; 52.2%), *Senecio* spp. (22 cases; 32.8%), *Brachiaria brizantha* (4 cases; 5.9%), *Phytolacca decandra* (3 cases; 4.4%), *Erythroxylum argentinum* (2 cases; 3%), and *Solanum pseudocapsicum* (1 case; 1.5%). Additionally, in 23 cases a presumptive diagnosis of toxic plant poisoning was made based on the epidemiology, clinical signs, and pathology, but the identification or the confirmation of the possible specific plant involved as a cause was not determined. These cases include 13 of hepatocellular centrilobular necrosis, five of hepatogenous photosensitization, five of acute renal tubular necrosis, and one of subacute skeletal muscle degeneration and necrosis. Five cases of *Senecio* spp. poisoning in this study were biopsies that were done in sheep that later either died or were euthanized and necropsied.

©CAB International 2011. *Poisoning by Plants, Mycotoxins, and Related Toxins* (eds F. Riet-Correa, J. Pfister, A.L. Schild, and T.L. Wierenga)

## Poisoning by *Nierembergia veitchii*

Sheep affected by systemic mineralization due to ingestion of *N. veitchii* presented with two different clinical diseases. Twenty-four had a history of chronic disease with weight loss, stiff gait, tucked abdomen, kyphosis, and recumbency. Eleven sheep died suddenly with white froth oozing from the nostrils. Grossly, all affected sheep had poor body condition and widespread mineralization of soft tissues, more prominent in blood vessels and heart valves. Mineralization was characterized by multiple, firm to hard, irregular, whitish and opaque plaques in the intimal surface of large elastic arteries and endocardium. In the kidneys, lungs, and in the ruminal, omasal, abomasal, and uterine serosae mineralization was characterized by multifocal, white and chalky streaks. Additionally, sheep with sudden death had moderate to severe pulmonary edema. Histologically, soft tissue mineralization was characterized by multifocal deposition of finely granular, basophilic granules in soft tissues (more prominent in the arteries, endocardium, and lungs). In the arteries there was mineralization of the tunica media (with occasional cartilaginous and osseous metaplasia) and intimal proliferation. Cartilaginous and osseous metaplasia was usually observed in the alveolar septa. In some cases it was possible to observe multinucleated giant cells surrounding these foci of metaplasia.

Systemic mineralization due to ingestion of *N. veitchii* has been described as a chronic disease with severe weight loss and ill thrift (Barros *et al.* 1970, 1992; Riet-Correa *et al.* 1987). Cases with chronic disease occurred in the spring and summer which is the epidemiological pattern described in cases of poisoning by *N. veitchii*. Usually sheep will ingest the plant in the flowering period (September and October) and clinical cases will be detected from these months through February. All cases with sudden death were seen in the fall and winter (Rissi *et al.* 2007b). In these cases the toxicosis had a chronic development with an acute outcome. Probably these sheep with acute clinical disease also ingested the plant in the flowering period, but not in sufficient amounts to cause clinical disease. Stress could have played a role in the debilitation and precipitation of death. Mineralized heart valves and/or vasculature could predispose to heart failure in acutely stressed animals, resulting in terminal pulmonary edema and death (Barros *et al.* 1992).

## Poisoning by *Senecio* spp.

Most of the cases of poisoning by *Senecio* spp. in sheep were previously reported (Ilha *et al.* 2004). Affected sheep presented with ulcerative or crusting lesions of photosensitization on the external aspect of the ears, periocular areas, and nostrils. Less frequently there was wool loss in the dorsal areas of the body. Lethargy, weight loss, depression, incoordination, aimless walking, wide base stance, recumbency, and dyspnea were also commonly observed. In these cases affected sheep either died due to hepatic insufficiency or were euthanized *in extremis*. Six sheep died or were euthanized in consequence of acute hemolysis due to hepatogenous secondary copper poisoning. In these cases affected sheep also presented with intense icterus and dark red to brown urine (confirmed as hemoglobinuria). Grossly, all affected sheep had a small and diffusely firm liver with numerous, well circumscribed, white to yellow nodules 1-3 mm in diameter (regeneration nodules) distributed throughout the natural and cut surface. A diffuse, white, and fine reticular pattern (fibrosis) could be observed on the cut surface. Other consistent findings included ascites and hydropericardium. In addition, sheep that presented with hemolytic anemia had icterus, diffuse tan discoloration in the liver, dark red and swollen kidneys, and dark urine. Variable degrees of hepatic fibrosis with portal or bridging deposition of connective tissue, bile duct hyperplasia, hepatomegalocytosis, hepatocellular necrosis (of small groups or isolated hepatocytes), and regenerative nodules were seen histologically in all affected sheep. Macrophages with intracytoplasmic, dark brown or gray pigment were observed predominantly in the portal areas but also in the sinusoids. This pigment was confirmed to be either ceroid or copper (through periodic acid-Schiff and rhodanine special stains, respectively). Hepatocellular pseudoinclusions and multifocal lymphoplasmacytic inflammatory infiltrate were also seen in the liver. Additionally, multifocal tubular renal necrosis with intraepithelial, granular, dark brown pigment was observed in the kidneys of sheep that developed hemolytic anemia. This pigment was confirmed to be hemosiderin through Pearls's special stain. Intratubular casts of glassy, tan pigment (hemoglobin) were also seen in these cases. Spongy degeneration (periaxonal vacuolation of the myelin sheets) was observed in the brain of nine affected sheep. Changes were more prominent in the subcortical white matter in the cerebral cortex, thalamus, midbrain and brainstem. Epithelial necrosis, with serocellular crust formation, hyperkeratosis, and diffuse dermal neutrophilic infiltrate was observed in the skin.

Sheep are much more resistant than cattle to the action of pyrrolizidine alkaloids (PA) present in species of *Senecio* spp. (Craig *et al.* 1992; Huan *et al.* 1998). This resistance has been used to control the plant in areas where the toxicosis is a problem in cattle. However, sheep can be affected by PA toxicosis under certain circumstances such as i) intense infestation of the pastures by *Senecio* spp. or other plants with PA; ii) low numbers of sheep grazing in infested pastures; iii) long periods of time in infested pastures; and iv) starvation (Ilha *et al.* 2004). In these cases, affected sheep can develop either chronic hepatic insufficiency with hepatic fibrosis due to ingestion of PA for a long period or massive acute hepatic necrosis due to ingestion of high amounts of PA in a short period (Kellerman *et al.* 2007). In six cases in this study affected sheep developed acute hemolysis due to hepatogenous copper toxicosis. In these cases copper is released from the hepatocytes due to chronic hepatic disease caused by the action of PA. The copper reaches the blood stream and causes intravascular hemolysis with consequent icterus, hemoglobinuria, and renal tubular necrosis. This is a common outcome in cases of chronic PA toxicosis in sheep since the diffuse liver damage overwhelms the capacity of copper storage by the hepatocytes.

## Poisoning by *Brachiaria brizantha*

Only one outbreak of poisoning by *B. brizantha* was observed during these 18 years. Affected sheep developed intense facial edema and ulcerative and crusting lesions on the ears, periocular skin, nostrils, and vulva. Grossly there was mild to moderate icterus and marked hepatic lobular pattern. The liver had a diffuse tan to golden discoloration. In one case severe pulmonary edema was observed. Histologically there were variable degrees of portal hepatic fibrosis, hepatomegalocytosis, lymphohistioplasmacytic cholangitis, bile duct hyperplasia, and intracellular and intracanalicular bilestasis. Numerous macrophages with abundant and foamy cytoplasm were observed scattered throughout the parenchyma. No birefringent crystals were observed in these cases. Fragments of liver were submitted to lectin immunohistochemistry using *Canavalia ensiformis* agglutinin (Con A), *Dolichos biflorus* agglutinin (DBA), *Glycine max* agglutinin (SBA), *Arachis hypogaea* agglutinin (PNA), *Ricinus communis* agglutinin-I (RCA), and *Triticum vulgaris* agglutinin (WGA). These tests were performed in the Department of Veterinary Clinical Pathology of

Universidade Federal do Rio Grande do Sul. Cytoplasmic positive staining was observed in hepatocytes, foamy macrophages, and biliary epithelium as a finely granular, brown material. Lectin histochemistry is useful mostly when it is difficult to define the presence of foamy macrophages or cells displaying cytoplasmic vacuolation in cases of poisoning by *Brachiaria* spp.

## Poisoning by *Phytolacca decandra*

One outbreak of poisoning by *P. decandra* was observed and was reported elsewhere (Peixoto *et al.* 1997). Affected sheep did not show any clinical signs and died acutely. Grossly there was mild reddening of the ruminal mucosa. Histological findings included epithelial ruminal coagulative necrosis. Food restriction was attributed as the main factor that induced the ingestion of the plant in these cases. Experiments were done in our laboratory on two occasions (Peixoto *et al.* 1997; Ecco *et al.* 2001). In these cases sheep were fed *P. decandra* and developed clinical signs characterized by apathy, dyspnea, abdominal pain, reduced ruminal activity, hyperthermia, difficulty to stand, diarrhea, incoordination, muscular twitching, head-pressing, hyperesthesia, and hind limb ataxia. Gross and histological findings were similar to those observed in the natural cases. Cases of poisoning by *P. decandra* shared similar clinical and pathological characteristics to those described in cases of poisoning by *Baccharis coridifolia* in sheep in Rio Grande do Sul (Rozza *et al.* 2006) and *Baccharidastrum triplinervum* (Langohr *et al.* 2005).

## Poisoning by *Erythroxylum argentinum*

Four outbreaks were seen and two necropsies were performed in cases of poisoning by *E. argentinum* (Barros *et al.* 2004). Affected sheep presented with lethargy, trembling, wide base stance, and reluctance to move. When forced to move they developed an uncoordinated gait and fell. Some sheep died after exercise, with marked salivation, dyspnea and cyanosis. No gross or histological changes were seen. A similar disease was reported in sheep from other areas of the state of Rio Grande do Sul (Colodel *et al.* 2004). These cases were attributed to ingestion of *E. deciduum*.

## Poisoning by *Solanum pseudocapsicum*

One single case of poisoning by *S. pseudocapsicum* was observed. The affected sheep had trembling, ataxia, and salivation. No gross or histological changes were seen. The diagnosis was made based on the evidence that the affected animals had eaten *S. pseudocapsicum*. Attempts to reproduce the disease in sheep and in a rabbit in our laboratory by administration of *S. pseudocapsicum* via oral gavage were unsuccessful.

## Cases with Presumptive Diagnosis of Plant Toxicosis

Sheep that presented with hepatic centrilobular necrosis either died suddenly or had clinical signs of lethargy, depression, recumbency, opisthotonus, and bloody diarrhea. Gross findings include hepatic swelling, with marked diffuse lobular pattern characterized by red, depressed areas surrounded by a clear halo. Additionally there was mesenteric and abomasal edema, and multifocal serosal hemorrhages. Diffuse, centrilobular to massive hepatocellular necrosis was observed histologically. Potential plants that could have caused the disease in these cases include *Xanthium cavanillesii*, *Cestrum* spp., *Vernonia* spp., *Sessea brasiliensis*, *Dodonea viscosa*, and *Trema micrantha* (Rissi *et al.* 2007a). Acute poisoning by *Senecio* spp. causing hepatocellular necrosis has also been reported (Kellerman *et al.* 2007), and although it has not been reported in Brazil, it cannot be excluded in these cases.

Sheep with hepatogenous photosensitization presented with crusting and ulcerative lesions in the face. Two sheep also developed mandibular subcutaneous edema. Grossly all affected sheep had a diffusely tan liver and variable degree of icterus. Histologically all affected sheep had diffuse bilestasis.

No clinical records were given for the cases of renal tubular necrosis and no gross changes were seen. Histologically there was multifocal tubular epithelial necrosis. Additionally in two cases there were birefringent intratubular crystals. Additionally to tubular necrosis and deposition of crystals in the tubules, two sheep which had been placed in a pasture infested with *Amaranthus* sp. also presented with multifocal ruminal epithelial necrosis with deposition of birefringent crystals in the epithelium and submucosa. It is known that *Amaranthus* spp. may contain oxalates (Marshall *et al.* 1967; Hill and Rawate 1982). Although it was clear that on that occasion affected sheep had ingested the plant, we were unable to reproduce the disease in sheep fed *Amaranthus* sp.

Cases of degenerative myopathy were seen on one occasion. These sheep had ingested a ration made from agricultural byproducts. Affected sheep presented with dyspnea, falling down, and death. Grossly there were multifocal to coalescing pale areas distributed in several muscles. Histologically there was multifocal myofiber degeneration, necrosis and regeneration. On that occasion, the veterinary practitioner reported that horses, cattle, chickens, and a dog developed a similar disease at the same period after the ingestion of the same ration that the affected sheep had eaten. Experimentally, sheep fed the same ration for one month did not show any clinical signs.

## Conclusions

During 1990 to 2007, 517 sheep were examined in our laboratory. Out of these animals, 67 cases of plant toxicosis were diagnosed. Additionally, 23 diagnoses of presumptive toxic plant toxicosis were made. Poisoning by *N. veitchii* and *Senecio* spp. were by far the most important cause of death in sheep associated with toxic plants in this period. Cases of toxicosis by *B. brizantha*, *Phytolacca* spp., *E. argentinum*, and *S. pseudocapsicum* were uncommon or even rare, indicating that those plants are not an important cause of death in sheep flocks in southern Brazil.

## References

Barros RR, Teixeira FR, Oliveira FN, Rissi DR, Rech RR, and Barros CSL (2004). Poisoning in sheep from the ingestion of fruits of *Erythroxylum argentinum*. *Veterinary and Human Toxicology* 46:173-175.

Barros SS, Pohlenz J, and Santiago C (1970). Zur kalzinose beim schaf. *Deutsche Tier Wochen* 77:321-356.

Barros SS, Driemeier D, Santos MN, and Guerreiro JAM (1992). Evolução clínica e reversibilidade das lesões da calcinose enzoótica dos ovinos induzida por *Nierembergia veitchii*. *Pesquisa Veterinária Brasileira* 12:5-10.

Colodel EM, Seitz AL, Schmitz M, Borba MR, Raymundo DL, and Driemeier D (2004). Intoxicação por *Erythroxylum deciduum* (Erythroxylaceae) em ovinos. *Pesquisa Veterinária Brasileira* 24:165-168.

Craig AM, Latham CJ, Blythe LL, Schmotzer WB, and O'Connor OA (1992). Metabolism of toxic pyrrolizidine alkaloids from tansy ragwort (*Senecio jacobaea*) in ovine ruminal fluid under anaerobic conditions. *Applied and Environmental Microbiology* 58:2730-2736.

Ecco R, Barros CSL, and Irigoyen LF (2001). Intoxicação experimental por *Phytolacca decandra* em ovinos. *Ciência Rural* 31:319-322.

Hill RM and Rawate PD (1982). Evaluation of food potential, some toxicological aspects, and preparation of a protein isolate from the aerial part of amaranth (pigweed). *Journal of Agricultural and Food Chemistry* 30:465-469.

Huan J, Miranda CL, Buhler DR, and Cheeke PR (1998). Species differences in the hepatic microsomal enzyme metabolism of the pyrrolizidine alkaloids. *Toxicology Letters* 99:127-137.

Ilha MRS, Loretti AP, Barros SS, and Barros CSL (2004). Intoxicação espontânea por *Senecio brasiliensis* (Asteraceae) em ovinos no Rio Grande do Sul. *Pesquisa Veterinária Brasileira* 21:123-138.

Kellerman TS, Coetzer JAW, Naude TW, and Botha CJ (2007). *Plant poisonings and mycotoxicoses of livestock in South Africa*, 310 pp. Oxford University Press, Cape Town.

Langohr IM, Gava A, and Barros CSL (2005). Intoxicação por *Baccharidastrum triplinervium* (Asteraceae) em bovines. *Pesquisa Veterinária Brasileira* 25:235-238.

Marshall VL, Buck WB, and Bell GL (1967). Pigweed (*Amaranthus retroflexus*): an oxalate-containing plant. *American Journal of Veterinary Research* 28:888-889.

Peixoto PV, Wouters F, Lemos RA, and Loretti AP (1997). *Phytolocca decandra* poisoning in sheep in southern Brazil. *Veterinary and Human Toxicology* 39:302-303.

Riet-Correa F, Schild AL, Mendez MC, Wasserman R, and Krook L (1987). Enzootic calcinosis in sheep caused by the ingestion of *Nierembergia veitchii* (Solanaceae). *Pesquisa Veterinária Brasileira* 7:85-95.

Rissi DR, Driemeier D, Silva MC, Barros RR, and Barros CSL (2007a). Poisonous plants producing acute hepatic disease in Brazilian cattle. In *Poisonous Plants: Global Research and Solutions* (Panter KE, Wierenga TL, Pfister JA, eds), pp. 72-76. CAB International, Wallingford.

Rissi DR, Rech RR, Pierezan F, Kommers GD, and Barros CSL (2007b). Intoxicação em ovinos por *Nierembergia veitchii*: observações em quarto surtos. *Ciência Rural* 37:1393-1398.

Rozza DB, Raymundo DL, Correa AMR, Leal J, Seitz AL, Driemeier D, and Colodel EM (2006). Intoxicação espontânea por *Baccharis coridifolia* (Compositae) em ovinos. *Pesquisa Veterinária Brasileira* 26:21-25.

# Chapter 9

# Toxic Plants of the State of Piauí, Northeastern Brazil

S.M.M.S. Silva[1], G.W. Mello[1], F.A.L. Costa[1], C.J.S. Carvalho[1], M.V.F.L. Cavalcante[1], D.M. Oliveira[2], and F. Riet-Correa[2]

[1]*Centro de Ciências Agrárias, Universidade Federal do Piauí, Campus Socopo, 64049-550 Teresina, Brazil;* [2]*Hospital Veterinário, Universidade Federal de Campina Grande, Patos, Paraíba, 58700-000, Brazil*

## Introduction

The state of Piauí is located in the northeastern region of Brazil, between 2°44' and 10° 52' S latitude and between 40° 25' and 45° 59' West longitude, with a surface area of 251,529,186 km$^2$ (IBGE 2002; Medeiros 2004). The climate is tropical with a well defined rainy season from December to April and a dry season from May to November; temperature varies between 19 and 36°C and relative humidity varies between 40% and 80% (Medeiros 2004). Vegetation is semi-deciduous forest including cerrado (savanna) and caatinga (white forest because of the bleached appearance during the dry season) (IBGE 2002).

In the State of Piauí, knowledge about plants poisonous to livestock began to emerge during the 1950s when many plants were recognized as toxic (Tokarnia *et al.* 2000). Presently, 16 plants are known to be toxic in the State, and four plants are considered toxic by farmers and veterinarians, but there is no experimental evidence of their toxicity (Mello *et al.* 2010a). In this paper the poisonous plants of Piauí are reviewed.

## Poisonous Plants of Piauí

### Plants causing acute cardiac failure

*Mascagnia rigida* (Malpighiceae) is the most important toxic plant in the Brazilian semiarid region (Tokarnia *et al.* 1990), including municipalities such as Elesbão Velloso, Oeiras, Conceição do Canindé, and others in Piauí (Tokarnia 1993, unpublished). Variation in the toxicity of this plant is reported, and the occurrence of poisoning varies between regions and farms in the same region. It causes sudden death and mainly affects cattle. Sheep and goats may also be poisoned (Vasconcelos *et al.* 2008). Clinical signs only occur when the animals are exposed to physical exercise, and these signs are instability, muscle tremors, pedaling movements, opisthotonos, loud vocalization, tachycardia with arrhythmia,

©CAB International 2011. *Poisoning by Plants, Mycotoxins, and Related Toxins* (eds F. Riet-Correa, J. Pfister, A.L. Schild, and T.L. Wierenga)

and engorged, pulsing jugular vein. The animals have difficulty in standing up and may recover if they are not forced to walk. Some animals died in minutes or hours (Riet-Correa *et al.* 2006). Fluoracetate is likely the toxic principle of the plant (Cunha *et al.* 2006). Macroscopic lesions are not observed. Histologically, hydropic degeneration and necrosis of epithelial cells from the renal distal convoluted tubules may be observed in some animals (Tokarnia *et al.* 1990). *Mascagnia* aff. *rigida*, another species causing sudden death (Tokarnia *et al.* 2000), is also found in the municipalities of Elesbão Velloso, Conceição do Canindé, and others (Tokarnia 1993, unpublished).

**Plants that cause digestive alterations**

*Stryphnodendron coriaceum* (Leg. Mimoseae) is a tree found in the areas of cerrado and poisoning affects cattle after the ingestion of pods that fall during the dry season (July to September). Animals from other regions are more frequently affected than those raised in areas where the plant exists. In northern Piauí this tree has been responsible for the highest economic losses of any toxic plant, but cases of poisoning are decreasing each year as the trees are being cut down (Mello *et al.* 2010a). Clinical signs are apathy, dry nose, ruminal atony, muscular tremors, drooling, ataxia, and diarrhea. Photosensitization or abortion occurs in animals that survive for longer periods of time. At necropsy lesions are mesenteric and intestinal wall edema, dry contents in the intestine and forestomachs, liquid content, ulcers and membranous inflammation in the intestine, yellowish liver, and pale kidneys. Seeds of the plant are observed in the forestomachs and abomasum. The most important histopathologic lesions are pseudomembranous inflammation and necrosis in the intestine. In kidneys there is accumulation of proteinaceous material within the Bowman's space, tubular dilatation, hyaline cylinders, interstitial edema, and hyaline degeneration in the tubular cells. In the liver, hepatocytes show tumefaction and necrosis (Tokarnia *et al.* 1991).

*Enterolobium contortisiliquum* (Leg. Mimoseae) causes intoxication in cattle and goats due to a large intake of pods (Riet-Correa *et al.* 2006) that fall from September to November (Mello *et al.* 2010a). The toxic principles are probably saponins (Mimaki *et al.* 2003). Clinical signs are anorexia, fetid diarrhea, dehydration (Tokarnia *et al.* 1960; Marques *et al.* 1974), and in some cases photosensitization (Riet-Correa *et al.* 2009). Abortions are commonly reported by farmers as one of the main clinical signs in cattle (Mello *et al.* 2010a). Abortions were induced in experiments with guinea pigs (Bonel-Raposo *et al.* 2008). The main gross lesion is hemorrhagic enteritis. The liver may be yellowish (Motta *et al.* 2000). Histopathologic lesions are enlarged hepatocytes with diffusely vacuolated cytoplasm (Riet-Correa *et al.* 2009).

*Luetzelburgia auriculata* is a tree found in northern Piauí and its pods are toxic to goats (Mello *et al.* 2010b). The leaves are not toxic to cattle (Tokarnia *et al.* 2000). Clinical signs are anorexia, apathy, regurgitation, decreased ruminal movements or ruminal atony, bloat, pasty to liquid feces, and dehydration. Experimentally goats ingesting 2.5 g/kg BW died in 2 to 5 days, and those ingesting 0.5-1 g/kg BW recovered in 1-4 days. At necropsy the mucosa of the forestomachs is reddish and detaches easily from the underlying tissues. Other lesions are diffuse reddening of the mucosa of the abomasum, liver and kidney congestion, and edema of the mesentery and renal pelvis. On histologic examination the mucosa of the forestomachs shows diffuse vacuolation and ballooning degeneration of keratinocytes, with necrosis and vesicle and pustule formation in the epithelium. In some areas sloughing of the ruminal epithelium is observed (Mello *et al.* 2010b).

## Nephrotoxic plants

Outbreaks of poisoning by *Thiloa glaucocarpa* (Combretaceae) occur at the beginning of the rainy season, 10-15 days after the first rains. The intoxication affects cattle, with a case fatality rate of nearly 75% (Tokarnia *et al.* 2000). Tannins are probably the toxic compounds of this plant (Itakura *et al.* 1987). The main clinical signs observed are subcutaneous edema in the buttocks, perineum, supra-mammary region, prepuce, scrotum, and lateral-inferior abdominal wall. In some cases there is no subcutaneous edema, but only accumulation of abdominal fluid. Anorexia, ruminal atony, dry feces and mucus with or without blood are also observed. Serum values of urea, creatinine, and bilirubin are increased, and albumin, bile salts, and hyaline cylinders are present in the urine (Silva 1987). Main lesions observed at necropsy are accumulation of fluid in the cavities and edema of the mesentery, mesocolon, and perirenal tissue. The kidneys are pale and hemorrhages are observed in the epicardium, endocardium, trachea mucosal, abomasum, and intestine (Tokarnia *et al.* 1981). Histologically, there is degeneration and necrosis of tubular epithelium, and tubular dilatation (Tokarnia *et al.* 1981).

## Plants affecting the nervous system

*Ipomoea asarifolia* (Convolvulaceae) is the most common plant intoxication in northern Piauí and probably in other semiarid regions of the state. It affects sheep, goats, and cattle during periods of food shortage (Araújo *et al.* 2008), mainly in extensive farming systems (Mello *et al.* 2010a). It is found on the banks of rivers and lakes, beaches, vacant lands, road margins, and near inhabited places (Barbosa *et al.* 2005). Goats and sheep ingesting *I. asarifolia* at daily doses of 5 g/kg BW collected during the dry season had clinical signs after 19-31 days (Araújo *et al.* 2008). Clinical signs are muscular tremors in the limbs and head, head shaking, and sensitivity to noise and movements (Araújo *et al.* 2008). Generally, there are neither macroscopic nor microscopic lesions, except in cases of long duration, which then lead to degenerative lesions of the Purkinje cells and axonal spheroids in the granular layer of cerebellum (Guedes *et al.* 2007).

*Ipomoea carnea* subsp. *fistulosa* (Convolvulaceae) remains green during the whole year. Animals, mainly goats, become habituated or addicted to the plant as they consume them in preference to other forages, and later, by social facilitation, they induce other animals to eat it (Tokarnia *et al.* 1960; Riet-Correa *et al.* 2006). It is toxic to cattle, sheep, and goats (Tokarnia *et al.* 1960; Antoniassi *et al.* 2007; Armién *et al.* 2007), and contains the indolizidine alkaloid swainsonine, a well known inhibitor of lysosomal $\alpha$-mannosidase and Golgi mannosidase II, causing glycoprotein storage diseases. Calystegines $B_1$, $B_2$, $B_3$, and $C_1$ are also present in the plant (Haraguchi *et al.* 2003). The animals have progressive weight loss, depression, lateral head movements, nystagmus, opisthotonos, ataxia, spastic paralysis, weakness, and abnormal postural reactions (Armién *et al.* 2007). Animals may recover if removed from the pasture at the beginning of the clinical signs; if they consume the plant for longer periods of time the lesions become irreversible (Riet-Correa *et al.* 2006). There are no macroscopic lesions. Vacuolation is microscopically observed in neurons, pancreatic epithelial cells, kidney, and thyroid follicular cells. The poisoning by *I. carnea* subsp. *fistulosa* is frequent in southern Piauí.

*Ricinus communis* contains ricinine in its leaves and seeds (Audi *et al.* 2005) and causes muscular alterations. The intoxication in cattle occurs during the drought period. The clinical signs are unbalanced walking, muscular tremors, drooling, chewing movements, and sometimes eructation. There are no macroscopic lesions. Microscopically, the liver has mild vacuolar degeneration (Tokarnia *et al.* 2000).

**Hepatotoxic plants**

*Brachiaria decumbens*, *B. brizantha*, and *B. humidicola* (Poaceae) affect cattle (Lemos *et al.* 1997), sheep (Lemos *et al.* 1996), goats (Lemos *et al.* 1998), and horses (Barbosa *et al.* 2006). Sheep and young animals (lambs and calves) are more susceptible. The poisoning is more common in animals introduced for the first time into the pasture after being raised in pastures without *Brachiaria* spp. (Riet-Correa *et al.* 2009). These plants contain lithogenic steroidal saponins that induce the formation of crystals in the biliary system (Brum *et al.* 2004). The clinical signs are anorexia, depression, reduced ruminal movements, signs of pain, anxiety, jaundice, dark-brown urine, and photosensitization (Riet-Correa *et al.* 2009). At necropsy there is a variable degree of jaundice, edema mainly in the limbs, and yellowish, occasionally hard liver. In the cases of photosensitization the histopathological lesions are pericholangitis, portal fibrosis, degeneration and necrosis of hepatocytes, presence of foamy macrophages, and crystals within the macrophages or in lumen of bile ducts (Riet-Correa *et al.* 2006).

**Plants causing primary photosensitization**

*Froelichia humboldtiana* (Amaranthaceae) is found in northern Piauí and causes photosensitization in horses, mules, sheep, cattle, and goats. It is considered good forage and the poisoning occurs during the rainy season, when large amounts of the plant are available in the pasture. Primary photosensitization associated with this plant has been observed in northeastern Brazil for many years (Tokarnia, unpublished data). The animals generally recover when they are removed from the pastures. It mainly affects white-skinned animals. The lesions are located on the face, ear, around the eye, neck, withers, and back (Macedo *et al.* 2006, Pimentel *et al.* 2007a). The absence of ocular lesions suggests that the plant contains naphtodianthrone derivatives or similar substances (Pimentel *et al.* 2007a).

**Cyanogenic plants**

*Manihot* spp. (Euphorbiaceae) is a shrub which causes intoxication in ruminants after the first rains (Tokarnia *et al.* 2000; Riet-Correa *et al.* 2006). *Piptadenia macrocarpa* (Leg. Papilionoidea) is a tree and the animals have access to it when it is cut down or its branches are broken (Tokarnia *et al.* 1994). A large quantity of the plant must be ingested to cause intoxication. The plant remains toxic for a long time, even after being cut. The toxic principles of *Manihot* spp. and *P. macrocarpa* are unknown cyanogenic compounds. Acute clinical signs are dyspnea, anxiety, sialorrhea, cyanotic mucous, mydriasis, opisthotonus, muscular tremors, and convulsions. There are no significant macroscopic or microscopic lesions (Riet-Correa *et al.* 2009). The sodium picrate paper test is used for the diagnosis of the poisoning (Tokarnia *et al.* 1994).

**Teratogenic plants**

*Mimosa tenuiflora* (Leg. Mimosidae) leaves are normally consumed by ruminants as forage. The intoxication affects mainly pregnant goats and sheep, and cattle less often (Pimentel *et al.* 2007b, Riet-Correa *et al.* 2009). The number of affected newborns with malformations is variable. In some herds, the malformations are sporadic, affecting 1-10% of the animals, but at times affecting other herds with a higher incidence that can reach 100% of newborns. The higher incidences have been observed in sheep and goats supplemented with grain or byproducts at the end of the dry season in areas invaded by *M. tenuiflora*. After a rain event, the plant may resprout, even though precipitation is insufficient to provoke growth in other forage plants. In this situation, the females come into heat in response to the supplementation, but with *M. tenuiflora* the only available green forage, animals consume large quantities at the beginning of gestation (Riet-Correa *et al.* 2009). The main malformations are arthrogryposis, craniofacial anomalies, micrognathia, palatosquisis, and spinal malformations (Riet-Correa *et al.* 2009).

### Other toxic plants

Some well known toxic plants including *Lantana camara*, *Tephrosia cinerea*, and *Prosopis juliflora* are present in the region, but outbreaks of poisoning by these species had not been reported in the State (Mello *et al.* 2010a).

## Plants Suspected of Being Poisonous

The fruits of the tree *Buchenavia tomentosa* have been associated with diarrhea, weakness, weight loss, dry nose, and sometimes abortion or birth of weak animals at the end of pregnancy in goats and cattle (Mello *et al.* 2010a). The fruits fall from August to October. The oral administration of a single dose of 40 g/kg BW in two doses provoked abortion in one of the animals (Bandeira 2006).

The ingestion of *Brunfelsia* sp. has been associated with nervous signs in horses, mules, donkeys, cattle, sheep, and goats. Donkeys are the most severely affected because, according to farmers, the plant is more palatable to them. Intoxications always occur during the first part of the rainy season, which varies from December to March. *Brunfelsia* sp. shows variability in toxicity, depending on maturity. In an experiment with a sheep in March, with a dose of 9 g/kg BW, the animal showed nervous signs and recovered (Riet-Correa 2008, unpublished data); however, a month later, in a different experiment with goats, using plants collected in the same place, no apparent toxicity was noted (Silva 2008, unpublished data).

The ingestion of *Hybanthus ipecaconha* was associated with clinical signs of diarrhea and disequilibrium followed by death in cattle and goats during the rainy season. Tokarnia (1993, unpublished data) observed anorexia, ruminal atony, enlarged abdomen, thirst, nasal discharge, marked tachycardia, labored breathing, restlessness and death in two calves after ingestion of 20 and 30 g/kg BW of the plant. The fruits of *Spondias luta* are highly palatable and are eaten by goats to the exclusion of other foods, causing diarrhea, weakness, and death (Mello *et al.* 2010a). Studies are needed of all the plants mentioned in this section in order to verify their toxicity.

## Acknowledgements

This work has been financially supported by the National Institute for Science and Technology for the Control of Toxic Plants (CNPq), grant number 573534/2008-0.

## References

Antoniassi NAB, Ferreira EV, Santos CEP, Arruda LP, Campos JLE, Nakazato L, and Colodel EM (2007). Intoxicação espontânea por *Ipomoea carnea* subsp. *fistulosa* (Convolvulaceae) em bovinos no Pantanal Matogrossense. *Pesquisa Veterinária Brasileira* 27:415-418.

Araújo JAS, Riet-Correa F, Medeiros RMT, Soares MP, Soares MP, Oliveira DM, and Carvalho FKL (2008). Intoxicação Experimental por *Ipomoea asarifolia* (Convovulaceae) em caprinos e ovinos. *Pesquisa Veterinária Brasileira* 28(10):488-494.

Armién AG, Tokarnia CH, Peixoto PV, and Frees K (2007). Spontaneous and experimental glycoprotein storage disease of goats induced by *Ipomoea carnea* subsp *fistulosa* (Convolvulaceae). *Veterinary Pathology* 44:170-184.

Audi J, Belson M, Patel M, Schier J, and Osterloh J (2005). Ricin poisoning: a comprehensive review. *Journal American Medical Association* 294:2342-2351.

Bandeira YCM (2006). *Plantas tóxicas e casos de intoxicação na Microrregião de Balsas, sul do Maranhão,* 53 pp. Monografia. Universidade Federal de Campina Grande, Campina Grande, PB.

Barbosa JD, Oliveira CMC, Duarte MC, Peixoto PV, and Tokarnia CH (2005). Intoxicação experimental e natural por *Ipomoea asarifolia* (Convolvulaceae) em búfalos e outros ruminantes. *Pesquisa Veterinária Brasileira* 25(4):231-234.

Barbosa JD, Oliveira CMC, Tokarnia CH, and Peixoto PV (2006). Fotossensibilização hepatógena em equinos pela ingestão de *Brachiaria humidicola* (Gramineae) no Estado do Pará. *Pesquisa Veterinária Brasileira* 26(3):147-153.

Bonel-Raposo J, Riet-Correa F, Guim TN, Schuch ID, Grecco FB, and Fernandes CG (2008). Intoxicação aguda e aborto em cobaias pelas favas de *Enterolobium contortisiquiluum* (Leg. Mimosoideae*). Pesquisa Veterinária Brasileira* 28(12):593-596.

Brum KB, Haraguchi M, Garutti MB, Nóbrega FM, Rosa B, and Fioravante MC (2004). Análise semiquantitativa da saponina protodioscina do ciclo vegetativo de *Brachiaria decumbens* e *B. Brizantha*. *Pesquisa Veterinária Brasileira* 24(supl):13-14.

Cunha JD, Haraguchi M, Gorniak SL, Xavier FG, Riet-Correa F, and Florio JC (2006). *Palicourea marcgravii* e *Mascagnia rigida* um estudo por cromatografia em camada delgada (CCD). *Encontro de iniciação Científica*. FMZV, USP (abstract).

Guedes KMR, Riet-Correa F, Dantas AFM, Simões SVD, Miranda Neto EG, Nobre VMT, and Medeiros RMT (2007). Doenças do sistema nervoso central em caprinos e ovinos no semi-árido. *Pesquisa Veterinária Brasileira* 27(1):29-38.

Haraguchi M, Gorniak SL, Ikeda K, Minami H, Kato A, Watson AA, Nash R, Molyneux RJ, and Asano N (2003). Alkaloidal components in the poisonous plant *Ipomoea carnea* (Convolvulaceae). *Journal Agriculture Food Chemistry* 51:4995-5000.

Instituto Brasileiro de Geografia e Estatística (IBGE) (2002). Mapa da Se'rie Brasil Geogra´fico, second edn. Brasília.

Itakura Y, Habermehl G, and Mebs D (1987). Tannins occurring in the toxic Brazilian plant *Thiloa glaucocarpa*. *Toxicon* 25(12):1291-1300.

Lemos RAA, Ferreira LCL, Silva SM, Nazakato L, and Salvador SC (1996). Fotossensibilização e colangiopatia associada a cristais em ovinos em pastagem com *Brachiaria decumbens*. *Ciência Rural* 26(1):109-113.

Lemos RAA, Salvador SC, and Nakazato L (1997). Photosensitization and crystal associated cholangiohepatopathy in cattle grazing *Brachiaria decumbens* in Brazil. *Veterinary Human Toxicology* 39:376-377.

Lemos RAA, Nakazato L, Herrero Jr GO, Silvaira AC, and Porfirio LC (1998). Fotossensibilização e colangiopatia associada a cristais em caprinos, mantidos em pastagens de *Brachiaria decumbens* no Mato Grosso do Sul. *Ciência Rural* 28:507-510.

Macedo MC, Bezerra MB, and Soto-Blanco B (2006). Fotossensibilização em animais de produção na região semi-árida do Rio Grande do Norte. *Arquivos Instituto Biológico* 73:251-254.

Marques DC, Santos ML, Couto ES, Mello MA, Ribeiro RPM, and Ferreira PM (1974). Intoxicação experimental pelo tamboril *Enterolobium contortisiliquum* (Vell) Morong em bovinos. *Arquivos da Escola Veterinária*, UFMG 26(3):283-286.

Medeiros RM (2004). Estudo agrometeorológico para o Estado do Piauí. Secretaria do Meio Ambiente e Recursos Hídricos do Piauí, Teresina, pp.88-112.

Mello GWS, Oliveira DM, Carvalho CJS, Costa FAL, Riet-Correa F, and Silva SMMS (2010a). Toxic plants for ruminants and equidae in Northern Piauí. *Pesquisa Veterinária Brasileira* 30(1):1-9.

Mello GWS, Oliveira DM, Carvalho CJS, Cavacalte MVFL, Costa FAL, Riet-Correa F, and Silva SMMS (2010b). Poisoning of goats by the pods of *Luetzelburgia auriculata. Toxicon* 55:1115-1118.

Mimaki Y, Harada H, Sakuma C, Haraguchi M, Yui S, Kudo T, Yamazaki M, and Sashida Y (2003). Enterolosaponin A and B, novel triterpene bisdesmosidis from *Enterolobium contortisiliquum*, and evaluation for their macrophage-oriented cytotoxic activity. *Bioorganic and Medicinal Chemistry Letters* 13:623-627.

Motta AC, Rivero JRC, Schild AL, Riet-Correa F, Mendez MC, and Ferreira JL (2000). Fotossensibilização hepatógena em bovinos no Sul do Rio do Grande do Sul. *Ciência Rural* 30(1):143-149.

Pimentel LA, Riet-Correa F, Guedes KM, Macêdo JTSA, Medeiros RMT, and Dantas AFM (2007a). Fotossensibilização primária em eqüídeos e ruminantes no semi-árido causada por *Froelichia humboldtiana* (Amaranthaceae). *Pesquisa Veterinária. Brasileira.* 27(2):45-50.

Pimentel LA, Riet-Correa F,Gardner D, Panter KE, Dantas AFM, Medeiros RMT, Mota RA, and Araújo JAS (2007b). *Mimosa tenuiflora* as a cause of malformations in ruminants in northeastern Brazilian semiarid rangelands. *Veterinary Pathology* 44:928-931.

Riet-Correa F, Medeiros RMT and Dantas AFM (2006). *Plantas Tóxicas da Paraíba.* 1st edn. Patos: Centro de Saúde e Tecnologia Rural: SEBRAE/PB. 54 pp.

Riet-Correa F, Medeiros RMT, Pfister J, Schild AL, and Dantas AFF (2009). *Poisonings by plants, mycotoxins and related substances in Brazilian livestock,* 246 pp. Pallotti, Santa Maria.

Silva SV (1987). Aspectos clínicos, laboratoriais e anátomo-histopatológicos na intoxicação experimental por sipaúba (*Thiloa glaucocarpa* Eichl) em bovinos no Estado do Piauí, 89 pp. Dissertação de Mestrado, Universidade Federal Rural de Pernambuco, Recife.

Tokarnia CH, Canella CFC, and Döbereiner J (1960). Intoxicação experimental pela fava da 'timbaúba' (*Enterolobium contortisiliquum* (Vell.) Morong.) em bovinos. *Arquivos Instituto Biologia Animal* 3:73-81.

Tokarnia CH, Dobereiner J, Canella CFC, Cordeiro JEM, Silva ACC, and Araújo FV (1981). Intoxicação de bovinos por *Thiloa glaucocarpa* (Combretaceae) no nordeste do Brasil. *Pesquisa Veterinária Brasileira* 1:111-132.

Tokarnia CH, Peixoto PV, and Döbereiner J (1990). Poisonous plants affecting heart function of cattle in Brazil. *Pesquisa Veterinária Brasileira* 10:1-10.

Tokarnia CH, Peixoto PV, Gava A, and Dobereiner J (1991). Intoxicação experimental por *Stryphnodendron coriaceum* (fam. Leg. Mimosoidae) em bovinos. *Pesquisa Veterinária Brasileira* 11(1/2):25-29.

Tokarnia CH, Peixoto PV, and Döbereiner J (1994). Intoxicação experimental por *Piptadenia macrocarpa* (Leg. Mimosideae) em bovinos. *Pesquisa Veterinária Brasileira* 14(2/3):57-63.

Tokarnia CH, Döbereiner J, and Peixoto PV (2000). *Plantas Tóxicas do Brasil,* 320 pp. Helinthus, Rio de Janeiro.

Vasconcelos JS, Riet-Correa F, Dantas AFM, Medeiros RMT, Galiza GJN, Oliveira DM, and Pessoa AFA (2008). Intoxicação por *Mascagnia rigida* (Malpighiaceae) em ovinos e caprinos. *Pesquisa Veterinária Brasileira* 28(10):521-526.

# Chapter 10

# Poisonous Plants Affecting Ruminants in Southern Brazil

N.A.B. Antoniassi, D.L. Raymundo, F.M. Boabaid, G.D. Juffo, P.M. Bandarra, P.M.O. Pedroso, C.E.F. da Cruz, and D. Driemeier

*Setor de Patologia Veterinária, Universidade Federal do Rio Grande do Sul, 91540-000, Porto Alegre, RS, Brazil*

## Introduction

Plant poisoning in livestock is quite common in Brazil, especially in ruminants. Poisoning from plants causes not only severe direct losses but also substantial indirect economic losses (Riet-Correa and Medeiros 2001). A number of factors may trigger the consumption of those plants by ruminants, but primary factors may be the lack of suitable forage, introduction of naïve animals to weed-infested areas, water deprivation of long duration, or palatability and availability of the toxic plants (Riet-Correa and Méndez 1993). There are numerous poisonous plants in Brazil, and although a considerable volume of research has been developed on the subject (Tokarnia *et al.* 1979; Riet-Correa *et al.* 1993; Tokarnia *et al.* 2000; Riet-Correa and Mendez 2008), information on the prevalence of poisoning cases remains scarce. The annual mortality due to the consumption of poisonous plants in Rio Grande do Sul (RS) is assumed to be 10 to 14% in cattle and 15 to 20% in sheep (Riet-Correa and Medeiros 2001). It is also assumed that approximately 60% of all the deaths caused by toxic plants in Brazil are associated with those plants that may cause sudden death, of which *Palicourea marcgravii* is most important. *Palicourea marcgravii* is responsible for significant losses in almost all regions of the country, except for southern Brazil, where *Mascagnia exotropica* is the most common cause of sudden death (Tokarnia *et al.* 1990; Gava *et al.* 1998; Riet-Correa and Méndez 2008).

*Pteridium aquilinum*, another major poisonous plant in the country, causes important losses in southern Brazil (Rissi *et al.* 2007; Anjos *et al.* 2008) and is the primary cause of cattle death in Santa Catarina (Riet-Correa and Medeiros 2001; Gava *et al.* 2002). *Senecio* spp. is the toxic plant most frequently associated with poisoning and death of cattle in Rio Grande do Sul (Riet-Correa and Medeiros 2001; Karam *et al.* 2004; Rissi *et al.* 2007).

Gastroenteric lesions from *Baccharis* sp. (Rozza *et al.* 2006), neurological disorders from *Erythroxylum deciduum* (Colodel *et al.* 2004) and *Sida carpinifolia* (Seitz 2003), and the enzootic calcinosis from *Nierembergia veitchii* (Riet-Correa and Medeiros 2001; Rissi *et al.* 2007) are the most important intoxications caused by consumption of poisonous plants in sheep. Sheep have also had episodes of intoxication associated with the

©CAB International 2011. *Poisoning by Plants, Mycotoxins, and Related Toxins* (eds F. Riet-Correa, J. Pfister, A.L. Schild, and T.L. Wierenga)

consumption of *Mascagnia exotropica* (Silva *et al.* 2008; Vasconcelos *et al.* 2008), *Brachiaria decumbens*, and *Cynodon* sp. (Riet-Correa and Méndez 2008).

The lysosomal storage disease due to ingestion of *Sida carpinifolia* by goats is the main cause of death in this species in RS (Colodel *et al.* 2002). Cyanide intoxication from the ingestion of *Prunus sphaerocarpa* and *Cynodon* spp. (Riet-Correa and Méndez 2008), and the acute hepatic necrosis linked to consumption of *Trema micrantha* (Traverso *et al.* 2003), are also important causes of losses in goats.

## Results

A retrospective study was conducted using the records of the Setor de Patologia Veterinária da Universidade Federal do Rio Grande do Sul (SPV–UFRGS) for the period of 2000-2008. Necropsy and submitted samples for histology on spontaneous plant poisoning cases involving cattle, sheep, and goats in RS were evaluated. Diagnoses were made by the analysis of clinical, pathological, and epidemiological findings (presence of plants and evidence of consumption), besides experimental induction for confirming toxicity. In a total of 5582 cases in the period, 84.9% (4738), 9.2% (516), and 5.9% (328) affected cattle, sheep, and goats, respectively. Of the 5582 cases, 255 (4.6%) were attributed to poisonous plants, of which 79.6% (203) affected cattle, 12.1% (31) sheep, and 8.2% (21) goats. The prevalence attributed to each toxic plant from poisonings in RS is presented in Table 1.

**Table 1.** Cases of plant poisoning in ruminants diagnosed at the Setor de Patologia Veterinária da UFRGS during the period 2000-2008.

| Poisonous plant* | Cattle (%) | Sheep (%) | Goats (%) |
|---|---|---|---|
| *Ateleia glazioviana* | 1.5 | - | - |
| *Baccharis* sp. | 2.0 | 19.0 | - |
| *Brachiaria decumbens* | 1.5 | 10.0 | 4.8 |
| *Cestrum intermedium* | 1.0 | - | - |
| *Cynodon* sp. | - | - | 4.8 |
| *Dodonea viscose* | 1.5 | - | - |
| *Erythroxylum deciduum* | - | 29.0 | - |
| *Mascagnia exotropica* | 5.0 | 6.5 | 9.5 |
| *Nerium oleander* | 1.5 | - | - |
| *Nierembergia veitchii* | - | 6.5 | - |
| *Prunus sphaerocarpa* | - | - | 9.5 |
| *Pteridium aquilinum* | 5.0 | - | - |
| *Senecio* sp. | 72.0 | 3.0 | - |
| *Senna occidentalis* | 1.0 | - | - |
| *Sida carpinifolia* | 2.5 | 26.0 | 62.0 |
| *Solanum fastigiatum* | 0.5 | - | - |
| *Trema micrantha* | 1.0 | - | 9.5 |
| *Vicia villosa* | 3.0 | - | - |
| *Xanthium cavanillesii* | 1.0 | - | - |

*In total, 255 cases were attributed to poisonous plants.

The main plants causing disease in ruminants in RS were *Senecio* spp. with the highest number, followed by *Sida carpinifolia*. In cattle, 71.9% (24 necropsies and 122 histopathological examinations) of the cases were linked to the chronic hepatic lesions caused by consumption of *Senecio* spp. In sheep, there were 22 necropsies and nine histopathological exams of cases of plant poisoning, most of which were associated with the neurological disorders due to the consumption of *Erythroxylum deciduum* (29%) and *S. carpinifolia* (25.8%). Finally, in goats the total of cases associated with plant poisoning was 6.4%, 62% of which were due to the lysosomal storage disorder caused by consumption of *S. carpinifolia.*

## Discussion and Conclusions

This examination of veterinary records indicates that there were substantial losses of ruminants from plant poisonings in the state of RS, as has been shown in other studies (Riet-Correa and Medeiros 2001; Rissi *et al.* 2007). There were minor differences between this study and previous work probably associated with regional particularities such as plant distribution. On the other hand, all of the studies have consistently shown the importance of *Senecio* spp. poisoning as the major cause of cattle loss associated with plant poisoning. Therefore, these results confirm the position of *Senecio* spp. poisoning as the second leading cause of adult cattle death in RS, after anaplasmosis (Karam *et al.* 2004). For small ruminants, however, *Sida carpinifolia* was the poisonous plant that caused the biggest impact.

Information on the prevalence of losses caused by poisonous plants in different ruminant species is fundamental in order to make appropriate management decisions for the control and prevention of diseases affecting livestock. Knowing the main causes of losses makes it easier to decide on priorities to reduce and prevent losses. Those data are also useful to determine differential diagnosis of diseases which may affect the different ruminant species in RS.

## References

Anjos BL, Irigoyen LF, Fighera RA, Gomes AD, Kommers GD, and Barros CSL (2008). Intoxicação aguda por samambaia (*Pteridium aquilinum*) em bovinos na Região Central do Rio Grande do Sul. *Pesquisa Veterinária Brasileira* 28(10):501-507.

Colodel EM, Driemeier D, Loretti AP, Gimeno EJ, Traverso SD, Seitz AL, and Zlotowski P (2002). Aspectos clínicos e patológicos da intoxicação por *Sida carpinifolia* (Malvaceae) em caprinos no Rio Grande do Sul. *Pesquisa Veterinária Brasileira* 22(2):51-57.

Colodel EM, Seitz AL, Schimitz M, Borba MR, Raymundo DL, and Driemeier D (2004). Intoxicação por *Erythroxylum deciduum* (Erythoxylaceae) em ovinos. *Pesquisa Veterinária Brasileira* 24(3):165-168.

Gava A, Cristani J, Branco JV, Neves DS, Mondadori AJ, and Sousa RS (1998). Mortes súbitas em bovinos causadas pela ingestão de *Mascagnia* sp. (Malpighiaceae), no Estado de Santa Catarina. *Pesquisa Veterinária Brasileira* 18(1):16-20.

Gava A, Neves DS, Gava D, Moura ST, Schild AL, and Riet-Correa F (2002). Bracken fern (*Pteridium aquilinum*) poisoning in cattle in southern Brazil. *Veterinary and Human Toxicology* 44(6):362-5.

Karam FSC, Soares MP, Haraguchi M, Riet-Correa F, Méndez MC, and Jarenkow JA (2004). Aspectos epidemiológicos da seneciose na região sul do Rio Grande do Sul. *Pesquisa Veterinária Brasileira* 24:191-198.

Riet-Correa F and Medeiros RMT (2001). Intoxicações por plantas em ruminantes no Brasil e no Uruguai: importância econômica, controle e riscos para a saúde pública. *Pesquisa Veterinária Brasileira* 21(1):38-42.

Riet-Correa F and Méndez MC (1993). Introdução ao estudo das plantas tóxicas. In *Intoxicações por plantas e micotoxicoses em animais domésticos*. (F Riet-Correa, MC Méndez and AL Schild, eds), pp.1-20. Hemisfério Sul, Pelotas.

Riet-Correa F, Méndez MC, and Schild AL, eds (1993). *Intoxicações por plantas e micotoxicoses em animais domésticos*, 340 pp. Hemisfério Sur, Montevideo.

Riet-Correa F and Méndez MC (2008). *Plantas Tóxicas e Micotoxicoses,* 298 pp. Editora Universitária/UFPel, Pelotas.

Rissi DR, Rech RR, Pierezan F, Gabriel AL, Trost ME, Brum JS, Kommers GD, and Barros CSL (2007). Intoxicações por plantas e micotoxinas associadas a plantas em bovinos no Rio Grande do Sul: 461 casos. *Pesquisa Veterinária Brasileira* 27(7):261-268.

Rozza DB, Raymundo DL, Corrêa AMR, Leal J, Seitz AL, Driemeier D, and Colodel EM (2006). Intoxicação espontânea por *Baccharis coridifolia* (Compositae) em ovinos. *Pesquisa Veterinária Brasileira* 26(1):21-25.

Seitz AL (2003). Doença do Armazenamento lisossomal induzida pelo consumo de *Sida carpinifolia* (Malvaceae) em ovinos, 62 pp. Dissertação de Mestrado em Ciências Veterinárias, Universidade Federal do Rio Grande do Sul, Porto Alegre, Rio Grande do Sul.

Silva IP, Lira RA, Barbosa RR, Batista JS, and Soto-Blanco B (2008). Intoxicação natural pelas folhas de *Mascagnia rigida* (Malpighiaceae) em ovinos. *Arquivos do Instituto Biológico* 75(2):229-233.

Tokarnia CH, Döbereiner J, and Silva MF (1979). *Plantas Tóxicas da Amazônia a Bovinos e Outros Herbívoros,* 95 pp. Instituto Nacional de Pesquisas da Amazônia, Manaus.

Tokarnia CH, Peixoto PV, and Döbereiner J (1990). Poisonous plants affecting heart function of cattle in Brazil. *Pesquisa Veterinária Brasileira* 10:1-10.

Tokarnia CH, Döbereiner J, and Peixoto PV (2000). *Plantas Tóxicas do Brasil,* 310 pp. Editora Helianthus, Rio de Janeiro.

Traverso SD, Colodel EM, Loretti AP, Correia AM, and Driemeier D (2003). Intoxicação natural por *Trema micrantha* em caprinos. *Ciência Rural* 33(1):133-136.

Vasconcelos JS, Riet-Correa F, Dantas AFM, Medeiros RMT, Galiza GJN, Oliveira DM, and Pessoa AFA (2008). Intoxicação por *Mascagnia rigida* (Malpighiaceae) em ovinos e caprinos. *Pesquisa Veterinária Brasileira* 28(10):521-526.

# Chapter 11

# Recently Diagnosed Poisonous Plants in the Cariri Region, State of Paraíba, Brazil

C.R.M. Pessoa, A.F.A. Pessoa, A.F.M. Dantas, R.M.T. Medeiros, and F. Riet-Correa

*Veterinary Hospital, CSTR, Federal University of Campina Grande, Campus of Patos, 58700-970, Patos, PB, Brazil*

## Introduction

The Cariri region is situated in the semiarid region of the state of Paraíba, Brazil. Its flora belongs to the caatinga biome, which is found only in the Brazilian semiarid (see Chapter 1 of this book). Several poisonous plants have been reported in ruminants and equidae in the region (Riet-Correa *et al.* 2006). This paper reports three previously unknown poisonous plants for ruminants and horses in the Cariri region.

## Poisoning by *Arrabidae corallina* in Goats

Diarrhea, increased intestinal movements, and depression were observed in goats during the dry season on a farm in the municipality of Boqueirão at the end of 2005. Fifty-six (10.2%) of a flock of 550 goats older than 1 year were affected. Six animals (1.1%) died; the others recovered in a period of 1-2 weeks after having been removed from the paddock. The pasture had low forage availability and large amounts of *Arrabidae corallina* (Bignoneaceae), which was the only green plant observed in the area. One adult goat was euthanized and necropsied. At necropsy the gut had liquid, fetid and blackish content, and catarrhal enteritis. Non-suppurative acute, diffuse, and moderate enteritis was observed histologically, with occasional presence of *Eimeria* spp. Non-significant macroscopic or histologic lesions were observed in other tissues. The disease was experimentally reproduced in four 6 to12-month-old Moxoto goats. Two other goats were used as control. The experimental goats received daily doses of 15 g of fresh plant per kg body weight. The plant was collected from the farm where the disease occurred and was kept refrigerated until administration within 1 week after collection. It was administered by putting small quantities in the mouth of the experimental animals. All animals were fed with concentrated commercial ration in amounts equivalent to 1% body weight, and *Cynodon dactylon* hay and water *ad libitum*. All the experimental goats had diarrhea with dark feces 3-4 days after the start of the ingestion and recovered 5-6 days after the end of the administration. The animals of the control group had no clinical signs (Pessoa *et al.* 2010b). It is concluded that *A. corallina* was responsible for the outbreak of diarrhea in goats, but that parasitic disease or malnutrition can be a concomitant factor to cause the death of the animals. Poisoning by *A. corallina* has to be differentiated from poisoning by *Enterolobium contortisiliquum* in goats, which is also a cause of diarrhea in the semiarid region (Benício *et al.* 2007). Parasitic diarrhea and *Salmonella* spp. infection should be consider also in the differential diagnosis. Until now only two species of *Arrabidaea*, *A. bilabiata* and *A. japurensis*, were reported as toxic, causing sudden death in cattle in the Brazilian Amazon region (Tokarnia *et al.* 2000). The withdrawal of the goats from paddocks invaded by *A. corallina* is recommended if during the dry season there is no availability of other forage in the area.

## Malformations Caused by *Mimosa ophthalmocentra*

*Mimosa tenuiflora* (Leguminosae) has been reported to cause malformations in goats, sheep, and cattle in the Brazilian semiarid region (Riet-Correa *et al.* 2006; Pimentel *et al.* 2007). Malformations were reproduced experimentally in goats which consumed the plant during the entire pregnancy (Pimentel *et al.* 2007), and in rats which consumed a ration containing 10% seeds of *M. tenuiflora* between the 6th and 21st day of pregnancy (Medeiros *et al.* 2007).

*M. ophthalmocentra* is a common plant in the Cariri region where malformations in goats and sheep are frequent and *M. tenuiflora* is not found or is found in lesser amounts than *M. ophthalmocentra.* This experiment was performed in rats to test the embryotoxicity and fetotoxicity of *M. ophthalmocentra.*

Twenty-four Wistar female rats were divided into two groups of 12 rats each. One group was fed with a ration contained 10% *M. ophthalmocentra* seeds between day 6 and day 21 of pregnancy, and the control group received only their normal ration. Water was given *ad libitum*. On the 21st day of pregnancy the rats were anesthetized by ether inhalation and the ovaries and uteri were removed by cesarean section. The number of corpora lutea in each ovary was recorded and the gravid uterus was weighed. The fetuses were removed from the uteri, dried of amniotic fluid, weighed and examined for conformation of the eyes, mouth, head, limbs, tail, and ears, and the presence of the anal perforation, with the aim to verify external abnormalities and malformations. The placentas and the live fetuses were weighed. The number of implantation sites and resorptions was recorded in both uterine horns. After being weighed the fetuses were euthanatized with ether, fixed in acetone for 24 hours, examined for cleft palate, and eviscerated. For examination of the skeleton, the fetuses were submersed in a solution of 0.8% potassium hydroxide with alizarin-red S, which was changed daily for 3-4 days (Staples and Schnell 1964). Then the fetuses were cleared in a solution of 40% ethylic alcohol, 40% glycerin, and 20% benzilic alcohol. The degree of fetal bone development was evaluated by counting the ossification centers in some fetal bones (phalanges of the fore limbs, metacarpus, metatarsus, sternebrae, and caudal vertebrae). The kidney, lung, and liver of the fetuses were weighed. The rats were euthanized and necropsied. Lung, liver, heart, and kidneys were weighed and samples of these tissues were fixed in 10% buffered formalin, embedded in paraffin, sectioned at 5 μm, and stained by hematoxylin and eosin for histologic examination. The software GraphPad Instat V2.01 (GraphPad 1993) was used for the statistical analysis. Food consumption, water ingestion, body weight gains, organ weights, and date of the offspring were analyzed by the Student 't' test. Frequency of skeletal abnormalities and malformations were evaluated by Fisher's exact test. The percentages of pre- and post-implantation losses and the degree of ossification were examined using the Mann-Whitney U test.

There were no significant differences between groups in weight gain, food and water consumption, and weight of liver, lung, heart, and kidneys. No significant histologic lesions were observed in these organs. These data suggest that *M. ophthalmocentra* was not toxic to the dam. In five rats of the experimental group all embryos were dead and in the other rats the fetuses were significantly lighter that the control fetuses. The placental weight of the treated rats was also significantly lower than that of the control rats. Bone malformations were observed in 66% of the fetuses of the experimentally group and 1.14% of the control group ($P > 0.001$) (Table 1).

**Table 1.** Type and frequency of bone malformations in rat fetuses from the control group and the experimental group treated with *Mimosa ophthalmocentra* seeds

| Malformation | Control group (*n*=68) | Experimental group (*n*=59) |
|---|---|---|
| Aplasia of one or more sternebraes | 5 | 6 |
| Hypoplastic sternebrae | 0 | 2 |
| Bifid sternebrae | 1 | 2 |
| Rudimentary sternebrae | 3 | 14** |
| Asymmetric sternebrae | 6 | 13* |
| Disarranged sternebrae | 0 | 3 |
| Hypoplasia of 13th rib | 0 | 3 |
| Aplasia of one or more ribs | 0 | 3 |
| Aplasia of supraoccipital | 2 | 12** |
| Hypoplasia of supraoccipital | 5 | 13* |
| Aplasia of interparietal | 0 | 1 |
| Hypoplasia of interparietal | 0 | 6** |
| Deformed occipital | 0 | 9*** |
| Cleft palate | 0 | 1 |

*$P < 0.05$, **$P < 0.01$, ***$P < 0.001$ in Exact Fisher's Test.

These results demonstrate that *M. ophthalmocentra* seeds are embryotoxic and fetotoxic in rats causing embryonic deaths, poor development of fetuses, and malformations. The malformations observed in this experiment are similar to those observed in rats that ingested seeds of *M. tenuiflora*, suggesting that *M. ophthalmocentra* also causes the malformations observed in the Brazilian semiarid. However, the experimental reproduction of malformations in ruminants is necessary to connect *M. ophthalmocentra* to spontaneous outbreaks of malformations in ruminants. Embryonic death was also observed in rats ingesting *M. ophthalmocentra.* Recently it was demonstrated experimentally in goats that *M. tenuiflora* is also an important cause of embryonic deaths (Dantas 2009).

## Tremorgenic Disease in Ruminants and Equidae

Eight outbreaks of a tremorgenic disease were observed in ruminants and equidae of different ages. In total, the disease affected 17 out of 29 horses, 1 out of 2 mules, 19 out of 72 sheep and 3 out of 40 bovines. Two horses and four sheep died. Seven outbreaks occurred in 2007, from July to December, with the highest frequency in September and October. Another outbreak was observed in February 2008. All outbreaks occurred during the dry period in paddocks with low forage availability. In six outbreaks in equidae the animals were grazing in cultures of *Opuntia ficus-indica* invaded by different grasses. In one outbreak in sheep and another in cattle the animals were grazing in deforested areas of caatinga. Clinical signs were staggering, hypermetria, ataxia, wide-based stance, and alertness. After being removed from pastures, the animals recovered in a period of 3-4 days to 2 weeks, however when returned to the pasture, clinical signs reappeared. One affected sheep was euthanized and necropsied and no gross or histological lesions were observed. In all outbreaks the pastures were mature and dry, thus it was not possible to identify the grass species. Two horses were fed *ad libitum* for 7 days with mature forage collected in pastures where the disease occurred. One horse showed mild signs of the disease on the fifth day of consumption, but recovered 1 day later. After the rainy period the farms were visited and the main grasses present in the paddocks were identified as *Chloris virgata*, *Chloris barbata*, *Enteropogon mollis,* and *Digitaria bicornis* (Pessoa *et al.* 2010a). Previous reports mentioned the occurrence of a similar disease, between 1956 and 1962, in the semiarid region of Pernambuco in pastures with *Chloris orthonothon* (Tokarnia 1962). A similar disease was also observed in 2005 in cattle in the state of Rio Grande do Norte (F. Riet-Correa, unpublished data).

Two weeds (*Herssantia crispa* and *Jacquemontia* sp.) observed in some paddocks were administered to experimental sheep and horses with negative results. Some farmers claimed that the disease was caused by *Passiflora foetida*; however the administration of this weed to goats caused cyanide poisoning but no tremorgenic disease (Carvalho 2009).

In the northeast of Brazil, the only known tremorgenic plant is *Ipomoea asarifolia* (Medeiros *et al.* 2003) which was not present in the paddocks where the disease occurred. We suggest that the disease was caused by a tremorgenic toxin produced by endophytic fungi probably infecting *Chloris* spp. pastures. The only recommendation to control the disease is to remove the animals from the paddocks after the first clinical signs of the disease.

## Acknowledgements

This work has been financially supported by the National Institute for Science and Technology for the Control of Toxic Plants (CNPq), grant number 573534/2008-0.

## References

Benício TMA, Nardelli MJ, Nogueira FRB, Araújo JAS, and Riet-Correa (2007). Intoxication by the pods of *Enterolobium contortisiliquum* in goats. In *Poisonous Plants: Global Research and Solutions* (KE Panter, TL Wierenga, and JA Pfister, eds), pp. 514-519. CABI Publishing, Wallingford, Oxon, UK.

Carvalho FKL (2009). Intoxicação experimental por *Passiflora* sp. em caprinos, 29 pp. Monografia de Graduação em Medicina Veterinária, Centro de Saúde e Tecnologia Rural, Universidade Federal de Campina Grande, Patos, PB.

Dantas AF (2009). Malformações e morte embrionária em ruminantes causadas pela ingestão de *Mimosa tenuiflora* (jurema preta), 68 pp. Tese de Doutorado, Programa de Pós-Graduação em Ciências Veterinárias, Universidade Federal Rural de Pernambuco, Recife, PE.

GraphPad Instat (1993). V2.01. San Diego: *GraphPad Software*.

Medeiros RMT, Barbosa RC, Riet-Correa F, Lima EF, Tabosa IM, Barros SS, Gardner DR, and Molyneux RJ (2003). Tremorgenic syndrome in goats caused by *Ipomoea asarifolia* in Northeastern Brazil. *Toxicon* 41:933-935.

Medeiros RMT, Figueiredo APM, Benício TMA, Dantas AFM, and Riet-Correa F (2007). Teratogenicity of *Mimosa tenuiflora* seeds to pregnant rats. *Toxicon* 51:316-319.

Pessoa CRM, Medeiros RMT, Dantas AF, Oliveira OF, and Riet-Correa F (2010a). Doença tremorgênica em ruminantes e equídeos no semiárido da Paraíba. *PesquisaVeterinária Brasileira* 30(7):541-546.

Pessoa CRM, Medeiros RMT, Pessoa AFA, Araújo JA, Dantas AFM, Silva-Castro MM, and Riet-Correa F (2010b). Diarreia em caprinos associada ao consumo de *Arrabidaea corallina* (Bignoniaceae). *Pesquisa Veterinária Brasileira* 30(7):547-550.

Pimentel LA, Riet-Correa F, Gardner D, Panter KE, Dantas AFM, Medeiros RMT, Mota RA, and Araújo JAS (2007). *Mimosa tenuiflora* as a cause of malformations in ruminants in the Northeastern Brazilian semiarid rangelands. *Veterinary Pathology* 44:928-931.

Riet-Correa F, Medeiros RMT, and Dantas AFM (2006). *Plantas Tóxicas da Paraíba*, pp. 49-50. Centro de Saúde e Tecnologia Rural, PB, SEBRAE/PB.

Staples RE and Schnell VL (1964). Refinements in rapid clearing technique in the KOH-alizarin red S method for fetal bone. *Stain Tecnology* 39:61-63.

Tokarnia CH (1962). *Relatório de viagem á região do agreste do Estado de Pernambuco*. Seção de Anatomia Patológica do Instituto de Biologia Animal, Rio de Janeiro, 09/02/1962. Non-published report.

Tokarnia CH, Döbereiner J, and Peixoto PV (2000). *Plantas tóxicas do Brasil*. pp. 30-35. Editora Helianthus, Rio de Janeiro.

# Chapter 12

# Poisonous Plants on Dairy Farms of the Caparaó Microregion, Espírito Santo State, Brazil

E.V. Oliveira[1], C.M. Scardua[2], M.D. Dórea[3], and L.C. Nunes[4]

*[1]Veterinary Medicine graduate student, Universidade Federal do Espírito Santo; [2]Veterinary practitioner, João Neiva, Espírito Santo, Brazil; [3]Veterinary Medicine postgraduate student, Universidade Federal do Espírito Santo; [4]Department of Veterinary Medicine, Universidade Federal do Espírito Santo, Alto Universitário, PO Box 16, Alegre, Espírito Santo, Brazil, 29500-000*

## Introduction

Dairies in the Caparaó microregion, in the southern part of Espírito Santo state, Brazil, are economically important, ranking third in milk production with 46,696,000 l of milk per year (IBGE 2007). There has been no estimate of losses on dairy farms in the region from toxic plants.

Losses due to plant poisonings can be direct or indirect. Direct losses cause the death of animals, decreased reproductive rates (abortion, infertility, and malformations), reduced productivity in surviving animals, transient or subclinical diseases, decreased milk, meat or wool production, and increased susceptibility to other diseases due to immune depression. Indirect losses include costs of controlling toxic plants in pastures, management measures to prevent poisoning including the use of fences and alternative grazing, reduction of the value of forage due to the delay in use, reduced land value, purchase of livestock to replace the dead animals, and expenses associated with the diagnosis of poisoning and treatment of affected animals (Riet-Correa and Méndez 1993; James 1994; Riet-Correa and Medeiros 2001). In Brazil very frequently losses due to toxic plants are erroneously attributed to diseases such as anthrax or snake venom poisoning (Tokarnia *et al.* 2000).

In the state of Rio Grande do Sul it is estimated that the annual mortality of cattle is around 5%. From 10 to 14% of these deaths are due to plant poisonings (Riet-Correa and Medeiros 2001). However, in other states, due to lack of data on the frequency of causes of mortality, it is difficult to estimate the death losses caused by toxic plants.

The number of plants known to be toxic to ruminants and horses is steadily increasing. There are at least 122 toxic plant species in Brazil, belonging to 71 genera, and an ever growing list of newly recognized toxic species are reported each year (Riet-Correa *et al.* 2009). Despite the large number of toxic species those causing important losses are few. *Senecio* spp., *Nierembergia veitchii*, *Ateleia glazioviana*, *Pteridium arachnoideum*,

©CAB International 2011. *Poisoning by Plants, Mycotoxins, and Related Toxins* (eds F. Riet-Correa, J. Pfister, A.L. Schild, and T.L. Wierenga)

*Palicourea marcgravii*, *Arrabidaea bilabiata*, *Arrabidaea japurensis*, *Mascagnia rigida*, *Cestrum laevigatum*, and *Brachiaria* spp. are reported as the main causes of poisoning in different Brazilian states (Riet-Correa and Medeiros 2001).

In the Caparaó microregion, cases of poisoning of dairy cattle by plants have been frequent, but there are no records to estimate the losses. The objective of this study was to estimate the occurrence of different plant species causing poisonings in the region between 2007 and 2008.

## Material and Methods

Fifty-one dairy cattle farms were visited from March 2007 to April 2008 in the Caparaó microregion. This microregion includes the counties of Iúna, Ibatiba, Ibitirama, Irupi, Dores do Rio Preto, Divino São Lourenço, Guaçuí, Muniz Freire, Alegre, and São José do Calçado. Five farms were chosen at random in each county from lists of farms registered with the Instituto de Defesa Agropecuária e Florestal do Espírito Santo (IDAF-ES) and with the County Agriculture Secretaries. A questionnaire was given to the farmers after their informed consent was obtained. The questionnaire requested general data for each farm, cattle characteristics, observation of clinical cases and deaths associated with poisonous plants, occurrence of toxic plants in the property, and treatments used in cases of plant poisoning.

After collecting the information from the farmers, the pastures were inspected for the presence of toxic plants. Mounted plant specimens were prepared (Fidalgo and Bononi 1989) and the suspected or known toxic plants sent for identification to the botanical laboratory of the Centro de Ciências Agrárias of the Universidade Federal do Espírito Santo.

## Results and Discussion

The data from the epidemiological questionnaire showed that 40% of farmers were not aware of the existence of toxic plants on their farms, 54% mentioned the existence of toxic plants on the farm, and 6% did not answer the questionnaire or did not identify any toxic plants. All farmers reported the presence of plants using common names, which may vary from region to region. According to Tokarnia *et al.* (2000) the popular names for toxic plants should be used with caution; the name 'erva-de-rato' for *Palicourea marcgravii*, for example, is also applied to other plants, especially of the family Rubiaceae, but most of them are not toxic. Also the words 'tingui' and 'timbó' are routinely used by farmers and handlers as synonyms for any plant that they believe is toxic.

Some farmers (33%) answered that most cattle deaths are caused by plant poisoning. These deaths occur mainly in the dry season (53%) and in most cases are diagnosed by the handlers (39.5%). Clinical signs observed before death were lack of appetite (37.2%), presence of blood in urine (23.2%), bloat (11.6%), lateral recumbence (11.6%), and others (25.6%). Treatment was not used by 34.9% of the farmers, and 20.4% treated the affected animals with antibiotics.

During inspection of the pastures, supposedly toxic plants were found in 40 (78.4%) of the 51 farms. Botanical identification revealed that 20.7% of the samples were *Pteridium arachnoideum*, 13.8% *Oxypetalum banksii*, 10.3% *Cestrum intermedium*, 10.35% *Dicksonia sellowiana*, 10.3% *Ipomoea* spp., 6.9% *Ricinus communis*, 3.4% *Ipomoea alba*,

3.4% *Hedychium coronarium*, 3.4% *Asclepias curassavica*, 3.4% *Senna occidentalis*, 3.4% *Lantana camara*, 3.4% *Lantana trifolia*, 3.4% *Mikania micrantha*, and 3.4% *Conium maculatum*. According to Kissmann and Groth (2000a), among the 14 plants identified, ten are considered toxic. Only four plants including *Oxypetalum banksi*i, *Hedychium coronarium*, *Dicksonia sellowiana*, and *Mikania micrantha* are not reported as toxic.

The implementation of botanical identification in studies of toxicity is extremely important for the distinction of the species of tree, shrub, and grass and conditioning factors related to the toxicity of each species (Kissmann and Groth 2000).

*Pteridium arachnoideum* is found in the family Polypodiaceae. *P. arachnoideum* has two subspecies: *aquilinum* and *caudatum*. In Brazil, *P. caudatum* subspecies occur in mountainous and hilly regions, growing better in cold areas, with good rainfall and well drained soils. The variety *P. arachnoideum* is found in the north of Rio Grande do Sul, Santa Catarina, Paraná, Sao Paulo, south, southeast, and central Minas Gerais, north of Rio de Janeiro, and southwest of the Espírito Santo. There are also small pockets in Acre, Amazonas, Mato Grosso, Pernambuco, and Bahia (Tokarnia *et al.* 2000). Since the end of the 19th century, the scientific literature contains references about the toxicity of bracken fern for livestock, but only in 1965 was the carcinogenicity of this plant demonstrated (Evans *et al.* 1982).

*Cestrum intermedium* is a tree of the Solanaceae family, popularly known as 'mata-boi', 'coerana', 'piloteira preta', and 'erva de tinta'. It is considered the most important poisonous plant in the west and northwest of Santa Catarina and in the west of Paraná (Gava 1993). Poisoning by this plant causes acute liver failure with centrilobular necrosis. Bandarra *et al.* (2009) reported mortality in cattle in Rio Grande do Sul caused by *C. intermedium* poisoning.

Worldwide there are about 600 to 700 species of the genera *Ipomoea* from the Convolvulaceae family, and several of them are considered toxic. Among the plants considered toxic, *Ipomoea carnea* causes poisoning during dry periods when it is consumed by livestock in the absence of other sources of food. Its consumption causes apathy, unbalanced gait, and progressive weight loss, which are non-reversible in most cases (Maia and Figueiredo 1992). *Ipomoea alba* is known as a weed (Kissmann and Groth 2000b).

*Conium maculatum* is reported by Cruz *et al.* (2001) as a plant with a mechanism of action similar to nicotine, causing mydriasis, weakness, immobility, respiratory depression, and teratogenicity. This plant causes birth defects in cattle and swine and teratogenic effects are observed when cows eat the nuts and plants in the first third of pregnancy (Burrows and Tyrl 2001).

*Asclepias curassavica*, *Ricinus communis*, *Senna occidentalis*, and *Lantana camara* are reported as toxic plants with well known clinical signs and pathology (Tokarnia *et al.* 2000). *Lantana trifolia* among several other species of *Lantana* is considered toxic, although the classification of this species is very difficult (Kissmann and Groth 2000c).

*Mikania micrantha* of the Asteraceae family is a popular medical plant used as a bronchodilator, anti-ulcerogenic, and anti-rheumatic (Bighetti 2004). *Dicksonia sellowiana*, belonging to Dicksoniaceae family, is commonly known as 'xaxim', and *Hedychium coronarium*, belonging to the Zingiberarceae family, is called 'lírio-do-brejo' and is known for its ornamental characteristics (Vieira and Pessoa 2001; Mielke 2002). In the literature consulted, there is no evidence of toxicity of these plants for cattle.

Around 100 species of *Oxypetalum* occur in the world. In Brazil the most cited is *O. banksii* Roen. & Schult., which is not considered toxic to cattle. Other species also have toxic compounds, but no detailed studies have been performed (Kissmann and Groth

2000c). Tokarnia *et al.* (2000) showed experimentally that there is no evidence of toxicity for this plant for cattle.

The plant most prevalent in the region was *Pteridium arachnoideum*. Bovine enzootic hematuria is also very frequent in the region, being present in 56.4% of the dairy cattle farms (Silva *et al.* 2009). Despite the well known toxicity of *Pteridium arachnoideum* in cattle (Tokarnia *et al.* 2000), many farmers in the region studied do not believe that the plant is toxic. This is an important barrier to the control of the plant or for the establishment of other measures to prevent the disease. According to Cruz *et al.* (2001), farmers have difficulty in accepting the toxicity of some plants.

In summary, poisonous plants were observed in all counties of the Caparaó microregion. However, the prevalence was higher in the municipalities of Iúna, Ibatiba, Ibitirama, Irupi, and Divino de São Lourenço. These data are similar to those reported by Silva *et al.* (unpublished data) in the same region, where 12 cases (54.6%) of chronic poisoning by *Pteridium arachnoideum* occurred in Ibitirama, seven cases (31.9%) in Ibatiba, one case (4.5%) in Iúna, one case (4.5%) in Guaçuí, and one case (4.5%) in Divino de São Lourenço. The high occurrence of *P. arachnoideum* poisoning in these counties may be related to the fact that, as mentioned by Amelot (1999), the geographical aspects of the region may also interfere with the toxicity of the plant.

## Conclusions

This research showed that the farmers of the Caparaó microregion do not know the toxicity of the plants found in the region. Nevertheless, ten of the 14 species of plants identified on the dairy farms have been reported as toxic. The plant responsible for the highest incidence of poisoning in the region was *Pteridium arachnoideum*. Great damage to animal health and economical losses to farmers are caused by this plant.

## Acknowledgements

This work had financial support from the Fundação de Apoio à Ciência e Tecnologia do Espírito Santo and support from the Centro de Ciências Agrárias of the Universidade Federal do Espírito Santo.

## References

Amelot A (1999). Bracken fern, animal and human health. *Revista de la Universidad de Agronomia, Universidad del Zulia* 16(5):528-547.

Bandarra PM, Bezerra Júnior PS, Corrêa AMR, Pedroso PMOC, Raymundo DL, and Driemeier D (2009). Intoxicação natural por *Cestrum intermedium* em bovinos no Rio Grande do Sul, Brasil. *Ciência Rural* 39:262-265.

Burrows GE and Tyrl RJ (2001). *Toxic Plants of North America*, pp.54-57. Iowa State University Press, Ames.

Cruz CMO, Guerreiro CIPD, and Reis TAFC (2001). *Substancias tóxicas ou anti-nutricionais dos alimentos para animais. Módulo de Nutrição*, Curso de Mestrado em Produção Animal, Universidade Técnica de Lisboa. Disponível em: http://www.alpetratimia.net/consulting/downloads/substoxicas.pdf. Accessed on May 1, 2009.

Evans IA, Prorok JH, Cole RC, Al-Salmani MH, Al-Samarrai AMH, Patel MC, and Smith RMM (1982). The carcinogenic, mutagenic and teratogenic toxicity of bracken. *Proceedings of the Royal Society of Edinburgh* 81:65-77.

Fidalgo O and Bononi VLR (1989). *Técnicas de coleta, preservação e herborização de material botânico*, 62 pp. Instituto de Botânica, São Paulo.

Gava A (1993). Intoxicação por *Cestrum intermedium*. In *Intoxicações por plantas e micotoxicoses em animais domésticos* (F Riet-Correa, MC Mendez, and AL Schild, eds), pp. 72-74. Hemisfério Sul, Pelotas.

IBGE (2007). *Instituto Brasileiro de Geografia e Estatística*. Disponível em: www.ibge.gov.br (Accessed on January 28, 2007).

IDAF/ES. *Instituto de Defesa Agropecuária e Florestal do Espírito Santo*. Disponível em: www.idaf.es.gov.br (Accessed on May 1, 2009).

James LF (1994). Solving poisonous plant problems by a team approach. In *Plant Associated Toxins* (Colegate SM and Dorling PR, eds), pp.1-6. CAB International, Wallingford.

Kissmann KG and Groth D (2000a). *Plantas infestantes e nocivas*, 608 pp. Tomo I, BASF, São Paulo.

Kissmann KG and Groth D (2000b). *Plantas infestantes e nocivas*, 978 pp. Tomo II, BASF, São Paulo.

Kissmann KG and Groth D (2000c). *Plantas infestantes e nocivas*, 726 pp. Tomo III, BASF, São Paulo.

Maia DC and Figueiredo NO (1992). Gênero *Ipomoea* L. (Convolvulaceae) na Ilha de São Luis, MA. *Flora do Estado do Maranhão* 1:1-104.

Mielke EJC (2002). Análise da cadeia produtiva e comercialização do xaxim, *Dicksonia sellowiana*, no estado do Paraná, 75 pp. Dissertação de Mestrado. Universidade Federal do Paraná, Pós-Graduação em Engenharia Florestal, Curitiba, PR.

Riet-Correa F and Medeiros RMT (2001). Intoxicações por plantas no Brasil e no Uruguai: importância econômica, controle e riscos para a saúde pública. *Pesquisa Veterinária Brasileira* 21:38-42.

Riet-Correa F and Méndez MC (1993). Introdução ao estudo das plantas tóxicas. In *Intoxicações por plantas e micotoxicoses em animais domésticos* (F Riet-Correa, MC Mendez and AL Schild, eds) pp. 1-19. Hemisfério Sur, Montevideo.

Riet-Correa F, Medeiros RMT, Pfister J, Schild AL, and Dantas AFM (2009). *Poisonings by plants, mycotoxins and related substances in Brazilian livestock*, 7 pp. Palotti, Santa Maria.

Silva MA, Scárdua CM, Dórea MD, Nunes LC, Martins IVF, and Donatele DM (2009). Prevalência de hematúria enzoótica bovina em rebanhos leiteiros na microrregião do Caparaó, Sul do Espírito Santo, entre 2007 e 2008. *Ciência Rural* 39:1847-1850.

Tokarnia CH, Döbereiner J, and Peixoto PV (2000). *Plantas Tóxicas do Brasil*, 310 pp. Helianthus, Rio de Janeiro.

Vieira CN and Pessoa SVA (2001). Estrutura e composição florística do extrato herbáceo-subarbustivo de um pasto abandonado na reserva biológica de poço das antas, município de Silva Jardim, RJ. *Revista Rodriguésia* 52:17-30.

# Chapter 13

# Ornamental Toxic Plant Species Sold in Campina Grande's Market, Paraíba, Brazil

R.G. Brito[1], R.L. Santos[2], G.P. Guimarães[2], T.N. Ponciano[1], I.C. Dantas[1], and D.C. Felismino[1]

*[1]Departamento de Biologia, Universidade Estadual da Paraíba, Brazil; [2]Departamento de Farmácia, Universidade Estadual da Paraíba, Brazil*

## Introduction

A poisonous plant is any plant which may damage human or animal health when ingested or handled in an erroneous way. In everyday life, it is common to find plants that can be toxic. However, these plants can damage human health leading to death (Schvartsman 1991).

The main function of ornamental plants is to embellish human living places. According to Oliveira *et al.* (1986), these plants play a prominent role in gardening and landscaping. Among the ornamentals, there are many species that have toxic active compounds that, when they come in contact with the body, may trigger reactions that may cause damage or even death (Schvartsman 1992). Intoxication can be acute, as may occur with an accidental overdose, or chronic, as a consequence of a continuous ingestion (Schvartsman 1979).

Poisonous plants are responsible for a large number of cases of human poisoning, especially in children. Because children may be especially susceptible, it is extremely important to present information about the most common toxic species to those that are most likely to come in contact with them. Furthermore, it is important to avoid confusion by using both the scientific name as well as the common name. A large number of common names are often used for the same species, which causes confusion to the lay man in distinguishing non-toxic ornamental plants from those that represent a great risk to human health. One of the most typical cases of human poisoning by plants reported by the Center for Toxicological Assistance (CEATOX / PB) involves the accidental ingestion of parts of leaves and stems of *Dieffenbachia* spp. (Dias and Araújo 1997).

The public or even some health professionals may not have an adequate understanding or knowledge about toxic plants (Pinillos *et al.* 2003). This lack of knowledge hampers efforts to prevent poisoning by plants, and may also reduce the effectiveness and speed of treatment in cases of poisoning.

Consequences of the poisoning by plants vary depending on the type of plant, the toxic compound, and also the amount ingested or the type of exposure. The active

©CAB International 2011. *Poisoning by Plants, Mycotoxins, and Related Toxins*
(eds F. Riet-Correa, J. Pfister, A.L. Schild, and T.L. Wierenga)

ingredients most likely responsible for poisoning caused by plants are alkaloids, glycosides, resins, phytotoxins, minerals, oxalates, essential oils, and photosensitizing compounds (Filho *et al.* 2001). It is known today that 70% of garden plants with toxic substances cause skin reactions (Winters 2005), such as the case of *Dieffenbachia* spp., which causes irritation of the mucous membrane, suffocation, and even death. According to Schvartsman (1992), the ingestion of any part of *Dieffenbachia* spp. or just the simple act of chewing results in severe irritation of the mucous membrane: swelling of lips, tongue, and palate with burning pain, drooling, dysphagia, abdominal cramps, nausea, and vomiting. If the milky sap comes in contact with eyes it causes congestion, tearing, photophobia, and eyelid edema.

The use of two plants in the *Apocynaceae* family, *Nerium oleander* L. and *Thevetia peruviana* Schum, has been related to some cases in which they were used as a weapon of both crime and suicide (Hoehne 1978).

According to Hoehne (1978), from a medical and toxicological point of view, Euphorbiaceae is seen as the most important plant family because of the large number of species of the genera including *Acalypha*, *Aleurities*, *Cnidosculus*, *Croton*, *Euphorbia*, *Jatropha*, *Ricinus*, *Manihot*, and *Sapium* that can be potentially toxic.

Dantas *et al.* (2007) warns about taking proper care with plants for ethnobotanical use which are also cultivated as ornamentals such as: *Argemone mexicana* L., which has hallucinogenic properties and may lead to miscarriage; *Euphorbia tirucalli* L., a poisonous plant that contains caustic compounds and causes severe irritation to eyes and mucous membranes; *Ricinus communis* L., with seeds containing the highest concentration of ricin, a heat-labile toxalbumin that causes nausea, vomiting, gastric bleeding, sore throat, diarrhea, abdominal cramps, hypothermia, tachycardia, dizziness, drowsiness, kidney and liver damage, convulsions, hypotension, depression, coma, as well as death.

Thus, some ornamental plants found in squares, parks, gardens, backyards, and vacant lands represent considerable toxicological risks for children and the elderly, or for any person that accidentally consumes or even mishandles them. Strategies to prevent poisoning should be based on informing the population about the toxic plants that are present in their local environment, and preventing exposure or misuse sufficient to cause health problems. Bochner (2006) states that the toxic plants should not be removed from the environment, rather the public should be made aware of the potential danger that these species represent.

In this context, this study aimed to conduct an inventory in the market of toxic ornamental plant species in Campina Grande, PB, and report their possible symptoms.

## Methodology

The study was carried out in Campina Grande, state of Paraíba, Brazil, during October and November 2008. Nine sellers of ornamental plants supplied us with the data that were collected in the city center and in the regional neighborhoods of Bodocongó, Catolé, São José, and Feira Central.

A questionnaire on ornamental plants was used as the research tool. Following the survey, a series of toxic ornamental plants was identified by the salesmen. Finally, a literature survey was carried out in order to investigate the family, scientific name, popular name, and symptoms of toxicity of such species.

## Results and Discussion

Twelve plant species were cited (Table 1), which according to the sellers have toxic effects. However, two of the salesmen identified the Papoula (*Hibiscus rosa-sinensis* L.) as being a toxic plant, but the literature states that this species is not toxic. *Calathea* sp. and *Rhipsalis capilliformes* were also mentioned as toxic, but no references were found confirming their toxic status.

**Table 1.** Ornamental plants listed by salesmen as being poisonous in Campina Grande.

| Popular name | Scientific name | No. of records | Percentage (%) |
|---|---|---|---|
| Comigo-ninguém-pode | *Dieffenbachia picta* Schott | 9 | 39.5 |
| Papoula | *Hibiscus rosa-sinensis* L. | 2 | 8.7 |
| Espirradeira | *Nerium oleander* L. | 3 | 13.1 |
| Agave | *Agave americana* L. | 1 | 4.3 |
| Coroa-de-cristo | *Euphorbia milii* Des Moul. | 1 | 4.3 |
| Calatéia | *Calathea* sp. | 1 | 4.3 |
| Palmeira rabo de peixe | *Caryota urens* L. | 1 | 4.3 |
| Espada de São Jorge | *Sansiviera trifasciata* var. *laurentii* (De Wild.) N. E. Br. | 1 | 4.3 |
| Lírio da paz | *Sphathiphyllum wallisii* Regel. | 1 | 4.3 |
| Ripsalis | *Rhipsalis capilliformes* Weber | 1 | 4.3 |
| Jasmim manga | *Plumeria rubra* L. | 1 | 4.3 |
| Avelós | *Euphorbia tirucalli* L. | 1 | 4.3 |
| TOTAL | | 23 | 100 |

In the 12 species reported in Table 1, 75% are toxic; in 55.5% the toxic compound is calcium oxalate. Other active compounds including glycosides, latex, indole alkaloids, and belladonna alkaloids are each present in 9% of the toxic species.

Among the species cited, the salesmen were unanimous in pointing out *Dieffenbachia* spp. as a toxic plant, which demonstrates its popularity. Considering the other species, most sellers were unable to identify the toxic plants correctly due to the large number of species of plants sold in the marketplace.

The types of exposure to toxic plants that result in poisoning were investigated, with the salesmen stating that ingestion (54%) was the most common way, followed by touching (23%) and chewing (23%). When the symptoms of poisoning were analyzed, itching (25%) was the most prominent, followed by nausea (19%) and a burning feeling (12.5%). Other symptoms like death, diarrhea, headache, and swelling lips each represented 6.2%.

With respect to toxic species that are cultivated in squares, parks, and gardens, plant sellers were unanimous in saying that all of the species mentioned can be grown in these areas. The exception is *Dieffenbachia* spp., which contains poisonous properties that make the species unfit for growing in areas open to public access. More specifically, the stem, leaves, and latex of this plant contain calcium oxalate crystals that cause irritation, manifested as swelling of one's lips and tongue, as well as pain and a burning sensation in one's mouth and esophagus, apart from abdominal pain. Also, the patient becomes unable to speak. The contact of these crystals with the eyes causes tearing, photophobia, and edema of one's eyelids (Schvartsman 1975).

## Conclusion

Based on the results, 12 species were identified by the sellers as poisonous plants, but only nine of these species are actually toxic. *Dieffenbachia* spp. was the most quoted among them, confirming its popularity. Calcium oxalate is present in 55.5% of the nine toxic plant species mentioned. Among the intoxication pathways reported, ingestion was the prevailing form. The most commonly observed symptoms were itching, nausea, and a burning sensation. Plant sellers have limited knowledge about toxic plants and the symptoms of intoxication.

## References

Bochner R (2006). Perfil das intoxicações em adolescentes no Brasil no período de 1999 a 2001. *Cadernos de Saúde Pública*, Rio de Janeiro, 22 (3): 587-595.

Dantas IC, Felismino DC, Dantas GDS, Dantas VS, and Chaves TP (2007). *O Raizeiro*. 1st edn, p. 540. Editora EDUEP, Campina Grande.

Dias EPF and Araújo RS (1997). *Toxinformes: a toxicologia ao alcance da comunidade,* pp.136-154. Editora Universitária, João Pessoa.

Filho AA, Campolina D, and Dias MB (2001). *Toxicologia na prática clínica*, 1st edn, p. 341. Editora Folium, Belo Horizonte.

Hoehne FC (1978). *Plantas e Substâncias Vegetais Tóxicas e Medicinais*, p. 355. Editora DBE, São Paulo.

Oliveira RF, Antunes IT, Alcantara J, Carneiro JBA, and Silva ZL (1986). *Atlas Escolar de Botânica*, 1st edn, pp. 107-109. Editora MEC/ FAE, Rio de Janeiro.

Pinillos MA, Gómez J, Elizalde J., and Duenas A (2003). Intoxicacion por alimentos, plantas, y setas. *Anales Sin San Navarra* 26(1):243-263.

Schvartsman S (1975). *Manual de envenenamentos: diagnóstico e tratamento*, p. 542. Editora Atheneu, São Paulo.

Schvartsman S (1979). *Plantas Venenosas*. p.176. 1st edn. Editora Sarvier, São Paulo.

Schvartsman S (1991). *Intoxicações agudas*. p. 155. 4th edn. Editora Sarvier, São Paulo.

Schvartsman S (1992). *Plantas Venenosas e Animais Peçonhentos*. p.288. 2nd edn. Editora Sarvier, São Paulo.

Winters GHM (2005). Plantas Ornamentais Tóxicas. *Apostila do Centro Paisagístico Gustaaf Winters*. Centro paisagístico Gustaaf Winters LTDA, São Paulo.

# Chapter 14

# Toxic Plants Grown in Gardens in Alto Branco, Campina Grande, Paraíba, Brazil

R.G. Brito[1], R.L. Santos[2], G.P. Guimarães[2], I.C. Dantas[1], and D.C. Felismino[1]

[1]*Departamento de Biologia, Universidade estadual da Paraíba, Brazil;* [2]*Departamento de Farmácia, Universidade Estadual da Paraíba, Avenida das Baraúnas 351, Campus Universitário - Bodocongó, 58109-753, Campina Grande-PB, Brazil*

## Introduction

The cultivation of plants in gardens is a very old practice. Such cultivation has not been given much study, but perhaps such gardens deserve more attention because a diverse array of plant species is found in such spaces, including those that are potentially toxic and represent a risk to human health.

The species that are used for ornamental purposes should ideally be lacking in toxic substances that may cause adverse reactions in human beings (Balensiefer and Wiecheteck 1987; Graziano 1994). However, many people do not realize that different plants grown in their home gardens have active ingredients like tannins and cardiotoxic glycosides that may be toxic. The analysis of presence or absence of toxic ingredients is important in the choice of species (Cavalcanti *et al.* 2003).

Toxic plants can produce a great variety of chemical substances. Some of these substances, such as proteins and lipids, are used in several metabolic processes within the plant. However, a large number of chemical compounds produced by plants have other functions. The pigments (flavonoids, anthocyanins, and betalains) and the essential oils (monoterpenes, sesquiterpenes, and phenylpropanoid) attract pollinators, while some other substances such as tannins, alkaloids, and the iridoids, produce an unpleasant taste which may deter consumption as well as being toxic and irritating to other organisms, thus protecting the plant against pathogens and predators (Sousa *et al.* 1991).

Some of these toxic substances can cause serious poisoning cases in humans or domestic animals when they are ingested or come in contact with their skin. However, the presence of these substances in a particular plant species may not be sufficient for classifying the plant as toxic because the concentration of these compounds may be too low to cause intoxication (Oliveira and Gokithi 2000).

The lack of knowledge displayed by many people about the toxic potential of garden or ornamental plant species is a serious concern because it increases the probability of accidental ingestion, especially by children, who are the main victims of accidents

©CAB International 2011. *Poisoning by Plants, Mycotoxins, and Related Toxins*
(eds F. Riet-Correa, J. Pfister, A.L. Schild, and T.L. Wierenga)

involving toxic species such as *Nerium oleander* or *Dieffenbachia picta.* These plants are typically found in many home gardens, and they contribute to a high number of incidents at hospital emergency care (Oliveira *et al.* 2003). The close proximity of humans and these toxic plants grown in gardens, and the ease of access, results in frequent accidents (Morgan 2003).

According to Oliveira *et al.* (2003) several circumstances can result in human intoxication from ornamental or garden plants, such as the inability to distinguish a non-toxic plant from a toxic one, the abusive use of medicinal plants, and the mistaken medicinal use of plants by laymen not including suicide and abortion attempts.

Poisoning can result in serious illness or death if the victim is not helped in time (Simões *et al.* 2003). The severity of poisoning depends on many factors such as the susceptibility of the individual, the amount and type of toxic agent ingested, and the time of exposure to the substance.

The aim of this study was to do a survey on toxic plant species present in gardens in Alto Branco, located in the city of Campina Grande, as well as to do a brief summary of the main characteristics related to each one of the species in order to have a greater understanding about toxic plant diversity in the gardens of the city.

## Methodology

This work was carried out in Campina Grande city, state of Paraíba, Brazil, from October to December 2008. Our data were collected in home gardens in the district of Alto Branco. The research had a qualitative and descriptive nature, and a total of 1080 gardens were examined. Toxicity of the ornamental species was used as a criterion for inclusion in the search. Subsequently, a literature research was done in order to investigate the common name, scientific name, family, and the toxic compounds of the plants found.

## Results and Discussion

After visiting all the gardens, 22 plants growing within some gardens were determined to be toxic (Table 1) in accordance with Schvartsman (1992). Six species belonged to the Euphorbiaceae family and seven to the Araceae family, showing that they are widely cultivated in the gardens of Campina Grande. One of the most important plants was *Dieffenbachia* spp. Accidental exposure is the major cause of poisoning by *Dieffenbachia picta*, which occurs due to the lack of knowledge of people about the plant toxicity (Silva and Takemura 2006).

Four plants belonged to the Apocynaceae family, which comprises 424 genera distributed in five subfamilies, 32 of which are found in the Amazon region (Rio *et al.* 2002; Pereira *et al.* 2007). One of the known toxic species found was *Allamanda cathartica*. The latex of this plant contains iridoids as active compounds (Nascimento *et al.* 2006). Its ingestion causes gastrointestinal disorders in humans.

Plants from other families, including Moraceae (two species), Araliaceae, Asteraceae, and Balsaminaceae (one species each) were also found (Table 1). These species are considered toxic because they contain such compounds as oxalates (Winters 2005).

The toxic compounds reported in the poisonous species are presented in Table 1. The exposure to toxicity can occur by ingestion, or by contact with eyes or skin. Symptoms of poisoning by these plants range from irritation of mucous membranes to suffocation and even death.

**Table 1.** Plants, toxic compounds, toxic parts, and number of gardens (1080 surveyed).

| Plants/ family | Toxic active ingredient | Toxic part | Gardens |
|---|---|---|---|
| Alamanda (*Allamanda cathartica* L.)/ Apocynaceae | Cardiotoxic glycosides | W | 24 |
| Avelós (*Euphorbia tirucalli* L.)/ Euphorbiaceae | Toxalbumine (4 deoxigenol) | Lx | 6 |
| Beijo (*Impatiens balsamina* L.)/ Balsaminaceae | Calcium oxalate | Sp | 14 |
| Bico-de-papagaio (*Euphorbia pulcherrima* Willd)/Euphorbiaceae | Toxalbumine | W | 8 |
| Boa-noite (*Catharanthus roseus* (L.) G. Don.)/Apocynaceae | Alkaloid, glycoside, vinceine | W | 70 |
| Café-de-salão (*Leea rubra* Blume & Spreng.)/Araceae | Oxalate | L, F, Sp | 96 |
| Candelabro (*Euphorbia lactea* Haw)/ Euphorbiaceae | Toxalbumine | Lx | 10 |
| Cardo-santo (*Argemone mexicana* L.)/Compositae | Alkaloid (Protopine and Berberine) | Sd | 3 |
| Chápeu-de-Napoleão (*Thevetia peruviana* (Pers.) Schum.)/ Apocynaceae | Cardiotoxic glycosides | W | 6 |
| Cheflera (*Brassaia actinophylla* (Endl.) Harms)/Araliaceae | Calcium oxalate | L, St | 45 |
| Comigo-ninguém-pode (*Dieffenbachia* spp.)/Araceae | Calcium oxalate, saponins | W | 59 |
| Coroa-de-Cristo (*Euphorbia milii* L.)/ Euphorbiaceae | Toxalbumine (5 deoxigenol) | Lx | 96 |
| Costela-de-Adão (*Monstera deliciosa* Liebm)/Araceae | Calcium oxalate | L | 18 |
| Espirradeira (*Nerium oleander* L. / Apocynaceae | Cardiotoxic glycosides | W | 18 |
| Ficus (*Ficus benjamina* L.)/ Moraceae | Furanocumarines | Lx | 71 |
| Hera (*Ficus pumila* L.)/Moraceae | Furanocumarines, Calcium oxalate | L, Lx | 16 |
| Lírio da paz (*Spathiphyllum wallisi* Regel.)/Araceae | Calcium oxalate | St, L | 12 |
| Mamona (*Ricinus communis* L. )/ Euphorbiaceae | Toxalbumine, ricin, hematoaglutinin | Sd | 3 |
| Pinhão-roxo (*Jatropha curcas* L.)/ Euphorbiaceae | Toxalbumine (curcin) | L, F | 23 |
| Taioba (*Colocasia antiquorum* Schott.)/Araceae | Calcium oxalate | W | 4 |
| Jibóia (*Scindapsus aureus* Engl.)/ Araceae | Calcium oxalate, toxalbumine | L, St, Lx | 12 |
| Tinhorão (*Caladium bicolor* Vent.)/ Araceae | Calcium oxalate | W | 2 |

W=entire plant; L=leaf; F=flowers; St=stem; Lx=latex; Sd=seed; Sp=Sap; F=fruit

## Conclusion

Ornamental plant gardens are a bond between man and nature in urban spaces and have brought many benefits not only for the esthetics of homes but also for the mental and physical health and wellbeing of its residents. It is in the public's interest to have a general understanding of the toxicity of some ornamental plants. Members of the public that are interested in growing ornamental plants should be educated about the toxicity of the 22 toxic species found in urban gardens in the city of Campina Grande.

## References

Balensiefer M and Wiecheteck M (1987). *Arborização das cidades*. Impresso pelo instituto de terras, cartografia e florestas; vinculado à secretaria de estado da agricultura e abastecimento, Curitiba.

Cavalcanti MLF, Dantas IC, Lira RS, Oliveira JMC, Albuquerque HN, and Albuquerque ICS (2003). Identificação dos vegetais tóxicos da cidade de Campina Grande-PB. Paraíba. *Revista de Biologia e Ciências da Terra* 3:1.

Graziano TT (1994). *Arborização de ruas*. Departamento de Horticultura – FCAVJ – UNESP. Notas de Aula.

Morgan R (2003). *Enciclopédia das Ervas e Plantas Medicinais*. 9th edn. 555 pp. Editora Hemus, São Paulo.

Nascimento CAA, Arévalo E, Afonso-Neto IS, Bessa ECA, and Soares GLG (2006). Efeito do extrato aquoso de folhas de *Allamanda cathartica* L. (Apocynaceae) sobre *Bradybaena similaris* (Férussac, 1821) (Mollusca, Bradybaenidae) em condições de laboratório. *Revista Brasileira de Zoociência* 8(1):77-82.

Oliveira F and Gokithi A (2000). *Fundamentos da farmacobotânica*. 2nd edn. pp. 147-155. Editora Atheneu, São Paulo.

Oliveira RB, Godoy SAP, and Costa FB (2003). *Plantas Tóxicas–conhecimento e Prevenção de acidentes*. 1st edn. p.9, 53. Holos Editora, São Paulo.

Pereira MM, Lisieux R, Jácome RP, Alcântara AFC, Alves RB, and Raslan DS (2007). Alcalóides indólicos isolados de espécies do gênero Aspidosperma (Apocynaceae). *Revista Química Nova* 30:4.

Rio MCS, Castro MM, and Kinoshita LS (2002). Distribuição e caracterização anatômica dos coléteres foliares de *Prestonia coalita* (Vell.) Woodson (Apocynaceae). *Revista Brasileira de Botânica* 25:3.

Schvartsman S (1992). *Plantas venenosas e animais peçonhentos*. 2nd edn. pp. 69-135. Editora Sarvier, São Paulo.

Silva IGR and Takemura OS (2006). Aspectos de intoxicações por *Dieffenbachia* spp. (Comigo-ninguém-pode) – Araceae. *Revista de Ciências Médicas e Biológicas* 5(2):151-159.

Simões CMO, Schenkel EP, Gosmann G, de Mello JCP, Mentz LA, and Petrovick PR (2003). *Farmacognosia da planta ao medicamento*. 5th edn. pp. 971, 973-978. Editora da UFRGS, Porto Alegre.

Sousa MP, Matos MEO, Matos FJA, Machado MIL, and Craveiro AA (1991). *Constituintes Químicos Ativos de Plantas Brasileiras*. 1st edn. p. 416. Edições UFC, Fortaleza.

Winters G. (2005). Plantas Ornamentais Tóxicas. *Apostila do Centro Paisagístico 'Gustaaf Winters'*.

# THE LIVER

# Chapter 15

# *Brachiaria* spp. Poisoning in Sheep in Brazil: Experimental and Epidemiological Findings

M.B. Castro[1], H.L. Santos Jr[1], V.S. Mustafa[1], C.V. Gracindo[1], A.C.R. Moscardini[1], H. Louvandini[1], G.R. Paludo[1], J.R.J. Borges[1], M. Haraguchi[2], M.B. Ferreira[3], and F. Riet-Correa[4]

[1]*Veterinary Medicine Course, University of Brasília, Brasília, DF, 70910-970, Brazil;* [2]*Biological Institute of São Paulo, SP 04014-002, Brazil ;* [3]*Pathology Department, UFMS, Campo Grande, MS 79070-900, Brazil;* [4]*Veterinary Hospital, CSTR, UFCG, Patos, PB, 58700-970, Brazil*

## Introduction

*Brachiaria* spp. poisoning in sheep is an important cause of economic loss in the Brazilian central-western region. *B. decumbens* (signal grass) was introduced in Brazil in 1952 by the Instituto de Pesquisas Experimentais Agropecuárias do Norte (IPEAN) (Nobre and Andrade 1976). Later, other species of *Brachiaria* including *B. brizantha*, *B. humidicola*, and *B. ruzziziensis* were introduced (Seiffert 1980). Because of its high production of dry matter, good adaptation in different soils, and ability to grow during most of the year, *Brachiaria* spp. is the most important forage species in the central-western, southeastern, and northern regions of Brazil. However, an important limiting factor for the use *Brachiaria* spp. as forage is its toxicity, causing hepatogenous photosensitization due to steroid lithogenic saponins (Graydon *et al.* 1991; Lajis *et al.* 1993; Smith and Miles 1993; Meagher *et al.* 1996; Lemos *et al.* 1997). *Brachiaria* spp. have been associated with photosensitization and death in cattle (Lemos *et al.* 1996, 1997; Tokarnia *et al.* 2000), sheep (Lemos *et al.* 1996; Brum *et al.* 2007), goats (Lemos *et al.* 1998), and buffalo (Rozza *et al.* 2004). The most characteristic histologic lesion of the intoxication is the presence of crystals in the biliary ducts and large vacuolated macrophages in the liver, lymph nodes, gut, and spleen. Ultrastructurally, foamy macrophages show negative images of crystals involved partially or totally by membranes. These images represent crystals absorbed from the food and carried to the lymphatic circulation (Driemeier *et al.* 1998). Brazilian samples of *Brachiaria* spp. contain steroidal saponins (Cruz *et al.* 2001; Siqueira-Souza *et al.* 2005; Brum *et al.* 2007), which are associated with photosensitization and deposition of crystalloid material within the biliary system (Cruz *et al.* 2000). Sheep are considered more susceptible than cattle to *Brachiaria* spp. poisoning, and young sheep are more susceptible than older animals (Hasiah *et al.* 2000; Riet-Correa and Mendez 2007). In the same flock, there are differences in susceptibility of sheep of the same breed and age. This difference in susceptibility seems to

©CAB International 2011. *Poisoning by Plants, Mycotoxins, and Related Toxins* (eds F. Riet-Correa, J. Pfister, A.L. Schild, and T.L. Wierenga)

be the reason for the continuous decrease in the frequency of the disease in cattle. Presently the poisoning is less frequent in cattle, possibly because susceptible animals have died or become resistant. *Brachiaria* spp. poisoning deters the introduction of sheep in central-western Brazil because numerous outbreaks occur when sheep from other regions are introduced to *Brachiaria* spp. pastures. The purpose of these experiments is to examine the aspects of *Brachiaria* spp. poisoning in sheep; epidemiological records from spontaneous outbreaks were also analyzed.

## Susceptibility of Sheep to Signal Grass Poisoning

The first experiment evaluated differences in susceptibility to signal grass poisoning between 20 crossbreed male sheep, 4-6 months old. A naïve group ($n$=10) was composed of animals from flocks that never grazed in *Brachiaria* spp. and a control group ($n$=10) originated from flocks that had been grazing for more than 3 years in *B. decumbens* paddocks. Both groups were introduced into a *B. decumbens* pasture for 11 weeks. Nine sheep (90%) in the naïve group presented clinical signs, mainly ocular discharge and edema of the face; serum activities of AST and GGT (Figure 1) increased 2-7 weeks after their introduction into the pasture. Two sheep died in this group. Numerous foamy macrophages, sometimes containing negative images of crystals, and cholestasis were observed histologically in the liver. Similar foamy macrophages were observed in the mesenteric and hepatic lymph nodes. Only one animal showed clinical signs and died in the control group, but serum activities of AST and GGT from control animals were also markedly increased 5-9 weeks after the start of the experiment despite the absence of clinical signs of intoxication. Saponin (protodioscin) concentrations of the pasture were 1.06% at the start of the experiment, and decreased to 0.52% at the end of the experiment when the pasture was senescing.

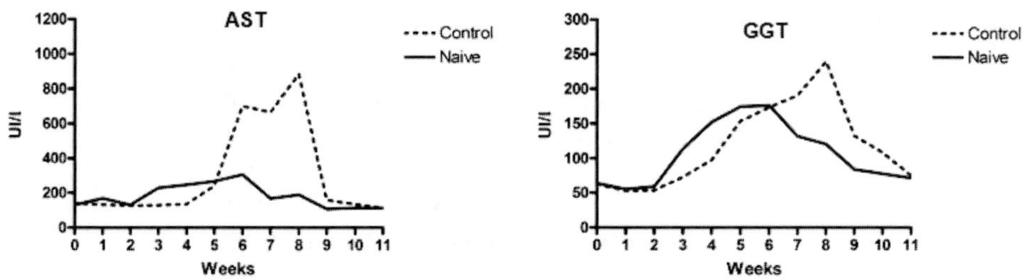


**Figure 1.** AST and GGT medium serum levels in control and naïve groups grazing *B. decumbens* during the experimental period.

## Effect of Signal Grass Maturity on Poisoning in Naïve Sheep

Three groups, each with seven young naïve sheep (4-6 months old), were kept on *B. decumbens* pastures at various levels of maturity: 15 (L group), 45 (M group), and 90 (H group) days of growth. There was an increase in serum levels of AST and GGT, and clinical signs began during the second week (Figure 2). The main clinical signs included apathy, photophobia, bilateral ocular secretions, mucosal hyperemia, jaundice, as well as facial and ear swelling. A hyperacute clinical manifestation of signal grass poisoning occurred just 7 to 10 days after the start of the experiment, which presented some differences from traditional acute and chronic manifestations (Fagliari *et al.* 1993). The hyperacute clinical form was characterized by severe depression, absence of jaundice, mucosal hyperemia, facial and ear swelling as well as rapid death of all animals. All the sheep that showed clinical signs of severe intoxication were removed from the paddocks, and ultrasound-guided liver biopsies were performed. Sheep that died during the experiment were necropsied. General morbidity was 57% with a fatality rate of 75%. Groups L, M, and H presented mortality rates of 44.44%, 33.33%, and 22.22%, respectively. The main macroscopic lesions in sheep that died were hepatomegaly, jaundice, and distension of the gall bladder. Upon histologic examination, the liver showed swelling and vacuolization of hepatocytes and foamy macrophages disseminated in sinusoids. Crystalloid negative images were observed in the macrophages and also within bile ducts. Rare and scattered foci of individual hepatocyte necrosis, inflammatory mononuclear multifocal infiltrate in the periportal region, and bile duct proliferation were also observed. Prominent hyperplasia of smooth endoplasmic reticulum associated with the presence of crystals in the cytoplasm of hepatocytes was visualized by electron microscopy in liver of sheep with and without clinical signs. *Pithomyces chartarum* spores were not observed in grass samples taken periodically during the entire experiment. Mean protodioscin levels of *B. decumbens* in paddocks L, M, and H during the experiment were 2.03%, 1.63% and 1.26%, respectively.

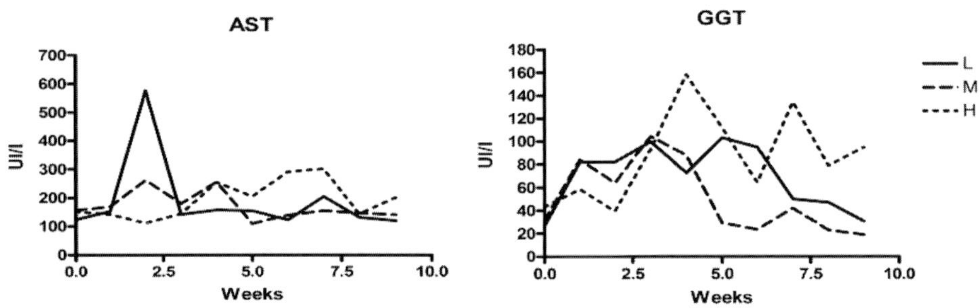


**Figure 2.** AST and GGT medium serum levels in groups L, M, and H grazing *B. decumbens* during the experimental period.

## Performance of Sheep Grazing *Brachiaria* spp. and Other Pastures

The objective of this experiment was to compare body weight gain in four groups of eight lambs (5-6 months old) each, born into flocks that had been grazing for more than 3 years in *Brachiaria* spp. pastures. The groups were introduced at the same time and stayed for 2 months in four paddocks of *B. decumbens*, *B. brizantha*, *Panicum maximum* var. Aires, and *Andropogon gayanus* var. Planaltina, respectively. Four naïve sheep were also introduced within each group in the paddocks as controls. No significant differences in weight gain between *Brachiaria* and *Panicum* groups were observed. Sheep in *Andropogon* pasture had lower weight gains than sheep in the other grasses (Figure 3). None of the lambs born to experienced mothers showed either clinical signs of intoxication or significant elevation of serum AST or GGT activity during the experiment. One naïve control sheep grazed in the *B. decumbens* paddock showed signs of poisoning and died.

Protodioscin levels in *B. decumbens*, *B. brizantha*, *P. maximum*, and *A. gayanus* paddocks during the experiment were 0.86%, 0.54%, 0%, and 0.12% respectively.

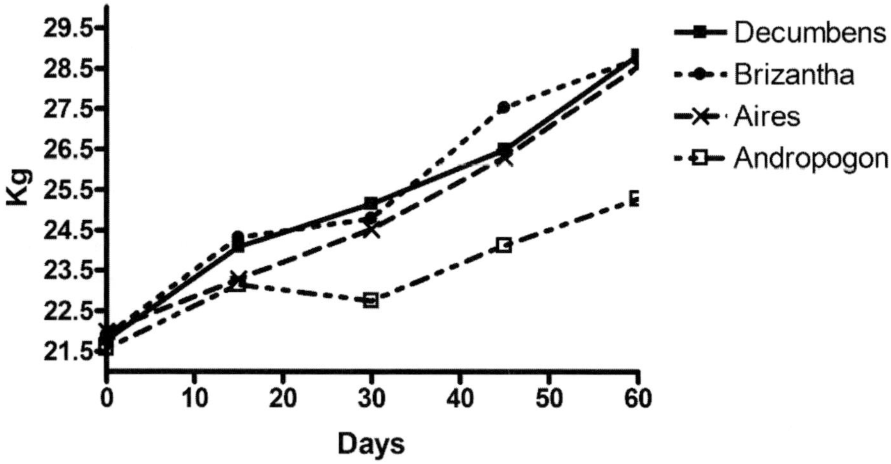


**Figure 3.** Performance of sheep groups (medium weight gain - kg) grazing *B. decumbens*, *B. brizantha*, *P. maximum* (var. Aires) and *A. gayanus* (var. Planaltina) pastures.

## Natural Outbreaks of *Brachiaria* spp. Poisoning

Epidemiological records involving 1272 animals from ten different sheep flocks with spontaneous outbreaks of *Brachiaria* spp. poisoning were evaluated. All outbreaks were diagnosed by characteristic clinical signs and macroscopic and histologic lesions. General morbidity (*n*=238), mortality (*n*=198), and fatality rates were 18.7% (0.42% to 57.5%), 15.5% (0% to 57.5%), and 83.2% (0% to 100%), respectively. Sheep younger than 1 year were most frequently intoxicated (90% of all poisoned animals). Clinical signs emerged 15 to 60 days after the introduction of the animals in the paddocks. The main signs observed included facial and ear swelling (70%), apathy/anorexia (30%), ocular secretion/mucosal hyperemia (30%), photosensitization (30%), jaundice (20%), weight loss (20%), and nasal discharge (20%). The major pathological findings included hepatomegaly, jaundice, distension of the gall bladder, presence of foamy macrophages disseminated in hepatic sinusoids, inflammatory mononuclear multifocal infiltrate in the periportal region, and crystalloid negative images within bile ducts. *B. decumbens* (70%), *B. brizantha* (20%), and mixed *B. decumbens*, *B. humidicola*, and *Andropogon* spp. (10%) were the grasses present in the paddocks where the outbreaks occurred. Initial and vigorous growth stages of *Brachiaria* spp. were present in 50% of the paddocks; 40% of the pastures were in seed, and 10% with seed shatter. *Pithomyces chartarum* spores were not observed in wash method counts in 80% of pastures. Spores (25,000/g of grass) were detected in only one outbreak. Protodioscin levels in the forage in five outbreaks were 0.30% to 2.56%.

## Discussion

All experimental and epidemiological findings suggest extreme variations in the susceptibility/resistance of sheep to *Brachiaria* spp. poisoning in the same flock and between flocks. Lambs from sheep grazing in *Brachiaria* spp. for years were more resistant to signal grass poisoning than naïve sheep. However, even apparently resistant animals had higher serum levels of AST and GGT than a naïve group in the absence of clinical signs of poisoning, suggesting that is not possible to associate AST and GGT serum levels with the severity and the prognosis of signal grass poisoning. These observations suggest that the pastures are toxic to all animals, but resistant sheep have a higher tolerance for liver damage from *B. decumbens*. The mechanisms involved in resistance to *Brachiaria* spp. poisoning are still unknown.

Sheep with signal grass poisoning demonstrated impairment in drug-metabolizing enzymes in the liver and kidney (Khairi *et al.* 2000). Animals dosed with phenobarbitone, a potent inducer of P450 isoenzymes and other mixed function oxidases, showed less severe clinical signs of signal grass poisoning (Hasiah *et al.* 2000) and this substance also increased the resistance of sheep to *Narthecium ossifragum* intoxication (Flaoyen and Jensen 1991). Moreover, prominent hyperplasia of smooth endoplasmic reticulum (SER) of hepatocytes was detected in sheep with and without clinical signs. Hyperplasia of SER in hepatocytes was previously described in sheep (Driemeier *et al.* 2002) and in cattle intoxicated with *Brachiaria* spp. (Driemeier *et al.* 1998). Induction or elevated microsomal enzyme activity could be speculated to be an important mechanism of resistance to signal grass poisoning, but possibly other factors should be included.

Records from natural outbreaks of *Brachiaria* spp. poisoning in the Brazilian central-western region demonstrated that young lambs are more susceptible than sheep over one year old as previously reported in Malaysia (Hasiah *et al.* 2000). Possibly adult sheep develop variable degrees of resistance to poisoning due to improved hepatic drug-metabolizing enzyme activity or other unknown mechanisms.

Ruminal metabolism degradation of steroidal saponins should be also considered as another possible form of resistance to poisoning. Some ruminal bacteria degrade dihydroxypyridine (DHP) produced from mimosine, and colonization of the rumen by bacteria that degrade DHP enables the successful utilization of *Leucaena leucocephala* as ruminant forage (Allison *et al.* 1990). Nevertheless, ruminal detoxification of *Brachiaria* spp. and other saponin-containing plants has not been demonstrated (Meagher *et al.* 1996; Flaoyen *et al.* 2001).

Heritability should also be considered in resistance/susceptibility to *Brachiaria* spp. poisoning. Young lambs from flocks adapted to graze signal grass pastures were more resistant to poisoning than animals from naïve herds. Some degree of epigenetic inheritance could be considered and may interact with other acquired unknown mechanisms of resistance. There is some evidence of genetic variation in the susceptibility of beef sires to fescue toxicosis (Gould and Hohenboken 1993).

*Brachiaria* spp. pastures should be considered as an option for grazing sheep in central-western Brazil. Lambs from resistant herds kept in *Brachiaria* spp. paddocks may have similar performance in weight gain when compared with animals grazed in standard *Panicum maximum* grass pastures. *Brachiaria* spp. is touted as the best option for livestock production in the central-western region due to dry matter production, persistence under heavy grazing, resistance to drought, and low cost maintenance (Costa *et al.* 2008). It represents more than 50% of all pastures in Brazil (Castro *et al.* 2007). The selection of sheep from adapted/resistant herds or management techniques to adapt ruminants to

*Brachiaria* spp. pastures are important strategies to reduce economical losses in the flocks. The selection for resistance to facial eczema caused by sporidesmin toxicosis has been demonstrated to be effective in sheep and cattle in New Zealand (Morris *et al.* 2004). Sheep may also develop short-term tolerance to *Narthecium ossifragum*, implicated in outbreaks of nephrotoxicity in cattle, when exposed to small to moderate amounts (Flaoyen *et al.* 2001). This finding suggests that it might be possible to develop some management protocols to adapt herds to graze saponin-containing plants. Experimental and epidemiological records showed that all *Brachiaria* spp. growth stages were potentially toxic to sheep, but paddocks with early and vigorous grass growth (15 days and 45 days of growth, respectively) and with higher protodioscin levels were more toxic. These observations and the absence (or presence of only small amounts) of *P. chartarum* spores in the grass indicate that the initial growth stages of signal grass are more toxic due to higher levels of protodioscin. Farmers and technicians in the central-western region frequently report *Brachiaria* spp. poisoning in sheep and cattle at the beginning of the rainy season which can be explained by the higher toxicity of the immature grass. However, data from experiments and spontaneous outbreaks demonstrated extreme variation in protodioscin concentrations in *Brachiaria* spp. and a threshold for toxicity is difficult to establish for saponin grass levels. Because of these observations and differences in sheep resistance/susceptibility to toxicity, it is very difficult to prevent the occurrence of outbreaks of *Brachiaria* spp. poisoning. Large variation in the concentration of saponins in other saponin-containing plants has been demonstrated in sheep with alveld in Norway (Flaoyen *et al.* 2004).

## Conclusions

Susceptibility or resistance to *Brachiaria* spp. poisoning in sheep is complex and is probably influenced by genetic and acquired resistance and factors determining the saponin concentration in grasses. Nevertheless, *Brachiaria* spp. is an important option for grazing sheep in central-western Brazil. Therefore, more research is necessary to develop management measures to prevent the intoxication.

## Acknowledgements

We give special thanks to the CNPq for financial support, INCT for the control of plant poisonings (grant 573534/2008-8) and Universal Edict, and to Dr Mauro Pereira Soares, College of Veterinary Medicine (DVM) from the Federal University of Pelotas (UFPel) for EM sample evaluation.

## References

Allison MJ, Hammond AC, and Jones RJ (1990). Detection of ruminal bacteria that degrade toxic dihydroxypyridine compounds produced from mimosine. *Applied And Environmental Microbiology* 56:590-594.

Brum KB, Haraguchi M, Lemos RAA, Riet-Correa F, and Fioravante MC (2007). Crystal associated cholangiopathy in sheep grazing *Brachiaria decumbens* containing the saponin protodioscin. *Pesquisa Veterinária Brasileira* 27:39-42.

Castro GHF, Graça DS, Gonçalves LC, Mauricio RM, Rodriguez NM, Borges I, and Tomich TR (2007). Cinética de degradação e fermentação ruminal da *Brachiaria brizantha* cv. Marandu colhida em diferentes idades ao corte. *Arquivos Brasileiros de Medicina Veterinária e Zootecnia* 59:1538-1544.

Costa KAP, Faquin V, Oliveira IP, Araújo JL, and Rodrigues RB (2008). Doses e fontes de nitrogênio em pastagem de capim-marandu: II - nutrição nitrogenada da planta. *Revista Brasileira de Ciências do Solo* 32:1601-1607.

Cruz C, Driemeier D, Pires VS, Colodel EM, Taketa ATC, and Schenkel EP (2000). Isolation of steroidal sapogenins implicated in experimentally induced cholangiopathy of sheep grazing *Brachiaria decumbens* in Brazil. *Veterinary and Human Toxicology* 42:142-145.

Cruz C, Driemeier D, Pires VS, and Schenkel EP (2001). Experimentally induced cholangiopathy by dosing sheep with fractionated extracts from *Brachiaria decumbens*. *Veterinary Diagnostic Investigation* 13:170-172.

Driemeier D, Barros SS, Peixoto PV, Tokarnia CH, Döbereiner J, and Brito MF (1998). Estudo histológico, histoquímico e ultra-estrutural de fígados e linfonodos de bovinos com presença de macrófagos espumosos ('foam cells'). *Pesquisa Veterinária Brasileira* 18:29-34.

Driemeier D, Colodel EM, Seitz AL, Barros SS, and Cruz CEF (2002). Study of experimentally induced lesions in sheep by grazing *Brachiaria decumbens*. *Toxicon* 40:1027-1031.

Fagliari JJ, Passipieri M, Kuchembuck MRG, and Curi PR (1993). Intoxicação natural de bovinos pela micotoxina esporodesmina. II. Aspectos clínicos. *Arquivos Brasileiros de Medicina Veterinária e Zootecnia* 45:275-282.

Flaoyen A and Jensen EG (1991). Microssomal enzymes in lambs and adult sheep, and their possible relationship to alveld. *Veterinary Research Communication* 15:271-278.

Flaoyen A, Hove K, and Wilkins AL (2001). Tolerance to the nephotoxic component of *Narthecium ossifragum* in sheep: the effects of repeated oral doses of plants extracts. *Veterinary Research Communication* 25:127-136.

Flaoyen A, Wilkins AL, di Menna ME, and Sandivik M (2004). The concentration of steroidal sapogenins in and degree of fungal infection on *Narthecium ossifragum* plants in More and Romsdal County, Norway. In *Poisonous Plants and Related Toxins* (T Acamovic, CS Stewart, and TW Pennycott, eds), pp. 79-83. CABI Publishing, Cambridge, Massachusetts.

Gould LS and Hohenboken WD (1993). Differences between progeny of beef sires in susceptibility to fescue toxicosis. *Journal of Animal Science* 71:3025-3032.

Graydon RI, Hamid H, and Zahari P (1991). Photosensitization and crystal associated cholangiohepatopathy in sheep grazing *Brachiaria decumbens*. *Australian Veterinary Journal* 68:234-236.

Hasiah AH, Elsheikh HA, Salam Abdullah A, Khairi HM, and Rajion MA (2000). Effect of phenobarbitone treatment against signal grass (*Brachiaria decumbens*) toxicity in sheep. *The Veterinary Journal* 160:267-272.

Khairi HM, Elsheikh HA, and Abdullah AS (2000). The effect of signal grass (*Brachiaria decumbens*) on drug-metabolizing enzymes in sheep and comparison with normal cells. *Veterinary and Human Toxicology* 42:193-195.

Lajis NH, Abdullah AS, Salim SJ, Bremner JB, and Khan MN (1993). Epi-sarsasapogenin and epi-smilagenin: two sapogenins isolated from the rumen content of sheep intoxicated by *Brachiaria decumbens*. *Steroids* 58:387-389.

Lemos RAA, Ferreira LCL, Silva SM, Nakazato L, and Salvador SC (1996). Fotossensibilização e colangiopatia associada a cristais em ovinos em pastagem com *Brachiaria decumbens*. *Ciência Rural* 26:109-113.

Lemos RAA, Salvador SC, and Nakazato L (1997). Photosensitization and crystal associated cholangiohepatopathy in cattle grazing *Brachiaria decumbens* in Brazil. *Veterinary and Human Toxicology* 39:376-377.

Lemos RAA, Nakazato L, Herrero Junior GO, Silveira AC, and Porfírio LC (1998). Fotossensibilização e colangiopatia associada a cristais em caprinos mantidos sob pastagens de *Brachiaria decumbens* no Mato Grosso do Sul. *Ciência Rural* 28(3):507-510.

Meagher LP, Wilkins AL, Miles CO, Collin RG, and Fagliari JJ (1996). Hepatogenous photosensitization of ruminants by *Brachiaria decumbens* and *Panicum dichotomiflorum* in the abscence of sporidesmim: lithogenic saponins may be responsible. *Veterinary and Human Toxicology* 38:271-273.

Morris CA, Towers NR, Hohenboken WD, Maqbool N, Smith BL, and Phua SH (2004). Inheritance of resistance to facial eczema: a review of research findings from sheep and cattle in New Zealand. *New Zealand Veterinary Journal* 52:205-215.

Nobre D and Andrade SO (1976). Relação entre fotossensibilização em bovinos jovens e gramínea *Brachiaria decumbens* Stapf. *Biológico* 42:249-258.

Riet-Correa F and Méndez MC (2007). Intoxicações por Plantas e Micotoxinas. In *Doenças de Ruminantes e Eqüídeos* (F Riet-Correa, AL Schild, RAA Lemos, and JRJ Borges, eds), pp. 99-219. Editora Pallotti, Santa Maria, RS.

Rozza DB, Seitz AL, Bandarra PM, Santos EO, and Driemeier D (2004). Fotossensibilização por *Brachiaria decumbens* em búfalo. *Pesquisa Veterinária Brasileira* 24 (supl.):55-56.

Seiffert NF (1980). *Gramíneas forrageiras do gênero Brachiaria,* n1, 83 pp. Technical Bulletin. EMBRAPA-CNPGC, Campo Grande.

Siqueira-Souza V, Sandrini CNM, Fioravanti MCS, and Haraguchi M (2005). Sazonal evaluation of saponins in the pastures of *Brachiaria* and *Andropogon* of the Brazilian Middle-West. *Arquivos do Instituto Biológico de São Paulo* 72:43.

Smith BL and Miles CO (1993). A role for *Brachiaria decumbens* in hepatogenous photosensitization of ruminants. *Veterinary Human Toxicology* 35:256-257.

Tokarnia CH, Döbereiner J, and Peixoto PV (2000). *Plantas Tóxicas do Brasil.* pp. 164-175. Editora Helianthus, Rio de Janeiro, RJ.

# Chapter 16

# Variation in Saponin Concentration in *Brachiaria brizantha* Leaves as a Function of Maturation: Preliminary Data

M.B. Ferreira[1], K.B. Brum[1], C.E. Fernandes[1], C.F. Martins[2], G.S. Pinto[2], V.S. Castro[2], K.G. Rezende[2], F. Riet-Correa[3], M. Haraguchi[4], H.L. Wysocki Jr..[4], and R.A.A. Lemos[5]

[1]*Departamento de Patologia (DPA/CCBS) UFMS, Campo Grande, MS 79070-900, Brazil;* [2]*Centro de Tecnologia em Ovinocultura (CTO) UNIDERP, Campo Grande, MS 79003-010, Brazil;* [3]*Centro de Saúde e Tecnologia Rural, UFCG, Campus de Patos, 58700-000 Patos, PB;*[4]*Instituto Biológico de São Paulo, São Paulo, SP 04014-002,Brazil;* [5]*Departamento de Medicina Veterinária (DMV/FAMEZ), UFMS, Campo Grande, MS 79070-900, Brazil*

## Introduction

*Brachiaria* species are important forages in tropical regions. In Brazil, approximately 51 million ha are cultivated with *Brachiaria* species, including *B. brizantha* (A. Rich.) Stapf (palisadegrass), *B. decumbens* Stapf (signalgrass), and *B. humidicola* (Rendle) Schweick (Koroniviagrass) (Macedo 2005). These forages are associated with outbreaks of hepatogenous photosensitization in cattle (Lemos *et al.* 1996b, 1997; Meagher *et al.* 1996; Fioravanti 1999), sheep (Graydon *et al.* 1991; Lemos *et al.* 1996a; Seitz *et al.* 2004), and goats (Lemos *et al.* 1998).

Ruminants affected by hepatogenous photosensitization due to *Brachiaria* spp. ingestion present histological lesions of cholangiohepatopathy with birefringent crystals in bile ducts and hepatocytes. These crystals have been reported as being insoluble salts of sapogenin glucuronides originating from steroidal saponins present in the plant (Meagher *et al.* 1996; Cruz *et al.* 2000, 2001). *B. brizantha* and *B. decumbens* contain a furostanol-like steroidal saponin known as 25R- and 25S- protodioscin isomers (Brum *et al.* 2009), which are associated with photosensitization of ruminants (Brum *et al.* 2007) and horses (Barbosa *et al.* 2006).

Saponin content in *Brachiaria* spp. can vary in the same rangeland, for example the content of protodioscin isomers in one pasture where there was an outbreak of poisoning by *B. decumbens* in sheep was 2.36% while the content in a neighboring paddock grazed by cattle was 1.63% (Brum *et al.* 2007). Data from experiments and outbreaks demonstrate a large variation in protodioscin concentrations in *Brachiaria* spp. thus making it difficult to establish a threshold for toxicity. The objective of this research was to determine how

©CAB International 2011. *Poisoning by Plants, Mycotoxins, and Related Toxins* (eds F. Riet-Correa, J. Pfister, A.L. Schild, and T.L. Wierenga)

saponin concentrations vary as a function of plant maturation and identify factors that may influence these changes.

Field observations in sheep grazing *B. brizantha* pastures with good forage availability showed that sheep were very selective in how and what they grazed. Sheep first grazed the fresh sprouting leaves followed by the mature fully expanded leaves. They never ate old senescent leaves or the whole plant. Thus, the objective of this preliminary study was to detect variation in protodioscin concentrations of *B. brizantha* leaves collected in various stages of maturation. We also compared saponin variation in relation to meteorological data such as insolation (hours of exposure to solar radiation during the day) and temperature.

## Material and Methods

Meteorological data were obtained from Centro Estadual de Monitoramento do Tempo Clima e Recursos Hídricos de Mato Grosso do Sul (CEMTEC), located 20 km from the studied pasture. We used data from 1 day before each leaf collection. Insolation records (h/day) were obtained using a sunshine recorder (Campbell-Stokes sunshine recorder), which burns a special tape when sunshine hits it (Ceballos *et al.* 1992; Varejão-Silva 2005).

Five collections were performed every other week in a 1 ha pasture of *B. brizantha* at the Centro de Tecnologia em Ovinocultura, Universidade para o Desenvolvimento do Estado e da Região do Pantanal (CTO/UNIDERP), Campo Grande, Mato Grosso do Sul State, Brazil, during a 60 day period at the end of the rainy season. Each collection of leaves, obtained on the same day, was separated into three phases of maturation: (i) young leaves (developing and sprout leaves); (ii) mature leaves (totally expanded leaves); and (iii) old leaves (senesced leaves) (Figure 1). After harvesting they were air dried, ground, and sent to the Laboratório de Química e Farmacologia de Produtos Naturais, Instituto Biológico, São Paulo State, Brazil. Analyses of protodioscin and its methylated derivatives were evaluated by HPLC/ELSD (Evaporative Light Scattering Detector) using a reversed-phase (RP-18) column and a water/acetonitrile gradient.

Statistical analysis was performed using Pearson correlation ($P < 0.05$), considering associations among meteorological variables and saponin concentrations during the experimental period.

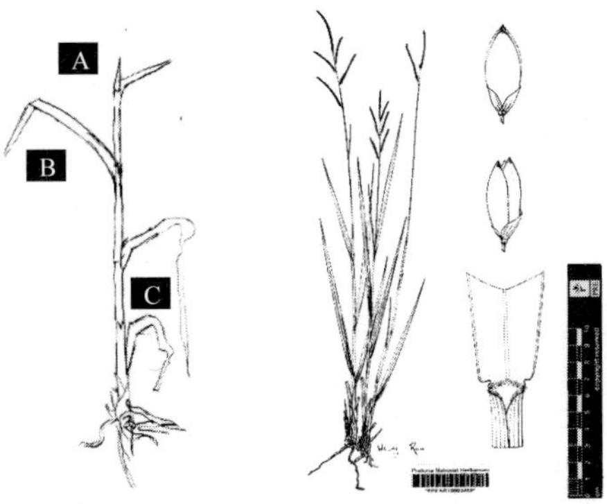


Adapted from www.aluka.org

**Figure 1.** Illustration of *Brachiaria brizantha* plant. The drawing on the left shows the collected leaves. A: young leaves (sprout); B: mature leaves; C: old leaves.

## Results

Humidity didn't have any correlation among saponin concentration and maturation of the leaves. Young leaves had higher saponin content (3.61 ± 1.12) than mature (1.94 ± 0.97 and old leaves (1.01 ± 0.79%) ($P < 0.01$) (Figure 2). Saponin variation in young leaves was strongly related ($r$=0.90; $P < 0.05$) to hours of solar radiation over the 60 day measurement period (Figure 3).

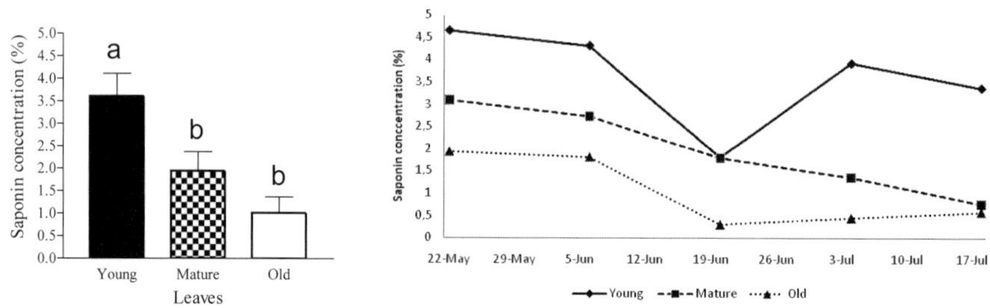


[a] differs significantly from b. One-Way ANOVA ($P < 0.05$), followed by Dunnett's Multiple Comparison Test.

**Figure 2.** Variation in saponin concentration (%) of *B. brizantha* leaves collected at different stages of maturity over 60 days. Left: Mean and standard error of saponin concentration of the leaves. Right: Levels (%) of saponin recorded during the collection period.

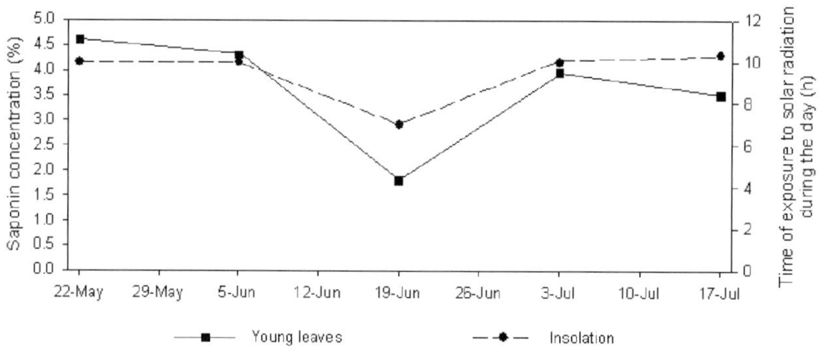


**Figure 3.** Saponin variation (%) in young *B. brizantha* leaves and solar insolation (h), evaluated over 60 days. Saponin concentrations are strongly associated ($r$=0.90; $P < 0.05$) with the time of exposure to solar radiation (h/day).

In the mature leaves, saponin content had a high positive correlation ($r$=0.97; $P < 0.01$) with average temperature (ºC) on the collection day (Figure 4).

## Discussion

Saponins are secondary metabolites of many plants, naturally occurring as sugar conjugates of triterpenoids or steroids, and possessing the properties of being able to form stable froth when shaken with water (Dewick 2002). Frequently, saponins are found in tissues that are most vulnerable to fungal or bacterial attack or insect predation. Therefore, one of their roles may be to act as a chemical barrier or shield in the defense system of the plant (Wina *et al.* 2005). In our work, the saponin concentrations were the greatest in young leaves of *B. brizantha* (1.80 to 4.65%) followed by mature leaves (0.75 to 3.10%) and lastly old leaves (0.30 to 1.03%). These data are consistent with Sen *et al.* (1998) who found higher content of saponins in immature plants than in mature plants of *Medicago lupulina*. Sprouts are the most fragile part of the grass and the higher concentrations of saponins in these young leaves are probably necessary for the plant to survive. In addition, the grazing habits of sheep in *B. brizantha* pastures result in exposure to the part of the forage with the highest concentrations of saponins.

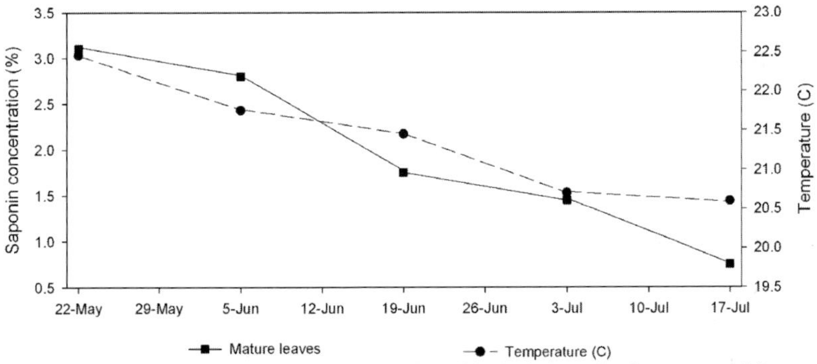


**Figure 4.** Saponin variation (%) in the mature *Brachiaria brizantha* leaves and its correlation between the average temperature (ºC), evaluated during 60 days. Saponin concentrations are strongest correlated ($r$=0.97; $P$ < 0.01) with temperature along the day.

Santos Jr (2008) studied *B. decumbens* at Goiás State in the cental-western region of Brazil and found elevated concentrations of saponins when the plant was sprouting which were associated with numerous cases of photosensitization in sheep. Outbreaks of the disease can occur at any time but primarily during the start of the rainy season (Riet-Correa and Méndez 2007). In central-western Brazil this period coincides with the end of winter and the beginning of spring (September) when insolation, humidity, and temperature are increasing, leading to growth of new leaves. We speculate that immature leaves have higher saponin concentrations as a defense against herbivory, rendering them more toxic to sheep.

The present work shows that saponin concentration is influenced by insolation and temperature during the end of rainy season, and it decreases when these meteorological data also decrease. The data could explain the reduced occurrence of outbreaks during this time, different from that observed during beginning of the rainy period when the insolation and temperature are increased and lead to larger amounts of sprouts and mature leaves available to animals that can result in outbreaks of photosensitization. New studies have been initiated to better understand how saponin concentrations vary in *B. brizantha* over the growing season and its relation with health of the ruminants feeding on that forage.

## Conclusions

Saponin concentrations vary due to climatic changes and plant maturity. Young leaves have greater concentrations of saponins followed by mature leaves, and lastly by old senesced leaves.

## Acknowledgements

This work was supported by Instituto Nacional de Ciência e Tecnologia Para o Controle das Intoxicações por Plantas – CNPq, MCT (Grant 573534/2008-0). The authors also thank Anhanguea-Uniderp that permits the studies in the CTO.

## References

Barbosa JD, Oliveira CMC, Tokarnia CH, and Peixoto PV (2006). Hepatogenous photosensitization in horses caused by *Brachiaria humidicola* (Gramineae) in the State of Pará. *Pesquisa Veterinária Brasileira* 26(3):147-153.

Brum KB, Haraguchi M, Lemos RAA, and Fioravanti CS (2007). Crystal-associated cholangiopathy in sheep grazing *Brachiaria decumbens* containing the saponin protodioscin. *Pesquisa Veterinária Brasileira* 27(1):39-42.

Brum KB, Haraguchi M, Garutti MB, Nóbrega FN, Rosa B, and Fioravanti MCS (2009). Steroidal saponin concentrations in *Brachiaria decumbens* and *B. brizantha* at different developmental stages. *Ciência Rural* 39(1):279-281.

Ceballos JC, Moura GB de A, Bezerra VF, and Farias JDA (1992). Desempenho de heliógrafos e actinógrafos na estimativa de insolação e fluxo direciomal. *Revista Brasileira de Meteorologia* 7(2):563-581.

Cruz C, Driemeier D, Pires VS, Colodel EM, Taketa ATC, and Schenkel EP (2000). Isolation of steroidal sapogenins implicated in experimentally induced cholangiopathy of sheep grazing *Brachiaria decumbens* in Brazil. *Veterinary and Human Toxicology* 42(3):142-145.

Cruz C, Driemeier D, Pires VS, and Schenkel EP (2001). Experimentally induced cholangiopathy by dosing sheep with fractionated extracts from *Brachiaria decumbens*. *Journal of Veterinary Diagnostic Investigation* 13:170-172.

Dewick PM (2002). *Medicinal Natural Products*, 2nd edn, 515 pp. John Wiley & Sons, West Sussex, England.

Fioravanti MCS (1999). Incidência, avaliações clínica, laboratorial e anatomopatológica da intoxicação subclínica por esporidesmina em bovinos. 256 pp. Tese (Doutorado em Medicina Veterinária) – Faculdade de Medicina Veterinária e Zootecnia, Universidade Estadual Paulista, Botucatu.

Graydon RJ, Hamid H, Zaha RIP, and Gardiner C (1991). Photosensitization and crystal-associated cholangiohepatopathy in sheep grazing *Brachiaria decumbens*. *Australian Veterinary Journal* 68(7): 234-236.

Lemos RAA, Ferreira LCL, Silva SM, Nakazato L, and Salvador (1996a). Fotossensibilização e colangiopatia associada a cristais em ovinos em pastagem com *Brachiaria decumbens*. *Ciência Rural* 26:109-113.

Lemos RAA, Osório ALAR, Rangel JMR, and Herrero Jr GO (1996b). Fotossensibilização e colangiopatia associada a cristais em bezerros ingerindo *Brachiaria brizantha*. *Arquivos do Instituto. Biológico* 63(Supl.):22.

Lemos RAA, Salvador SC, and Nakazato L (1997). Photosensitization and crystal associated cholangiohepatopathy in cattle grazing *Brachiaria decumbens* in Brazil. *Veterinary and Human Toxicology* 39(6):376-377.

Lemos RAA, Nakazato L, Herrero Jr GO, Silveira AC, and Porfírio LC (1998). Fotossensibilização e colangiopatia associada a cristais em caprinos mantidos sob pastagens de *Brachiaria decumbens* no Mato Grosso do Sul. *Ciência Rural* 28(3):507-510.

Macedo MCM (2005). Pastagens no Ecossistema Cerrados: Evolução das pesquisas para o desenvolvimento sustentável. In *42ª Reunião Anual da Sociedade Brasileira de Zootecnia – A produção animal e o foco no agronegócio*, 56-84. Abstracts, Universidade Federal de Goiás, Goiânia.

Meagher LP, Miles CO, and Fagliari JJ (1996). Hepatogenous photosensitization of ruminants by *Brachiaria decumbens* and *Panicum dichotomiflorum* in the absence of sporidesmin: lithogenic saponins may be responsible. *Veterinary and Human Toxicology* 38(4):271-274.

Riet-Correa F and Méndez MC(2007). Intoxicações por Plantas e Micotoxinas. In: Riet-Correa F, AL Schild, RAA, Lemos, and Borges JRJ (eds) *Doenças de Ruminantes e Eqüídeos*. Editora Pallotti, Santa Maria, RS, Brazil. V2:99-219.

Santos Jr HL (2008). Estudo da toxicidade de diferentes estágios de crescimento da *Brachiaria decumbens* em ovinos. Dissertação (Mestrado) – Universidade de Brasília/Faculdade de Agronomia e Medicina Veterinária, Brasília, p. 65.

Seitz AL, Rozza DB, Feltrin C, Travesso SD, Colodel EM, and Driemeier D (2004). Fotossensibilização por *Brachiaria decumbens* em ovinos no Rio Grande do Sul. *Pesquisa Veterinária Brasileira* 24(supl.):67.

Sen S, Makkar HPS, and Beckerplant K (1998). Alfalfa saponins and their implication in animal nutrition. *Journal of Agricultural and Food Chemistry* 46(1):131-140.

Varejão-Silva MA (2005). *Meteorologia e Climatologia.* Versão Digital, Recife, 2005. p.21. Available at: http://d.yimg.com/kq/groups/21945308/1524981508/name/Livro.

Wina E, Muetzel S, and Becker K (2005). The impact of saponins or saponin-containing plant materials on ruminant production – a review. *Journal of Agricultural and Food Chemistry* 53:8093-8105.

# Chapter 17

# Lectin Histochemistry on Sections of Liver and Hepatic Lymph Nodes from Sheep Grazing on *Brachiaria* spp.

F.M. Boabaid[1], N.A.B. Antoniassi[1], C.A. Pescador[2], M.A. Souza[2], N.D. Gasparetto[2], C.E.F. Cruz[1], P.S. Bezerra Júnior[3], D. Driemeier[1], and E.M. Colodel[2]

*[1]Setor de Patologia Veterinária, Universidade Federal do Rio Grande do Sul, Porto Alegre, RS, 91540-000, Brazil; [2]Laboratório de Patologia Veterinária, Universidade Federal de Mato Grosso, Cuiabá, MT, 78068-900, Brazil; [3]Laboratório de Patologia Veterinária, Universidade Federal de Lavras, Lavras, MG, 37200-000, Brazil*

## Introduction

*Brachiaria* spp. (Gramineae) are important sources of forage for ruminants in Brazil, especially on the 'cerrado' areas from the midwest and southeast regions (Brum *et al.* 2007). Sporadic outbreaks of photosensitization in ruminants grazing on *Brachiaria* spp. have been reported. The condition has been associated with steroidal saponins in the plant, which through ruminant metabolism form biliary crystals that may deposit within hepatocytes, macrophages, biliary ducts, and bile causing cholestasis, hepatic lesions, and secondary photosensitization (Driemeier *et al.* 1998; Cruz *et al.* 2000; Brum *et al.* 2007). Animals ingesting *Brachiaria* spp. may not always show clinical signs (Driemeier *et al.* 1999; Gomar *et al.* 2005). The disease is found in cattle (Lemos *et al.* 1997; Ecco *et al.* 2004), sheep (Lemos *et al.* 1996; Brum *et al.* 2004; Albernaz *et al.* 2008), goats (Lemos *et al.* 1998), and buffalo (Rozza *et al.* 2004). There have been chronic cases of *B. decumbens* poisoning in cattle showing progressive weight loss (Riet-Correa *et al.* 2002).

Besides the dermatitis lesions, macroscopic changes linked to the disease include subcutaneous edema, jaundice (Tokarnia *et al.* 2000; Saturnino *et al.* 2008), enhanced and yellowish liver, enhanced lobular pattern, increased liver consistency (Lemos *et al.* 1996, 1998; Brum *et al.* 2004), and distended gall bladder (Lemos *et al.* 1996). Enhanced hepatic and mesenteric lymph nodes displaying whitish foci surrounded by reddened halos on their cut surface were also reported (Driemeier *et al.* 1998; Riet-Correa *et al.* 2002). Main microscopic changes seen on the liver sections were tumefaction, vacuolation, and individual necrosis of hepatocytes (Lemos *et al.* 1996; Santos *et al.* 2008). Periportal regions may present mononuclear infiltrates, fibrosis, pericholangitis, bile duct proliferation, and crystals within biliary duct and canaliculi (Lemos *et al.* 1996, 1998; Brum *et al.* 2007). These crystalline structures are rarely seen on conventional microscopy, since

©CAB International 2011. *Poisoning by Plants, Mycotoxins, and Related Toxins* (eds F. Riet-Correa, J. Pfister, A.L. Schild, and T.L. Wierenga)

they may be dissolved by ethanol during routine histological processes (Driemeier *et al.* 1998). Macrophages displaying cellular enlargement and fine cytoplasmic vacuolation (foamy macrophages), sometimes forming giant cells, are consistent hepatic findings (Driemeier *et al.* 1998, 1999; Gomar 2002). These foamy macrophages may also be observed in hepatic and mesenteric lymph nodes (Lemos *et al.* 1996; Riet-Correa *et al.* 2002; Gomar *et al.* 2005), spleen (Driemeier *et al.* 1998, 1999; Riet-Correa *et al.* 2002), and small intestine (Riet-Correa *et al.* 2002).

Lectins are carbohydrates binding proteins of non-immune origin that are purified through chromatography and recombinant DNA techniques. Lectins recognize and bind to carbohydrates present on the surface of, or inside cells and are used for the study of both normal and altered cells (Brooks *et al.* 1997; Lis and Sharon 1998). We report the changes in liver and hepatic lymph nodes of sheep that grazed on *Brachiaria* spp. and describe the carbohydrate residues stored within the foamy cells through lectin histochemistry.

## Material and Methods

Liver and hepatic lymph nodes from 19 female Santa Inês sheep kept exclusively on *Brachiaria* spp. pastures and slaughtered in the Mato Grosso state, Brazil were grossly evaluated, collected, fixed in buffered 10% formalin, and submitted to the Laboratório de Patologia Veterinária da Universidade Federal de Mato Grosso (LPV-UFMT). Fragments were routinely processed and stained by hematoxylin and eosin (Prophet *et al.* 1992) and histological sections were submitted to the Setor de Patologia Veterinária da Universidade Federal do Rio Grande do Sul (SPV-UFRGS), where the histochemical study of lectins was performed as described by Brooks *et al.* (1997). Fragments from tissues of sheep fed exclusively lucerne (alfafa) hay and concentrate served as controls. Eight biotinylated lectins were used (Table 1). After deparaffination, the sections were incubated in 0.3% hydrogen peroxide for 30 min at room temperature, and then incubated with citrate buffer pH 6.0 at 100°C for 15 min. Sections were subsequently treated with 5% fat-free milk (Molico®) in distilled water for 30 min and then incubated with biotinylated lectins at 5 μg/ml dilution (except for ConA and RCA, which were diluted at 0.5 μg/ml and 3 μg/ml, respectively) at 4°C overnight, followed by incubation with streptavidin for 20 min at room temperature. Diaminobenzidina (DAB) was the chromogen and Harris's hematoxylin was the counterstain. Negative control was PBS.

**Table 1.** Biotinylated lectins used to examine carbohydrate residues stored within foamy cells from sheep grazing *Brachiaria* spp. pastures.

| Lectins | Acronyms | Carbohydrate specificity[a] |
|---|---|---|
| *Canavalia ensiformis* | ConA | α-D-Man; α-D-Glc; GlcNAc |
| *Arachis hypogaea* | PNA | β-D-Gal(1-3)-D-GalNAc |
| *Ricinus communis* | RCA | β-D-Gal; D-GalNAc |
| *Glycine max* | SBA | D-GalNAc; Gal |
| *Triticum vulgaris* | WGA | β-D-GlcNAc; β-D-GlcNAc NeuAc, GalNAc |
| *Lens culinares* | LCA | α-D-Man; D-Glc |
| *Phaseolus vulgaris* | PHA-L | Complex carbohydrates |
| *Pisum sativum* | PSA | α-D-Man; α-D-Glc |

Brookes *et al.* 1997. [a]Gal = galactose, GalNac = N-acetyl-galactosamine, Glc = glucose, GlcNAc = N-acetyl-glucosamine, Man = mannose, NeuAc = N-acetyl-neuraminic acid.

## Results

Grossly, livers had occasionally normal aspect but often had enhanced lobular pattern, light to intense diffuse yellow coloration, mildly increased consistency, and multifocal capsular thickening. Those areas were irregular, scattered, whitish, firm, depressed, and mostly confined to the capsule at cut surface. Lymph nodes had no gross changes. Microscopically, there were hepatocellular tumefaction and vacuolation, individual necrosis of hepatocytes, mononuclear pericholangitis, and fibrosis and bile duct proliferation in the portal triads. Infiltration of foamy cells was a frequent finding in the lobule, but predominantly surrounding centrolobular veins or forming agglomerates, which sometimes joined in giant multinucleated cells. In samples from three sheep, there were no foamy cells in the liver, but they were present in the corresponding hepatic lymph nodes. Occasionally, there was cholestasis, inflammatory infiltrates, and negative images of crystals within biliary ducts, hepatocytes, and foamy macrophages. The most severe microscopic lesions corresponded to the whitish areas seen grossly. In total, 14 of the 16 hepatic lymph nodes evaluated had foamy cells that were more consistent and in higher numbers than in the correspondent livers. These cells were distributed mainly in the cortical region and surrounding the lymphoid follicles and sometimes had crystals. The foamy macrophages were marked by lectins in both liver and lymph node samples. In liver samples, the lectins that showed higher intense staining in the foamy macrophages were ConA, RCA, WGA, LCA, and PHA-L. PNA had mild but specific staining in foamy macrophages. RCA, LCA, and PHA-L also showed evident staining in the vascular endothelium and Kupffer cells in the livers. SBA had consistent staining in the biliary ducts, mainly in hyperplasic ducts. In lymph node samples, foamy macrophages were consistently marked by ConA, LCA, PHA-L, and PSA. PNA also had mild but specific staining in the foamy cells. All the lectins marked foamy macrophages in the tissues studied and made those cells more evident, especially when they were isolated.

## Discussion

Gross and microscopic changes seen in the livers and hepatic lymph nodes of these sheep were similar to those described in ruminants and associated with the consumption of *Brachiaria* spp. (Lemos *et al.* 1996; Driemeier *et al.* 1999; Cruz 2000; Seitz *et al.* 2004). Cattle kept on *Brachiaria* spp. pastures may show lesions in hepatic and mesenteric lymph nodes and in livers even when they are clinically healthy (Driemeier *et al.* 1998). Sheep from this study were also apparently healthy; however, chronic changes were found in their livers and hepatic lymph nodes and associated with the consumption of *Brachiaria* spp. In those samples, there were infiltrates of foamy macrophages and negative images or cholesterol clefts associated with crystals in the lumen of the bile ducts, hepatocytes, and foamy macrophages of the livers and lymph nodes. These were the main findings associated with chronic insult caused by steroidal saponins of *Brachiaria* spp. (Driemeier *et al.* 1998). Sporadic outbreaks of wasting disease of cattle kept on *Brachiaria decumbens* pastures in Mato Grosso state still have uncertain etiopathogeny; however, it is possible that foamy cells present in intestinal submucosa may be involved (Riet-Correa *et al.* 2002).

Lectin histochemistry has demonstrated the accumulation of glycoconjugates within cytoplasm of foamy macrophages. These cells were consistently marked in the liver and lymph node sections by the lectins ConA, RCA, WGA, LCA, PHA-L, and PSA, indicating the accumulation of mannose, glucose, N-acetyl-glucosamine, galactose, N-acetylgalactosamine, N-acetyl-neuraminic acid, besides other complex carbohydrates (Brooks *et al.* 1997). Although at differentiated intensities, this study also showed PNA, SBA, and WGA staining of foamy cells as seen previously (Gomar *et al.* 2005). A remarkable finding also seen previously (Gomar *et al.* 2005) is the selective pattern of staining of foamy macrophages by PNA. The specificity of PNA by macrophages has also been highlighted in humans (Ree and Kadin 1985). These foamy macrophages in livers and associated lymph nodes may be linked to the phagocytosis of necrotic hepatocytes affected by the steroidal saponins from *Brachiaria* spp. or the direct phagocytosis from this substance absorbed in the digestive tract or present in the enterohepatic cycle (Driemeier *et al.* 2002). Morphological and histochemical findings described here suggest that the inhibition of a lysosomal lipase enzyme also could be involved in the formation of the foamy macrophages, leading to the intracellular storage of glycoconjugates in cells of sheep grazing on *Brachiaria* spp. pastures.

## References

Albernaz TT, Silveira JAS, Reis ASB, Oliveira CHS, Oliveira CMC, Duarte MD, Cerqueira VD, Riet-Correa G, and Barbosa JD (2008). Fotossensibilização em ovinos associada à ingestão de *Brachiaria brizantha* no Pará. *Anais do Encontro Nacional de Diagnóstico Veterinário,* pp. 73-74. Campo Grande, Mato Grosso do Sul, Brasil.

Brooks SA, Leathem AJC, and Schurmacher U (1997). *Lectin Histochemistry – a concise practical handbook,* 177 pp. Bios Scientific Publishers, Oxford.

Brum KB, Haraguchi M, Lemos RAA, and Fioravanti MCS (2004). Colangiopatia associada a cristais em ovinos mantidos em pastagens de *Brachiaria decumbens*. *Pesquisa Veterinária Brasileira* 24 (Supl.):14-15.

Brum KB, Haraguchi M, Lemos RAA, Riet-Correa F, and Fioravanti MCS (2007). Crystal-associated cholangiopathy in sheep grazing *Brachiaria decumbens* containing the saponin protodioscin. *Pesquisa Veterinária Brasileira* 27(1):39-42.

Cruz C, Driemeier D, and Pires VS (2000). Isolation of steroidal sapogenins implicated in experimentally induced cholangiopathy of sheep grazing *Brachiaria decumbens* in Brazil. *Veterinary and Human Toxicology* 42(3):142-145.

Cruz CEF (2000). Contribuição ao estudo da etiopatogenia das lesões hepáticas em ovinos associadas ao consumo de *Brachiaria decumbens*, 66 pp. Dissertação de mestrado em Ciências Veterinárias, Universidade Federal do Rio Grande do Sul, Porto Alegre, Rio Grande do Sul, Brasil.

Driemeier D, Barros SS, Peixoto PV, Tokarnia CH, Döbereiner J, and Brito MF (1998). Estudos histológicos, histoquímicos e ultra-estrutural de fígados e linfonodos de bovinos com presença de macrófagos espumosos ('foam cells'). *Pesquisa Veterinária Brasileira* 18(1):29-34.

Driemeier D, Döbereiner J, Peixoto PV, and Brito MF (1999). Relação entre macrófagos espumosos ('foam cells') no fígado de bovinos e ingestão de *Brachiaria* spp. no Brasil. *Pesquisa Veterinária Brasileira* 19(2):79-83.

Driemeier D, Colodel EM, Seitz AL, Barros SS, and Cruz CEF (2002). Study of experimentally induced lesions in sheep by grazing *Bracharia decumbens*. *Toxicon* 40:1027-1031.

Ecco R, Santos Jr HL, Túry E, and Jacobina GC (2004). Intoxicação crônica por *Brachiaria* spp. em bovinos. *Pesquisa Veterinária Brasileira* 24(Supl.):19-20.

Gomar MS (2002). Características das células espumosas no fígado, linfonodos mesentéricos e intestino de bovinos associados ao consumo do *Brachiaria* spp., 62 pp. Dissertação de Mestrado em Ciências Veterinárias, Universidade Federal do Rio Grande do Sul, Porto Alegre, Rio Grande do Sul, Brasil.

Gomar MS, Driemeier D, Colodel EM, and Gimeno J (2005). Lectin histochemistry of foam cells in tissues of cattle grazing *Brachiaria* spp. *Journal of Veterinary Medicine* 52:18-21.

Lemos RAA, Ferreira LCL, Silva SM, Nakazato L, and Salvador SC (1996). Fotossensibilização e colangiopatia associada a cristais em ovinos em pastagem com *Brachiaria decumbens*. *Ciência Rural* 26(1):109-113.

Lemos RAA, Salvador CS, and Nakazato L (1997). Photosensitization and crystal-associated cholangiohepatopathy in cattle grazing *Brachiaria decumbens* in Brazil. *Veterinary and Human Toxicology* 39(6):376-377.

Lemos RAA, Nakazato L, Herrero Jr GO, Silveira AC, and Porfírio LC (1998). Fotossensibilização e colangiopatia associada a cristais em caprinos mantidos sob pastagens de *Brachiaria decumbens* no Mato Grosso do Sul. *Ciência Rural* 28(3):507-510.

Lis H and Sharon (1998). Lectins: Carbohydrate-specific proteins that mediate cellular recognition. *Chemistry Revision* 98:637-674.

Prophet EB, Mills B, Arrington JB, and Sobin LH (1992). *Laboratory Methods in Histotechnology* 279 pp. Armed Forces Institute of Pathology, Washington, DC.

Ree HJ and Kadin ME (1985). Macrophage-histiocytes in Hodkin's disease. The relation of peanut-agglutinin-binding macrophage-histiocytes to clinicopathologic presentantion and course of disease. *Cancer* 56(2):333-338.

Riet-Correa G, Riet-Correa F, Schild AL, and Driemeier D (2002). Wasting and death in cattle associated with chronic grazing of *Brachiaria decumbens*. *Veterinary and Human Toxicology* 44(3):179-180.

Rozza DB, Seitz AL, Bandarra PM, Santos EO, and Driemeier D (2004). Fotossensibilização por *Brachiaria decumbens* em búfalo. *Pesquisa Veterinária Brasileira* 24(Supl.):55-56.

Santos JCA, Riet-Correa F, Simões SVD, and Barros CSL (2008). Patogênese, sinais clínicos e patologia das doenças causadas por plantas hepatotóxicas em ruminantes e eqüinos no Brasil. *Pesquisa Veterinária Brasileira* 28(1):1-14.

Saturnino KC, Mariani TM, and Lemos RAA (2008). Intoxicação experimental por *Brachiaria decumbens* em ovinos. *Anais do Encontro Nacional de Diagnóstico Veterinário,* pp. 215-216. Campo Grande, Mato Grosso do Sul, Brasil.

Seitz AL, Rozza DB, Feltrin C, Traverso SD, Colodel EM, and Driemeier D (2004). Fotossensibilização por *Brachiaria decumbens* em ovinos no Rio Grande do Sul, Brazil. *Pesquisa Veterinária Brasileira* 24 (Supl.):67-68.

Tokarnia CH, Döbereiner J, and Peixoto PV (2000). *Plantas tóxicas do Brasil*, 310 pp. Editora Helianthus, Rio de Janeiro, Brasil.

# Chapter 18

# *Brachiaria* spp. Poisoning in Ruminants in Mato Grosso do Sul, Brazil

R.A.A. de Lemos[1], A.P.A. Nogueira[2], R.I.C. Souza[2], B.S. Santos[2], N.M. Carvalho[3], A.C.M. Aniz[4], and P.C. de Freitas[5]

*[1]Departamento de Medicina Veterinária (DMV/FAMEZ), UFMS, Campo Grande, MS 79070-900, Brazil; [2]Programa de Mestrado em Ciência Animal, FAMEZ, UFMS, Campo Grande, MS 79070-900, Brazil; [3]Médico Veterinário. FAMEZ, UFMS, Campo Grande, MS 79070-900, Brazil; [4]Médica Veterinária autônoma.Campo Grande, MS, Brazil; [5]Aluno de graduação do curso de Medicina veterinária, FAMEZ*

## Introduction

*Brachiaria* spp. are important forage for bovines and ovines in midwestern Brazil, but their ingestion may cause hepatogenous photosensitization in cattle, sheep, goats, buffalo, and horses (Riet-Correa and Méndez 2007). Progressive weight loss due to ingestion of *B. decumbens* is also reported (Riet-Correa *et al.* 2002). At first, outbreaks from *Brachiaria* spp. in bovines and ovines were thought to be contamination by *Pithomyces chartarum* spores in the forage. Later reports from Indonesia (Graydon *et al.* 1991), Malaysia (Abas-Mazni *et al.* 1985), Africa (Opasina 1985), and Brazil described *Brachiaria* poisoning in sheep and goats without *P. chartarum* spores in forage and with histologic findings similar to those reported in poisoning by saponin-containing plants (Riet-Correa and Méndez 2007). The objective of this paper is to report the epidemiology, clinical signs, and pathology of *Brachiaria* spp. poisoning in ruminants in the state of Mato Grosso do Sul.

## Material and Methods

The survey was conducted during research on spontaneous outbreaks of *Brachiaria* spp. poisoning in ruminants in Mato Grosso do Sul. Records of the Veterinary Pathology Laboratory at the Federal University of Mato Grosso do Sul from 1996 to 2008 were reviewed and epidemiology (period of year, species affected, age, morbidity, lethality, and pasture vegetative stage), clinical signs, and gross and histology findings were recorded.

©CAB International 2011. *Poisoning by Plants, Mycotoxins, and Related Toxins* (eds F. Riet-Correa, J. Pfister, A.L. Schild, and T.L. Wierenga)

## Results and Discussion

During this time period, 32 outbreaks were diagnosed in bovines and 21 in ovines. Outbreaks in cattle represented 1.23% of the 2600 diagnoses performed during the period. In contrast the 21 sheep outbreaks represent 10.5% of the 207 diagnoses in sheep, demonstrating that the disease is much more frequent in sheep than cattle. Two outbreaks were observed in goats.

Epidemiologic data referring to the time of the year the outbreaks occurred and age of the affected animals are given in Figures 1 and 2.

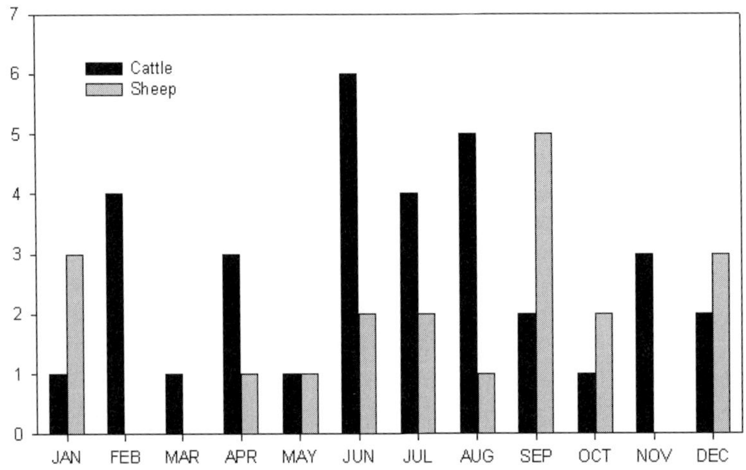


**Figure 1.** Monthly distribution of outbreaks of *Brachiaria* spp. poisoning.

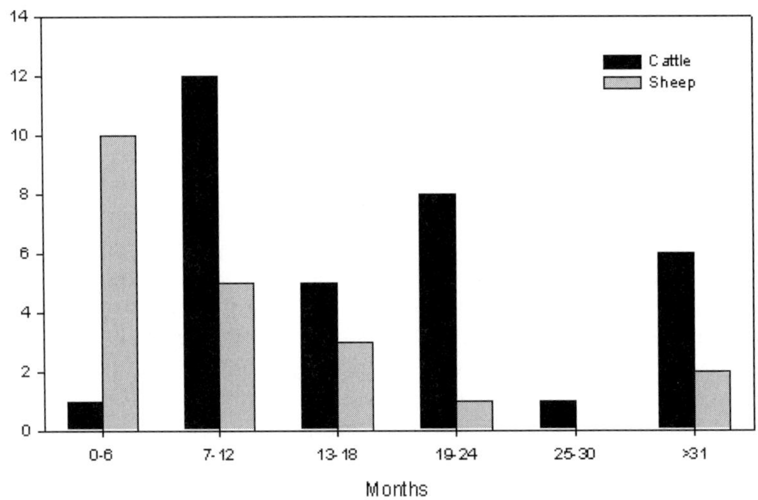


**Figure 2.** Age in months of sheep and cattle affected by *Brachiaria* spp. poisoning.

The poisoning affected cattle and sheep of different ages and occurred in all periods of the year. Previous reports mentioned that sheep of any age and cattle up to 2 years old are affected (Tokarnia *et al.* 2000). In two outbreaks in cattle the animals were introduced into a pasture that regrew after being burned in the previous year. This condition was not observed in other outbreaks.

In cattle, morbidity varied from 0.08% to 7.9% and lethality from 10% to 100%. In sheep, morbidity varied between 4.6% and 60% and lethality was 50% to 100%. Despite the similar frequency of the poisoning in cattle and sheep, the disease is more severe in sheep than cattle. These results are similar to other reports that mention a greater susceptibility in sheep than cattle (Riet-Correa and Méndez 2007).

In cattle the clinical and pathological findings in the majority of the outbreaks were hepatogenous photosensitization. First signs were ear edema, jaundice, restlessness, and seeking shade, followed by cutaneous lesions with dermatitis, crust formation, and loss of the epidermis mainly in the perineal and flank regions. Ulcerations of the ventral side of the tongue were also observed. A frequent finding was thickness and healing retraction of the ears. Six outbreaks of a progressive weight loss syndrome without photosensitization were observed in cattle. Other clinical signs were apathy, anorexia, and cachexia. This clinical syndrome was reported earlier (Riet-Correa *et al.* 2002).

In sheep, the predominant clinical sign is facial eczema. Severe jaundice is also frequent in animals with prolonged clinical evolution, and crusts on the face and ears are observed in the survivors.

Gross and histological findings are similar in cattle, sheep, and goats. Gross lesions are several stages of jaundice and enlarged and yellow liver with evident lobular pattern. The gall bladder is frequently full and distended. Microscopically, the main lesions are in the liver, characterized by swelling and vacuolization of hepatocytes, biliary retention, infiltration of macrophages with peripheral nucleus and vesicular cytoplasm, and the presence of optically active refractile crystals or negative crystals images in the lumen of canaliculi and bile ducts. Cholangitis, pericholangitis, and proliferation of bile duct epithelial cells are observed. These alterations are similar to those in other reports (Cruz *et al.* 2001). Chronic cases can show proliferation of fibrous tissue mainly in the intermediary region. Macrophages with foamy cytoplasm occasionally in groups are observed mainly in the periacinar region. These alterations were previously described by Driemeier *et al.* (1998). Foamy macrophages and fibrosis are very common in cases of *Brachiaria* spp. poisoning and can be found in the liver of healthy ruminants grazing the same plant (Lemos and Purisco 2002).

Because factors that determine the occurrence of the intoxication are not completely known, control and preventative measures are currently lacking. The existence of resistant (Saturnino 2009) and susceptible sheep (Aniz 2008) to the intoxication was recently demonstrated. In sheep, adaption to the consumption of the plant did not prevent the intoxication. After the introduction to *Brachiaria* spp. pastures, no significant differences in prevalence of the intoxication were observed between sheep raised in *Brachiaria* spp. and sheep raised in other pastures (Aniz 2008) and later introduced to *Brachiaria* pastures. Complementary research is necessary to establish the relationship between the saponin concentrations in the forage and the occurrence of the intoxication and to investigate the conditions that determine this concentration. Further work should determine if animal resistance to *Brachiaria* spp. poisoning is genetic in origin or acquired. This knowledge will help to establish effective control measures by aiding in pasture management, selection of resistant animals, and selection of *Brachiaria* varieties with low saponin levels.

## Acknowledgements

This work has been financially supported by the National Institute for Science and Technology for the Control of Toxic Plants (CNPq), grant number 573534/2008-0.

## References

Abas-Mazni O, Sharif H, and Khusahry M (1985). Photosensitization in goats grazed on *Brachiaria decumbens*. *Mardi Research Bulletin* 13(2):203-206.

Aniz ACM (2008). Efeito da adaptação ao consumo de *Brachiaria decumbens* e a existência de resistência ou suscetibilidade individual em ovinos à intoxicação, 29 pp. Course Conclusion Monograph, Universidade Federal do Mato Grosso do Sul, Campo Grande.

Cruz C, Driemeier D, Pires VS, and Schenkel EP (2001). Experimentally induced cholangiopathy of sheep grazing *Brachiaria decumbens* in Brazil. *Veterinary and Human Toxicology* 42:142-145.

Driemeier D, Barros SS, Peixoto PV, Tokarnia CH, Döbereiner J, and Brito MF (1998). Estudos histológico, histoquímico e ultra-estrutural de fígados e linfonodos de bovinos com presença de macrófagos espumosos ('foam cells'). *Pesquisa Veterinária Brasileira* 18(1):29-34.

Graydon RJ, Hamid H, Zaha RIP, and Gardiner C (1991). Photosensitization and crystal-associated cholangiohepatopathy in sheep grazing *Brachiaria decumbens*. *Australian Veterinary Journal* 68(7):234-236.

Lemos RAA and Purisco E (2002). Plantas que causam fotossensibilização hepatógena. In *Enfermidades de interesse econômico em bovinos de corte: perguntas e respostas* (RAA Lemos, N Barros, and KB Brum, eds), 292 pp. UFMS, Campo Grande.

Opasina BA (1985). Photosensitization jaundice syndrome in West African dwarf goats and sheep. *Tropical Grasslands* 19:120-123.

Riet-Correa F and Méndez MDC (2007). Intoxicações por plantas e micotoxinas. In *Doenças de Ruminantes e Eqüídeos* (F Riet-Correa, AL Schild, RAA Lemos, and JRJ Borges, eds), pp. 114-122. Editora Palloti, Santa Maria.

Riet-Correa G, Riet-Correa F, Schild AL, and Driemeier D (2002). Wasting and death in cattle associated with chronic grazing of *Brachiaria decumbens*. *Veterinary and Human Toxicology* 44:179-180.

Saturnino KC (2009). Intoxicação experimental por *Brachiaria decumbens* em ovinos, 36 pp. Master Dissertation, Universidade Federal do Mato Grosso do Sul, Campo Grande.

Tokarnia CH, Döbereiner J, and Peixoto PV (2000). *Plantas Tóxicas do Brasil,* 310 pp. Helianthus, Rio de Janeiro.

# Chapter 19

# Practical Rules for the Differentiation between *Brachiaria* spp. Poisoning and Pithomycotoxicosis

P.V. Peixoto[1], J.N. Seixas[2], T.N. França[3], C.H. Tokarnia[1], J. Döbereiner[4], and B.L. Smith[5]

*[1]Departamento de Nutrição Animal e Pastagem, Universidade Federal Rural do Rio de Janeiro (UFRRJ), Seropédica, RJ 23890-000, Brazil; [2]Doutorado em Ciências Veterinárias, UFRRJ, Seropédica, RJ 23890-000, Brazil; [3]Departamento de Epidemiologia e Saúde Pública, UFRRJ, Seropédica, RJ 23890-00, Brazil; [4]Projeto Sanidade Animal, Embrapa-CNPAB/PSA, Seropédica, RJ 23851-970, Brazil; [5]56 Orchard Avenue, Hamilton, New Zealand*

## Introduction

*Brachiaria* spp. are an important source of food for ruminants under range conditions in many countries (Crispim and Branco 2002). In some tropical countries like Brazil, these grasses cause significant economic losses due to outbreaks of photosensitization that occur yearly in cattle, sheep, and sometimes in goats, buffalo, and horses. The first reports on photosensitization in cattle kept on *B. decumbens* pastures in Brazil were attributed to the toxicity of the sporidesmin-containing spores of *Pithomyces chartarum* (Camargo *et al.* 1976; Döbereiner *et al.* 1976; Nobre and Andrade 1976). Later research showed that *Brachiaria* spp. containing saponins are responsible for liver lesions and consequent photosensitization (Salam Abdullah *et al.* 1992; Wilkins *et al.* 1994; Lemos *et al.* 1996, 1997a, b; Driemeier *et al.* 1997). Although there is some information about differentiation between those conditions (Smith and Miles 1993), in Brazil procedures to minimize the problem are not adopted or are wrong; prophylactic measures against the fungus, like zinc sulfate administration with a salt mix, are recommended. There is a similar situation in Panama (Peixoto and França 2003, unpublished). Herein are practical rules for an easy differentiation between *Brachiaria* poisoning and pithomycotoxicosis based on epidemiological, toxicological, clinical, and pathological aspects.

## Epidemiological Differentiation

Epidemiological conditions which differentiate *Brachiaria* poisoning from pithomycotoxicosis are distinct (Table 1). Lush green pasture mainly at the beginning of the

©CAB International 2011. *Poisoning by Plants, Mycotoxins, and Related Toxins* (eds F. Riet-Correa, J. Pfister, A.L. Schild, and T.L. Wierenga)

rainy season is associated only with outbreaks of *Brachiaria* poisoning. Outbreaks of pithomycotoxicosis occur when there is dead plant litter (moldy grass) and the number of *Pithomyces chartarum* spores reaches more than 50,000/g, while only few or no spores are found in *Brachiaria* pastures during outbreaks of photosensitization.

**Table 1.** Epidemiological differentiation .

| *Brachiaria* spp. poisoning | Pithomycotoxicosis |
|---|---|
| OCURRENCE OF NATURAL OUTBREAKS | |
| Sheep [6,15,26,32,47,53,57], cattle [15,17,25,31,35,45,50,67] goats [34,38,47,48], horses [3,46,56] (*B. humidicola*), buffalo [2,21,30,66] | Sheep [4,14,16,28,36,49], cattle [8,49], goats [58] |
| SUSCEPTIBILITY ACCORDING TO ANIMAL SPECIES | |
| Sheep +++**, cattle ++, goats +, horses + [6,26,47,54], buffaloes + [1] | Sheep+++, cattle ++, goats+, horses [58,42,62] |
| AGE | |
| Sheep: any age; lambs more sensitive [6,15,32,57]<br>Cattle: almost all younger than 2 years [15,35,45] | Sheep: any age [29], mostly in lambs [14,28]<br>Cattle: lactating dairy cows +++ [42] |
| PERIOD OF THE YEAR | |
| Throughout the year; for cattle mostly in the rainy season [32,34,35,39,45,56] | End of summer and autumm [4,29,49,59] |
| WEATHER | |
| For cattle, mostly after the beginning of the rainy season [23,32,35] | >12.2°C/>72 h, RH>98%<br>Rainfall > 3.8 mm [14,24,28,36,49] |
| PLANT/GRASS | |
| *Brachiaria* spp. usually green grass, lush in appearance [6, 39] | *Lolium perenne* (mostly) – dead plant litter moldy grass, sometimes black color [8,28,29] |
| START OF THE SYMPTOMS | |
| Cattle: 10 days-months [6,26,35,45]<br>Sheep: 5-89 days [6,13,18,20] | Cattle: 10-24 days [29,42]<br>Sheep: 2-24 days [29,36,42,58] |
| CLINICAL COURSE | |
| Few days [6,47] to 120 days [26] | Few days to months [28,42] |
| MORBIDITY | |
| Cattle: 1-20% [35] or more<br>Sheep: 2.5-58% [6,32,47,69] or more | Cattle: 1-90% [8,49]<br>Sheep: 10-100% [4,14,28,36,49] |
| MORTALITY | |
| Cattle: 0-50% [35,45]<br>sheep: 0-50% [6,32] or more | Cattle: 0-16% [8,49]<br>Sheep: 0-20% [14,36,49] |
| PRESENCE OF SPORIDESMIN IN THE PASTURE | |
| Few or no spores of *P. chartarum* [12,13,20,32,34,35,39,60] | Normally 50,000 to 300,000 spores/g [29,49,62] or more [8,28] |
| SPORIDESMIN-PRODUCING STRAINS | |
| Almost only non-toxigenic strains (Brazil, Texas, Colombia, North America) [5,9,27,39,40,60] | Toxigenic strains (New Zealand, South Africa, Australia, France, Portugal, Spain, Uruguay, Argentina) [9,36,49,60,62] |

The numbers refer to the reference superscripts.
Susceptibility + = less, ++ = moderate, +++ = more, - = not susceptible, RH = % relative humidity.

*P. chartarum* strains are almost not toxigenic in South and North America (Carillo *et al.* 1980; Hansen *et al.* 1994; Kellerman *et al.* 2005). The toxin sporidesmin has never been isolated from *P. chartarum* collected from pastures where photosensitization occurred in Brazil. Adequate weather conditions for the proliferation of the fungus are present only in very restricted areas of southern Brazil, where pastures usually are not formed by *Brachiaria* spp. (Seixas 2009). Although adult cattle can sometimes be slightly affected, the occurrence of photosensitivity outbreaks in under 2 years old (Nunes 1976; Lemos *et al.* 1997b; Lemos and Brum 2001; Torres and Coelho 2001) indicates *Brachiaria* poisoning. Possibly, older cattle may develop a progressive capacity to metabolize the *Brachiaria* saponins. Horses are not susceptible to sporidesmin and normally do not graze *Brachiaria* spp. because of its low palatability, but do develop photosensitization from ingestion of the palatable *B. humidicola.* We could not find references on pithomycotoxicosis in buffalo, but we have observed rare *B. brizantha* poisoning in this species.

## Clinical Differentiation

The differentiation can be difficult if based only on clinical aspects, but photosensitization, associated with cystitis, hemoglobinuria and hemolysis which are alterations of pithomycotoxicosis (Connor 1977; Smith and Embling 1991), does not occur in ruminants poisoned by *Brachiaria* (Table 2).

**Table 2.** Clinical differentiation.

| Clinical findings | *Brachiaria* spp. poisoning | Pithomycotoxicosis |
|---|---|---|
| Skin lesions | Photosensitization [3,12,13,21,26,32,34,47,53,56] (No differences) | Photosensitization [4,8,16,28,36,42,49,58,59] (No differences) |
| Icterus | Slight to moderate [12,26,31,34,45,47,53] | Moderate to marked [4,8,16,36,42,49,58] |
| Diarrhoea transient | Absent | Present [36,42,49,58,61] |
| Hemoglobinuria | Absent | Sometimes present [10,59] |
| Hemolysis | Absent | Sometimes present [10,42] |
| Signs of cystitis | Absent | Sometimes present [29,36,42,58] |
| Anemia | Absent | Mild [10,36,42] |
| Neurologic disorders | Rare [31,47,53] | Rare [65] |

The numbers correspond to the reference superscripts.

## Pathological Differentiation

At necropsy, the diagnosis may be relatively easy (Table 3). The macroscopic aspects of the liver and gall bladder differ in both conditions and allow differentiation, mainly in chronic cases of pithomycotoxicosis with liver fibrosis and atrophy, enlarged bile ducts and nodular hyperplasia. In *Brachiaria* poisoned sheep and cattle the liver is usually yellowish or orange without other important morphological alterations, but sometimes is slightly harder. Bladder lesions are described only in animals affected by pithomycotoxicosis. Also the histological appearance of the liver is very characteristic and differs for each condition (Table 4).

**Table 3.** Macroscopic differentiation.

| *Brachiaria* spp. poisoning | Pithomycotoxicosis |
|---|---|
| LIVER | |
| Yellow or orange, usually not hard 17,25,35,50,57<br>Rarely multiples white foci 12,13,18,25 | Swollen liver, yellow-to-green AD;8,28,29,49,58<br>Reduced size, firm CD;8,16,28<br>Thick bile ducts CD; 16,28,29,58<br>Hepatic cirrhosis/atrophy CD;4,16,29,36,49, the left lobe is particularly affected CD; 16,29,42<br>Regeneration nodules CD;4,16,29,36 |
| GALL BLADDER | |
| No lesions | Enlarged AD; 16,49, hemorrhages, edema, ulceration 28,29,36, 58 |
| URINARY BLADDER | |
| No lesions | Hemorrhages, edema, ulceration, sometimes inflammation and necrosis 16,29,42,58,61 |
| HEPATIC AND MESENTERIC LYMPH NODES | |
| Sometimes with white foci or lines 12,13,17,18,25,50 | Enlargement 16<br>Mostly no lesions |

The number corresponds to the reference superscripts.
AD = acute disease, CD = chronic disease

**Table 4.** Histopathological differentiation.

| Organ | *Brachiaria* spp. poisoning | Pithomycotoxicos |
|---|---|---|
| LIVER | | |
| Presence of foamy cells | Always (or sometimes) present (cattle) 17,19,23,35,50,67 sometimes present (sheep) 6,18,26,32,57 absent (horse) 3 | Always absent |
| Birefringent crystals in bile ducts | Sometimes present 6,18,20,26,32,35,57,67 | Always absent |
| Pericholangitis | Sometimes present 12,13,18,20,23,25 but slight | Sometimes present 8,36,42,49,61 |
| Bile stasis (major ducts) | Normally absent | Present 4,8,16,49,58 |
| Necrosis/ mineralization of bile ducts | Rare 26,32 | Present 8,16,28,36,49,58 |
| Proliferation of bile ducts | Slight 2,12,13,18,25,32 | Moderate to marked CD; 8,16,28,36,49 |
| Fibrosis in portal area | Normally slight[12.13,23,26,32] | Moderate to marked CD; 4,8,16,28,36,49,58 |
| Regeneration nodules | Absent | Present CD;4,16,36 |
| LYMPH NODES | | |
| Presence of foamy cells | Sometimes present 17,23,26,32,50 | Absent |
| SPLEEN | | |
| Presence of foamy cells | Rare 17,19,50 | Absent |
| BRAIN | | |
| Status spongiosus | Rare 53 | Rare 65 |

The number corresponds to the reference superscripts.
CD = chronic disease

The presence of many foamy cells associated with severe diffuse hepatocyte swelling (Driemeier *et al.* 1999), with or without birefringent crystals in hepatocytes and/or bile ducts, indicates steroidal saponin poisoning (Miles 1993; Miles *et al.* 1994; Driemeier *et al.* 1998) caused by *Brachiaria* spp. However, the foamy cells may not be present in the liver of ruminants poisoned by *Brachiaria* grasses: if the animal died after a very short clinical course, due to liver insufficiency caused by saponin storage, the foamy cells are not formed, because they probably appear only after 20 days or more of the poisoning. Foamy cells sometimes are also present in hepatic and mesenteric lymph nodes or even in the spleen of some animals. Those cells are more prominent in cattle than in small ruminants, but are not described in horses poisoned by *B. humidicola*. The microscopic lesions seen in the liver of horses poisoned by this grass species are different from those observed in cattle, as hepatocytes are swollen ('ground-glass hepatocytes'), the cellular limits are distinct, nuclei are vesicular and some are bizarre, and many of them are bi- or trinucleated (amitosis). Recently, we observed similar lesions in sheep that died after the ingestion of *Brachiaria decumbens* with high saponin content. Severe portal fibroplasia, bile duct proliferation and mineralization, and regeneration nodules – lesions which characterize a cirrhotic condition – are described only in animals with pithomycotoxicosis (Marasas *et al.* 1972; Smith and Embling 1991; Kellerman *et al.* 2005; Pinto *et al.* 2005). Ultrastructurally foam cells and also hepatocytes of cattle show negative images of crystals involved in part or entirely by membranes (Driemeier *et al.* 1998).

## Conclusions

*Brachiaria* spp. and *Pithomyces chartarum* poisoning are very different from each other. The differential diagnosis should not be difficult if the epidemiological, clinical, and pathological aspects are all considered.

## References

Barbosa JD[1], Oliveira CMC, Duarte MD, and Silveira JAS (2005). *Doenças de búfalos na Amazônia*. II Simpósio Mineiro de Buiatria, Belo Horizonte, Minas Gerais, Brazil.

Barbosa JD[2], Oliveira CMC, Tokarnia CH, and Peixoto PV (2006). Fotossensibilização hepatógena em eqüinos pela ingestão de *Brachiaria humidicola* (Gramineae) no Estado do Pará. *Pesquisa Veterinária Brasileira* 26(3):147-153.

Benzille P[3], Braun JP, and Bars JL. (1984). Première identification de l'eczéma facial chez les ovins en Europe: Aspects épidémiologiques, cliniques et biologiques. *Recueil Médicine Véterinaire* 160(4):339-347.

Brewer D[4], Russe DW, Melo Amaral RE, and Aycardi ER (1989). An examination of North and South America isolates of *Pithomyces chartarum* for production of sporidesmin and sporidesmolides. *Proceedings of the Nova Scotia Institute of Science* 38:73-81.

Brum KB[5], Haraguchi M, Lemos RAA, Riet-Correa F, and Fioravanti MC (2007). Crystal associated cholangiopathy in sheep grazing *Brachiaria decumbens* containing the saponin protodioscin. *Pesquisa Veterinária Brasileira* 27(1):39-42.

Camargo WVA[6], Nazário W, Fernandes NS, and Amaral REM (1976). Fotossensibilização em bovinos de corte: provável participação do fungo *Pithomyces chartarum* na etiologia do processo (Comunicado). *O Biológico,* São Paulo, 42(11/12):259-261.

Carrillo BM[7], Carcagno C, Corbellini CN, Duffy SJ, Miquet JM, and Miguel MS (1980). Fotosensibilización por *Pithomyces chartarum* en bovinos en la República Argentina. *Revista de Investigaciones Agropecuarias* 15(3):527-537.

Collin RG[8], Odriozola E, and Towers NR (1998). Sporidesmin production by *Pithomyces chartarum* isolates from Australia, Brazil, New Zealand and Uruguay. Part 2. *Mycological Research* 102:163-166.

Connor HE[9] (1977). *Poisonous Plants in New Zealand,* 247 pp. Keating, Govt Printer, Wellington, NZ.

Crispim SMA[10] and Branco OD (2002). Aspectos gerais das braquiárias e suas características na sub-região da Nhecolândia, Pantanal, MS. *Boletim de Pesquisa e Desenvolvimento 33, Embrapa-Centro de Pesquisa Agropecuária do Pantanal,* 25 pp., Corumbá, MS.

Cruz C[11], Driemeier D, Pires VS, Colodel EM, Taketa ATC, and Schenkel EP (2000). Isolation of steroidal sapogenins implicated in experimentally induced cholangiopathy of sheep grazing *Brachiaria decumbens* in Brazil. *Veterinary and Human Toxicology* 42(3):142-145.

Cruz C[12], Driemeier D, Pires VS, and Schenkel EP (2001). Experimentally induced cholangiohepatopathy by dosing sheep with fractionated extracts from *Brachiaria decumbens*. *Journal of Veterinary Diagnostic Investigation* 13(2):170-172.

De Wet JAL[13] and Eramus JA (1984). Suspected facial eczema in sheep in the central Orange Free State. *Journal of the South African Veterinary Association* 55(4):199-200.

Döbereiner J[14], Tokarnia CH, Monteiro MCC, Cruz LCH, Carvalho EG, and Primo AT (1976). Intoxicação de bovinos e ovinos em pastos de *Brachiaria decumbens* contaminados por *Pithomyces chartarum*. *Pesquisa Agropecuária Brasileira, Série Veterinária* 11:87-94.

Done J[15], Mortimer PH, and Taylor A (1960). Some observations on field cases of eczema facial: liver pathology and determinations of serum bilirubin, cholesterol, transaminase and alkaline phosphatase. *Research in Veterinary Science* 1:76-83.

Driemeier D[19], Kreimeier RD, Colodel EM, Fries C, Seitz AL, Oliveira RT, and Schmitt AC (1997). Colangiopatia experimental induzida por alimentação de ovinos com *Brachiaria decumbens*. *VIII Encontro Nacional de Patologia Veterinária* (Resumo PE-16). Pirassununga, SP.

Driemeier D[16], Barros SS, Peixoto PV, Tokarnia CH, Döbereiner J, and Brito MF (1998). Estudos histológico, histoquímico e ultra-estrutural de fígados e linfonodos de bovinos com presença de macrófagos espumosos (foam cells). *Pesquisa Veterinária Brasileira* 18(1):29-34.

Driemeier D[18], Döbereiner J, Peixoto PV, and Brito MF (1999). Relação entre macrófagos espumosos ('foam cells') no fígado de bovinos e ingestão de *Brachiaria* spp no Brasil. *Pesquisa Veterinária Brasileira* 19(2):79-83.

Driemeier D[17], Colodel EM, Seitz AL, Barros SS, and Cruz CEF (2002). Study of experimentally induced lesions in sheep by grazing *Brachiaria descumbens. Toxicon* 40:1027-1031.

Fagliari JJ[20], Marques LC, and Alessi AC (2000a). Relato sobre a ocorrência de fotossensibilização em búfalos mantidos em pastagem de *Brachiaria decumbens*. *XXVII Congresso Brasileiro de Medicina Veterinária* (Resumo 50). Águas de Lindóia, SP.

Fagliari JJ[21], Marques LC, Alessi AC, Martins F, Cadioli FA, and Silva SL (2000b). Ocorrência de eczema facial em ovinos e caprinos no Estado de São Paulo. *XXVII Congresso Brasileiro de Medicina Veterinária*, Águas de Lindóia, SP. (Resumo 51)

Fioravanti MCS[22], Trindade BR, Brum KB, Carneiro RB, Menezes LB, França AFS, Orsini GF, and Silva LAF (2003). Estudo histopatológico do fígado, linfonodos mesentéricos e aorta de bovinos alimentados com *Brachiaria brizantha*. *XI Encontro Nacional de Patologia Veterinária,* 34 pp. Botucatu, SP.

Flaoyen A[23] and Froslie A (1997). Photosensitization disorders. *Handbook of plant and fungal toxicant*. (JP Felix D´Mello, ed.), 368 pp. CRC Press, Boca Raton, FL.

Gomar MS[24], Driemeier D, Colodel EM, and Gimeno EJ (2005). Lectin histochemistry of foam cells in tissues of cattle grazing *Brachiaria* spp. *Journal of Veterinary Medicine A* 52(1):18-21.

Graydon RJ[25], Hamid H, Zahari P, and Gardiner C (1991). Photosensitization and crystal-associated cholangiohepatopathy in sheep grazing *Brachiaria decumbens*. *Australian Veterinary Journal* 68(7):234-236.

Halder CA[26], Taber RA, and Camp BJ (1981). Absence of sporidesmin production by 12 Texas isolates of *Pithomyces* spp. *Applied and Environmental Microbiology* 41(1):212-215.

Hansen DE[27], McCoy RD, Hedstrom OR, Snyder SP, and Ballerstedt PB (1994). Photosensitization associated with exposure to *Pithomyces chartarum* in lambs. *Journal of the American Veterinary Medical Association* 204(10):1668-1671.

Kellerman TS[28], Coetzer JAW, Naudé TW, and Botha CJ (2005). *Plant Poisonings and Mycotoxicoses of Livestock in Southern Africa*, 310 pp. Oxford University Press, Cape Town.

Láu HD[29] (1990). Efeitos tóxicos de *Lantana camara* e de *Pithomyces chartarum* em búfalos. *Documento 54, Embrapa-CPATU*, 18 pp. Belém, PA.

Lemos RAA[30] and Brum KB (2001). Intoxicação por *Brachiaria decumbens* em bovinos. *X Encontro Nacional de Patologia Veterinária*, p. 222. Pirassununga, SP.

Lemos RAA[31], Ferreira LCL, Silva SM, Nakazato L, and Salvador SC (1996). Fotossensibilização e colangiopatia associada a cristais em ovinos em pastagem com *Brachiaria decumbens*. *Ciência Rural* 26(1):109-113.

Lemos RAA[32], Nakazato L, and Salvador SC (1997a). Alterações histológicas no fígado de ruminantes com fotossensibilização hepatógena associada à ingestão de *Brachiaria decumbens* e *Brachiaria brizantha*. *VIII Encontro Nacional de Patologia Veterinária*, Pirassununga, SP. (Resumo PA-59)

Lemos RAA[34], Salvador SC, and Nakazato L (1997b). Photosensitization and crystal-associated cholangiohepatopathy in cattle grazing *Brachiaria decumbens* in Brazil. *Veterinary and Human Toxicology* 39(6):376-377.

Lemos RAA[33], Nakazato L, Herrero Júnior GO, Silveira AC, and Porfírio LC (1998). Fotossensibilização e colangiopatia associada a cristais em caprinos mantidos sob pastagens de *Brachiaria decumbens* no Mato Grosso do Sul. *Ciência Rural* 28(3):507-510.

Marasas WF[35], Adelaar TF, Kellerman TS, Minné JA, Van Rensburg IB, and Burroughs GW (1972). First report of facial eczema in sheep in South Africa. *Onderstepoort Journal of Veterinary Research* 39(2):107-112.

Martha Júnior G[36] and Corsi M (2001). Pastagens no Brasil: situação atual e perspectivas. *Preços Agrícolas* pp. 3-6.

Mazni OA[37], Sharif H, Khusahry MYM, and Vance HN (1985). Photosensitization in goats grazed on *Brachiaria decumbens*. *Mardi Research Bulletin* 13:203-206.

Meagher LP[38], Wilkins AL, Miles CO, Collin RG, and Fagliari JJ (1996). Hepatogenous photosensitization of ruminants by *Brachiaria decumbens* and *Panicum*

*dichotomiflorum* in the absence of sporidesmin: Lithogenic saponins may be responsible. *Veterinary and Human Toxicology* 38(4):271-274.
Miles CO[39] (1993). A role for steroidal saponins in hepatogenous photosensitization. *New Zealand Veterinary Journal* 41:221.
Miles CO, Wilkins AL, Erasmus GL, and Kellerman TS (1994). Photosensitivity in South Africa. III. Ovine metabolism of *Tribulus terrestris* saponins during experimentally induced geeldikkop. *Onderstepoort Journal of Veterinary Research* 61:351-359.
Moreira CN[40], Bankys VL, Rosa BC, Pinto AS, Silva LAF, Haraguchi M, and Fioravanti MCS (2007). Bovinos alimentados com capim *Brachiaria* e *Andropogon*: desempenho, avaliação da quantidade de esporos do fungo *Pithomyces chartarum* e teor de saponina das pastagens. *VII Congresso Brasileiro de Buiatria, Curitiba, PR.* (Resumo 100)
Mortimer PH[41] and Ronaldson JW (1983). Fungal toxin induced photosensitization. In *Handbook of Natural Toxins: Plant and Fungal Toxins* (RF Keeler and AT Tu, eds), pp. 361-419. Vol 1. Marcel Dekker, New York.
Nascimento TC[42], Morais M, Carvalho TF, Rabelo RE, Bannys VL, and Sandrini CNM (2005). Relato de caso de bovino com fotossensibilização com avaliação de parâmetros hematológicos e bioquímicos. *2º Congresso de Pesquisa, Ensino e Extensão da Universidade Federal de Goiás.* (Anais eletrônicos do XIII Seminário de Iniciação Cientifica [CD-ROM], UFG, Goiânia, n.p.)
Nobre D[43] and Andrade SO (1976). Relação entre fotossensibilização em bovinos jovens e a gramínea *Brachiaria decumbens* Stapf. *O Biológico,* São Paulo 42(11/12):249-258.
Nunes LP[44] (1976). Fotossensibilização: o problema pode estar no capim. *Ruralidade,* Goiânia 19:64-65.
Nunes SG[45], Silva JM, and Schenk JAP (1990). Problemas com cavalos em pastagens de humidicola. *Comunicado Técnico nº 37, Embrapa-CNPGC,* 4 pp. Campo Grande, MS.
Opasina BA[46] (1985). Photosensitization jaundice syndrome in West African dwarf sheep and goats grazed on *Brachiaria decumbens*. *Tropical Grasslands* 19(3):120-123.
Peixoto PV[47] and França TN (2003). Personal communication (Universidade Federal Rural do Rio de Janeiro, Seropédica, RJ, Brazil).
Pinto C[48], Santos VM, Dinis J, Peleteiro MC, Fitzgerald JM, Hawkes AD, and Smith BL (2005). Pithomycotoxicosis (facial eczema) in ruminants in the Azores, Portugal. *Veterinary Record* 157(25):805-810.
Riet-Correa G[49], Riet-Correa F, Schild AL, and Driemeier D (2002). Wasting and death in cattle associated with chronic grazing of *Brachiaria decumbens*. *Veterinary and Human Toxicology* 44(3):179-180.
Russomanno OMR[50], Portugal MASC, Coutinho LN, Calil BEM, and Figueiredo MB (2003). *Leptosphaerulina chartarum* (=*Pithomyces chartarum*) e seu envovimento no eczema facial.: artigo de revisão. *Arquivos do Instituto Biológico, São Paulo* 70(3):385-390.
Salam Abdullah A[52], Noordin MM, and Rajion MA (1989). Neurological disorders in sheep during signal grass (*Brachiaria decumbens*) toxicity. *Veterinary and Human Toxicology* 31(2):128-129.
Salam Abdullah A[51], Lajis NH, Bremner JB, Davies NW, Mustapha W, and Rajion MA (1992). Hepatotoxic constituents in the rumen of *Brachiaria decumbens* intoxicated sheep. *Veterinary and Human Toxicology* 34(2):154-155.
Santos JCA[53], Riet-Correa F, Simões SVD, and Barros CSL (2008). Patogênese, sinais clínicos e patologia das doenças causadas por plantas hepatotóxicas em ruminantes e eqüinos no Brasil. *Pesquisa Veterinária Brasileira* 28(1):1-14.

Schenk MAM[54] and Schenk JAP (1983). Aspectos gerais da fotossensibilização hepatógena de bovinos. *Comunicado Técnico nº 19, Embrapa-CNPGC,* 7 pp. Campo Grande, MS.

Schenk MAM[55], Nunes SG, and Silva JM (1991). Ocorrência de fotossensibilização em eqüinos mantidos em pastagem de *Brachiaria humidicola. Comunicado Técnico nº 40, Embrapa-CNPGC,* 4 pp. Campo Grande, MS.

Seixas JN (2009). Diferenciação das intoxições por *Brachiaria* spp e *Pithomyces chartarum* através dos aspectos epidemiológicos, clínico-patológicos e toxicológicos, 245 pp. Tese de Doutorado, Instituto de Veterinária, UFRRJ.

Shons S[56], Bonel-Raposo J, Schild AL, Gevehr-Fernandes C, and Soares MP (2005). Intoxicação por *Brachiaria brizantha* em ovinos no Rio Grande do Sul. *Arquivo Brasileiro de Medicina Veterinária e Zootecnia* 57(1):35.

Smith BL[61] (2000). Facial eczema, a mycotoxic hepatogenous photosensitization, in New Zealand. *I Taller Internacional de Toxicoses por Plantas en Animales y Humanos,* 14 pp. La Havana, Cuba.

Smith BL[57] and Embling PP (1991). Facial eczema in goats: The toxicity of sporidesmin in goats and its pathology. *New Zealand Veterinary Journal* 39:18-22.

Smith BL[59] and Miles CO (1993). A letter to the editor: A role for *Brachiaria decumbens* in hepatogenous photosensitization of ruminants? *Veterinary and Human Toxicology* 35(3):256-257.

Smith BL[58] and O'Hara PJ (1978). Bovine photosensitization in New Zealand. *New Zealand Veterinary Journal* 26:2-5.

Smith BL[60] and Towers NR (2002). Mycotoxicoses of grazing animals in New Zealand. *New Zealand Veterinary Journal* 25:124-127.

Soares Filho CV[62] (1994). Recomendações de espécies e variedades de *Brachiaria* para diferentes condições. *Anais do Simpósio sobre Manejo da Pastagem,* pp.25-48, Piracicaba, SP.

Soares PC[63], Mota RA, Teixeira MN, and Santos NVM (2000). Aspectos epidemiológicos e clínicos da intoxicação por *Pithomyces chartarum* em ovinos da raça Santa Inês, no município de Gravatá- PE. *Revista Brasileira de Ciências Veterinárias* 7(2):78-82.

Thompson KG[64], Lake DE, and Cordes DO (1979). Hepatic encephalopathy associated with chronic facial eczema. *New Zealand Veterinary Journal* 27:221-223.

Tokarnia CH[65] and Langenegger J (1983). *Relatório de viagem realizada no período de 1-10.2.83 para estudar doença de etiologia obscura em búfalos na UEPAE de Manaus, Embrapa,* 8 pp. (Datilografado)

Torres MBAM[66] and Coelho KIR (2001). Macrófagos espumosos em fígados de bovinos alimentados com feno de *Brachiaria brizantha. Anais X Encontro Nacional de Patologia Veterinária,* 174 pp. Pirassununga, SP.

Wilkins AL[67], Miles CO, Smith B, Meagher LP, and Ede R (1994). GC/MS method for the analysis of plant and animal samples associated with the ovine photosensitization. In *Poisonous Plants of the World: Agricultural, phytochemical and ecological aspects* (SM Colegate and PR Dorling, eds), pp. 263-268. CAB International, Wallingford, UK.

Zamri-Saad M[68], Sharif H, and Mazni OA (1987). Pathological changes in indigenous sheep of Malaysia following grazing on *Brachiaria decumbens. Kajian Veterinary* 19(1):9-12.

# Chapter 20

# Measurement of Steroidal Saponins in *Panicum* and *Brachiaria* Grasses in the USA and Brazil

S.T. Lee[1], R.B. Mitchell[2], D.R. Gardner[1], C.H. Tokarnia[3], and F. Riet-Correa[4]

*[1]USDA-ARS Poisonous Plant Research Laboratory, Logan, Utah 84341, USA; [2] Grain, Forage, and Bioenergy Research Unit, Agricultural Research Service, United States Department of Agriculture, 314 Biochemistry Hall, UNL, East Campus, Lincoln, Nebraska 68583-0737, USA; [3]Department of Animal Nutrition and Pastures, Federal Rural University of Rio de Janeiro, Serpedica, RJ 23890-000, Brazil; [4]Rural Center for Health and Technology, Federal University of Campina Grande Patos, Paraiba 58700-000, Brazil*

## Introduction

Several grasses in the *Panicum* genus have been reported to cause hepatogenous photosensitization in animals throughout the world (Flaoyen 2000; Riet-Correa *et al.* 2009). In the USA, switchgrass (*P. virgatum* L.) has been reported to cause hepatogenous photosensitization in lambs and horses (Puoli *et al.* 1992; Lee *et al.* 2001, 2009; Stegelmeier *et al.* 2007). In Brazil, cultivars of *P. maximum* have been implicated in toxicity and severe colic in horses and mules (Cerqueira *et al.* 2009). *Brachiaria decumbens* has been reported to cause hepatogenous photosensitization in cattle (Meagher *et al.* 1996; Lemos *et al.* 1997), sheep (Lemos *et al.* 1996a), and goats (Lemos *et al.* 1998) while *B. brizantha* has been reported to cause the same disease in cattle (Lemos *et al.* 1996b) and sheep (Brum *et al.* 2007). Glycosidic steroidal saponins have been found in some species of the *Panicum* genus and in *B. decumbens* and *B. brizantha* and these compounds have been suggested as the primary agents causing hepatogenous photosensitization in animals grazing these grasses (Patamalai *et al.* 1990; Holland *et al.* 1991; Miles *et al.* 1992; Munday *et al.* 1993; Lee *et al.* 2001, 2009; Siqueira-Souza *et al.* 2005; Brum *et al.* 2007). In the USA, switchgrass has been identified for development into an efficient and environment friendly biomass energy crop. A 5-year study demonstrated that switchgrass grown for biofuel production produced 540% more energy than what is needed to grow, harvest, and process it into cellulosic ethanol (Schmer *et al.* 2008). If switchgrass is grown on a scale useful for a bio-energy source, some of the material could be used by livestock as hay or pasture. The chemical structures of the three major saponins–dichotomin, protodioscin and Saponin B–present in switchgrass were determined and are shown in Figure 1 (Lee *et al.* 2009). A simple extraction and rapid reversed phase HPLC-ESI-MS method was developed for quantifying the major saponins in *Panicum* and

©CAB International 2011. *Poisoning by Plants, Mycotoxins, and Related Toxins* (eds F. Riet-Correa, J. Pfister, A.L. Schild, and T.L. Wierenga)

*Brachiaria* samples. Differences in the relative concentration of different saponins were observed between the different *Panicum* and *Brachiaria* samples.

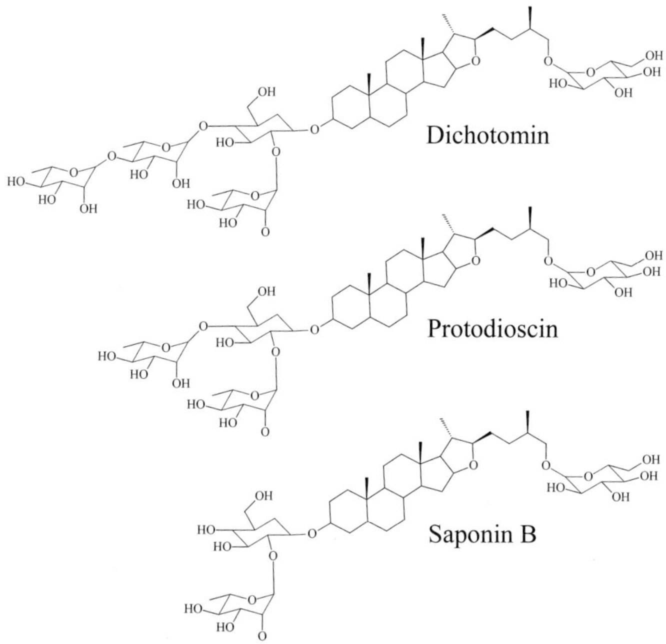


**Figure 1.** Chemical structures of dichotomin, protodioscin, and saponin B.

## Material and Methods

### Plant material

Four switchgrass cultivars (Trailblazer, Cave-in-Rock, Summer, and Kanlow) were grown at the University of Nebraska Agricultural Research and Development Center near Mead, Nebraska, USA. Each cultivar was hand-harvested to a 10 cm stubble height on July 3, 2002. Grass samples were transported to the lab and separated into leaf and stem components. The leaf and stem components for each switchgrass cultivar were dried at 55°C, ground to pass a 2 mm screen in a Wiley mill, and sent to the Poisonous Plant Research Laboratory for analysis. Five *P. maximum* samples were collected on February 4, 2005 in the municipality of Xinguara, state of Pará, Brazil. Sample 1 is variety Mombaça collected from the Santa Rosa Farm, Samples 2 and 3 are variety Tanzania and were both collected from the Brasil Verde Farm, Sample 4 is variety Mombaça collected from the Rio Vermelho Farm, and Sample 5 is variety Tobiata from the Rio Verde Farm.

Two additional *P. maximum* samples were collected. Sample 6 is variety Massai collected in the municipality of Cumaru do Norte in the state of Pará. Sample 7 is variety Mombaça collected from the Brasil Verde Farm, municipality of Xingaura, state of Pará. Two *B. humidicola* samples (Samples 8, 9) were collected from the municipality of

Castanhal, state of Pará. Two *B. decumbens* samples (samples 10, 11) were collected in the municipality of Conceição do Mato Dentro in the state of Minas Gerais, and a *B. brizantha* sample (Sample 12) was collected in the municipality of Curvelo, state of Minas Gerais. All samples collected in Brazil were ground and sent to the Poisonous Plant Research Laboratory for analysis.

## Extraction

Extraction of plant material for saponin analysis was accomplished by weighing 100 mg of ground plant material into a 13 ml screw top test tube equipped with Teflon-lined caps (Pierce, Rockford, IL, USA). Methanol (MeOH, 5 ml) was added to each test tube and placed in a mechanical shaker for 30 min, then centrifuged to separate the plant residue and MeOH extract. The MeOH extract was transferred to a clean 20 ml test tube. The switchgrass residue was extracted two more times with 5 ml MeOH for 30 min and all MeOH extracts combined for a total of 15 ml. A 0.5 ml aliquot was transferred to a 7 ml vial and evaporated to dryness on a heat block at 65°C under a gentle flow of nitrogen. The dried aliquot was then reconstituted in 1.0 ml 0.1% formic acid:acetonitrile (90:10). The sample was then passed through a 0.20 µm syringe filter (National Scientific, Rockwood, TN, USA) and 0.5 ml was then transferred to a 1 ml autosample vial for analysis.

## HPLC-ESI-MS

Samples were injected (5 µl) onto a Betasil C-8 reversed phase column (100 2.1 mm i.d.) (Thermo Electron Corporation, Waltham, MA, USA) protected by a guard column of the same phase. The saponins were eluted from the column with an isocratic flow (0.500 ml/min) of 72:28 (0.1% formic acid:acetonitrile) mobile phase. The total HPLC run time was 3 min. Flow from the column was connected directly to a Thermo Finnigan (San Jose, CA USA) LCQ ion trap mass spectrometer via an electrospray ionization (ESI) source. Full scan mass data were collected for a mass range of 300-1300 amu. MS/MS product ion spectra were collected after isolation of a selected precursor ion (± 5 amu) and the relative collision energy manually adjusted to observe significant fragmentation of the selected ion. Protodioscin (ChromaDex Irvine, CA USA) was used as a standard to quantify Saponins A, B, and C. Saponin A standard was prepared in a solution of 0.1% formic acid:acetonitrile (90:10) to give a ten point standard curve over the range of 0.0977 µg/ml to 50.0 µg/ml. Peak areas of the individual saponins (dichotomin, protodioscin, and Saponin B) were determined from reconstructed ion chromatograms of the respective $MH^+$-$H_2O$ ions ($m/z$ = 1177, 885, and 1031). Figure 2 shows reconstructed ion chromatograms ($m/z$ = 1177, 1031, and 885) from *Panicum virgatum* L. cv. Kanlow plant material and *Brachiaria brizantha* plant material.

A correction to the peak areas measured for protodioscin in the plant samples was needed because dichotomin and protodioscin co-elute, and because the ion at $m/z$ 1031 may result from either protodioscin ($MH^+$ - $H_2O$) or a fragment ion of dichotomin ($MH^+$ - $H_2O$-146). The contribution of the latter would be subtracted from the peak area determined for protodioscin. The relative average percentage peak area of the 1031 ion was 7.8 % compared to the peak area for dichotomin (1177 ion) when six different concentrations over the range of 1.56-50 µg/ml of previously isolated dichotomin was run. For the correction, when dichotomin (1177 ion) was present in the samples, the relative average percentage area plus two standard deviations was 10.5% and was thus subtracted from the integrated 1031 peak areas to obtain corrected 1031 peak (protodioscin) areas (Lee *et al.* 2009).

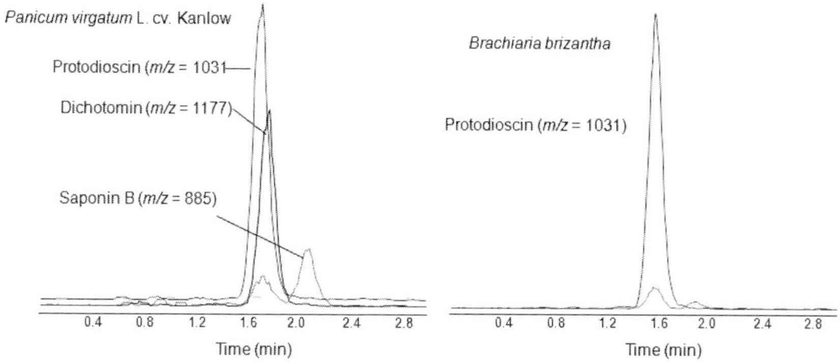


**Figure 2.** Reconstructed ion chromatograms (*m/z* = 1177, 1031, and 885) from *P. virgatum* L. cv. Kanlow plant material and *B. brizantha* plant material.

## Saponin Concentrations in *Panicum* and *Brachiaria* Cultivars

The total saponin content in the switchgrass samples was consistent across the different varieties ranging from a low of 0.108 % for the Cave-in-Rock variety to a high of 0.124% for the Summer variety. In comparing switchgrass cultivars the concentrations of dichotomin, protodioscin, and saponin B in the Summer Trailblazer and Cave-in-Rock cultivars are very similar with dichotomin as the major saponin (Table 1). Among the four switchgrass cultivars analyzed, Kanlow was the most distinct in its saponin profile. In Kanlow, the concentrations of protodioscin and saponin B are both higher than dichotomin. Switchgrass has two distinct ecotypes, lowland and upland. Lowland ecotypes are found on flood plains and other areas that receive run-on water, whereas upland ecotypes occur in upland areas that are not subject to inundation (Vogel 2004). The differences in saponins across cultivars may be explained by differences in ecotype. Summer, Trailblazer, and Cave-In-Rock are upland ecotypes and had similar saponin profiles. Kanlow, a lowland ecotype, had a dissimilar saponin profile to the upland ecotypes.

None of the *P. maximum* varieties analyzed in this study contained dichotomin, protodioscin, or saponin B. However, hepatic disease was reported in the pastures where Samples 1, 2, and 4 were collected, while severe colic occurred in horses and mules in the pastures where Samples 6 and 7 were collected.

One *B. humidicola* sample, Sample 9, had a low level (< 0.01%) of protodioscin while Sample 8 did not have any measurable protodioscin, dichotomin, or saponin B. Both of these samples are from pastures where photosensitization occured in horses.

The two *B. decumbens* samples had saponin levels of 1.55% and 1.15%. The *B. brizantha* sample had a total saponin level of 0.628%. These *Brachiaria* samples are from pastures where photosensitization occurred in sheep and had high levels of total protodioscin and saponins when compared to the *Panicum* samples.

## Conclusions

We developed a simple extraction and rapid analysis method to quantify the saponins in plant material. We examined four *P. virgatum* cultivars and found that the saponin profile of Kanlow was unique compared to the other three varieties. Steroidal saponins were not found in any of the *P. maximum* samples analyzed and was found at low levels in only one of the *B. humidicola* samples. Conversely, high levels of total saponins and in particular protodioscin were found in the *B. decumbens* and *B. brizantha* samples. The quantitation of saponins in these grasses is important considering the current hypothesis that saponins are a primary agent causing hepatic photosensitization in animals grazing these grasses. If switchgrass is grown on a scale that is useful for a bio-energy source, the risk of disease associated with animals using switchgrass as feed should be known.

**Table 1.** Measured amounts of protodioscin, dichotomin and saponin B in *Panicum* and *Brachiaria* samples.

| Sample | | Protodioscin (μg/mg) | Dichotomin (μg/mg) | Saponin B (μg/mg) |
|---|---|---|---|---|
| US | *P. virgatum* L. cv. Kanlow | 0.607 | 0.439 | 0.139 |
| US | *P. virgatum* L. cv. Summer | 0.0532 | 1.14 | 0.0442 |
| US | *P. virgatum* L. cv. Trailblazer | 0.0334 | 1.12 | 0.0655 |
| US | *P. virgatum* L. cv. Cave-in-Rock | 0 | 1.01 | 0.0681 |
| 1 | *P. maximum* var. Mombaça | 0 | 0 | 0 |
| 2 | *P. maximum* var. Tanzania | 0 | 0 | 0 |
| 3 | *P. maximum* var. Tanzania | 0 | 0 | 0 |
| 4 | *P. maximum* var. Mombaça | 0 | 0 | 0 |
| 5 | *P. maximum* var. Tobiata | 0 | 0 | 0 |
| 6 | *P. maximum* var. Massai | 0 | 0 | 0 |
| 7 | *P. maximum* var. Mombaça | 0 | 0 | 0 |
| 8 | *B. humidicola* | 0 | 0 | 0 |
| 9 | *B. humidicola* | 0.196 | 0 | 0 |
| 10 | *B. decumbens* | 15.2 | 0 | 0.311 |
| 11 | *B. decumbens* | 11.3 | 0 | 0.240 |
| 12 | *B. brizantha* | 6.09 | 0 | 0.191 |

## References


Brum KB, Haraguchi M, Lemos RAA, Riet-Correa F, and Fioravanti MCS (2007). Crystal-associated cholangiopathy in sheep grazing *Brachiaria decumbens* containing the saponin protodioscin. *Pesquisa Veterinária Brasileira* 27:39-42.

Cerqueira VD, Riet-Correa G, Barbosa JD, Duarte MD, Oliveira MC, de Oliveira CA, Tokarnia C, Lee ST, and Riet-Correa F (2009). Colic caused by *Panicum maximum* in equidae in northern Brazil. *Journal of Veterinary Diagnostic Investigation* 21:882-888.

Flaoyen A (2000). Plant-Associated Hepatogenous Photosensitization Diseases. In: *Natural and Selected Synthetic Toxins: Biological Implications* (AT Tu and W Gaffield, eds), pp.204-219. ACS Symposium Series 745: Washington DC.

Holland PT, Miles CO, Mortimer PH, Wilkins AL, Hawkes AD, and Smith BL (1991). Isolation of steroidal sapogenin epismilagenin from the bile of sheep affected by *Panicum dichotomiflorum* toxicosis. *Journal of Agricultural and Food Chemistry* 39:1963-1965.

Lee ST, Stegelmeier BL, Gardner DR, and Vogel KP (2001). The isolation and identification of steroidal sapogenins in switchgrass. *Journal of Natural Toxins* 10:273-281.

Lee ST, Mitchell RB, Wang Z, Heiss C, Gardner DR, and Azadi P (2009). The isolation, characterization, and quantification of steroidal saponins in switchgrass (*Panicum virgatum* L.). *Journal of Agricultural and Food Chemistry* 57:2599-2604.

Lemos RAA, Ferreira LCL, Silva SM, Nakazato L, and Salvador SC (1996a). Fotossensibilizção e colangiopatia associada à cristais em ovinos em pastagem com *Brachiaria decumbens*. *Ciência Rural* 26:109-113.

Lemos RAA, Osorio ALAR, Rangel JMR, and Herrero Jr GO (1996b). Fotossensibilização e colangiopatia associada a cristais em bezerros ingerindo *Brachiaria brizantha*. *Arquivos Insitute Biologico*, Sao Paulo, 63:22.

Lemos RAA, Salvador SC, and Nakazato L (1997). Photosensitization and crystal associated cholangiohepatopatia in cattle grazing *Brachiaria decumbens* in Brazil. *Veterinary and Human Toxicology* 39:376-377.

Lemos RAA, Nakazato L, Herrero Jr GO, Silveira AC, and Porfirio LC (1998). Fotossensibilizacao e colangiopatia associada a cristais em caprinos mantidos sob pastagens de *Brachiaria decumbens* no Mato Grossodo sul. *Ciência Rural* 28:507-510.

Meagher LP, Miles CO, and Fagliari JJ (1996). Hepatogenous photosensitization of ruminants by *Brachiaria decumbens* and *Panicum dichotomiflorum* in the absence of sporidesmin: lithogenic saponins may be responsible. *Veterinary and Human Toxicology* 38:271-274.

Miles CO, Wilkens AL, Munday SC, Holland PT, Smith BL, Lancaster MJ, and Embling PP (1992). Identification of the calcium salt of epismilagenin β-D-glucuronide in the bile crystals of sheep affected by *Panicum dichotomiflorum* and *Panicum schinzii* toxicoses. *Journal of Agricutural and Food Chemistry* 40:1606-1609.

Munday SL, Wilkins AL, Miles CO, and Holland PT (1993). Isolation and structure elucidation of dichotomin, a furostanol saponin implicated in hepatogenous photosensitization of sheep grazing *Panicum dichotomiforum*. *Journal of Agricultural and Food Chemistry* 41:267-271.

Patamalai B, Hejtmancik E, Bridges CH, Hill DW, and Camp BJ (1990). The isolation and identification of steroidal sapogenins in kleingrass. *Veterinary and Human Toxicology* 32:314-318.

Puoli JR, Reid RL, and Belesky DP (1992). Photosensitization in lambs grazing switchgrass. *Agronomy Journal* 84:1077-1080.

Riet-Correa F, Haraguchi M, Dantas AFM, Burakovas RG, Yokosuka A, Mimaki Y, Medeiros RMT, and deMatos PF (2009). Sheep poisoning by *Panicum dichotomiflorum* in northeastern Brazil. *Pesquisa Veterinária Brasileira* 29:94-98.

Schmer MR, Vogel KP, Mitchell RB, and Perrin RK (2008). Net energy of cellulosic ethanol from switchgrass. *Proceedings of the National Academy of Sciences* 10:464-469.

Siqueira-Souza V, Sandrini, CNM, Fioravanti MCS, and Haraguchi M (2005). Sazonal evaluation of saponins in the pastures of *Bracharia* and *Andropogon* of the Brazilian Middle-West. *Arquivos de Insituto Biológico*, Sao Paulo, 71:43.

Stegelmeier BL, Elmore SA, Lee ST, James LF, Gardner DR, Panter KE, Ralphs MH, and Pfister JA (2007). Switchgrass (*Panicum virgatum*) Toxicity in Sheep, Goats and Horses. In *Poisonous Plants: Global Research and Solutions* (KE Panter, TL Wierenga, and JA Pfister eds), pp. 113-117. CAB International, Wallingford, UK.

Vogel KP (2004). Switchgrass. In *Warm-season ($C_4$) grasses* (LE Moser, L Sollenberger, and B Burson, eds), pp. 561-588. ASA-CSSA-SSSA Monograph No. 45, Madison, WI.

# Chapter 21

# Acute Poisoning by *Crotalaria spectabilis* Seeds in Pigs of Mato Grosso State, Brazil

F.M. Boabaid, R.L. Alberton, D.G. Ubiali, R.A.S. Cruz, M.I.V. Silva, C.A. Pescador, M.A. Souza, and E.M. Colodel

*Veterinary Pathology Laboratory, Federal University of Mato Grosso, Cuiabá, MT 78068-900, Brazil*

## Introduction

The genus *Crotalaria* (Fabaceae), with about 600 species, contains shrubs which are distributed worldwide (Williams and Molyneux 1987). They are popularly known in Brazil as 'guizo or chocalho-de-cascavel', 'chocalho-de-cobra', 'feijão-de-guizo' or 'xique-xique' (Riet-Correa *et al.* 2003), common names recalling the sound produced by the rattle of dried beans (Boghossian *et al.* 2007). Some species are known to contain toxic pyrrolizidine alkaloids (PAs) as the active principle (Tokarnia and Döbereiner 1982). The main PA found in *C. spectabilis* is monocrotaline, detected in all parts of the plant including seeds (Neal *et al.* 1935). Pigs are more sensitive to the effects of PAs followed by chickens, cattle, horses, sheep, and goats (Torres *et al.* 1997).

Some *Crotalaria* spp. are used in crop rotation systems as green manure, especially in soybean, maize, and sorghum (Souza *et al.* 1997). In commercial pig farms the poisoning occurs when toxic plant seeds are mixed with the grains used in the formulation of feed (Calegari *et al.* 1992; Timm and Riet-Correa 1997).

In Brazil spontaneous outbreaks of poisoning were caused by *C. juncea* in horses and *C. mucronata* in cattle in Minas Gerais State (Nobre *et al.* 1994; Lemos *et al.* 1997) and by *C. retusa* in horses, cattle, and sheep in Paraíba State (Riet-Correa *et al.* 2003; Dantas *et al.* 2004; Nobre *et al.* 2004, 2005). Poisoning was reproduced with *C. mucronata* and *C. anagiroides* in cattle (Tokarnia and Döbereiner 1982, 1983; Boghossian *et al.* 2007), with *C. retusa* in sheep, horses, and goats (Nobre *et al.* 2004, 2005; Lucena *et al.* 2007), and with *C. spectabilis* in pigs (Souza *et al.* 1997; Torres *et al.* 1997).

The main clinical signs described in pigs poisoned by *C. spectabilis* are related to liver failure (Souza *et al.* 1997; Torres *et al.* 1997) associated with hepatocellular necrosis or liver fibrosis (Torres *et al.* 1997). The aim of this paper is to report an outbreak of acute intoxication by seeds of *C. spectabilis* in pigs in Mato Grosso State due to contamination of sorghum grain, describing the epidemiology, clinical signs, pathological findings, and the experimental reproduction of the poisoning in pigs with seeds of this plant.

©CAB International 2011. *Poisoning by Plants, Mycotoxins, and Related Toxins* (eds F. Riet-Correa, J. Pfister, A.L. Schild, and T.L. Wierenga)

## Material and Methods

### Natural poisons

Epidemiological and clinical history was obtained from the veterinarian and owners of the affected farms. All four properties were in Juscimeira city, State of Mato Grosso (MT), Brazil. Two out of four farms were visited by veterinarians from the Laboratório de Patologia Veterinária, Universidade Federal do Mato Grosso (LPV-UFMT). Sorghum grains contaminated with seeds of *C. spectabilis* and swine feeds that were contaminated were collected in the factory and on the properties with problems. The property in the municipality of Santo Antônio do Leveger, MT, Brazil that provided contaminated sorghum was also visited to collect samples of *C. spectabilis* for identification. The plant samples were pressed and sent to the Central Herbarium of UFMT for botanical classification. The percentage of *C. spectabilis* seed contamination was determined in ten 100 g samples of grain sorghum by manually separating all seeds in the samples.

### Experimental poisoning

Pigs used were clinically healthy and crossbred. Pig 1 was male and Pig 2 was female, weighing 12 and 10 kg, respectively. The pigs received a single dose of 1g/kg (Pig 1) or 5g/kg (Pig 2) of *C. spectabilis* seeds separated from the sorghum grains collected in the animal feed factory. The whole seeds were administered by gavage. The pigs were kept in individual cages and received a ration for growing pigs acquired in the swine sector of Federal University of Mato Gross and water *ad libitum*.

### Pathologic study

Five pigs that had been naturally intoxicated in the city of Juscimeira-MT and the two experimental pigs were necropsied. Liver, lung, kidney, heart, spleen, lymph nodes, tonsils, stomach, intestine, and central nervous system tissues were collected and fixed in a 10% formaldehyde solution. The samples were routinely processed, stained with hematoxylin and eosin (Prophet *et al.* 1992), and examined microscopically. The liver and lung slices were selected and stained by the Masson trichrome (Prophet *et al.* 1992).

## Results

### Natural poisonings

Four small properties in the municipality of Juscimeira, MT, Brazil, showed mortality of pigs in August 2008. All properties have rudimentary pig breeding and use feed that contained sorghum grain, which was later noted to have been contaminated with seeds of *C. spectabilis* by both the factory owner and the responsible technician from the feed factory.

At property A, the manufacturer reported that the animals were fed with the contaminated rations and also with residues from a restaurant. The lot of 96 pigs received contaminated feed for a day. The use was discontinued after the owner of the feed plant was alerted to the contamination of the feed. Clinical signs started 36 h after consumption of contaminated feed and were characterized by depression, lethargy, lack of appetite, vomiting, and increased abdominal volume, followed by death between 48 to 60 h after the onset of clinical signs. On a visit to Farm A, it was noted that the animals were confined regardless of age, gender, or class. Thirty-seven pigs died, and necropsies were performed on five: four that died naturally and one that was euthanized after suffering severe depression and anorexia.

Property B pigs were reared outdoors without separation by category, and some were crossed with a wild boar (*Sus scrofa*). According to the owner, the animals were fed with sorghum grain purchased at the feed plant and mixed with soybean hulls at a proportion of 50%. Clinical signs, starting 24-48 h after ingestion of this mixture, were depression, anorexia, lethargy, and sternal recumbence, followed by death. Two pigs were necropsied: one that was dead and another showing apathy, lateral recumbence, jaundiced mucous membranes, seizures, and periodic pedal movements, which died during the clinical examination. The owner reported that the contaminated sorghum was also supplied to 100 chickens, 20 of which died.

According to the technical report of the animal food factory in properties C and D, 12 and 25 pigs died, respectively, between 24 to 48 hours after ingestion of the ration produced in the factory. Clinical signs reported were depression, lethargy, lack of appetite, dyspnea, ascites, vomiting, and epistaxis. Property D belonged to the owner of the animal food factory who reported that, after notification of the death of pigs by owners, the lots of sorghum and contaminated feed were collected and stored on his property, and given accidentally to pigs, which died with a clinical picture similar to that of the other farms.

It was found that contamination of sorghum occurred accidentally during the processing and storage of seed at the production property of sorghum grain. While visiting this property, the use of *Crotalaria* as green manure in a crop rotation system in sorghum and soybean was observed. The sample sent for botanical identification was classified as *C. spectabilis*. The average contamination of grain sorghum with seeds of *C. spectabilis* was estimated at 1.97% and the commercial diet was approximately 0.53%. The factory technician responsible for the ration reported that a property had been provided bovine feed for cattle reared under confinement, and 48 h later two cattle had died. The cattle were not examined clinically or pathologically.

## Experimental intoxication

Seeds of *C. spectabilis* collected in the animal feed factory, were separated from the residue of sorghum and given to two pigs. A single dose of 1g/kg (Pig 1) caused no clinical signs. This pig was euthanized 33 days after ingestion of the seeds. In Pig 2 clinical signs started 16 h after administration of 5g/kg of seeds and were characterized by depression, lack of appetite, and vomiting. There was marked postural instability, apathy, anorexia, and slightly jaundiced mucous membranes 48 h after dosing. The pig remained in supine position for long periods and showed a slight increase in the abdomen with moderate dyspnea and abdominal breathing. After 72 h, the animal was euthanized and necropsied.

## Pathological findings

The pathological changes were similar in animals spontaneously poisoned and in Pig 2. The main changes were found in the liver, which showed protein filamentous material in the capsular surface and increased lobular pattern characterized by red areas surrounded by lighter areas or streaky reddish brown areas on the capsular and cutting surfaces. There was also edema and hemorrhage of the gall bladder. Sometimes the pigs showed severe jaundice and liquid in the abdominal cavity. Submandibular edema, hydropericardium, and petechial hemorrhages in the peritoneum, epicardium, and pleural surface were occasionally observed. Microscopically, the lesion consisted of hepatic central lobular necrosis associated with congestion and hemorrhage and disruption of hepatic cords. Occasionally, in the periphery of the necrotic areas, vacuolated hepatocytes, bilestase, and a few hepatocytes with nuclear megalocytosis were noted. Proliferation of connective tissue was not seen. In the brain, there were multifocal areas of vacuolization of the white matter, mainly located in the region of the cerebellar peduncles, medulla oblongata and thalamus.

Pig 1, slaughtered 33 days after ingestion of seeds of *C. spectabilis*, showed no macroscopic and histologic changes.

## Discussion

The diagnosis of poisoning by the seeds of *C. spectabilis* was based on epidemiological, clinical, and pathological features similar to those described in cases of acute poisoning by seeds of *Crotalaria* spp. (Figueredo *et al.* 1987; Nobre *et al.* 2005) and in the experimental reproduction of the intoxication. Unlike the number of reports of chronic poisoning by seeds of *Crotalaria* spp. the hepatocellular necrosis associated with consumption of this plant is uncommon, probably due to low palatability of the plant; ingestion of high doses in a short period of time is rare (Kelly 1993). Ingestion of *Crotalaria* spp. typically occurs when there is a shortage of pasture in areas invaded by the plant or contamination of food or feed (Radostits *et al.* 2000; Riet-Correa *et al.* 2006).

In this outbreak, the contamination of feed with *C. spectabilis* was approximately 0.53%. In experimental studies described by Torres *et al.* (1997) pigs developed chronic poisoning ingesting contaminated with seeds 0.3%, 0.5%, and 1%. Only one animal developed acute signs after receiving feed contaminated with 1% of *C. spectabilis*.

The lesion was reproduced in a pig by oral administration of 5 g/kg body weight in a single dose of seeds of *C. spectabilis*. In sheep, acute signs were reproduced using the same dosage of seeds of *C. retusa* (Nobre *et al.* 2005). The sheep showed no clinical signs at a dose of 2.5 g/kg. There were no clinical and pathological changes in the pig which ingested *C. spectabilis* seeds at a dose of 1 g/kg.

The toxicity of PAs varies depending on biotransformation into pyrroles, on the capacity of detoxification of the animal, and on the reactivity of pyrroles with the target cell. These factors vary according to species, age, and individual metabolism (Kelly 1993; Santos *et al.* 2008).

In Brazil, cases of poisoning by *C. retusa* occur throughout the year in horses and seasonally in sheep and cattle (Riet-Correa *et al.* 2006). Poisoning by other plants of the genus *Senecio* and *Echium plantagineum*, which also have PAs as the main compound, are reported in Brazil (Tokarnia *et al.* 2000; Riet-Correa and Méndez 2007; Santos *et al.* 2008). The genus *Senecio* is of great importance in Rio Grande do Sul, being a major cause of economic losses in cattle (Pilati and Barros 2007). Additional studies on the acute toxicity of *C. spectabilis* in pigs are in development in the Federal University of Mato Grosso.

## References

Boghossian MR, Peixoto PV, Brito MF, and Tokarnia CH (2007). Aspectos clínico-patológicos da intoxicação experimental pelas sementes de *Crotalaria mucronata* (Fabaceae) em bovinos. *Pesquisa Veterinária Brasileira* 27(4):149-156.

Calegari A, Alcantara PB, Miyasaka S, and Amado TJC (1992). Caracterização das principais espécies de adubo verde. In *Adubação verde no sul do Brasil* (MBB Costa, ed.), pp. 209-327. AS-PTA, Rio de Janeiro, Brasil.

Dantas AFM, Nobre VMT, Riet-Correa F, Tabosa IM, Junior GS, Medeiros JM, Silva RMN, Silva EMN, Anjos BL, and Medeiros JKD (2004). Intoxicação crônica espontânea por *Crotalaria retusa* (Fabaceae) em ovinos na região do semi-árido paraibano, Brasil. *Pesquisa Veterinária Brasileira* 24(Supl.):18-19.

Figueredo MLA, Rodriguez J, and Alfonso HA (1987). Patomorfologia de la intoxicación experimental aguda por *Crotalaria retusa* y *C. spectabilis* em pollos. *Revista Cubana Ciencia Veterinaria* 18(1/2):63-71.

Kelly WR (1993). The liver and biliary system. In *Pathology of domestic animals*. (KVF Jubb, PC Kennedy, and N Palmer, eds), v 2, pp. 319-406, 4th edn. Academic Press, San Diego.

Lemos RAA, Dutra IS, Souza GF, Nakazato L, and Barros CSL (1997). Intoxicação espontânea por *Crotalaria mucronata* em bovinos em Minas Gerais. *Arquivos do Instituto Biológico* 64(Supl.), resumo 46.

Lucena RB, Nobre VMT, Dantas AFM, Maia LA, and Riet-Correa F (2007). Intoxicação experimental aguda por *Crotalaria retusa* (Fabaceae) em caprinos. In: *Anais do XIII Encontro de Patologia Veterinária*. Campo Grande, Mato Grosso do Sul, Brasil.

Neal WM, Rusoff LL, and Ahmann CF (1935). The isolation and some properties of an alkaloid from *Crotalaria spectabili* roth. *Journal of the American Chemical Society* 57(12):2560-2561.

Nobre D, Dagli MLZ, and Haraguchi M (1994). *Crotalaria juncea* intoxication in horses. *Veterinary and Human Toxicology* 36(5):445-448.

Nobre VMT, Riet-Correa F, Barbosa Filho JM, Dantas AFM, Tabosa IM, and Vasconcelos JS (2004). Intoxicação por *Crotalaria retusa* (Fabaceae) em eqüídeos no semi-árido da Paraíba. *Pesquisa Veterinária Brasileira* 24(3):132-143.

Nobre VMT, Dantas AFM, Riet-Correa F, Barbosa Filho JM, Tabosa IM, and Vasconcelos JM (2005). Acute intoxication by *Crotalaria retusa* in sheep. *Toxicon* 45:347-352.

Pilati C and Barros CSL (2007). Intoxicação experimental por *Senecio brasiliensis* (Asteraceae) em eqüinos. *Pesquisa Veterinária Brasileira* 27(7):287-296.

Prophet EB, Mills B, Arrington JB, and Sobin LH (1992). *Laboratory Methods in Histotechnology,* 279 pp. Armed Forces Institute of Pathology, Washington, DC.

Radostits OM, Gay CC, Blood DC, and Hinchcliff KW (2000). Diseases caused by toxins in plants: pyrrolizidine alkaloid poisoning. *Veterinary Medicine: a textbook of the diseases of cattle, sheep, pigs, goats and horse,* pp. 1661-1665, 9th edn. W.B. Saunders, London.

Riet-Correa F and Méndez MDC (2007). Intoxicações por plantas e micotoxinas – Plantas Hepatotóxicas. In *Doenças de ruminantes e eqüídeos* (F. Riet-Correa, AL Schild, RAA Lemos, and JRJ Borges, eds). 3rd edn., pp. 99-114. Pallotti, Santa Maria.

Riet-Correa F, Tabosa IM, Azevedo EO, Medeiros RMT, Simões SVD, Dantas AFM, Alves CJ, Nobre VMT, Athayde ACR, Gomes AA, and Lima EF (2003). Intoxicação por *Crotalaria retusa*. Doenças de ruminantes e equinos no semi-árido da Paraíba. *Semi-árido em foco* 1(1):63-68.

Riet-Correa F, Medeiros RMT, and Dantas AFM (2006). *Plantas Tóxicas da Paraíba,* 58 pp. Centro de Saúde e Tecnologia Rural, SEBRAE, João Pessoa, Paraíba.

Santos JCA, Riet-Correa F, Simões SVD, and Barros CSL (2008). Patogênese, sinais clínicos e patologia das doenças causadas por plantas hepatotóxicas em ruminantes e eqüinos no Brasil. *Pesquisa Veteterinária Brasileira* 28(1):1-14.

Souza AC, Hatayde MR, and Bechara GH (1997). Aspectos patológicos da intoxicação de suínos por sementes de *Crotalaria spectabilis* (Fabaceae). *Pesquisa Veterinária Brasileira* 17(1):12-18.

Timm CD and Riet-Correa F (1997). Plantas tóxicas para suínos. *Ciência Rural* 27(3):521-528.

Tokarnia CH and Döbereiner J (1982). Intoxicação experimental por *Crotalaria mucronata* (Leg. Papilionoideae) em bovinos. *Pesquisa Veterinária Brasileira* 2(2):77-85.

Tokarnia CH and Döbereiner J (1983). Intoxicação experimental por *Crotalaria anagyroides* (Leg. Papilionoideae) em Bovinos. *Pesquisa Veterinária Brasileira* 3(4):115-123.

Tokarnia CH, Döbereiner J, and Peixoto PV (2000). *Plantas tóxicas do Brasil,* 309 pp. Editora Helianthus, Rio de Janeiro, Brasil.

Torres MBAM, Salles MWS, Headley AS, and Barros CSL (1997). Intoxicação experimental por sementes de *Crotalaria spectabilis* (Leguminosae) em Suínos. *Ciência Rural* 27(2):307-312.

Williams MC and Molyneux RJ (1987). Occurrence, concentration, and toxicity of pyrrolizidine alkaloids in *Crotalaria* seeds. *Weed Science* 35(4):476-481.

# Chapter 22

# Possible Association between Precipitation and Incidence of *Senecio* spp. Poisoning in Cattle in Southern Brazil

F.B. Grecco[1], A.L. Schild[1], P. Estima-Silva[1], C. Marcolongo-Pereira[1], M.P.S. Soares[1], and E.S.V. Sallis[2]

[1] *Laboratório Regional de Diagnóstico, Faculdade de Medicina Veterinária, Universidade Federal de Pelotas, Campus universitário s/n, Pelotas-RS, Brazil;* [2]*Departamento de Patologia Animal Faculdade de Veterinária, Universidade Federal de Pelotas, Campus Universitário s/n, Pelotas, RS 96010-900, Brazil*

## Introduction

*Senecio* spp. is the most important poisonous plant of southern Brazil, causing 7% of all cattle deaths in this region (Riet-Correa and Medeiros 2001; Riet-Correa and Mendez 2007; Rissi *et al.* 2007). Much research has been done on both spontaneous and experimental poisoning and reporting the epidemiological, clinical, and pathological characteristics of the intoxication and the different factors that induce the ingestion of *Senecio* spp. by cattle (Barros *et al.* 1987, 1992; Méndez *et al.* 1987, 1990; Liddell *et al.* 1992; Méndez and Riet-Correa 1993; Karam *et al.* 2004). Environmental conditions such as drought and high temperatures increase the concentration of pyrrolizidine alkaloids (Radostits *et al.* 2002). Poisoning is generally chronic in cattle and characterized by diffuse hepatic fibrosis with regenerative and hyperplastic nodules (Driemeier *et al.* 1991; Karam *et al.* 2004). The objective of this chapter is to report an increase in the number of outbreaks of *Senecio* spp. poisoning in southern Brazil in 2006-2008, probably related to low rainfalls during 2004-2006.

## Material and Methods

A retrospective study was conducted of cases of *Senecio* spp. poisoning in cattle submitted to the Regional Diagnostic Laboratory (LRD) of Pelotas University (UFPel) between 2000 and 2008. The diagnosis of *Senecio* spp. poisoning was confirmed by the macroscopic and histological lesions observed. Data including epidemiology, clinical signs, and gross and histologic lesions were collected by careful review of necropsy files and microscopic slide archives. These data were then associated with precipitation data from the

©CAB International 2011. *Poisoning by Plants, Mycotoxins, and Related Toxins* (eds F. Riet-Correa, J. Pfister, A.L. Schild, and T.L. Wierenga)

Agroclimatological Institute of Meteorology (INMET) at EMBRAPA/UFPel, Pelotas, Rio Grande do Sul, Southern Brasil.

## Results and Discussion

In the period studied 86 outbreaks of *Senecio* spp. poisoning were diagnosed by the Regional Diagnostic Laboratory (Table 1). The highest incidence of poisoning occurred in the years following prolonged drought in Rio Grande do Sul, especially in the year 2007 in which the number of outbreaks (21) was the highest throughout the study period (Figure 1), representing 15.9% of the cattle cases received during the year. There was also a higher frequency of outbreaks affecting animals less than 3-years-old (10 outbreaks), a category that is not usually the most affected. In previous years the poisoning affected mainly adult cows (Barros *et al.* 1992; Karam *et al.* 2004; Rissi *et al.* 2007).

**Table 1.** Outbreaks of *Senecio* spp. distributed monthly from 2000 to 2008.

| Year | Jan | Feb | Mar | Apr | May | Jun | Jul | Aug | Sep | Oct | Nov | Dec | Total |
|---|---|---|---|---|---|---|---|---|---|---|---|---|---|
| 2000 | | 1 | 1 | | | | 1 | 1 | 2 | 2 | 2 | 1 | 11 |
| 2001 | 2 | 1 | | 1 | | 1 | 2 | 2 | | 1 | 1 | 1 | 12 |
| 2002 | 1 | | | | | 1 | | 2 | 1 | | 1 | 1 | 7 |
| 2003 | | | 2 | | | | | | | | | | 2 |
| 2004 | | | 1 | 2 | | | | | | | | | 3 |
| 2005 | | | | 2 | | | | | | 1 | 1 | | 4 |
| 2006 | | 1 | 3 | | 2 | 1 | 1 | | 1 | | 4 | 1 | 14 |
| 2007 | | 4 | 3 | 3 | 3 | 2 | | 1 | 1 | 4 | | | 21 |
| 2008 | | | 2 | 3 | 2 | | 1 | 2 | 2 | | | | 12 |
| **Total** | 3 | 7 | 12 | 11 | 7 | 5 | 5 | 8 | 7 | 8 | 9 | 4 | 86 |

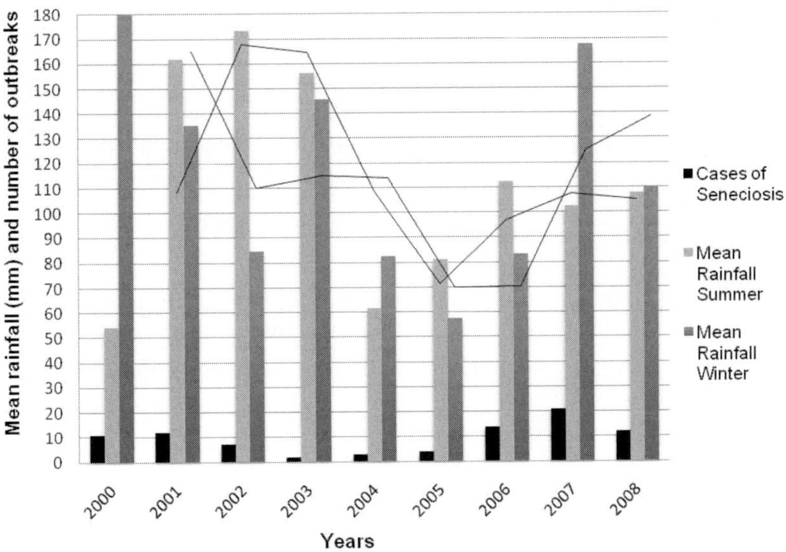


**Figure 1.** Number of *Senecio* spp. outbreaks between 2000-2008 and the mean rainfall (mm) in winter and summer in the same period.

The climate of Rio Grande do Sul is temperate, with an even distribution of rainfall during all seasons. In summer even when there is regular rainfall, evaporation is higher due to the high temperatures and greater sunlight. During winter, there is less evaporation balancing the accumulated rainfall during the entire year (Julio R.Q. Marques 2009, personal communication). Climate records show that the accumulated rainfall decreased from 2004 to 2006, both during the winter and summer. During these dry years in Rio Grande do Sul, there was a consequent shortage of forage, probably associated with a continuous increase in the amounts of *Senecio* spp. in the pastures. In Rio Grande do Sul *Senecio* spp. can sprout during the whole year, and the lack of grass cover promotes *Senecio* spp. seed germination (Karam *et al.* 2002).

It is probable that the increased amount of *Senecio* spp. associated with increased ingestion by cattle from scarcity of forage and the potentially long lag time between ingestion and appearance of clinical signs (Riet-Correa *et al.* 2009) are responsible for the higher frequency of the poisoning in 2007. It is also probable that after 3 years of drought the amount of *Senecio* spp. in pastures was considerably increased in 2007 in relation to previous years.

Driemeier *et al.* (1991) reported an increased number of outbreaks in years following droughts in the state of Rio Grande do Sul. Drought may have altered the concentration of pyrrolizidine alkaloids in the populations of *Senecio* spp., rendering the plants more toxic, as it has been shown that both drought and high temperatures increase the concentration of pyrrolizidine alkaloids in plants (Tokarnia *et al.* 2000; Radostits *et al.* 2002).

Typical chronic lesions of *Senecio* poisoning were observed in most cases. However, in two cases sub-acute histological lesions were observed. In both cases the affected animals lost weight and developed diarrhea and nervous signs with a clinical manifestation period of 7 days. These animals had ascites and edema of the intestinal mesentery and omentum. The livers were firm and dark with yellow spots or whitish areas mainly in the capsular surface. Histologically these animals had hepatocellular swelling with vacuolation and prominent enlarged, vacuolated nuclei (megalocytosis). Binucleated hepatocytes were common. Less prominent changes included hepatocyte necrosis with minimal fibrosis, proliferation of biliary epithelium and hemorrhages. The occurrence of sub acute poisoning with hemorrhages, necrosis, and moderate fibrosis may also be associated with variation in the amount of plant ingested, and the longer exposures relating to plant density and palatability. More research is needed to determine if plant populations, plant toxicity, and the subsequent type and frequency of poisoning truly correlate with rainfall.

## Acknowledgements

This work has been financially supported by the National Institute for Science and Technology for the Control of Toxic Plants (CNPq), grant number 573534/2008-0.

## References

Barros CSL, Metzdorf LL, and Peixoto PV (1987). Ocorrência de surtos de intoxicação por *Senecio* spp. (Compositae) em bovinos no Rio Grande do Sul. *Pesquisa Veterinária Brasileira* 7:101-107.

Barros CL, Driemeier D, Pilati C, and Barros SS (1992). *Senecio* spp. poisoning in cattle in Southern Brazil. *Veterinary and Human Toxicology* 34(3):241-246.

Driemeier D, Barros CSL, and Pilati C (1991). Seneciose em bovinos. *A Hora Veterinária* 10:23-30.

Karam FSC, Méndez MC, Jarenkow JA, and Riet-Correa F (2002). Fenologia de quatro espécies tóxicas de *Senecio* (Asteraceae) na região Sul do Rio Grande do Sul. *Pesquisa Veterinária Brasileira* 22:33-39.

Karam FSC, Soares MP, Haraguchi M, Riet-Correa F, Méndez MC, and Jarenkow JA (2004). Aspectos epidemiológicos da seneciose na região sul do Rio Grande do Sul. *Pesquisa Veterinária Brasileira* 24:191-198.

Liddell JR, Stermitz FR, and Barros CS (1992). Pyrrolizidine alkaloids from *Senecio oxyphyllus*, a Brazilian poisonous plant. *Biochemical Systematics and Ecology* 20: 393.

Méndez MC and Riet-Correa F (1993). Intoxication by *Senecio tweediei* in cattle in southern Brazil. *Veterinary and Human Toxicology* 35:55.

Méndez MC, Riet-Correa F, and Schild AL (1987). Intoxicação por *Senecio* spp. (Compositae) em bovinos no Rio Grande do Sul. *Pesquisa Veterinária Brasileira* 7(2): 51-56.

Méndez MC, Riet-Correa F, Schild AL, and Martz W (1990). Intoxicação experimental por cinco espécies de *Senecio* em bovinos e aves. *Pesquisa Veterinária Brasileira* 10:63-69.

Radostits OM, Gay CC, Blood DC, and Hinchcliff KW (2002). *Clínica Veterinária: um tratado de doenças dos bovinos, ovinos, suínos, caprinos e eqüinos,* 1881 pp. Guanabara, Koogan.

Riet-Correa F and Medeiros RMT (2001). Intoxicação por plantas em ruminantes no Brasil e no Uruguai: importância econômica, controle e riscos para a saúde pública. *Pesquisa Veterinária Brasileira* 21:38-42.

Riet-Correa F and Méndez MC (2007). Intoxicações por plantas e micotoxinas. In *Doenças de Ruminantes e Eqüídeos* (F Riet-Correa, AL Schild, RAA Lemos, and JR Borges, eds), vol. 2, pp. 99-219. Ed. Palotti, Santa Maria, RS, Brazil.

Riet-Correa F, Medeiros RMT, Pfister JA, Schild AL, and Dantas AFM (2009). *Poisoning by plants, mycotoxins and related substances in Brazilian livestock.* Pallotti, Santa Maria vol. 1. 246 pp.

Rissi DR, Rech RR, Pierezan F, Gabriel AL, Trost ME, Brum JS, Kommers GD, and Barros CSL (2007). Intoxicações por plantas e micotoxinas associadas a plantas em bovinos no Rio Grande do Sul: 461 casos. *Pesquisa Veterinária Brasileira* 27:261-268.

Tokarnia CH, Döbereiner J, and Peixoto PV (2000). *Plantas tóxicas do Brasil,* 309 pp. Editora Helianthus, Rio de Janeiro, Brasil.

# Chapter 23

# Phenology of *Senecio* spp. and Vegetation Cover in Rio Grande do Sul State, Southern Brazil

F.S.C. Karam[1] and J.A. Jarenkow[2]

[1]*Desidério Finamor Veterinary Research Institute – FEPAGRO: Estrada do Conde, 6000. Eldorado do Sul, RS, Brazil, 92990-000;* [2]*Department of Botanics, Universidade Federal do Rio Grande do Sul, Av. Bento Gonçalves, 9500, Porto Alegre, RS, Brazil, 91501-970*

## Introduction

Phenology, which consists of the observation of repetitive biological events and their relationship with changes of the biotic and abiotic environment (Galetti *et al.* 2003), is regarded as an excellent technique for ecosystem characterization (Morellato 1992). Each distinguishable stage of the life cycle of a species is known as phenophase, and the phenological differences across species can pinpoint ecological and physiological problems (Fournier 1976).

The phenological study of four toxic *Senecio* species reviewed the relationship between their occurrence and levels of vegetation cover in native grasslands in the southern region of Rio Grande do Sul, Brazil. The phenology of *S. brasiliensis*, *S. oxyphyllus*, *S. heterotrichius*, and *S. selloi* shows that their vegetative phases occur throughout the year whereas the reproductive phases are concentrated between September and December, exhibiting an annual and monocarpic behavior. *S. oxyphyllus* was the least persistent in the environment, and *S. heterotrichius* the most persistent, behaving like a psammophilic species (Karam *et al.* 2002). Both humid and dry soils seem to be favorable to these species, as they show a positive photoblastic and anemochorous behavior with vegetative propagation capability (Matzenbacher 1998).

## Environmental Aspects

In the southwestern region of the state of Rio Grande do Sul, environmental factors such as cold temperatures in the winter or water deficits in the summer and inappropriate management (fires and excessive animal load, among others) reduce the grazing activity on natural grasslands (Crawshaw *et al.* 2007; Overbeck *et al.* 2007). Patches of bare soil interspersed with the remaining vegetation cover allow for the growth of undesirable plants,

©CAB International 2011. *Poisoning by Plants, Mycotoxins, and Related Toxins* (eds F. Riet-Correa, J. Pfister, A.L. Schild, and T.L. Wierenga)

changing the appearance of grasslands and limiting the supply of good quality pasture (Gonçalves and Girardi-Deiro 1986; Girardi-Deiro *et al.* 1992).

The fall-winter period is generally considered to be more favorable to seedling growth than the summer, when high temperatures and evapotranspiration are sources of stress (Castellani *et al.* 1999). According to Beskow (1995), seedlings that germinate in the fall have a longer time to develop deep roots before enduring water stress in the summer, especially in drier regions, thus increasing their chances of survival and establishment. Conversely, seedling deaths are not observed in regions with more frequent rains during the summer. Heavy grazing in the winter, or even during the rainy season, contributes towards reducing the vegetation cover layer, augmenting the incidence of light upon the soil and, consequently, temperature, which predisposes the germination of seeds (Thompson *et al.* 1977; Coombs *et al.* 1991; Beskow 1995). Seed germination responds to daily oscillations in temperature and this response varies according to the magnitude of these oscillations and to the presence or absence of light (Thompson *et al.* 1977). In studies involving *S. jacobaea* in Oregon (USA), McClements *et al.* (1998) and McEvoy *et al.* (1991) attach special importance to the soil seed bank, where seeds are invulnerable to natural enemies due to the depth at which they are buried. Maia *et al.* (2003, 2004) conducted studies in Rio Grande do Sul and established a positive correlation between vegetation cover (chiefly for perennial species) and the soil seed bank, showing that this could play a key role in the dynamics of natural grassland vegetation. The viability of seeds in the soil varies according to temperature, light, and moisture (van der Meijden and van der Waals-Kooi 1979).

For the emergence and establishment of seedlings, the lack of protection on the soil surface implies damage that varies according to local conditions. Both microorganisms and plants require mild temperatures and sufficient moisture, as excessive heating of the soil kills microorganisms and reduces oxygen solubility. Water absorption comes to a halt at temperatures greater than 32°C and plants cease to grow, increase their respiration, and use more photosynthesized substances. Every 0.5°C rise in soil temperature above 36°C causes a 2% decrease in carbohydrate reserves, and roots build up fewer reserves (Guevara 1993).

Beskow (1995), while investigating *S. jacobaea* in New Zealand, found that seedling emergence is higher in a bare soil than at sites with dense vegetation cover, and also that it is higher in humid soil in the winter, when temperatures are low. Seed germination is inhibited in the summer because of poor moisture and at any time of the year due to the presence of pasture cover. Seedling emergence is not directly influenced by trampling. Trampling often stimulates germination through the damage it causes to vegetation cover and is a determining factor for the development of bare soils in the winter, when plants are more severely damaged. Emergence of *S. jacobaea* was minimal where pasture cover was undisturbed. McClements *et al.* (1998) observed that the incidence of *S. jacobaea* and *S. aquaticus* increases as vegetation cover decreases. They found that soils that are richer in phosphorus are less infested by *S. jacobaea* while low soil pH had increased growth of *S. jacobaea*. Low pH and low phosphorus content are common in soils in several regions of Rio Grande do Sul, including those soils in the region analyzed (Macedo 1984). Therefore, these factors should also be considered when assessing the occurrence of *Senecio* spp. in Rio Grande do Sul.

## Vegetative and Reproductive Phenological Aspects

In grassland areas, the seedlings of some species can remain for several months without apparent growth or exhibit extremely low growth. These plants can vary considerably in age and produce multiple branches if they suffer some kind of damage. The root system is a vigorous vegetative propagation agent. Roots of seedlings in the cotyledon stage may produce sprouts. Roots and rosettes form sprouts more readily than flowering plants, and root sprouts may emerge when the plant naturally dies after having blossomed (Beskow 1995). According to Castellani *et al.* (1999), resprouting occurs under favorable moisture conditions which usually occur during fall to spring in perennial species found in coastal dunes and is positively correlated with rainfall. In a study of the herbaceous layer of forest ecosystems in the southern Brazilian Plateau, Cestaro (1984) observed that many species which produce leaves all year round are constantly attacked by herbivores.

The reproduction of various species relies mainly on photoperiod and temperature in order to trigger the flowering process (Cestaro 1984). In *S. grisebachii* (found in the province of Buenos Aires, Argentina), reproductive phenophases coincided with an increase in sunlight exposure and also with the use of available water in the soil (Madanes *et al.* 1996). Flowering stages often begin when the first rains fall, with subsequent fruiting and seeding. This phenomenon takes place when the photoperiod is longer and when the temperature rises. These events are synchronized for seed dispersal, allowing for maintenance of the species. Anemochorous species are aided by fruit dehiscence due to high temperatures, low relative air humidity, and more frequent winds (Miranda 1995; Morellato and Leitão-Filho 1996; Machado *et al.* 1997). McEvoy (1984) mentions that the pappus, an anemochorous dispersal structure, is found mainly in the central florets of the capitulum of *S. jacobaea*, which constitute most of the achenes. These achenes are released shortly after their maturation. Ray achenes do not contain this dispersal structure, and therefore dispersal occurs by animals; they may be retained for months after maturation. According to Matzenbacher (1998), the diaspores (achenes) of *Senecio* species containing a pappus with very thin hairs are carried by the wind and grassland species, but the woods are a natural barrier to their dispersal and thus they end up restricted to the margin of these woods. In *S. jacobaea*, seeds can also be dispersed by water, birds, man, and the droppings of sheep that fed on plants which were in the fruiting stage and whose seeds were not damaged in the digestive tract (Harper and Wood 1957).

Among the analyzed species, *S. brasiliensis* is a perennial plant whilst *S. oxyphyllus*, *S. heterotrichius*, and *S. selloi* are annuals (Matzenbacher 1998). These species behave like annual and monocarpic plants but with some individual variation. Depending on the damage, they can behave like annual, biennial, or even perennial plants. If damage is severe and/or frequent, some of them will have a biennial cycle, and most of them will need 2 or more years before they can flower. If growth conditions are always favorable, some plants can flower in the first year, behaving like annuals. According to van der Meijden and van der Waals-Kooi (1979), biennials show better growth strategies as they exploit the environmental resources that are available only in intermittent periods.

The vegetative strength of the plants depends on the environment in which they live and is determined by sunlight exposure (Borgignon and Piccolo 1981). *S. brasiliensis* and *S. heterotrichius* are heliophilous species (Cabrera and Klein 1975) like *S. selloi* and *S. oxyphyllus* (Matzenbacher 1998). The poor vitality of these species in shaded areas indicates that this is not the most favorable environment for their growth.

## Conclusions

The phenophases of *Senecio* spp. vary according to several factors, including air and soil temperatures, photoperiod, and biotic damage caused by inadequate management, suggesting that they are amenable to anthropogenic activities. In a balanced ecosystem, there is not usually excessive proliferation of undesirable plants such as *Senecio* spp. Dense vegetation cover prevents excessive sunlight, thereby decreasing the germination of seeds present in the soil. This is in agreement with the findings obtained for the *Senecio* species analyzed, which show a positive photoblastic behavior.

## References


Beskow WB (1995). A study of the factors influencing the emergence and establishment of ragwort (*Senecio jacobaea* L.) seedlings in pastures, 116 pp. Master Thesis, Massey University, New Zealand.

Borgignon OJ and Piccolo ALG (1981). Fenologia de *Hydrocotyle leucocephala* Cham. *Rodriguésia* 33(56):91-99.

Cabrera AL and Klein RM (1975). Compostas, 2. Tribo: Senecioneae. In *Flora ilustrada catarinense* (R Reitz, ed.), pp. 126-222. Herbário Barbosa Rodrigues, Itajaí, SC.

Castellani TT, Caus CA, and Vieira S (1999). Fenologia de uma comunidade de duna frontal no sul do Brasil. *Acta Botanica Brasílica* 13(1):99-114.

Cestaro LA (1984). Ecologia do estrato herbáceo da mata de araucária da Estação Ecológica de Aracuri, Esmeralda, Rio Grande do Sul, 110 pp. Dissertação de Mestrado em Ecologia, Instituto de Biociências, UFRGS, Porto Alegre.

Coombs EM, Bedell TE, and McEvoy PB (1991). Tansy ragwort (*Senecio jacobaea*): importance, distribution and control in Oregon. In *Noxious Range Weeds* (LF James, JO Evans, MH Ralphs, and RD Child, eds), pp. 419-428. Westview Press, Boulder, CO.

Crawshaw D, Dall'Agnol M, Cordeiro JLP, and Hasenack H (2007). Caracterização dos campos sul-rio-grandenses: uma perspective da ecologia da paisagem. *Boletim Gaúcho de Geografia* 33:233-252.

Fournier LA (1976). El dendrofenograma, una representación gráfica del comportamiento fenológico de los árboles. *Turrialba* 26(1):96-97.

Galetti M, Pizo MA, and Morellato PC (2003). Fenologia, frugivoria e dispersão de sementes. In *Métodos de estudos em biologia da conservação e manejo da vida silvestre* (L Cullen Jr, R Rudran, C Valladares-Pádua, eds), pp. 395-422. UFPR/Fundação Boticário de Proteção à Natureza, Curitiba.

Girardi-Deiro AM, Gonçalves JON, and Gonzaga SS (1992). Campos naturais ocorrentes nos diferentes tipos de solos no município de Bagé, RS. 2. Fisionomia e composição florística. *Iheringia* (Sér. Bot.) 42:55-79.

Gonçalves JON and Girardi-Deiro AM (1986). Efeito de três cargas animais sobre a vegetação de pastagem natural. *Pesquisa Agropecuária Brasileira* 21(5):547-554.

Guevara GJ (1993). *Plagas y Cómplices*, 116 pp. Orientación Gráfica Editora S. R. L., Buenos Aires.

Harper JL and Wood WA (1957). Biological flora of the British Isles: *Senecio jacobaea* L. *Journal of Ecology* 45:617-637.

Karam FSC, Méndez MC, Jarenkow JA, and Riet-Correa F (2002). Fenologia de quatro espécies tóxicas de *Senecio* (Asteraceae) na região Sul do Rio Grande do Sul. *Pesquisa Veterinária Brasileira* 22(1):33-39.

Macedo W (1984). Levantamento de reconhecimento dos solos do Município de Bagé. Bagé: Embrapa/UEPAE de Bagé. Documentos, 69 pp.

Machado ICS, Barros LM, and Sampaio EVSB (1997). Phenology of caatinga species in Serra Talhada, PE, Northeastern Brazil. *Biotropica* 29(1):57-68.

Madanes N, Vicari R, and Bonaventura SM (1996). Fenologia de las especies de los bordes de caminos en agroecosistemas y su relación con los parámetros climáticos. *Parodiana* 9(1-2):149-158.

Maia FC, Medeiros RB, Pillar VP, Chollet DMS, and Olmedo MOM (2003). Composição, riqueza e padrão de variação do banco de sementes do solo em função da vegetação de um ecossistema de pastagem natural. *Iheringia* (Sér. Bot.) 58:61-80.

Maia FC, Medeiros RB, Pillar VP, and Focht T (2004). Soil seed bank variation patterns according to environmental factors in a natural grassland. *Revista Brasileira de Sementes* 26(2):126-137.

Matzenbacher NI (1998). O complexo 'Senecionoide' (Asteraceae-Senecioneae) no Rio Grande do Sul – Brasil. Tese de Doutorado em Botânica, 274 pp. Instituto de Biociências, UFRGS, Porto Alegre.

McClements I, Courtney AD, and Malone FE (1998). Management and edaphic factors related with the incidence of marsh ragwort. In *Toxic plants and other natural toxicants* (T Garland and AC Barr, eds), pp. 40-44. Biddles Ltd Guildford and King's Lynn, UK.

McEvoy PB (1984). Dormancy and dispersal in dimorphic achenes of tansy ragwort, *Senecio jacobaea* L. (Compositae). *Oecologia* 61:160-168.

McEvoy P, Cox C, and Coombs E (1991). Successful biological control of ragwort, *Senecio jacobaea*, by introduced insects in Oregon. *Ecological Applications* 1(4):430-442.

Miranda IS (1995). Fenologia do estrato arbóreo de uma comunidade de cerrado em Alter-do-Chão, PA. *Revista Brasileira de Botânica* 18(2):235-240.

Morellato LPC (1992). Sazonalidade e dinâmica de ecossistemas florestais na Serra do Japi. In *História natural da Serra do Japi: ecologia e preservação de uma área florestal no Sudeste do Brasil.* (LPC Morellato, ed.), pp. 98-111. UNICAMP/FAPESP, Campinas.

Morellato PC and Leitão-Filho HF (1996). Reproductive phenology of climbers in a Southeastern Brazilian forest. *Biotropica* 28(2):180-191.

Overbeck GE, Müller SC, Fidelis A, Pfadenhauer J, Pillar VD, Blanco C, Boldrini II, Both R, and Forneck ED (2007). Brazil's neglected biome: the South Brazilian Campos. *Perspectives in Plant Ecology Evolution and Systematics* 9:101-116.

Thompson K, Grime JP, and Mason G (1977). Seed germination in response to diurnal fluctuations of temperature. *Nature* 267:147-149.

van der Meijden E and van der Waals-Kooi RE (1979). The population ecology of *Senecio jacobaea* in a sand dune system. *Journal of Ecology* 67:131-153.

# Chapter 24

# Nutritional Implications of Pyrrolizidine Alkaloid Toxicosis

P.R. Cheeke

*Department of Animal Sciences, Oregon State University, Corvallis, OR 97331*

## Introduction

The pyrrolizidine alkaloids (PA) constitute one of the most important classes of toxicants of plant origin. Although as many as 6000 plant species contain PA (Smith and Culvenor 1981), only a few are of major agricultural or toxicological importance. Livestock poisoning problems are associated mainly with the PA occurring in three plant families, the Boraginaceae, Compositae, and Leguminosae. Of particular significance are PA in *Senecio* spp., with *S. jacobaea* (tansy ragwort) the major plant involved.

The studies summarized in this paper have been conducted mainly in the author's laboratory and represent a review of a 30 year research program on the toxicologic and nutritional effects of *Senecio* PA.

## PA Interactions with Protein and Amino Acids

The toxicity of PA is influenced by dietary protein concentration. Cheeke and Garman (1974) found that the severity of PA toxicosis was increased with low levels of dietary protein. No gross signs of PA toxicosis were noted in *S. jacobaea*-fed rats receiving 25% dietary protein, whereas with 8% dietary protein, severe signs of toxicosis were observed, even though *S. jacobaea* intake was only 68% of that of the high protein group. The protective effect was associated with sulfur amino acids. Subsequent studies by Buckmaster *et al.* (1976) indicated that the protective effect of sulfur amino acids in rats was due primarily to cysteine; methionine had little activity. Feeding 1% cysteine doubled the survival time of rats fed *S. jacobaea*, and total plant intake (thus PA intake) was 300% greater with the cysteine treatment. Dietary or injected cysteine has been shown in numerous studies to partially protect against the toxicity of consumed or injected PA (Hayashi and Lalich 1967; Buckmaster *et al.* 1976; Miranda *et al.* 1981, 1982; Garrett and Cheeke 1984). Cysteine probably exerts its effects as a constituent of glutathione, which conjugates PA and PA metabolites for excretion as mercaptans (Robertson *et al.* 1977). The protective activity of dietary cysteine is enhanced by simultaneous administration of synthetic antioxidants such as ethoxyquin (Miranda *et al.* 1981a), butylated hydroxyanisole (BHA), and butylated hydroxytoluene (BHT) (Miranda *et al.* 1981b, 1982; Garrett and

©CAB International 2011. *Poisoning by Plants, Mycotoxins, and Related Toxins* (eds F. Riet-Correa, J. Pfister, A.L. Schild, and T.L. Wierenga)

Cheeke 1984). Synthetic antioxidants produce a marked increase in glutathione-S-transferase activity, stimulating the conjugation of PA with glutathione (Miranda *et al.* 1981a, 1982).

The favorable results with sulfur amino acids and synthetic antioxidants were obtained with laboratory animals. Experiments with large animals have been disappointing. Dietary supplementation with 1% cysteine and 0.75% BHA did not have protective effects against the toxicity of *S. jacobaea* in horses (Garrett *et al.* 1984). Similarly, a trial with beef cattle fed a source of rumen nondegradable sulfur amino acid (methionine hydroxy analog) and ethoxyquin did not affect PA toxicosis induced with *S. jacobaea* (Cheeke *et al.* 1985).

The branched chain amino acids (BCAA) isoleucine, leucine, and valine were suggested to have protective activity against PA toxicosis (Rogers *et al.* 1979). The rationale of this treatment is that liver dysfunction alters the plasma amino acid pattern. Because the BCAA are metabolized in peripheral tissues to a greater extent than other amino acids, the plasma BCAA levels relative to other amino acids decrease with liver disease. The plasma ratio of BCAA to phenylalanine and tyrosine correlates well with degree of liver damage or encephalopathy (Rogers *et al.* 1979; Gulick *et al.* 1980). Rogers *et al.* (1979) suggested that restoring this ratio to normal by supplementation with BCAA improves the clinical state of PA-intoxicated animals. With rats, Garrett *et al.* (1984) observed that supplementation of the diet with a mixture of BCAA did not influence the toxicity of *S. jacobaea.*

Although high dietary protein levels may be helpful in reducing the susceptibility of animals to PA toxicosis, high protein intakes intensify the pathological effects in animals already afflicted with PA toxicity signs (McGinness 1980). This is particularly true in horses which develop pronounced neurological signs in PA toxicosis (Giles 1983; Giesecke 1986). These signs develop because of elevated levels of blood ammonia, resulting from impaired hepatic metabolism of amino acids and the inability to adequately convert ammonia from amino acid degradation to urea (Hintz *et al.* 1970). Hyperammonemia causes spongy degeneration of the central nervous system (Hooper *et al.* 1974). Biochemically, ammonia poisoning of the central nervous system is associated with the detoxification of ammonia in the brain by formation of glutamine from glutamate, which in turn is derived from α-ketoglutarate. Depletion of α-ketoglutarate results in impaired citric acid cycle activity and thus ATP deficit and neurological signs (depression, head pressing, and coma). These reactions have been reviewed previously (Cheeke 1989). Animals suffering from sublethal chronic PA toxicosis are susceptible to stresses from other sources besides high protein intake. For example, horses that appeared clinically normal 14 months after intoxication by *S. vulgaris* were unable to tolerate exercise stress (Lessard *et al.* 1986) and developed signs of depression, edema, and anorexia when started in a training program.

Ascites and edema are characteristic signs of chronic PA toxicity. Ascites is attributed to low serum albumin, a hepatically synthesized blood protein with an important role in regulation of osmotic relationships and fluid balance. Serum albumin is depressed in animals consuming PA (Cheeke and Garman 1974). High dietary protein levels protected against ascites in rats fed *S. jacobaea* (Cheeke and Garman 1974), possibly by helping to maintain hepatic albumin synthesis.

## PA Interactions with Minerals

Interrelationships of dietary PA with mineral metabolism are well known, arising from observations in Australia on copper poisoning of sheep (Bull *et al.* 1956). Enzootic copper poisoning had been observed for many years in areas where the forage copper content was in the normal range. Bull *et al.* (1956) demonstrated that the consumption of PA-containing plants such as *Heliotropium europaeum* and *Echium lycopsis* (*plantagineum*) predisposed sheep to excessive liver uptake of copper, ultimately leading to the hemolytic crisis of chronic copper toxicity. They postulated that PA increased the avidity of liver cells for copper. Copper toxicosis in sheep associated with PA consumption is now well documented (Bull *et al.* 1956; St. George-Grambauer and Rac 1962; Seaman 1985, 1987). Breed differences in sheep are noted; the British breeds are more susceptible to PA-induced copper toxicosis than Merinos (Culvenor *et al.* 1984; Seaman 1987)

Increased liver copper levels in PA-poisoned animals is not always found. Swick *et al.* (1983) and White *et al.* (1984) did not find increased liver copper levels in sheep consuming *S. jacobaea*. Liver copper concentrations were normal in cattle with chronic heliotrope poisoning (Bull 1961). Horses consuming *S. jacobaea* had liver copper levels almost six-fold high than controls (Garrett *et al.* 1984). Rabbits fed a diet with 5% *S. jacobaea* and 100 ppm Cu had a 2.6-fold increase in liver copper over the appropriate controls (Swick *et al.* 1982a). Australian researchers have differentiated two PA-induced syndromes in sheep: PA poisoning and hepatogenous chronic copper poisoning (Seaman 1987). The PA poisoning is typical PA toxicosis with shrunken fibrotic liver, ascites, and histopathologic liver lesions characteristic of PA toxicity. Hepatogenous chronic copper toxicity, also known as the yellows, toxemic jaundice, and hemolytic jaundice, characteristically involves a hemolytic crisis, with depression, jaundice, hemoglobinuria, and elevated liver copper (Seaman 1987). In Australia, induced copper toxicity is the more common of the PA-related conditions (Seaman 1987). As sheep are quite resistant to PA toxicosis, exposure is usually on a chronic basis. The chronic copper poisoning syndrome is mainly associated with *E. plantagineum* consumption over a period of 2 or more years, while PA toxicity occurs mainly with sheep consuming *H. europaeum* or a mixture of both echium and heliotrope. The PA in echium have a low hepatoxic potency (Culvenor *et al.* 1984), so it appears that the chronic copper toxicity in sheep is associated with prolonged exposure to PA of low potency, whereas PA toxicosis is due to consumption of a higher dosage of hepatotoxins.

The degree of copper accumulation in the liver is related both to the intakes of copper and PA and probably also to the pattern of PA intake. Swick *et al.* (1982b) found that elevated liver copper levels in rats fed *S. jacobaea* occurred only in the presence of high (50 and 250 ppm) copper levels, and with low PA intakes, copper accumulation occurred only with the highest (250 ppm) dietary copper level used. Similarly, Australian workers (Howell *et al.* 1991) reported that heliotrope consumption caused elevated levels of tissue copper in sheep only when copper was also administered. These workers noted that the combined administration of PA and copper intensified the hepatic histologic lesions and resulted in clinical signs of toxicosis, absent in animals receiving only heliotrope (Howell *et al.* 1988). In pair-fed rats having the same feed and copper intakes, rats fed 10% *S. Jacobaea* had liver copper levels of 585 μg/g liver (dry weight), whereas controls had 77 μg/g, a 7.6-fold difference (Swick *et al.* 1982c). In the control rat livers, 37.5 and 30.6% of the copper was in the nuclei and cytosol fractions, respectively, while in the PA-fed animals corresponding values were 53.4 and 16.5%. This suggests that cytosolic proteins, such as metallothionein, copper chelatin, and superoxide dismutase, were not the fractions accumulating copper. Elevated copper levels in the nuclei and debris fractions suggested an impairment of normal subcellular excretory mechanisms, such as a lysosomal defect. In chronic copper toxicity, the copper initially accumulates in all subcellular fractions and then concentrates in the nuclei and debris (Gooneratne *et al.* 1979; Helman *et al.* 1983). As these fractions become saturated with copper, the lysosomes take up the element. Because of their increased density, they then separate with the nuclei and debris fraction during centrifugation. Rupture of the lysosomes *in vivo* releases hydrolases and the copper, causing cell death and copper-induced hemolysis (Gooneratne *et al.* 1979). An alternative viewpoint (Fuentealba and Haywood 1988) is that copper-induced hepatic damage is a consequence of nuclear degeneration caused by the movement of copper into the nucleus, rather than from lysosomal damage.

In rats pair-fed control and *S. jacobaea*-containing diets, serum ceruloplasmin activity was significantly increased in the animals receiving PA (Swick *et al.* 1982c). During PA intoxication, induction of ceruloplasmin synthesis may be an attempt by the liver to excrete excess copper. Alternatively, the elevated ceruloplasmin may be involved in transfer of iron to transferrin in the serum, associated with metabolism of excess iron due to the block in hematopoesis that occurs in PA toxicosis (Swick *et al.* 1982b).

Serum copper is elevated and zinc depressed in rats intoxicated by *S. jacobaea* (Swick *et al.* 1982b, c). Elevated serum copper was also noted in rats treated with monocrotaline pyrrole (Ganey and Roth 1987); the increase was attributed to pulmonary hypertension.

In studies with chicks, Huan *et al.* (1992) reported that both serum and liver copper were markedly increased in birds fed diets with 5% *S. jacobaea* and 250 ppm copper, while serum and liver zinc concentrations were decreased. Liver iron was increased in chicks fed *S. jacobaea*, while liver selenium was not affected. In similar trials with Japanese quail, Huan and Cheeke (Chapter 32, this volume) found no increase in liver copper in birds fed *S. jacobaea*. Japanese quail are totally resistant to the hepatoxic effects of PA (Buckmaster *et al.* 1977). The lack of effect of *S. jacobaea* on tissue copper in this PA-resistant species suggests that some degree of hepatotoxicity is necessary to induce the changes in tissue copper concentrations seen in PA-susceptible species.

Farrington and Gallagher (1960) noted that copper formed complexes with PA and their necic acids. Although this observation has not been pursued further, such complexes could have biological significance.

Copper, as a cofactor of enzymes involved in melanin synthesis, is necessary for normal hair pigmentation. An interesting observation is that black-haired pigs have been reported to lose their coloration when fed grain suspected of being contaminated with *Crotalaria* seeds (Gibbons 1967). This could suggest an impairment of copper utilization by dietary PA.

Interrelationships between copper and molybdenum metabolism in ruminants are well known. Molybdenum helps protect against copper toxicity by promoting its excretion. Sheep consuming *Echium* and *Heliotropium* spp. in Australia accumulate high levels of liver copper and many develop hemolytic jaundice, as previously described. Molybdenum supplementation would appear to offer potential as a means of reducing the copper accumulation. Contrary results, however, were obtained in a study by White *et al.* (1984) who observed that liver copper levels in sheep fed *S. jacobaea* were not reduced when molybdenum supplementation was provided. The copper levels were slightly high in sheep receiving molybdenum and survival time of animals fed molybdenum was reduced, indicating a possible negative effect of molybdenum rather than a beneficial one.

Besides effects of PA on copper metabolism, other minerals are apparently influenced by PA. Accumulation of liver copper in animals consuming PA is accompanied by depressed zinc levels (Swick *et al.* 1982b, c); copper has a higher affinity for metallothionein than zinc and may displace it. Thus the change in tissue zinc levels with PA exposure is probably an indirect one.

Zinc supplementation protects animals against various hepatotoxins, including carbon tetrachloride (Cagen and Klaassen 1979; Clarke and Lui 1986) and phomopsins (Allen and Master 1980). Zinc induces synthesis of metallothionein, a class of low molecular weight, cysteine-rich proteins with a high concentration of reactive sulfhydryl groups. Because PA metabolites react with sulfhydryl groups, a tenable hypothesis is that zinc might have protective effects against PA toxicosis, by 'soaking up' pyrroles with metallothionein. Miranda *et al.* (1982) have shown protective effects of zinc supplementation against *S. jacobaea* toxicity in rats.

Hematopoiesis is markedly impaired by PA consumption (Swick *et al.* 1982a), resulting in anemia and changes in tissue iron distribution. McLean (1970) reviewed several studies showing a loss of hematopoietic tissue and bone marrow lesions in PA-poisoned rats. Rats fed *S. jacobaea* as a source of PA show characteristic changes in gross appearance of the tissues. In the early stages of PA exposure, the liver is very dark in color due to the accumulation of iron. Shortly thereafter, the liver becomes light in color as the iron deposits are shifted to the spleen. The spleen becomes enlarged with a high iron content (Swick *et al.* 1982b). These changes are reflected in organ weights and tissue-mineral concentrations. Incorporation of $^{59}Fe$ into erythrocytes is markedly impaired in rats following PA exposure. Other workers have also reported effects of PA consumption on tissue iron. In vervet monkeys administered retrorsine, liver iron values were 10.8 to 992 ppm in control animals and 202 to 22,707 ppm in those given retrorsine, with mean values of 43.3 and 449.0 ppm, respectively (Van der Watt *et al.* 1972). These authors, from South Africa, suggested that PA-induced iron accumulation might contribute to a human health problem in Zimbabwe, Malawi, Mozambique, and South Africa. A significant incidence of siderosis (excessive accumulation of iron in hemosiderin deposits) occurs in the Bantu people (Bantu siderosis). Many of these people consume herbs known to contain PA (Van der Watt *et al.* 1972) which, in conjunction with the use of iron cooking pots, might lead to siderosis.

A likely explanation for these effects on iron metabolism is impaired protein synthesis. A major action of PA metabolites is to cause cross-linking of DNA strands, thus inhibiting cell replication and protein synthesis. By these effects, pyrrolic metabolites may inhibit heme biosynthesis in the liver and other tissues as a result of alkylation of DNA. Because iron then cannot be used for hemoglobin synthesis, the excess iron accumulates as hemosiderin deposits in liver and spleen. Two iron storage compounds are ferritin and hemosiderin. Ferritin consists of iron and apoferritin, an iron-free protein. Hemosiderin is essentially a protein-free aggregate (Morris 1987). In the early stages of PA toxicosis, iron is probably stored as ferritin, while in later stages, when liver protien synthesis is impaired, hemosiderin is probably the main storage form. This area is one in which further studies are needed to fully elucidate the mode of action of PA in affecting tissue iron distribution and hematopoiesis.

Hepatotoxicity is induced by high liver concentrations of copper (Kumaratilake and Howell 1986) and iron (Bacon and Britton 1989). Thus the increased liver levels of these elements with exposure to PA suggests that the hepatotoxic effects could be associated not only with the PA but with the metals as well. Miranda *et al.* (1981c) provided evidence that high dietary copper levels enhance the hepatoxicity of PA. This provides another dimension to PA-mineral interrelationships.

There has been little work on the interrelationships of PA with other minerals. Shull *et al.* (1977) found that *in vitro* metabolism of monocrotaline and a *Senecio* PA mixture was not affected by severe selenium deficiency in rats (verified by very low glutathione peroxidase levels), although it was observed that the ability of phenobarbital to induce PA-metabolizing enzymes was reduced in selenium-deficient rats (Shull *et al.* 1979). In other work (Burguera *et al.* 1983), it was noted that selenium supplementation of the diet of turkey poults did not influence their susceptibility to *Crotalaria* intoxication.

Finally, it is of interest that cattle losses to *S. jacobaea* toxicosis have been linked to mineral deficiency. Palfrey *et al.* (1967) noted that farms in Nova Scotia, Canada, having cattle mortality due to ragwort consumption had forage with low phosphorus and trace mineral contents, and the ragwort in the pastures had higher concentrations of these elements than the forage. They suggested that animals consumed the ragwort as a source of supplementary minerals. The animal losses were higher on farms where no mineral supplement was provided than on farms where minerals were fed. Attempts to reduce PA toxicity by feeding increased levels of minerals have not been successful (Johnson 1982). However, since mineral-deficient animals often have depraved appetites, it is not unlikely that mineral deficiency could be a predisposing factor to consumption of PA-containing plants, which are generally unpalatable.

## PA Interactions with Vitamins

Hepatotoxic agents such as PA might be expected to affect the metabolism of nutrients for which the liver is a major site of storage and/or metabolism. This is particularly true for nutrients which are transported or stored in association with proteins synthesized in the liver. Hence, it is not surprising that PA have a marked effect on vitamin A (Vit A) metabolism (Moghaddam and Cheeke 1989). In rats fed *S. jacobaea*, both plasma and liver Vit A levels were markedly depressed. Significant reductions in plasma Vit A occurred within 10 days after initial PA consumption, indicating that an influence on Vit A distribution occurs early in PA toxicosis. The much lower liver Vit A levels in PA-fed animals compared to controls is interesting because PA damage occurs primarily in hepatocytes, while Vit A is stored mainly in stellate (Ito) cells (Ong 1985). Possible explanations for the depressed liver and blood Vit A levels include: (i) PA may inhibit the synthesis of proteins involved in Vit A transport and storage; (ii) PA may inhibit Vit A absorption; and (iii) PA damage may impair the ability of the liver to take up Vit A. Because of the pronounced inhibitory effects of PA on hepatic protein synthesis, it is likely that synthesis of retinol-binding proteins (RBP) and other proteins involved in Vit A metabolism is impaired (Huan *et al.* 1993). Biliary hyperplasia and impaired bile secretion are characteristic of PA toxicosis. Bile is necessary for absorption of fat-soluble vitamins; thus depressed tissue Vit A levels may reflect diminished absorption. Another factor which may influence Vit A absorption is that severe intestinal lesions, including inhibition of crypt cell mitosis and villus atrophy, occur in PA toxicosis (Hooper 1975). Further work is needed to elucidate the mechanisms by which PA depress tissue Vit A levels.

Other hepatotoxins are known to affect Vit A metabolism. Dietary DDT inhibits Vit A storage (Azais-Braesco *et al.* 1989), as do other organochlorines, organophosphates, and polychlorinated dibenzo-*p*-dioxins and dibenzofurans (Hakansson and Hanberg 1989). The dioxin TCDD inhibits storage of Vit A in stellate cells (Hakansson and Hanberg 1989). TCDD inhibits the storage of newly administered Vit A (Hakansson and Ahlborg 1989) and increases the mobilization of stored Vit A. Thus toxins such as TCDD affect the Vit A storage system. Vitamin A is first taken up by the parenchymal cells (hepatocytes) and within a few hours most is transferred to the stellate cells for storage. These transport mechanisms are not fully understood. Since PA metabolites specifically damage the hepatocytes, it is possible that PA exposure interferes with transfer of Vit A to the stellate cells. The fate of absorbed Vit A that does not become stored in the liver presumably is to be excreted. Huan *et al.* (1992) observed in chickens that PA inhibit the mobilization of previously stored Vit A from the liver, probably by inhibiting hepatic synthesis of retinol-binding protein.

Moore *et al.* (1972) noted that while copper and Vit A behave similarly in being stored preferentially in the liver and being transported by blood proteins, the tissue concentrations of these nutrients often behave in an inverse manner. Factors that increase the concentration of one usually decrease the concentration of the other. The effect of PA on blood and liver copper and Vit A levels follows this pattern.

During investigation of effects of *Senecio* PA on Vit A, Moghaddam and Cheeke (1989) noted that the red blood cells of PA-intoxicated rats were susceptible to *in vitro* hemolysis. This observation could have implications in the pathology of PA effects. *In vitro* hemolysis is characteristic of vitamin E (Vit E) deficiency. Since the absorption and tissue distribution of the fat-soluble vitamins A and E share some similarities, alterations in tissue Vit E levels similar to those observed for Vit A might be anticipated. Reduction in tissue Vit E concentrations in chicks fed *S. jacobaea* was observed by Lulay *et al.* (2007). Vit E functions *in vivo* as an antioxidant. The PA metabolites, including pyrroles and reactive aldehydes, may act as oxidizing agents. Thus, in PA toxicosis the pathology induced by the alkaloids may be intensified by an induced deficiency of cellular antioxidant (Vit E). This could also explain the protective effects of synthetic antioxidants against PA toxicosis (Miranda *et al.* 1981a,b; Miranda *et al.* 1982; Garrett and Cheeke 1984). Interrelationships with fat-soluble vitamins could have important implications in PA toxicosis. There may also be a copper–Vit A–Vit E–PA interaction. Copper increases lipid peroxidation which could increase Vit E requirements and increase Vit A destruction. Copper has an involvement in synthesis of Vit A transport proteins (Rachman *et al.* 1987). Vit A has a role in the regulation of ceruloplasmin synthesis (Barber and Cousins 1987). Injection of rats with retinoic acid increases serum ceruloplasmin activity; this increase does not occur in copper-deficient rats unless copper is also given (Barber and Cousins 1987). Barber and Cousins (1987) suggested that because ceruloplasmin functions as a free radical scavenger, part of the role of Vit A in increasing resistance of animals to stress and infections could arise through its effect on ceruloplasmin. The increase in ceruloplasmin activity in rats fed PA (Swick *et al.* 1982c) may relate to the lipid peroxidation effects of PA metabolites and the role of ceruloplasmin in protection against peroxidation. Thus the elucidation of these interactions between copper and Vit A is a fertile area for further research.

As with copper and iron, high intakes of Vit A are hepatotoxic (Jacques *et al.* 1979). Furthermore, synthetic antioxidants such as BHT, which protect against PA toxicosis (Miranda *et al.* 1981a; Garrett and Cheeke 1984) potentiate Vit A hepatoxicity (McCormick *et al.* 1987). This potentiation is particularly interesting because it occurs in the presence of decreased, rather than increased, Vit A levels in the liver (McCormick *et al.* 1987), which is the situation induced by PA intake (Moghaddam and Cheeke 1989).

A few other relationships between PA and vitamins have been reported. Vit $B_{12}$ has been implicated by Australian workers in the ruminal detoxification of PA (Dick *et al.* 1963), but does not seem to have been followed up with further work. Garrett and Cheeke (1984) hypothesized that since folic acid and Vit $B_{12}$ have roles in hematopoesis, they might have protective effects against the depressed erythrocyte formation characteristic of PA toxicosis (Swick *et al.* 1982b). After 12 weeks of consumption of a diet with 5% *S. jacobaea*, 100% of rats receiving Vit $B_{12}$ and folic acid were alive, whereas there was 50% mortality in those not receiving extra vitamins. However, overall survival time was not prolonged by inclusion of the vitamins. A supplement containing these vitamins was ineffective as a protective agent when fed to ponies (Garrett *et al.* 1984) and cattle (Cheeke *et al.* 1985) fed *S. jacobaea.*

## Conclusions

PA constitute one of the main groups of natural toxicants in foods and feeds. This review has concentrated on nutritional interactions of PA. Their toxicity is influenced by dietary protein level and various amino acids including cysteine and the branched chain amino acids. The PA inhibit protein synthesis in the liver via alkylation and cross-linking of DNA. Impaired protein synthesis may be involved in other nutritional interactions. The PA have pronounced effects on mineral metabolism. Liver copper concentrations and blood levels of copper and ceruloplasmin are elevated in PA toxicosis. Hematopoesis is greatly impaired, probably because of inhibited heme synthesis. As a result, iron concentrations of various tissues such as liver and spleen are elevated, from storage of excess iron that cannot be used in hemoglobin synthesis. There is a pronounced effect of PA on tissue Vit A and Vit E concentrations, with marked reductions in both plasma and liver concentrations of both vitamins. These effects may also be a reflection of impaired hepatic synthesis of proteins involved in Vit A metabolism such as retinol-binding protein and tocopherol-binding proteins. Thus there are numerous nutritional interactions involving PA.

## Acknowledgements

The participation of Dr Peter Cheeke to the 8th International Symposium on Poisonous Plants was financially supported by Conselho Nacional de Desenvolvimento Científico e Tecnológico (CNPq), grant 454084/2008-0, and by Coordenação de Aprefeiçoamento de Pessoal de Nível Superior (CAPES), grant 0017/09-4.

## References

Allen JG and Master HG (1980). Prevention of ovine lupinosis by the oral administration of zinc sulphate and the effect of such therapy on liver and pancreas zinc and liver copper. *Australian Veterinary Journal* 56:168-171.

Azais-Braesco V, Pascal G, Mareschi JP, Fayet Y, Degiuli A, Letoublon R, Got R, and Frot-Coutaz J (1989). Influence of vitamin A status and DDT on vitamin A-dependent protein mannosylation in rat liver. *Chemico-Biological Interactions* 69:259-267.

Bacon BR and Britton RS (1989). Hepatic injury in chronic iron overload. Role of lipid peroxidation. *Chemico-Biological Interactions* 70:183-226.

Barber EF and Cousins RJ (1987). Induction of ceruloplasmin synthesis by retinoic acid in rats. Influence of dietary copper and vitamin A status. *Journal of Nutrition* 117:1615-1622.

Buckmaster GW, Cheeke PR, and Shull LR (1976). Pyrrolizidine alkaloid poisoning in rats: protective effects of dietary cysteine. *Journal of Animal Science* 43:464-473.

Buckmaster GW, Cheeke PR, Arscott GH, Dickinson EO, Pierson ML, and Shull LR (1977). Response of Japanese quail to dietary and injected pyrrolizidine (*Senecio*) alkaloid. *Journal of Animal Science* 45:1322-1325.

Bull LB (1961). Heliotropium poisoning in cattle. *Australian Veterinary Journal* 37:37-43.

Bull LB, Dick AT, Keast JC, and Edgar G (1956). An experimental investigation of the hepatotoxic and other effects on sheep of consumption of *Heliotropium europaeum* L.: heliotrope poisoning of sheep. *Australian Journal of Agricultural Research* 7:281-332.

Burguera JA, Edds GT, and Osuna O (1983). Influence of selenium on aflatoxin $B_1$ or crotalaria toxicity in turkey poults. *American Journal of Veterinary Research* 44:1714-1717.

Cagen SZ and Klaassen CD (1979). Protection of carbon tetrachloride-induced hepatotoxicity by zinc: role of metallothionein. *Toxicology and Applied Pharmacology* 51:107-116.

Cheeke PR (1989). Pyrrolizidine alkaloid toxicity and metabolism in laboratory animals and livestock. In *Toxicants of Plant Origin.* pp. 1-22, vol. 1. CRC Press, Boca Raton, Florida.

Cheeke PR and Garman GR (1974). Influence of dietary protein and sulfur amino acid levels on the toxicity of *Senecio jacobaea* (tansy ragwort) to rats. *Nutrition Reports International* 9:197-207.

Cheeke PR, Schmitz JA, Lassen ED, and Pearson EG (1985). Effects of dietary supplementation with ethoxyquin, magnesium oxide, methionine hydroxyl analog, and B vitamins on tansy ragwort, (*Senecio jacobaea*) toxicosis in beef cattle. *American Journal of Veterinary Research* 46:2179-2183.

Clarke IS and Lui EMK (1986). Interaction of metallothionein and carbon tetrachloride on the protective effect of zinc on hepatotoxicity. *Canadian Journal of Physiological Pharmacology* 64:1104-1110.

Culvenor CCJ, Jago MV, Peterson JE, Smith LW, Payne AL, Campbell DG, Edgar JA, and Frahn JL (1984). Toxicity of *Echium plantagineum* (Paterson's Curse). I. Marginal toxic effects in Merino wethers from long-term feeding. *Australian Journal of Agricultural Research* 35:293-304.

Dick AT, Dann AT, and Bull LB (1963). Vitamin $B_{12}$ and the detoxification of hepatotoxic pyrrolizidine alkaloids in rumen liquor. *Nature* 197:207-208.

Farrington KJ and Gallagher CH (1960). Complexes of copper with some pyrrolizidine alkaloids and with some of their esterifying acids. *Australian Journal of Biological Science* 13:600-603.

Fuentealba IC and Haywood S (1988). Subcellular changes and metal mobilization in the livers of copper loaded rats. In *Trace Elements in Man and Animals 6* (LS Hurley, CL Keen, B Lonnderdal, and RB Rucker, eds), pp. 179-180. Plenum Press, New York.

Ganey PE and Roth RA (1987). Elevated serum copper concentration in monocrotaline pyrrole treated rats with pulmonary hypertension. *Biochemical Pharmacology* 36:3535-3537.

Garrett BJ and Cheeke PR (1984). Evaluation of amino acids, B vitamins and butylated hydroxyanisole as protective agents against pyrrolizidine alkaloid toxicity in rats. *Journal of Animal Science* 58:138-144.

Garrett BJ, Holtan DW, Cheeke PR, Schmitz JA, and Rogers QR (1984). Effects of dietary supplementation with butylated hydroxyanisole, cysteine, and vitamins B on tansy ragwort *Senecio jacobaea* toxicosis in ponies. *American Journal of Veterinary Research* 45:459-464.

Gibbons WJ (1967). Decoloration of hogs (questions and answers). *Modern Veterinary Practice* 48:52.

Giesecke PR (1986). Serum biochemistry in horses with *Echium* poisoning. *Australian Veterinary Journal* 63:90-91.

Giles CJ (1983). Outbreak of ragwort *Senecio jacobea* poisoning in horses. *Equine Veterinary Journal* 15:248-250.

Gooneratne SR, Howell M, and Gawthorne J (1979). Intracellular distribution of copper in the liver of normal and copper loaded sheep. *Research in Veterinary Science* 27:30-37.

Gulick BA, Lui IKM, Qualls CW Jr, Gribble DH, and Rogers QR (1980). Effect of pyrrolizidine alkaloid-induced hepatic disease on plasma amino acid patterns in the horse. *American Journal of Veterinary Research* 41:1894-1898.

Hakansson H and Ahlborg UG (1985). The effect of 2,3,7,8-tetrachlorodibenzo-p-dioxin (TCDD) on the uptake, distribution and excretion of a single oral dose of [11,12-$^{3}$H]retinylacetate and on the vitamin A status in the rat. *Journal of Nutrition* 115:759-771.

Hakansson H and Hanberg A (1989). The distribution of [$^{14}$C]-2,3,7,8-tetrachlorodibenzo-p-dioxin (TCDD) and its effect on vitamin A content in parenchymal and stellate cells of rat liver. *Journal of Nutrition* 119:573-580.

Hayashi L and Lalich JJ (1967). Renal and pulmonary alterations induced in rats by a single injection of monocrotaline. *Proceedings of the Society for Experimental Biological Medicine* 124:392-396.

Helman RG, Adams LG, Pierce KR, Bridges CH, and Bailey EM. (1983). The role of lysosomes in the pathogenesis of copper-induced hepatotoxicity. *Toxicological Applied Pharmacology* 67:238-245.

Hintz HF, Lowe JE, Clifford AJ, and Visek WJ (1970). Ammonia intoxication resulting from urea ingestion by ponies. *Journal of American Veterinary Medical Association* 157:963-966.

Hooper PT (1975). Experimental acute gastrointestinal disease caused by the pyrrolizidine alkaloid, lasiocarpine. *Journal of Comparative Pathology* 85:341.

Hooper PT, Best SM, and Murray DR (1974). Hyperammonaemia and spongy degeneration of the brain in sheep affected with hepatic necrosis. *Research into Veterinary Science* 16:216-222.

Howell J McC, Patel H, and Dorling P (1988). Heliotrope alkaloids and copper. In *Trace Elements in Man and Animals 6* (LS Hurley, CL Keen, B Lonnerdal, and RB Rucker, eds), pp. 319-320. Plenum Press, New York.

Howell J McC, Deol HS, and Dorling P (1991). Experimental copper and *Heliotropium europeaum* intoxication in sheep: Clinical syndromes and trace element concentrations. *Australian Journal of Agricultural Research* 42:979-992.

Huan J, Cheeke PR, Lowry RR, Nakaue HS, Snyder SP, and Whanger PD (1992). Dietary pyrrolizidine (*Senecio*) alkaloids and tissue distribution of copper and vitamin A in broiler chickens. *Toxicology Letters* 62:139-153.

Huan JY, Cheeke PR, and Blaner WS (1993). Modification of vitamin A metabolism by dietary pyrrolizidine (*Senecio*) alkaloids in rats. *Proceedings, Western Section, American Society of Animal Science* 44:275-278.

Jacques EA, Buschmann RJ, and Layden TJ (1979). The histopathologic progression of vitamin A-induced hepatic injury. *Gastroenterology* 76:599-602.

Johnson AE (1982). Failure of mineral-vitamin supplements to prevent tansy ragwort (*Senecio jacobaea*) toxicosis in cattle. *American Journal of Veterinary Research* 43:718-723.

Kumaratilake JS and Howell J McC (1986). Histochemical study of the accumulation of copper in the liver of sheep. *Research into Veterinary Science* 42:73-81.

Lessard P, Wilson WD, Olander HJ, Rogers QR, and Mendel VE (1986). Clinicopathologic study of horses surviving pyrrolizidine alkaloid *Senecio vulgaris* toxicosis. *American Journal of Veterinary Research* 47:1776-1780.

Lulay AL, Leonard SW, Traber MG, Keller MR, and Cheeke PR (2007). Effects of dietary pyrrolizidine (*Senecio*) alkaloids on plasma and liver vitamin E distribution in broiler chickens. In *Poisonous Plants:Global Research and Solutions* (KE Panter, TL Wierenga, and JA Pfister, eds), p. 77. CABI Publishing, Wallingford, UK.

McCormick DL, Hultin TA, and Detrisac CJ (1987). Potentiation of vitamin A hepatotoxicity by butylated hydroxytoluene. *Toxicology and Applied Pharmacology* 90:1-9.

McGinness JP (1980). *Senecio jacobaea* as a cause of hepatic encephalopathy. *California Veterinarian* 34:20.

McLean EK (1970). The toxic actions of pyrrolizidine *Senecio* alkaloids. *Physiology Review* 22:429-483.

Miranda CL, Carpenter HM, Cheeke PR, and Buhler DR (1981a). Effect of ethoxyquin on the toxicity of the pyrrolizidine alkaloid monocrotaline and on hepatic drug metabolism in mice. *Chemical Biological Interactions* 37:95.

Miranda CL, Reed RL, Cheeke PR, and Buhler DR (1981b). Protective effects of butylated hydroxyanisole against the acute toxicity of monocrotaline in mice. *Toxicology and Applied Pharmacology* 59:424-430.

Miranda CL, Henderson MC, and Buhler DR (1981c). Dietary copper enhances the hepatotoxicity of *Senecio jacobaea* in rats. *Toxicology and Applied Pharmacology* 60:418-423.

Miranda CL, Buhler DR, Ramsdell HS, Cheeke PR, and Schmitz JA (1982). Modifications of chronic hepatoxicity of pyrrolizidine *Senecio* alkaloids by butylated hydroxyanisole and cysteine. *Toxicology Letters* 10:177-182.

Moghaddam MF and Cheeke PR (1989). Effects of dietary pyrrolizidine (*Senecio*) alkaloids on vitamin A metabolism in rats. *Toxicology Letters* 45:149-156.

Moore T, Sharman IM, Todd JR, and Thompson RH (1972). Copper and vitamin A concentrations in the blood of normal and Cu-poisoned sheep. *British Journal of Nutrition* 28:23-30.

Morris ER (1987). Iron. In *Trace Elements in Human and Animal Nutrition* (W Mertz, ed.), chap. 4. Academic Press, San Diego.

Ong DE (1985). Vitamin A-binding proteins. *Nutrition Reviews* 43:225-232.

Palfrey GD, MacLean KS, and Langille WM (1967). Correlation between incidence of ragwort (*Senecio jacobaea* L) poisoning and lack of mineral in cattle. *Weed Research* 7:171-175.

Rachman FI, Conjat F, Carreau JP, Bleiberg-Daniel F, and Amedee-Manesme O (1987). Modification of vitamin A metabolism in rats fed a copper-deficient diet. *International Journal of Vitamin Nutrition Research* 57:247-252.

Robertson KA, Seymour JL, Hsia M-T, and Allen JR (1977). Covalent interaction of dehydroretronecine, a carcinogenic metabolite of the pyrrolizidine alkaloid monocrotaline, with cysteine and glutathione. *Cancer Research* 37:3141-3144.

Rogers QR, Knight HD, and Gulick BA (1970). Proposed method of diagnosis and treatment of pyrrolizidine alkaloid poisoning in horses. In *Symposium on Pyrrolizidine (*Senecio*) alkaloids: Toxicity, Metabolism and Poisonous Plant Control Measures* (PR Cheeke, ed.), p. 145. Nutrition Research Institute, Oregon State University, Corvallis.

Seaman JT (1985). Hepatogenous chronic copper poisoning in sheep associated with grazing *Echium plantagineum*. *Australian Veterinary Journal* 62:247.

Seaman JT (1987). Pyrrolizidine alkaloid poisoning of sheep in New South Wales. *Australian Veterinary Journal* 64:164-167.

Shull LR, Buckmaster GW, and Cheeke PR (1977). Dietary selenium status and pyrrolizidine alkaloid metabolism *in vitro* by rat liver microsomes. *Research Communications in Chemical Pathology and Pharmacology* 17:337-340.

Shull LR, Buckmaster GW, and Cheeke PR (1979). Effect of dietary selenium status on *in vitro* hepatic mixed-function oxidase enzymes of rats. *Journal of Environmental Pathology and Toxicology* 2:1127-1138.

Smith LW and Culvenor CCJ (1981). Plant sources of hepatotoxic pyrrolizidine alkaloids. *Journal of Natural Products* 44:129-152.

St George-Grambauer TD and Rac R (1962). Hepatogenous chronic copper poisoning in sheep in South Australia due to consumption of *Echium plantagineum* L. (Salvation Jane). *Australian Veterinary Journal* 38:288-293.

Swick RA, Cheeke PR, Patton NM, and Buhler DR (1982a). Absorption and excretion of pyrrolizidine *Senecio* alkaloids and their effects on mineral metabolism in rabbits. *Journal of Animal Science* 55:1417-1424.

Swick RA, Cheeke PR, Miranda CL, and Buhler DR (1982b). The effect of consumption of the pyrrolizidine alkaloid-containing plant *Senecio jacobaea* on iron and copper metabolism in the rat. *Journal of Toxicology and Environmental Health* 10:757-768.

Swick RA, Cheeke PR, and Buhler DR (1982c). Subcellular distribution of hepatic copper, zinc and iron and serum ceruloplasmin in rats intoxicated by oral pyrrolizidine *Senecio* alkaloids. *Journal of Animal Science* 55:1425-1430.

Swick RA, Cheeke PR, Ramsdell HS, and Buhler DR (1983). Effect of sheep rumen fermentation and methane inhibition on the toxicity of *Senecio jacobaea. Journal of Animal Science* 56:645-651.

Van der Watt JJ, Purchase IFH, and Tustin RC (1972). The chronic toxicity of retrorsine, a pyrrolizidine alkaloid, in vervet monkeys. *Journal of Pathology* 107:279-287.

White RD, Swick RA, and Cheeke PR (1984). Effects of dietary copper and molybdenum on tansy ragwort (*Senecio jacobaea*) toxicity in sheep. *American Journal of Veterinary Research* 45:159-161.

# Chapter 26

# Seasonal Variation in Pyrrolizidine Alkaloid Concentration and Plant Development in *Senecio madagascariensis* Poir. (Asteraceae) in Brazil

F.S.C. Karam[1], M. Haraguchi[2], and D.R. Gardner[3]

*[1]Desidério Finamor Veterinary Research Institute – FEPAGRO: Estrada do Conde, 6000, Eldorado do Sul, RS, Brazil, 92.990-000; [2]Center for Animal Health, Biological Institute, S. Paulo, SP, Brazil, 04014-002; [3]USDA-ARS Poisonous Plant Research Laboratory, Logan, Utah 84341, USA*

## Introduction

Intoxication by *Senecio* spp. is among the principal causes of adult cattle death in the state of Rio Grande do Sul (RS), Brazil (Pedroso *et al.* 2007; Rissi *et al.* 2007; Grecco *et al.* 2008). Among the most common species are *S. brasiliensis*, *S. selloi*, *S. oxyphyllus*, and *S. heterotrichius* (Karam *et al.* 2004). In addition to these native species, the introduced species *S. madagascariensis*, a native of Madagascar and South Africa (Gardner *et al.* 2006), has been spreading in RS (Matzenbacher 1998). The current study measured the pyrrolizidine alkaloid (PA) concentrations of *S. madagascariensis* plant material in different plant phenological growth stages throughout the year. In addition, observations were recorded concerning the phenological variation in plants during the year.

## Material and Methods

### Plant material

The aerial parts including leaves, flowers, and stems were collected in the area of Desidério Finamor Veterinary Research Institute, municipality of Eldorado do Sul, State of Rio Grande do Sul, Brazil, in July and October 2007 and January and May 2008. A herbarium voucher specimen was deposited in the herbarium of the Universidade Federal do Rio Grande do Sul under number ICN 150755 and identified by Nelson Ivo Matzenbacher. Plant materials were dried in an oven at 45°C and then ground in a mill.

©CAB International 2011. *Poisoning by Plants, Mycotoxins, and Related Toxins* (eds F. Riet-Correa, J. Pfister, A.L. Schild, and T.L. Wierenga)

### Extraction

Each powdered sample weighing 100 mg was placed in a 15 ml screw cap test tube and 4 ml of 1*M* HCl and 4 ml chloroform was added and the samples extracted for 1 h with agitation (mechanical rotation). After extraction, the samples were centrifuged and the upper aqueous acid layer removed to a second test tube. The remaining sample was re-extracted with 2 ml 1*M* HCl for 5 min followed by centrifugation and the acid extract was removed and added to the first acid extract. The combined acid extract was reduced with zinc dust for 30 min, centrifuged, and decanted into a third test tube to which concentrated ammonium hydroxide (28%) was added drop wise with a Pasteur pipette until pH 9-10. This sample was extracted twice with chloroform using 4 ml and 2 ml, respectively, for 5 min with agitation, centrifuged, and the chloroform layer removed. The combined chloroform extract was filtered through anhydrous sodium sulfate into a clean 8 ml vial and concentrated under a flow of nitrogen at 60°C to dryness to give the crude alkaloid extract. The samples were stored until analysis.

### LC/MS analysis

The alkaloid extract of each sample was prepared for analysis by the addition of 1.0 ml of 50% MeOH containing atropine (50 μg) as internal reference standard. It was analyzed by liquid chromatography/mass spectrometry (LC-MS) using an Aquasil $C_{18}$ (Thermo Fisher, 3 μg, 100 × 2.1 mm) column, a mobile phase of 0.1% formic acid and acetonitrile (ACN) at a flow rate of 0.200 ml/min. The programmed gradient was: 5% ACN (0-5 min); 5-70% ACN linear gradient (5-15 min); 70-5% ACN (15-16 min); 5% ACN (16-25 min), a modification of the method from Colegate *et al.* (2005). Detection was by electrospray ionization (ESI) using the LCQ-Advantage Max MS. Individual pyrrolizidine alkaloids or alkaloid groups were identified based on the resulting ions $[M+H]^+$ and correlation to previously identified alkaloids by GC/MS (Gardner *et al.* 2006).

### Vegetative and reproductive phenological aspects

During the collection periods observations were recorded about phenological stages: sprouts, young and adult leaves (vegetative phenophases), flower buds, flowers, unripe and ripe fruits, seed dispersal (reproductive phenophases), and death of leaves.

## Results and Discussion

The aerial plant parts containing stems, leaves, and flowers of *S. madagascariensis*, collected in the south of Brazil during July and October (2007), January and May (2008) corresponding respectively to winter, spring, summer, and autumn, were extracted and the concentrations of pyrrolizidine alkaloids were measured by LC-MS.

### Pyrrolizidine alkaloids of *Senecio madagascariensis*

In previous work, a total of 12 different PA were detected in *S. madagascariensis* from Australia and Hawaii after alkaloid extraction and analysis by GC-MS (Gardner *et al.* 2006). Samples of *S. madagascariensis* collected in Rio Grande do Sul were found to contain the same chromatographic profile (GC/MS and LC/MS) as those samples from Australia and Hawaii. The alkaloids are macrocyclic diesters of retronecine (**1**-**6**) and otonecine (**7**-**12**) bases (Figure 1). Upon prior analysis it was determined that the alkaloids based on retronecine in the plant material were present almost entirely as *N*-oxides. Before analysis, the *N*-oxide form of the alkaloids was reduced to the free base by addition of Zn dust to the aqueous acid extract. The identification of each alkaloid was made by simple correlation of the mass of the protonated molecule ($MH^+$) and the previously reported molecular weight of known *S. madagascariensis* alkaloids (Gardner *et al.* 2006).

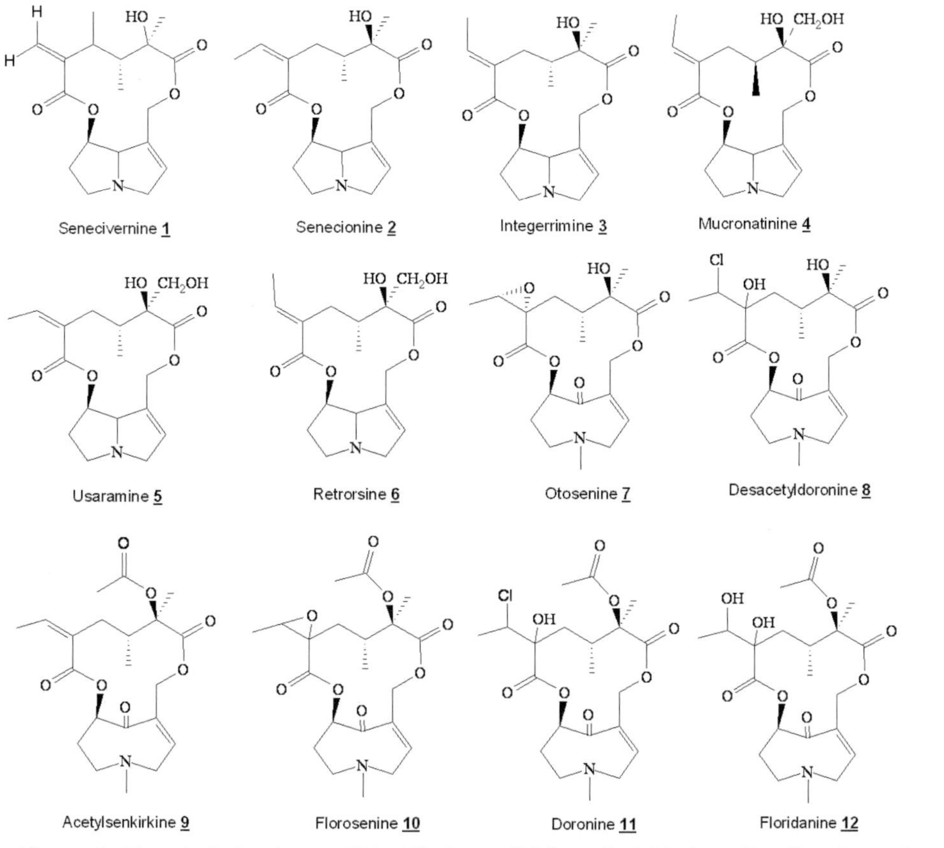


**Figure 1.** Chemical structures of identified pyrrolizidine alkaloids from Brazilian *Senecio madagascariensis*.

All previously identified alkaloids could be accounted for in the LC-MS chromatographic profiles (Figure 2). The alkaloid senkirkine ($MH^+$ 366) was only detected in trace concentrations and was thus eliminated from quantitative measurements. In addition to those previously reported known alkaloids an unknown alkaloid ($MH^+$ 442) was detected from analysis of selected ion chromatograms. Based on the retention time and molecular weight it is proposed that this alkaloid is floridanine which is simply the dihydroxy derivative (hydrolysis of the epoxide ring) of florosenine. The corresponding derivative of otosenine, known as onetine ($MH^+$ 400), is similarly present but only at trace concentrations. Under the chromatographic conditions isomeric compounds were not resolved. For example senecivernine, senecionine, and integerrimine, all of molecular weight 335, were found to elute in the same peak (Figure 2).

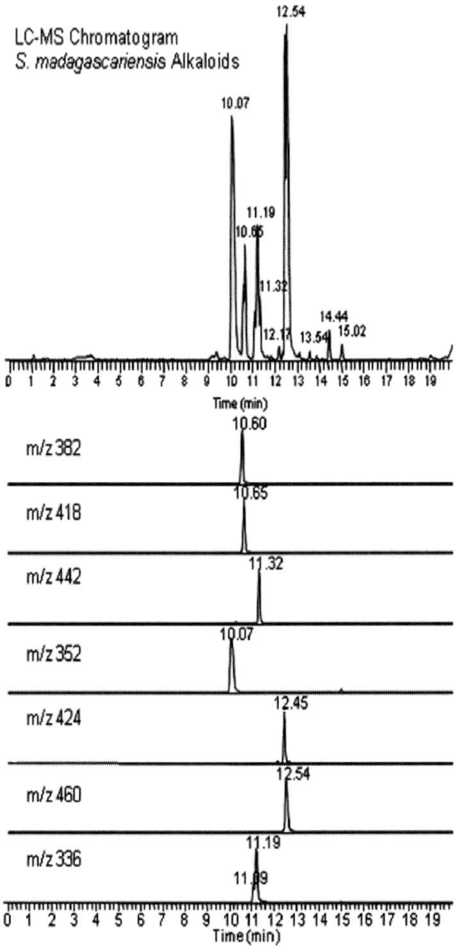


**Figure 2.** LC-MS chromatogram and selected reconstructed ion chromatograms for the different pyrrolizidine alkaloids $[M+H]^+$ found in *Senecio madagascariensis*. Alkaloids included the following: *m/z* 336 (senecivernine **1**, senecionine **2**, integerrimine **3**); 352 (mucronatinine **4**, usaramine **5**, retrorsine **6**); 382 (otosenine **7**); 418 (desacetyldoronine **8**); 424 (acetylsenkirkine **9**, florosenine **10**); 442 (floridanine **12**); 460 (doronine **11**).

## Seasonal Variation in Pyrrolizidine Alkaloid Concentration

The flowers of *S. madagascariensis* collected in Rio Grande do Sul contained the highest total PA concentration in all seasons (0.18-0.35%, dry matter basis) but more so in the spring (0.35%). The combined aerial plant parts (stems, flowers, and leaves) also had the highest concentration of total PA in the spring at 0.17%. In contrast, the lowest PA concentration (0.017%) was measured during the summer collection (Table 1 and Figure 3).

**Table 1**. Concentration (µg/g) of PA in *S. madagascariensis* throughout a year in RS, Brazil.

| Plant Part | Season | Alkaloid $[M+H]^+$ (µg/g) | | | | | | | |
|---|---|---|---|---|---|---|---|---|---|
| | | 336 | 352 | 382 | 418 | 424 | 442 | 460 | Total |
| Aerial | Winter | 211 | 138 | 22 | 52 | 74 | 30 | 238 | 764 |
| | Spring | 244 | 366 | 34 | 38 | 188 | 39 | 304 | 1782 |
| | Summer | 118 | 403 | 100 | 156 | 297 | 97 | 700 | 735 |
| | Autumn | 221 | 406 | 118 | 107 | 287 | 56 | 412 | 966 |
| Leaves | Winter | 102 | 81 | 20 | 24 | 178 | 48 | 283 | 735 |
| | Spring | 175 | 187 | 24 | 35 | 189 | 43 | 316 | 968 |
| | Summer | 44 | 208 | 55 | 113 | 212 | 85 | 632 | 1348 |
| | Autumn | 149 | 283 | 94 | 86 | 380 | 66 | 613 | 1671 |
| Stems | Winter | 132 | 112 | 49 | 51 | 235 | 48 | 340 | 966 |
| | Spring | 172 | 220 | 126 | 127 | 184 | 35 | 191 | 1055 |
| | Summer | 68 | 208 | 89 | 89 | 192 | 42 | 251 | 940 |
| | Autumn | 300 | 341 | 153 | 161 | 397 | 74 | 665 | 2091 |
| Flowers | Winter | 653 | 825 | 19 | 18 | 100 | 27 | 140 | 1782 |
| | Spring | 1324 | 1622 | 23 | 14 | 204 | 42 | 261 | 3490 |
| | Summer | 547 | 1334 | 37 | 43 | 118 | 65 | 332 | 2477 |
| | Autumn | 634 | 1370 | 47 | 10 | 237 | 41 | 167 | 2506 |

Alkaloids $[M+H]^+$ included the following: 336 (senecivernine **1**, senecionine **2**, integerrimine **3**); 352 (mucronatinine **4**, usaramine **5**, retrorsine **6**); 382 (otosenine **7**); 418 (desacetyldoronine **8**); 424 (acetylsenkirkine **9**, florosenine **10**); 442 (floridanine **12**); 460 (doronine **11**).

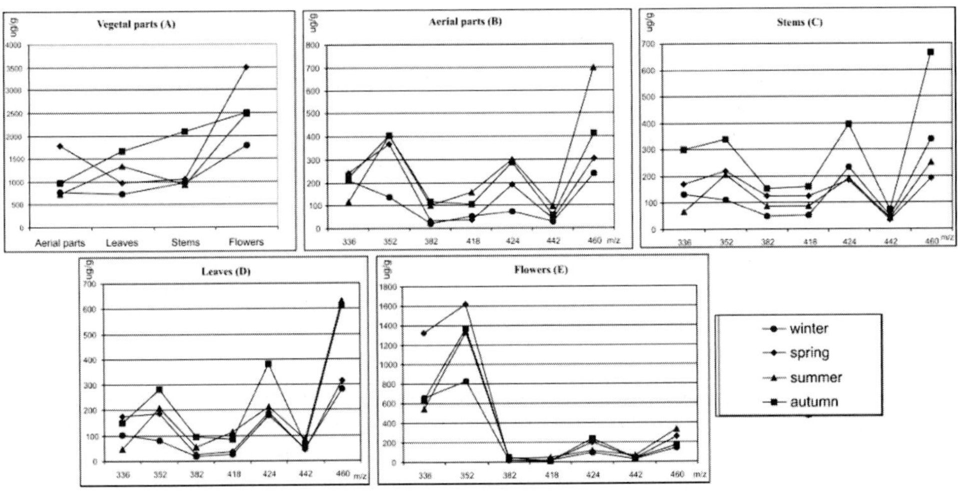


**Figure 3.** Concentration of total PA in the vegetal parts (A), in the aerial parts (B), stems (C), leaves (D), and flowers (E) from *S. madagascariensis* during the seasons.

Among the PA, the macrocyclic diester alkaloids identified in *S. madagascariensis* are the most toxic types (Mattocks 1986) such as senecionine (**2**), integerrimine (**3**), and retrorsine (**6**) with concentrations varying from 0.01% to 0.04% in aerial parts during the year. Flowers had the highest concentration of retronecine-based alkaloids (0.05% to 0.16%), followed by stems (0.006% to 0.03%) and then leaves (0.004% to 0.03%).

Of the macrocyclic diesters of otonecine (**7**-**12**) bases, otosenine (**7**) and desacetyldoronine (**8**) were found at concentrations lower than those of the retronecine bases. Concentrations increased in autumn relative to the other times of the year. Doronine (**11**) was detected at the highest concentrations in aerial parts of *S. madagascariensis* mainly during the summer.

### Observed phenological stages

During each season of the year, there were some plants in all phenological stages from vegetative to reproductive as well as dead leaves. The most active growth occurred during the spring, similar to other *Senecio* species that we have observed (Karam *et al.* 2002). Plant vigor, determined by the Braun-Blanquet scale (Mueller-Dombois and Ellenberg 1974), was reduced only during drought periods. Although *Senecio* leaves were continually senescing, senescent leaves were not predominant in the vegetation. Most plants continued to maintain growth and vigor year-round, which greatly favors their persistence and spread in the vegetation community.

## Conclusions

Twelve PA were detected from the aerial parts (stems, leaves, and flowers) of *S. madagascariensis* collected in southern Brazil. The alkaloid profile was similar to that reported from plants in Australia and Hawaii. The alkaloids were macrocyclic diesters of retronecine and otonecine bases and would be presumed to cause poisoning in cattle. The flowers contained the highest total PA in spring. Since PA were detected in *S. madagascariensis*, we conclude that this plant should be included with those *Senecio* spp. able to produce seneciosis in livestock, especially in cattle. Therefore, control measures should be implemented by the local livestock industry to prevent or diminish loss of livestock by these plants.

## References

Colegate SM, Edgar JA, Knill AM, and Lee ST (2005). Solid-phase extraction and HPLC-MS Profiling of pyrrolizidine alkaloids and their N-oxides: a case study of *Echium plantagineum*. *Phytochemical Analysis* 16:108-119.

Gardner DR, Thorne MS, Molyneux RJ, Pfister JA, and Seawright AA (2006). Pyrrolizidine alkaloids in *Senecio madagascariensis* from Australia and Hawaii and assessment of possible livestock poisoning. *Biochemical Systematics and Ecology* 34:736-744.

Grecco FB, Fiss L, Soares MP, Marcolongo-Pereira C, Assis Brasil N, Quevedo P, and Schild AL (2008). Influência dos fatores climáticos na prevalência da intoxicação por *Senecio* spp. em bovinos na região Sul do Rio Grande do Sul no período de 2000-2007. *Boletim do Laboratório Regional de Diagnóstico*, 28:27-30. Editora Universitária, Universidade Federal de Pelotas, Pelotas.

Karam FSC, Méndez MC, Jarenkow JA, and Riet-Correa F (2002). Fenologia de quatro espécies tóxicas de *Senecio* (Asteraceae) na região Sul do Rio Grande do Sul. *Pesquisa Veterinária Brasileira* 22(1):33-39.

Karam FSC, Soares MP, Haraguchi M, Riet-Correa F, Méndez MC, and Jarenkow JA (2004). Aspectos epidemiológicos da seneciose na região sul do Rio Grande do Sul. *Pesquisa Veterinária Brasileira* 24(4):191-198.

Mattocks AR (1986). Toxicology of Pyrrolizidine Alkaloids in Animals. In *Chemistry and Toxicology of Pyrrolizidine Alkaloids*, pp.191-219. Academic Press Inc., London.

Matzenbacher NI (1998). O complexo "Senecionoide" (Asteraceae-Senecioneae) no Rio Grande do Sul – Brasil, 274 pp. Tese de Doutorado em Botânica, Instituto de Biociências, UFRGS, Porto Alegre, RS, Brasil.

Mueller-Dombois D and Ellenberg H (1974). *Aims and Methods of Vegetation Ecology*, 547 pp. John Wiley, New York.

Pedroso PMO, Pescador CA, Oliveira EC, Sonne L, Bandarra PM, Raymundo DL, and Driemeier D (2007). Intoxicações naturais por plantas em ruminantes diagnosticadas no Setor de Patologia Veterinária. *Acta Scientiae Veterinariae* 35:213-218.

Rissi DR, Rech RR, Pierezan F, Gabriel AL, Trost ME, Brum JS, Kommers GC, and Barros CSL (2007). Intoxicações por plantas e micotoxinas associadas a plantas em bovinos no Rio Grande do Sul: 461 casos. *Pesquisa Veterinária Brasileira* 27:261-268.

# Chapter 27

# Buffalo Calves Intoxicated with *Ageratum houstonianum* Mill.

P.B. Pal[1], D.K. Singh[1], and B.L. Stegelmeier[2]

[1]*Institute of Agriculture and Animal Science, Tribhuvan University, Nepal;* [2]*USDA-ARS Poisonous Plant Research Laboratory, Logan, Utah 84341, USA*

## Introduction

*Ageratum houstonianum* Mill is a drought tolerant flowering annual of the Asteraceae family. Originally a Central American plant, it now is widely distributed in many of the tropical and subtropical regions worldwide. It commonly invades pastures, irrigation canals, and marginal lands and it can be harvested and included in prepared feeds. Animals generally do not consume *A. houstonianum* unless other forages are lacking or if it is included in prepared forages. Livestock poisoning has been reported in Cuba, Nepal, and Mexico (Alfonso *et al.* 1989; Dhakal 1989; Noa *et al.* 2004). The *Ageratum* genus has been sporadically associated with both human and livestock poisoning. Most recently, studies suggest that poisoning in Ethiopia is likely attributed to *Ageratum* species that contaminate feeds and food (The Oasis Foundation 2009).

*A. houstonianum* poisoned animals develop liver disease with secondary or hepatogeneous photosensitization or more acute hemorrhagic disease. The liver disease is poorly described with icterus, hyperbilirubenia, and light-induced photosensitivity. The hemorrhagic disease is also poorly understood with prolonged coagulation times and mucosal hemorrhages with little understanding of the cause. It has been suggested *A. houstonianum* coumarins may alter vitamin K-dependent synthesis of coagulation proteins; however, this has not been confirmed and the clinical hemorrhage could equally be produced by inadequate hepatic protein synthesis as is seen in extensive hepatic disease and failure (Alfonso *et al.* 1989; Noa *et al.* 2004). Other liver diseases including those caused by pyrrolizidine alkaloid (PA)-containing plants have also been documented to produce hemorrhagic disease (Stegelmeier *et al.* 1994; Sanchez-Campos *et al.* 2004).

Poisoning is sporadic and although it has been partially reproduced, the causative toxin has not been definitively identified (Alfonso *et al.* 1989). Several different toxins including flavonoids, phytosterols, and coumarins were originally suggested as the cause of toxicity. Later, long chain hydrocarbons were also isolated from *A. houstonianum* and they were shown to produce hemorrhagic disease in rats (Alfonso *et al.* 1989; Garcia *et al.* 1999; Noa *et al.* 2004). However, Wiedenfeld and Andrade-Cetto (2001) isolated four PA and as these have been shown to be toxic, speculate that they contribute to plant toxicity. All of

©CAB International 2011. *Poisoning by Plants, Mycotoxins, and Related Toxins* (eds F. Riet-Correa, J. Pfister, A.L. Schild, and T.L. Wierenga)

these findings suggest the presence of multiple toxins and more information is needed to determine their involvement or combined involvement in subsequent poisoning.

The purpose of this study is to document the toxicity of *A. houstonianum* from Nepal, better characterize the chemistry of toxic *A. houstonianum*, and describe the clinical and pathologic changes of poisoning in livestock.

## Materials and Methods

Multiple samples of full blooming *A. houstonianum* were collected near Rampur, Chitwan, Nepal. The samples were dried, finely ground, and shipped for analysis. Composite samples were made, extracted, and analyzed for PA following previously described methods (Molyneux *et al.* 1979). Later in a pilot study to verify plant toxicity, six preconditioned, crossbred, 12-month-old buffalo calves were fed fresh *A. houstonianum* Mill *ad libitum* daily until they became clinically poisoned. The calves were examined daily and blood and serum were collected for whole blood counts and serum analysis of alanine aminotransferase (ALT), aspartate aminotransferase (AST), total protein (TP), albumin (ALB), bilirubin (BIL), and glucose (GLU) using routine laboratory techniques. When animals developed disease they were euthanized and necropsied. All lesions and routine tissues were collected, fixed in formalin, and processed for microscopic examination. The significance of hematologic and serum biochemical parameters were determined with the Student's 't' test.

## Results and Discussion

Chemical analysis of the composite *A. houstonianum* samples contained lycopsamine and several related isomers (Figure 1). The concentration was estimated to be 0.56 mg/kg dry weight.

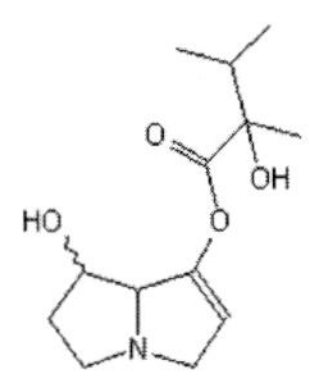


**Figure 1.** Structure of lycopsamine-type PA. Several of these alkaloids were isolated from *A. houstonianum* from Rampur Chitwan. Additional work is needed to definitively identify the structures of these alkaloids and to better characterize their involvement in poisoning.

After ingesting fresh plant for several days several of the calves developed anorexia, hypothermia, erratic respiration, low pulse rate, vomiting, and icterus (acute cases). These animals had swollen livers with histologic changes of severe hepatocellular necrosis with collapse of hepatic cords and hemorrhage. The remaining calves ingested plant for several weeks longer before they developed a chronic disease which included facial edema, alopecia, secondary photosensitization, and wasting. The liver from these animals had moderate hepatocellular necrosis with periportal fibrosis and mild biliary proliferation.

Many clinical signs observed in this study, e.g. anorexia, vomiting, icterus, and photosensitivity, are typical of PA-associated liver disease (Stegelmeier *et al.* 1999). These clinically affected calves developed hematologic changes of anemia characterized by reduced mean erythrocyte counts, hematocrits and hemoglobin concentrations when compared to pretreatment values ($P < 0.05$). As hemorrhagic disease was not a component of this clinical disease, this anemia seems to be a result of the hemorrhage associated with the hepatic necrosis. Of course there may be other yet-unidentified toxins that might impair hematopoiesis. For example, plant saponins have been shown to produce similar anemia (Adedapo *et al.* 2004). Increased ALT and AST activities, hypoalbuminemia, decreased glucose concentrations, and hyperbilirubinemia ($P < 0.05$) have all been associated with PA-induced hepatic dysfunction and necrosis (Stegelmeier *et al.* 1996, 1999) (Table 1).

**Table 1.** Hematology and serum biochemistry results of buffalo calves fed *A. houstonianum.*

| | Initial Values | Clinical Values |
|---|---|---|
| Erythrocyte Counts (millions) | 4.7±0.3 | 3.6±0.3* |
| Hematocrit (%) | 27.8±3.3 | 17±2* |
| Hemoglobin (g/dl) | 8.7±0.8 | 5.7±0.5* |
| AST (IU) | 42±2 | 87±10* |
| ALT (IU) | 19±1 | 38±5* |
| Total Protein (mg/dl) | 7.6±0.4 | 9.2±0.6 |
| Albumin (mg/dl) | 3.4±0.1 | 2.4±0.5* |
| Bilirubin (mg/dl) | 0.4±0.3 | 5.4±0.8* |
| Glucose (mg/dl) | 51±6 | 33±9* |

* Values that were significantly different from the initial values ($P < 0.05$).

## Conclusions

We have shown that fresh *A. houstonianum* fed to buffalo calves produces hepatic disease similar to that seen in spontaneous poisonings. Chemical analysis suggests that this toxicity may be largely due to PA. Additional work is needed to better characterize *A. houstonianum* toxins, define the toxic dose, characterize *A. houstonianum*-induced lesions, and to determine when *A. houstonianum* is likely to cause livestock poisoning.

## Acknowledgements

We thank Dr Peetamber Kushwaha and Dr Subir Singh for their constructive suggestions. We also thank Dr Dale Gardner for analysis of the plant material. This research was conducted by the approval and supervision of the Directorate of Research and Publication, Institute of Agriculture and Animal Sciences, Tribhuvan University, Nepal.

## References

Adedapo AA, Matthew O, and Olufunso O (2004). Toxic effects of some plants in the genus *Euphorbia* on haematological and biochemical parameters of rats. *Veternarski Archives* 74:53-62.

Alfonso HA, Rivera M, Aparicio JM, Ancisar J, Marrero E, and Cabrera JM (1989). Natural and experimental poisoning of cattle with *Ageratum houstonianum* (blue celestine). *Revista Cubana de Ciencias Veterinarias* 20:113-119.

Dhakal IP (1989). Major problems of livestock in Chitwan district. *Journal of Agriculture and Animal Sciences* 85:16-19.

Garcia T, Aparicio J, Alfonso HA, Perez C, and Sanchez L (1999). Constituyentes hidrocarbonados de *Ageratum houstonianum* MILL. *Revista de Salud Animal* 21:97-100.

Molyneux RJ, Johnson AE, Roitman JN, and Benson ME (1979). Chemistry of toxic range plants. Determination of pyrrolizidine alkaloid content and composition in *Senecio* species by nuclear magnetic resonance spectroscopy. *Journal of Agricultural and Food Chemistry* 27:494-499.

Noa M, Sanchez LM, and Durand R (2004). *Ageratum houstonianum* toxicosis in zebu cattle. *Veterinary and Human Toxicology* 46:193-194.

Sanchez-Campos S, Alvarez M, Culebras JM, Gonzalez-Gallego J, and Tunon MJ (2004). Pathogenic molecular mechanisms in an animal model of fulminant hepatic failure: rabbit hemorrhagic viral disease. *Journal Laboratory Clinical Medicine* 144:215-222.

Stegelmeier BL, Gardner DR, Molyneux RJ, and James LF (1994). *Cynoglossum officinale* (houndstongue) poisoning in horses. *Veterinary Pathology* 31:621.

Stegelmeier BL, Gardner DR, James LF, and Molyneux RJ (1996). Pyrrole detection and the pathologic progression of *Cynoglossum officinale* (houndstongue) poisoning in horses. *Journal of Veterinary Diagnostic Investigation* 8:81-90.

Stegelmeier BL, Edgar JA, Colegate SM, Gardner DR, Schoch TK, Coulombe RA, and Molyneux RJ (1999). Pyrrolizidine alkaloid plants, metabolism and toxicity. *Journal of Natural Toxins* 8:95-116.

The Oasis Foundation (2009). *The Oasis Foundation Grace Village Ethiopia*. Available at: http://www.oasisfoundationethiopia.org/kelakil_update_jul_12_2008.htm.

Wiedenfeld H and Andrade-Cetto A (2001). Pyrrolizidine alkaloids from *Ageratum houstonianum* Mill. *Phytochemistry* 57:1269-1271.

# Chapter 28

# Evaluation of Immunotoxic Properties of *Senecio brasiliensis*: Study of Toxicity in Rats

F. Elias[1], I.M. Hueza[1], M. Haraguchi[2], and S.L. Górniak[1]

[1]*Research Center of Veterinary Toxicology – Department of Pathology, School of Veterinary Medicine and Animal Science, University of São Paulo, SP 13635-900, Brazil;* [2]*Biological Institute of São Paulo, SP 04014-002, Brazil*

## Introduction

More than 1200 species of plants from genus *Senecio* are cataloged worldwide, and *S. brasiliensis* is the most important species in Brazil. Plants from this genus have frequently been associated with poisoning in livestock such as cattle, horses (Tokarnia *et al.* 2000), sheep (Ilha *et al.* 2001), and Murrah buffalo (Corrêa *et al.* 2008). About 50% of all deaths caused by toxic plants in cattle in southern Brazil and Uruguay are due to the consumption of *Senecio* spp., therefore this plant causes large economic losses for farms and ranches in the region (Karam *et al.* 2004).

Several cases of poisoning by *Senecio* spp. in humans are also associated with the use of their leaves to make tea (Cheeke 1998; Prakash *et al.* 1999). The toxic active ingredients in this plant have been found to contaminate human food sources such as wheat, honey, herbal medicines, and herbal teas, and this may potentially cause widespread human health problems. It is also possible that the toxins from this plant can contaminate other food products such as milk (Goeger *et al.* 1982).

Acute poisoning with this plant causes massive hepatotoxicity with hemorrhagic necrosis. Chronic poisoning takes place mainly in liver, leading to hepatocyte enlargement (megalocytosis), veno-occlusion in liver and lungs, fatty degeneration, nuclei enlargement with increasing nuclear chromatin, loss of metabolic function, inhibition of mitosis, fatty degeneration, proliferation of biliary tract epithelium, liver cirrhosis, nodular hyperplasia, and adenomas or carcinomas (Cheeke 1998; Fu *et al.* 2002; Barros *et al.* 2007; Corrêa *et al.* 2008). Also it causes individual hepatocyte necrosis, apoptosis, and nuclear inclusions (Torres and Coelho 2008).

*S. brasiliensis* contains a mixture of pyrrolizidine alkaloids (PA): senecionine, integerrimine, retrorsine, usaramine, and seneciphylline (Toma *et al.* 2004; Silva *et al.* 2006). The mechanism of hepatotoxicity induced by these alkaloids has been extensively investigated, and it is well established that PA must be activated by microsomal liver enzymes into pyrrolic compounds to be toxic (Cheeke 1998, Fu *et al.* 2002) as follows: after absorption and distribution, PA are first oxidized (dehydrogenation) by monooxygenases of the cytochrome P-450 and the pyrrole compounds thus generated are

©CAB International 2011. *Poisoning by Plants, Mycotoxins, and Related Toxins* (eds F. Riet-Correa, J. Pfister, A.L. Schild, and T.L. Wierenga)

reactive and undergo spontaneous conversion, leading to eletrophilic compounds that react with cellular nucleophiles.

Thus, the pyrrole derivatives can react with nucleic acids (DNA) and vital macromolecules as proteins, forming adducts which can lead to mutagenicity (Fu *et al.* 2002), teratogenic effects (Peterson and Jago 1980), genotoxicity (Petry *et al.* 1984), and carcinogenicity (Mattocks and Cabral 1982). It is clear that pyrrolic compounds can damage DNA resulting in cellular malfunction, thus we raise the possibility that systems that require a high rate of proliferative activity may be compromised by PA toxicosis.

The immune system is dependent on cell-cell signaling between membrane proteins and receptors to induce lymphocyte activation, proliferation, and finally, establishment of a strong functional immune response. However, it is well known that the immune system is sensitive to disruption by different noxious stimuli such as nutritional deficiencies (Cunningham-Rundles *et al.* 2005), or hormonal status (Reichlin 1993). Our first objective was to determine the dose of *S. brasiliensis* extract that could cause toxicity in male rats. Secondly, we wished to establish if there were lower doses that did not induce substantial toxicity that may, however, induce immunotoxic effects from *S. brasiliensis*, as this knowledge would be useful for subsequent experiments.

## Materials and Methods

Five hundred grams of dry leaves of *S. brasiliensis* were collected in Pelotas, state of Rio Grande do Sul, Brazil, and identified as *S. brasiliensis* (Spreng.) Less. var. *brasiliensis*, voucher number 24592. Finely ground *S. brasiliensis* leaves, defatted with hexane, were exhaustively extracted using 92% ethanol. The extract was suspended in ethanol:water (1:1) and applied to a column containing resin Amberlite IR-120B. It was subsequently eluted with 0.5*M* ammonium hydroxide solution and concentrated under vacuum to obtain the alkaloidal crude residue (ACR) (Mattocks 1968).

Forty male Wistar rats (10 weeks-of-age) were randomly divided into four equal groups and treated by gavage with 0.0, 0.3, 1.0, and 3.0 mg/kg of ACR for 28 days. Food intake and body weight gain were measured every other day. On the 29th day of the experiment rats were euthanized in order to collect lymphoid organs (thymus and spleen) to evaluate their relative organ weight and cellularity. Blood samples were also collected for biochemical analysis and to evaluate blood parameters. Tissue samples were obtained for histopathology (thymus, spleen, liver, heart, lungs, and kidney).

Data were analyzed by one-way analysis of variance (ANOVA), with post-hoc analysis using Dunnett's test. Differences between the control and the experimental groups were considered to be statistically significant when $P < 0.05$.

## Results

No statistical differences were observed in body weight gain and food intake among treatment groups and controls. In addition, the analysis of lymphoid organs and blood parameters from rats treated with *S. brasiliensis* did not show any alterations when compared with controls. Moreover, there were no morphological alterations in animals treated with the alkaloid fraction of *S. brasiliensis*.

## Conclusions

Typically one of the first parameters affected when toxic compounds are dosed in animal models is reduced food consumption and related body weight changes (Tamborini *et al.* 1990; Hoffman *et al.* 2002). These results indicate that the doses employed did not have deleterious effects on these or any other parameters evaluated in this study.

It is well known that damage to the immune system can generally be produced with lower doses of xenobiotics than those which induce overt toxic effects in other systems (Sjoblad 1988). However, in this study we found no toxic effects from the alkaloid extract, and we also did not find any indication that, at these doses, *S. brasiliensis* compromised the immune system in any way. For this reason, additional experiments are being conducted in our laboratory to determine the toxic dose of *S. brasiliensis* extract to rats. Once a toxic threshold is determined, it will be possible to use lower doses to determine if *S. brasiliensis* will have immunomodulatory effects on animals exposed to this plant.

## Acknowledgements

This study was supported by grants from Fundação de Amparo à Pesquisa do Estado de São Paulo – FAPESP, Brazil (Proc. No. 06/60397-8) and CAPES.

## References

Barros CSL, Castilhos LML, Rissi DR, Kommers GD, and Rech RR (2007). Biópsia hepática no diagnóstico da intoxicação por *Senecio brasiliensis* (Asteraceae) em bovinos. *Pesquisa Veterinária Brasileira* 27(1):53-60.

Cheeke PR (1998). *Natural Toxicants in Feeds, Forages and Poisonous Plants*. 2nd edn, 479 pp. Interstate Publishers, Danville.

Corrêa AMR, Bezerra PSJ, Pavarini SP, Santos AS, Sonne L, Zlotowski P, Gomes G, and Driemeier D (2008). *Senecio brasiliensis* (Asteraceae) poisoning in Murrah buffaloes in Rio Grande do Sul. *Pesquisa Veterinária Brasileira* 28(3):187-189.

Cunningham-Rundles S, McNeeley DF, and Moon A (2005). Mechanisms of nutrient modulation of the immune response. *Journal of Allergy and Clinical Immunology* 115(6):1119-1128.

Fu PP, Qingsu X, Ge L, and Ming W (2002). Genotoxic pyrrolizidine alkaloids – Mecanisms leading to DNA adduct formation and tumorigenicity. *International Journal of Molecular Sciences* 3:948 – 964.

Goeger DE, Cheeke PR, Schmitz JA, and Buhler DR (1982). Effect of feeding milk from goats fed tansy ragwort (*Senecio jacobaea*) to rats and calves. *American Journal of Veterinary Research* 43:1631-1633.

Hoffman WP, Ness DK, and Van Lier RBL (2002). Analysis of rodent growth data in toxicology studies. *Toxicological Sciences* 66:313-319.

Ilha MRS, Loretti AP, Barros SS, and Barros CSL (2001). Intoxicação espontânea por *Senecio brasiliensis* (Asteraceae) em ovinos no Rio Grande do Sul. *Pesquisa Veterinária Brasisleira* 21(3):123-138.

Karam FSC, Soares MP, Haraguchi M, Riet-Correa F, Méndez MC, and Jarenkow JA (2004). Aspectos epidemiológicos da seneciose na região sul do Rio Grande do Sul. *Pesquisa Veterinária Brasileira* 24(4):191-198.

Mattocks AR (1968). Toxicity of pyrrolizidine alkaloids. *Nature* 217:723-728.

Mattocks AR and Cabral JRO (1982). Carcinogenicity of some pyrrolic pyrrolizidine alkaloid metabolites and analogues. *Cancer Letters* 17:61-66.

Peterson JE and Jago MV (1980). Comparison of the toxic effects of dehydroheliotridine and heliotrine in pregnant rats and their embryos. *Journal of Pathology* 131:339-355.

Petry TW, Bowden GT, Huxtable RJ, and Sipes IG (1984). Characterization of hepatic DNA damage induced in rats by the pyrrolizidine alkaloid monocrotalina. *Cancer Research* 44:1505-1509.

Prakash AS, Pereira TN, Reilly PEB, and Seawright AA (1999). Pyrrolizidine alkaloids in human diet. *Mutation Research* 443:53–67.

Reichlin S (1993). Neuroendocrine-immune interactions. *The New England Journal of Medicine* 329(17):1246-1253.

Silva CM, Bolzan AA, and Heinzmann BM (2006). Alcalóides pirrolizidínicos em espécie do gênero *Senecio*. *Química Nova* 29(5):1047-1053.

Sjoblad RB (1988). Potential future requirements for immunotoxicology testing of pesticides. *Toxicology and Industrial Health* 4(3):391-395.

Tamborini P, Sigg H, and Zbinden G (1990). Acute toxicity testing in the nonlethal dose range: a new approach. *Regulatory Toxicology and Pharmacology* 12:69-87.

Tokarnia CH, Dobereiner J, and Peixoto PV (2000). *Plantas tóxicas do Brasil,* 297 pp. Helianthus, Rio de Janeiro.

Toma W, Trigo JR, De Paula ACB, and Brito ARMS (2004). Preventive activity of pyrrolizidine alkaloids from *Senecio brasiliensis* (Asteraceae) on gastric and duodenal induced ulcer on mice and rats. *Journal of Ethnopharmacology* 95:345-351.

Torres MBAM and Coelho KIR (2008). Experimental poisoning by *Senecio brasiliensis* in calves: quantitative and semi-quantitative study on changes in the hepatic extracellular matrix and sinusoidal cells. *Pesquisa Veterinária Brasileira* 28(1):43-50.

# Chapter 29

# Hepatic Biopsy as a Diagnostic Tool for Detecting *Senecio* spp. Poisoning in Live Cattle

K.L. Takeuti[1], P.M. Bandarra[1], J.S. Brum[2], K.S. Carvalho[3], A.G.C. Dalto[1], D.L. Raymundo[1], C.E.F. Cruz[1], and D. Driemeier[1]

[1]*Setor de Patologia Veterinária, Universidade Federal do Rio Grande do Sul, Avenida Bento Gonçalves, 9090, CEP 91540-000, Porto Alegre, RS, Brazil;* [2]*Departamento de Patologia, Universidade Federal de Santa Maria, CEP 97105-900, Santa Maria, RS, Brazil;* [3]*Universidade Federal de Campina Grande, CEP 58429-900, Campina Grande, PB, Brazil*

## Introduction

The annual mortality of cattle in Rio Grande do Sul (RS) is about 5%, of which 10 to 14% is caused by the consumption of poisonous plants. Seneciosis and anaplasmosis are the two principal causes of death in adult cattle in RS, 7% of which is attributed to the former (Riet-Correa and Medeiros 2001). *Senecio* spp. are generally unpalatable and the intoxication occurs mainly from May to August, when there is a shortage of alternative forage (Riet-Correa and Méndez 2007). *Senecio* poisoning occurs when animals ingest toxic doses of pyrrolizidine alkaloids (PA) that induce chronic, progressive, and irreversible hepatic disease (Riet-Correa and Méndez 2007). This condition is clinically characterized by tenesmus, apathy, progressive emaciation, dried feces, ascites, and neurological signs, which may develop for several months after the ingestion of the plant. Characteristic histological lesions affect the liver and include megalocytosis, biliary ductal hyperplasia, fibrosis, degeneration and hepatocyte necrosis, and nodular regeneration. As chemical detection of PA metabolites is difficult, liver biopsies are excellent means to reveal these characteristic histological changes in clinically affected animals as well as subclinical cases. This communication describes the liver biopsy as a useful diagnostic method for detecting *Senecio* poisoning in live cattle (Riet-Correa and Méndez 2007).

## Materials and Methods

On a farm located in the municipality of Rio Pardo, Rio Grande do Sul, 50 out of 300 cattle died from *Senecio* poisoning. A request was made to perform liver biopsies in the remaining herd. The objective of the procedure was to determine the magnitude of the problem in order to minimize future losses. Blood samples were collected simultaneously from animals to compare biopsy results with GGT findings. Biopsies were done by

©CAB International 2011. *Poisoning by Plants, Mycotoxins, and Related Toxins* (eds F. Riet-Correa, J. Pfister, A.L. Schild, and T.L. Wierenga)

introducing percutaneously a Menghini needle in the transthoracic intersection of the 11th right intercostal space, nearly 20 cm under the dorsum line, in an intersection of an imaginary line between the iliac tuberosity and the scapula, and another perpendicular line at the 11th intercostal space. This point corresponds to the topographic position of the right hepatic lobule (Barros *et al.* 2007). Biopsies were performed in 227 Aberdeen Angus cows. The biopsies were fixed in buffered 10% formalin, routinely processed for histology, and stained with hematoxylin and eosin and Masson's trichrome staining for histologic evaluation and scoring. Hepatic sections were scored as positive or Group 1 when there were significant changes of biliary ductal proliferation and fibrosis. In this group megalocytosis may be present, especially near the portal zones. Liver biopsies that showed only megalocytosis or no histologic change at all were negative (Group 2). Sera from 31 animals of these same animals were analyzed for GGT activity since an increase in activity may indicate hepatic disorder (Kaneko 1989). Nineteen of these 31 samples were taken from animals in Group 1, and 12 samples were taken from animals in Group 2. These GGT activities were analyzed to determine if they were associated with the histologic scores or degree of megalocytosis.

## Results

During the farm visit, the *Senecio* species was identified as *S. brasiliensis*, and the grazed pastures were found to be severely infested with this plant. Animals showed wasting, diarrhea, and enhanced volume of the lower abdomen (ascites). Death affected approximately 17% of the herd, which motivated the owner to request the biopsy examinations. The biopsy results revealed that 55 (24%) out of 227 animals were positive (Group1) and 172 animals (76%), negative (Group 2).

Both groups had increased GGT activity when compared to a normal GGT range of 6.1-17.4 U/l (Kaneko 1989). In Group 1, 11 and 8 samples had mild and moderate megalocytosis, respectively. In Group 2, 6 samples had mild megalocytosis, 4 moderate, and 2 showed no alteration (Table 1). No significant association was observed between degree of microscopic changes and GGT levels (Tukey test, confidence interval 95-98%), whereas mild and moderate lesions showed enzymatic values above the normal levels for the species.

## Discussion and Conclusions

Clinical, epidemiological, and histological findings from liver biopsies confirmed the *Senecio* spp. poisoning. Cows with positive biopsies showed severe and advanced lesions of seneciosis (biliary ductal hyperplasia and fibrosis), which are associated with imminent risk of death. Therefore, this group was sent to slaughter to minimize economic losses to the farmer. Megalocytosis may be the initial lesion seen in animals that ingested *Senecio* spp. (Thorpe and Ford 1968), and in this study, animals with only that specific lesion were regarded as negative biopsies. There was no significant difference between both groups of cows with positive and negative biopsies (both had increased values of GGT), and there was no association between degree of megalocytosis and GGT levels. Since all of the cows with mild and severe changes had increased GGT levels, this enzymatic assay was not adequate to determine the severity of seneciosis in cattle under the conditions of this study.

Not all animals with high GGT levels had a clinically imminent risk of death; therefore they did not need to be slaughtered.

**Table 1.** Evaluation of *S. brasiliensis* poisoning in cattle using gamma-glutamyl transferase (GGT) activity from 31 cows, with 19 animals from Group 1 (categorized as positive biopsies[1]) and 12 animals from Group 2 (categorized as negative biopsies), and associated degree of megalocytosis. Animals were assigned to groups based on lesions in material taken from liver biopsies.

| Group | Cattle | GGT[2] | Degree of megalocytosis | Group | Cattle | GGT[2] | Degree of megalocytosis |
|---|---|---|---|---|---|---|---|
| 1 | 1 | 41 | mild | 2 | 1 | 34 | mild |
| | 2 | 57 | mild | | 2 | 36 | mild |
| | 3 | 63 | mild | | 3 | 37 | mild |
| | 4 | 65 | mild | | 4 | 43 | mild |
| | 5 | 67 | mild | | 5 | 44 | mild |
| | 6 | 68 | mild | | 6 | 52 | mild |
| | 7 | 68 | mild | | 7 | 37 | moderate |
| | 8 | 70 | mild | | 8 | 55 | moderate |
| | 9 | 78 | mild | | 9 | 57 | moderate |
| | 10 | 101 | mild | | 10 | 80 | moderate |
| | 11 | 107 | mild | | 11 | 41 | no change |
| | 12 | 39 | moderate | | 12 | 47 | no change |
| | 13 | 42 | moderate | | | | |
| | 14 | 47 | moderate | | | | |
| | 15 | 50 | moderate | | | | |
| | 16 | 70 | moderate | | | | |
| | 17 | 70 | moderate | | | | |
| | 18 | 84 | moderate | | | | |
| | 19 | 90 | moderate | | | | |
| | Mean: 7.2 U/l | | | | | Mean: 46.9 U/l | |

[1]Hepatic sections were considered positive (Group 1) when displaying biliary ductal proliferation associated with fibrosis, especially around portal spaces. Liver sections that showed only megalocytosis or no change at all were negative (Group 2).
[2]The normal range for GGT in cattle is 6.1-17.4 U/l.

Liver biopsies allowed a reasonably fast, precise, and efficient detection of animals with imminent risk of death due to seneciosis, since lesions attributed to pyrrolizidine alkaloids were characteristic and the degree of severity of lesions is often indicative of the stage of development. Further, all diseased animals, including the subclinical cases, may be detected with this method. The technique is quite simple, easily applied in the field, and associated with low risks to the animal (Braga *et al.* 1985). The early diagnosis of *Senecio* poisoning by liver biopsy may allow owners of *Senecio* spp. affected animals to send them to slaughter before greater death losses occur.

## References

Barros CSL, Castilhos LML, Rissi DR, Kommers GD, and Rech RR (2007). Biópsia hepática no diagnóstico da intoxicação por *Senecio brasiliensis* (Asteraceae) em bovinos. *Pesquisa Veterinária Brasileira* 27(1):53-60.

Braga MM, Castilhos LML, and Santos MN (1985). Biópsia hepática em bovinos: proposta de nova técnica. *Revista Centro de Ciências Rurais* 15(1):79-88.

Kaneko JJ (1989). Liver function. In *Clinical Biochemistry of Domestic Animals* (CE Cornelius, ed.), 4th edn, pp. 385. Academic Press, San Diego/Oxford University Press, Cape Town, South Africa.

Riet-Correa F and Medeiros RMT (2001). Intoxicações por plantas em ruminantes no Brasil e no Uruguai: importância econômica, controle e riscos para a saúde pública. *Pesquisa Veterinária Brasileira* 21(1):38-42.

Riet-Correa F and Méndez MDC (2007). Intoxicações por plantas e micotoxinas. In *Doenças de Ruminantes e Eqüideos* (F Riet-Correa, AL Schild, RAA Lemos, and JRJ Borges, eds), vol. 2, pp. 99-219. Pallotti, Santa Maria.

Thorpe E and Ford EJH (1968). Development of hepatic lesions in calves fed with ragwort (*Senecio jacobaea*). *Journal of Comparative Pathology* 78:195-205).

# Chapter 30

# Poisoning of Cattle by *Senecio* spp. in Uruguay

M. Preliasco[1] and R. Rivero[2]

[1]*División de Laboratorios Veterinarios (DILAVE) 'Miguel C. Rubino', Laboratorio Central, Montevideo, Uruguay;* [2]*DILAVE 'Miguel C. Rubino', Laboratorio Regional Noroeste, Paysandú, Uruguay*

## Introduction

The genus *Senecio* (Compositae family Asteraceae) is composed of more than 1200 species distributed worldwide with the exception of the Pacific Islands, Antarctica, and the Amazon Forest (Podestá *et al.* 1976; Lombardo 1984; Riet-Correa *et al.* 1993). Several species are abundant in South American countries and their geographical distribution is directly related to climate and topography. Mountainous countries such as Chile and Argentina have abundant species of *Senecio* (about 210 and 300, respectively), while countries like Brazil, Uruguay, and Paraguay have fewer species (128, 25, and 7, respectively) (Gallo 1987; Tokarnia *et al.* 2000). *Senecio* plants are known by common names like 'spring field' and 'yellow flower' in Argentina, 'Maria Mole', 'souls blossom', and 'lancet weed' in Brazil, and 'Spring weed' in Uruguay (Podestá *et al.* 1976; Gallo 1987; Riet-Correa *et al.* 1993; Tokarnia *et al.* 2000; Romero *et al.* 2002).

*Senecio* spp. cause significant losses in livestock production, mainly due to the toxic (pyrrolizidine alkaloids) and invasive nature of the plants. Pyrrolizidine alkaloids (PA) are also found in other botanical genera such as *Erectites* and *Eupatorium* (Compositae), *Crotalaria* (Leguminosae), *Echium plantagineum*, and *Heliotropium* spp. (Boraginaceae), among others (Garner and Papwort 1970; Lombardo 1984; Gallo 1987; Kelly 1990; Tokarnia *et al.* 2000; Santos *et al.* 2008). *Senecio* species belonging to the *Paucifolii* group are considered particularly toxic (Stöber *et al.* 2005). The first description of *Senecio* poisoning was carried out by Gilruth in 1903, who demonstrated the association between intake of *S. jacobaea* plants by horses and cattle in New Zealand, with the subsequent development of liver cirrhosis. After this first report, concern about the toxicity of these plants was increased (Bull 1955; Podestá *et al.* 1976).

## Epidemiological Situation in Uruguay

### Distribution and habitat

The first report about the presence of *Senecio* in Uruguay was made in the 1970s (mainly *S. brasiliensis* and *S. selloi*) (Podestá *et al.* 1976). *S. madagascariensis* is listed as

©CAB International 2011. *Poisoning by Plants, Mycotoxins, and Related Toxins* (eds F. Riet-Correa, J. Pfister, A.L. Schild, and T.L. Wierenga)

the newest species in the country, being identified in the late 1990s by farmers of southwestern coastal departments (Colonia, San José, Canelones, and Montevideo). It is now recognized that the distribution of the plant is much broader than originally identified, with plants being found in northern and eastern Uruguay (Ferreira Chaves and Fumerol 2008). It is assumed that the introduction of *S. madagascariensis* into the country was from the importation of contaminated seeds (Rivero 2008, personal communication).

Approximately 25 different species of *Senecio* have been identified in Uruguay. *S. selloi*, *S. madagascariensis*, *S. brasiliensis*, and *S. grisebachii* are considered the most important species because of the economic losses they cause, either due to their high toxicity or their invasiveness. These species are not evenly distributed in the country, as some regions have varying populations of different species (Marzocca *et al.* 1976; Gallo 1987). According to information provided by the database of East and West Regional Laboratories of the DILAVE 'Miguel C. Rubino', *Senecio* poisoning in cattle is the leading cause of death in eastern Uruguay (mainly extensive type production systems), and the second leading cause of death in the northwest coast of Uruguay (Dutra *et al.* 2009; Rivero *et al.* 2009).

Morphologically these plants represent shrubs of upright and leafy stems, which can be up to 5 feet tall. Leaves are lanceolate or oblong-lanceolate, arranged in chapters alone or in small amounts at the end of the stems. They are long and acute, narrowed towards the base, margins irregularly serrate, sessile or slightly petiolate (lower leaves) (Cabrera 1953; Marzocca *et al.* 1976; Podestá *et al.* 1976; Lombardo 1984; Gallo 1987; Teibler *et al.* 1999; Villar and Ortiz 2006). The flowers are typical of this genre: similar to daisies, grouped in chapters which in turn are arranged in dense corymbs at the apex of the branches. Generally they have 13 yellow petals (characteristic of the genre) 12 mm in length, with 20 to 24 bracts at the involucre (Cabrera 1953; Marzocca *et al.* 1976; Podestá *et al.* 1976; Lombardo 1984; Gallo 1987).

Each species has unique characteristics that allow its differentiation from the rest. *S. brasiliensis* plants may reach 1.5 m in height, with erect and branched stems that have serrated edged leaves, alternate distributed. *S. grisebachii* has light green leaves of gray underside (hence its name) and with hairs on their surface. *S. selloi* presents rounded fleshy leaves of sticky texture (hence it is known as 'sticky senecio'). *S. madagascariensis* are smaller plants (generally not exceed 60 cm in height), with fewer branches than the other species of *Senecio* and bright green leaves usually devoid of hairs (Cabrera 1953; Marzocca *et al.* 1976; Lombardo 1984; Villalba and Fernández 2007).

In biennial species (*S. grisebachii*, *S. selloi*), flowering occurs in spring (October and November). In perennial species (*S. brasiliensis*, *S. madagascariensis*) plants can flower throughout the year (Cabrera 1953; Marzocca *et al.* 1976; Podestá *et al.* 1976; Lombardo 1984; Gallo 1987; Castillos Karam *et al.* 2004; Villar and Ortiz 2006).

The plants of this genus are highly invasive. This is attributed to having two types of reproduction: sexually by seeds and asexually through stolons. This feature is exacerbated in the short-cycle species (*S. madagascariensis*) (Cabrera 1953; Marzocca *et al.* 1976; Lombardo 1984).

A mature plant can produce from 50,000 to 150,000 seeds. These are small (2 mm long) and equipped with a crown of white hair that aids its dispersal by wind. After reaching the ground, the seeds take 10 days to mature, being able to germinate at any time of year if appropriate conditions are provided (moderate temperature and high humidity) (Marzocca *et al.* 1976; Gallo 1987; Castillos Karam *et al.* 2004).

One of the main peculiarities of *Senecio* plants is the poor palatability for animals, typically being consumed only under conditions of scarce forage (Garner and Papwort

1970; Gallo 1987; Moraes and Rivero 1991; Blood *et al.* 2002; Stöber *et al.* 2005; Villar and Ortiz 2006). Concentrations of the toxic alkaloids are not substantially diminished by drying, thus animals can be poisoned from hay, contaminated grain, or silage.

In 1975 the first *Senecio* poisoning was experimentally reproduced in Uruguay (Podesta *et al.* 1976). These authors were able to demonstrate the toxicity of *S. brasiliensis* var. *tripartirtus*, a species that had been suspected as a cause of major losses since 1970. Many years later other researchers tested the toxicity of two *Senecio* species, *S. grisebachii* (Preliasco and Monroy 2008) and *S. madagascariensis* (Ferreira Chaves and Fumero, 2008; Arrospide 2009, personal communication) experimentally in cattle. *S. grisebachii* was found to be highly toxic for cattle, killing all the animals that were dosed (Preliasco and Monroy 2008). Ferreira Chaves and Fumero (2008) failed to reproduce *S. madagascariensis* poisoning, but later research established the toxicity of this species in cattle at higher doses (Arrospide 2009, personal communication).

## Sensitive animal species

*Senecio* poisoning has been described in horses, cattle, sheep, goats, swine, chickens, quails, pigeons, and humans (Garner and Papwort 1970; Riet-Correa *et al.* 1987; Tokarnia *et al.* 2000). In Uruguay, cases of poisoning have been reported in cattle and horses (Dutra *et al.* 2009; Rivero *et al.* 2009). Two outbreaks of *Senecio* poisoning have been detected in horses, one in 2007 in eastern Uruguay and another in 2009 in the Department of Paysandú (Dutra and Rivero 2009, personal communication).

Sheep and goats are less susceptible to the action of PA. There are several theories to explain this phenomenon of resistance. Some suggest that the main reason is liver metabolism of the alkaloids. The toxic pyrrole metabolites are synthesized in smaller quantities while the majority of the alkaloids react with the enzyme glutathione peroxidase allowing increased biliary elimination of these substances (da Silva *et al.* 2006; Brambilla *et al.* 2007; Santos *et al.* 2008). A second theory suggests that rumen bacterial activity has the ability to metabolize the PA, resulting in less enteric absorption and their subsequent hepatic metabolism, and that this activity is about 8 times higher in goats and 4.5 times higher in sheep compared to cattle (Teibler *et al.* 1999; Santos *et al.* 2008). However, sheep are not entirely resistant to the action of PA and are capable of being poisoning by *Senecio* (Ilha *et al.* 2001). The increased presence of *Senecio* plants in the country has been linked in part to the depopulation of sheep in the fields of Uruguay, which were considered the traditional way of controlling this weed (Riet Correa *et al.* 1987, 1993; Tokarnia *et al.* 2000; García y Santos *et al.* 2003; Castillos Karam *et al.* 2004).

## General conditions

Two factors are typically present in outbreaks of *Senecio* poisoning: shortage of fodder and presence of the plants in the fields (or contaminated food). In Uruguay the largest outbreaks were in 1988 and 2007, years in which drought resulted in a scarcity of forage (Rivero *et al.* 1989; Dutra and Rivero 2009, personal communication).

Management conditions generally determine the characteristics of the outbreaks. The literature suggests that males and young animals are more susceptible to the action of PA, but in Uruguay the most affected animals are more often adult females which are confined to areas with scarce forage (Podestá *et al.* 1976; Rivero *et al.* 1989; Maclachlan and Cullen 1995; Castillos Karam *et al.* 2004; Dutra *et al.* 2009). The type of production system is also important. There is no indication that breeds of cattle show differences of susceptibility to the action of PA, however in Uruguay a higher frequency of outbreaks in meat breeds is observed. This is most likely the result of different management conditions, particularly extensive grazing for beef cattle (pastoral-based systems) in relation to dairy systems (Podestá *et al.* 1976; García y Santos *et al.* 2003; Dutra *et al.* 2009; Rivero *et al.* 2009).

Most of the outbreaks are reported in late spring and early summer (October to December) after the critical period with forage shortages has ended and pastures have recovered (Podestá *et al.* 1976; Rivero *et al.* 1989; Castillos Karam *et al.* 2004; Dutra *et al.* 2009; Rivero *et al.* 2009). Cases are detected weeks to months after the animals have eaten the toxic plants; even when the populations of *Senecio* plants in the fields are low, animals may still become intoxicated after a long period of grazing (Podestá *et al.* 1976; Araya and Fuentealba 1984; Riet-Correa *et al.* 1987; Kelly 1990, 2002; Araya 1991; Tokarnia *et al.* 2000; Blood *et al.* 2002; Stöber *et al.* 2005; Villar and Ortiz 2006; Santos *et al.* 2008). This chronic liver poisoning from PA begins to be manifested clinically when the hepatic functional reserve has been exhausted (when 70 to 75% of the liver has been damaged). Toxicity is dependent on the specific types and concentration of alkaloids present in the plants. Thus, the PA determine the development of a fatal chronic liver disease that may develop from days to months after the ingestion of the plants (Kelly 1990; Riet-Correa *et al.* 1993; Tokarnia *et al.* 2000; García y Santos *et al.* 2003; Villar and Ortiz 2006; Santos *et al.* 2008).

## Toxic compounds

Pyrrolizidine alkaloids are considered to be plant secondary metabolites. They are generally considered to be bitter and represent a chemical mechanism of plant defense against herbivores (Araya 1990; da Silva *et al.* 2006; Brambilla *et al.* 2007). Chemically, most of the PA are esters of amino alcohols (necine, heliotridine, retronecine), with a pyrrolizidinic core (necine) and aliphatic acids (necic acids) that may occur in the form of monoesters, non-cyclic diesters (open), and cyclic diesters (in increasing order of toxicity). The necine base structure consists of two rings of five carbon atoms each of which share a nitrogen atom (Blood *et al.* 2002; da Silva *et al.* 2006; Brambilla *et al.* 2007; Santos *et al.* 2008). There are about ten necines and still more necic acids. These can be combined in many different ways and thus give rise to the more than 250 PA that have been identified and characterized so far (Garner and Papwort 1970; Maclachlan and Cullen 1995; Brambilla *et al.* 2007).

Pyrrolizidine alkaloids are chemically very stable and conventional forage conservation processes do not inactivate them. Animals can be poisoned by consuming contaminated hay and silage (Gallo 1987; Araya 1990; Riet-Correa *et al.* 1993; Tokarnia *et al.* 2000; Blood *et al.* 2002; García y Santos *et al.* 2003; Villar and Ortiz 2006). Under experimental conditions, *Senecio* plants are most often administered in their dry form to animals (Méndez *et al.* 1990; Preliasco *et al.* 2009). Another important aspect in outbreaks is the *Senecio* spp. present in the fields and their stage of growth. Many studies have shown that both quantity and type of alkaloids present in the plants are different between *Senecio* spp. Riet Alvariza *et al.* (1983) found a concentration of 0.16% in *S. brasiliensis* var. *tripartitus*. These plants contained three alkaloids (senecionine, anacrotine, and retrorsine), of which senecionine represented 90% (Riet Alvariza *et al.* 1983). Significant variations were found in the concentrations of PA between different species in Rio Grande do Sul: *S. brasiliensis* (0.31%), *S. selloi* (0.022%), and *S. leptolobus* (0.005%) (Riet-Correa *et al.* 1987).

In collaboration with the Department of Pharmacognosy of the Faculty of Chemistry, we found a PA concentration of 0.25% in samples of *S. grisebachii*. Six different alkaloids were detected, but only senecionine and retrorsine were identified. Retrorsine was found in higher proportion (23.9%), with low levels of senecionine (7.9%) (Preliasco *et al.* 2009). In a further collaboration with the Poisonous Plant Research Laboratory, (USDA, ARS, Logan, Utah, USA) the PA content in *S. grisebachii* (samples from Uruguay), showed a concentration of free base alkaloids equal to 0.29%; senecionine (46.2%), seneciophiline (29%), and retrorsine (24.8%) were identified. The samples were reduced with zinc to detect the presence PA-*N*-oxides; concentration of *N*-oxides was 0.09% resulting in a total PA (free base + *N*-oxide) concentration of 0.37%. At this institute, *S. madagascariensis* from two locations in Uruguay were examined. The results were compared with similar studies on *S. madagascariensis* from Australia and Hawaii. The samples from Uruguay showed PA concentrations of 0.073% and 0.10%, comparable to those found in plants from Australia and Hawaii (0.02% to 0.19%) (Gardner *et al.* 2006). In Uruguay we found *S. grisebachii* contains a higher concentration of PA than *S. madagascariensis*. Experimental research in Uruguay showed that *S. grisebachii* was more toxic than *S. madagascariensis* (Ferreira Chaves and Fumerol 2008; Preliasco *et al.* 2009; Arrospide 2009, pers. comm.). Although numerous studies have shown that shoots, seeds, and flowers of *Senecio* have higher concentrations of PA, it is not known which factors affect these concentrations throughout the lifecycle of the plants (Gallo 1987; Castillos Karam *et al.* 2004; Villar and Ortiz 2006).

*Senecio* spp. in Uruguay have shown great variability in toxicity among the different species. Experimental research was performed in calves with *S. grisebachii* and *S. madagascariensis*. Both studies used similar doses of dry plant, getting totally opposite results. *S. grisebachii* was highly toxic at doses of 45, 24, and 15 g dry plant/kg BW, leading to the death of all animals (Preliasco *et al.* 2008). *S. madagascariensis* caused no alteration in the health of animals at doses of 49, 65, and 80 g/kg BW (Ferreira Chaves and Fumerol 2008).

*S. madagascariensis* was tested in three calves at total accumulative doses of 61.93, 81.88, and 163.88 g/kg BW for 13, 15, and 21 days, respectively (Arrospide *et al.* 2009, personal communication). These researchers succeeded on reproducing the poisoning in the animal treated with the highest dose. Although the other two animals did not develop the clinical disease, they showed reduced feed intake and weight gain in comparison to the control calves, indicating the effect that the plant has upon productivity of the animals. In Brazil experiments with *S. brasiliensis* demonstrated clinical signs at 22.5 g/kg BW, but only caused death at a dose of 90 g/kg BW (Mendez *et al.* 1990).

## Clinical patterns

Only chronic *Senecio* poisoning is seen in Uruguay (Garner and Papwort 1970; Riet Alvariza *et al.* 1983; Riet-Correa *et al.* 1987; Kelly 1990; Dutra and Rivero 2009, personal communication). The main clinical signs are progressive weight loss, sudden drop in milk production, depression, anorexia, salivation, diarrhea, tenesmus, nervous signs, recumbency, and death (Rivero *et al.* 1989). Jaundice and photosensitization were only reported in severe advanced natural cases and not seen in experimental animal intoxication (Ferreira Chaves and Fumerol 2008; Preliasco *et al.* 2009; Arrospide *et al.* 2009, personal communication). The main manifestation of nervous signs in cattle from Uruguay is aggressiveness and frequently a lack of coordination and ataxia (Dutra and Rivero 2009, personal communication). The mortality rate ranges from 0.22% to 64% (Dutra and Rivero

2009, personal communication). The intensity and development of clinical signs depends on the toxic dose ingested, the time of ingestion, the susceptibility of animals, and secondary factors (such as stress) that promote the development or exacerbation of the clinical patterns (Araya 1990; Riet-Correa *et al.* 1993; Stalker and Hayes 2007).

## Necropsy findings

Macroscopic findings seen in experimental reproductions and spontaneous cases in Uruguay agree with those reported in the literature: generalized edema, especially edema of the digestive tract (mesenteric edema and petechiation). Abomasal edema is particularly common. Edema is caused by increased portal pressure caused by the interruption of blood circulation at the portal veins due to atrophy and diffuse liver fibrosis combined with hypoproteinemia caused by liver dysfunction (Podestá *et al.* 1976; Rivero *et al.* 1989; Kelly 1990; Stalker and Hayes 2007; Santos *et al.* 2008; Preliasco *et al.* 2009).

The liver is usually visibly decreased in size and presents a hardened cutting surface. The gall bladder is edematous, enlarged, with bleeding walls and altered biliary content. Other possible findings are hemorrhages, edema, and congestion in the heart and lungs, diarrhea, and rectal prolapse (Podestá *et al.* 1976; Rivero *et al.* 1989; Kelly 1990; Stalker and Hayes 2007; Santos *et al.* 2008; Preliasco *et al.* 2009).

## Histopathology

Histopathological examination is the main diagnostic method for this disease. The characteristic lesions produced by these types of alkaloids are known as 'end stage liver disease' formerly called hepatic cirrhosis. The main findings are loss of hepatic structure with detrabeculitation, megalocytosis, hepatocyte vacuolization, periportal fibrosis, proliferation of fibroblasts and epithelial cells of the bile ducts from portal tracts and centrilobular areas. Hepatic regenerative nodules surrounded by fibrous tissue, with absence of hepatocytes and mononuclear cell infiltrate, are often present. Spongy degeneration of the white matter of the brain is characteristic of hepatic encephalopathy (Podestá *et al.* 1976; Rivero *et al.* 1989; Kelly 1990; Méndez *et al.* 1990; Tokarnia *et al.* 2000; Stöber *et al.* 2005; Villar and Ortiz 2006; Stalker and Hayes 2007; Santos *et al.* 2008; Preliasco *et al.* 2009). These lesions are not unique to this condition. Other poisonings such as aflatoxins, nitrosamines, or other plants present in the country that contain PA (*Echium plantagineum*, *Erechtites hieracifolia*), can produce very similar pathological changes (Kelly 1990; Tokarnia *et al.* 2000; Kelly 2002; Stalker and Hayes 2007).

## Diagnosis

The diagnosis of this condition is based upon epidemiological data, necropsy findings, and histopathological studies. It has been shown that certain tests like liver biopsy and liver function tests may be helpful in the diagnosis. While Podesta *et al.* (1976) concluded that liver function tests have little diagnostic significance, other authors suggest that serum levels of certain enzymes (serum aspartate aminotransferase, alanine aminotransferase, and lactate dehydrogenase) and changes in enzyme activity over time indicate liver disease and may be useful for diagnosis (Araya 1991; Ferreira Chaves and Fumerol 2008; Preliasco and Monroy 2008).

## Treatment

Because this is a chronic condition, irreversible, and usually fatal, there is no effective therapy to reverse the lesions produced in the liver and other affected organs (Blood *et al.* 2002; Kelly 2002; García y Santos *et al.* 2003; Castillos Karam *et al.* 2004; Stöber *et al.* 2005). Animals should be removed from the source of the contaminated feed (pasture, hay, or silage). Symptomatic treatment is usually uneconomical and is generally applied to satisfy the desires of the producer (Blood *et al.* 2002; Stöber *et al.* 2005).

## Control Methods

*Senecio* causes significant production losses related to decreased production in animals and decreased use of the land. Major efforts should be directed to the control of these weeds. *Senecio* plants are extremely difficult to control. The plants are drought tolerant, have a dual mode of reproduction (sexual and asexual), are prolific seed producers with efficient seed distribution mechanisms, and have annual or biennial phenological cycles (Cabrera 1953; Marzocca *et al.* 1976; Lombardo 1984; Ferreira Chaves and Fumero 2008).

Control methods can be classified as physical, chemical, and biological. In Uruguay the most common methods of control integrate physical and chemical categories. The traditional method for the control of *Senecio* is sheep grazing. Since sheep are remarkably resistant to the action of PA, grazing with sheep at high stocking rates has been an effective measure for many years. However, sheep populations countrywide have declined in recent years, and the trend is partially responsible for the re-invasion of *Senecio* into pastures, a situation similar to that observed in the state of Rio Grande do Sul in Brazil (Riet-Correa *et al.* 1987; Moraes and Rivero 1991; García y Santos *et al.* 2003; Castillos Karam *et al.* 2004). Further, rapid changes experienced by forestry and agriculture in recent years have resulted in the relocation of livestock to less fertile areas which were previously assigned to sheep. In these areas plant competition for light and soil nutrients is reduced, which encourages the invasion of weeds and therefore, a greater exposure of livestock to weedy species (Riet-Correa *et al.* 1987).

Physical methods such as mechanically cutting the plants in the fields before flowering will not only prevent consumption by animals but prevent dispersal of the seeds as well. It is a relatively efficient method for species with short growth cycles in which several phenological stages can be found growing simultaneously, and the period between emergence and flowering is brief (e.g. *S. madagascariensis*). As new sprouts may occur from cut or damaged stems, it is essential to pull up the plants by their roots. Plants can also be extracted manually or by using tools like the hoe. After collection the harvested plants should be correctly managed so that the seeds cannot re-contaminate the soil. For these measures to be effective, *Senecio* plants should be controlled also on neighboring fields in order to avoid reintroductions from wind-dispersed seeds (Marzocca *et al.* 1976; Moraes and Rivero 1991; Villalba and Fumerol 2007; Ferreira Chaves and Fernández 2008).

Chemical control methods have the disadvantage of being more expensive yet are more practical, especially in cases of major infestations. The doses of chemicals to be applied vary with the type and mechanism of application, infestation rate, and physiological status of the weeds. The best time for application is before flowering. In fallow or old plants from the previous year, the use of Glyphosate controls 95 to 100% at application rates of 4 l/ha. Using specialized application equipment (rope, carpet, or rollers), differentiated adult plants taller than desirable pasture plants can be controlled by applying Picloram or Glyphosate at 10-20%. Flumetsulam alone or in mixtures with 4-(2,4-dichlorophenoxy) butyric acid gives good results for young and adult plants in reproductive state of growth with minimum regrowth 1 year after application (Marzocca *et al.* 1976; Villalba and Fernández 2007; Ferreira Chaves and Fumero 2008; Zanoniani 2006, personal communication).

There are some natural diseases and insects that are able to kill *Senecio* plants. *S. jacobea* was successfully controlled in the USA by the action of three insects: *Tyria jacobeae*, *Longitarsus jacobeae*, and *Pegohylemia seneciella* (Coombs *et al.* 1997). There are currently no research efforts in the area of biological control methods in Uruguay. One must consider the potential danger the introduction of alien species (arthropods, mollusks, plants) may represent to the ecosystem balance (Ferreira Chaves and Fumero 2008).

Prevention is also important and is based upon the use of clean seeds, fence row maintenance, and the permanent elimination of plants. An integrated approach that includes prevention, control, and monitoring work would be successful if carried out using effective, consistent, and systematical methods (Ferreira Chaves and Fumero 2008).

## References


Araya O (1990). Seneciosis en caballos. *Monografías de Medicina Veterinaria.* Instituto Ciencias Clínicas Veterinarias. Universidad Austral de Chile. http://www.monografias veterinaria.uchile.cl/CDA/mon_vet_seccion/0,1419,SCID%253D14002%2526ISID%25 3D420,00.html.

Araya O (1991). Manifestaciones clínicas de insuficiencia hepática en bovinos: diagnóstico y tratamiento. In *XIX Jornadas Uruguayas de Buiatría* (CMVP ed), pp. I1-I12. Paysandú, Uruguay.

Araya O and Fuentealba I (1984). Alteración hepática en terneros debido al consumo de *Senecio erraticus* en dos años consecutivos. In *XII Jornadas Uruguayas de Buiatría* (CMVP ed.), (2):c.c.10.1-c.c.10.2. Paysandú, Uruguay.

Blood DC, Radostits OM, Gay CC, and Hinchcliff KW (2002). Enfermedades causadas por toxinas vegetales, de hongos, cianofitos, clavibacterias y venenos de garrapatas y animales vertebrados. In *Medicina Veterinaria* (DC Blood, OM Radostits, CC Gay, KW Hinchcliff, eds), (2):1939-2029. Interamericana ed, México.

Brambilla G, Epifane M, Fumeo L, and Pontiggia R (2007). Alcaloides. *Revista de Facultades de Ciencias Exactas y Naturales, y Salud.* Universidad de Belgrano. http://www.ub.edu.ar/revistas_digitales/Ciencias/A2Num5/articulos.htm.

Bull LB (1955). The histological evidence of liver damage from pyrrolizidine alkaloids: Megalocytosis of the liver cells and inclusion globules. *The Australian Veterinary Journal* 31:33-41.

Cabrera AL (1953). *Manual de la flora de los alrededores de Buenos Aires*, 589 pp. Acme ed, Buenos Aires, Argentina.

Castillos Karam FS, Pereira Soares M, Haraguchi M, Riet-Correa F, Méndez MC, and Jarenkow JA (2004). Aspectos epidemiológicos da seneciose na região sul do Rio Grande do Sul. *Pesquisa Veterinária Brasileira* 24:191-198.

Coombs E, Mallory-Smith LC, Burril RH, Calliha R, Parker & Radtke H (1997). Tansy ragwort *Senecio jacobea* L. *Pacific Northwest Extension Publication* 157:1-7.

da Silva C, Abati A, and Heinzmann BM (2006). Alcaloides pirrolizidínicos em espécies do gênero *Senecio*. *Quim. Nova* 29:1047-1053.

Dutra F, Matto C, and Rivero R (2009). Descriptive statistics and spatiotemporal analysis of bovine hepatotoxic diseases diagnosed in Uruguay, 1998-2008. In *8th International Symposium on Poisonous Plants*, p. 11. João Pessoa, Paraíba, Brasil.

Ferreira Chaves S and Fumero R (2008). Investigación sobre la toxicidad de *Senecio madagascariensis* en bovinos del Uruguay, 69 pp. MVD Dissertation, Universidad de la República. Montevideo, Uruguay.

Gallo G (1987). *Plantas tóxicas para el ganado en el cono sur de América*, 213 pp. Hemisferio sur ed, Buenos Aires, Argentina.

García y Santos C, Elias F, Ramos A, Soares MP, and Schild AL (2003). Intoxicaciones diagnosticadas en bovinos por el Laboratorio Regional de Diagnóstico (UFPel) entre 1990 y 2002. In *XXXI Jornadas Uruguayas de Buiatría* (CMVP ed), pp. 141-143. Paysandú, Uruguay.

Gardner DR, Thorne MS, Molyneux RJ, Pfister JA, and Seawright AA (2006). Pyrrolizidine alkaloids in *Senecio madagascariensis* from Australia and Hawaii and assessment of possible livestock poisoning. *Biochem. Syst. Ecol.* 34:736-744.

Garner RJ and Papwort DS (1970). *Toxicología Veterinaria*, 470 pp. Acribia ed, Zaragoza, Spain.

Ilha MRS, Loretti AP, Barros SS, and Barros CSL (2001). Intoxicação espontânea por *Senecio brasiliensis* (Asteraceae) em ovinos no Rio Grande do Sul. *Pesquisa Veterinária Brasileira* 21:123-138.

Kelly WR (1990). El hígado y sistema biliar. In *Patología de los animales domésticos* (KVF Jubb, PC Kennedy, N Palmer, eds), (2):277-360. Hemisferio Sur ed, Montevideo, Uruguay.

Kelly WR (2002). Enfermedad del hígado en grandes y pequeños rumiantes. In *XXX Jornadas Uruguayas de Buiatría* (CMVP ed.), pp. 1-6. Paysandú, Uruguay.

Lombardo A (1984). *Flora montevidensis.* (2):465 pp. Intendencia Municipal de Montevideo ed, Montevideo, Uruguay.

Maclachlan NJ and Cullen JM (1995). Liver, Biliary System, and Exocrine Pancreas. In *Thomson's Special Veterinary Pathology* (W Carlton, MD McGavin, eds), pp. 81-115. Mosby ed, St Louis, EEUU.

Marzocca A, Marisco OJ, and Del Puerto O (1976). Guía de identificación de las principales malezas. In *Manual de malezas* (A Marzocca, OJ Marisco, O Del Puerto, eds), pp. 137-507. Hemisferio Sur ed, Buenos Aires, Argentina.

Méndez MC, Riet-Correa F, Schild A, and Martz W (1990). Intoxicaçaõ experimental por cinco espécies de *Senecio* em bovinos e aves. *Pesquisa Veterinária Brasileira* 10:63-69.

Moraes J and Rivero R (1991). La seneciosis en bovinos como limitante productiva. In *II Jornadas Técnicas de la Facultad de Veterinaria*, p. 36. Montevideo, Uruguay.

Podestá M, Tórtora JL, Moyna P, Izaguirre PR, Arrillaga B, and Altamirano J (1976). Seneciosis en bovinos. Su comprobación en el Uruguay. In *IV Jornadas Uruguayas de Buiatría* (CMVP ed), pp. iii/1-iii/18. Paysandú, Uruguay.

Preliasco M and Monroy IN (2008). Investigación sobre la toxicidad de *Senecio grisebachii* en bovinos del Uruguay, 75 pp. MVD Dissertation, Universidad de la República. Montevideo, Uruguay.

Preliasco M, Monroy IN, Horvath F, Vázquez A, Moraes J, and Rivero R (2009). Toxicity of *Senecio grisebachii* in Uruguay. In *8th International Symposium on Poisonous Plants*, p. 101. João Pessoa, Paraíba, Brasil.

Riet Alvariza F, Perdomo E, Rodríguez J, Duran J, Paullier C, Moyna P, Del Puerto O, Manta E, and Gilmet J (1983). Seneciosis en bovinos. Detección química de los alcaloides de *Senecio brasiliensis* var. *tripartitus* en la planta. In *I Jornadas Técnicas de Facultad de Veterinaria,* pp. 1-2. Montevideo, Uruguay.

Riet-Correa F, Riet Alvariza F, Schild AL, and Méndez MC (1987). Plantas tóxicas para bovinos en el Uruguay y Río Grande del Sur. In *XV Jornadas Uruguayas de Buiatría* (CMVP ed.), pp. G2-G3. Paysandú, Uruguay.

Riet-Correa F, Méndez MC, and Schild AL (1993). *Intoxicações por plantas e micotoxicoses em animais domésticos*, 340 pp. Agropecuaria Hemisferio Sur ed, Montevideo, Uruguay.

Rivero R, Quintana S, Ferola R, and Haedo F (1989). Principales enfermedades diagnosticadas en el área de influencia del Laboratorio de Diagnóstico Regional Noroeste de CIVET – Miguel C. Rubino. In *XVII Jornadas Uruguayas de Buiatría* (CMVP ed.), pp. 1-73. Paysandú, Uruguay.

Rivero R, Matto C, Dutra F, and Riet-Correa F (2009). Toxic plants affecting cattle and sheep in Uruguay. In *8th International Symposium on Poisonous Plants*, p. 1. João Pessoa, Paraíba, Brasil.

Romero A, Zeinsteger P, Teibler P, Montenegro M, Ruiz de Torrent R, Ríos E, and Acosta de Pérez O (2002). Lesiones hepáticas inducidas por componentes volátiles de *Senecio grisebachii* (margarita del campo o primavera) en ratones. *Revista Veterinaria UNNE* 12/13:1-2.

Santos JC, Riet-Correa F, Simões S, and Barros C (2008). Patogênese, sinais clínicos e patologia das doenças causadas por plantas hepatotóxicas em ruminantes e eqüinos no Brasil. *Pesquisa Veterinária Brasileira* 28:1-14.

Stalker MJ and Hayes MA (2007). Liver and biliary system. In *Pathology of Domestic Animals* (KVF Jubb, PC Kennedy, N Palmer, eds), (2):297-388. Elsevier Saunders, London, UK.

Stöber M, Martig J, Renner E, and Laiblin C (2005). Enfermedades alimentarias, metabólicas, carenciales y tóxicas con la participación de varios sistemas orgánicos. In *Medicina Interna y Cirugía del Bovino* (G Dirksen, HD Gründer, M Stöber, eds), (2):1125-1157. Interamericana ed, Buenos Aires, Argentina.

Teibler P, Rios E, Amarilla O, Ciotti E, and Acosta de Pérez O (1999). Resistencia del ovino a la intoxicación con *Senecio grisebachii* (Margarita Del Campo). *Revista de Investigaciones Agropecuarias*, N° 30 – INTA, Argentina. Available at: http://www.agroparlamento.com.ar/agroparlamento/notas.asp?n=0794.

Tokarnia CH, Döbereiner J, and Peixoto PV (2000). *Plantas tóxicas do Brasil*, 297 pp. Helianthus ed, Río de Janeiro, Brasil.

Villalba J and Fernández G (2007). *Senecio madagascariensis* poir. In *Seminario de actualización técnica en control y manejo de malezas de campo sucio. Serie Técnica N° 164* (INIA), pp. 23-28, Uruguay.

Villar D and Ortiz JJ (2006). *Plantas tóxicas de interés veterinario: Casos clínicos*, 179 pp. Masson ed, Barcelona, Spain.

# Chapter 31

# Risks from Plants Containing Pyrrolizidine Alkaloids for Livestock and Meat Quality in Northern Australia

M.T. Fletcher, R.A. McKenzie, K.G. Reichmann, and B.J. Blaney

*Department of Employment, Economic Development and Innovation, Health and Food Sciences Precinct, PO Box 156, Archerfield Qld 4108*

## Introduction

Plants containing pyrrolizidine alkaloids (PA) are widespread across the rangelands of northern Australia including Queensland (Qld), the Northern Territory (NT) and the northern half of Western Australia (WA). Livestock exposed to these plants are occasionally poisoned but overall impact of these plants on productivity, while negative, is unquantified. To better assess these impacts, all sources of PA needed to be identified, exposure quantified, and pharmacokinetics of PA metabolism in livestock clarified.

Common known sources of PA in the study region include plants within the genera *Crotalaria* (rattlepods), *Senecio* (fireweeds), *Heliotropium* (heliotropes), *Trichodesma zeylanicum* (cattle bush), and *Ageratum* spp. However, many taxa within these genera had not previously been assayed for PA. Consequently, we collected several hundred samples of these plants across the region and assayed them by standard GCMS procedures. This allowed compilation of mass spectral libraries, comparison with published data, and characterization of several new alkaloids (Fletcher *et al.* 2009). This report highlights some of our findings by reference to known poisoning scenarios that occur in northern Australia.

In addition to effects on livestock, worldwide concern over the hepatotoxic properties of PA in food has raised questions over the likelihood of PA residues occurring in meat of ruminants (WHO 1988; ANZFA 2001). PA and their *N*-oxides in plants are rapidly metabolised after ingestion, forming reactive pyrroles that bind with protein and DNA in the liver and other tissues. The resultant adducts persist in tissue and have been used as a diagnostic test for previous PA ingestion by livestock (Winter *et al.* 1990; Seawright *et al.* 1991). The significance of these adducts for human health is uncertain but their possible lability and release following consumption of meat have been questioned by toxicologists (Seawright 1994; Colegate *et al.* 1998). We have been investigating these risks and will briefly describe our preliminary findings as they relate to consumption of PA-containing plants by cattle and horses in rangelands of northern Australia.

©The State of Queensland (through the Department of Employement, Economic Development and Innovation) 2011. *Poisoning by Plants, Mycotoxins, and Related Toxins* (eds F. Riet-Correa, J. Pfister, A.L. Schild, and T.L. Wierenga)

## *Crotalaria* spp.

In the Kimberley region of WA and across the NT and northern Qld, notable PA poisonings are associated with horses consuming *Crotalaria* species. The disease is called 'walkabout disease' or Kimberley horse disease since it was first investigated in the Kimberley region (Gardiner *et al.* 1965). It should be noted that cattle are also affected, but the health of horses tends to be more closely monitored by property managers because they are used for mustering. One of the main plants originally identified as a causal agent in the Kimberley region was *C. crispata*, but this species has since been botanically divided into two, *C. crispata* and *C. ramosissima*, raising uncertainty over the relative risks. From our observations, both are sprawling, branched herbs up to 30 cm high with grey-green foliage and prominent flower pods (1-2 cm diameter), but *C. ramosissima* tends to be denser and larger. We have found these to have different alkaloid contents and profiles – *C. crispata* contains roughly equal amounts of fulvine, monocrotaline, and crispatine, while *C. ramosissima* contains very high concentrations of fulvine, lesser amounts of monocrotaline, and only traces of crispatine (Fletcher *et al.* 2009). *C. ramosissima* also contained the highest concentrations of PA of all *Crotalaria* taxa examined in this study (up to 6% by dry weight). Based on PA content and plant prevalence, both species present a high risk of livestock poisoning.

*C. novae-hollandiae* has also been associated with Kimberley horse disease; it is the most widespread and common *Crotalaria* species across northern Australia. Botanically, three subspecies are recognised in northern Australia: *C. novae-hollandiae* subsp. *novae-hollandiae*; *C. novae-hollandiae* subsp. *crassipes*; and *C. novae-hollandiae* subsp. *lasiophylla* (Holland 2002). From our GCMS analysis, we can expand this differentiation on the basis of PA profiles, such that the subspecies *novae-hollandiae* can be split into three distinct 'chemotypes'. Our data on alkaloid composition of these five taxa are given in Table 1 (data from Fletcher *et al.* 2009). In *C. novae-hollandiae* subsp. *novae-hollandiae* Chemotype 1 the otonecine alkaloid retusamine predominates and the profile is not unlike *C. novae-hollandiae* subsp. *crassipes*. In Chemotype 2 retronecine alkaloids monocrotaline and pumiline A (tentative) predominate and are present as free alkaloids. In addition to the two major profiles, three samples of *C. novae-hollandiae* subsp. *novae-hollandiae* Chemotype 3 from WA had an alkaloid composition intermediate between these profiles, with both monocrotaline and retusamine present. The risk associated with consumption of these three chemotypes would be markedly different, with considerably higher risk associated with the higher level of free retronecine alkaloids present in Chemotype 2, mainly collected in northern Qld.

We fed *C. novae-hollandiae* subsp. *novae-hollandiae* (approximately 15% of a basic maintenance diet) to weaned calves (110-120 kg) for 6 weeks at a rate to supply 5.5 mg PA/kg BW/day. Alkaloids detected by GCMS and LCMS were typical of Chemotype 2: the retronecine cyclic diesters monocrotaline, pumiline A, and trichodesmine (predominantly as the free alkaloid rather than *N*-oxide) and the otonecine cyclic diester crosemperine. This intake of plant was deliberately chosen to be below the level required to induce toxicity and produced no clinical signs, histopathological changes, nor significant variations in biochemical or hematological parameters in calves. Total PA in blood generally plateaued around days 7-28 with levels up to 150 μg/kg before decreasing. Muscle and liver total PA generally paralleled this trend with maximum levels up to 250 μg/kg and 2500 μg/kg, respectively. This plateau effect is consistent with activation of rumen bacteria and/or microsomal enzymes brought about by the prolonged exposure to PA as seen in other PA-containing plants (Craig *et al.* 2005). All PA present in the plant were detected in tissues but at varying levels which did not mirror relative levels of alkaloids in the plant material. For example, levels of monocrotaline (the predominant plant alkaloid) were at best comparable in tissue analysis with those of trichodesmine, which is a minor plant constituent. This reflects the differing rates at which these alkaloids are metabolized by hydrolysis or oxidation and excreted, and it is established that the more toxic forms are most reactive with the shortest half-life in tissues. PA-adducts were detectable from the first blood samples taken after 7 days of feeding and there was an apparent trend for PA-adduct levels to increase to a maxima at 14-21 days and then decrease towards the end of the feeding trial. PA-adducts were detected in all tissues taken at autopsy in the order liver >kidney ≈ heart > muscle.

**Table 1.** Pyrrolizidine alkaloid content within *C. novae-hollandiae* taxa.

| *C. novae-hollandiae* subspecies | PA content range (mean) (mg/g) | PA present (bold = major components) | Sample location (number of samples) |
|---|---|---|---|
| *crassipes* | 2.3 (2.3) | **Retusamine**<br>Monocrotaline<br>Crosemperine[a]<br>Croaegyptine | WA (2) |
| *lasiophylla* | 0 - 0.6 (0.2) | Retusamine<br>Crosemperine | NT (4) |
| *novae-hollandiae* Chemotype 1 | 0.1 - 1.4 (0.6) | **Retusamine**<br>Monocrotaline<br>Crosemperine | WA (2); NT (4) |
| *novae-hollandiae* Chemotype 2 | 0.2 - 23 (6.0) | **Monocrotaline**[b]<br>**Pumiline A**[b,c]<br>Crispatine[b]<br>Crosemperine<br>Trichodesmine[a,b] | Qld (9); NT (1) |
| *novae-hollandiae* Chemotype 3 | 0.2 - 0.7 (0.4) | Retusamine<br>Monocrotaline[b]<br>Crosemperine<br>Pumiline A[b,c] | WA (3) |

[a]Two stereiosomers, [b]present as free alkaloid, [c]tentative identification

## *Heliotropium* spp.

There is a significant history of PA poisoning of livestock by exotic naturalized *Heliotropium* spp. in Australia, particularly *H. europaeum* (Bull *et al.* 1961; Jones *et al.* 1981) and *H. amplexicaule* (Ketterer 1987). *H. indicum* has been reported to cause poisoning elsewhere (van Weeren *et al.* 1999) and is suspect in Australia. In addition, there are a large number of native heliotropes in northern Australia that constitute a potential PA risk which we are investigating (unpublished results). Creeper *et al.* (1999) first drew attention to PA in Australian native heliotropes when they ascribed cases of Kimberley horse disease to *H. ovalifolium.*

*H. amplexicaule* (blue heliotrope), a native of South America which is naturalized in southeast Qld, presents the greatest known risk of livestock poisoning from heliotropes in northern Australia. This plant (prostrate perennial up to 30 cm high with a deep taproot, multi-branched stems, and blue flowers with a yellow throat) is widespread across south-eastern Qld and is extending north and west. It flowers for much of the year and can be much faster to regenerate than pasture when spring rain follows a customary dry winter. Spread can also be favored by soil cultivation and fertilization and by disturbance of pasture in rangelands. Several cases of cattle poisoning occurring in the previous 8 years are described by Ketterer *et al.* (1987) and subsequent diagnostic records show blue heliotrope poisoning in a further seven cattle herds and one horse herd during 1987-2003. Ketterer *et al.* (1987) noted that young cattle were observed to regularly consume the plant despite the presence of alternative feed contrary to the accepted view that PA-containing plants are unpalatable.

We fed *H. amplexicaule* (approximately 15% of diet) to weaned calves for 6 weeks at a rate to supply 15 mg PA/kg BW/day. Alkaloids present were identified by GCMS as indicine (major) and heliospathine (minor), both present predominantly as *N*-oxides. Again, this intake of plant produced no clinical signs, histopathological changes, nor significant variations in biochemical or hematological parameters in calves. PA were assayed by LCMSMS in a range of tissues, but these were at the limit of quantitation (1 μg/kg) or less. PA-adducts were detectable in the first blood samples taken after 7 days of feeding and tended to increase throughout the feeding trial. PA-adduct levels also increased in biopsied muscle and liver samples taken throughout the trial. PA-adducts were detected in all tissues taken at autopsy in the order liver > kidney ≈ heart ≈ muscle.

Blue heliotrope is the target of biological control programs (Briese and Walker 2002), and we have worked with property owners involved in this program. These producers consider the plant a serious pest, but once again the actual effect on productivity was difficult to quantify. We tested blood samples of cattle grazing among blue heliotrope on six such properties. The cattle appeared clinically normal, and there was no evidence of liver damage from the blood clinical profile. Free PA (indicine and heliospathine) were detected at trace levels, 1-3 μg/kg in whole blood from four of ten animals on one of the six properties. Indicine *N*-oxide was detected (2 μg/kg) in whole blood from only one animal on a different property. PA-adducts were detected in almost all of these blood samples.

We also conducted an abattoir survey of 50 cattle from ten properties within the area where property owners and/or our departmental advisors believed blue heliotrope to be a serious pest. PA-adducts were detected at trace levels that were only about 1% of levels detected in our feeding trial, in liver samples of nine out of ten animals from one property and one out of one from a second property, but were not detected in livers of animals from the remaining eight properties. Despite widespread exposure, other factors such as herd behavioral patterns restrict consumption. Additionally, in areas where animals are continually exposed to blue heliotrope, it is very likely that there will be some adaptation by the animal, for example, an increased destruction of toxin in the rumen and liver.

## *Senecio* spp.

The native fireweed *S. brigalowensis* (previously classified as *S. lautus* (Thomson 2005)) is a regular cause of cattle poisoning in the Callide-Dawson region of central Qld. The plant is about 30 cm high with a yellow daisy-type flower. While the prevailing climate in northern Australia is for summer-dominant rainfall, this fireweed grows most extensively in seasons when unusual wet winter rainfall follows a dry summer when pasture is depleted, when it can form a continuous mass of blooms across paddocks. Noble *et al.* (1994) reported serious incidents in ten cattle herds during 1988-1992 involving mortalities ranging from 2-58%. More recently, extensive growth occurred in 2007 and in 2008, but there was no increase in PA-poisoning cases (over the average of about five cases of PA-poisoning/year in Qld) submitted to our diagnostic laboratories. Although there are widespread perceptions that such seasons present a high risk of lost cattle productivity, it is very difficult to estimate the real impact: poor performance as well as random cattle deaths are generally under-reported to veterinarians and diagnostic laboratories, while common anecdotal reports of lost production due to fireweed could easily be over-estimated due to the highly visible and extensive nature of blooms combined with knowledge that it has potential to kill stock.

We fed *S. brigalowensis* (approximately 15% of diet) to weaned calves for 6 weeks at a rate to supply 2.5 mg PA/kg BW/day. Alkaloids present were identified by GCMS and LCMS as the retronecine cyclic diester sceleratine (predominantly present as *N*-oxide) with a number of otonecine alkaloids, namely senkirkine, otosenine, desacetyldorinine, florosenine, and dorinine. Again, this intake of plant produced no clinical signs, histopathological changes, nor significant variations in biochemical or hematological parameters in calves. PA were assayed by LCMSMS in a range of tissues. Free PA were detected in blood and liver, tending to plateau after 2 to 3 weeks with levels up to 90 µg/kg in blood and 400 µg/kg in liver but then decrease to 30 and 40 µg/kg, respectively, by the end of the trial. Muscle levels followed a similar trend. The PA detected were all of the otonecine type with neither scleratine nor its *N*-oxide being detected in tissue, although this was the main PA in the plant. Uptake of sceleratine *N*-oxide like all PA *N*-oxides is dependent on rumen reduction to the free PA before absorption (Mattocks 1986). The requirement for this additional transformation compared to the otonecine alkaloids, which can be directly absorbed, may be responsible for the differential detection of these alkaloids in tissues. PA-adducts were detectable from the first blood samples taken after 7 days of feeding and there was an apparent trend for PA-adduct levels to increase to a maxima at 35 days and then decrease towards the end of the feeding trial. PA-adducts were detected in all tissues taken at necropsy in the order liver > kidney > heart ≈ muscle.

We conducted an abattoir survey for PA residues in meat from cattle originating on properties in the fireweed-affected areas in the few months after the 2007 bloom in central Qld. Low concentrations of PA-adducts were detected in livers of 80% of the 189 animals assayed at levels approximately 1-10% of that measured in livers of calves in our feeding trial. We conclude that PA-adducts will occasionally be present in some animals in these regions and seasons, but the overall prevalence of adducts in meat is likely to be much lower than expected on a purely exposure basis. In both our fireweed and *Crotalaria* feeding trials, both free alkaloid and PA-adduct residue levels decreased with prolonged feeding. It has been hypothesized that this decrease may relate to increased levels of PA-metabolizing rumen bacteria and/or liver enzymes induced by the prolonged exposure. This observation would suggest that animals exposed to PA plants for lengthy periods could develop a form of resistance to PA similar to that seen in sheep (Craig *et al.* 1992; Hovermale and Craig 2002).

## Risks for Meat Quality of PA Residues

Although continued debate over safe levels of PA should be expected, the Australia New Zealand Food Safety Authority has set a provisional tolerable daily intake (PTDI) of 1 µg PA/kg BW/day (ANZFA 2001). The basic premise for setting this PTDI in Australia and New Zealand was the link between PA and veno-occlusive disease in humans, in the absence of evidence of PA causing human liver cancer (ANZFA 2001). Grain and some herbal remedies are indisputably the major sources of human dietary exposure to PA, but honey, eggs, and meat (mainly offal) were also considered to make minor contributions. In respect to meat and offal, the key questions are: the levels of PA present and their prevalence; the chemical nature of those residues and their relative toxicity; and whether or not they could make a significant contribution to the PTDI. It is worth noting that differing toxicities of PA are not yet accounted for in the PTDI.

From our present studies, we estimate that the risk to health of persons consuming liver (or other tissues) from stock exposed to PA-containing plants is negligible. Even in the regions and seasons where exposure to PA is maximal, the likelihood of occurrence of free PA residues is extremely low and would be confined to less-reactive forms. Adducts might occasionally be present in meat, but the low concentrations detected in maximum exposure situations, combined with their undisputed much lower toxicity compared to free PA, leads us to conclude that they would make no significant contribution to the PTDI of PAs for the human consumer.

## Acknowledgements

Meat and Livestock Australia provided financial support for this work. All feeding trials were approved and overseen by the ARI Animal Ethics Committee (approval numbers ARI044/2004; ARI009/2004; SA 2005/09/46).

## References

ANZFA (2001). Pyrrolizidine alkaloids in food: A toxicological review and risk assessment. Australian New Zealand Food Authority, *Technical Report Series No. 2.*

Briese DT and Walker A (2002). A new perspective on the selection of test plants for evaluating the host-specificity of weed biological control agents: the case of *Deuterocampta quadrijuga*, a potential insect control agent of *Heliotropium amplexicaule*. *Biological Control* 25:273-287.

Bull LB, Rogers ES, Keast JC, and Dick AT (1961). Heliotropium poisoning in cattle. *Australian Veterinary Journal* 37:37-43.

Colegate SM, Edgar JA, and Stegelmeier BL (1998). Plant-associated toxins in the Human Food Supply. In *Environmental Toxicology: Current Developments* (Environmental Topics Volume 7) pp. 317-344, Gordon and Breach Science Publishers, Amsterdam, the Netherlands.

Craig AM, Latham CJ, Blythe LL, Schmotzer WB, and O'Connor OA (1992). Metabolism of toxic pyrrolizidine alkaloids from tansy ragwort (*Senecio jacobaea*) in ovine ruminal fluid under anaerobic conditions. *Applied and Environmental Microbiology* 58:2730-2736.

Craig AM, Duringer JM, and Blythe LL (2005). An Overview of Pyrrolizidine Alkaloid Toxicity in Livestock: Microbial and Metabolic Perspectives. In *Poisonous Plants: Global Research and Solutions* (K Panter, TL Wierenga, and JA Pfister, eds), pp. 99-106. Cromwell Press, Trowbridge.

Creeper JH, Mitchell AA, Jubb TF, and Colegate SM (1999). Pyrrolizidine alkaloid poisoning of horses grazing a native heliotrope (*Heliotropium ovalifolium*). *Australian Veterinary Journal* 77:401-402.

Fletcher MT, McKenzie RA, Blaney BJ, and Reichmann KG (2009). Pyrrolizidine alkaloids in *Crotalaria* taxa from northern Australia: risk to grazing livestock. *Journal of Agricultural and Food Chemistry* 57:311-319.

Gardiner MR, Royce R, and Bokor A (1965). Studies on *Crotalaria crispata*, a newly recognised cause of Kimberley Horse Disease. *Journal of Pathology and Bacteriology* 89:43-55.

Holland AE (2002). A review of *Crotalaria* L. (Fabaceae: Crotalarieae) in Australia. *Austrobaileya* 6:293-324.

Hovermale JT and Craig AM (2002). Metabolism of pyrrolizidine alkaloids by *Peptostreptococcus heliotrinreducens* and a mixed culture derived from ovine ruminal fluid. *Biophysical Chemistry* 101-102:387-399.

Jones RT, Drummond GR, and Chatham RO (1981). *Heliotropium europaeum* poisoning of pigs. *Australian Veterinary Journal* 57:396.

Ketterer PJ, Glover PE, and Smith LW (1987). Blue heliotrope (*Heliotropium amplexicaule*) poisoning in cattle. *Australian Veterinary Journal* 64:115-116.

Mattocks AR (1986). *Chemistry and Toxicology of Pyrrolizidine Alkaloids.* Academic Press, London.

Noble JW, Crossley JdeB, Hill BD, Pierce RJ, McKenzie RA, Debritz M, and Morley AA (1994). Pyrrolizidine alkaloidosis of cattle associated with *Senecio lautus*. *Australian Veterinary Journal* 71:196-200.

Seawright AA (1994). Toxic plant residues in meat. In *Plant-associated Toxins: Agricultural, Phytochemical and Ecological Aspects* (SM Colegate, PR Dorling, eds), pp. 77-82. CAB International, Wallingford, UK.

Seawright AA, Kelly WR, Hrdlicka J, McMahon P, Mattocks AR, and Jukes R (1991). Pyrrolizidine alkaloidosis in cattle due to *Senecio* species in Australia. *Veterinary Record* 129:198-199.

Thomson IR (2005). Taxonomic studies of Australian *Senecio* (Asteraceae): 5. The *S. pinnatifolius*/*S. lautus* complex. *Muelleria* 21:23-76.

van Weeren PR, Morales JA, Rodriguez LL, Cedeno H, Villalobos J, and Poveda LJ (1999). Mortality supposedly due to intoxication by pyrrolizidine alkaloids from *Heliotropium indicum* in a horse population in Costa Rica: a case report. *Veterinary Quarterly* 21:59-62.

WHO (1988). Pyrrolizidine Alkaloids, Environmental Health Criteria 80. World Health Organisation, Geneva.

Winter H, Seawright AA, Mattocks AR, Jukes R, Tshewang U, and Gurung BJ (1990). Pyrrolizidine alkaloid poisoning in yaks. First report and confirmation by identification of sulphur bound pyrrolic metabolites of the alkaloids in preserved liver tissue. *Australian Veterinary Journal* 67:411-412.

# Chapter 32

# Effects of Dietary Pyrrolizidine (*Senecio*) Alkaloids on Copper and Vitamin A Tissue Concentrations in Japanese Quail

J. Huan and P.R. Cheeke

*Department of Animal Sciences, Oregon State University, Corvallis, OR 97331*

## Introduction

The pyrrolizidine alkaloids (PA) are a large and important family of natural toxicants produced by a variety of plant species. Most PA-containing plants which produce toxic effects in livestock and humans are in the genera *Senecio*, *Crotalaria*, *Heliotropium*, and *Echium* (Cheeke 1998).

The hepatotoxicity and metabolism routes of PA are well known (Mattocks 1986; Cheeke 1998). They also have an important interaction with nutrient metabolism. An interaction between PA toxicosis and minerals (primarily copper) has been noted. The evidence arose from the observation that consumption of *E. plantagineum* and *H. europaeum* by sheep led to excessive liver copper concentrations followed by the hemolytic crisis of copper toxicity (Bull *et al.* 1956). Since then, other workers have noted that PA exposure causes elevated liver copper and iron and decreased zinc in horses (Garrett *et al.* 1984), rabbits (Swick *et al.* 1982a), and rats (Swick *et al.* 1982b, c). However, elevated liver copper levels in PA-poisoned animals do not always occur. Swick *et al.* (1983) and White *et al.* (1984) did not find elevated liver copper levels in sheep consuming *S. jacobaea*. Liver copper concentrations were normal in cattle with chronic heliotrope poisoning (Bull 1961). Species differences in susceptibility to PA toxicosis are well known (Cheeke 1998). Both PA-susceptible species such as horses (Garrett *et al.* 1984) and rats (Swick *et al.* 1982b, c), and PA-resistant animals such as sheep (Bull *et al.* 1956; Seaman 1987) and rabbits (Swick *et al.* 1982a) have been observed to have increased hepatic copper levels when fed sources of PA. However, Swick *et al.* (1982b) found that elevated liver copper levels in rats fed *S. jacobaea* occurred only in the presence of high levels of copper (50 and 250 ppm) and with high PA intakes. Similar results have been reported by Howell *et al.* (1991) who observed that heliotrope consumption caused elevated levels of liver copper in sheep only when copper was also administered. Is the occurrence of hepatoxic effects of PA necessary to influence copper metabolism? This question has yet to be conclusively answered.

Moghaddam and Cheeke (1989) observed that liver and blood Vit A concentrations are depressed in PA-intoxicated rats; similar results were seen in chicks fed *S. jacobaea* (Huan *et al.* 1992). Liver concentrations of copper and Vit A have an inverse relationship:

©CAB International 2011. *Poisoning by Plants, Mycotoxins, and Related Toxins*
(eds F. Riet-Correa, J. Pfister, A.L. Schild, and T.L. Wierenga)

factors that increase the concentration of one usually decrease the concentration of the other (Moore *et al.* 1972). Rachman *et al.* (1987) reported that in copper deficient rats, retinol and retinyl esters increase significantly in the liver and decrease in the serum. Sklan *et al.* (1987) reported that high levels of Vit A in the diet result in increased copper concentration in the liver and decreased copper concentration in the serum.

The objectives of this experiment were to study the effects of PA on copper and Vit A concentrations in serum and liver in Japanese quail, a species that is highly resistant to the hepatoxic effects of PA (Buckmaster *et al.* 1977), and to determine if there is an association between the PA-induced change in tissue mineral and Vit A concentrations and hepatic damage. Tansy ragwort (*S. jacobaea*) flowers were used as the PA source.

## Materials and Methods

### Animals and feed

Eight-eight Japanese quail of mixed sex, 2-weeks-old of age, were housed in electrically heated brooder batteries with experimental feed and water provided *ad libitum*. They were randomly assigned to eight groups and maintained on 24 h artificial light at 30 ± 2°C. Birds were wing banded and body weights were obtained initially and at the end of each week. The experiment lasted 6 weeks.

Eight different test diets were used, with a maize-soy basal diet. Dietary concentrations of PA, copper, and Vit A can be seen in Table 1. *S. jacobaea* was collected in the bloom stage near Corvallis, Oregon, air dried, ground through a 1 mm screen in a Wiley mill, and incorporated into the test diets, replacing maize, in order to make up 0 and 10% of diet. All-*trans*-retinol palmitate with a potency of 500,000 IU/g was used as the Vit A supplement. Cupric sulfate (anhydrous powder) was used as the copper supplement.

### Sample preparation and analysis

At the end of the experiment, six birds from each group were killed with injection of T-61 euthanasia solution (Hoechst-Roussel Agri-Vet Company). Blood samples were immediately taken by cardiac puncture from each bird, and serum was separated in an automatic refrigerated centrifuge at 1032 g, 4°C for 10 min. The serum samples were divided into three portions and then frozen at −80°C for future analysis. Whole liver tissue was removed, trimmed, and weighed. The right lobe of the liver was cut to provide consistent liver samples in order to minimize interlobular variability in Vit A and mineral distribution. Liver samples were immediately frozen at −80°C for later analysis. The methods for Vit A, copper, zinc, and iron analyses were previously described (Huan *et al.* 1992).

### Statistics

Statistical analysis was performed using the statistical software base SAS (SAS Institute Inc.). Data were assessed for homogeneity of variance using Analysis of Variance procedure with means compared by Student-Newman-Keuls (SNK) test at $P < 0.05$; uneven number of replications were analyzed by using the General Linear Models procedure with SNK test from SAS.

## Results

The inclusion of 10% tansy ragwort (TR) in the diet fed to Japanese quail did not cause significant depression in body weight gain (Table 1). Average weekly gains and final body weights were slightly lower with no significant differences ($P > 0.05$) between TR-fed groups compared to the control groups (Table 1). Adding copper and Vit A to the diet had no influence on the growth rate. The liver weights were not different in TR-containing groups compared to the controls. Similar results indicating high resistance to PA toxicosis were obtained in a previous Japanese quail trial (Buckmaster *et al.* 1977). In the TR-containing groups, the gross examination did not show any differences between TR-fed and the control groups. There was no mortality attributable to diet in any treatments.

**Table 1.** Growth and feed efficiency in Japanese quail.

| Dietary treatment | | | Avg. initial wt (g) | Avg. final wt (g) | Avg. weekly gain (g) | Feed efficiency (F/G)* |
|---|---|---|---|---|---|---|
| % Dietary TR | Cu (ppm) | Vit A (IU/kg) | | | | |
| 0 | 0 | 0 | 22.13±1.11 | 131.98±5.76 | 18.31±0.89[a] | 5.340 |
| 10 | 0 | 0 | 22.11±1.03 | 113.94±3.64 | 15.31±0.56[ab] | 6.750 |
| 0 | 250 | 0 | 22.08±0.90 | 121.02±5.23 | 16.49±0.80[ab] | 6.260 |
| 10 | 250 | 0 | 22.69±1.04 | 118.04±3.73 | 15.89±0.65[ab] | 6.500 |
| 0 | 0 | 25,000 | 22.20±0.98 | 126.29±4.85 | 17.35±0.83[ab] | 5.520 |
| 10 | 0 | 25,000 | 22.49±0.85 | 118.33±3.98 | 15.97±0.65[ab] | 6.683 |
| 0 | 250 | 25,000 | 22.43±1.11 | 129.98±5.28 | 17.93±0.83[ab] | 5.420 |
| 10 | 250 | 25,000 | 22.91±1.28 | 112.33±4.92 | 14.76±0.53[b] | 6.340 |

Means ± SE in the same column followed by different superscripts are different ($P < 0.05$).
*Feed intake (g)/body weight (g).

Concentrations of copper, zinc, and iron in serum and liver are summarized in Tables 2 and 3. The levels of copper in serum were not different among treatment groups ($P > 0.05$). Liver copper concentration was significantly decreased ($P < 0.05$) in TR-containing groups compared to the control groups, but there was no significance when individual groups were compared with each other. Adding high levels of Vit A supplement with TR caused a significant depression on liver copper concentration compared to the control group. Serum zinc concentrations were not different among treatment groups with the exception of the comparison between Vit A supplement alone and Vit A supplement plus copper (Table 3). TR did not affect liver zinc concentration, while copper supplement in the diet caused a significant depression in the zinc concentration. There were no significant differences in the iron concentrations of serum and liver among treatment groups (Table 3). Ceruloplasmin activity was not significantly different among treatment groups (Table 2).

There were no significant differences in serum Vit A concentration ($P > 0.05$) among treatment groups (Table 4). The liver Vit A concentrations were also not significantly different among treatment groups with exception of the TR alone group (Table 4).

## Discussion

Consumption of TR did not affect the growth rate of Japanese quail. Performance data showing no adverse effects of 10% TR were in agreement with previous results

(Buckmaster *et al.* 1977). The feed efficiency for the TR-fed groups was slightly lower than for controls, probably due to the lower energy content in the TR-containing diet. Similar results were also seen in previous work (Buckmaster *et al.* 1977).

**Table 2.** Copper concentration and ceruloplasmin in serum and liver of Japanese quail.

| Dietary treatment | | | Serum copper (ppm) | Liver copper (ppm) (wet tissue) | Ceruloplasmin (U/ml) |
|---|---|---|---|---|---|
| % Dietary TR | Cu (ppm) | Vit A (IU/kg) | | | |
| 0 | 0 | 0 | 0.340±0.04 | 6.49±0.74[a] | 24.6±3.3 |
| 10 | 0 | 0 | 0.297±0.03 | 4.62±0.44[ab] | 28.1±4.3 |
| 0 | 250 | 0 | 0.305±0.04 | 5.75±0.60[ab] | 22.3±5.4 |
| 10 | 250 | 0 | 0.318±0.01 | 5.75±0.91[ab] | 23.0±7.5 |
| 0 | 0 | 25,000 | 0.292±0.01 | 5.62±0.72[ab] | 24.4±2.1 |
| 10 | 0 | 25,000 | 0.298±0.03 | 3.71±0.34[b] | 24.0±3.9 |
| 0 | 250 | 25,000 | 0.335±0.02 | 5.30±0.48[ab] | 22.2±1.8 |
| 10 | 250 | 25,000 | 0.310±0.02 | 4.06±0.12[ab] | 26.7±6.7 |

Means ± SE in the same column followed by different superscripts are different ($P < 0.05$).

**Table 3.** Zinc and iron concentrations in serum and liver (wet tissue) of Japanese quail.

| Dietary treatment | | | Serum zinc (ppm) | Liver zinc (ppm) | Serum iron (ppm) | Liver iron (ppm) |
|---|---|---|---|---|---|---|
| % Dietary tansy ragwort | Cu (ppm) | Vitamin A (IU/kg) | | | | |
| 0 | 0 | 0 | 2.80±0.45[ab] | 31.16±1.81[a] | 7.87±0.63 | 236.01±25.78 |
| 10 | 0 | 0 | 3.38±0.59[ab] | 27.84±1.66[ab] | 8.83±1.08 | 253.72±25.40 |
| 0 | 250 | 0 | 3.43±0.59[ab] | 21.18±1.80[b] | 9.08±0.98 | 169.81±32.67 |
| 10 | 250 | 0 | 2.72±0.40[ab] | 23.94±2.76[ab] | 7.35±0.48 | 181.22±31.11 |
| 0 | 0 | 25,000 | 2.23±0.17[b] | 26.40±3.53[ab] | 8.82±0.80 | 156.81±15.66 |
| 10 | 0 | 25,000 | 3.35±0.73[ab] | 25.29±1.78[ab] | 9.45±1.40 | 248.26±25.94 |
| 0 | 250 | 25,000 | 4.55±0.38[a] | 21.20±1.51[b] | 7.47±0.63 | 169.91±19.80 |
| 10 | 250 | 25,000 | 3.52±0.40[ab] | 21.45±0.63[b] | 8.17±0.59 | 237.27±26.94 |

Means + SE in the same column followed by different superscripts are different ($P < 0.05$).

**Table 4.** Liver weight and Vit A concentration in serum and livers of Japanese quail.

| Dietary treatment | | | Liver weight (g) | Serum Vit A (IU/ml) | Liver Vit A* (IU/g) (wet tissue) |
|---|---|---|---|---|---|
| % Dietary TR | Cu (ppm) | Vitamin A (IU/kg) | | | |
| 0 | 0 | 0 | 3.22±0.40 | 4.306±0.424[c] | 160.62±36.32[a] |
| 10 | 0 | 0 | 2.83±0.17 | 3.948±0.638[c] | 524.40±80.85[b] |
| 0 | 250 | 0 | 3.55±0.48 | 5.121±0.253[bc] | 148.27±25.47[a] |
| 10 | 250 | 0 | 3.08±0.38 | 4.206±0.399[c] | 171.09±40.71[a] |
| 0 | 0 | 25,000 | 3.13±0.26 | 7.858±0.665[ab] | 3235.63±576.26[c] |
| 10 | 0 | 25,000 | 3.02±0.24 | 7.258±0.458[abc] | 3466.54±584.59[c] |
| 0 | 250 | 25,000 | 3.52±0.52 | 9.836±1.68[a] | 4266.84±733.99[c] |
| 10 | 250 | 25,000 | 3.45±0.31 | 9.797±1.08[a] | 4901.76±206.45[c] |

Means ± SE in the same column followed by different superscripts are different ($P < 0.05$).
*comparison was made in log values.

There were no significant differences in serum copper concentrations among treatment groups, but liver copper levels were decreased overall with TR treatment. This contrasts with the marked increase in liver copper concentration in chicks fed TR (Huan *et al.* 1992). In work with Merino sheep, there were only slight hepatoxic effects with a long-term exposure to *E. plantagineum* and no elevation in the liver copper concentration was noted (Culvenor *et al.* 1984). Bull (1961) reported that liver copper concentrations were normal in cattle suffering chronic heliotrope poisoning. These results suggest that the hepatoxic effects of PA are necessary to influence copper metabolism. Japanese quail are very resistant to the PA and may consume over 2000% of body weight of TR with no pathological signs (Buckmaster *et al.* 1977). In this experiment, the total TR intake per bird was only 55% of body weight. This TR intake may be not enough to cause liver damage. The concentrations of zinc and iron in the serum and liver also were not affected during PA exposure in this experiment. An antagonism between copper and zinc in their metabolism may be one of the reasons for this (Mertz 1986). The higher affinity of copper for metallothionine compared to zinc may allow it to displace zinc in the tissues (Mertz 1986).

The serum Vit A levels were not affected by PA consumption (Table 4). The concentration of liver Vit A remained normal with the exception of the group fed TR without added copper or Vit A. In this group the liver Vit A content was elevated (Table 4). Buckmaster *et al.* (1977) showed that Japanese quail have a very low rate of *in vitro* pyrrole production. In Japanese quail, the liver may have detoxification enzymes against PA toxicosis. The results obtained from the TR without supplement group may be due to inhibition of PA on the mobilization of Vit A.

The results suggest that some degree of hepatotoxicity is necessary to cause the PA-induced changes in tissue copper and Vit A concentrations that occur in PA-susceptible species.

## Acknowledgements

The participation of Dr Peter Cheeke to the 8th International Symposium on Poisonous Plants was financially supported by Conselho Nacional de Desenvolvimento Científico e Tecnológico (CNPq), grant 454084/2008-0, and by Coordenação de Aprefeiçoamento de Pessoal de Nível Superior (CAPES), grant 0017/09-4.

## References

Buckmaster GW, Cheeke PR, Arscott GH, Dickinson EO, Pierson ML, and Shull LR (1977). Response of Japanese quail to dietary and injected pyrrolizidine (*Senecio*) alkaloid. *Journal of Animal Science* 45:1322-1325.

Bull LB (1961). *Heliotropium* poisoning in cattle. *Australian Veterinary Journal* 37:37-43.

Bull LB, Dick AT, Keast JC, and Edgar G (1956). An experimental investigation of the hepatoxic and other effects on sheep of consumption of *Heliotropium europaeum* L.: heliotrope poisoning of sheep. *Australian Journal of Agricultural Research* 7:281-332.

Cheeke PR (1998). *Natural Toxicants in Feeds, Forages and Poisonous Plants.* Prentice-Hall, Upper Saddle River, New Jersey, USA.

Culvenor CCJ, Jago MV, Peterson JE, Smith LW, Payne AL, Campbell DG, Edgar JA, and Frahn JL (1984). Toxicity of *Echium plantagineum* (Paterson's curse) I. Marginal toxic effects in Merino wethers from long-term feeding. *Australian Journal of Agricultural Research* 35:293-304.

Garrett BJ, Holtan DW, Cheeke PR, Schmitz JA, and Rogers QR (1984). Effects of dietary supplementation with butylated hydroxyanisole, cysteine, and vitamins B on tansy ragwort (*Senecio jacobaea*) toxicosis in ponies. *American Journal of Veterinary Research* 45:459-464.

Howell JMcC, Deol HS, and Dorling PR (1991). Experimental copper and *Heliotropium europeaum* intoxication in sheep: Clinical syndromes and trace element concentrations. *Australian Journal of Agricultural Research* 42:979-992.

Huan J, Cheeke PR, Lowry RR, Nakaue HS, Snyder SP, and Whanger PD (1992). Dietary pyrrolizidine (*Senecio*) alkaloids and tissue distribution of copper and vitamin A in broiler chickens. *Toxicology Letters* 62:139-153.

Mattocks AR (1986). *Chemistry and Toxicology of Pyrrolizidine Alkaloids.* Academic Press, Orlando, Florida.

Mertz W (1986). *Trace Elements in Human and Animal Nutrition*, pp. 301-364, 5th edn, Vol. 1. Academic Press, San Diego, California.

Moghaddam MF and Cheeke PR (1989). Effects of dietary pyrrolizidine (*Senecio*) alkaloids on vitamin A metabolism in rats. *Toxicology Letters* 45:149-156.

Moore T, Sharman IM, Todd JR, and Thompson RH (1972). Copper and vitamin A concentrations in the blood of normal and Cu-poisoned sheep. *British Journal of Nutrition* 28:23-30.

Rachman F, Conjat F, Carreau JP, Bleiberg-Daniel F, and Amedee-Manesme O (1987). Modification of vitamin A metabolism in rats fed a copper-deficient diet. *International Journal of Vitamin Nutrition Research* 57:247-252.

Seaman JT (1987). Pyrrolizidine alkaloid poisoning of sheep in New South Wales. *Australian Veterinary Journal* 64:164-167.

Sklan D, Halevy O, and Donoghue S (1987). The effect of different dietary levels of vitamin A on metabolism of copper, iron and zinc in the chick. *International Journal of Vitamin Nutrition Research* 57:11-18.

Swick RA, Cheeke PR, Patton NM, and Buhler DR (1982a). Absorption and excretion of pyrrolizidine (*Senecio*) alkaloids and their effects on mineral metabolism in rabbits. *Journal of Animal Science* 55:1417-1424.

Swick RA, Cheeke PR, Miranda CL, and Buhler DR (1982b). The effect of consumption of the pyrrolizidine alkaloid-containing plant *Senecio jacobaea* on iron and copper metabolism in the rat. *Journal of Toxicology and Environmental Health* 10:757-768.

Swick RA, Cheeke PR, and Buhler DR (1982c). Subcellular distribution of hepatic copper, zinc and iron and serum ceruloplasmin in rats intoxicated by oral pyrrolizidine *Senecio* alkaoids. *Journal of Animal Science* 55:1425-1430.

Swick RA, Miranda CL, Cheeke PR, and Buhler DR (1983). Effect of phenobarbital on toxicity of pyrrolizidine (*Senecio*) alkaloids in sheep. *Journal of Animal Science* 56:887-894.

White RD, Swick RA, and Cheeke PR (1984). Effects of dietary copper and molybdenum on tansy ragwort (*Senecio jacobaea*) toxicity in sheep. *American Journal of Veterinary Research* 45:159-161.

# Chapter 33

# Poisoning by *Cycas revoluta* in Dogs in Brazil

B.M. Cunha[1], T.N. França[2], M.S.F. Pinto[1], M.A. Esteves[1], E.M. Yamasaki[3], and P.V. Peixoto[4]

[1]*Medicina Veterinária, UNESA, Rio de Janeiro, RJ 22783-320, Brazil;* [2]*Instituto de Veterinária, UFRRJ, RJ 23890-000, Brazil;* [3]*Curso de Pós-graduação, UFRRJ, RJ 23890-000, Brazil;* [4]*Instituto de Zootecnica, UFRRJ, Seropédica, RJ 23890-000, Brazil*

## Introduction

Poisoning in humans caused by the ingestion of 'nuts' from palms of the order Cycadales dates from the 18th century (Reagor *et al.* 1986). This order includes plants of the families Cycadaceae (with one genus, *Cycas*), Stangeriaceae (with one genus and one species, *Stangeria eriopus*), and Zamiaceae (with eight genera, among them *Encephalartos*) (Tustin 1983; Botha *et al.* 1991), whose species are found in tropical and subtropical areas. Among these, *C. revoluta* and *C. circinalis* are the most frequently used as ornamental or residential plants (Hooper 1978; Tustin 1983; Botha *et al.* 1991). Cases of poisoning in humans determined by ingestion of Cycadales seeds occur due to their use as an alternative source of food, a prophylactic measure against cancer, growth promoters, cosmetic use, or even supposedly therapeutic purposes (Chang *et al.* 2004). The first report of poisoning by this plant in humans and pigs dates from 1770 in Australia (Reagor *et al.* 1986; Hall 1987). Reports on the toxicity of Cycadales were also described in soldiers during the Boer War (1899-1902) (Reitz 1929; Botha *et al.* 1991). Inhabitants of Guam, a Micronesia island, used *C. circinalis* seeds as food after eliminating the toxic substances from the seeds (Ziemer 1997). Also in Australia, *Macrozamia riedlei* seeds are an important source of food after detoxification by roasting and drying (Mills *et al.* 1996).

Reports of diseases caused by ingestion of plants of the order Cycadales have been described in various animal species (Reagor *et al.* 1986; Albretsen *et al.* 1998). In Australia, cases of poisoning by *Cycas* and *Zamia* have been described in cattle and sheep (Botha *et al.* 1991). Poisonings have also been reported in New Guinea, Puerto Rico, Mexico, and the Dominican Republic (Senior *et al.* 1985). Ruminants exhibit acute gastroenteritis, hind limb paresis (determined by myelin degeneration in the spinal cord), ataxia, and hepatic necrosis (Hall 1987). Poisoned rats and primates are affected by the carcinogenic and teratogenic effects (Senior *et al.* 1985; Reagor *et al.* 1986), with formation of intestinal, renal, and hepatic tumors (Senior *et al.* 1985). Reports of the toxic effects of *C. revoluta* were described in dogs after ingestion of the stem in South Africa (Botha *et al.* 1991) and Texas (Reagor *et al.* 1986). In Florida, seeds of *Zamia floridiana* have been shown to be toxic for dogs (Senior *et al.* 1985). In the same state, poisoning of a dog by ingestion of *Dioon edule* seeds was observed (Morton 1967). In Australia, roasted seeds of

©CAB International 2011. *Poisoning by Plants, Mycotoxins, and Related Toxins* (eds F. Riet-Correa, J. Pfister, A.L. Schild, and T.L. Wierenga)

*Macrozamia riedlei* were provided as food and determined a clinical picture of poisoning in a Dachshund breed dog (Mills *et al.* 1996).

Dogs poisoned by Cycadales suffer from gastrointestinal and hepatic disorders (Reagor *et al.* 1986; Albretsen *et al.* 1998; Gfeller and Messonnier 2006), with vomiting (Botha *et al.* 1991; Gfeller and Messonnier 2006), hematemesis, diarrhea with or without blood, abdominal pain (Albretsen *et al.* 1998; Gfeller and Messonnier 2006) accompanied or not by a neurological syndrome with depression (Senior *et al.* 1985; Botha *et al.* 1991; Mills *et al.* 1996), 'rigid state', ataxia, and seizures (Reagor *et al.* 1986; Albretsen e*t al.* 1998). Hemorrhage, ascites, anuria (Reagor *et al.* 1986; Albretsen *et al.* 1998; Gfeller and Messonnier 2006), and jaundice (Senior *et al.* 1985; Reagor *et al.* 1986; Botha *et al.* 1991; Mills *et al.* 1996; Gfeller and Messonnier 2006) were also observed in dogs poisoned by these plants. Generally, the laboratory tests showed lymphopenia, leukocytosis, thrombocytopenia, increase in hepatic enzymes (ALT and AST) (Botha *et al.* 1991; Gfeller and Messonnier 2006), hyperbilirubinemia, hypoproteinemia, and hypocalcemia (Ziemer 1997; Albretsen *et al.* 1998; Gfeller and Messonnier 2006).

The necropsy findings consist of hemorrhages in the gastrointestinal tract, enlarged and congested liver, kidney tumefaction, jaundice, generalized ecchymotic hemorrhages (Reagor *et al.* 1986), ascites (Albretsen *et al.* 1998; Gfeller and Messonnier 2006), gall bladder wall edema (Mills *et al.* 1996), and even cirrhosis (Gfeller and Messonnier 2006). Microscopically, there is acute toxic hepatitis characterized by diffuse (Reagor *et al.* 1986), centrilobular or 'midzonal' hepatocellular necrosis (Albretsen *et al.* 1998), fatty degeneration of hepatocytes, thrombi in liver sinusoids, biliary retention in hepatocytes, canaliculi, and biliary ducts, and passive congestion (Reagor *et al.* 1986). There are biliary thrombi inside renal tubules and biliary pigment in the cytoplasm of tubular epithelial cells (Reagor *et al.* 1986). Cardiomyocyte degeneration, slight degeneration of cerebellar tracts (Senior *et al.* 1985), renal tubular necrosis, and a decrease in the population of granulocytes in the bone marrow (Mills *et al.* 1996) may be observed in poisoning caused by ingestion of *Cycas* palms. This report describes two cases of poisoning by *C. revoluta* in dogs.

## Case Report

### History and clinical picture

The owner reported that a *C. revoluta* palm tree, approximately two and a half years old, had been repeatedly attacked by a young female dog (Dog 2), which would ingest the buds in small portions through the gaps in a protective fence. On September 22, 2008, 2 days after the palm had been dug up for transportation, the owner found vomit with a material similar to 'moist sawdust' and bloody diarrhea at around 6 pm. She also found several pieces of leaves and remains of roots of *C. revoluta*. On that occasion, she saw an adult female dog (Dog 1) showing restless behavior. As time went on during the night, the animal showed prostration, lowering and paralysis of hind limbs, stiffness of the whole body especially of the forelimbs, generalized tremors, bulging eyes, and opisthotonus. The animal appeared to be totally disconnected from the environment, not reacting to any stimuli. Dog 1 was taken to a veterinarian, who claimed the animal was suffering from 'abdominal colic.' After administration of atropine and hyoscine butylbromide (Buscopan), the animal was taken home and exhibited convulsive seizures. When taken again to the clinic, it was medicated with benzodiazepine compound (Diazepam) and fluid therapy. The animal regained response to auditory stimuli but only showed movements of the eyes. With the stable yet critical clinical picture, the owner took the animal to a hospital in the morning on September 23. Marked prostration, ocular mucosa hyperemia, cold extremities in the hind limbs, and anuria were seen. Supportive therapy was administered with fluid infusion for approximately 12 h; the animal did not urinate. The laboratory exams for this animal revealed 77 mg/dl of urea, 450 IU/l of ALT (GPT), 599 IU/l of AST (GOT), 4.4 g/dl of total serum proteins, and 2.1 g/dl of albumin, a platelet count of 96,000/$mm^3$, and lymphopenia. Death occurred at the end of the day; the time from onset of symptoms until death was approximately 30 h. This animal did not undergo necropsy.

Also on September 22, the owner saw that the young female dog (Dog 2) vomited small amounts of contents similar to 'sawdust,' then vomited a viscous liquid of yellowish color and exhibited diarrhea with streaks of blood at around 6 pm. On September 23, Dog 2 was taken to the hospital with dehydration, prostration, emesis in the form of gushes, and oliguria. On September 25, the clinical picture progressed to bloody diarrhea, vomiting, and convulsive seizures. Saline solution and hepatic protectors were administered. On September 30 the animal exhibited jaundice.

The laboratory findings of Dog 2 consisted of 30 mg/dl of urea on September 25th progressing to 59 mg/dl on October 7th, 519 IU/I of ALT (GPT) regressing to 366 IU/I in the same period, 4.4 g/dl of total serum proteins on the 25th to 4 g/dl on the 7th, a decrease in albumin in the same period from 3.1 g/dl to 1 g/dl. Values for total, direct, and indirect bilirubin were, respectively, 0.18-0.55 mg/dl, 0.14-0.43 mg/dl, and 0.04-0.12 mg/dl from September 25 to October 7, 2008. Platelet counts were 129,000/$mm^3$ and there was lymphopenia on the 24th. Death occurred during the night of October 7.

### Necropsy and histology

The necropsy of Dog 2 was performed with restrictions and revealed generalized marked jaundice, ascites (approximately 1.5 l of orange-reddish liquid content), and liver with a perceptible lobular pattern with a green-yellowish color interspersed with red spots. In some regions close to the hepatic borders under the capsule, there were tortuous linear structures of several diameters disposed as a disorganized web, sometimes whitish, sometimes reddish (dilated and congested lymphatic and blood vessels, respectively). Kidneys exhibited green-orangish color with streaks radiating perpendicular to the cortex, and a red-orangish tonality was observed in the medulla.

Microscopic examination revealed that most hepatocytes were increased in volume (megalocytosis), sometimes with one or more cytoplasmic vacuoles of various sizes and large, vesicular nucleus. Hepatocytes with two, three, or even more nuclei were frequent. Necrotic hepatocytes with marked cytoplasmic eosinophilia, pycnosis, karyorrhexis, or karyolysis were distributed randomly. Marked biliary retention was also seen (biliary thrombi and bile inside hepatocytes and Kupffer cells). A discrete inflammatory infiltration, predominantly mononuclear, was seen in portal spaces and between hepatic cords. In some areas, there were macrophages, phagocyting cell detritus, and bile. Bile was seen inside tubules and in the cytoplasm of tubular epithelial cells and in the kidneys, as well as tumefaction and vacuolation of the epithelium, mainly on the distal convoluted tubules.

## Discussion

The diagnosis of poisoning by *C. revoluta* was based on the clinical picture, necropsy findings, and histological lesions compatible with those described for dogs, which confirmed the evidence reported by the owner that the dogs had eaten the plant. As described by Reagor *et al.* (1986), stems of *C. revoluta* when ingested can cause poisoning in dogs. Likewise our report shows that inadvertent exposure of dogs to *C. revoluta* that had been dug up to be transported can cause poisoning. Care should be taken in avoiding exposure of animals to *C. revoluta* when it is being dug up and transported.

The gastrointestinal disorders and hepatic lesions observed in the dogs of this study are probably related to a direct action of methylazoxymethanol (MAM), an aglycone derived from the metabolism of azoglycosides present in several parts of plants of the order Cycadales (Botha *et al.* 1991). Methylazoxymethanol is known to be responsible for the digestive tract lesion (Hooper 1978; Botha *et al.* 1991; Albretsen *et al.* 1998) and for hepatic necrosis (Botha *et al.* 1991; Mills *et al.* 1996; Ziemer 1997; Albretsen *et al.* 1998) in dogs and ruminants poisoned by these plants.

Besides these signs, we saw that Dog 1 showed a symptomatology already described as a 'rigid state,' a condition probably related to a 'semicomatose' state (Reagor *et al.* 1986). The acute clinical evolution, cold extremities, congested ocular mucosa, prostration, and anuria observed in this animal seem to be compatible with the establishment of a peripheral tissue capillary perfusion deficiency and venous stasis, which suggests shock.

It has been discussed that the neurological signs in animals poisoned by Cycadales are secondary to hepatic encephalopathy or caused by direct action of a neurotoxin on the central nervous system (Albretsen *et al.* 1998). Some have speculated that this neurotoxic effect may be mediated by the azoglycoside (Mills *et al.* 1996) and/or β-methylamino-L-alanine amino acid on the central nervous system (Spencer *et al.* 1987). Depression is the most frequent neurological symptom in dogs poisoned by Cycadales (Albretsen *et al.* 1998), as seen in the cases of dogs poisoned by *Zamia floridiana* (Senior *et al.* 1985) and *Marcozamia riedlei* (Mills *et al.* 1996). The dogs in our study also exhibited seizures. Besides this sign, Dog 1 exhibited hind limb paralysis, a sign also described in cases of poisoning by *Cycas* and *Zamia* in cattle in South Africa (Botha *et al.* 1991). However, in this study it was not possible to determine the nature of the lesions responsible for the neurological clinical picture since central nervous system tissue was not collected.

Though incomplete, the necropsy of Dog 2 revealed hepatic lesions compatible with those described in cases of poisoning by *C. revoluta* (Reagor *et al.* 1986) and other Cycadales (Senior *et al.* 1985; Mills *et al.* 1996) in dogs.

Histologically, the hepatic necrosis lesions and cholestasis observed in Dog 2, although less severe and of random distribution, are compatible with those described in other cases of dog poisoning by *C. revoluta* (Reagor *et al.* 1986). No staining was performed to confirm if the micro and macrovacuolations of hepatocytes present in this case were indeed fatty degeneration, as mentioned in cases described by Reagor and colleagues (1986). The megalocytosis of hepatocytes is noteworthy since it has not been described in cases of poisoning by *C. revoluta* (Reagor *et al.* 1986) and other Cycadales (Morton 1967; Senior *et al.* 1985; Mills *et al.* 1996; Ziemer 1997) in dogs. This finding can indicate a chronic and cryptic hepatic injury in accordance with the report from the owner of a continuous ingestion of small amounts of the palm buds by Dog 2. The presence of hepatocytes with two or more nuclei may indicate an attempt by the liver to regenerate tissue after the toxic insult.

The anuria detected at clinical examination in Dog 1 is probably of prerenal origin, which corroborates the circumstantial evidence of shock. Although the urea levels in the urine of this dog were above normal values, azotemia was not present. The histological exam of the kidneys was not performed for this animal. The oliguria seen in Dog 2 also seems to be of prerenal origin, resulting from marked transient dehydration, establishment of a relative hypovolemic picture, and decrease in renal perfusion. In addition, there were no significant histological lesions as nephritis in the kidneys of Dog 2. The vacuolization and biliary pigments in the cytoplasm of epithelial cells of the convoluted tubules as well as bile plugs in the lumen of renal tubules may have partly contributed to the decrease in formation and flow of the renal ultrafiltrate. There is no evidence of azotemia in the laboratory data of this animal.

The increase in the levels of hepatic enzymes and hypoproteinemia detected in the laboratory exams of both animals, described in other cases of poisoning by *C. revoluta* (Botha *et al.* 1991) and Cycadales (Senior *et al.* 1985; Mills *et al.* 1996), reinforce the findings of hepatic lesion and ascites, respectively. Hyperbilirubinemia and jaundice are also frequent in cases of poisoning by Cycadales in dogs (Senior *et al.* 1985; Mills *et al.* 1996). Although the laboratory data have revealed thrombocytopenia in Dogs 1 and 2, we did not observe clinical or macroscopic signs of coagulopathy.

Poisoning by *C. revoluta* in dogs should be considered in the differential diagnosis of hemorrhagic diseases that primarily affect the digestive tract and cause acute toxic hepatic lesions.

## References


Albretsen JC, Khan AS, and Richardson JA (1998). Cycad palm toxicosis in dogs: 60 cases (1987-1997). *Journal of the American Medical Veterinary Association* 213:99-101.

Botha CJ, Naude TW, Swan GE, and Asthon MM (1991). Suspected cycad (*Cycas revoluta*) intoxication in dogs. *Journal of the South African Veterinary Association* 62:189-190.

Chang SS, Chan YL, Wu ML, Deng JF, Chiu TF, Chen JC, Wang FL, and Tseng CP (2004). Acute *Cycas* seed poisoning in Taiwan. *Journal of Toxicology-Clinical Toxicology* 42:49-54.

Gfeller RW and Messonnier SP (2006). *Manual de Toxicologia e Envenenamentos em Pequenos Animais*, p. 284. Roca, São Paulo, Brasil.

Hall WTK (1987). Cycad (*Zamia*) poisoning in Australia. *Australian Veterinary Journal* 64:149-151.

Hooper PT (1978). Cycad poisoning in Australia – etiology and pathology. In *Effects of poisonous plants on livestock* (Keeler RF, VanKampen KR, James LF, eds), pp. 337-347. Academic Press Inc, New York.

Mills JN, Lawley MJ, and Thomas J (1996). Macrozamia toxicosis in a dog. *Australian Veterinary Journal* 73:69-72.

Morton JF (1967). Some notes on cycad uses and hazards. In *Proceedings 5th Conference on Cycad Toxicity*, pp. 24-25.

Reagor JC, Ray AC, Dubuisson L, and Jones LP (1986). Sago Palm (*Cycas-revoluta*) poisoning in the canine. *Southwestern Veterinarian* 37:20.

Reitz D (1929). *Commando: A Journal of the Boer War*. Second revised ed., reprinted 1969, p. 240. Faber & Faber, London.

Senior DF, Sundlof SF, Buergelt CD, Hines AS, O'Neil-Foil CS, and Meyer DJ (1985). Cycad intoxication in the dog. *The Journal of the American Veterinary Hospital Association* 21:103-109.

Spencer PS, Nunn PB, Hugon J, Ludolph AC, Ross SM, Roy DN, and Robertson RC (1987). Guam amyotrophic lateral sclerosis-parkinsonism-dementia linked to a plant excitant neurotoxin. *Science* 237:517-522.

Tustin RC (1983). Notes on the toxicity and carcinogenicity of some South African cycad species with special reference to that of *Encephalartos lanatus*. *Journal of the South African Veterinary Association* 54:33-42.

Ziemer P (1997). Durch Pflanzen und pflanzliche Materialien verursachte Vergiftungen bei Haustieren unter besonderer Berücksichtigung der Kleintiere Eine Literaturübersicht, 275 pp. PhD Dissertation, University of Hannover, Germany.

# Chapter 34

# Natural and Experimental Poisoning of Bovines by *Cestrum corymbosum* Schltdl. in the State of Minas Gerais, Brazil

M.S. Varaschin[1], F. Wouters[1], I. Petta[2], P.S. Bezerra Jr[1], and A.T.B. Wouters[3]

[1]*Universidade Federal de Lavras, DMV, Caixa, Postal 3037, Lavras MG, Brazil, CEP 37200-000;* [2]*Practitioner Veterinarian, Estiva, MG, Brazil;* [3]*Universidade José do Rosário Vellano, Alfenas, MG, Brazil*

## Introduction

Cases of death with postmortem findings of acute hepatic insufficiency have been observed during the last few years in bovines from the Estiva region in the state of Minas Gerais, Brazil. A plant commonly found in the pasture was identified as *Cestrum corymbosum* Schltdl., a shrub belonging to the Solanaceae family. In Brazil there is only one study of spontaneous and experimental poisoning by *C. corymbosum* var. *hirsutum*, associated with acute hepatic necrosis in bovines in Santa Catarina state (Gava *et al.* 1991), and one case report of spontaneous poisoning by *C. corymbosum* in Minas Gerais state (Petta *et al.* 2001). In this report we describe epidemiological, clinical, and pathological aspects of spontaneous and experimental cases of poisoning by *C. corymbosum* from Estiva County in the state of Minas Gerais and to discuss the importance of the plant in the region.

## Natural and Experimental Cases

The spontaneous poisonings by *C. corymbosum* in Minas Gerais were associated with lack of forage during drought periods. The plant is found growing in cultivated fields and roadsides. Two out of five bovines spontaneously poisoned by *C. corymbosum* between August and November were necropsied.

Clinical manifestations were anorexia, apathy, ruminal atony, dry feces covered by mucus, sunken eyes, muscular tremors, and staggering gait. The main necropsy findings were marked lobular patterns of the liver, dried contents of colon and rectum, and hemorrhages in the heart and serosal layer of the rumen. Histologically, there were centrilobular and midzonal coagulative necrosis and hemorrhages and mild cytoplasmic vacuolation in the hepatocytes from the periportal areas.

©CAB International 2011. *Poisoning by Plants, Mycotoxins, and Related Toxins*
(eds F. Riet-Correa, J. Pfister, A.L. Schild, and T.L. Wierenga)

The disease was experimentally reproduced by oral administration of fresh leaves of *C. corymbosum*, collected from a farm where field mortalities had been diagnosed, to 18 cattle during a period of 3 years (between 2000 and 2003), with single doses varying from 17.5 g/kg to 2.5 g/kg BW. The plant was collected during sprouting and flowering vegetative stages. Eleven animals showed clinical signs between 6 and 21 h after plant administration, and four of them died between 19 and 91 h after the ingestion of single doses of 5, 7.5, 15, and 17.5 g of fresh leaves/kg BW. Some cattle that received the same doses did not have clinical signs or recovered after the clinical disease. Bovine 9 was euthanized without clinical signs of poisoning 124 h after the clinical recovery due to a metatarsal fracture. In the animals that did not die, the recovery occurred in a period from 25 to 106 h after plant administration (Table 1). No clinical signs were observed in seven bovines.

**Table 1.** Experimental poisoning of cattle by *C. corymbosum*.

| Bovine | Body weight (kg) | Dose (g/kg) | Total dose (g) | Time/year | Onset of clinical signs* | Outcome* |
|---|---|---|---|---|---|---|
| 1 | 114 | 17.5 | 2000 | 15/11/00 | 12h 30min | 19 h died |
| 2 | 128 | 7.5 | 980 | 04/12/00 | 20 h | 76 h died |
| 3 | 62.8 | 5.0 | 314 | 14/12/00 | 6 h | 24 h died |
| 4 | 127 | 2.5 | 317.5 | 20/12/00 | 8 h | 48 h recovered |
| 5 | 89.8 | 2.5 | 224.5 | 06/02/01 | no clinical signs | - |
| 6 | 150 | 5.0 | 750 | 20/02/01 | 15 h | 25 h recovered |
| 7 | 160 | 5.0 | 800 | 10/03/01 | no clinical signs | - |
| 8 | 142 | 5.0 | 710 | 28/08/01 | no clinical signs | - |
| 9* | 140 | 7.5 | 1050 | 02/10/01 | 8 h | 72 h recovered 196 h euthanized |
| 10 | 140 | 7.5 | 1050 | 22/10/01 | 24 h | 48 h recovered |
| 11 | 100 | 10.0 | 1000 | 05/11/01 | 22 h | 48 h recovered |
| 12 | 100 | 10.0 | 1000 | 03/12/01 | 21 h | 31 h recovered |
| 13 | 65 | 15.0 | 975 | 15/04/02 | no clinical signs | - |
| 14 | 107 | 15.0 | 1605 | 19/08/02 | 21 h | 91 h died |
| 15 | 105 | 10.0 | 1050 | 18/08/03 | 14 h | 106 h recovered |
| 16 | 123 | 7.5 | 923 | 18/08/03 | no clinical signs | - |
| 17 | 67 | 15.0 | 1010 | 30/08/03 | no clinical signs | - |
| 18 | 83 | 15.0 | 1245 | 30/08/03 | no clinical signs | - |

* after plant administration

The clinical manifestations were apathy (Bovines 1, 3, 14, and 15) and/or excitement (Bovines 2, 4, 6, 9, 10, 14, and 15), anorexia (Bovines 1, 2, 12, and 14), sunken eyes (Bovines 2, 9, 10, 11, and 14), muscular tremors (Bovines 1, 2, 3, 4, 6, 9, 10, 12, and 14) more accentuated in the head region (Bovines 1, 9, and 10), reluctance to walk and to stand up (Bovines 1, 2, 3, 14, and 15), staggering gait (Bovines 1, 2, 3, 4, 10, 11, 14, and 15), ruminal atony (Bovines 1, 2, 3, and 14), dried feces covered by mucus (Bovines 1, 2, 3, 4, 6, 9, 10, 14, and 15), sternal recumbence with the head kept down (Bovines 1, 2, 3, 4, 9, and 14), and lateral recumbence (Bovines 1, 2, 3, and 14). Bovine 14 with a longer clinical manifestation period walked in circles, and Bovine 3 had abdominal distention. The clinical manifestations were mild (Bovines 4, 6, 11, and 12), moderate (Bovines 9 and 10), or intense (Bovines 1, 2, 3, 14 and 15).

Postmortem findings were a nutmeg appearance of the liver, where the lobulation was slightly accentuated (Bovines 1 and 3) or markedly accentuated (Bovines 2 and 14), subcapsular petechiae in the liver (Bovines 1 and 3), dried feces in the colon and rectum covered by mucus in the rectum (Bovines 1, 2, 3, and 14) and sometimes streaked with blood (Bovines 1 and 2). Other lesions were petechiae and ecchymoses at the coronary sulcus (Bovines 2, 3, and 14), petechiae, ecchymoses and suffusive hemorrhages in the epicardial and subepicardial region (Bovines 1 and 14), petechiae and echymoses in the subendocardial (Bovines 1, 3, and 14) and myocardial (Bovine 1) region. Petechiae and ecchymoses in the omentum (Bovine 14), serosal layer of the rumen (Bovines 1 and 14), kidney fat (Bovine 3), and urinary bladder mucosa (Bovine 1) were seen. There were edema and hemorrhages of the wall of the gall bladder (Bovine 3). The brain of Bovine 14 had swollen and flattened gyri and herniation of the cerebellar vermis through the foramen magnum. Coning of the cerebellum is described in one bovine associated with *Cestrum parqui* poisoning (McLennan and Kelly 1984), but histopathological description of it was not given. Edema of the brain was described with *C. laevigatum* poisoning in cattle (Van Der Lugt *et al.* 1991), but the pathogenesis was not discussed in these two cases.

In the necropsy of Bovine 3 with abdominal distension, we also observed congestion of the visible mucous membranes and dilation of the abomasum which contained gases, clotted milk, and liquid material. Abomasal dilation is related to atony and/or hypomotility of the organ or to a diet that promotes the accumulation of gases (McGavin and Zachary 2007). In this case dilatation was probably related to the ingestion of the plant along with milk consumption. Histologically, livers had marked centrilobular and midzonal coagulative necrosis and hemorrhages (Bovines 1, 2, and 3). In some cases the necrotic areas joined one another (central bridging necrosis) (Bovines 2 and 14). The necrotic cells had karyopyknosis, karyorrhexis, and increased cytoplasmic eosinophilia. Also, the hepatocytes from the periportal and midzonal areas showed accentuated cytoplasmic vacuolation (Bovines 2 and 14). The cytoplasmic vacuolation was more accentuated in Bovine 14 which had a longer clinical evolution. Bovine 14 also had moderate hemorrhages, necrotic cells, and decrease of cellularity in the centrilobular area. The gall bladder showed hemorrhages mainly in the muscularis and edema of the submucosa, associated with occasional necrosis focuses of the mucous acinus (Bovine 3). Bovine 14 had cerebral edema involving the gray matter with widening of the perivascular and perineuronal spaces. The astrocytes of the gray matter were enlarged and vesicular, some in pairs and surrounded by a clear space similar to the Alzheimer's type II astrocytes. Alzheimer's type II astrocytes are seen in animals with hepatic encephalopathy (McGavin and Zachary 2007). Lesions of mild to moderate edema of the brain involving particularly the cerebral gray matter are described in cattle poisoning by *C. laevigatum* (Van Der Lugt *et al.* 1991) and *Trema micrantha* (Traverso *et al.* 2004). No microscopic lesions were seen in the animal euthanized (Bovine 9).

Several other plants that cause acute hepatotoxicosis in Brazil must be included in the differential diagnosis of intoxication by *C. corymbosum*, such as *C. laevigatum*, *C. parqui*, *C. intermedium*, *Sessea brasiliensis*, *Vernonia mollissima*, *V. rubricaulis*, *Xanthium* spp., *Myoporum laetum* (Tokarnia *et al.* 2000), and *Trema micrantha* (Traverso *et al.* 2004).

*C. corymbosum* studied in Minas Gerais had a lower toxic dose (5-17.5 g/kg BW) than *C. corymbosum* var. *hirsutum* (35 and 39 g/kg) tested in Santa Catarina (Gava *et al.* 1991). Even though *C. corymbosum* has not been classified as a variety in Santa Catarina, some botanists believe it is the same plant and suggest that they should be considered as synonyms (Márcia Vignoli, Federal University of Rio Grande do Sul, personal communication). The variations of toxic doses among the tested plants in both states can be related to variations in the toxicity of the plant or to different susceptibility of the animals.

## Conclusions

There were no other hepatotoxic plants in the fields where the intoxications took place, thus the results of this study demonstrate that *C. corymbosum* is the cause of a disease characterized by acute hepatic insufficiency in the Estiva region, south of Minas Gerais state. Clinical signs, necropsy findings of nutmeg liver and hemorrhages in several tissues, and histological findings of necrosis and acute hepatic hemorrhages are characteristics of the intoxication by *C. corymbosum*. Similar signs and lesions are caused by other hepatotoxic plants with acute action found in Brazil.

## Acknowledgements

This research was supported by FAPEMIG (Fundação de Amparo à Pesquisa do Estado de Minas Gerais). We thank botanist João R. Stehmann (UFMG) for the plant identification.

## References

Gava A, Stolf L, Pilati C, Neves DS, and Viganó L (1991). Intoxicação por *Cestrum corymbosum* var. *hirsutum* (Solanaceae) em bovinos no Estado de Santa Catarina. *Pesquisa Veterinária Brasileira* 11(3/4):71-74.

McGavin MD and Zachary JF (2007). *Pathologic Basis of Veterinary Disease*, 4th edn, 1488 pp. Elsevier, St Louis.

McLennan MW and Kelly WR (1984). *Cestrum parqui* (green cestrum) poisoning in cattle. *Australian Veterinary Journal* 61(9):289-291.

Petta I, Varaschin MS, Wouters F, and Della Lucia MT (2001). Intoxicação natural por *Cestrum corymbosum* Schlecht em bovino no Estado de Minas Gerais – Relato de caso. In *Anais do* X *Encontro Nacional de Patologia Veterinária*, p. 218. Funep, Jaboticabal.

Traverso SD, Corrêa AMR, Schmitz M, Colodel EM, and Driemeier D (2004). Intoxicação experimental por *Trema micrantha* (Ulmaceae) em bovinos. *Pesquisa Veterinária Brasileira* 24 (4):211-216.

Tokarnia CH, Döbereiner J, and Peixoto PV (2000). *Plantas tóxicas do Brasil*, 310 pp. Helianthus, Rio de Janeiro.

Van Der Lugt JJ, Nel PW, and Kitchin JP (1991). The pathology of *Cestrum laevigatum* (Schlechtd.) poisoning in cattle. *Onderstepoort Journal of Veterinary Research* 58:211-221.

# Chapter 35

# *Trema micrantha* Poisoning in Domestic Herbivores

P.M. Bandarra[1], S.D. Traverso[2], D.L. Raymundo[1], S.P. Pavarini[1], L. Sonne[1], C.E.F. Cruz[1], and D. Driemeier[1]

[1]*Setor de Patologia Veterinária, Universidade Federal do Rio Grande do Sul, Porto Alegre, RS, Brazil;* [2]*Laboratório de Patologia Animal, Universidade Estadual de Santa Catarina, Lages, SC, Brazil*

## Introduction

*Trema micrantha* (Ulmaceae) is a fast growing perennial tree that may reach 5 to 15 m high and is widely distributed in South America. The plant is abundant in woodland habitats and as secondary vegetation on abandoned areas. It has also been used as a pioneer in reforestation, especially for the recovery of burned or degraded soils (Castellani and Aguiar 1998; Kissmann 1999). *T. micrantha* leaves are palatable and readily consumed by cattle and other ruminants, especially during drought (Kissmann 1999). However, it has been associated with outbreaks of acute hepatic insufficiency due to hepatocellular necrosis in goats (Traverso *et al.* 2004). *T. tomentosa*, a related species from Australia, has similarly been described as a cause of spontaneous poisoning in horses (Hill *et al.* 1985), camels (Trueman and Powell 1991), cattle, goats, sheep (Mulhearn 1942), and deer (McKenzie *et al.* 1985). This chapter describes clinical and pathological aspects of *T. micrantha* poisoning in domestic herbivores in the state of Rio Grande do Sul, Brazil.

## Epidemiology

Data regarding natural and experimental cases of *T. micrantha* poisoning were retrospectively retrieved from farmers and the records of the Veterinary Pathology Service of the Federal University of Rio Grande do Sul during the period of 2000-2008. In the indicated period, there were 7 goats, 2 horses, 6 cattle, and 9 rabbits poisoned by the plant. While spontaneous cases affected goats and horses, experimental reproduction was induced in goats, cattle, and rabbits. Natural poisoning occurred after intentional (plant was added to the diet) or accidental (falling branches or trees) consumption of green leaves of the plant by the animals. Experimentally, *T. micrantha* has been toxic to goats, rabbits, and cattle at 30, 35, and 54g/kg BW, respectively (Traverso and Driemeier 2000; Traverso *et al.* 2002, 2004).

©CAB International 2011. *Poisoning by Plants, Mycotoxins, and Related Toxins* (eds F. Riet-Correa, J. Pfister, A.L. Schild, and T.L. Wierenga)

## Clinical Signs

Main clinical signs seen in *T. micrantha* poisoning of cattle, goats, and horses include changes in fecal consistency (varying between liquid and pasty), anorexia, apathy, progressive weakness, sialorrhea, muscular tremors (especially of the anterior members), aggressive behavior, hypermetria, dysphagia (animals hold forage in their mouth without masticating or swallowing), abnormal posture, reluctance to movement, jaundice, sternal or lateral recumbence, paddling, coma, and death. Experimentally, in cattle first signs were seen 16 h after consumption of a toxic dose and death occurred between 67 and 153 h. In goats, first signs and death occurred at 48 h and 4 days after the end of ingestion (Traverso *et al.* 2002, 2004).

## Macroscopic Changes

The most significant gross lesions are observed in the livers, which are yellowish, friable, and with pronounced lobular pattern. Their cut surfaces show reddened and depressed areas alternated with whitish ones. Petechial hemorrhages in the subcutaneous tissues, in the region between the chest and scapula, in the epicardium, mediastinum, and serosal membranes of the abdominal organs are also observed. Dried feces covered by mucus and blood have also been consistent findings.

## Microscopic Changes

The main histological changes consist of coagulative centrilobular to massive hepatic necrosis, sometimes associated with congestion, hemorrhages, and degenerative changes in adjacent hepatocytes. Additional microscopic lesions include vacuolation, degeneration, and necrosis in the neurons of the brain stem, cortex, hippocampus, Purkinje cells, and gray matter of the medulla.

## Discussion

Centrolobular necrosis and hepatocyte degeneration are particularly common in hepatotoxic diseases due to sanguineous irrigation particularities and pronounced enzymatic activity of mixed-function oxidases that occur in hepatocytes from this area. Those enzymes may transform inactive compounds into toxic metabolites (Cullen 2007). Hemorrhages are secondary to the damaged liver due to both the excessive consumption in necrotic insults and inability to further synthesize coagulation factors and platelets (Stalker and Hayes 2007). Neurological signs resultant of hepatic encephalopathy are caused by the systemic accumulation of ammonia, short chain fatty acids, and mercaptanes, besides a decrease in neurotransmissor and glucose levels. Clinical signs and pathological changes seen in cases of *T. micrantha* poisoning are typical of intoxications associated with acute hepatic necrosis. Therefore, this condition must be differentiated from those caused by other hepatotoxic agents. The general clinical and pathological picture described here may also be seen in poisoning caused by other plants such as *Dodonea viscosa* (Colodel *et al.* 2003), *Xanthium cavallinesii* (Méndez *et al.* 1998), *Myoporum laetum* (Raposo *et al.* 1998), *Cestrum* spp. (Riet-Correa *et al.* 1986; Gava *et al.* 1991; Peixoto *et al.* 2000), *Vernonia* spp.

(Döbereiner *et al.* 1976; Tokarnia and Döbereiner 1982), *Sessea brasiliensis* (Canella *et al.* 1968), and *Crotalaria retusa* (Nobre *et al.* 2005), all of which also may cause acute hepatic damage in ruminants in Brazil. Information presented here emphasizes the importance of *T. micrantha* as a cause of acute hepatopathy in domestic herbivores managed or kept in areas where the plant occurs.

## References

Canella FCC, Tokarnia CH, and Döbereiner J (1968). Intoxicação por *Sessea brasiliensis* Toledo em bovinos. *Pesquisa Agropecuária Brasileira* 3:333-340.

Castellani ED and Aguiar IB (1998). Preliminary conditions for germination of *Trema micrantha* (L.) Blume seeds. *Revista Brasileira de Engenharia Agrícola e Ambiental* 2(1):80-83.

Colodel EM, Traverso SD, Seitz AL, Correa A, Oliveira FN, Driemeier D, and Gava A (2003). Spontaneous poisoning by *Dodonea viscosa* (Sapindaceae) in cattle. *Veterinary and Human Toxicology* 45(3):147-148.

Döbereiner J, Tokarnia CH, and Purisco E (1976). *Vernonia mollissima*, planta tóxica responsável por mortandades de bovinos no sul de Mato Grosso. *Pesquisa Agropecuária Brasileira* 11:49-58.

Gava A, Stolf L, Pilati C, Neves DS, and Viganó L (1991). Intoxicação por *Cestrum corymbosum* var. *hirsutum* (Solanaceae) em bovinos no estado de Santa Catarina. *Pesquisa Veterinária Brasileira* 11(3/4):71-74.

Hill BD, Wills LD, and Dowling RM (1985). Suspected poisoning of horses by *Trema aspera* (poison peach). *Australian Veterinary Journal* 62(3):107-108.

Kissmann KG (1999). *Plantas Infestantes e Nocivas*, pp. 643-644. BASF, São Paulo.

McKenzie RA, Green PE, Thornton AM, Chung YS, Mackenzie AR, Cybinski DH, and George TD (1985). Diseases of deer in south eastern Queensland. *Australian Veterinary Journal* 62(12):424.

Méndez MC, dos Santos RC, and Riet-Correa F (1998). Intoxication by *Xanthium cavanillesii* in cattle and sheep in southern Brazil. *Veterinary and Human Toxicology* 40(3):144-147.

Mulhearn CR (1942). Poison peach (*Trema aspera*): a plant poisonous to stock. *Australian Veterinary Journal* 18:68-72.

Nobre VMT, Dantas AFM, Riet-Correa F, Barbosa JM, Tabosa IM, and Vasconcelos JS (2005). Acute intoxication by *Crotalaria retusa* in sheep. *Toxicon* 45(3):347-352.

Peixoto PV, Brust LC, Duarte MD, Franca TN, Duarte VC, and Barros CS (2000). *Cestrum laevigatum* poisoning in goats in southeastern Brazil. *Veterinary and Human Toxicology* 42(1):13-14.

Raposo JB, Mendez MC, de Andrade GB, and Riet-Correa F (1998). Experimental intoxication by *Myoporum laetum* in cattle. *Veterinary and Human Toxicology* 40 (5):275-277.

Riet-Correa F, Schild AL, and Méndez MC (1986). Intoxicação por *Cestrum parqui* (Solanaceae) em bovinos no Rio Grande do Sul. *Pesquisa Veterinária Brasileira* 6(4):111-115.

Stalker MJ and Hayes MA (2007). Liver and biliary system. In *Jubb, Kennedy, and Palmer's Pathology of Domestic Animals* (MG Maxie, ed.), pp. 297-381. Elsevier, Philadelphia, Pennsylvania.

Tokarnia CH and Döbereiner J (1982). Intoxicação de bovinos por *Vernonia rubricaulis* (Compositae) em Mato Grosso. *Pesquisa Veterinária Brasileira* 2(4):143-147.

Traverso SD and Driemeier D (2000). Experimental *Trema micrantha* (Ulmaceae) poisoning in rabbits. *Veterinary and Human Toxicology* 42(5):301-302.

Traverso SD, Corrêa AMR, Pescador CA, Colodel EM, Cruz CEF, and Driemeier D (2002). Intoxicação experimental por *Trema micrantha* (Ulmaceae) em caprinos. *Pesquisa Veterinária Brasileira* 22(4):141-147.

Traverso SD, Correa AMR, Schmitz M, Colodel EM, and Driemeier D (2004). Intoxicação experimental por *Trema micrantha* (Ulmaceae) em bovinos. *Pesquisa Veterinária Brasileira* 24 (4):211-216.

Trueman KF and Powell MW (1991). Suspected poisoning of camels by *Trema tomentosa* (poison peach). *Australian Veterinary Journal* 68(6):213-214.

# REPRODUCTIVE SYSTEM

# Chapter 36

# Plants Teratogenic to Livestock in the United States

K.E. Panter, K.D. Welch, S.T. Lee, D.R. Gardner, B.L. Stegelmeier, M.H. Ralphs, T.Z. Davis, B.T. Green, J.A. Pfister, and D. Cook

*USDA-ARS Poisonous Plant Research Laboratory, Logan, Utah 84341, USA*

## Introduction

Teratology, as a scientific discipline, is relatively new and recognition of poisonous plants that cause birth defects in livestock only came to the forefront in the 1950s and 1960s. Prior to this time many of the congenital deformities in livestock were assumed to be of genetic origin and because of the negative connotations related to the quality of the livestock producer's gene pool, the resultant offspring were destroyed with little or no recognition or follow up investigation. *Veratrum*-induced 'monkey faced' lamb syndrome and lupine-induced 'crooked calf disease,' both studied extensively at the Poisonous Plant Research Laboratory (PPRL), were two very significant discoveries that elevated the importance of plant-induced birth defects in livestock and also advanced recognition of teratology as an important and relevant scientific discipline. The clinical and gross manifestations of many of these birth defects in animals have their counterparts in certain human disease conditions. Thus, the study of plant-induced malformations in animals, where research can be humanely and readily conducted, provides applicable and relevant research information to related human conditions. For example, one of the teratogenic alkaloids in *Veratrum*, cyclopamine (identified and characterized at the PPRL), is now being investigated for cancer chemotherapy and derivatives of cyclopamine are currently in human clinical trials. A Spanish goat cleft palate model developed to study the mechanism of lupine-induced crooked calf disease is being utilized to develop new biomedical tools and improve methods of treatment for cleft palate in children.

Since the discovery of the teratogenic effects of lupine and *Veratrum* other plants have been added to the list known to cause birth defects in animals. The ensuing research efforts at the PPRL and other institutions have characterized many specific teratogenic chemical compounds, determined mechanisms of action, described chemical structure–activity relationships, and provided the impetus to develop management strategies to improve understanding of the underlying causes and reduce losses to livestock producers. In addition to *Veratrum* and lupines, poison hemlock, *Nicotiana* spp., locoweeds, *Lathyrus*, *Solanum* spp., cyanide-containing plants, and others contribute to the overall losses to the livestock industry in the USA from poisonous plants.

©CAB International 2011. *Poisoning by Plants, Mycotoxins, and Related Toxins* (eds F. Riet-Correa, J. Pfister, A.L. Schild, and T.L. Wierenga)

## *Veratrum*

There are over 11 species of *Veratrum* in the lily family distributed across the USA and Canada. *V. californicum* is the most notable because of its early history of large losses to the sheep industry from extensive and lethal craniofacial birth defects. During the mid-20th century, sheep flocks in central Idaho that grazed certain *Veratrum*-infested ranges experienced a rate of birth defects in their ewes of over 25%, and when early embryonic loss was included, the losses were even greater (Binns *et al.* 1965; James 1999). The birth defects reported included a range of malformations from gross craniofacial anomalies (cyclopia) to less severe deformities of the upper and lower jaws or limbs. The Basque shepherds called the cyclopic defect 'chatto' which translates to 'monkey-faced' lamb. While losses from *Veratrum* have long been reduced or eliminated on these ranges because of research and subsequent management strategies, the cyclopic defect is still reported. In recent years isolated cases of congenital cyclops have been reported in a flock of sheep in Utah and in a group of Alpacas in Big Lake, Alaska (personal communications). *V. californicum* is the species most associated with these birth defects, however *V. album* is suspected in Alaska and other species are also believed to be teratogenic.

The most notable malformations are a multitude of craniofacial anomalies including synophthalmia (cyclopia). Pregnant ewes fed *Veratrum* on gestation day 13 or 14 produced malformed lambs while those dosed on day 15 produced normal lambs and ewes dosed for 3 consecutive days (13-15) resorbed their fetuses (Welch *et al.* 2009). High embryonic losses have been reported when *Veratrum* ingestion occurred during any stage of gestation from days 13-19, and limb reductions and tracheal stenosis resulted when *Veratrum* ingestion included gestation days 28-33 (Keeler *et al.* 1986).

*V. viride* is the most widespread species of this genera and grows in the northwestern USA, Canada, into Alaska and across the northeastern USA; *V. insolitum* grows in a relatively small region of the northwest in northern California and southern Oregon; *V. parviflorum* grows in the central southeastern states; and *V. woodii* grows in the midwestern and southern states (Burrows and Tyrl 2001). *V. californicum* grows in high mountain ranges of the western USA and is found in open alpine meadows, open woodlands, along marshes, swamps or lakes (Knight and Walter 2001). Three additional species are common in other countries, i.e. *V. japonicum* in Korea and *V. album* and *V. nigrum* in Europe. All *Veratrum* species grow in similar habitats of moist meadows and woodlands where adequate soil moisture and growing conditions support populations. Plant description is similar among all species with coarse erect stalks 1-2.5 m tall. The leaves are large (up to 30 cm long and 15 cm wide), smooth, oval or lanceolate with parallel veins. The inflorescence is a panicle of numerous, small, white or greenish white, star-shaped flowers. Seeds are three-chambered.

The teratogens responsible for the malformations are steroidal alkaloids including cyclopamine, cycloposine, and jervine (Keeler 1984). In recent biomedical research these alkaloids have been used as probes or tools providing a basic understanding of a multitude of biological developmental processes in mammals (Gaffield *et al.* 2000). The most significant finding is the powerful and selective inhibition by cyclopamine of the Sonic hedgehog signaling pathway, an important feature in the research of complex biochemical mechanisms of birth defects in humans and cancers wherein this pathway is integral.

## Nightshades

The nightshade family is large and comprises over 2300 species worldwide. While most contain toxic alkaloids only a few genera including *Datura*, *Solanum*, and *Nicotiana* have been associated with birth defects in the USA. *Nicotiana* spp. will not be discussed here because the mechanism of teratogenesis is similar to that of lupine and poison-hemlock and will be included below.

Spirosolane alkaloids found widely in the *Solanum* genus are structurally related to the teratogenic *Veratrum* alkaloids and contribute to the list of toxins with known teratogenic activity. While solasodine was teratogenic in a hamster model, neither tomatidine nor the form of solasodine which lacks a nitrogen atom (diosgenin) showed any teratogenic activity. The solanidane alkaloids in potatoes and other plants are less closely related to the *Veratrum* alkaloids but research studies demonstrated that the alkaloid glycosides alpha-solanine and alpha-chaconine and the aglycone epimers, solanidine and demissidine, were teratogenic in hamsters, although at a reduced level of activity (Keeler *et al.* 1993).

## Cyanogenic Plants

While the implication of malformations induced by plants containing cyanogenic glycosides lacks experimental substantiation, reports of skeletal defects in pigs, calves, and foals associated with maternal ingestion of cyanogenic plant species is cause for further investigation. Limb contractures in calves and foals have been reported with a known history of maternal ingestion of sudan or sorghum (Van Kampen 1970; Seaman *et al.* 1981). Other signs of toxicoses and pathology in mares such as posterior ataxia, cystitis, and myelomalacia indicate that a potential teratogenic effect may exist. Similar contracture malformations in pigs have been historically implicated with consumption of wild black cherries by pregnant sows (Selby *et al.* 1971). While cyanogenic glycosides are believed to be the cause, confirmation of a specific compound or group of compounds is lacking, although fetal anoxia from HCN has been speculated.

## Lupines, Poison-hemlock, and *Nicotiana* spp.

These three genera are combined for this discussion because the gross descriptions of the malformations are essentially the same and the mechanism of action is the same (Panter *et al.* 1990; Weinzweig *et al.* 2008). The teratogenic alkaloids include anagyrine, ammodendrine, and ammodendrine derivatives present in some lupines, coniine and coniine derivatives in poison-hemlock, and anabasine and anabasine derivatives in *Nicotiana* spp. All of these teratogenic alkaloids inhibit fetal activity and when this occurs during susceptible stages of pregnancy the fetus is born with multiple congenital defects including arthrogryposis, scoliosis, kyphosis, torticollis, or lordosis and cleft palate or any combination of these contractures.

### Lupines

There are over 150 lupines including annual, perennial, or woody species. Most wild species contain toxic or teratogenic alkaloids. Lupines belong to the legume family with alternate palmately compound leaves. Flowers are pea-like and can be blue, purple, white, yellow or reddish. Seeds are flattened in legume-like pods. Many lupines are difficult to identify taxonomically, and chemical analysis is required to determine toxic or teratogenic risk.

Lupines are considered toxic to all livestock species, but overt poisoning generally occurs in small ruminants while the major problem in the USA is 'crooked calf syndrome' (Panter *et al.* 2009). This results when pregnant cows graze lupines containing the teratogenic alkaloids during gestation days 40-100. Cleft palate occurs occasionally when ingestion includes 40-50 days gestation and severity and extent of the skeletal contractures are dependent on duration of ingestion, amount consumed, and stage of pregnancy when the fetus is exposed to teratogenic alkaloids (Panter *et al.* 1999).

**Poison-hemlock**

Few species of poison-hemlock occur worldwide and only one is described in the USA, *Conium maculatum*. A biennial plant 1-3 m tall, poison-hemlock has stems that are stout, rigid, smooth, and hollow except at the nodes. Purple splotches are a distinguishing characteristic as is the carrot-like single white taproot. Leaves are large, triangular, carrot-like, and alternate on the stem. Flowers are small, white, and form into umbellate clusters. Seeds are grayish-brown, with wavy, knotted ridges. The plant frequently grows along fences and in waste areas but will encroach into hay fields and pastures. Geographically, it grows throughout the USA and has adapted to most climates.

There are at least eight known piperidine alkaloids, three of which are believed to be teratogenic: coniine, gamma-coniceine, and *N*-methyl coniine. Historically, teratogenic effects were most significant in pigs but cattle, sheep, goats, and horses have also been reported to be affected (Panter *et al.* 1999). Other species including wildlife and birds have also been poisoned. The susceptible period of gestation in pigs, sheep, and goats is 30-60 days with comparable periods in cattle and horses. Cleft palate is also reported if poison-hemlock ingestion includes gestation days 30-40 in pigs, sheep, and goats (Panter *et al.* 1999).

***Nicotiana* spp.**

About 60 species of *Nicotiana* are known throughout the world. While only two species, *Nicotiana tabacum* and *N. glauca* have been implicated in the induction of teratogenic effects in livestock in the USA, others are suspected to contain alkaloids with teratogenic activity, and further research is needed.

*N. tabacum* (burley tobacco) is an annual plant with erect stalks and branches containing ovate or lanceolate leaves. Flowers are cream-colored, trumpet-like and seeds are kidney-shaped, brown, and fluted or ribbed. *N. glauca* (wild tree tobacco) is a shrub or tree depending on the habitat; stalks or branches are woody, leaves are ovate, blue-green in color with a waxy appearance. Flowers are yellow, tubular, and appear throughout the seasons. Seeds are small and dark reddish-brown.

*N. tabacum* was first reported to cause skeletal birth defects and cleft palate in pigs. This occurred when tobacco stalks were fed to pregnant sows in the midwestern and southern states (Menges *et al.* 1970). It was determined that anabasine was the teratogenic alkaloid, not nicotine (Keeler *et al.* 1984). This was further supported when *N. glauca*, containing exclusively anabasine, was experimentally fed to pregnant sows, sheep, goats, and cows causing the same contracture skeletal defects and cleft palate as reported in pigs and the same as described in lupine-induced 'crooked calf syndrome' in cattle.

## Locoweeds

Locoweeds are those species of the *Astragalus* and *Oxytropis* genera that contain the indolizidine alkaloid swainsonine. There are approximately 24 *Astragalus* and *Oxytropis* species known to contain swainsonine, and while neurological and reproductive disorders are the most common disease conditions reported with locoweed poisoning, occasional birth defects occur in sheep. The mechanism of action is not known and whether the teratogen is swainsonine, some other compound, or a combination is speculative.

Locoweeds are distributed worldwide and their toxic effects are known in many parts of the world. Locoweed toxicoses throughout the world occur sporadically because of the cyclic nature of plant populations due to changing climatic conditions. The timing and amount of precipitation influences the cyclic nature of locoweed populations and subsequent outbreaks of poisoning.

The congenital malformations reported in sheep are characterized by excessive flexure of the carpal joints or contracted tendons, or both (James *et al.* 1967). Some animals have anterior flexure and hypermobility in the hock joints. Ingestion of locoweed by pregnant ewes at almost any period during gestation may cause the contractures. The nature of these deformities would suggest that inhibition of fetal movement *in utero* as shown in lupine-induced crooked calf syndrome may be a contributing factor. Obviously, there are substantial differences and the mechanism of action has not been determined nor has the specific teratogen been isolated. Similar malformations associated with plants containing swainsonine have occurred in sheep and goats in South America. However, other similar indolizidine alkaloids are also present and it is suspected that a combination of toxins may be responsible for these plant-induced malformations.

## *Lathyrus* and *Vicia*

Certain members of the *Lathyrus* and *Vicia* genera contain compounds called osteolathyrogens which are teratogenic, causing congenital skeletal defects in offspring. The malformations are characterized as contracture or flexure of the pastern and carpal joints, lateral rotation of the forelimbs, scoliosis, kyphosis, torticollis, and front limb abductions. The extent to which these two genera contribute to congenital malformations in livestock grazing on native ranges in the USA is unknown but believed to be minimal. The malformations have been reproduced by experimental feeding of the synthetic osteolathyrogen aminoacetonitrile for as few as 10 days anytime during gestation days 20-129 in sheep (Keeler and James 1971). The natural lathyrogen believed to be the teratogen is β-aminopropionitrile.

## *Leucaena* and Related Plants

*Leucaena* is a tropical plant used for forage in tropical states including the US Virgin Islands. Mimosine is considered toxic; however *Leucaena* is considered a good source of forage protein for ruminants if ingestion is limited to less than 50% of their diet. Malformations have been reported in rats and swine but are apparently of little significance to grazing livestock. However, in South American countries *Mimosa tenuiflora* is responsible for various skeletal and craniofacial malformations in livestock and the first trimester is believed to be the most susceptible gestational period (Riet-Correa *et al.* 2009).

While mimosine is suspected as a teratogen in these cases, the putative toxin remains unknown.

## Discussion

Reported estimates are that 5% of all grazing livestock in the USA encounter poisonous plants with some form of negative effects and that 1-2% result in death or are otherwise lost to production (Keeler 1979; Keeler *et al.* 1993). The incidence of livestock congenital malformations has been estimated at 1-3% of all births (Dennis and Leipold 1979) and it has been speculated that 33% of all congenital malformations in livestock are induced by poisonous plants. Most of the relevant data on plant teratogenesis in livestock have been obtained through field observations or from studies on domestic livestock. There are obvious limitations with research on large livestock species, particularly when using isolated and purified toxins. Some of the limitations include size of the animal, length of gestation, large dosage requirements, housing limitations, etc. In some cases, rodent models may provide relevant data, however subsequent research is required in the target animal species for confirmation. Research at the PPRL will continue to identify teratogenic plants, identify and characterize putative teratogens, define mechanisms of teratogenesis, and develop management strategies to prevent livestock losses.

## References

Binns W, Shupe JL, Keeler RF, and James LF (1965). Chronological evaluation of teratogenicity in sheep fed *Veratrum californicum. Journal of the American Veterinary Medical Association* 147:839-842.

Burrows GE and Tyrl RJ (2001). *Toxic Plants of North America*, Iowa State Press, Ames, 1342 pp.

Dennis SM and Leipold HW (1979). Ovine congenital defects. *Veterinary Bulletin* 49:233-239.

Gaffield W, Incardona JP, Kapur RP, and Henk R (2000). Mechanistic investigation of *Veratrum* alkaloid-induced mammalian teratogenesis. In *Natural and Selected Synthetic Toxins: Biological Implications* (Tu AT and Gaffield W, eds), pp. 173-187. American Chemical Society, Washington DC.

James LF (1999). Teratological research at the USDA-ARS Poisonous Plant Research Laboratory. *Journal of Natural Toxins* 8:63-80.

James LF, Shupe JL, Binns W, and Keeler RF (1967). Abortive and teratogenic effects of locoweed on sheep and cattle. *American Journal of Veterinary Research* 28:1379-1388.

Keeler, RF (1979). Toxins and teratogens of the Solanaceae and Liliaceae. In *Toxic Plants* (AD Kinghorn, ed.), pp. 59-82. Columbia University Press, Irvington-Hudson, New York.

Keeler RF (1984). Teratogens in plants. *Journal of Animal Science* 58:1029-1039.

Keeler RF and James LF (1971). Experimental teratogenic lathyrism in sheep and further comparative aspects with teratogenic locoism. *Canadian Journal of Comparative Medicine* 35:332-341.

Keeler RF, Crowe MW, and Lambert EA (1984). Teratogenicity in swine of the tobacco alkaloid anabasine isolated from *Nicotiana glauca. Teratology* 30:61-69.

Keeler RF, Stuart LD, and Young S (1986). When ewes ingest poisonous plants: the teratogenic effects. *Veterinary Medicine* 81:449-454.

Keeler RF, Gaffield W, and Panter KE (1993). Natural products and congenital malformation: structure-activity relationships. In *Dietary Factors and Birth Defects* (RP Sharma, ed.), pp. 310-331. California Academy of Sciences, San Francisco, California.

Knight AP and Walter RG (2001). *A Guide to Plant Poisoning of Animals in North America*, 367 pp. Teton NewMedia, Jackson, Wyoming.

Menges RS, Selby LA, Marienfeld CJ, Aue WA, and Greer DL (1970). A tobacco related epidemic of congenital limb deformities in swine. *Environmental Research* 3:285-290.

Panter KE, Bunch TD, Keeler RF, Sisson DV, and Callan RJ (1990). Multiple congenital contractures (MCC) and cleft palate induced in goats by ingestion of piperidine alkaloid-containing plants: Reduction in fetal movement as the probable cause. *Clinical Toxicology* 28:69-83.

Panter KE, James LF, and Gardner DR (1999). Lupines, poison-hemlock and *Nicotiana* spp: Toxicity and teratogenicity in livestock. *Journal of Natural Toxins* 8:117-134.

Panter KE, Motteram E, Cook D, Lee ST, Ralphs MH, Platt TE, and Gay CC (2009). Crooked calf syndrome: Managing lupines on the rangelands of the Channel Scablands of east-central Washington State. *Rangelands* 31:10-15.

Riet-Correa F, Medeiros RMT, Pfister J, Schild AL, and Dantas AFM (2009). *Poisonings by plants, mycotoxins and related substances in Brazilian livestock,* pp. 170-174. Sociedade Vicente Pallotti-Editora, Santa Maria, RS.

Seaman JT, Smeal MG, and Wright JC (1981). The possible association of a sorghum (*Sorghum sudanese*) hybrid as a cause of developmental defects in calves. *Australian Veterinary Journal* 57:351-352.

Selby LA, Manges RW, Houser EC, Glatt RE, and Case AA (1971). Outbreak of swine malformations associated with the wild black cherry, *Prunus serotina*. *Archives of Environmental Health* 22:496-501.

Van Kampen KR (1970). Sudan grass and sorghum poisoning of horses: a possible lathyrogenic disease. *Journal of the American Veterinary Medical Association* 156:629-630.

Weinzweig J, Panter KE, Jagruti P, Smith DM, Spangenberger A, and Freeman MB (2008). The fetal cleft palate: V. Elucidation of the mechanism of palatal clefting in the congenital caprine model. *Plastic and Reconstructive Surgery* 121:1328-1334.

Welch KD, Panter KE, Lee ST, Gardner DR, Stegelmeier BL, and Cook D (2009). Cyclopamine-induced synophthalmia in sheep: Defining a critical window and toxicokinetic evaluation. *Journal of Applied Toxicology* 29:414-421.

# Chapter 37

# Dose-Response Evaluation of *Veratrum californicum* in Sheep

K.D. Welch, S.T. Lee, D.R. Gardner, K.E. Panter, B.L. Stegelmeier, and D. Cook

*USDA-ARS Poisonous Plant Research Laboratory, Logan, Utah 84341, USA*

## Introduction

*Veratrum* belongs to the Liliaceae (Lily) family and is comprised of at least five species in North America. *V. viride* is the most widespread species and grows throughout the northwestern USA, western Canada, Alaska, and the northeastern USA. Two other common species are *V. album* and *V. californicum*. Common names for *Veratrum* species include western false hellebore, hellebore, skunk cabbage, corn lily, Indian poke, wolfsbane, etc. Most *Veratrum* species are found in similar habitats of moist, open alpine meadows or open woodlands, marshes, along waterways, in swamps or bogs, and along lake edges in high mountain ranges (Kingsbury 1964). Most species grow at higher elevations. All species are similar with coarse, erect plants about 1-2.5 m tall with short perennial rootstalks. The leaves are smooth, alternate, parallel veined, broadly oval to lanceolate, up to 30 cm long, 15 cm wide, in three ranks, and sheathed at the base (Burrows and Tyrl 2001). All species should be considered poisonous and capable of causing acute intoxication. Over 50 complex steroidal alkaloids have been indentified from *Veratrum* species. Alkaloid concentrations are highest in the leaves from June through early July and then decline in August, when the roots attain their highest concentrations. The stems appear to be intermediate between leaves and roots (Keeler and Binns 1971).

Over 50 years ago scientists at the Poisonous Plant Research Laboratory demonstrated that holoprosencephaly and the related craniofacial deformities (called 'monkey face lamb disease') were produced in lamb fetuses when pregnant ewes grazed *V. californicum* early in gestation (Binns *et al.* 1962, 1963, 1965). Early field poisonings reported incidences as high as 25% of the lambs in large flocks of sheep (flocks of 5000-10,000 ewes) (Binns *et al.* 1963). Further studies demonstrated that maternal *Veratrum* ingestion produced a variety of congenital defects including tracheal stenosis (Keeler *et al.* 1985), carpal and tarsal shortening (Keeler and Stuart 1987), and early embryonic death and resorption (Keeler 1990). Retrospectively early embryonic death and resorption were later associated with low reproductive rates when sheep were grazed in areas with abundant *V. californicum* (Binns *et al.* 1963; Van Kampen *et al.* 1969). Management schemes were developed and implemented to avoid maternal *Veratrum* exposure during susceptible times and the

©CAB International 2011. *Poisoning by Plants, Mycotoxins, and Related Toxins* (eds F. Riet-Correa, J. Pfister, A.L. Schild, and T.L. Wierenga)

incidence of *Veratrum*-induced birth defects in range animals is now negligible. However, sporadic cases of *Veratrum*-induced malformations in sheep and other species such as llamas and alpacas have been reported to our laboratory (personal communications). The alkaloids responsible for terata induction in *V. californicum* have been identified as jervine, 11-deoxojervine (which has been named cyclopamine), and cycloposine (the glycoside of cyclopamine) (Keeler and Binns 1968).

The mechanism of cyclopamine-induced birth defects has been shown to result from the inhibition of the Sonic Hedgehog signal transduction pathway ( Cooper *et al.* 1998; Incardona *et al.* 1998). Further research demonstrated that cyclopamine antagonized Hedgehog signaling by binding directly to Smoothened, a key factor in the Hedgehog signaling pathway (Chen *et al.* 2002). The Hedgehog signaling pathway plays an integral role in cell growth and differentiation, including embryonic development of the eyes (Rubenstein and Beachy 1998; Lum and Beachy 2004).

Early evaluation of the chronology of teratogenicity of *V. californicum* in sheep indicated that the plant must be ingested between gestation days (GD) 10-15 in order to cause synophthalmia malformations (Binns *et al.* 1963). It was later reported that GD 14 is the critical day for synophthalmia malformations to occur (Binns *et al.* 1965), as it was observed that ewes dosed with *V. californicum* on GD 11-13 and 15-16 all had normally developed fetuses. These data suggest that the critical window for synophthalmia formation is approximately 1 day. Recent work demonstrated that the elimination rate of cyclopamine is very rapid (approximately 1.1 h) (Welch *et al.* 2009). The rapid clearance of cyclopamine indicates that ingestion of *V. californicum* must occur during a very narrow window for synophthalmia formation to occur.

The clinical signs of *Veratrum* intoxication are typically limited to excess salivation with froth around the mouth, weakness, trembling, incoordination, limb paresis, and recumbence. These signs may be accompanied by slow respiration, irregular heart rate and rhythm, and cyanosis. Signs may begin 2-3 h post-ingestion and persist for several hours in the case of mild intoxications or for 1-4 days for more severe poisonings (Binns *et al.* 1965; Keeler 1990). The objective of this study was to determine the maximum tolerated dose of ground *V. californicum* root and the corresponding dose of cyclopamine that would not severely incapacitate the ewes or cause fetal death, but still cause craniofacial malformations in the lambs.

## Materials and Methods

### Plant material

Root material from *V. californicum* plants was the source of cyclopamine for the oral dosing experiments in this study. Both the aerial and root/rhizome portions of the plant contain the teratogen cyclopamine (Keeler and Binns 1966a), and both can induce 'monkey face lamb' defects (Binns *et al.* 1965). However, the concentration of cyclopamine is 5-10 times higher in root material (Keeler and Binns 1966a, b, 1971). The plant material was collected in Muldoon Canyon at the headwaters of the Lost River Drainage in Idaho (PPRL collections number 85-18). Plant material was transported to our laboratory, dried in sunlight, finely chopped, and stored in an enclosed shed at ambient temperature.

Extraction of *V. californicum* for cyclopamine analysis was accomplished by weighing 100 mg of ground *Veratrum* root material into a 10 ml screw top test tube equipped with Teflon lined caps. Then 4 ml $CH_2Cl_2$ and 200 μl concentrated $NH_4OH$ were added and the test tubes placed in a mechanical shaker for 16 h, then centrifuged to separate the plant residue and the $CH_2Cl_2/NH_4OH$ extract. The extract was transferred to a clean 10 ml screw top test tube. This extraction was repeated 2 times for 3 h and all extracts combined. The combined extracts were evaporated to dryness under a stream of $N_2$ at 65°C. The samples were reconstituted in 1ml MeOH and then diluted 1/100 by addition of 10 µl of sample into 990 µl MeOH. Cyclopamine standards of 1.0, 0.50, 0.25, 0.13, and 0.063 µg/ml were prepared by serial dilution. Colchicine (250 µl of a 100 µg/ml MeOH solution) was added to each standard and sample as an internal standard. The samples were evaporated to dryness under a stream of $N_2$ at 65°C. The samples were then reconstituted with 2 ml of a 50:50 MeOH:0.1% formic acid solution and 1 ml transferred to a 1.5 ml auto sample vial. Quantitative analysis of the alkaloids was made using a Surveyor HPLC and autosampler system coupled to a ThermoFinnigan LCQ Advantage Max mass spectrometer (Thermo Finnigan, San Jose, CA). Samples (10 µl) were injected onto a Betasil C-18 reversed phase column (100 2.1 mm i.d.) (Thermo Electron Corporation, Waltham, MA USA) protected by a guard column of the same adsorbent. The column was eluted with a gradient flow (0.250 ml/min.) of 0.1% formic acid:methanol (A:B). Mobile phase B was increased from 50 to 80% over 10 min, followed by a second linear gradient to 100% B over 10 to 13 min of the run. Retention times for cyclopamine and cycloposine under these conditions were 3.9 and 1.7 min, respectively. Flow from the HPLC column was connected directly to the electrospray source of the mass spectrometer (MS). The MS was operated in full scan MS mode. Selected ion chromatograms for *m/z* 412.3 and 574.3 were used for detection and quantitation of cyclopamine and cycloposine, respectively. Both cyclopamine and cycloposine were quantified against a five-point standard curve of cyclopamine over the range of 0.63 to 1.0 µg/ml.

## Animal studies

Before conducting teratogenic experiments, a simple dose finding experiment was performed in order to determine the maximum dose of plant material that would not severely incapacitate the animals. Four sheep (one animal per dose) were dosed twice (7 am and 3 pm) at doses of 0.75, 1.0, 1.25, and 1.5 g *V. californicum*/kg BW. Sheep were monitored for clinical signs of poisoning for 48 h after treatment.

Twenty Western white-faced ewes weighing 77±8 kg were synchronized in estrus using intravaginal sponges impregnated with fluorgestone acetate (Intervet International B.V., Netherlands). Each ewe was hand mated to Suffolk rams three times a day for 3 days following removal of the intravaginal pessaries; the last day that each ewe exhibited standing estrus was considered GD 0. Each ewe was dosed at 7 am and 3 pm on the specified day(s) of gestation, with ground plant material (0.88 g *V. californicum*/kg BW), which corresponds to a dose of 0.88 mg cyclopamine/kg BW. On GD 60 all the ewes were checked for pregnancy and fetal lambs were evaluated for malformations via ultrasound examinations. The ewes were examined transabdominally using an Aloka SSD-900V scanner fitted with a 5 MHz convex electronic transducer (Wallingford, CT). The ewes were restrained on their backs to facilitate access to the hairless areas of the abdominal wall just in front of the udder. After parturition the lambs were assessed for malformations.

## Results

A mass spectrum of steroidal alkaloid extract of root material from *V. californicum* is shown in Figure 1. The extract contained many of the commonly found alkaloids including veratramine ($MH^+$ = 410), cyclopamine ($MH^+$ = 412), muldamine ($MH^+$ = 458), and cycloposine ($MH^+$ = 574), the glycoside of cyclopamine. The plant material contained 1 mg cyclopamine/g plant material and approximately 6 mg cycloposine/g plant material. No jervine was detected in this collection of plant material.

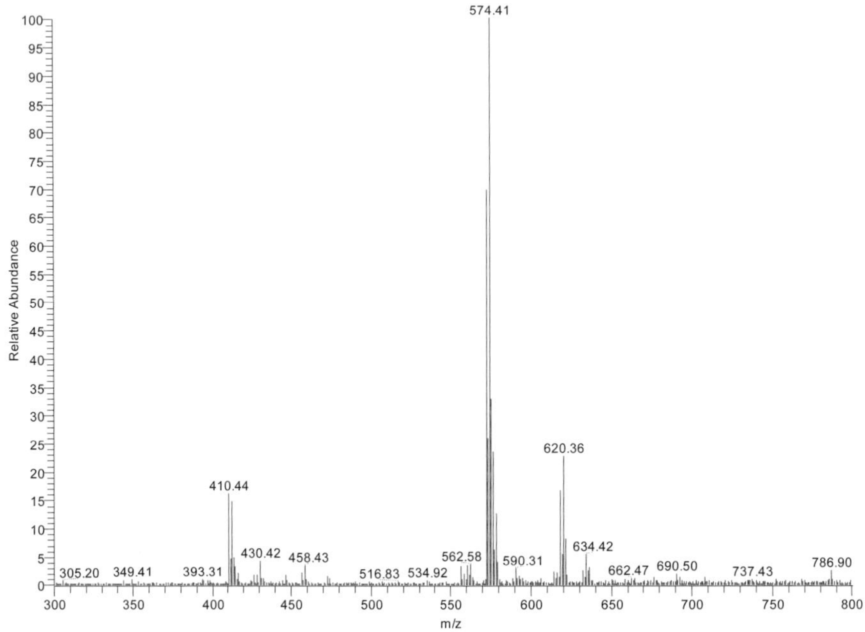


**Figure 1.** Electrospray mass spectrum of a total alkaloid extract of root material from *V. californicum*. The alkaloids veratramine ($MH^+$ = 410), cyclopamine ($MH^+$ = 412), muldamine ($MH^+$ = 458), the glycoside of veratramine ($MH^+$ = 572), cycloposine ($MH^+$ = 574) the glycoside of cyclopamine, and the glycoside of muldamine ($MH^+$ = 620) were identified.

A simple dose finding experiment was performed in order to determine the maximum dose of plant material that would not severely incapacitate the animals. Sheep were dosed twice (7 am and 3 pm) at doses of 0.75, 1.0, 1.25, and 1.5 g/kg BW. Five hours after the first dose no animals showed clinical signs. Two hours after the second dose (10 h after the first dose), the wether receiving the 0.75 g/kg dose had minor salivation but was otherwise normal while the wether receiving the 1.0 g/kg dose was recumbent and weak but could stand when prompted. The wether receiving the 1.25 g/kg dose was also weak and recumbent and could not stand when prompted. At this time the dose of 1.5 g/kg dose was found to be lethal. Twenty-four hours after the first dose, the low dose animal showed no clinical signs, the 1.0 g/kg dosed animal was still weak and trembling but was up walking around. The 1.25 g/kg dosed animal was still recumbent and unable to stand. Forty-eight hours after the first dose, both animals receiving the lower doses showed no clinical signs, while the animal receiving the 1.25 g/kg dosc was still recumbent. This animal remained recumbent for the next 2 days whereupon it was euthanized. From the dose-response experiment, it was determined that a dose of 0.88 g/kg would be used for treating pregnant ewes, which corresponds to a dose of 0.88 mg cyclopamine/kg BW.

For the teratogenic experiments, 20 ewes were randomly divided into four groups of five ewes that were dosed twice on GD 13, 14, or 15, and one group that was dosed twice on all three days. The only clinical signs observed in the ewes that received two doses of plant material were minor salivation and minor muscle weakness. The sheep that received six doses of plant material all showed salivation and general muscle twitching and weakness. One animal from this group was excluded from the study due to severe reactions to treatment. A second ewe was injured 24 days after completion of the dosing regimen (GD38). This ewe was euthanized and necropsied. It was determined that the injury was not due to treatment.

Four of the five ewes treated on GD 13 were pregnant, all five of the ewes treated on GD 14 were pregnant, three out of five ewes treated on GD 15 were pregnant, while none of the ewes treated on GD 13-15 were pregnant at GD 60 (Table 1). Four of the five ewes diagnosed by ultrasonography with abnormal lambs on GD 60 gave birth to malformed lambs. Two of the seven ewes diagnosed by ultrasonography with normal lambs at GD 60 gave birth to malformed lambs. The malformations consisted of many craniofacial malformations ranging from mild maxillary hypoplasia to severe cranial doming, abnormal proboscis formation, to cyclopia and maxillary aplasia.

**Table 1.** Diagnosis of ewes for pregnancy and their lambs for malformations.

| | Assessment at GD 60 | | Assessment at Parturition | | | |
|---|---|---|---|---|---|---|
| Group | Pregnant | Malformed | Lambs | Normal lambs | Cyclops lambs | Lambs with other craniofacial malformations |
| GD 13 | 4 of 5 | 2 of 4 | 7 | 2 | 2 | 3 |
| GD 14 | 5 of 5 | 3 of 5 | 5 | 0 | 3 | 2 |
| GD 15 | 3 of 5 | 0 of 3 | 3 | 3 | 0 | 0 |
| GD 13-15 | 0 of 5 | | 0 | | | |

## Discussion

Several *Veratrum* species in North America, including *V. viride*, *V. album*, and *V. californicum*, produce numerous toxic alkaloids that are distributed in all parts of the plant. Over 50 complex steroidal alkaloids have been identified from *Veratrum* species. Five classes have been characterized: veratrines, cevanines, jervanines, solanidines, and cholestanes. The veratrines and cevanines are of considerable interest in toxicology as they are neurological toxins and hypotensive agents. The primary effect on the heart is to cause a repetitive response to a single stimulus resulting from prolongation of the sodium current (Jaffe *et al.* 1990). Clinical signs of poisoning are most likely caused by the neurotoxic cevanine alkaloids present in most species of *Veratrum*. Typical signs begin 2-3 h post-ingestion with excess salivation with froth around the mouth, slobbering, and vomiting progressing to ataxia, collapse, and death.

The objective of this study was to determine the maximum tolerated dose of ground *V. californicum* root that would not severely incapacitate the ewes and still cause craniofacial malformations in the lambs. The results from the dose finding study indicated that doses of 1.25 mg/kg and above were too toxic as the animals had severe muscle weakness and dyspnea and even died. Conversely, the sheep dosed with 0.75 mg/kg showed little clinical signs indicating that this dose would most likely not be effective. The sheep dosed with 1.0 mg/kg showed signs of poisoning but seemed to recover without problem. However, this sheep was only dosed twice, and due to the fact that one group was to receive two doses per day for 3 days, we decided to use a dose of 0.88 mg/kg for the teratogenic studies. This amount of plant material corresponded to a dose of 0.88 mg cyclopamine/kg BW. The results from the teratogenic study indicate that this dose was very effective in causing craniofacial malformations when administered twice. However, when this dose was administered six times, none of the ewes were still pregnant at GD 60. One of these ewes was necropsied and the observation was made that it had been pregnant but the embryo had died and had been reabsorbed. Consequently the assumption was made that a dosing regimen of 0.88 mg/kg twice a day for 3 days leads to embryonic death.

The development processes of the embryo occur very quickly (DeSesso 2006). Consequently the critical window for birth defects can be fairly narrow depending upon the species and the deformity (Wilson 1973; DeSesso 2006). In this study, the ewes were bred multiple times with the assumption that the time of conception would be within 12 h of the last time of breeding (Jainudeen *et al.* 2000). We confirmed in this study that the window for craniofacial deformities is short, with malformations observed from ewes treated on GD 13 and 14. However, with the techniques employed here, the exact timing of conception was still unknown. In the future, *in vitro* fertilization or embryonic transfer studies could be useful to more accurately determine the timing of conception and more closely relate the time of exposure to cyclopamine with specific terata. This information would be used to determine if the timing of exposure to cyclopamine dictates the type and extent of cyclopamine-induced craniofacial malformations.

Another important aspect of this study is that the plant material used for this study contained cycloposine, the glycoside of cyclopamine. The concentration of cycloposine was approximately six times that of cyclopamine. Cycloposine has also been shown to be teratogenic, causing similar terata as cyclopamine (Keeler and Binns 1968; Keeler 1969, 1970). Cycloposine effectively increases the concentration of teratogenic compounds available from this plant material. However, it is unknown if cycloposine itself is teratogenic, or if the glycoside is first cleaved and then it is cyclopamine that is the teratogen. Initial *in vitro* experiments indicate that the glycoside is cleaved from cycloposine very rapidly in the rumen (data not shown).

In conclusion, the results from this study indicate that a dosing regimen of 0.88 mg/kg of ground *V. californicum* root material administered twice in one day is sufficient to cause minor clinical signs of poisoning but not enough to severely incapacitate the animal. This dose corresponded to a dose of 0.88 mg/kg of cyclopamine for the plant material used in this study. This dose of cyclopamine caused craniofacial malformations in the majority of the lambs from ewes dosed on either GD 13 or 14. This supports previous findings that the critical window for synophthalmia formation is short. However, even though we found that ewes dosed on GD 13 in addition to GD 14 are susceptible to synophthalmia formation, the exact time of conception is still inaccurate and consequently with the methods employed here we cannot more accurately define the critical window. Toxicokinetic analysis demonstrated that the elimination of cyclopamine from sheep is very quick indicating that the plant must be consumed in sufficient quantities during the very narrow critical window for teratogenesis to occur.

## Acknowledgements

We acknowledge the technical assistance of Kendra Dewey, Andrea Dolbear, and Scott Larsen. We also acknowledge Al Maciulus, Rex Probst, and Danny Hansen for their help with the care and handling of the animals.

## References

Binns W, James LF, Shupe JL, and Thacker EJ (1962). Cyclopian-type malformation in lambs. *Archives of Environmental Health* 5:106-108.
Binns W, James LF, Shupe JL, and Everett G (1963). A congenital cyclopian-type malformation in lambs induced by maternal ingestion of a range plant, *Veratrum californicum*. *American Journal of Veterinary Research* 24:1164-1175.
Binns W, Shupe JL, Keeler RF, and James LF (1965). Chronologic evaluation of teratogenicity in sheep fed *Veratrum californicum*. *Journal of the American Veterinary Medical Association* 147:839-842.
Burrows GE and Tyrl RJ (2001). *Toxic Plants of North America*, 1342 pp., 1st edn. Iowa State University Press, Ames, Iowa.
Chen JK, Taipale J, Cooper MK, and Beachy PA (2002). Inhibition of Hedgehog signaling by direct binding of cyclopamine to Smoothened. *Genes & Development* 16:2743-2748.
Cooper MK, Porter JA, Young KE, and Beachy PA (1998). Teratogen-mediated inhibition of target tissue response to Shh signaling. *Science* 280:1603-1607.
DeSesso JM. (2006). Compartive Features of Vertebrate Embryology. In *Developmental and Reproductive Toxicology* (RD Hood, ed.), pp. 147-198. CRC Press Taylor and Francis, Boca Raton, Florida.
Incardona JP, Gaffield W, Kapur RP, and Roelink H (1998). The teratogenic *Veratrum* alkaloid cyclopamine inhibits sonic hedgehog signal transduction. *Development* 125:3553-3562.
Jaffe AM, Gephardt D, and Courtemanche L (1990). Poisoning due to ingestion of *Veratrum viride* (false hellebore). *The Journal of Emergency Medicine* 8:161-167.
Jainudeen MR, Wahid H, and Hafez ESE (2000). *Reproduction in Farm Animals*, p. 509, 7th edn. Blackwell Publishing, Philadelphia, Pennsylvania.
Keeler RF (1969). Teratogenic compounds of *Veratrum californicum* (Durand). VII. The structure of the glycosidic alkaloid cycloposine. *Steroids* 13:579-588.
Keeler RF (1970). Teratogenic compounds in *Veratrum californicum* (Durand) IX. Structure-activity relation. *Teratology* 3:169-173.
Keeler RF (1990). Early embryonic death in lambs induced by *Veratrum californicum*. *Cornell Veterinarian* 80:203-207.
Keeler RF and Binns W (1966a). Teratogenic compounds of *Veratrum californicum* (Durand). I. Preparation and characterization of fractions and alkaloids for biologic testing. *Canadian Journal of Biochemistry* 44:819-828.
Keeler RF and Binns W (1966b). Teratogenic compounds of *Veratrum californicum* (Durand). II. Production of ovine fetal cyclopia by fractions and alkaloid preparations. *Canadian Journal of Biochemistry* 44:829-838.

Keeler RF and Binns W (1968). Teratogenic compounds of *Veratrum californicum* (Durand). V. Comparison of cyclopian effects of steroidal alkaloids from the plant and structurally related compounds from other sources. *Teratology* 1:5-10.

Keeler RF and Binns W. (1971). Teratogenic compounds of *Veratrum californicum* as a function of plant part, stage, and site of growth. *Phytochemistry* 10:1765-1769.

Keeler RF and Stuart LD (1987). The nature of congenital limb defects induced in lambs by maternal ingestion of *Veratrum californicum. Journal of Toxicology – Clinical Toxicology* 25:273-286.

Keeler RF, Young S, and Smart R (1985). Congenital tracheal stenosis in lambs induced by maternal ingestion of *Veratrum californicum. Teratology* 31:83-88.

Kingsbury JM (1964). *Poisonous Plants of the United States and Canada*, 626 p., 1 ed. Prentice-Hall, Inc., Englewood Cliffs, New Jersey.

Lum L and Beachy PA (2004). The Hedgehog response network: sensors, switches, and routers. *Science* 304:1755-1759.

Rubenstein JL and Beachy PA (1998). Patterning of the embryonic forebrain. *Current Opinion in Neurobiology* 8:18-26.

Van Kampen KR, Binns W, James LF, and Balls LD. (1969). Early embryonic death in ewes given *Veratrum californicum. American Journal of Veterinary Research* 30:517-519.

Welch KD, Panter KE, Lee ST, Gardner DR, Stegelmeier BL, and Cook D. (2009). Cyclopamine-induced synophthalmia in sheep: defining a critical window and toxicokinetic evaluation. *Journal of Applied Toxicology* 29:414-421.

Wilson JG (1973). *Environment and Birth Defects*. Academic Press, New York, New York.

# Chapter 38

# Toxic Effects of *Ipomoea carnea* on Placental Tissue of Rats

L.L. Lippi, F.M. Santos, C.Q. Moreira, and S.L. Górniak

*Dept. of Pathology, School of Veterinary Medicine and Animal Sciences, University of São Paulo, Av. Prof. Dr. Orlando Marques Paiva 87, 05508-270, Brazil*

## Introduction

*Ipomoea carnea* Jacq. spp. *fistulosa* Choisy (Convolvulaceae) is a toxic plant widely distributed in Brazil (Tokarnia *et al.* 2000) and other tropical countries (Austin and Huáman 1996). During periods of drought, animals graze this plant which grows even under adverse climatic conditions (Keeler 1988). After prolonged periods of plant intake the animals exhibit a variety of clinical signs such as depression, general weakness, loss of body weight, staggering gait, muscle tremors, ataxia, posterior paresis, and paralysis (Idris *et al.* 1973; Damir *et al.* 1987; De Balogh *et al.* 1999; Tokarnia *et al.* 2000).

Two kinds of toxic principles have been isolated from the plant, the nortropane alkaloids calystegines B1, B2, B3, and C1 and the indolizidine alkaloid swainsonine (De Balogh *et al.* 1999; Haraguchi *et al.* 2003). The latter alkaloid has a known mechanism of action as a potent inhibitor of two distinct intracellular enzymes, the lysosomal α-mannosidase and the Golgi mannosidase II. Inhibition of α-mannosidase results in lysosomal accumulation of incompletely processed oligosaccharides rich in α-mannosyl and β-*N*-acetyl glucosamine moieties inside vacuoles, which progresses to loss of cellular function and ultimately to cell death (Tulsiani *et al.* 1988). Histologically, cellular vacuolization of Purkinje cells, thyroid follicles, exocrine pancreas, liver, and kidney cells has been observed.

Swainsonine inhibition of the Golgi mannosidase II enzyme causes alteration of the *N*-linked glycoprotein process (Elbein 1989), producing increased numbers of high-mannose, hybrid, or complex types of oligosaccharide structures that participate in hormones, cytokines, membrane receptors, and adhesion molecules (Stegelmeier *et al.* 1998). These molecular effects can alter hormonal and endocrine function (Stegelmeier *et al.* 1995) and cause abnormal gastrointestinal (Pan *et al.* 1993), immunological (Karasuno *et al.* 1992), and reproductive function (Nelson *et al.* 1980).

Recently, many studies in our laboratory have shown that *I. carnea* has teratogenic effects in rats (Hueza *et al.* 2007), goats (Shumaher-Henrique *et al.* 2003), and rabbits. However, it is not known if the malformations observed in the fetuses are due to alterations in the placenta or if they can be directly related to the transplacental transfer of the active

©CAB International 2011. *Poisoning by Plants, Mycotoxins, and Related Toxins* (eds F. Riet-Correa, J. Pfister, A.L. Schild, and T.L. Wierenga)

principle. The present study was performed to evaluate the effects of *I. carnea* in the placental tissue and in the litters of treated female rats.

*I. carnea* was collected in May, 2004 from plants cultivated at the Research Center for Veterinary Toxicology (CEPTOX), University of São Paulo (USP), Pirassununga, Brazil. Plants were first frozen, then ground; ground plant material was extracted with ethyl alcohol (96%). After total solvent evaporation under reduced pressure at 50°C a dark green extract was obtained which was suspended in water to remove the waxy residue and consecutively fractionated with n-butanol saturated with water. This procedure yielded the aqueous fraction (AF) that was stored at −20°C.

Male and female Wistar rats from the Department of Pathology (School of Veterinary Medicine, University of São Paulo) weighing 180-200 g and approximately 90 days of age were used. For breeding, males were housed overnight with females (1 male:2 females) and females were checked for sperm-positive vaginal smears the next morning which was designated as GD0. The experiments were carried out in accordance with the ethical principles for animal research adopted by the Bioethics Committee of the School of Veterinary Medicine and Zootechny, University of São Paulo.

Pregnant rats ($n$=50) on GD0 were weighed and kept in separate cages. The dams were divided into five groups (one control, one peer-feeding, and three experimental groups). The 3 experimental groups were treated orally by gavage once a day from GD6 to GD19 with (A) 1, (B) 3, and (C) 7 g/kg of *I. carnea* AF. The control (Co) and peer-feeding (PF) group received tap water by gavage. The peer-feeding group received the same amount of food as group C to control for effects of nutrient intake. Total body weight gain and water and food consumption were measured on 3 days during the experimental period. On GD20 the dams were euthanized, the uterus was removed, and the number of corpora lutea, implantations, resorptions, and live and dead fetuses were counted. The fetuses were individually weighed and examined for macroscopic external malformations and the placentas were weighed and analyzed histopathologically. Fragments of kidney, liver, spleen, and central nervous system (CNS) were removed, weighed, and submitted for histopathological analysis; blood was collected for serum biochemistry analysis. The percentage of pre-implantation loss was calculated as number of corpora lutea – number of implantations × 100/number of corpora lutea. Percentage of post-implantation loss was number of implantations – number of live fetuses × 100/number of implantations.

For analysis of the data the statistical program GraphPad Prism 5.00® (GraphPad Software, Inc., San Diego, CA, USA) was used. Bartlett's test was used to determine data homogeneity. One-way ANOVA followed by the Dunnett's test was used to analyze the parametric data. The nonparametric data were analyzed by the Kruskal–Wallis test followed by the Dunn test for multiple comparisons. In all cases, results were considered to be significant when $P < 0.05$. The data are expressed as means ± SEM and the percentage data are expressed as medians (with minimum and maximum).

## Results

There were no significant differences in water and food consumption between the pregnant rats treated with *I. carnea* AF at 1 or 3 g/kg BW and control females. However, weight gains were significantly ($P < 0.01$) lower in females treated with 7 g/kg of *I. carnea* AF compared to controls (Table 1).

The reproductive performance of the dams treated during organogenesis was similar to that of control animals (Table 2). There were no significant differences in the percentages of the implantation sites, live fetuses per litter, pre-implantation and post-implantation losses, fetal weight, placental weight, maternal weight, and maternal weight at term minus uterus weight among control and experimental groups. Significant differences were observed in fetal length and uterus weight at term between the PF and Co groups; in addition there were significant differences observed in the number of live fetuses per litter between the B and PF group compared to controls.

**Table 1.** Total water and food consumption and body weight gain in female rats receiving different treatments during gestation.

| Groups[1] | Water consumption (ml) | Food consumption (g) | Weight gain (g) |
|---|---|---|---|
| Co (*n*=10) | 546±13.6 | 256±8.1 | 29±4.4 |
| A (*n*=12) | 532±30.6 | 257±8 | 27.2±2.7 |
| B (*n*=11) | 553±42.8 | 225±15.7 | 25.8±2.4 |
| C (*n*=11) | 550±42.8 | 225±12.3 | 8.8±2.4** |
| PF (*n*=5) | 528±20.5 | – | 27.5±4.4 |

***P* < 0.01 versus control group Co.
[1]Co = Controls; A = 1 g/kg of *I. carnea* AF; B = 3 g/kg; C = 7 g/kg; PF = peer-feeding.

**Table 2.** Reproductive performance[1] of rats treated with 1, 3, and 7 g/kg of *I. carnea* AF or controls from gestation day 6 to day 19.

| | Treatments[2] | | | | |
|---|---|---|---|---|---|
| Variable of Interest | Co (*n*=10) | A (*n*=12) | B (*n*=11) | C (*n*=11) | PF (*n*=5) |
| Maternal weight (g) | 300.6±8.62 | 296.7±7.6 | 303.4±5.2 | 283±7.53 | 287.7±8.73 |
| Uterus weight at term (g) | 60.3±3.04 | 58.8±2.73 | 57.8±4.67 | 64.5±2.62 | 41.4±3.51* |
| Maternal weight at term minus uterus weight (g) | 240.3±7.32 | 238±5.86 | 239.2±3.92 | 220.4±5.96 | 246±7.74 |
| Placental weight per litter (g) | 0.44±0.016 | 0.45±0.013 | 0.44±0.01 | 0.44±0.024 | 0.46±0.029 |
| Fetal body weight per litter (g) | 3.52±0.084 | 3.51±0.067 | 3.57±0.069 | 3.38±0.076 | 3.55±0.173 |
| Fetal length per litter (cm) | 3.66±0.016 | 3.63±0.026 | 3.62±0.026 | 3.56±0.028 | 3.43±0.062** |
| Number of live fetuses per litter | 10.4±0.47 | 10.58±0.41 | 11.45±0.54 | 11.63±0.51 | 7.8±1.11* |
| Implantation sites per litter | 100 (69.2-100) | 100 (83.4-100) | 100 (73.3-100) | 100 (80-100) | 80 (30.77-100) |
| Live fetuses per litter | 89.9 (75-100) | 90.91 (71.4-100) | 93.3 (81.8-100) | 91.6 (76.9-100) | 100 (88.9-100) |
| Pre-implantation loss per litter | 0 (0-30.76) | 0 (0-16.6) | 0 (0-26.6) | 0 (0-20) | 20 (0-69.23) |
| Post-implantation loss per litter | 10.1 (0-25) | 9.09 (0-28.57) | 6.6 (0-18.18) | 8.33 (0-23.07) | 0 (0-11.1) |

[1]Data are expressed either as mean ± SEM or as median percentages (%) with the minimum and maximum in parentheses. Means were analyzed by ANOVA followed by the Dunnett's test, whereas percentages were analyzed by Kruskal–Wallis test followed by the Dunn's Multiple Comparisons Test.
[2]Treatment designations: Co = Controls; A = 1 g/kg of *I. carnea* AF; B = 3 g/kg; C = 7 g/kg; and PF = peer-feeding. *Differences are statistically significant (**P* < 0.05 or ***P* < 0.01) between PF group compared to the Co group.

The relative weight of the organs of the females did not show significant alterations between the experimental and control groups. Animals treated with the higher dose of *I. carnea* presented histological alterations on the kidney which were characterized by intense intracellular vacuolization of proximal tubular cells, increased glomerular space, presence of hyaline cylinders, and cellular death. Serum biochemistries showed a reduction in the levels of total protein in the animals from the B and C groups ($P < 0.0001$) and albumin in animals from group C ($P < 0.01$) when they were compared to the Co group. In the placental tissue a thickening of labyrinth zone and reduction of the thickness of the junctional zone were observed in the dams from the C group.

## Conclusions

The primary result obtained in this study was the alteration in the labyrinth zone in placental tissue from treatment with the highest dose of the aqueous fraction from *Ipomoea*. This maternal tissue is located at the fetal interface and contains the maternal sinusoids and fetal vessels, and this locale is responsible for the maternal-fetal exchange, thus alterations of this area may compromise fetal development.

## References

Austin DF and Huáman Z (1996). A synopsis of *Ipomoea* (Convolvulaceae) in the Americas. *Taxon* 45:3-38.

Damir HA, Adam SEI, and Tartour G (1987). Effects of *Ipomoea carnea* on goats and sheep. *Veterinary and Human Toxicology* 29: 316-319.

De Balogh KIM, Dimande AP, Van Der Lugt JJ, Molyneux RJ, Naudé TW, and Welman WG (1999). A lysosomal storage disease induced by *Ipomoea carnea* in goats in Mozambique. *Journal Veterinary Diagnostic Investigation* 11:266-273.

Elbein AD (1989). The effects of plant indolizidine alkaloids and related compounds in glycoprotein processing. In *Swainsonine and related glycosidase inhibitors* (LF James, AD Elbein, RJ Molyneux, and CD Warren, eds), pp. 87-155. Ames University Press, Iowa.

Haraguchi M, Górniak SL, Ikeda K, Minami H, Kato A, Watson AA, Nash R, Molyneux RJ, and Asano N (2003). Alkaloidal components in the poisonous plant, *Ipomoea carnea* (Convolvulaceae). *Journal of Agriculture and Food Chemistry* 51:4995-5000.

Hueza IM, Guerra JL, Haraguchi M, Gardner DR, Asano N, Ikeda K, and Górniak SL (2007). Assessment of the perinatal effects of maternal ingestion of *Ipomoea carnea* in rats. *Experimental and Toxicologic Pathology* 58:439-446.

Idris OF, Adam SEI, and Tartour G (1973). Toxicity to goats of *Ipomoea carnea. Tropical Animal Health and Production* 5: 119-123.

Karasuno T, Kanayama Y, Nishiura T, Nakao H, Yonezawa T, and Tarui S (1992). Glycosidase inhibitors (castanospermine and swainsonine) and neuraminidase inhibit pokeweed mitogen-induced B-cell maturation. *European Journal of Immunology* 22: 2003-8.

Keeler RF (1988). Livestock models of human birth defects, reviewed in relation to poisonous plants. *Journal Animal Science* 66:2414-27.

Nelson BK, James LF, Sharma RP, and Cheney CD (1980). Locoweed embryotoxicity in rats. *Clinical Toxicology* 16:149-166.

Pan YT, Ghidoni J, and Elbein AD (1993). The effects of castanospermine and swainsonine on the activity and synthesis of intestinal sucrase. *Archives of Biochemistry and Biophysics* 303:134-144.

Shumaher-Henrique B, Górniak SL, Dagli MLZ, and Spinosa HS (2003). Toxicity of long-term administration of *Ipomoea carnea* to growing goats: clinical, biochemical, haematological and pathological alterations. *Veterinary Research Communication* 27:311-319.

Stegelmeier BL, Molyneux RJ, Elbein AD, and James LF (1995). The lesions of locoweed (*Astragalus mollissimus*), swainsonine, and castanospermine in rats. *Veterinary Pathology* 32:289-298.

Stegelmeier BL, Snyder PD, James LF, Panter KE, Molyneux RJ, Ralphs MH, and Pfister JA (1998). The immunologic and toxic effects of chronic locoweed (*Astragalus lentiginosus*) intoxication in cattle. In *Toxic Plants and Other Natural Toxicants* (T Garland, AC Barr, eds) p. 285-290. CAB International, Wallingford, UK.

Tokarnia CH, Dobereiner J, and Peixoto PV (2000). *Plantas tóxicas do Brasil*, pp. 120-123. Editora Helianthus, Rio de Janeiro.

Tulsiani DR, Broquist HP, James LF, and Touster O (1988). Production of hybrid glycoproteins and accumulation of oligosaccharides in the brain of sheep and pigs administered swainsonine or locoweed. *Archives of Biochemistry and Biophysics* 264:607-617.

# Chapter 39

# Chronic Heart Failure and Abortion Caused by *Tetrapterys* spp. in Cattle in Brazil

P.V. Peixoto[1], S.A. Caldas[2], T.N. França[3], T.C. Peixoto[4], and C.H. Tokarnia[1]

[1]*Instituto de Zootecnia, Universidade Federal Rural do Rio de Janeiro (UFRRJ), Seropédica, RJ 23890-000, Brazil;* [2]*Pós-graduação em Ciências Veterinárias, UFRRJ, Seropédica, RJ;* [3]*Instituto de Veterinária, UFRRJ, Seropédica, RJ;* [4]*Pós-graduação em Medicina Veterinária, UFRRJ, Seropédica, RJ*

## Introduction

Several cardiotoxic plants in the world cause different clinical-pathological syndromes due to numerous active principles. In Brazil, there are many plants that cause sudden death (Brazilian sudden death causing plants; BSDCP) which may have sodium monofluoroacetate (MF) as a toxic principle (Tokarnia *et al.* 2000; Peixoto *et al.* unpublished data). This group of plants is responsible for the death of at least 600,000 cattle per year in Brazil (Tokarnia 2009, personal communication). In general, the animals poisoned by BSDCP do not develop heart lesions and the clinical evolution usually lasts only a few minutes.

In Brazil there are two other genera of cardiotoxic plants whose toxic agents cause fibrosis and regressive alterations of the myocardium of cattle with subacute to chronic evolution: *Tetrapterys* (*T. multiglandulosa* and *T. acutifolia*) and *Ateleia* (*A. glazioviana*). *A. glazioviana* was recently shown to induce abortion (Caldas 2008). Cattle poisoned by *A. glazioviana* can additionally exhibit nervous signs (Gava *et al.* 2001). Although the epidemiological and clinical-pathological aspects of poisoning by *Tetrapterys* spp. have been extensively studied, the pathogenesis remains obscure and the toxic principle is still unknown. Further, spongy degeneration lesions of the CNS can eventually occur in poisoned animals (Tokarnia *et al.* 1989; Carvalho *et al.* 2006). The aim of this study was to collect the most important data and discuss the aspects that are still obscure about natural and experimental poisoning by *Tetrapterys* spp.

## General Aspects, Distribution, and Habitat

*T. multiglandulosa* Adr. Juss. and *T. acutifolia* Cav. (Malpghiaceae) are vines or climbing shrubs widely distributed in the southeast region of Brazil, mainly in the states of

©CAB International 2011. *Poisoning by Plants, Mycotoxins, and Related Toxins* (eds F. Riet-Correa, J. Pfister, A.L. Schild, and T.L. Wierenga)

São Paulo, Minas Gerais, Rio de Janeiro, and Espírito Santo, and in other places of the country. Both species grow on the upper part of hillsides (Tokarnia *et al.* 2000).

Recently in the municipalities of Barra do Piraí and Valença, state of Rio de Janeiro, another plant from the *Tetrapterys* genus was determined to cause the same disease as other *Tetrapterys* spp. (Caldas 2008); however, it is morphologically different from the species mentioned above. Specimens of this plant have been submitted to a botanist specializing in Malpighiaceae but have not yet been identified.

## History

Since the late 1960s, the former Animal Biology Institute (ABI), Rio de Janeiro, has records of the occurrence of abortions in cows in the northwest part of the state of São Paulo. Microbiological studies did not reveal the involvement of *Brucella*, *Campylobacter*, or *Tritrichomonas* in these abortions and the few plants tested, especially *Rynchosia pyramidalis*, did not cause abortion. On the other hand, there was an anecdotal report that Borges had reproduced abortion in cows by oral administration of leaves of *T. acutifolia* (Tokarnia *et al.* 1989), but this work was not published. In the mid 1980s, the opportunity arose to study the disease on site. It was reported that in the same areas where the abortions occurred in cattle, numerous cases of subacute to chronic heart failure (SCHF) also occurred during the year (Tokarnia *et al.* 1989). In this study, the heart failure of adult animals was reproduced experimentally.

## The Natural Disease

Under natural conditions, poisoning by *Tetrapterys* spp. is described only in cattle (Tokarnia *et al.* 1989; Carvalho *et al.* 2006). In the majority of cases of natural poisoning, the clinical evolution is subacute and on occasion tending to chronic (Tokarnia *et al.* 2000). Apparently, this plant is palatable because cattle ingest the sprouts promptly when offered the plant (Caldas 2008). The plant becomes less toxic as the leaves mature (Tokarnia *et al.* 1989). The morbidity rate varies from 6% to 28% and the fatality rate is close to 100% (Tokarnia *et al.* 1989).

### Typical picture of subacute to chronic heart failure (SCHF)

Unlike some of the histories suggested, SCHF occurs throughout the year (Tokarnia *et al.* 1989); however, the incidence of the disease is higher during the dry period when cattle ingest the aerial parts of the plant (Tokarnia *et al.* 2000).

Well-defined clinical signs of heart failure are generally observed. The signs include subcutaneous edema of the lower part of the dewlap sternal regions, engorgement of the jugular vein, positive venous pulse, and cardiac arrhythmia; lethargy, weakness, anorexia, muscle tremors, and slight dyspnea are also seen (Tokarnia *et al.* 1989).

The main necropsy findings are observed in the heart as clear areas that are visible through the epicardium; on cut surfaces there are distinct whitish stains and clusters throughout most of the myocardium which, in some cases, was firmer. There was also concentric hypertrophy. In most cases, the liver exhibited marked lobulation or 'nutmeg appearance' (Tokarnia *et al.* 1989). Most histological findings were restricted to the heart and liver. Interstitial fibrosis, lysis of fibers, individual or group necrosis of myocytes, mononuclear inflammatory infiltrate, and atrophy of cardiac fibers were seen on the heart. In a few cases, extensive areas of necrosis of the myocardium surrounded by fibrosis were seen. In the liver there was centrilobular congestion, slight vacuolization and lysis of hepatocytes as well as fibrosis (Tokarnia *et al.* 2000).

### Abortion

Abortions occur during all phases of gestation (Riet-Correa and Méndez 2007); however, most reports indicate that there is a higher rate of incidence between the 6th and 9th months of gestation in cattle (Tokarnia and Peixoto, unpublished data). Tokarnia *et al.* (1989) studied the macroscopic and microscopic alterations in aborted fetuses or newborns of cows with suspected poisoning by *Tetrapterys* spp. In this study, the main macroscopic findings seen in a fetus aborted at 8 months of gestation consisted of subcutaneous edema, hydroperitoneum, hydrothorax, and hydropericardium. In a calf that died 24 h after birth, areas significantly lighter and firmer on the myocardium were seen, as well as hydroperitoneum. Histopathology was performed on some organs of five aborted fetuses in areas where the disease occurred. The most constant histological alterations observed were in the heart and liver. In the majority of the cases the heart alterations were fibrosis (5/6), necrosis (3/6), atrophy (2/6), and intracellular (3/6) and extracellular (3/6) edema of cardiac fibers. In the liver there was fibrosis (6/7), congestion (4/7), and vacuolation of hepatocytes (1/7) (Tokarnia *et al.* 1989).

Recently, an outbreak of natural poisoning by *T. multiglandulosa* was described in cattle in the state of Mato Grosso do Sul in which 230 cows (79.3%), out of a herd of 290, aborted or had stillborn or weak calves which died within a few days after birth (Carvalho *et al.* 2006).

## Experimental Reproduction

### Subacute to chronic heart failure (SCHF)

Experimentally, cattle (Tokarnia *et al.* 1989; Riet-Correa *et al.* 2005; Caldas 2008), sheep (Riet-Correa *et al.* 2005, 2009; Carvalho *et al.* 2006), and rabbits (Tokarnia and Peixoto, unpublished data) have been poisoned by the oral route.

SCHF was reproduced experimentally in cattle by oral administration of fresh sprouts at the following dosages (g/kg BW): 5.0 g/kg/day for 60 days (two animals); 10 g/kg/day for 13 to 41 days (three animals), and 20 g/kg/day for 10 days (three animals). The daily dose of 2.5 g/kg/day administered for 130 days (one animal) only caused mild signs and cardiac lesions (Tokarnia *et al.* 1989). A large single dose (100 g/kg) administered to one animal did not reproduce the disease (Tokarnia *et al.* 1989), which indicates that under natural conditions the disease most likely results from repeated ingestion for prolonged periods.

Tokarnia *et al.* (1989) saw sudden death occurring after exercise in only one bovine naturally poisoned by *Tetrapterys* spp., and this animal was already exhibiting previous signs of heart failure for weeks. In this study, the clinical and pathological aspects of experimental and natural poisoning were qualitatively identical; however, the tissue and cavitary edema as well as the hepatic macroscopic lesions and the intensity of cardiac microscopic lesions were more marked in the cases of natural poisoning. Three experimentally poisoned animals exhibited marked apathy (Tokarnia *et al.* 1989). This difference is related to the difference in frequency and intensity of exercise to which the animals are submitted in the regions where the natural disease occurs. The increase in exercise likely overloads the cardiovascular system.

### Abortion

Caldas (2008) experimentally reproduced the abortifacient effect of *Tetrapterys* spp. in cattle by daily administration of sprouts and young leaves in doses of 2.5, 5, and 10 g/kg BW for 23 to 76 days. The stage of gestation at beginning of the experiment was 6 to 8 months. Two animals ingested the fresh plant which was given to one animal in a trough and to the other by oral gavage. For two other cows it was necessary to administer the ground plant mixed with grass. In this study, abortion occurred after 246 to 266 days of gestation (Caldas 2008). Additionally, in one cow, typical signs of SCHF were seen and death occurred 36 days after abortion of the fetus. At necropsy the fetuses revealed hydrothorax, hydropericardium, hydroperitoneum, petechiae and ecchymosis on the epicardium, and hepatic congestion; on cut surfaces there were pale areas on the myocardium. Histology showed that the heart of the fetuses exhibited interstitial edema with incipient fibrosis (Caldas 2008). It is important to emphasize that in this study, gynecology exams were performed as well as serology tests for brucellosis, bovine viral diarrhea (BVD), infectious bovine rhinotracheitis (IBR), leptospirosis, and exams to detect campylobacteriosis and trichomoniasis. The results of all these exams were negative which allowed us to discard the involvement of such agents in the pathogenesis of abortion.

In experiments performed in pregnant sheep, it was shown that daily administration of dried leaves of *T. multiglandulosa* in doses of 3.68 g/kg BW for 35 days and 6.93 g/kg for 29 days can result in macerated and edematous fetuses, respectively (Carvalho *et al.* 2006). Recently, Riet-Correa *et al.* (2009) experimentally reproduced abortion and neonatal mortality in pregnant sheep by daily administration of dried leaves of *T. multiglandulosa.* These authors concluded that the occurrence of neonatal mortality or abortion depended on the dose ingested and phase of gestation of the sheep when ingesting the plant. The pathological alterations found by Riet-Correa *et al.* (2009) in the aborted fetuses and newborn lambs were similar to those described in cattle, which in general consisted of anasarca, pale right ventricle, cardiac dilation, pale and firm areas on the surface of the myocardium, and hepatic congestion. Histology revealed mainly multifocal areas of fibrosis on the heart associated with mononuclear infiltrates and necrosis of muscle fibers.

## Discussion

Even though it is evident that death of adult animals poisoned by *Tetrapterys* spp. occurs due to cardiac insufficiency, the pathogenesis of the cardiac lesions in adults as well as in fetuses remains unknown. In the beginning we thought the primary lesions were degenerative-necrotic in nature which would then initiate events of proliferative reactions (fibroblasts and collagen). It is possible, however, as suggested by Barros (2009, personal communication), that degenerative-necrotic lesions (large areas of coagulative necrosis, necrosis of isolated myocytes) can be secondary to ischemia caused by a difficulty in irrigation associated with marked interstitial fibrosis and in consequence of scar retraction (an aspect that prevails in the histological picture). In fact, only 2 out of 14 adult cattle naturally poisoned by *Tetrapterys* spp. exhibited large areas of coagulative necrosis in the myocardium while 13 out of 14 exhibited interstitial fibrosis (Tokarnia *et al.* 1989). Similar findings were also seen in aborted fetuses of cows naturally poisoned by *Tetrapterys* spp., in which necrosis of the myocardium was seen in 3 out of 6 fetuses and interstitial fibrosis was seen in almost all cases examined (5/6) (Tokarnia *et al.* 1989). Interstitial fibrosis, in turn, can be secondary to interstitial edema, that is, proliferation of connective tissue and deposition of collagen can be a consequence of serum protein in the interstitium, as suggested by Jones *et al.* (2000) for the pathogenesis of fibrosis of the myocardium.

A similar histological picture has been described in rats chronically stimulated by isoproterenol, a drug derived from noradrenaline and a potent β-adrenergic agonist (Hoffman and Lefkowitz 1996) that induces cardiac hypertrophy (Zierhut and Zimmer 1989), myocardial fibrosis, and progressive dysfunction which culminates in marked cardiac insufficiency (Shubeita *et al.* 1992). In these cases, giant cells are also observed, as described by Tokarnia *et al.* (1989) in cattle poisoned by *Tetrapterys* spp. These authors termed them 'myogenic giant cells' (bizarre) and interpreted them as unproductive attempts of myocardial regeneration based on observations by Stünzi and Teuscher (1970). However, the exact pathogenesis of the cardiac lesion caused by isoproterenol still has not been fully clarified (Grimm *et al.* 1998). Some authors have the opinion that, in these cases, the marked proliferation of collagen tissue in animals that develop myocardial hypertrophy results from reparative processes from ischemic muscular necrosis induced by the drug (Tang and Taylor 1996; Lin 1973) due to the abrupt and intense increase in cardiac work without the necessary oxygen supply through the coronary circulation (Grimm *et al.* 1998).

The cause of death of the aborted fetuses is the same seen for adult animals that ingest *Tetrapterys* spp.: heart failure caused by direct damage of the toxic principle of the plant to the myocardium, which crosses the placental barrier. Necroscopic and histopathological evidence of heart failure in the fetus and the absence of placental lesions in cattle poisoned by *Tetrapterys* spp. (Caldas 2008) and sheep poisoned by *T. multiglandulosa* (Riet-Correa *et al.* 2009) support this hypothesis. These results differ from those described in goats experimentally poisoned by *T. multiglandulosa* (Melo *et al.* 2001); in this study, abortion was attributed to placental lesions. It is also believed that the pathogenesis of abortion caused by *T. multiglandulosa* and *A. glazioviana* can be the same (Carvalho *et al.* 2006) and that the active principle of these plants may be similar (Riet-Correa *et al.* 2009).

Poisoning by *Tetrapterys* spp. in cattle and sheep is comparable to the poisoning caused by several species of Rubiaceae (*Pachystigma pygmaeum*, *P. thamnus*, *P. latifolium*, *Pavetta harborii*, *P. schumanniana*, and *Fadogia homblei*) that occur in South Africa (Hunter *et al.* 1972; Fourie *et al.* 1989; Kellerman *et al.* 2005) and cause the so-called 'gousiekte' disease of cattle poisoned by plants that contain pavetamine (Schultz *et al.* 2004). A more careful observation indicates that the severe heart lesions of regressive-proliferative nature found in 'gousiekte' (Kellerman *et al.* 2005) share resemblances to those described in poisoning by *Tetrapterys* spp. To cause heart lesions in cattle, these plants also need to be ingested in large amounts for prolonged periods. Another similarity seen in clinical cases of poisoning by plants that cause 'gousiekte' and *T. multiglandulosa* was recently described by Carvalho *et al.* (2006), who observed clinical signs and cardiac lesions in cattle that had been removed for 2 months from the pasture where there was the plant. This latency period (4 to 8 weeks after ingestion of the plant) is a characteristic of poisoning by the African plants (Hunter *et al.* 1972; Fourie *et al.* 1989; Kellerman *et al.* 2005). The rare influence of exercise as well as the subacute to chronic clinical evolution in poisoning by *Tetrapterys* spp. differ from the African plants which in general cause death in a peracute way (acute heart failure and sudden death), especially when the animals are moved and, less frequently, determine chronic congestive heart failure. However, the African plants do not cause abortion nor central nervous system lesions (status spongiosus) as described in cattle poisoned by *Tetrapterys* spp. and *A. glazioviana* (Tokarnia *et al.* 1989; Gava and Barros 2001; Gava *et al.* 2001; Stigger *et al.* 2001).

Spongiosis of the CNS similar to that caused by *T. multiglandulosa* in cattle (Tokarnia *et al.* 1989) has been recently described in sheep (Riet-Correa *et al.* 2005, 2009; Carvalho *et al.* 2006). At the time when it was described, Tokarnia *et al.* (1989) did not think the lesion was important. Similar lesions, however, were also seen in cattle and sheep poisoned by *A. glazioviana* (Gava and Barros 2001; Gava *et al.* 2001; Stigger *et al.* 2001; Raffi *et al.* 2004, 2006). From the histological point of view, this lesion is similar to that seen in cases of hepatic encephalopathy (Riet-Correa *et al.* 2005). However, the spongy lesions in the CNS have no apparent correlation with the liver chronic stasis. Gava *et al.* (2001) believe that the lesions found in the brain of cattle poisoned by *A. glazioviana* are probably related to the direct effect of the active principle of the plant on the nervous system. In any case, similar primary lesions have been described in the brains of cattle, sheep, and goats poisoned by plants such as *Helichrysum argyrosphaerum*, *Ornithogalum saudersiae*, and *O. prasinum* (Basson *et al.* 1975; Van der Lugt *et al.* 1996, 2002).

Recent studies using electron microscopy proved that spongiosis of the nervous system is caused by intramyelinic edema (Riet-Correa *et al.* 2005, 2009) similar to that occurring in poisoning by *A. glazioviana* (Raffi *et al.* 2006). Thus, it appears that there are still many obscure aspects related to the pathogenesis of poisoning by *Tetrapterys* spp. in cattle. Even though rabbits are susceptible to poisoning by plants of this genus, the myocardium of these animals do not exhibit chronic lesions like those seen in cattle (Tokarnia and Peixoto, unpublished data).

## References

Basson PA, Kellerman TS, Albl P, Maltitz LJF, von Miller ES, and Welman WG (1975). Blindness and encephalopathy caused by *Helichrysum argyrosphaerum* D.C. (Compositae) in sheep and cattle. *Onderstepoort Journal of Veterinary Research* 42:135-148.

Caldas SA (2008). Abortos em bovinos determinados pela ingestão natural e experimental de *Tetrapterys* sp. no Estado do Rio de Janeiro. Tese de doutorado, Universidade Federal Rural do Rio de Janeiro, Seropédica, RJ (in progress).

Carvalho NM, Alonso LA, Cunha TG, Ravedutti J, Barros CSL, and Lemos AA (2006). Intoxicação de bovinos por *Tetrapterys multiglandulosa* (Malpighiaceae) em Mato Grosso do Sul. *Pesquisa Veterinária Brasileira* 26:139-146.

Fourie N, Schultz RA, Prozesky L, Kellerman TS, and Labuschagne L (1989). Clinical pathological changes in gousiekte, a plant-induced cardiotoxicosis of ruminants. *Onderstepoort Journal of Veterinary Research* 56:73-80.

Gava A and Barros CSL (2001). Field observations of *Ateleia glazioviana* poisoning in cattle in southern Brazil. *Veterinary and Human Toxicology* 43:37-41.

Gava A, Barros CSL, Pilati C, Barros SS, and Mori AM (2001). Intoxicação por *Ateleia glazioviana* (Leg. Papilionoideae) em bovinos. *Pesquisa Veterinária Brasileira* 21:49-59.

Grimm D, Cameron D, Griese DP, Riegger GA, and Kromer EP (1998). Differential effects of growth hormone on cardiomyocyte and extracellular matrix protein remodeling following experimental myocardial infarction. *Cardiovascular Research* 40:297-306.

Hoffman BB and Lefkowitz RJ (1996). Catecolaminas, drogas simpaticomiméticas e antagonistas dos receptores adrenérgicos. In *Goodman and Gilman's The pharmacological basis of therapeutics* (JG Hardman, LE Limbird, PB Molinoff, RW Ruddon, and AG Gilman, eds), 9th edn, pp. 146-182. McGraw-Hill, New York.

Hunter LR, Naudé TW, Adelaar TF, Smit JD, and Codd LE (1972). Ingestion of the plant *Fadogia monticule* Robins as an additional cause of gousiekte in ruminants. *Onderstepoort Journal of Veterinary Research* 39:71-83.

Jones TC, Hunt RD, and King NW (2000). *Patologia Veterinária,* 1415 pp., 6th edn. Manole, São Paulo.

Kellerman TS, Coetzer JAW, Naudé TW, and Botha CJ (2005). *Plant Poisonings and Mycotoxicosis of Livestock in Southern Africa,* 310 pp., 2nd edn. Oxford University Press, Cape Town.

Lin YC (1973). Hemodynamics in the rat with isoproterenol induced cardiac hypertrophy. *Research Communications in Chemical Pathology and Pharmacology* 6:213-220.

Melo MM, Vasconcelos AC, Dantas GC, Serakides R, and Alzamora Filho F (2001). Experimental intoxication of pregnant goats with *Tetrapterys multiglandulosa* A. juss. (Malpighiaceae). *Arquivo Brasileiro de Medicina Veterinária e Zootecnia* 53:58-65.

Raffi MB, Barros RR, Bragança JFM, Rech RR, Oliveira FN, and Barros CSL (2004). The pathogenesis of reproductive failure induced in sheep by the ingestion of *Ateleia glazioviana. Veterinary and Human Toxicology* 46:233-238.

Raffi MB, Rech RR, Sallis ESV, Oliveira FN, Barros SS, and Barros CSL (2006). Chronic cardiomyopathy and cerebral spongy changes in sheep experimentally fed with *Ateleia glazioviana. Ciência Rural* 36:1860-1866.

Riet-Correa F and Méndez MC (2007). Intoxicações por plantas e Micotoxinas. In *Doenças de Ruminantes e Eqüídeos* (F Riet-Correa, AL Schild, RAA Lemos, and JRJ Borges, eds), pp. 99-221, vol. 2. Gráfica Editora Pallotti, Santa Maria.

Riet-Correa G, Terra FF, Schild AL, Riet-Correa F, and Barros SS (2005). Intoxicação experimental por *Tetrapterys multiglandulosa* (Malpighiaceae) em ovelhas. *Pesquisa Veterinária Brasileira* 25:91-96.

Riet-Correa G, Riet-Correa F, Schild AL, Barros SS, and Soares MP (2009). Abortion and neonatal mortality in sheep poisoned with *Tetrapterys multiglandulosa. Veterinary Pathology* 46:960-965.

Schultz RA, Fourie N, Bode ML, Basson KM, Labuschagne L, Snyman LD, and Prozesky L (2004). Pavetamine: an inhibitor of protein synthesis in the heart. In: *Poisonous Plants and Related Toxins* (Acamovic T, Stewart CS, and Pennycott TW, eds), pp. 408-411. CABI Publishing, Wallingford, UK.

Shubeita HE, Martinson EA, Van Bilsen M, Chien KR, and Brow JH (1992). Transcriptional activation of the cardiac myosin light chain 2 and atrial natriuretic factor genes by protein kinase C in neonatal rat ventricular myocytes. *Proceedings of the National Academy of Sciences* 89:1305-1309.

Stigger AL, Barros CSL, Langohr IM, and Barros SS (2001). Intoxicação experimental por *Ateleia glazioviana* (Leg. Papilionoideae) em ovinos. *Pesquisa Veterinária Brasileira* 21:98-108.

Stünzi H and Teuscher E (1970). Herzmuskulatur (Myocardium). In *Ernst Joest Handbuch der Speziellen Pathologischen Anatomie der Haustiere* (J Dobberstein, G Pallaske, and H Stünzi, eds), pp. 93-96, vol. 3. Paul Parey, Berlin.

Tang Q and Taylor PB (1996). Altered contractile function in isoproterenol-induced hypertrophied rat heart. *Journal of Hypertension* 14:751-757.

Tokarnia CH, Peixoto PV, Döbereiner J, Consorte LB, and Gava A (1989). *Tetrapterys* sp. (Malpighiaceae), a causa de mortalidades em bovinos caracterizadas por alterações cardíacas. *Pesquisa Veterinária Brasileira* 9:23-44.

Tokarnia CH, Döbereiner J, and Peixoto PV (2000). *Plantas Tóxicas do Brasil*, 310 pp. Editora Helianthus, Rio de Janeiro.

Van der Lugt JJ, Oliver J, and Jordaan P (1996). Status spongiosis, optic neuropathy and retinal degeneration in *Helichrysum argyrosphaerum* poisoning in sheep and a goat. *Veterinary Pathology* 33:495-502.

Van der Lugt JJ, Kellerman TS, Fourie N, and Coetzer JAW (2002). Diarrhoea, blindness and cerebral spongy changes in cattle caused by *Ornithogalum saudersiae* and *O. prasinum*. In *The Clinicopathology and Pathology of Selective Toxicoses and Storage Diseases of Nervous System of Ruminants in Southern Africa* (JJ Van der Lugt, ed.), pp. 99-114. Tese de Doutorado, University of Utrecht, the Netherlands.

Zierhut W and Zimmer HG (1989). Significance of myocardial alpha- and beta-adrenoreceptors in catecholamine-induced cardiac hypertrophy. *Circulation Research* 65:1417-1425.

# Chapter 40

# Effects of *Senna occidentalis* Seeds Ingested during Gestation on Kid Behavior

M. Barbosa-Ferreira[1], J.A. Pfister[2], A.T. Gotardo[1], P.C.F. Raspantini[1], and S.L. Górniak[1]

[1]*Centro de Pesquisa em Toxicologia Veterinária (CEPTOX), Faculdade de Medicina Veterinária e Zootecnia, Universidade de São Paulo, SP 13635-900, Brasil;* [2]*USDA-ARS Poisonous Plant Research Laboratory, Logan, Utah 84341, USA*

## Introduction

Plant toxins may cause death and loss of productivity in farm animals as well as losses that are subtle and not easily noticed by livestock producers. The weed *Senna occidentalis* (L.) Link, Fabaceae-Caesalpinoideae family (formerly *Cassia occidentalis*), known as fedegoso or coffee senna is distributed throughout the tropical and subtropical regions of the world. The entire plant is toxic (Dollahite and Henson 1965) with seeds the most toxic; when eaten in large amounts, *Senna* causes death (Henson *et al.* 1965; Pierce and O'Hara 1967; Barros *et al.* 1990). When eaten in small amounts for an extended duration, it may reduce animal productivity (Barbosa-Ferreira 2003). The toxic principle is dianthrone, an anthraquinone derivative (Haraguchi *et al.* 1996). Dianthrone may cause disruption of mitochondrial oxidative phosphorylation (Cavaliere *et al.* 1997) leading to death of the organelle, resulting in tissue necrosis in skeletal and cardiac muscles (Herbert *et al.* 1983; Barros *et al.* 1990; Cavaliere *et al.* 1997; Haraguchi *et al.* 1998), liver, kidneys (Tasaka *et al.* 2000; Barbosa-Ferreira *et al.* 2005), and in the central and peripheral nervous system (Barbosa-Ferreira 2003; Barbosa-Ferreira *et al.* 2005). Despite the clinical and histopathological knowledge about *S. occidentalis* intoxication, little is known about the effects of this plant when eaten during the perinatal period. Thus, the purpose of this work was to evaluate some behavioral effects on the development of goat kids that were intoxicated by seeds during gestation. Additionally, we sought to refine a previously proposed model for toxicity risk assessment in ruminants.

## Material and Methods

Ripe *S. occidentalis* seeds were collected from a cultivated garden at the Centro de Pesquisa em Toxicología Veterinária (CEPTOX), Departamento de Patologia Veterinária, Faculdade de Medicina Veterinária e Zootecnia, Universidade de São Paulo (FMVZ/USP), Campus of Pirassununga, SP. *S. occidentalis* was identified at the species level at the Maria

©CAB International 2011. *Poisoning by Plants, Mycotoxins, and Related Toxins* (eds F. Riet-Correa, J. Pfister, A.L. Schild, and T.L. Wierenga)

Eneida Fidalgo Herbarium of the Botanical Institute of São Paulo, São Paulo state, Brazil. The herbarium specimen was deposited in the Botanical Institute of São Paulo (SP) as voucher number SP-363817. After harvesting, the seeds were quickly frozen, immediately finely ground to a powder, and mixed in different concentrations in the animal rations.

The protocol employed met the guidelines of the Bioethics Committee of Faculdade de Medicina Veterinária e Zootecnia da Universidade de São Paulo, São Paulo, Brazil. The protocol was conducted under veterinary supervision and all efforts were made to minimize animal pain and distress.

Ten Saanen mixed-breed female goats (30-33 kg BW) 9-14 months old were hand-mated to one Alpine buck. All females were negative for reproductive diseases such as brucellosis, toxoplasmosis, leptospirosis, and mycoplasmosis. Estrus was synchronized using D-(+)-cloprostenol (Veteglan, Serono). The day of breeding was defined as day 1 of gestation. Once pregnancies were confirmed by ultrasound (US) (Scanner 100 Vet, Pie Medical Lineal probe, 5.0 / 7.0 MHz) on day 27 (Traldi 1994), the goats began to receive the experimental rations. Pregnant goats were randomly allocated into two treatment groups and given the following percentages by weight (as fed basis) of *S. occidentalis* seeds in the ration: 0 (control) and 3% (SO3 group) from the 27th day of gestation to parturition (about day 150). The doses of *S. occidentalis* seeds used in the present experiment were chosen based on data from an earlier study conducted in our laboratory using young goats fed rations having a low concentration of *S. occidentalis* seeds (Barbosa-Ferreira 2003).

At birth, each neonate's body weight was immediately recorded and gender determined; each kid was then examined carefully for gross abnormalities (Szabo 1989). Each kid was weighed weekly from birth to 120 days old. There were eight animals in the control group and seven in the SO3 group. Evaluation of the offspring was done using various tests after birth:

(A) Reaching and discriminating mother and mother vs. alien dam (12 and 24 h postpartum) (Figure 1A): kids had to leave a start box and reach their mother, using her vocalizations as motivation. In a second test the kid had to leave a start box and discriminate its mother from an alien dam. Each test was repeated 3 times and we recorded time to leave the start box, time to reach a female, and whether the choice was correct or not in the discrimination trial.

(B) Navigating a progressive maze (2, 4, 6 days postpartum) (Figure1B): kids were given progressively more difficult barriers to navigate on test days. First, kids had to move from behind a barrier and make their way to their mother, with the mother's vocalizations as a guide. On the first test day the maze had only one barrier, and another barrier was added each test day. On each test day each kid was given three trials. Time spent leaving the starting point and time to reach the mother was recorded.

(C) T maze (6, 8, 10 weeks postpartum) (Figure 1C): kids had to reach their mother, and using her vocalizations, choose an arm of a T maze with the mother at one end. The mother was randomly placed in one arm, and moved to the other arm for the second trial. Kids were run for three trials, and the mother was in a different arm for each to increase the difficulty. Time spent to leave the starting point, and time to reach their own mother or the alien mother was recorded.

(D) Lashley Maze (6, 8, 10 weeks postpartum) (Figure 1D): kids had to reach their mother by navigating in a maze with a zigzag shape. Each kid was tested three times per day. Time spent to leave the starting point, time to enter into different arms and time to reach the mother was recorded.

(E) Hebb-Williams Maze (6, 8, 10 weeks postpartum) (Figure 1E): kids had to reach their mother, navigating in a more elaborate maze. Each kid was tested three times per day.

Time spent to leave the starting point, time to enter into different arms and time to reach the mother was recorded.

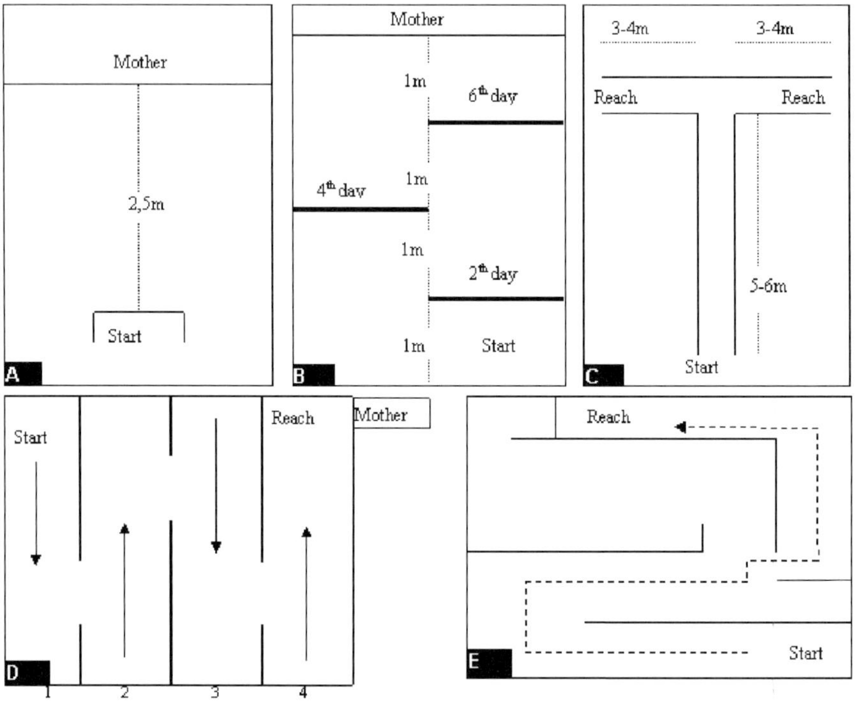


**Figure 1.** Designs of behavioral tests used to evaluate offspring. **A**-reaching and discriminating mother vs alien dam; **B**-progressive maze; **C**-T maze; **D**-Lashley maze; **E**-Hebb-Williams maze. The configuration of the maze was altered over time so that it became more difficult to reach the end of the maze.

Variables were analyzed statistically using a linear mixed model with treatments, individual animals nested within treatments, and repeated measurements over time (Proc Mixed of SAS), with main effects as treatment, gender, and time with all possible interactions. Significant main effects ($P < 0.1$) were followed by mean separation using the PDIFF option in SAS. Data are expressed as least squares means (LS means and SE of the LS means).

## Results

During Test A, kids from control and SO3 group didn't leave the starting point of the maze leading to mother. During Test B, there were differences ($P < 0.01$) between trials. Thus, on the first try, *Senna*-treated kids spent more time leaving the start point and to reach their mothers but this latency diminished during subsequent trials.

In the T maze (Test C), kids from the SO3 group were slower to reach their mother when an alien goat was added and placed in the arm where the mother formerly had been (Figure 2).

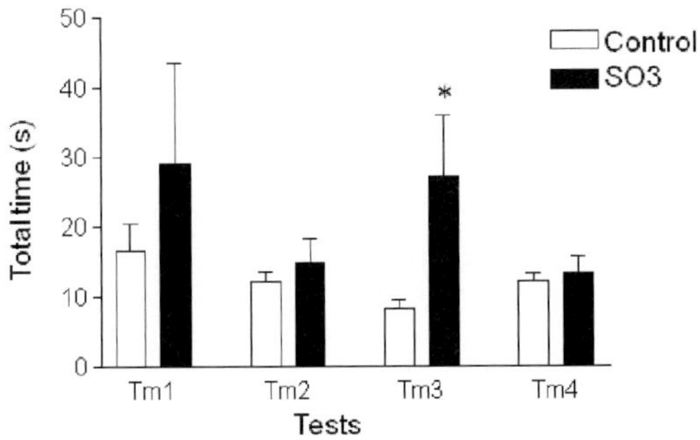


Tm1 = mother alone in left arm; Tm2 = mother alone in right arm; Tm3 = mother in left arm, alien goat in right; Tm4 = mother in right arm, alien goat in left arm.

**Figure 2.** Behavioral tests used to evaluate offspring. Shown here is the total time (sec) to reach mother in the T maze. Animals from the SO3 group differed (*P* < 0.1) from controls in their time to reach mother in Tm3.

In the Hebb-Williams maze (Test E), intoxicated kids were slower (P<0.1) than kids from control group at the 10th week test and took more time to reach the end point; however, treated kids were faster than control kids during the 6th week at the beginning of the tests (Figure 3).

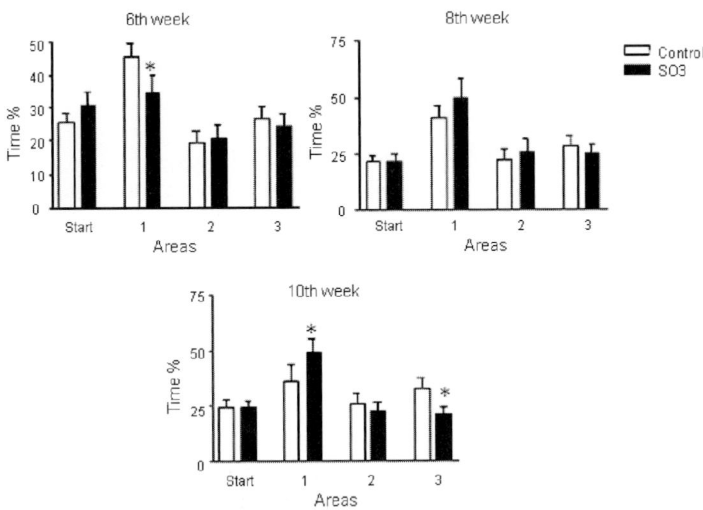


**Figure 3.** Behavioral tests used to evaluate offspring. Shown is the percentage of total time spent in each area of the maze before reaching the end point in the Hebb-Williams maze (Test E). Note that treated animals increased time (**P* < 0.10) spent relative to controls during the week 10 tests.

## Discussion

This study evaluated the reproductive and developmental toxicity of *S. occidentalis* in goats and will help improve the ruminant model for teratology studies developed in our laboratory for the study of potential fetotoxic and developmental effects not only of plant toxins but for a number of chemicals to which domestic ruminants are frequently exposed.

These results show that the toxic compounds in *S. occidentalis* can reach the fetus when the plant is eaten during gestation. Some agents given to mothers during gestation can lead to toxic effects in the offspring, as shown by the use of Heptachlor that causes death only on the 6th day after birth or cataracts after the 21st day of birth (Narotsky and Kavlock 1995). Developmental delay and retardation in neonates could result in offspring that have great difficulty handling environmental challenges postpartum with negative effects on kid survival and economic losses in animal production (Wechsler and Lea 2007).

The present work shows that treated kids had more difficulty when presented with stressing situations, such as when conditions in the tests were changed as observed in the T maze. Intoxicated kids apparently had difficulty in discriminating their mother from an alien dam when their mother was in the opposite side of the arm.

The delay in reaching the end point on the 10th week in Test E reveals a learning deficit. Treated animals were not different initially from control animals in the first trials. The capability to reach the end point in later tests depends on information acquired in the first attempts. We found that treated kids had diminished capacity to achieve success in reaching the end point during later tests whereas control animals had no problems. This study suggests that *S. occidentalis* ingested during gestation results in slower learning in treated kids compared to controls.

These results show that the active principle of the plant passes through the placental barrier and interferes with fetal development. Typically toxic plants are considered teratogenic if they result in neonate weight loss and malformations. These were not seen in the *S. occidentalis* intoxication. More subtle signs of toxicity that we noted with changes in animal behavior suggest that treated kids were developmentally delayed. These results indicate that livestock owners should avoid grazing pregnant animals in areas with infestations of *S. occidentalis*. Further, *S. occidentalis* is used as a medicinal plant and we recommend that pregnant humans also not ingest the plant.

## Conclusions

*S. occidentalis* is teratogenic because it causes behavioral alterations in offspring development and behavior. The use of this plant should be avoided during gestation. The protocol used here is suitable to study developmental toxicity caused by plant toxins.

## Acknowledgements

The authors thank Leonila E.R. Raspantini for technical assistance and Estevão Belloni, Marco A. Faustino, and Adilson Baladore for animal care and handling.

This work is part of the PhD of Marcos Barbosa-Ferreira at the Departamento de Patologia Veterinária, Faculdade de Medicina Veterinária e Zootecnia, Universidade de São Paulo, Brazil and is supported by Coordenação de Aperfeiçoamento de Pessoal de Nível Superior – CAPES.

## References

Barbosa-Ferreira M (2003). Estudo dos efeitos tóxicos produzidos pela administração prolongada de sementes de *Senna occidentalis*. Avaliação em ratos e caprinos, 138 pp. Dissertação de Mestrado, Universidade de São Paulo, São Paulo.

Barbosa-Ferreira M, Dagli MLZ, Maiorka PC, and Górniak SL (2005). Sub-acute intoxication by *Senna occidentalis* seeds in rats. *Food and Chemical Toxicology* 43(4):497-503.

Barros CSL, Pilati C, Andujar, MB, Graça DL, Irigoyen LF, Lopes ST, and Santos CF (1990). Intoxicação por *Cassia occidentalis* (Leg. Caes.) em bovinos. *Pesquisa Veterinária Brasileira* 10(3-4):47-48.

Cavaliere MJ, Calore EE, Haraguchi M, Górrniak SL, Dagli MLZ, Raspantini PCF, Calore NMP, and Weg R (1997). Mitochondrial myopathy in *Senna occidentalis*-seed-fed chicken. *Ecotoxicology and Environmental Safety* 37:181-185.

Dollahite JW and Henson JB (1965). Toxic plants as the etiologic agent of myopathies in animals. *American Journal Veterinary Research* 112(26):749-752.

Haraguchi M, Górniak SL, Dagli MLZ, and Raspantini PCF (1996). Determinação dos constituintes químicos das frações tóxicas de fedegoso, (*Senna occidentalis* L.). In *Anais do XIX Reunião Anual da Sociedade Brasileira de Química*, p. 96. Poços de Caldas, SP.

Haraguchi M, Górniak SL, Calore EE, Cavaliere MJ, Raspantini PCF, Calore NM, and Dagli MLZ (1998). Muscle degeneration in chicks caused by *Senna occidentalis* seeds. *Avian Pathology* 27:346-351.

Henson JB, Dollahite JW, Bridges CH, and Rao RR (1965). Myodegeneration in cattle grazing *Cassia* species. *Journal of the American Veterinary Medical Association* 147:142-145.

Herbert CD, Flory W, Seger C, and Blanchard RE (1983). Preliminary isolation of a myodegenerative toxic principle from *Cassia occidentalis*. *American Journal of Veterinary Research* 44(7):1370-1374.

Narotsky MG and Kavlock RJ (1995). A multidisciplinary approach to toxicological screening. II. Developmental toxicity. *Journal of Toxicological and Environmental Health* 45:145-171.

Pierce KR and O'Hara PJ (1967). Toxic myopathy in Texas cattle. *The Southwestern Veterinarian* 20:179-184.

SAS - Statistical Analysis System (2004). *SAS user's guide: statistics*. SAS Inst. Inc., Cary, NC; Version 9.1 for Windows.

Szabo KT (1989). *Congenital Malformations in Laboratory and Farm Animals*. Academic Press, San Diego, 313 pp.

Tasaka AC, Weg R, Calore EE, Sinhorini IL, Dagli MLZ, Haraguchi M, and Górniak SL (2000). Toxicity of *Senna occidentalis* seed in rabbits. *Veterinary Research Communications* 24:573-582.

Tasaka AC, Sinhorini IL, Dagli MLZ, Haraguchi M, and Górniak SL (2004). Perinatal study of *Senna occidentalis*. Intoxication in rabbits. In *Poisonous Plants and Related Toxins* (T Acamovic, CS Stewart, TW Pennycott, eds), pp. 459-464. CABI Publishing, Wallingford.

Traldi AS (1994). *Tópicos em Reprodução de Caprinos.*.Departamento de Reprodução da Faculdade de Medicina Veterinária Zootecnia da Universidade de São Paulo, SP, p.84.

Wechsler B and Lea SEG (2007). Adaptation by learning: Its significance for farm animal husbandry. *Applied Animal Behavior Science* 108:197–214.

# Chapter 41

# Evaluation of the Abortifacient Effect of *Luffa acutangula* Roxb. in Rats

L.C.B. Fernandes, L.A.V. Cordeiro, and B. Soto-Blanco

*Department of Animal Sciences, Universidade Federal Rural do Semi-Árido (UFERSA), BR 110 Km 47, Mossoró, RN, 59625-900, Brazil*

## Introduction

Plants of the Cucurbitaceae family are characteristically but not exclusively producers of cucurbitacins, highly oxygenated triterpenes. Cucurbitacins isolated from *Luffa* spp. have various pharmacological properties including anti-inflammatory, antimicrobial, anti-tumoral, and hepatocurative effects and are cytotoxic (Miró 1995). Other compounds isolated from *Luffa* spp. include luffaculins, which are ribosome-inactivating proteins (Ng *et al.* 1992; Chan *et al.* 1994; Lin *et al.* 2002; Wang and Ng 2002; Hou *et al.* 2007), and a lectin that is specific for chito-oligosaccharides (Anantharam *et al.* 1986).

Several farmers from the northeastern region of Brazil have reported abortions in ruminants that had ingested fruits of *L. acutangula* (Silva *et al.* 2006). Tea made from this plant has been widely used by women for induction of abortion and as a purgative (Lorenzi and Matos 2002). However, studies of the effects of this plant on reproduction are needed. The present work aimed to study the abortifacient effect of *L. acutangula* tea in rats.

## Methodology

Mature fruits of *L. acutangula* Roxb. were collected at Catolé do Rocha municipality, Paraíba state, Brazil. Botanical identification was done by Prof Odaci Fernandes de Olivera, a retired professor from Universidade Federal Rural do Semi-Árido (UFERSA), Mossoró, RN, Brazil. A total of 50 g of dried powdered fruits were added to 100 ml of distilled water. The mixture was heated until it began boiling and it was then filtered. The solution obtained was kept refrigerated for up to 7 days.

Adult male and virgin female Wistar rats were obtained at the age of about 12 weeks from the Animal Sciences Department, Universidade Federal Rural do Semi-Árido, Mossoró, RN, Brazil. Commercial food ration (Labina, Purina, Paulínia, SP, Brazil) and tap water were provided *ad libitum*. For mating, two females were placed together with one male. Females that showed evidence of mating (a vaginal plug, or a vaginal smear showing sperm cells) were assigned in rotation to each dose group until the required 12 females had

©CAB International 2011. *Poisoning by Plants, Mycotoxins, and Related Toxins* (eds F. Riet-Correa, J. Pfister, A.L. Schild, and T.L. Wierenga)

been allotted to each group. The day on which there was evidence of mating was recorded as day 0 of gestation. During gestation, the female rats were housed individually in plastic cages measuring 40×50×20 cm, which were covered with metal lids. The animal room was maintained at 22-24°C and on a 12 h light/dark cycle.

Eleven pregnant Wistar female rats were used for the experiment. On the 15th day of gestation, six rats were dosed by gavage with 10 ml/kg BW of the infusion of *L. acutangula*, and the other five rats were dosed with 10 ml/kg BW of saline solution. On the 21st day of gestation, all the rats were submitted to Cesarean section. The number of implantation sites on the uterine horns, live and dead fetuses, corpora lutea, and number of fetal resorptions were recorded. Fetuses and placentas were weighed individually and the fetuses were evaluated carefully for the presence of gross malformations. The placentas were fixed in 10% buffered formalin and paraffin-embedded sections were stained with hematoxylin and eosin for pathological studies. Data are reported as the mean ± standard deviation, and they were compared using the unpaired *t* test for parametric data or the Mann–Whitney test for non-parametric data using GraphPad Prism v.4 for Mac. The level of significance was set at $P < 0.05$.

## Results and Discussion

The average body weight and rate of body weight gain in rats dosed with *L. acutangula* were not significantly ($P > 0.05$) different from controls (Table 1). Furthermore, no clinical signs of maternal toxicity were observed. It is well known that maternal toxicity may be responsible for abnormal fetal development, because maternal–fetal interactions are affected (Khera 1984; Chernoff *et al.* 1989). In the present study there was no evidence that aberrant fetal development was due to maternal toxicity.

**Table 1.** Body weight and gain in body weight (in g) of pregnant rats dosed with *L. acutangula* or saline solution (control) on the 15th day of pregnancy. The data are presented as means followed by the respective standard deviation.

| | | Control | Treated |
|---|---|---|---|
| Body weight | 1st day | 184.8±17.0 | 182.8±7.19 |
| | 15th day | 244.5±19.4 | 245.0±13.6 |
| | 21st day | 303.0±14.4 | 300.7±27.6 |
| Body weight gain | 1st to 15th day | 59.8±7.18 | 62.2±19.1 |
| | 15th to 21st day | 53.8±10.2 | 62.8±17.6 |

The number of implantation sites, live and dead fetuses, corpora lutea, and number of resorptions did not differ ($P > 0.05$) between groups; however, reduced ($P < 0.05$) fetal weight was significant in the group treated with *L. acutangula* (Table 2). Fetal weight is an important parameter in studies of developmental toxicity (Manson 1989). Thus, we conclude that *L. acutangula* is fetotoxic under the conditions of this experiment.

*L. acutangula* is historically used by women to interrupt pregnancy (Lorenzi and Matos 2002). Furthermore, several farmers from the northeastern region of Brazil have reported abortions in ruminants that had ingested fruits of *L. acutangula* (Silva *et al.* 2006). Several studies have identified proteins from *L. acutangula* and other *Luffa* species that have abortifacient action (Schilling and Heiser 1981; Ng *et al.* 1992; Chan *et al.* 1994; Lin *et al.* 2002; Wang and Ng 2002), including the proteins luffin b and lufaculin (Chan *et al.*

1994). These proteins promote the inactivation of ribosomes (Ng *et al.* 1992; Chan *et al.* 1994; Wang and Ng 2002; Hou *et al.* 2007), which affects the synthesis of cellular proteins and potentially interferes with embryogenesis and fetal development. In the present study, the post-implantation loss in rats dosed with *L. acutangula* was 21.5%, which is much higher than that of the control group (5%), but this result was not statistically significant.

**Table 2.** Reproductive performance (mean±SD) of pregnant rats dosed with *Luffa acutangula* or saline solution (control) on the 15th day of pregnancy (*=significant).

| | Control | Treated |
|---|---|---|
| Number of corpora lutea | 11.2±1.92 | 11.5±0.84 |
| Number of implantations | 10.4±1.82 | 10.7±1.37 |
| Number of live fetuses | 10.0±2.55 | 8.17±2.04 |
| Pre-implantation loss (%) | 6.82±7.03 | 7.20±9.79 |
| Post-implantation loss (%) | 5.00±11.2 | 21.5±23.5 |
| Fetal weight (g) | 3.72±0.22 | 3.22±0.38* |
| Placental weight (g) | 0.46±0.04 | 0.51±0.10 |

The search for external malformations revealed a fetus with a cleft palate from a female rat treated with *L. acutangula*. While this is the first time that a gross malformation has been attributed to *L. acutangula*, it has been reported that *L. operculata* was responsible for significant alterations in the palatal epithelium of the frog (Menon-Miyake *et al.* 2005). There are several chemicals known to induce cleft palate (DePass and Weaver 1982; Bonner 1984; Hood and Ottley 1985; Alsdorf and Wyszynski 2005; Bock and Köhle 2006) and multiple mechanisms have been proposed. However, proposing *L. acutangula* as the cause of the single cleft palate in this study is somewhat speculative and further research is needed.

The placenta is a critical organ for normal embryogenesis and fetal development and it performs a number of different and specialized functions in pregnancy. Several chemicals may induce damage to the placenta subsequently affecting embryonic and fetal development (Goodman *et al.* 1982). In the present work, histological evaluation of placentas of treated rats revealed no lesions. While lack of histological lesions would suggest that *L. acutangula* does not cause fetotoxicity, biochemical changes could still alter placental function and requires further research investigation.

In conclusion, the ingestion of *L. acutangula* during pregnancy may inhibit normal development in exposed rat pups as shown by reduced fetal weight and the occurrence of a single cleft palate. Further research into the potential fetotoxic effects of *L. acutangula* is necessary to fully understand the field reports of abortion in small ruminants and the preliminary data presented here.

## References

Alsdorf R and Wyszynski DF (2005). Teratogenicity of sodium valproate. *Expert Opinion on Drug Safety* 4:345-353.

Anantharam V, Patanjali SR, Swamy MJ, Sanadi AR, Goldstein IJ, and Surolia A (1986). Isolation, macromolecular properties, and combining site of a chito-oligosaccharide-specific lectin from the exudate of ridge gourd (*Luffa acutangula*). *The Journal of Biological Chemistry* 261:14621-14627.

Bock KW and Köhle C (2006). Ah receptor: dioxin-mediated toxic responses as hints to deregulated physiologic functions. *Biochemical Pharmacology* 72:393-404.

Bonner JJ (1984). The H-2 genetic complex, dexamethasone-induced cleft palate, and other craniofacial anomalies. *Current Topics on Developmental Biology* 19:193-215.

Chan WY, Ng TB, and Yeung HW (1994). Differential abilities of the ribosome inactivating proteins luffaculin, luffins and momorcochin to induce abnormalities in developing mouse embryos *in vitro*. *General Pharmacology* 25:363-367.

Chernoff N, Rogers JM, and Kavlock RJ (1989). An overview of maternal toxicity and prenatal development: considerations for developmental toxicity hazard assessments. *Toxicology* 59:111-125.

DePass LR and Weaver EV (1982). Comparison of teratogenic effects of aspirin and hydroxyurea in the Fischer 344 and Wistar strains. *Journal of Toxicology and Environmental Health* 10:297-305.

Goodman DR, James RC, and Harbison RD (1982). Placental toxicology. *Food and Chemical Toxicology* 20:123-128.

Hood RD and Ottley MS (1985). Developmental effects associated with exposure to xylene: a review. *Drug and Chemical Toxicology* 8(4):281-297.

Hou X, Chen M, Chen L, Meehan EJ, Xie J, and Huang M (2007). X-ray sequence and crystal structure of luffaculin I, a novel type I ribosome-inactivating protein. *BMC Structural Biology* 7:29-38.

Khera KS (1984). Maternal toxicity – a possible factor in fetal malformations in mice. *Teratology* 29:411-416.

Lin JK, Chen MH, Xie JM, Zhao R, Ye XM, Shi XL, and Wang ZR (2002). Purification and characterization of two luffaculins, ribosome-inactivating proteins from seeds of *Luffa acutangula*. *Chinese Journal of Biochemistry and Molecular Biology* 18:609-613.

Lorenzi H and Matos FJA (2002). *Plantas Medicinais do Brasil: nativas e exóticas*, pp.259-260. Instituto Plantarum, Nova Odessa, SP, Brazil.

Manson JM (1989). Test methods for assessing female reproductive and developmental toxicology. In *Principles and Methods of Toxicology* (Hayes AW, ed.), pp. 311-360. Raven Press, New York.

Menon-Miyake MA, Saldiva PH, Lorenzi-Filho G, Ferreira MA, Butugan O, and Oliveira RC (2005). Efeitos da *Luffa operculata* sobre o epitélio do palato de rã: aspectos histológicos. *Revista Brasileira de Otorrinolaringologia* 71:132-138.

Miró M (1995). Cucurbitacins and their pharmacological effects. *Phytotherapy Research* 9:159-168.

Ng TB, Chan WY, and Yeung HW (1992). Proteins with abortifacient, ribosome inactivating, immunomodulatory, antitumor and anti-AIDS activities from Cucurbitaceae plants. *General Pharmacology* 23:579-590.

Schilling EE and Heiser Jr CB (1981). Flavonoids and the systematics of *Luffa*. *Biochemical Systematics and Ecology* 9:263-265.

Silva DM, Riet-Correa F, Medeiros RMT, and Oliveira OF (2006). Plantas tóxicas para ruminantes e eqüídeos no Seridó Ocidental e Oriental do Rio Grande do Norte. *Pesquisa Veterinária Brasileira* 26:223-236.

Wang H and Ng TB (2002). Luffangulin, a novel ribosome inactivating peptide from ridge gourd (*Luffa acutangula*) seeds. *Life Science* 70:899-906.

# Chapter 42

# Experimental Studies of Poisoning by *Aspidosperma pyrifolium*

M.C.J.S. Lima[1] and B. Soto-Blanco[2]

[1]*Post-Graduate Program of Animal Science, Universidade Federal Rural do Semi-Árido (UFERSA), BR 110 Km 47, Mossoró, RN, 59625-900, Brazil;* [2]*Department of Animal Sciences, Universidade Federal Rural do Semi-Árido (UFERSA), BR 110 Km 47, Mossoró, RN, 59625-900, Brazil*

## Introduction

Poisonous plants may affect animal reproduction; they promote dysfunctions that include abortions, infertility, and teratogenesis. In Brazil, poisonous plants that promote abortions and neonatal loss include *Aspidosperma pyrifolium* (Medeiros *et al.* 2004), *Ateleia glazioviana* (Stolf *et al.* 1994), *Tetrapterys acutifolia*, *T. multiglandulosa* (Tokarnia *et al.* 1989), and *Stryphnodendron obovatum* (Tokarnia *et al.* 1998). Abortions induced by *A. pyrifolium* are frequently observed and have been demonstrated experimentally in goats (Medeiros *et al.* 2004). This plant is one of the most important toxic species in the caatinga, the forest of the semiarid region of Brazil (Silva *et al.* 2006). Despite its economic impact, knowledge about this poisonous plant is very limited. This study (i) describes spontaneous cases of poisoning by *A. pyrifolium*; (ii) characterizes the toxicity of this plant in rats; and (iii) investigates the cytotoxicity by *in vitro* tests.

## Materials and Methods

### Field cases

Six farms, three in the municipality of Mossoró and three in the municipality of Angicos located in the state of Rio Grande do Norte, Brazil, that had a history of cases of abortions in small ruminants attributed to *A. pyrifolium*, were visited to determine the epidemiology of the disease and for investigation of the paddocks.

### Plant material for experimental studies

Leaves from *A. pyrifolium* were collected near Mossoró city, northeastern Brazil, (5°11'15"S and 37°20'39"W) at an altitude of 16 m above sea level. The climate in the

©CAB International 2011. *Poisoning by Plants, Mycotoxins, and Related Toxins* (eds F. Riet-Correa, J. Pfister, A.L. Schild, and T.L. Wierenga)

region is characterized as semiarid. The mean annual temperature is 27.4°C, while the mean annual rainfall and mean relative humidity are 674 mm and 68.9%, respectively. Fresh leaves were extracted with 100% ethanol, filtered, and the solvent was removed by rotary evaporation. The ethanol residue was dissolved in distilled water and filtered. The aqueous solution of the ethanol extract was used for the assays performed *in vivo* and *in vitro*.

### *In vivo* toxicity assays in rats

Adult male and virgin female Wistar rats were used. Food and tap water were provided *ad libitum*. For mating, two females were placed together with one male. The day on which there was evidence of mating was recorded as day 0 of gestation. During gestation, the female rats were housed individually in plastic cages measuring 40×50×20 cm and covered with metal lids. The animal room was maintained at 22-24°C and with a 12 h light/dark cycle.

Twenty-four female Wistar rats were dosed by gavage with 0 (control), 10, or 20 g/kg BW of *A. pyrifolium* in an aqueous solution from the ethanolic extract of the plant on the 15th gestational day, or with 20 g/kg BW from the 15th to the 17th gestational day. On day 21 of gestation, the females were killed with ether anesthesia and the ovaries and uteri were removed by cesarean section. The number of corpora lutea in each ovary was recorded and the gravid uteri were weighed. The fetuses were removed from the uteri, dried of amniotic fluid, weighed, and examined for gross abnormalities. The placenta of the live fetuses was also weighed. The number of implantation sites and resorptions in both uterine horns was recorded.

An aqueous solution of the ethanol extract was injected intraperitoneally into rats at concentrations corresponding to 10, 20, 30, and 60 g of plant/kg body weight. Three male and three female rats were used for each dose. The rats were monitored closely for up to 48 h after dosing.

### Osmotic fragility assay and lethality for *Artemia salina* larvae

The evaluation of the osmotic fragility of red blood cells was performed on a fresh blood sample collected from a healthy goat using EDTA as anticoagulant. The blood sample was gently mixed with 0.90% NaCl solution and the mixture was centrifuged (1500 rpm, 15 min). The supernatant was discarded and the procedure was repeated. The concentrated red blood cells (10 μl) were added to 1.0 ml of plant extract at different concentrations (corresponding to 0, 25, 50, 100, and 200 mg of plant/ml in 0.9% NaCl solution) and incubated for 60 min at room temperature with two repetitions for each concentration. An extract was also prepared at a concentration of 200 mg/ml containing different concentrations of NaCl (0 up to 0.9%) and incubated with blood samples. The hemolytic percentage was determined by measurement of hemoglobin in the supernatants using a commercial kit in a spectrophotometer at 540 nm.

The toxicity of the extract of *A. pyrifolium* to *Artemia salina* (brine shrimp) was tested at concentrations corresponding to 10, 15, 20 and 30 mg of plant/ml in 10 ml of sea-water solution. Ten 1-day-old larvae were used in each test, and the survivors counted after 24 h. Three replications were used for each concentration. A parallel series of blank controls were conducted. The mortality after 24 h of exposure was measured.

### Statistical analysis

The data are reported as mean ± SEM and were compared using analysis of variance (ANOVA) with separation of means by Duncan's method (GraphPad Prism v.4 for Mac). The level of significance was set at $P < 0.05$.

## Results

### Field cases

Naturally occurring cases of poisoning were observed on six farms in the region of the cities of Mossoró and Angico, RN, Brazil. On all the evaluated farms, the ingestion of the plant and the cases of abortion occurred exclusively in goats and did not affect sheep. The majority of the abortions occurred early in the dry season and early in the rainy season. There had been no acquisition of new animals on two of the farms from Mossoró during the last 3 years, and the incidence of abortions had decreased to almost zero. The third farm from Mossoró had abortions in goats and also cases of malformation in sheep, goats, and cattle, characterized as permanent flexure of the front legs, cranial deformities, and brachygnathia. On this farm there was intense exploitation of wood for construction and fuel, mainly of the species *Mimosa tenuiflora*. Most of the *M. tenuiflora* was blossoming and was avidly ingested by the animals.

### *In vivo* assays of toxicity in rats

The administration of the extract of *A. pyrifolium* caused the death of a pregnant rat treated with 20 g/kg 1 day after administration. A rat treated with 20 g/kg for 2 days, and another treated with the same dose for 3 days also died. At postmortem examination, no macroscopic changes were observed. Histological analysis was not performed because the tissues showed autolysis.

The data on reproduction are summarized in Table 1. No statistically significant differences were observed in the numbers of corpora lutea, implantations, live and dead fetuses, and placental weights. However, the fetal weights were lower ($P < 0.05$) in groups that received the extract of *A. pyrifolium* at all doses evaluated.

**Table 1.** Reproductive performance (mean±SD) of pregnant female rats treated by gavage with *A. pyrifolium* extract at doses corresponding to 0 (control), 10, or 20 g of plant/kg BW on the 15th gestational day or 20 g/kg from the 15th to the 17th gestational day.

| | Group | | | |
|---|---|---|---|---|
| | Control | 10 g/kg | 20 g/kg | 20 g/kg for 3 days |
| Number of corpora lutea | 11.3±1.75 | 9.67±1.34 | 9.75±0.96 | 8.00±2.45 |
| Number of implantations | 10.7±1.75 | 8.00±2.19 | 9.00±1.41 | 8.00±2.45 |
| Number of live fetuses | 10±2.28 | 7.50±1.97 | 6.75±3.69 | 7.25±2.22 |
| Number of dead fetuses | 0 | 0 | 0 | 0 |
| Fetal weight | 3.81±0.30 [a] | 3.02±0.35 [b] | 3.07±0.27 [b] | 2.96±0.46 [b] |
| Placental weight | 0.44±0.06 | 0.45±0.09 | 0.44±0.04 | 0.48±0.10 |

[a,b] Data for each parameter followed by different letters are significantly different ($P < 0.05$, ANOVA followed by Bonferroni's multiple comparisons test)

After intraperitoneal injection of the plant extract, the female rats treated with 20 g/kg of extract presented with paralysis of the hind-limbs for 1 h and one of the three rats died. The administration of 30 and 60 g/kg was fatal to all female rats. Plant extract at 10 g/kg did not produce any sign of poisoning in female rats. Administration of the extract at concentrations up to 30 g/kg to male rats did not produce disturbances, but treatment with 60 g/kg produced paralysis of the hind-limbs for 1 h in all male rats.

### Assays of osmotic fragility and lethality for *Artemia salina* larvae

The mean percentage of erythrocyte lysis was 0% at 0 mg/ml, 18.1% at 25 mg/ml, 36.4% at 50 mg/ml, 66.7% at 100 mg/ml, and 100% at 200 mg/ml. The extract at a concentration of 200 mg/ml prepared with different concentrations of NaCl (0 to 0.9%) did not show reduced hemolytic activity.

The extract was lethal to 11.5% of larvae of *Artemia salina* at 10 mg/ml, 26.9% at 15 mg/ml, 34.6% at 20 mg/ml, and 46.2% at 30 mg/ml. No deaths were observed in those treated with the blank control.

## Discussion

In the present study, abortions induced by *A. pyrifolium* were observed exclusively in goats. Most cases were reported to occur early in the dry season (usually from July to September) and early in the rainy season (usually from December to January). Silva *et al.* (2006) also found a higher incidence of abortion during the early dry season. On one farm cases of malformations in sheep, goats, and cattle were found. The alterations observed were permanent flexure of the front legs, cranial deformities, and brachygnathia; these are characteristic signs of the teratogenic effect of *Mimosa tenuiflora* (Medeiros *et al.* 2007; Pimentel *et al.* 2007). In fact, this farm grew large amounts of *M. tenuiflora* and it was avidly ingested by the animals. Thus, none of the malformations could be attributed definitively to *A. pyrifolium*. An interesting observation was that the incidence of abortions had decreased to almost zero on two farms that had purchased no new animals recently. This is an indication that the plant may promote a natural aversion reaction to its ingestion in experienced goats. It may be possible to prevent poisoning by *A. pyrifolium* by avoiding the purchase of new females or by promoting a conditioned aversion to the plant in goats, similar to that used for *Mascagnia rigida* (Barbosa *et al.* 2008). Alternatively, goats could be adapted to the plant as a result of increased toxin degradation by liver metabolism or rumen degradation.

In the present work, the administration of *A. pyrifolium* to pregnant rats promoted reduction of fetal weight. This is a significant effect because fetal weight is one of the most important parameters in studies of developmental toxicity (Manson 1989). However, pregnant rats dosed with an extract of *A. pyrifolium* showed strong evidence of maternal toxicity, and it was lethal to some of them. It is well known that maternal toxicity is responsible for disturbances of fetal development as maternal–fetal interactions are affected (Chernoff *et al.* 1989). It was not possible to determine if the reduced fetal weight observed was the consequence of maternal toxicity or of a direct action of the plant on the fetus.

*In vitro* tests of cytotoxicity were performed by investigation of erythrocyte lysis and of lethality in *Artemia salina* larvae. Chemicals cause hemolysis by damaging the integrity of the membrane; this may involve direct effects on specific ion transport pathways, induction of oxidative damage of the cell membrane, disturbance of the structure of the lipid layer, and/or unpredicted toxic effects on processes that control cell volume, leading to cell swelling (Mineo and Hara 2007; Watts and Handy 2007). In this study, the ability of *A. pyrifolium* to promote hemolysis is indicative of cytotoxicity by damage to the plasma membrane.

The brine shrimp (*Artemia*) is one of the best known aquatic organisms. Toxicological bioassays using *Artemia* nauplii are very useful and are characterized by low cost, short time to obtain results, and no requirement for sophisticated apparatus, which permits their use in a wide variety of studies (Sleet and Brendel 1985; Nunes *et al.* 2006). The results presented here showed that *A. pyrifolium* is lethal to *A. salina*, which confirms its cytotoxicity. The assays of lysis of erythrocytes and lethality in *Artemia salina* may be used as alternative tests for the identification of the toxic compound(s) present in *A. pyrifolium*.

## References


Barbosa RR, Pacífico da Silva I, and Soto-Blanco B (2008). Development of conditioned taste aversion to *Mascagnia rigida* in goats. *Pesquisa Veterinária Brasileira* 28:571-574.

Chernoff N, Rogers JM, and Kavlock RJ (1989). An overview of maternal toxicity and prenatal development: considerations for developmental toxicity hazard assessments. *Toxicology* 59:111-125.

Manson JM (1989). Test methods for assessing female reproductive and developmental toxicology. In *Principles and Methods of Toxicology* (Hayes AW, ed.), pp. 311-360. Raven Press, New York.

Medeiros RMT, Neto SAG, Riet-Correa F, Schild AL, and Sousa NL (2004). Mortalidade embrionária e abortos em caprinos causados por *Aspidosperma pyrifolium*. *Pesquisa Veterinária Brasileira* 24(supl.):42-43.

Medeiros RMT, Figueiredo APM, Benício TMA, Dantas FPM, and Riet-Correa F (2007). Teratogenicity of *Mimosa tenuiflora* seeds to pregnant rats. *Toxicon* 51:316-319.

Mineo H and Hara H (2007). Chemical specificity in short-chain fatty acids and their analogues in increasing osmotic fragility in rat erythrocytes *in vitro*. *Biochimica et Biophysica Acta – Biomembranes* 1768:1448-1453.

Nunes BS, Carvalho FD, Guilhermino LM, and Van Stappen G (2006). Use of the genus *Artemia* in ecotoxicity testing. *Environmental Pollution* 144:453-462.

Pimentel LA, Riet-Correa F, Gardner D, Panter KE, Dantas AFM, Medeiros RMT, Mota RA, and Araújo JAS (2007). *Mimosa tenuiflora* as a cause of malformations in ruminants in the northeastern Brazilian semiarid rangelands. *Veterinary Pathology* 44:928-931.

Silva DM, Riet-Correa F, Medeiros RMT, and Oliveira OF (2006). Toxic plants for livestock in the western and eastern Seridó, state of Rio Grande do Norte, in the Brazilian semiarid. *Pesquisa Veterinária Brasileira* 26:223-236.

Sleet RB and Brendel K (1985). Homogeneous populations of *Artemia* nauplii and their potential use for *in vitro* testing in developmental toxicology. *Teratogenesis Carcinogenesis Mutagenesis* 5:41-54.

Stolf L, Gava A, Varaschin MS, Neves DS, Mondadori AJ, and Scolari LS (1994). Aborto em bovinos causado pela ingestão de *Ateleia glazioviana* (Leg. Papilionoideae). *Pesquisa Veterinária Brasileira* 14:15-18.

Tokarnia CH, Peixoto PV, Döbereiner J, Consorte LB, and Gava A (1989). *Tetrapterys* spp. (Malpighiaceae), a causa de mortandades em bovinos caracterizadas por alterações cardíacas. *Pesquisa Veterinária Brasileira* 9:23-44.

Tokarnia CH, Brito MF, Driemeier D, Costa JBD, and Camargo AJR (1998). Aborto em vacas na intoxicação experimental pelas favas de *Stryphnodendron obovatum* (Leg. Mimosoideae). *Pesquisa Veterinária Brasileira* 18:35-38.

Watts TJ and Handy RD (2007). The haemolytic effect of verapamil on erythrocytes exposed to varying osmolarity. *Toxicology in Vitro* 21:835-839.

# Chapter 43

# Determination of Teratogenic Effects of *Aspidosperma pyrifolium* Ethanolic Extract in Rats

A.P.M. Figueiredo[1], F.P.M. Dantas[1], R.M.T. Medeiros[1], J.M. Barbosa Filho[2], and F. Riet-Correa[1]

[1]*Centro de Saúde e Tecnologia Rural, Universidade Federal de Campina Grande, Campus de Patos, Paraíba, PB 58700-970, Brazil;* [2]*Laboratório de Tecnologias Farmacêuticas/UFPB, João Pessoa, PB 58000-000, Brazil*

## Introduction

*Aspidosperma pyrifolium* (Apocynaceae) causes abortion and embryonic death in goats (Medeiros *et al.* 2004; Lima and Soto-Blanco 2009) and probably in sheep and cattle (Silva *et al.* 2006). Goats abort when they eat *A. pyrifolium* at different stages of gestation. Abortions occur primarily during the dry season when forage becomes scarce from lack of rain and *A. pyrifolium*, which maintains green foliage, is the main available forage. Abortions also occur during the dry season because after a rain event the plant resprouts rapidly and is often eaten by pregnant goats (Riet-Correa *et al.* 2006) or when the flock is introduced in areas with large amounts of *A. pyrifolium*. The toxic compound of *A. pyrifolium* has not been determined, but the plant contains monoterpenoid indole alkaloids aspidofractinine, 15-demethoxypyrifoline, and *N*-formylaspidofractine (Araújo *et al.* 2007).

In the semiarid Brazilian northeastern region, malformations are frequent in goats, sheep, and cattle (Medeiros *et al.* 2005; Nóbrega *et al.* 2005; Riet-Correa *et al.* 2006). The main malformations are permanent flexure of the forelimbs (arthrogryposis) which may also be shortened or twisted, malformations of the bones of the head and face including micrognathia, primary cleft lip that occurs with hypoplasia, or unilateral or bilateral aplasia of the incisive bone, secondary cleft palate (palatoschisis), and malformations of the spine (kyphosis, scoliosis, torticollis, or hyperlordosis). Some animals are born blind with varying degrees of opacity of the cornea and/or microphthalmia, others with ocular dermoids. Other malformations include acephaly, bicephaly, hydranencephaly, hypoplasia of the tongue, meningocele, and syringocele. Some animals show variations of these malformations, and may be termed monsters. The majority of animals with malformations of the head and spine die, but many that only have flexion of the forelimbs survive with the defect. Some of these malformations were experimentally produced by the administration of *Mimosa tenuiflora* to experimental goats (Pimentel *et al.* 2007). Experiments in rats

©CAB International 2011. *Poisoning by Plants, Mycotoxins, and Related Toxins* (eds F. Riet-Correa, J. Pfister, A.L. Schild, and T.L. Wierenga)

showed that the ingestion of 10% of *Mimosa tenuiflora* seeds in the diet from days 6-21 of pregnancy causes fetal bone malformation (Medeiros *et al.* 2008). Seeds of *M. ophtalmocentra* mixed at 10% in the diet and fed days 6-21 of pregnancy were also teratogenic for rats (Pessôa 2007).

Before the experimental demonstration that *M. tenuiflora* causes malformations in goats and sheep, many farmers claimed that the cause of those malformations was the ingestion of *A. pyrifolium*. Experiments conducted in goats caused abortion and embryonic deaths but not malformations (Medeiros *et al.* 2004). The objective of this experiment was to study the effect of an ethanolic extract of *A. pyrifolium* on rat reproduction.

## Material and Methods

Wistar rats (*Rattus novergicus*), males and females, 12 weeks old, were used. The animals were housed in 40×50×20cm plastic cage, under controlled temperature (22-26°C) in a natural cycle of light. Twenty-four pregnant rats were used, divided into two groups: one control and one experimental.

*A. pyrifolium* leaves were collected, dried in shade, and then ground to prepare the ethanolic extract. The leaves were steeped in ethanol for 7 days. Later, this solution was filtered and the solvent was evaporated by rotoevaporation.

For mating, one male and two females were allotted to each cage for 12 h from evening to the following morning. Females with evidence of mating (vaginal plug or vaginal smear with sperm cells) were assigned to one of the two groups. Females that failed to become pregnant for two consecutive times were rejected. The females received water and food *ad libitum*. Water ingestion, food consumption, and weight gain were measured every 4 days, beginning on the 6th day of pregnancy. From days 6-21 of pregnancy, the *A. pyrifolium* ethanolic extract was administrated daily by gavage in a dose of 0.06 g diluted in 2 ml of soya oil. The animals from the control group received only 2 ml of soya oil, also by gavage.

On the 21st day of pregnancy the rats were anesthetized by ether inhalation and the ovaries and uteri were removed by cesarean section. The number of corpora lutea in each ovary was recorded and the gravid uterus was weighed. The fetuses were removed from the uteri, dried of amniotic fluid, weighed, and examined for conformation of the eyes, mouth, head, limbs, tail, and ears, and the presence of the anal perforation to verify external abnormalities and malformations. The placentas and the live fetuses were weighed. The number of implantation sites and resorptions was recorded for both uterine horns. After being weighed the fetuses were euthanatized with ether, fixed in acetone for 24 h, examined for cleft palate, and eviscerated. For examination of the skeleton the fetuses were submersed in a solution of 0.8% potassium hydroxide with alizarin-red S which was changed daily for 3-4 days (Staples and Schenell 1964). Then the fetuses were cleared in a solution of 40% ethyl alcohol, 40% glycerin, and 20% benzilic alcohol. The degree of fetal bone development was evaluated by counting the ossification centers in some fetal bones (phalanges of the fore limbs, metacarpus, metatarsus, sternebrae, and caudal vertebrae). The kidney, lung, and liver of the fetuses were weighed. The rats were euthanized and necropsied. Lung, liver, heart, and kidneys were weighed and samples of these tissues were fixed in 10% buffered formalin, embedded in paraffin, sectioned at 5 μm, and stained by hematoxylin and eosin for histologic examination. The software GraphPad Instat V2.01 (GraphPad 1993) was used for the statistical analysis. Food consumption, water ingestion, body weight gains, organ weight, and data from the offspring were analyzed by the

Students 't' test. Frequency of skeletal abnormalities and malformations were evaluated by Fisher's exact test. The percentages of pre- and post-implantation losses and the degree of ossification were evaluated by the Mann-Whitney U test.

## Results

Reproductive performance is summarized in Table 1. All fetuses were alive during the cesarean. There was a significant ($P < 0.001$) decrease in the fetal and placental weights of the experimental group compared with the control group, and a significant ($P < 0.005$) increase in the number of resorptions in the control group compared with experimental group. The other reproductive parameters showed no statistical differences between groups.

**Table 1.** Reproductive performance (mean ± SD ) of dams that received 0.06 g *A. pyrifolium* ethanolic extract from days 6-20 of pregnancy and control dams.

| Reproductive parameters | Control Group $\bar{x} \pm$ SD | Experimental Group $\bar{x} \pm$ SD |
|---|---|---|
| Gravid uterus weight (g) | 48.13 ± 16.96 | 56.00 ± 12.43 |
| Number of fetuses | 8.33 ± 2.93 | 10.33 ± 2.42 |
| Weight of fetuses (g) | 3.56 ± 0.54 | 3.32 ± 0.38** |
| Weight of placenta (g) | 0.47 ± 0.07 | 0.43 ± 0.06** |
| Length of fetuses (cm) | 4.40 ± 0.33 | 4.39 ± 0.26 |
| Number of corpora lutea | 12.58 ± 2.57 | 13.58 ± 2.64 |
| Number of resorptions | 1.25 ± 1.35 | 0.33 ± 0.65* |

* Significantly different ($P < 0.05$) from the control group; ** ($P < 0.01$); ***$P < 0.001$

The only malformation observed in the control group was aplasia of two sternebrae in one fetus. In the experimental offspring there were 22 different malformations (Table 2), with cleft palate, scoliosis, and aplasia of one or two sternebrae the most frequent (Table 2). The frequency of these four specific malformations in the treated group was significantly higher ($P < 0.01$ or less) compared with controls as determined by the Fisher test.

The means and standard deviation of ossification centers are presented in Table 3. The Mann-Whitney U test indicated a significant ($P < 0.05$) decrease in the number of ossification centers in the sternebrae and caudal vertebrae of the experimental offspring when compared with the controls.

## Discussion

These results demonstrated that *A. pyrifolium* is another native plant from the caatinga that can adversely affect reproduction, causing embryonic death and malformations. Embryonic death in goats (Dantas 2009) and malformations in goats (Pimentel *et al.* 2007) and rats (Medeiros *et al.* 2008) were induced by the administration of *Mimosa tenuiflora*. The latter is associated with spontaneous malformations in goats, sheep, and cattle in the Brazilian semiarid (Riet-Correa *et al.* 2006). Despite the teratogenic effects demonstrated in rats by the administration of *A. pyrifolium*, the experimental administration of this plant to goats caused abortion and embryonic death, but failed to cause malformations (Medeiros *et al.* 2004). Nevertheless, the involvement of *A. pyrifolium* as a potential cause of malformations in ruminants in the Brazilian semiarid cannot be definitively ruled out.

**Table 2.** Skeletal malformations (number and percentage) in the offspring of dams that received 0.06 g of *A. pyrifolium* ethanolic extract from days 6-20 of pregnancy.

| Malformations | Control Group (% affected) (*n*=100) | Experimental Group (% affected) (*n*=124) |
|---|---|---|
| Cleft palate | 0 | 35 (28.22%) *** |
| Scoliosis | 0 | 21 (16.93%)*** |
| Lordosis | 0 | 4 (3.22%) |
| Bifid sternum | 0 | 1 (0.80%) |
| Aplasia of one sternebrae | 0 | 22 (17.74%)*** |
| Aplasia of two sternebrae | 1 (2.32%) | 10 (8.06%)* |
| Aplasia of three sternebrae | 0 | 3 (2.41%) |
| Aplasia of four sternebrae | 0 | 1 (0.80% ) |
| Aplasia of sternum | 0 | 1 (0.80%) |
| Disarranged ribs | 0 | 2 (1.61%) |
| Aplasia of one rib | 0 | 2 (1.61%) |
| Aplasia of two ribs | 0 | 1 (0.80%) |
| Aplasia of three ribs | 0 | 3 (2.41%) |
| Deviation on lumbar vertebrae | 0 | 1 (0.80%) |
| Aplasia of three lumbar vertebrae | 0 | 1 (0.80%) |
| Aplasia of five lumbar vertebrae | 0 | 1 (0.80%) |
| Aplasia of one thoracic vertebra | 0 | 1 (0.80%) |
| Aplasia of parietal bone | 0 | 2 (1.61%) |
| Aplasia of interparietal bone | 0 | 1 (0.80%) |
| Total aplasia of caudal vertebrae | 0 | 1 (0.80%) |
| Aplasia of two caudal vertebrae | 0 | 2 (1.61%) |
| Aplasia of one caudal vertebra | 0 | 3 (2.41%) |

* Significantly different ($P < 0.05$) from the control group; ** ($P < 0.01$); ***$P < 0.001$
*n*=number of animals

**Table 3.** Number of ossification centers in the offspring (mean ± SD) of dams that received 0.06 g of *A. pyrifolium* ethanolic extract from days 6-20 of pregnancy or from control dams.

| Ossification Centers | Control Group $\bar{x} \pm$ SD | Experimental Group $\bar{x} \pm$ SD |
|---|---|---|
| Phalanges of the forelimbs | 0 ± 0 | 0 ± 0 |
| Metacarpus | 3 ± 0 | 3 ± 0 |
| Metatarsus | 4 ± 0 | 4 ± 0 |
| Sternebrae | 5.98 ± 0.20 | 5.51 ± 0.95* |
| Caudal vertebrae | 3.74 ± 0.44 | 3.34 ± 0.72* |
| Total | 16.72 ± 0.64 | 15.85 ± 1.67 |

* Significantly different ($P < 0.05$) from the control group

## Acknowledgements

This work has been financially supported by the National Institute for Science and Technology for the Control of Toxic Plants (CNPq), grant number 573534/2008-0.

## References

Araújo Jr C, Antheaume RCP, Trindade RCP, Schmitt M, Bourguignon JJ, and Sant'ana AEG (2007). Isolation and characterization of the monoterpenoid indole alkaloids of *Aspidosperma pyrifolium*. *Phytochemical Reviews* 6:183–188.

Dantas AF (2009). Malformações e morte embrionária em ruminantes causadas pela ingestão de *Mimosa tenuiflora* (jurema preta), 68 pp. Tese de Doutorado, Programa de Pós-Graduação em Ciências Veterinárias, Universidade Federal Rural de Pernambuco. Recife, Pernambuco.

GraphPad Instat (1993). V2.01. San Diego: *GraphPad Software*. (1 computer disk, 3½ in, IBM).

Lima MCGS and Soto-Blanco B (2009). Poisoning in goats by *Aspidosperma pyrifolium* Mart.: Biological and cytotoxic effects. *Toxicon* 55:320-324.

Medeiros RMT, Neto SAG, Riet-Correa F, Schild AL, and Sousa NL (2004). Mortalidade embrionária e abortos em caprinos causados por *Aspidosperma pyrifolium*. *Pesquisa Veterinária Brasileira* 24:(Supl.):42-43.

Medeiros JM, Tabosa IM, Simões SVD, Nóbrega JR, Vasconcelos JS, and Riet-Correa F (2005). Mortalidade perinatal em cabritos no semi-árido da Paraíba. *Pesquisa Veterinária Brasileira* 25:201-206.

Medeiros RMT, Figueiredo APM, Benício TMA, Dantas FPM, and Riet-Correa F (2008). Teratogenicity of *Mimosa tenuiflora* seeds to pregnant rats. *Toxicon* 51:316-319.

Nóbrega JR, Riet-Correa F, Nóbrega RS, Medeiros JM, Vasconcelos JS, Simões SVD, and Tabosa IM (2005). Mortalidade perinatal de cordeiros no semi-árido da Paraíba. *Pesquisa Veterinária Brasileira* 25:171-178.

Pessôa CRM (2007). Embriofetotoxicidade da *Mimosa ophtalmocentra* em ratas. Patos. 28 pp. Monografia de Graduação, Universidade Federal de Campina Grande.

Pimentel LA, Riet-Correa F, Gardner D, Panter KE, Dantas AFM, Medeiros RMT, Mota RA, and Araújo JAS (2007). *Mimosa tenuiflora* as a cause of malformations in ruminants in the northeastern Brazilian semiarid rangelands. *Veterinary Pathology* 44:928-931.

Riet-Correa F, Medeiros RMT, and Dantas AFM (2006). *Plantas Tóxicas da Paraíba.* 58 pp. SEBRAE, João Pessoa.

Silva DM, Riet-Correa F, Medeiros RMT, and Oliveira OF (2006). Plantas tóxicas para ruminantes e eqüídeos no Seridó Ocidental e Oriental do Rio Grande do Norte. *Pesquisa Veterinária Brasileira* 26:223-236.

Staples RE and Schenell VL (1964). Refinements in rapid clearing technique in the KOH-alizarin red S method for fetal bone. *Stain Technology* 39:61-63.

# Chapter 44

# Effects of Gossypol Present in Cottonseed Cake on Spermatogenesis in Sheep

F. Guedes and B. Soto-Blanco

*Department of Animal Sciences, Universidade Federal Rural do Semi-Árido (UFERSA), BR 110 Km 47, Mossoró, RN, 59625-900, Brazil*

## Introduction

Cottonseed is used as an alternative to soy because of its low cost and accessibility in areas in which it is grown. However, cotton seeds present a substance with toxic potential in their composition, gossypol. Gossypol is a phenolic yellow pigment produced by pigment glands found in cotton roots, branches, leaves, and seeds (Cheeke 1998).

Multiple factors influence the presence of gossypol in the plant. Weather conditions play a significant role. Gossypol concentration is positively correlated with rainfall and negatively correlated with temperature. Variation between cotton species is another important factor: *Gossypium barbadense* contains a higher concentration than *G. hirsutum.* Storage of cotton has little influence on gossypol content (Cheeke 1998).

Gossypol is a highly reactive compound that binds rapidly to different substances, including minerals and amino acids. Iron binds to gossypol, forming a gossypol-iron complex. Iron bound to gossypol is inaccessible and iron deficiency may occur affecting hematopoiesis. In addition, the presence of this complex in the yolk of eggs causes the formation of a green color (Patton *et al.* 1985; Kerr 1989; Cheeke 1998).

Since the concentrations of gossypol in cotton are not high enough to cause acute intoxication, the natural intoxication by gossypol occurs through prolonged ingestion of the plant. The effects of gossypol are cumulative and may appear suddenly after a variable period of ingestion (Eagle 1950; Patton *et al.* 1985; Kerr 1989; Cheeke 1998).

In males, gossypol promotes reduction of motility and spermatozoid concentration; testosterone level and testicular morphology remain unaltered (Qian and Wang 1984). In non-ruminant females, exposure to gossypol is associated with the interruption of estrous cycle, pregnancy, and early embryo development; females from ruminant species are less sensitive (Randel *et al.* 1992). The present work evaluated spermatogenesis in male sheep fed cottonseed cake.

©CAB International 2011. *Poisoning by Plants, Mycotoxins, and Related Toxins* (eds F. Riet-Correa, J. Pfister, A.L. Schild, and T.L. Wierenga)

## Methodology

Twelve male adult purebred Santa Inês sheep were used. They were separated into two groups; the first was fed daily with 0.5 kg per animal cottonseed cake (treated group) and the second was fed daily with 0.5 kg per animal corn meal (control group), both for 120 consecutive days. Cottonseed cake (torta de algodão tangará, tangará, RN, Brazil) or corn meal was offered moiled.

At the end of the experimental period, samples of semen were collected for laboratorial determination of quality, including density and spermatic pathologies. Sample collection was accomplished by the use of an artificial vagina. Semen ejaculates were collected directly into a graduated tube, the volume recorded, and samples were immediately examined. Motility was estimated by examining a fresh drop of semen on a slide without cover slip using a light microscope at 100× magnification. Motility was scored as: 1: little or no individual spermatozoa motion with no wave; 2: slow motion with no swirl; 3: rapid motion with slow swirl and eddies; and 4: vigorous progressive motion with rapid swirls and eddies. From each ejaculate, 10 μl of semen were suspended in 2 ml of 10% buffered formol saline solution and spermatozoids were counted in a Neubauer chamber using a light microscope. Total spermatozoid counts were obtained by multiplication of spermatozoids concentration by semen volume. The percentage of abnormal sperm was calculated for a total of 200 sperm from each sample with a microscope under 1000× magnification.

Data are reported as mean ± standard deviation and were compared using unpaired *t* test (GraphPad Prism v.4 for Mac). The level of significance was set at $P < 0.05$.

## Results and Discussion

None of the experimental animals presented any clinical alterations during the evaluation period. This fact is relevant because it indicates that both groups had similar nutritional metabolic states, especially related to similar gain of weight. Therefore, no clinical alterations were observed indicative of diseases that could affect sperm production and the sperm analysis.

Gossypol is a non-specific enzymatic inhibitor (Hervé *et al.* 1996) and forms chemical complexes with cations including iron (Abou-Donia 1976). The administration of gossypol to rats can induce diarrhea (Bender *et al.* 1988; Silva *et al.* 2002), which was attributed to the possible inhibition of pepsinogen and/or other digestive enzymes (Bender *et al.* 1988). In the present study, no diarrhea or any other disturbance of the digestive tract was found. This can be attributed to differences in the gossypol concentration that the animals were exposed to and/or to differences in the susceptibility between species. The results of the analysis of semen samples are shown in Table 1. No statistically significant differences were found between the two groups.

Gossypol has a proven deleterious action on sperm mobility (Chongthammakun *et al.* 1986; Hong *et al.* 1989), blocking the production, release, and use of ATP in these cells (Ueno *et al.* 1988). Abnormal spermatozoids are formed in animals exposed to gossypol with ultrastructural abnormalities mainly in the mitochondrial membranes (Haffer 1983). In this study, no morphological or mobility abnormalities were found in sheep fed cottonseed.

Naturally occurring gossypol can be found in the free form or bound to proteins. In intact cotton seeds, gossypol is mainly present in its free form. During the process of oil extraction, the binding of gossypol to proteins from the seeds occurs, probably to the radical epsilon-amine from lysine (Calhoun *et al.* 1995).

**Table 1.** Sperm analysis from sheep treated with 0.5 kg/animal/day corn meal (control group) or 0.5 kg/animal/day cottonseed cake (treated group) for 120 consecutive days. Data are shown as mean ± SD (*n*=6).

| Parameters | Control | Treated |
|---|---|---|
| Semen volume (ml) | 1.62±0.19 | 1.64±0.18 |
| Spermatozoids concentration x$10^9$/ml) | 4.11±0.53 | 3.98±0.59 |
| Total spermatozoids (x$10^9$) | 6.66±0.62 | 6.53±0.74 |
| Motility | 4.0±0.0 | 4.0±0.0 |
| Abnormal spermatozoids (%) | 4.72±0.87 | 5.18±0.96 |

Gossypol bound to proteins is not absorbed by the gastrointestinal tract of ruminants, and thus this form is considered non-toxic. Ruminal microbes of developed animals are able to detoxify gossypol by binding it to proteins (Calhoun *et al.* 1995). In this study, one possible explanation for the absence of deleterious effects in sheep is that the free gossypol concentration in cottonseed cake is low because of the heating treatment performed during the oil extraction process.

The ruminal microbial action could also have contributed to the reduction of the amount of free form of gossypol. However, the binding of gossypol to proteins can be broken during digestion, releasing the toxin (Blackwelder *et al.* 1998; Noftsger *et al.* 2000). Therefore, further studies are necessary to determine the residual amounts of gossypol, both free and bound forms, in cotton residues.

We conclude that, based on our experimental conditions, cottonseed cake can be administered, at the concentration and time evaluated in this study, to adult male sheep without compromising spermatogenesis.

## References

Abou-Donia M (1976). Physiological effects and metabolism of gossypol. *Residue Revue* 61:125-160.

Bender HS, Saunders GK, and Misra HP (1988). A histopathologic study of the effects of gossypol on the female rat. *Contraception* 38:585-592.

Blackwelder JT, Hopkins BA, Diaz DE, Whitlow LW, and Brownie C (1998). Milk production and plasma gossypol of cows fed cottonseed and oilseed meals with or without rumen-undegradable protein. *Journal of Dairy Science* 81:2934-2941.

Calhoun MC, Kuhlmann SW, and Baldwin BC (1995). Assessing the gossypol status of cattle fed cottonseed products. *Proceedings of the Pacific Northwest Animal Nutrition Conference*, pp. 147A-157A. Portland, Oregon.

Cheeke PR (1998). *Natural Toxicants in Feeds, Forages, and Poisonous Plants*, 2nd edn, 479 pp. Interstate Publishers, Danville.

Chongthammakun S, Ekavipat C, Sanitwongse B, and Pavasuthipaisit K (1986). Effects of gossypol on human and monkey sperm motility *in vitro*. *Contraception* 34:323-331.

Eagle E (1950). Effect of repeated doses of gossypol on the dog. *Archives of Biochemistry* 26:68-71.

Haffer AP (1983). Effects of gossypol on the seminiferous epithelium in the rat: a light and electron microscope study. *Biology of Reproduction* 28:1000-1003.

Hervé JC, Pluciennik F, Bastide B, Cronier L, Verrecchia F, Malassiné A, Joffre M, and Délezè J (1996). Contraceptive gossypol blocks cell-to-cell communication in human and rat cells. *European Journal of Pharmacology* 313:243-255.

Hong CY, Huang JJ, and Wu P (1989). The inhibitory effect of gossypol on human sperm motility: relationship with time, temperature and concentration. *Human Toxicology* 8:49-51.

Kerr LA (1989). Gossypol toxicosis in cattle. *Compendium on Continuous Education for Practicing Veterinaries* 11:1139-1146.

Noftsger SM, Hopkins BA, Diaz DE, Brownie C, and Whitlow LW (2000). Effect of whole and expanded-expelled cottonseed on milk yield and blood gossypol. *Journal of Dairy Science* 83:2539-2547.

Patton CS, Legendre AM, Gompf RE, and Walker MA (1985). Heart failure caused by gossypol poisoning in two dogs. *Journal of the American Veterinary Medical Association* 187:625-627.

Qian SZ and Wang ZG (1984). Gossypol: a potential antifertility agent for males. *Annual Reviews on Pharmacology and Toxicology* 24:329-360.

Randel RD, Chase Jr CC, and Wyse SJ (1992). Effects of gossypol and cottonseed products on reproduction of mammals. *Journal of Animal Science* 70:1628-1638.

Silva MA, Kozicki LE, and Dalsenter PR (2002). Toxicidade do gossipol na gestação e na lactação de ratas (*Rattus rattus norvegicus*). *Archives of Veterinary Science* 7:87-89.

Ueno H, Sahni MK, Segal SJ, and Koide SS (1988). Interaction of gossypol with sperm macromolecules and enzymes. *Contraception* 3:333-341.

# NERVOUS

# SYSTEM

# Chapter 45

# Poisonous Plants Affecting the Nervous System of Horses in Brazil

E.F. Lima[1], B. Riet-Correa[2], F. Riet-Correa[3], R.M.T. Medeiros[3], D.R. Gardner[4], and G. Riet-Correa[5]

[1]*Universidade do Estado do Amazonas, Escola Superior de Saúde, Av. Carvalho Leal, 1777, Cachoeirinha, Manaus, AM, 69065-001 and Escola Superior Batista do Amazonas, R. Leonor Teles, 153, Adrianópolis, Manaus, AM, 69057-510, Brazil;* [2]*Universidade Federal do Pará, Campus de Castanhal, Central de Diagnóstico Veterinário, Maximino Porpino da Silva, 1000, Pirapora, Castanhal, 68740-080, Brazil;* [3]*Hospital Veterinário, CSTR, UFCG, Patos, PB, 58700-000, Brazil;* [4]*USDA-ARS Poisonous Plant Research Laboratory, Logan, Utah 84341, USA;* [5]*Faculdade de Medicina Veterinária, Universidade Federal do Pará (UFPA),Rua Maximino Porpino da Silva 1000, Castanhal, PA, 68743-080, Brazil*

## Introduction

Well known diseases of the nervous system of horses in Brazil are rabies, equine eastern, Venezuelan, and western virus encephalomyelitis, and leukoencephalomalacia. However, the differential diagnosis of those diseases with other diseases affecting the nervous system of horses, including plant poisonings, is important. This chapter reports poisonous plants affecting the nervous system in horses in Brazil.

## Poisoning by *Indigofera lespedezioides*

Poisoning by *Indigofera lespedezioides* (= *Indigofera pascuori*) occurs in horses in at least five counties (Amajarí, Alto Alegre, Normandia, Cantá, and Bom Fim) in the northern region of the state of Roraima, northern Brazil. In this region the rainy season is from May to August/September and most cases occur in April at the end of the dry season when *I. lespedezioides* is nearly the only green vegetation available. Typically up to 10% of horses can be affected, but in one case a farmer reported 100% mortality in a herd of 30 horses. Cattle and sheep are not affected by the poisoning. Main clinical signs are anorexia, sleepiness, unsteady gait, severe ataxia, weakness, stumbling, and progressive weight loss. Gait alterations are more marked in the hind limbs while the hooves are dragged, causing excessive wear of the toes. Eye discharge and blindness are also observed. Some farmers have reported corneal opacity. Horses of all ages are affected. If the animals are disturbed or forced to move, nervous signs increase and the animals can fall. Abortion is commonly observed in mares. The clinical manifestation period is approximately 2-4 months. If the plant consumption is interrupted some animals may recover. In an experimental case the first clinical signs appeared after 44 days from the start of grazing in a small paddock invaded by the plant. The animal was euthanized on day 59. Clinical signs were weight loss, ataxia, and sleepiness. The only histologic lesion observed in a spontaneous and an experimental case was Wallerian degeneration in some brain stem tracts. Lipofuscinosis associated with neuronal and axonal degeneration was observed using electron microscopy of a spontaneous case.

The disease is very similar to Birdsville disease caused by *I. linnaei* in Australia (Caroll and Swain 1983) and *I. spicata* in the USA (Morton 1989). *I. linnaei* contains indospicine and nitro-compounds, but it has not been proven if any of these toxins is responsible for the clinical syndrome. Samples from northern Brazil of *I. lespedezioides* were analyzed and found to be positive for indospicine but negative for nitro-compounds (Dale Gardner, 2010, unpublished data).

## Swainsonine Poisoning

*Turbina cordata* in northeastern Brazil and *Sida carpinifolia* in southern Brazil, which contain swainsonine, cause mannosidosis in horses. The poisoning by *S. carpinifolia* in ponies was diagnosed in Rio Grande do Sul (Loretti *et al.* 2003), but it also occurs in goats, sheep, and cattle in the states of Santa Catarina, Rio de Janeiro, and São Paulo. *T. cordata* causes intoxication in goats and with less frequency in cattle and horses in the states of Bahia and Pernambuco (Dantas *et al.* 2007; Assis *et al.* 2010).

The chronic poisoning is characterized by rough hair coat, depression, progressive weight loss ataxia, intention tremors, wide-based stance, reluctance to walk, signs of abdominal pain manifested by kicking at the belly, rolling, falling and moaning, recumbence, and death (Loretti *et al.* 2003; Assis *et al.* 2010). In ponies in Rio Grande do Sul death occurred 15-20 days after the animals were introduced into the area invaded by *S. carpinifolia*. In the state of Bahia horses showed clinical signs for more than 1 year (Assis *et al.* 2010). There are no significant gross lesions. Histology revealed swollen neurons with multiple cytoplasmic vacuoles in the brain, cerebellum, spinal cord, autonomic trigeminal and celiac ganglia, and submucosal and myenteric plexuses of the intestines. These changes were constant in neurons throughout the central nervous system. In the kidneys, there was marked vacuolation of the proximal convoluted tubular cells. Stored material was not evident in other organs (Loretti *et al.* 2003).

## Hepatic Encephalopathy Caused by *Senecio* spp. and *Crotalaria retusa*

*Crotalaria retusa* is the most important toxic plant for equidae in the semiarid region of northeastern Brazil. Poisonings by *Senecio* spp. are reported in Rio Grande do Sul and Santa Catarina, in farms where *S. brasiliensis* and *S. selloi* are found. The pyrrolizidine alkaloids most frequently found are integerrimine, retrorsine, and senecionine in *Senecio* spp. and monocrotaline in *C. retusa*. Clinical signs in horses are weight loss, sometimes for 3-4 months, depression, jaundice, and nervous signs, including dullness, hyperexcitability, head pressing, compulsive walking, circling, ataxia, dysmetria, faulty food prehension, dysphasia, blindness, convulsions, and terminal coma. Some horses show frenzied behavior and are violent with uncontrollable galloping. Anorexia and occasionally diarrhea and photosensitization are also observed. Horses with nervous signs have a clinical manifestation period of 7-15 days. Most horses lose weight for a period of 30-60 days before the nervous signs are observed (Gava and Barros 1997; Nobre *et al.* 2004).

At necropsy, the liver is hard and whitish or yellowish with increased lobular pattern. Edema of the mesentery and fluid in the body cavities are common findings. Mild jaundice, ascites, hydropericardium, and hydrothorax are also observed. Histologic lesions are megalocytosis, bile duct cell proliferation, cholestasis, and fibrosis, mainly periportal. More acute cases show centrilobular hemorrhagic necrosis (Gava and Barros 1997; Nobre *et al.* 2004). Lesions in the central nervous system are characterized by the presence of enlarged astrocytes with vesicular nuclei, named Alzheimer type II astrocytes, found mainly in the cerebral cortex and basal nuclei (Gava and Barros 1997; Nobre *et al.* 2004).

The initial diagnosis is based on epidemiologic information, clinical signs, and gross lesions, and a definitive diagnosis is made from the characteristic histologic lesions of the liver. The time lag between ingestion and appearance of clinical signs must be considered, as death can occur weeks or months after exposure has ended when there are no more plants in the field or after animals have been moved to a location without the toxic plants. The determination of liver enzymes, mainly γ-glutamyl transferase (GGT), and hepatic biopsies can help in the clinical diagnosis. PA poisoning should be differentiated from other diseases causing nervous signs, diarrhea, weight loss, edema, ascites, or photosensitization.

*Crotalaria* spp. should be controlled by the use of herbicides or other methods. In areas invaded by the plant it is important to avoid grazing by horses. Sheep cannot be introduced into areas that are severely invaded by *Crotalaria*, especially when the plant is setting seed at the end of the rainy season in northeastern Brazil.

## Poisoning by *Bambusa vulgaris* f. *vulgaris*

Poisoning by *Bambusa vulgaris* f. *vulgaris* was diagnosed in horses of different ages in northeastern Pará in areas where the plant is cultivated for shade. Horses ingest the plant when other forage is scarce during the dry season or when it grows in *Brachiaria brizantha* or *B. decumbens* pastures, which are not very palatable to horses. The plant is found in other regions of Brazil but cases of intoxication have not been reported (Barbosa *et al.* 2006).

The main clinical signs are somnolence and severe ataxia with horses standing with abducted limbs and having difficulty turning around. Signs of impairment of cranial nerves are also observed such as paresis of the tongue, difficulty in prehending, chewing, and swallowing of food, and decreased palatal and labial reflexes. Cutaneous, anal, and flexor reflexes are depressed. Blindness and head pressing are occasionally observed. The clinical course is subacute or chronic and most horses recover after being removed from the pastures. Gross lesions are not observed. Histologically only slight edema and axonal degeneration had been reported in a few axons, mainly in the medulla oblongata (Barbosa *et al.* 2006).

The intoxication was reproduced experimentally by the administration of *B. vulgaris* leaves at daily doses of 10-31 g/kg BW for 6 to 60 days. First signs appeared 24-72 h after dosing, but clinical signs were less severe than those in the spontaneous intoxication (Barbosa *et al.* 2006). The active principle is unknown. The diagnosis is based on the characteristic clinical signs in horses grazing in areas with *B. vulgaris* and in the regression of clinical signs after consumption ceases. The disease should be differentiated from myeloencephalitis caused by *Sarcosystis neurona*, and from the poisonings by *Crotalaria* spp., *Equisetum* spp., and *Indigofera lespedezioides*. To prevent the intoxication horses should be removed from pastures with *B. vulgaris*, mainly during the dry season. Pastures of *Brachiaria brizantha* or *B. decumbens* are not good grazing lands for horses, especially if there are bamboo plants in the pastures.

## Stringhalt Caused by *Hypochaeris radicata*

*Hypochaeris radicata* causes stringhalt (high stepping with hyperflexion of the hind limb) in horses in different countries. In Brazil the disease has been reported from the states of Rio Grande do Sul and Paraná. It is observed during winter and spring (July to December). Typically, one or two horses are affected, but up to 50% of horses have been affected in some cases. Horses from the age of 1 to 13 years old have all been affected (Rodrigues *et al.* 2007; Araújo *et al.* 2008).

Clinical signs are characterized by abnormal gait with involuntary flexing of the hocks of one or both hind legs, impaired ambulation, and bunny hop-type of gait. In some horses the hyperflexion is so marked that the abdomen is kicked when walking. Affected horses have difficulty in stepping backward or circling. Left laryngeal hemiplegia (roaring) can be associated with stringhalt. Muscular atrophy can be observed in the hind limbs. Involvement of the forelimbs is also seen occasionally, taking the form of stumbling, toe-scuffing, and knuckling at the carpus. Axonal degeneration in peripheral nerves with reduction or absence of myelinated fibers and muscular atrophy are observed histologically. Ultrastructural findings included signs of demyelination, regeneration, and remyelination of peripheral nerves. When removed from pastures invaded by *H. radicata*, most animals recover without treatment over a period of time that can last several months, but some horses do not recover and even after 17 months show clinical signs (Rodrigues *et al.* 2007; Araújo *et al.* 2008).

The disease named Australian stringhalt is different from classical stringhalt. The disease caused by *H. radicata* is more severe, usually bilateral, occurs in outbreaks, is seasonal, and most animals recover spontaneously. Classical stringhalt is a sporadic disease of unknown cause that has to be treated surgically because there is no spontaneous recovery (Rodrigues *et al.* 2007; Araújo *et al.* 2008).

The disease was reproduced experimentally in a 6-month-old horse weighing 250 kg by the daily administration of 9.8 kg of fresh plant for 50 days. Clinical signs first appeared 19 days after dosing began (Araújo *et al.* 2008).

Treatment with phenytoin or other anticonvulsants can be of benefit. Grazing should be avoided in areas severely infested by the plant in order to prevent the disease. Horses have to be removed from the pastures immediately after clinical signs are first observed.

## Poisoning by *Equisetum* spp.

Poisoning by *Equisetum* spp. was reported in the 1940s in Minas Gerais, but new outbreaks have not been reported since then. Plant ingestion occurs during the dry season when the plants are still green or when fed with contaminated hay. Clinical signs characterized by weight loss, lethargy, staggers, unsteady gait, and ataxia are observed 3-6 weeks after the start of ingestion. No macroscopic or histologic lesions have been reported (Alvim 1948).

The plant contains a thiaminase. The horses recovered if treated with daily administration of 100 mg of thiamine, but when the animal is emaciated and recumbent treatment can be ineffective.

Other chronic nervous diseases such as equine protozoal myeloencephalitis from *Sarcosystis neurona*, and hepatic encephalopathy due to pyrrolizidine alkaloid-containing plants have similar clinical signs, but unlike *Equisetum* toxicity, these conditions have characteristic histologic lesions.

## Acknowledgements

This work has been financially supported by the National Institute for Science and Technology for the Control of Toxic Plants (CNPq), grant number 573534/2008-0.

## References

Alvim PT (1948). Envenenamento de cavalos por *Equisetum* spp. (cavalinha). *Revista Ceres*, Viçosa, 8(43):32-36.

Araújo JAS, Curcio B, Alda J, Medeiros RMT, and Riet-Correa F (2008). Stringhalt in Brazilian horses caused by *Hypochaeris radicata*. *Toxicon* 52:190-193.

Assis TS, Medeiros RMT, Riet-Correa F, Galiza GJN, Dantas AFM, and Oliveira MD (2010). Intoxicações por plantas diagnosticadas em ruminantes e equinos e estimativa das perdas econômicas na Paraíba. *Pesquisa Veterinária Brasileira* 30(1):13-20.

Barbosa D, Oliveira CM, Duarte MD, Riet-Correa G, Peixoto PV, and Tokarnia CH (2006). Poisoning of horses by bamboo, *Bambusa vulgaris*. *Journal of Equine Veterinary Science* 26(9):393-398.

Carroll AG and Swain BJ (1983). Birdsville disease in the Central high-land area of Queensland. *Australian Veterinary Journal* 60:316-317.

Dantas AFM, Riet-Correa F, Gardner DR, Medeiros RMT, Barros SS, Anjos BL, and Lucena RB (2007). Swainsonine-induced lysosomal storage disease in goats caused by the ingestion of *Turbina cordata* in Northeastern Brazil. *Toxicon* 49:111-116.

Gava A and Barros CSL (1997). *Senecio* spp. poisoning of horses in southern Brazil. *Pesquisa Veterinária Brasileira* 17: 36-40.

Loretti AP, Colodel EM, Gimeno EJ, and Driemeier D (2003). Lysosomal storage disease in *Sida carpinifolia* toxicosis: an induced mannosidosis in horses. *Equine Veterinary Journal* 35: 434-438.

Morton JF (1989). Creeping indigo (*Indigofera spicata* Forsk.) (Fabaceae)–A hazard to herbivores in Florida. *Economic Botany* 43(3):314-327

Nobre VMT, Riet-Correa F, Barbosa Filho JM, Tabosa IM, and Vasconcelos JS (2004). Intoxicação por *Crotalaria retusa* (Fabaceae) em eqüídeos no semiárido da Paraíba. *Pesquisa Veterinária Brasileira* 24:132-143.

Rodrigues A, De La Corte FD, Graça DL, Rissi DR, Schild AL, Kommers GD, and Barros CSL (2007). Harpejamento em eqüinos no Rio Grande do Sul. *Pesquisa Veterinária Brasileira* 48:23-28.

# Chapter 46

# Rational Uses of Mesquite (*Prosopis juliflora*) and the Importance of Spontaneous Poisoning by the Pods in Ruminants from Pernambuco, Northeastern Brazil

A.C.L. Câmara[1], J.A.B. Afonso[2], and F. Riet-Correa[3]

[1] *Hospital Escola de Grandes Animais, Universidade de Brasília. Galpão 4, Granja do Torto, 70636-200, Brasília, Distrito Federal, Brazil;* [2] *Clínica de Bovinos, Campus Garanhuns, Universidade Federal Rural de Pernambuco, PO Box 152, 55292-901, Garanhuns, Pernambuco, Brazil;* [3]*Hospital Veterinário, Centro de Saúde e Tecnologia Rural, Universidade Federal de Campina Grande, 58700-000, Patos, Paraíba, Brazil*

## Introduction

The genus *Prosopis* of the family *Leguminosae* (Fabaceae), subfamily Mimosoideae is native to the Americas, Africa, and Asia, and comprises 44 species. Several species from South and Central America, especially the subtropical *P. chilensis*, *P. glandulosa*, and *P. velutina* and the tropical *P. juliflora* and *P. pallida* have been distributed around the world over the last 200 years and are now widespread in dry parts of Sahelian and East Africa, South Africa, Pakistan, India, Australia, and Brazil (Pasiecznik *et al.* 2001).

*P. juliflora* is a tree that produces flattened, multi-seeded, and curved pods with hard pericarp. Pod production per tree can vary from a few kg to over 400 kg and is highly dependent on moisture availability to the plant (Riveros 1992). Due to the association of low costs, high palatability, and nutritional value, the pods or its 'bran' are largely used for feeding cattle (Tabosa *et al.* 2006; Câmara *et al.* 2009), sheep (Ravilaka *et al.* 1995; Mahgoub *et al.* 2005a; Obeidat *et al.* 2008), goats (Mahgoub *et al.* 2005b), swine (Silva *et al.* 1989), quail (Silva *et al.* 2002a), chickens (Silva *et al.* 2002b), and horses (Stein *et al.* 2005). Farmers designate as bran the dry and ground pods which are mixed with a small amount of maize or wheat bran to facilitate grinding. Grinding the pods to feed to livestock is important to destroy the seeds to avoid the ingestion of unground pods, which results in large quantities of germinating seeds in the livestock feces.

The pods are also ground into flour for human consumption as bread, cakes, biscuits, spirits, jellies, or a rich porridge (Tabosa *et al.* 2004; Chogel *et al.* 2007). Coffee substitute (Azevedo-Rocha 1987) and liquor (Silva *et al.* 2003) have also been made from *P. juliflora* in Brazil. The aim of the present paper is to review some rational uses of mesquite and the importance of spontaneous poisoning by the pods in ruminants in Pernambuco, northeastern Brazil.

©CAB International 2011. *Poisoning by Plants, Mycotoxins, and Related Toxins* (eds F. Riet-Correa, J. Pfister, A.L. Schild, and T.L. Wierenga)

## Epidemiology

Earlier reports had already associated the ingestion of the pods with the disease called jaw and tongue trouble in cattle in the USA (Dollahite 1964; Kingsbury 1964) and 'coquera' in goats from Peru (Bacca *et al.* 1966). In northeastern Brazil poisoning by the pods has been recognized as a disease affecting cattle popularly called 'cara torta' (twisted face) due to the head tilting during chewing as an attempt to keep the food inside the mouth (Tokarnia *et al.* 2000). Poisoning also occurs with the 'bran' (Tabosa *et al.* 2006).

In Brazil spontaneous poisoning was reported in cattle in the semiarid regions of the states of Paraíba (Dantas and Menezes 1994), Pernambuco (Dantas and Menezes 1994; Câmara *et al.* 2009), and Rio Grande do Norte (Silva *et al.* 2006) and in goats in Paraíba (Lima *et al.* 2004). Experimental intoxication was produced in goats (Tabosa *et al.* 2000) and cattle (Tabosa *et al.* 2006). Goats need to ingest the pods for longer periods than cattle to become intoxicated (Tabosa *et al.* 2004). Sheep seem to be resistant to the poisoning; spontaneous poisoning in this species had not been reported, and experimentally sheep were not affected even after 1 year of ingesting food containing 75% *P. juliflora* pods (Lima *et al.* 2004).

## Clinical Signs

Clinical signs, which are more prominent during eating and rumination, are characterized by masseter muscle atrophy, protrusion of the tongue, dropped (slack) mandible, tilting of the head during chewing, continuous licking of the nostrils, profuse salivation, yawning, swallowing impairment, and dysphagia (Figueiredo *et al.* 1996; Tabosa *et al.* 2004, 2006; Câmara *et al.* 2009). Some cattle, when they are not chewing, remain with their mouths slightly open and the tongue protruding 2-3 cm (Câmara *et al.* 2009); the saliva hangs continuously from the mouth in long strings (Tabosa *et al.* 2004, 2006). Dehydration, ruminal hypomotility or atony, inability to stand, hypothermia, gradual emaciation, anemia, and hypoproteinemia are frequent findings (Figueiredo *et al.* 1996; Tokarnia *et al.* 2000; Câmara *et al.* 2009). Goats can also develop increased rumination period, salivation, weight loss, and tremors of the lips, mandible, and head mainly during chewing, which are associated with muscular debility of the masticator muscles (Bacca *et al.* 1966; Lima *et al.* 2004). The clinical signs are consequences of cranial nerve impairment affecting mainly the trigeminal nuclei; however, impaired function of cranial nerves IX, X, and XII are often seen (Tabosa *et al.* 2004, 2006).

## Experimental and Spontaneous Poisoning

Experimental poisoning in cattle was achieved in the USA and Brazil. Suggestive signs of the intoxication appeared 30 days after the ingestion of pods and hay and 60-90 days after the ingestion of pods as the only food (Dollahite and Anthony 1957). In Brazil, the disease was produced experimentally in cattle ingesting food containing 50% and 100% pods after 3 months of administration (Figueiredo *et al.* 1996), whilst Tabosa *et al.* (2006) observed first signs of the toxicosis after 45-75 days of ingestion of diets containing 50% and 75% pods. If the animals had been affected for no more than 60 days, they recovered fully after the withdrawal of the pods; but in animals affected for more time the clinical signs were not reversible (Dollahite 1964). In cattle severely affected the denervation atrophy of the masseter and other muscles is irreversible and clinical signs remain after the withdrawal of the plant (Tabosa *et al.* 2004). Recovery of some cases in the initial stages of the disease can be attributed to the reversion of neuronal degeneration (mitochondrial dilatation) before neuronal death and the stabilization and recovery of muscular atrophy before fiber losses and replacement by fibrous and fatty tissue (Câmara *et al.* 2009).

Goats were experimentally intoxicated in Peru (Bacca *et al.* 1966) and Brazil (Tabosa *et al.* 2004). Bacca *et al.* (1966) observed clinical signs 9-11 months after the ingestion *ad libitum* of a concentrate containing 80% pods, plus chopped maize stalks and leaves. Another group of goats that received mesquite beans pods *ad libitum* and small amounts of dried whole maize plants died of emaciation after 43-102 days of ingestion without showing other clinical signs (Bacca *et al.* 1996). In Brazil, goats were fed with a ration containing 30%, 60%, and 90%, on a dry matter base, of *P. juliflora* pods. At 210 days after the start of the experiment goats receiving 60% and 90% pods showed signs characteristic of cranial nerve impairment and gradual weight loss (Tabosa *et al.* 2004).

Spontaneous poisoning is reported as affecting one or a few cattle in each farm, although morbidity can be as high as 50% in some properties (Dantas and Menezes 1994). Spontaneous poisoning is increasing in the semiarid region of the state of Pernambuco, northeastern Brazil. Câmara *et al.* (2009) reported three outbreaks of poisoning by *P. juliflora* pods in cattle grazing in fields invaded by the plant or ingesting mesquite beans as a concentrate food. In two farms the disease occurred sporadically, affecting one crossbred bovine (Holstein × Gir) in each farm. In another, 112 (9.28%) cattle were affected, 84 (6.96%) died due to emaciation, and 28 (2.32%) gained weight after the pods had been withdrawn from the feed. This major outbreak occurred in a beef farm of 1206 cattle (845 Nelore and 361 Nelore × Holstein crossbred cattle) affecting only lactating females with ages ranging from 4 to 12 years. A case-control study showed an odds ratio of 2.76 meaning that Nelore cattle have 2.76 times more chance to manifest the disease compared to the crossbred cattle in the same environmental and management condition. Because the age of each animal of the herd was not determined, it was not conclusive if the higher frequency of the intoxication in Nelore cattle was due to a higher susceptibility of this breed or because Nelore cows were probably older than the others. In the same areas goats have been intoxicated by the pods since 2002 with about 50 affected goats from a total of 380. Sheep were the most abundant species with 2015 animals and none of them were poisoned (Câmara *et al.* 2009). This fact confirms earlier reports mentioning that sheep can eat diets containing 70-100% of the pods for over a year without becoming intoxicated (Lima *et al.* 2004). A case of spontaneous poisoning in a goat ingesting leaves and pods of *P. glandulosa* has been reported in the USA (Washburn *et al.* 2002).

## Pathology

Gross lesions in cattle are emaciation and reduction in size of the masseter and other masticatory muscles which appear yellowish due to muscular atrophy (Tabosa *et al.* 2004). The primary lesions in *P. juliflora* intoxication are the damage of the mitochondria in neurons of the trigeminal and other cranial nerve nuclei causing fine vacuolation of the perikaryon of neurons which present a granular or spongy appearance. Cranial nerve degeneration and denervation atrophy of the muscles occurs as a consequence of the neuronal lesion. Dark distended nuclei, sometimes displaced to the margin of the perikaryon, are observed in some neurons. Occasionally ghost neurons, characterized by a pale perikaryon with dissolution of the Nissl substance and undefined borders, or round cavities bordered by eosinophilic material suggest neuronal loss. Axonal spheroids are rarely observed. Reactive astrocytes with vesicular dilated nuclei and scant eosinophilic cytoplasm are observed in low numbers within the trigeminal motor nuclei. Similar but milder lesions are present occasionally in neurons of the facial, hypoglossal, and oculomotor nuclei. Wallerian-type degeneration characterized by short chains of two to five vacuoles side by side and occasionally containing eosinophilic myelin residues or some macrophages are observed in the intracranial roots of facial, hypoglossal, and oculomotor nerves. Wallerian-like degeneration of variable severity is observed also in the maxillary, mandibular, hypoglossal, facial, and lingual nerves (Tabosa *et al.* 2004, 2006; Câmara *et al.* 2009).

In goats there is also vacuolation of motor neurons of the motor trigeminal, facial, and hypoglossal nuclei. Motor neurons of the spinal cord and trigeminal ganglia are also vacuolated (Lima *et al.* 2004; Tabosa *et al.* 2004).

Denervation atrophy characterized by fiber size variation with fibers of decreased size and some angular fibers, abundant internal nuclei, and occasional vacuolated fibers are observed mainly in the masticatory muscles. In advanced cases the myofibers are substituted by fibrous or fatty tissue (Tabosa *et al.* 2000, 2004, 2006; Câmara *et al.* 2009).

Under electron microscopy the neurons of the trigeminal nuclei have markedly swollen mitochondria corresponding to the vacuoles observed at light microscopy. The mitochondrial cristae are displaced peripherally and are disoriented and disintegrating. In severely affected mitochondria, the cristae are extremely shortened or absent. Intramitochondrial dense granules are absent. There is an increase in the number of lysosomes, consistent with secondary lysosomes with membranous to granular/amorphous electron-dense residual bodies (Tabosa *et al.* 2006).

## Rational Uses

*P. juliflora* is an important noxious weed in the semiarid region of northeastern Brazil. In the more humid areas its aggressive growth rapidly leads to woodland monocultures as it excludes other species by shading (Instituto Hórus de Desenvolvimento e Conservação Ambiental 2008) and allelopathy (Nakano *et al.* 2004). It also reduces moisture available for herbaceous understory growth. The loss of grass cover under canopies of this tree may promote soil erosion in areas invaded by the plant; farmers have to decide whether to eradicate the plant or attempt rational uses. Besides the proper use of mesquite pods to feed livestock and other species including humans (Tabosa *et al.* 2004; Chogel *et al.* 2007), the tree can also be used in several other ways including: decontamination of heavy metals contaminated soils (Senthilkumar *et al.* 2005); a source of lumber, fuelwood (Ramos *et al.* 2008), biofuel (Felker *et al.* 1981), or charcoal (Câmara *et al.* 2009); a form to increase organic matter and nitrogen in the soil (Herrera-Arreola *et al.* 2007); or even the use of juliflorine as a leading candidate for Alzheimer's disease therapy (Choudhary *et al.* 2005). Care must be taken when using mesquite trees to decontaminate soils since the foliage and pods are used as fodder for livestock; in view of heavy metal accumulation, such a practice should be avoided as otherwise it could pave the way for biomagnifications (Senthilkumar *et al.* 2005).

## Conclusions

Despite knowledge about the toxicity of *P. juliflora*, poisoning by pods of this species is still an important disease of cattle and goats in northeastern Brazil. To prevent poisoning it is necessary to avoid the use of concentrates with mesquite bean pods or bran in a percentage higher than 30% of the diet for no more than 6 months. Grinding the pods to feed to livestock is important to destroy the seeds because ingestion of unground pods results in large quantities of germinating seeds in the livestock feces.

In areas invaded by the plant we recommend harvesting and storing the pods for rational use since extensively raised cattle in those areas can manifest the toxicosis after a grazing period of 30-60 days. Besides livestock feed, *P. juliflora* has other potentially economical and rational uses. However, recommendations for rational uses or eradication of the plant need more research to determine the cost-benefit ratio of these options. Sheep are valuable for their wool production and can reduce the number of plants by grazing the young plants, obtaining more productivity and avoiding the invasion of the paddocks by the tree. After cutting the plant, regrowth can be avoided by the use of herbicides (clopyralid or triclopyr). Good progress was also achieved in eliminating resprouts from harvested *Prosopis* stumps by combinations of kerosene (20-40 ml per tree) applications followed by burning.

## Acknowledgements

This work has been financially supported by the National Institute for Science and Technology for the Control of Toxic Plants (CNPq), grant number 573534/2008-0.

## References

Azevedo-Rocha RG (1987). Algaroba na alimentação e farmacopéia do homem rural norte-riograndense. *Revista da Associação Brasileira de Algaroba,* Mossoró, 1:67-96.

Bacca SF, Vallenas A, Novoa C, Ochoa J, and Cueva S (1966). Estudio experimental de la ‘coquera’ em caprinos. *Revista Faculdad Medicina Veterinaria*, Lima, 18/20:131-159.

Câmara ACL, Costa NA, Riet-Correa F, Afonso JAB, Dantas AFM, Mendonça CL, and Souza MI (2009). Intoxicação espontânea por vagens de *Prosopis juliflora* (Leg. Mimosoideae) em bovinos em Pernambuco. *Pesquisa Veterinária Brasileira* 29(3):233-240.

Chogel SK, Pasiecznik NM, Harvey M, Wright J, Awan SZ, and Harris PJC (2007). *Prosopis* pods as human food, with special reference to Kenya. *Water SA* 33(3):419-424.

Choudhary MI, Nawaz SA, Haq Z-U, Azim MK, Ghayur MN, Lodhi MA, Jalil S, Khalid A, Ahmed A, Rode BM, Rahman A-U, Gilani A-U, and Ahmad VU (2005). Juliflorine: A potent natural peripheral anionic-site-binding inhibitor of acetylcholinesterase with calcium-channel blocking potential, a leading candidate for Alzheimer’s disease therapy. *Biochemical and Biophysical Research Communications* 332(4):1171-1179.

Dantas JRF and Menezes RV (1994). UFPB, UFBA e USP estudam ‘cara torta, doença que acomete bovinos na Paraíba, Pernambuco e Rio Grande do Norte. *Boletim Informativo do CRMV-PB*, jan/fev.

Dollahite JW (1964). Management of the disease produced in cattle on an unbalanced diet of mesquite beans. *Agriculture Experimental Station, Texas, USA. Progress Report* 2314, pp. 293-296.

Dollahite JW and Anthony WV (1957). Malnutrition in cattle on an unbalanced diet of mesquite beans. *Agriculture Experimental Station, Texas, USA, Progress Report* 193, 32 pp.

Felker P, Clark PR, Cannell GH, and Osborn JF (1981). Screening *Prosopis* (mesquite or algarrobo) for biofuel production on semiarid lands. In *Symposium on Dynamics and Management of Mediterranean-type Ecosystem*, pp. 179-185.

Figueiredo LJC, Ferreira MM, Távora JPF, Simões SDV, and Dantas JR (1996). Estudo clínico e anatomopatológico da doença 'cara torta' em bovinos do nordeste brasileiro. *Arquivos Escola Medicina Veterinária UFBA* 18:175-183.

Herrera-Arreola G, Herrerac Y, Reyes-Reyesd BG, and Dendooven L (2007). Mesquite (*Prosopis juliflora* (Sw.) DC.), huisache (*Acacia farnesiana* (L.) Willd.) and catclaw (*Mimosa biuncifera* Benth.) and their effect on dynamics of carbon and nitrogen in soils of the semi-arid highlands of Durango Mexico. *Journal of Arid Environments* 69(4):583-598.

Instituto Hórus de Desenvolvimento e Conservação Ambiental (2008). http://www.institutohorus.org.br/download/fichas/Prosopis_juliflora.htm. Accessed 10/30/2008.

Kingsbury JM (1964). *Poisonous Plants of the United States and Canada*, pp. 349-351. Prentice-Hall Inc., Englewood Cliffs, New Jersey.

Lima E, Riet-Correa F, Amorin SL, and Sucupira Júnior G (2004). Intoxicação por favas de *Prosopis juliflora* (algaroba) em caprinos no Nordeste do Brasil. *Pesquisa Veterinária Brasileira* 24(Supl.):36-37.

Mahgoub O, Kadim IT, Johnson EH, Srikandakumar A, Al-Saqri NM, Al-Abri AS, and Ritchie A (2005a). The use of a concentrate containing Meskit (*Prosopis juliflora*) pods and date palm byproducts to replace commercial concentrate in diets of Omani sheep. *Animal Feed Science and Technology* 120(1-2):33-41.

Mahgoub O, Kadim IS, Forsberg NE, Al-Ajmi DS, Al-Saqry NM, Al-Abri AS, and Annamalai K (2005b). Evaluation of Meskit (*Prosopis juliflora*) pods as a feed for goats. *Animal Feed Science and Technology* 121(3-4):319-327.

Nakano H, Nakajima E, Hiradate S, Fujii Y, Yamada K, Shigemori H, and Hasegawa K (2004). Growth inhibitory alkaloids from mesquite (*Prosopis juliflora* (Sw.) DC.) leaves. *Phytochemistry* 65(5):587-591.

Obeidat BS, Abdullah AY, and Al-Lataifeh FA (2008). The effect of partial replacement of barley grains by *Prosopis juliflora* pods on growth performance, nutrient intake, digestibility, and carcass characteristics of Awassi lambs fed finishing diets. *Animal Feed Science and Technology* 146(1-2):42-54.

Pasiecznik NM, Felker P, Harris PJC, Harsh LW, Cruz G, Tewari JC, Cadoret K, and Maldonado LJ (2001). *The Prosopis juliflora – Prosopis pallida Complex: A Monograph.* HDRA, Coventry, UK, 172 pp.

Ramos MA, Medeiros PM, Almeida ALS, Feliciano ALP, and Albuquerque UP (2008). Use and knowledge of fuelwood in an area of Caatinga vegetation in NE Brazil. *Biomass and Bioenergy* 32(6):510-517.

Ravikala K, Patel AM, Murthy KS, and Wadhwani KN (1995). Growth efficiency in feedlot lambs on *Prosopis juliflora* based diets. *Small Ruminant Research* 16(3):227-231.

Riveros F (1992). The genus *Prosopis* and its potential to improve livestock production in arid and semiarid regions. In *Legume trees and other fodder trees as protein sources for*

*Livestock* (A Speedy and P Pugliesi, eds), pp. 257-276. FAO Animal Production and Health Paper 102.

Senthilkumar P, Prince WSPM, Sivakumar S, and Subbhuraam CV (2005). *Prosopis juliflora* – A green solution to decontaminate heavy metal (Cu and Cd) contaminated soils. *Chemosphere* 60(10):1493-1496.

Silva AMA, Pereira JA, Costa PMA, and Melo HV (1989). Utilização da algaroba (*Prosopis juliflora*) na alimentação de suínos. *Revista Sociedade Brasileira de Zootecnia* 18:179-183.

Silva JHV, Oliveira JNC, Silva EL, Jordão Filho J, and Ribeiro MLG (2002a). Uso da farinha integral de vagem de algaroba (*Prosopis juliflora* (Sw) D.C.) na alimentação de codornas japonesas. *Revista Brasileira de Zootecnia* 31(3):1789-1795.

Silva JHV, Silva EL, Jordão Filho J, Toledo RS, Albino LFT, Ribeiro MLG, and Couto HP (2002b). Valores energéticos e efeitos da inclusão de farinha integral de vagem de algaroba (*Prosopis juliflora* (Sw.) D.C.) em rações de poedeiras comerciais. *Revista Brasileira de Zootecnia* 31(6):2255-2264.

Silva CG, Cavalcanti-Mata MERM, Braga MED, and Queiroz VS (2003). Extração e fermentação do caldo de algaroba (*Prosopis juliflora* (Sw.) DC) para obtenção de aguardente. *Revista Brasileira de Produtos Agroindustriais* 5(1):51-56.

Silva DM, Riet-Correa F, Medeiros RMT, and Oliveira OD (2006). Plantas tóxicas para ruminantes e eqüídeos no Seridó Ocidental e Oriental do Rio Grande do Norte. *Pesquisa Veterinária Brasileira* 26(4):223-236.

Stein RBS, Toledo LRA, Almeida FQ, Arnaut AC, Patitucci LT, Soares Neto J, and Costa VTM (2005). Uso do farelo de vagem de algaroba (*Prosopis juliflora* (Swartz) D.C.) em dietas para eqüinos. *Revista Brasileira de Zootecnia* 34(4):1240-1247.

Tabosa IM, Souza JC, Graça DL, Barbosa-Filho JM, almeida RN, and Riet-Correa F (2000). Neuronal vaculolation of the trigeminal nuclei in goats caused by ingestion of *Prosopis juliflora* pods (mesquite beans). *Veterinary and Human Toxicology* 42(3):155-158.

Tabosa IM, Riet-Correa F, Simões SVD, Medeiros RMT, and Nobre VMT (2004). Intoxication by *Prosopis juliflora* pods (mesquite beans) in cattle and goats in Northeastern Brazil. In *Toxic Plants and other Natural Toxicants* (T Acamovic, CS Stewart, and TW Pennycott, eds), pp. 341-346. CAB International Publishing, Wallingford, UK.

Tabosa IM, Riet-Correa F, Barros SS, Summers BA, Simões SVD, Medeiros RMT, and Nobre VMT (2006). Neurohistologic and ultrastructural lesions in cattle experimentally intoxicated with the plant *Prosopis juliflora*. *Veterinary Pathology* 43(5):695-701.

Tokarnia CH, Döbereiner J, and Peixoto PV (2000). *Plantas Tóxicas do Brasil*, 310 pp. Editora Helianthus, Rio de Janeiro.

Washburn KE, Breshears MA, Ritchey JW, Morgan SE, and Streeter RN (2002). Honey mesquite toxicosis in a goat. *Journal of the American Veterinary Medical Association* 220(12):1837-1839.

Chapter 47

# Neonate Behavior in Goats is Affected by Maternal Ingestion of *Ipomoea carnea*

A.T. Gotardo[1], J.A. Pfister[2], M. Barbosa-Ferreira[1], and S.L. Górniak[1]

[1]*Research Center of Veterinary Toxicology (CEPTOX), School of Veterinary Medicine and Animal Science, University of São Paulo, SP 13635-900, Brazil;* [2]*USDA-ARS Poisonous Plant Research Laboratory, Logan, Utah 84341, USA*

## Introduction

Toxic plants on rangelands often adversely affect animal production and fatally intoxicate animals that ingest such plants (James *et al.* 1992). However, many plant toxins also have more subtle effects on animal production and behavior that may not be immediately obvious to producers (Pfister *et al.* 2006). One such plant is *Ipomoea carnea*, a shrubby plant that grows in much of South America. *I. carnea* contains the indolizidine alkaloid, swainsonine, as well as toxic calystegines (Haraguchi *et al.* 2003). Related plants worldwide are *Astragalus* and *Oxytropis* species (so-called locoweeds); however, these two genera do not contain the additional toxic calystegines (Elbein and Molyneux 2004). Swainsonine and calystegines cause cellular accumulation of oligosaccharides due to inhibition of several important enzymes, resulting in cellular vacuolization and death in the central nervous system and in other body systems (De Balogh *et al.* 1999; Schumaher Henrique *et al.* 2003).

The CNS effects of *I. carnea* may have a profound effect on animal behavior similar to effects from ingestion of *Astragalus* and *Oxytropis* species (Molyneux *et al.* 1995). There is little information available on the effects of *I. carnea* species on animal behavior. This study evaluated the behavioral effects on dams and kids of prenatal ingestion of this plant. Offspring behavior was examined during 2 h postpartum. Kid behavior was further examined using various tests up to 6 days postpartum to determine if there were any behavioral alterations in early post-natal development.

## Materials and Methods

### Experimental subjects

The study was conducted at the University of São Paulo Experimental Station, Pirassununga, São Paulo state, Brazil (S21°58', W47°27'). All animal care and handling was done by experienced personnel under veterinary supervision.

©CAB International 2011. *Poisoning by Plants, Mycotoxins, and Related Toxins* (eds F. Riet-Correa, J. Pfister, A.L. Schild, and T.L. Wierenga)

Female adult (15 months) goats of Alpine breed weighed an average of 40 kg. Breeding was synchronized using standard methods with vaginal pessaries, and all does were bred twice by the same fertile male of the same breed. Pregnancy was verified using radio-ultrasound on day 32 of gestation. All animals were provided with a commercial ration: ground maize (60.6% on DM basis), soybean extract (36%), and 3.4% mineral salt at 200 g/day for each animal. Fresh water was available free choice. Animals were randomly divided into two treatment groups, controls (*n*=5) and treatment (*n*=9; 10 g/kg BW of *I. carnea* fresh plant material). Fresh plant material contains 19.3% dry matter, thus does consumed 1.93 g dry *I. carnea*/kg BW. Treatment animals were given freshly harvested and chopped (2.5 cm screen) *I. carnea* during the 5th to 16th weeks of gestation (day 35 to 112). Additionally, after animals had consumed both the commercial ration, and *I. carnea* for treated animals, they were given *ad libitum* access to chopped sugarcane residue (*Sacharum officinarum* L.) sufficient for overnight feeding. Residual sugarcane was removed before feeding commenced the next day.

**Behavior tests**

Kid behavior was evaluated for 2 h post partum. Further evaluation of the offspring was done using two tests after birth: (i) reaching and discriminating mother vs. alien dam (12 and 24 h postpartum), and (ii) navigating a progressive maze (2, 4, 6 days postpartum).

Pregnant goats were observed closely as their parturition date approached, and personnel were on hand when they gave birth. Immediately after birth, the kid weight and sex was recorded. The mothers were given 5 min to bond with their offspring (Poindron *et al.* 2007), and then moved to a separate pen for the 2 h observation period. The 5 × 3 m pen had a 1-m grid marked on the concrete floor for reference. This 2 h period was videotaped and various observations were made from these taped sessions. The postpartum session was divided into the following time periods: 0-15, 15-30, 30-60, and 60-120 minutes. The following kid behaviors were recorded for frequency: (i) head up: first attempt by the kid to lift its head off the ground any distance; (ii) crawl: movement of the legs as if attempting to stand but unable to get legs under the body and not able to lift body off the ground; (iii) attempt to stand: legs under body and able to lift body off the ground any distance; (iv) stand up: first successful attempt to stand on all legs regardless of the length of time; (v) nuzzle mother's front half: any nuzzling on the front half of the mother; (vi) nuzzle mother's rear half: any nuzzling on the rear half of the mother; and (vii) suckle: first successful grasping of the mother's teat with suckling for any length of time; each successful grasp of a teat with suckling was considered a bout of suckling.

Kids were tested for ability to discriminate their own mother from an alien dam at 12 h after birth. The alien dam used in all tests was one that had recently (within 5 days) given birth. Kids were not allowed to nurse for 120 min before the test, then placed into a test pen with the mother on one side and the alien dam on the other side. Kids were allowed a maximum of 5 min to complete this test. Several variables were recorded during this test: (i) time to exit start box; (ii) time when kid began a direct approach to the dam(s); and (iii) arrival (i.e. physical touching) at either mother or alien dam. Positions were switched for the dams and the test repeated immediately. The test was done at 12 h postpartum. If kids did not successfully leave the start box or did not complete the test by moving towards one dam or the other, the test was repeated at 36 h after birth.

Kids were tested at 2, 4, and 6 days in a progressive maze with incrementally increased difficulty. At 48 h after parturition, kids were placed into a simple maze with the dam secured at the exit of the maze at a distance of 1 m. The 2-day maze had only one barrier, the 4-day test had two barriers, and the 6-day maze had three barriers that the kids needed to navigate for successful completion. The test lasted for 5 min and variables recorded were i) time to leave the starting position; and ii) time to traverse the maze and reach the mother.

### Statistical analysis

Kid birth weights were tested using *t*-tests. Chi square tests were used to evaluate frequency of behaviors postpartum. Two statistical tests were used to examine kids' choice of mother or alien dam in the discrimination test. A binomial test was done using a probability of 0.5 to test if choices were different from random selection; additionally a 2 × 2 chi square test of independence was also done to determine if control and *I. carnea* kids differed in their number of incorrect choices. The times for kids in the two treatments to move towards their mothers or alien dam were evaluated using a linear mixed model (Proc Mixed) in SAS (Version 9.2, SAS Institute, Inc., Cary, N.C.) to evaluate kid times through the progressive maze on days 2, 4, and 6. The model included treatment, animals nested with treatment, and days and runs within days as repeated measures.

## Results

### Neonate mortality and weight at birth

Post-natal ($n$=2) and fetal ($n$=2) mortality were observed in the treated group but not in the control group. Pathological examination revealed that aborted and stillborn kids had lesions typical of cellular vacuolization (Hartley and James 1975). Body weight of treated kids was negatively affected by exposure to *I. carnea*. Treated kids weighed 3.42 ± 0.22 kg and control animals weighed 3.98 ± 0.17 kg ($P < 0.05$).

### Kid behavior 2 h postpartum

Treated kids were either unable to stand within 120 min (5/7) or slow to stand (2/7) compared to controls, which all stood up within 5-10 min postpartum. Similarly, control kids all nursed within 30-50 min postpartum, whereas only one treated kid nursed within 120 min.

Within the first 15 min control kids made 6.8 attempts/15 min to stand, compared to 1.2 attempts/15 min for treated kids ($P = 0.08$). During the 15-30 min period, control kids made 4.3 attempts to stand/15 min period, compared to treated kids with 0.4 attempts/15 min ($P = 0.004$). Likewise, control kids had a frequency to nuzzle their mothers front with 35 attempts during the 15-30 min period compared to 1.2 for treated kids ($P = 0.01$). In a similar manner, control kids nuzzled their mothers rear 8.3 times in this same 15-30 min period compared to 2.0 times for treated kids ($P = 0.04$) (Table 1).

Treated kids had a higher frequency (4.6 attempts/30 min; $P = 0.02$) of raising their heads during the 30-60 min period compared to control kids, all of whom had already raised their heads for the first time in an earlier period. In addition, control kids nuzzled their mothers front half at a much higher frequency (26.1 times/30 min) compared to treated kids (1.5 times/30 min; $P = 0.06$). Similarly, control kids nuzzled their mothers rear half at a higher frequency (13.0 times/30 min) compared to treated kids (0.8 times/30 min; $P = 0.01$) (Table 1).

**Table 1.** Kid behavior variables (frequency/time period) in the first 2 h after birth.

| Variables[1] | 0-15 min | | | 15-30 min | | | 30-60 min | | | 60-120 min | | |
|---|---|---|---|---|---|---|---|---|---|---|---|---|
| | C | T | P | C | T | P | C | T | P | C | T | P |
| Crawl | 8.5 | 10.8 | 0.58 | 1.8 | 8.4 | 0.05 | 0.3 | 4.0 | 0.14 | 0.5 | 14.2 | 0.01* |
| Head up | 1.3 | 7.2 | 0.31 | 0.3 | 2.8 | 0.12 | 0.0 | 4.6 | 0.02* | 0.5 | 16.0 | 0.01* |
| Attempt to stand | 6.8 | 1.2 | 0.08 | 1.5 | 1.8 | 0.86 | 0.5 | 0.8 | 0.71 | 0.8 | 4.3 | 0.38 |
| Stand up | 4.5 | 1.2 | 0.20 | 4.3 | 0.4 | 0.004* | 5.7 | 1.8 | 0.11 | 4.3 | 1.6 | 0.28 |
| Nuzzle front of dam | 12.8 | 3.6 | 0.19 | 35.0 | 1.2 | 0.01* | 26.1 | 1.5 | 0.06 | 17.5 | 1.0 | 0.12 |
| Nuzzle rear of dam | 9.5 | 1.6 | 0.27 | 8.3 | 2.0 | 0.04* | 13.0 | 0.8 | 0.01* | 14.1 | 0.6 | 0.05 |
| Suckle | 1.5 | 0.0 | 0.29 | 2.3 | 0.8 | 0.42 | 8.3 | 0.2 | 0.05 | 4.0 | 0.0 | 0.13 |

[1] Kid behavior was videotaped with a low-light camcorder and observations recorded for the various periods. Definitions are provided in the text.
* Statistical significance $P < 0.05$

During the final hour of observation, treated kids had a higher frequency of crawling ($P = 0.01$) compared to control kids (14.2 vs. 0.5 occurrences for treated and control kids, respectively). Similarly, the frequency of head raising was higher ($P = 0.01$) for treated kids (16.0 events/60 min) compared to controls kids (0.5 events/60 min). Control animals continued to have a higher ($P = 0.05$) frequency of nuzzling mothers front half (14.1 events/60 min) compared to treated kids (0.6 events/60 min) (Table 1).

## Mother-alien ewe discrimination test

Treated kids made many more errors (7/9) in choosing their own mother in the discrimination test than did control kids (0/10; $P = 0.01$). Choices by treated kids did not differ ($P = 0.18$ in binomial test) from randomness whereas control kids did not choose randomly ($P = 0.002$). Treated and control kids differed in their time to exit the start box and also differed in their approach time to mother. There was a tendency for treated kids to take more total time to arrive at their mother compared to control kids (Table 2).

**Table 2.** Discrimination of kids for own or alien mother. Kids were given two tests with the alien and own dam reversed in position from test 1 to test 2.

| Variable | Control | Treated | Probability |
|---|---|---|---|
| Exit[1] | 72.8 ± 13.0 | 137.4 ± 45.5 | 0.02* |
| Approach[2] | 87.9 ± 13.2 | 159.0 ± 42.1 | 0.08 |
| Arrival[3] | 105.8 ± 16.9 | 183.0 ± 38.5 | 0.08 |
| Incorrect choices[4] | 0/10 | 7/9 | 0.001* |

[1] Exit was time (seconds) for kids to leave the start box.
[2] Approach was time (seconds) for kids to begin a direct approach to the dams.
[3] Arrival was time (seconds) for kids to arrive at and touch either their own dam or the alien dam.
[4] Incorrect choices of alien dam / number of attempts by kids.
* Statistical significance $P < 0.05$.

### Maze test on days 2, 4, 6 postpartum

There was a treatment × day × run interaction for leaving the starting point of the maze ($P = 0.05$) and for arrival at the end ($P = 0.003$) therefore results are shown in detail for days and runs in Table 3. There were large numerical differences in kid response, but small sample size and high variability precluded finding some differences significant. During the test on day 2, treated and control kids did not differ ($P > 0.16$) in their times to leave the starting point or to arrive at the end of the maze. On day 4, control kids were faster to leave the starting point on runs 2 and 3 ($P < 0.1$) and tended ($P < 0.15$) to be quicker to reach the end of the maze compared to treated kids on all runs. Control kids left the starting point quicker on run 1 on day 6 ($P < 0.1$) and had a tendency to leave the starting point quicker than treated kids on the other runs. Control kids arrived more quickly at the end of the maze than did treated kids only during the first run on day 6 (Table 3).

**Table 3.** Kid performance in a progressive maze (seconds ± SE) on postpartum days 2, 4, and 6. Each kid was run for 3 consecutive runs and the maximum time for each run was 3 min. The maze had 1 barrier on day 2, 2 barriers on day 4, and 3 barriers on day 6. For this portion of the experiment, *P* values ≤ 0.10 were considered significant (alpha = 0.10) and those *P* values < 0.15 trending towards significance.

| Day/Run | Control | | Treated | | Probability | |
|---|---|---|---|---|---|---|
| | Leave[1] | Arrive[2] | Leave | Arrive | Leave | Arrive |
| Day 2 | | | | | | |
| Run 1 | 49.3 ± 24.7 | 72.0 ± 28.6 | 65.0 ± 31.6 | 78.4 ± 31.6 | 0.73 | 0.96 |
| Run 2 | 5.2 ± 1.4 | 19.7 ± 12.3 | 42.8 ± 34.4 | 49.4 ± 32.9 | 0.001* | 0.08 |
| Run 3 | 4.2 ± 1.5 | 10.2 ± 3.8 | 40.94 ± 34.9 | 56.4 ± 33.4 | 0.001* | 0.001* |
| Day 4 | | | | | | |
| Run 1 | 43.7 ± 18.0 | 52.8 ± 17.9 | 77.0 ± 42.1 | 109.6 ± 38.3 | 0.12 | 0.18 |
| Run 2 | 14.7 ± 7.7 | 30.0 ± 7.6 | 73.6 ± 43.4 | 48.2 ± 41.6 | 0.002* | 0.004* |
| Run 3 | 13.0 ± 3.6 | 23.7 ± 4.1 | 73.0 ± 43.7 | 76.4 ± 42.3 | 0.001* | 0.001* |
| Day 6 | | | | | | |
| Run 1 | 8.5 ± 1.1 | 49.5 ± 4.2 | 52.6 ± 32.6 | 135.6 ± 20.9 | 0.001* | 0.005* |
| Run 2 | 7.2 ± 2.4 | 28.0 ± 5.6 | 40.2 ± 35.0 | 47.0 ± 33.3 | 0.001* | 0.002* |
| Run 3 | 6.3 ± 2.1 | 23.0 ± 7.2 | 40.4 ± 34.9 | 53.0 ± 31.9 | 0.001* | 0.009* |

[1] Time (seconds) for kids to leave the starting point in the maze.
[2] Time (seconds) for kids to complete the maze.
* Statistical significance *P* < 0.05

## Discussion

The toxins in *I. carnea* are indolizidine alkaloids and calystegines that inhibit several important enzymes throughout the body and particularly in the CNS (Molyneux *et al.* 1995; Haraguchi *et al.* 2003). This inhibition leads to cellular death, thus disrupting many aspects of normal metabolism. In goats, swainsonine promotes serious lesions in the CNS, particularly in Purkinje cells and the cerebellum. These lesions are also found in offspring from mothers that have ingested swainsonine-containing plants for considerable periods of time during gestation (Hartley and James 1975).

This study has clearly shown that ingestion of *I. carnea* during gestation affected kid behavior, particularly in the first 2 h after birth. These observations indicate that the alterations in behavior are life-threatening to kids as treated animals were not able to stand

and suckle normally. Thus, these affected kids would be deprived of nourishment in the form of milk. Further, their passive immunity would be greatly diminished because of the lack of colostrum in the first hours after birth (Rajala and Castrén 1995).

Treated kids survived through human assistance and thus the kids participated in additional behavior tests. Treated kids were able to move well towards their own and an alien mother but the majority of treated kids (7/9) did not choose their own mother in the discrimination test, whereas control kids made no errors. Under normal conditions, if a kid approaches the wrong mother, the alien mother often butts the young goat with considerable force (Rutter 2002). For the weak kids from the treatment group in this study, that butting could cause serious injury and further complicate their ability to survive.

There was no treatment effect on kid performance in the progressive maze on the first day of testing (postpartum day 2). Thereafter treated kids were generally slower than control kids to leave the starting point (Table 3). On some of the runs within days, treated kids were also slower to complete the maze compared to controls. These results were not conclusive because of the high variability particularly amongst the treated kids. Two treated kids were unable or unwilling to leave the starting point whereas others started and completed the maze with little difficulty. Sheep tested in mazes can retain spatial memories for many days (Lee *et al.*, 2006), and the controls in our study also appeared to retain memories of the maze as their performance improved over time; however, treated kids' performance, particularly during the first run, actually deteriorated over time on average (Table 3). Similar work with sheep and swainsonine has shown that treated lambs had impaired maze performance compared to control lambs and treated lambs did not improve their maze performance over the course of several days postpartum (Pfister *et al.* 2006). These results with kids have implications for growth and development as treated kids may be handicapped in their ability to learn from mother and cohorts about foraging, location of food and water, and other aspects important for survival (Mendl 1999).

## Conclusions

Taken together, these results indicate that the behavior of kids from mothers treated with *I. carnea* during gestation would lead to poor kid survival at birth and later poor performance if the kid survived. Livestock producers with pastures that contain swainsonine-containing plants such as *I. carnea* must be aware of the potential effects of the plant on pregnant females and ensure that pregnant animals are limited in their exposure to such plants. The cost of exposure to toxic plants may be high (James *et al.* 1992). In addition to the loss of production such as meat and milk and reductions in animal health in both does and kids, there would likely be concomitant increased costs for veterinary care and treatment and additional opportunity costs to purchase other pastures to replace those infested with the toxic plants.

## Acknowledgements

This work was financially supported by the Fundação de Amparo à Pesquisa do Estado de São Paulo – FAPESP, Brazil (Proc nº 2006/58729-2).

We thank Paulo Cesar Fabricio Raspantini, Leonila Ester Reinert Raspantini, Estevão Belloni, Marco Antonio Faustino dos Santos and Adilson Baladore for valuable assistance with the study and for animal care.

## References

De Balogh KIM, Dimande AP, Van Der Lugt JJ, Molyneux RJ, Naudé TW, and Welman WG (1999). A lysosomal storage disease induced by *Ipomoea carnea* in goats in Mozambique. *Journal of Veterinary Diagnostic Investigation* 11:266-273.

Elbein AD and Molyneux RJ (2004). Inhibitors of Glycoprotein Processing. In *Iminosugars as Glycosidase Inhibitors* (AE Stütz, ed.), pp. 216-252. Wiley-VCH Verlag GmbH, Germany.

Haraguchi M, Gorniak SL, Ikeda K, Minami Y, Kato A, Watson AA, Nash RJ, Molyneux RJ, and Asano N (2003). Alkaloidal components in the poisonous plant, *Ipomoea carnea* (Convolvulaceae). *Journal of Agricultural and Food Chemistry* 51:4995-5000.

Hartley WJ and James LF (1975). Fetal and maternal lesions in pregnant ewes ingesting locoweed (*Astragalus lentiginosus*). *American Journal Veterinary Research* 36:825-826.

James LF, Nielsen DB, and Panter KE (1992). Impact of poisonous plants on the livestock industry. *Journal of Range Management* 45:3-8.

Lee C, Colegate S, and Fisher AD (2006). Development of a maze test and its application to assess spatial learning and memory in Merino sheep. *Applied Animal Behaviour Science* 96:43-51.

Mendl M (1999). Performing under pressure: stress and cognitive function. *Applied Animal Behaviour Science* 65:221-244.

Molyneux RJ, McKenzie RA, O'Sullivan BM, and Elbein AD (1995). Identification of the glycosidase inhibitors swainsonine and calystegine B2 in Weir vine (*Ipomoea sp.* Q6 [aff. calobra]) and correlation with toxicity. *Journal of Natural Products* 58:878-886.

Pfister JA, Astorga JB, Panter BL, Stegelmeier RJ, and Molyneux RJ (2006). Maternal ingestion of locoweed I. Effects on ewe-lamb bonding and behavior. *Small Ruminant Research* 65:51-53.

Poindron P, Lévy F, and Keller M (2007). Maternal responsiveness and maternal selectivity in domestic sheep and goats: The two facets of maternal attachment. *Developmental Psychobiology* 49:54-70.

Rajala P and Castrén H (1995). Serum immunoglobulin concentrations and health of dairy calves in two management sysems from birth to 12 weeks of age. *Journal of Dairy Science* 78:2737-2743.

Rutter SM (2002). Behavior of sheep and goats. In *Ethology of Domestic Animals* (P Jensen, ed.), pp. 145-158. CABI Publishing, Wallingford, Oxfordshire, UK.

Schumaher-Henrique B, Górniak SL, Dagli MLZ, and Spinosa HS (2003). Toxicity of long term administration of *Ipomoea carnea* to growing goats: clinical, biochemical, haematological and pathological alterations. *Veterinary Research Communications* 27:311-319.

# Chapter 48

# The Comparative Pathology of Locoweed Poisoning in Horses and Other Livestock

B.L. Stegelmeier, T.Z. Davis, K.D. Welch, B.T. Green, D.R. Gardner, S.T. Lee, M.H. Ralphs, J.A. Pfister, D. Cook, and K.E. Panter

*USDA-ARS Poisonous Plant Research Laboratory, Logan, Utah 84341, USA*

## Introduction

We have shown that some animal species are more likely to develop both clinical locoism and locoweed-induced histologic lesions (James and Van Kampen 1971, 1976; Van Kampen and James 1972; James *et al.* 1981; Stegelmeier *et al.* 2007). Mice and deer are relatively resistant and require long exposures and high doses to develop neurologic lesions (Stegelmeier *et al.* 1994, 2005). Sheep and cattle are more susceptible as we have shown that doses of 0.25 mg/kg BW for 30-45 days will produce irreversible neurologic damage (Stegelmeier *et al.* 1999). Horses appear to be highly sensitive to poisoning but no controlled dose response studies of the response of horses to locoweed poisoning are available. In this study we identify the minimal doses that produce clinical locoism and better describe the lesions of locoweed poisoning in horses.

## Materials and Methods

Sixteen mature mares were randomly divided into four groups of four animals. The horses were individually penned and fed a mixed locoweed ration twice a day to obtain swainsonine doses of 0.0, 0.2, 0.6, and 1.8 mg swainsonine/kg body weight/day. All groups were fed for 45 days after which the left ovary was surgically biopsied. After an additional 45 days two mares from each group were necropsied and tissues were collected for chemical, histology, and ultrastructural studies. The remaining two animals from each group were kept on control ration for an additional 45 days after which they were euthanized and necropsied. Repeated samples including weight, hematology, and serum biochemistry data were compared using a mixed model for repeated samples.

## Results and Discussion

Within 10 days of treatment the mares from the 0.6 and 1.8 mg groups were depressed and took longer to ingest their ration. These animals were hesitant to move and had subtle intention tremors when moving. These clinical signs became more severe and after 45 days five of the six mares had abnormal estrus cycles. When the ovaries were biopsied from this group, several were cystic. After 60 days of treatment the 0.2 mg/kg group also became depressed and started to lose weight. The clinical condition of the animals in the higher dose groups continued to deteriorate. Two animals became anorexic and dangerous to handle and these animals were euthanized and necropsied early. At necropsy the treated animals were thin with lack of visceral and subcutaneous adipose tissue. Histologically the 0.6 and 1.8 mg swainsonine groups all had severe vacuolation of nearly all neuronal and visceral tissues. The 0.2 mg swainsonine group had mild vacuolation of the Purkinje cells and renal tubular cells. The mares that were allowed to recover began to have normal estrus cycles and near the end of the recovery period began to eat better and gain weight. Histologically these animals had some neuronal vacuolation with loss of neurons and axonal dystrophy (spheroids). Morphometric and ultrastructural studies of these mares continue.

## Conclusions

These preliminary results suggest that horses are much more sensitive to locoweed poisoning than other livestock. Horses develop more severe clinical signs quicker and they are slower to recover than other species. Though previously poisoned mares appear to recover reproductive function, they still have neurologic damage and are likely to be a risk if they are ridden or worked.

## References

James LF and Van Kampen KR (1971). Acute and residual lesions of locoweed poisoning in cattle and horses. *Journal of the American Veterinary Medical Association* 158:614-618.

James LF and Van Kampen KR (1976). Effects of locoweed toxin on rats. *American Journal of Veterinary Research* 37:845-850.

James LF, Hartley WJ, and Van-Kampen KR (1981). Syndromes of *Astragalus* poisoning in livestock. *Journal of the American Veterinary Medical Association* 178:146-150.

Stegelmeier BL, Molyneux RJ, and James LF (1994). The pathology of swainsonine and locoweed (*Astragalus mollissimus*) in rodents. *Veterinary Pathology* 31:620.

Stegelmeier BL, James LF, Panter KE, Gardner DR, Pfister JA, Ralphs MH, and Molyneux RJ (1999). Dose response of sheep poisoned with locoweed (*Oxytropis sericea*). *Journal Veterinary Diagnostic Investigation* 11:448-456.

Stegelmeier BL, James LF, Gardner DR, Panter KE, Lee ST, Ralphs MH, Pfister JA, and Spraker TR (2005). Locoweed (*Oxytropis sericea*)-induced lesions in mule deer (*Odocoileius hemionus*). *Veterinary Pathology* 42:566-578.

Stegelmeier BL, Lee ST, James LF, Gardner DR, Panter KE, Ralphs MH, and Pfister JA (2007). The comparative pathology of locoweed poisoning in livestock, wildlife and rodents. In *Poisonous Plants Global Research and Solutions* (KE Panter, TL Wierenga, and JA Pfister, eds) pp. 359-365. CABI Publishing, Cambridge, Massachusetts.

Van Kampen KR and James LF (1972). Sequential development of the lesions in locoweed poisoning. *Clinical Toxicology* 5:575-580.

Chapter 49

# *Sida carpinifolia* (Malvaceae) Poisoning in Herbivores in Rio Grande do Sul

P.M.O. Pedroso[1], E.M. Colodel[2], P.M. Bandarra[1], D.L. Raymundo[1], A.L. Seitz[1], C.E.F. Cruz[1], and D. Driemeier[1]

[1]*Setor de Patologia Veterinária da Universidade Federal do Rio Grande do Sul, Porto Alegre, RS, Brazil, 91540-000;* [2]*Laboratório de Patologia Veterinária, Universidade Federal de Mato Grosso, Cuiabá, MT, Brazil, 78068-900*

## Introduction

*Sida carpinifolia* (Malvaceae) is an erect perennial scrub 30 to 70 cm tall that frequently invades humid and shady areas. Although native to tropical America, it has spread throughout the tropics and subtropics (Lorenzi 2000). The plant contains the indolizidine alkaloid swainsonine (Colodel *et al.* 2002) that is an inhibitor of the lysosomal enzyme α-mannosidase and induces the storage of mannose-containing oligosaccharides in lysosomes of several cells (Driemeier *et al.* 2000), especially neurons, hepatocytes, and acinar pancreatic cells (Agamanolis 1995; Stegelmeier *et al.* 1995). Spontaneous *S. carpinifolia* poisoning has been reported in goats (Driemeier *et al.* 2000), horses (Loretti *et al.* 2003), sheep (Seitz *et al.* 2005), and fallow deer (*Dama dama*) (Pedroso *et al.* 2009) in Rio Grande do Sul, cattle in Santa Catarina (Furlan *et al.* 2009), and goats (Godoy *et al.* 2005) and sheep (Paganini Filho e*t al.* 2008) in São Paulo state. Other swainsonine-containing plants linked to similar conditions are *Astragalus* and *Oxytropis* in the USA (Molyneaux and James 1982; James and Nielsen 1990; Stegelmeier *et al.* 1995), *Swainsona* in Australia (Colegate *et al.* 1979), *Ipomoea* in Mozambique (Balogh *et al.* 1999) and Brazil (Barbosa *et al.* 2006; Antoniassi *et al.* 2007; Armién *et al.* 2007), and *Turbina cordata* (Dantas *et al.* 2007) in Brazil.

Main clinical signs include general incoordination, ataxia, muscular tremors (especially head and neck), falls, hypermetria, difficulty in standing-up, recumbence, and abnormal behavior and posture (Driemeier *et al.* 2000; Loretti *et al.* 2003). There are no significant changes at necropsy (Colodel *et al.* 2002; Seitz *et al.* 2005; Furlan *et al.* 2009). Remarkable histological changes are swelling and multifocal cytoplasmic vacuolation in the Purkinje cells, pancreatic acinar epithelium, follicular cells in thyroid, tubules in kidneys, hepatocytes, and macrophages in lymphoid organs (Colodel *et al.* 2002; Furlan *et al.* 2009). Light and electronic microscopy are used to observe the presence of lysosomal storage abnormalities (Glew *et al.* 1985), and the detection of urinary oligosaccharides has been presented as a peripheral marker for *S. carpinifolia* exposure or poisoning (Bedin *et al.* 2009) and may serve as a useful diagnostic tool. This communication reports

©CAB International 2011. *Poisoning by Plants, Mycotoxins, and Related Toxins* (eds F. Riet-Correa, J. Pfister, A.L. Schild, and T.L. Wierenga)

retrospective cases of spontaneous poisoning by *S. carpinifolia* in herbivores in Rio Grande do Sul, Brazil.

## Materials and Methods

Data on history, clinical signs, and gross and microscopic lesions from cases occurring between 1997-2008 were taken from the records of the Setor de Patologia Veterinária da Universidade Federal do Rio Grande do Sul (SPV-UFRGS). At necropsies, fragments of tissues were collected, fixed in buffered 10% formalin, and subsequently processed under routine histology and stained by hematoxylin and eosin.

## Results

In total, 5 cattle, 14 goats, 6 sheep, and 1 horse have been affected by *S. carpinifolia* poisoning. In all cases, the history included paddocks being invaded by the plant that in most situations was the predominant vegetation. Clinical signs were very similar between species and included emaciation, incoordination, muscular tremors, and frequent falls but also included hypermetria, difficulty in standing-up, abnormal behavior and posture, recumbence, paddling, and death. There were no significant gross changes. The main microscopic finding was cytoplasmic vacuolation of multiple tissues but especially neurons, Purkinje cells in cerebellum, hepatocytes, follicular thyroid cells, renal tubular epithelial cells, and acinar pancreatic cells.

## Discussion and Conclusions

The diagnoses were based on epidemiological, clinical, and pathological findings. In most of these cases, *S. carpinifolia* was the predominant vegetation in the paddocks where animals were grazing and no supplemental feed was added to their diet. Both clinical and pathological findings were similar between animal species (Driemeier *et al.* 2000; Seitz *et al.* 2005; Furlan *et al.* 2009) and characterized by a lysosomal storage disorder. Since most cases included in this report were associated with *S. carpinifolia* infestation associated with lack of both suitable forage and additional food supplementation, the conditions related to these poisoning epidodes were linked to improper nutritional management. Since the plant is widely spread in Brazil and impacts many animals, intoxication by *S. carpinifolia* must be considered in the differential diagnosis of neurological conditions affecting herbivores.

## References

Agamolis DP (1995). The pathology lysosomal storage diseases. *Pathology Annual* 30:247-285.

Antoniassi NAB, Ferreira EV, Santos CEP, Arruda LP, Campos JLE, Nakazato L, and Colodel EM (2007). Intoxicação espontânea por *Ipomoea carnea* subsp. *fistulosa* (Convolvulaceae) em bovinos no Pantanal Matogrossense. *Pesquisa Veterinária Brasileira* 27:415-418.

Armién AG, Tokarnia CH, Peixoto PV, and Frese K (2007). Spontaneous and experimental glycoprotein storage disease of goats induced by *Ipomoea carnea* subsp *fistulosa* (Convolvulaceae). *Veterinary Pathology* 44:170-184.

Balogh KIM, Dimande AP, Van der Lugt JJ, Molyneux RJ, Naudé TW, and Welman WG (1999). A lysosomal storage disease induced by *Ipomoea carnea* in goats in Mozambique. *Journal of Veterinary Diagnostic Investigation* 11:266-273.

Barbosa RC, Riet-Correa F, Medeiros RMT, Lima EF, Barros SS, Gimeno EJ, Molyneux RJ, and Gardner DR (2006). Intoxication by *Ipomoea sericophylla* and *Ipomoea riedelli* in goats in the state of Paraíba, Northeastern Brazil. *Toxicon* 47:371-379.

Bedin M, Colodel EM, Giugliani R, Zlotowski P, Cruz CEF, and Driemeier D (2009). Urinary oligosaccharides: A peripheral marker for *Sida carpinifolia* exposure or poisoning. *Toxicon* 53:591-594.

Colegate SM, Dorling PR, and Hustable CR (1979). A spectroscopic investigation of swainsonina: an alfa-manosidase inhibitor isolated from *Swainsona canescens*. *Australian Journal Chemical* 32:2257-2264.

Colodel EM, Gardner DR, Zlotowski P, and Driemeier D (2002). Identification of Swainsonina as a glycoside inhibitor responsible for *Sida carpinifolia* poisoning. *Veterinary and Human Toxicology* 44:177-178.

Dantas AFM, Riet-Correa F, Gardner DR, Medeiros RMT, Barros SS, Anjos BL, and Lucena RB (2007). Swainsonine-induced lysosomal storage disease in goats caused by the ingestion of *Turbina cordata* in Northeastern Brazil. *Toxicon* 49:111-116.

Driemeier D, Colodel EM, Gimeno EJ, and Barros SS (2000). Lysosomal storage disease caused by *Sida carpinifolia* in goats. *Veterinary Patholoolgy* 37:153-159.

Furlan FH, Lucioli J, Veronezi LO, Medeiros A, Barros SS, Traverso SD, and Gava A (2009). Spontaneous lysosomal storage disease caused by *Sida carpinifolia* (Malvaceae) poisoning in cattle. *Veterinary Pathology* 46:343-347.

Glew RH, Basu A, Prence EM, and Remaley AT (1985). Lysosomal storage disease. *Laboratory Investigation* 53(3):250-269.

Godoy GS, Castro Netto A, Momo C, Dune ACG, Ávila LG, Alessi AC, Marques LC, and Castro MB (2005). Intoxicação natural por *Sida carpinifolia* (Malvaceae) em caprinos no estado de São Paulo. *Proceedings XII Encontro Nacional de Patologia Veterinária*, p.25. Belo Horizonte, MG.

James LF and Nielsen D (1990). Locoweeds assessment of the problem on western U.S. rangelands. In *The Ecology and Economic Impact of Poisonous Plants on Livestock Production* (LF James, MH Ralphs, and D Nielsen, eds), p. 428. Westview Press, Boulder, Colorado.

Lorenzi H (2000). *Plantas daninhas do Brasil: Terrestres, aquáticas, parasitas, tóxicas e medicinais*, 471 pp. Plantarum Ltda, Nova Odessa, São Paulo.

Loretti ALP, Colodel EM, Gimeno EJ, and Driemeier D (2003). Lysosomal storage disease in *Sida carpinifolia* toxicosis: an induced mannosidosis in horses. *Equine Veterinary Journal* 35:434-438.

Molyneaux RJ and James LF (1982). Loco intoxication: indolizidine alkaloids of spotted locoweed (*Astragalus lentiginosus*). *Science* 216:190-191.

Paganini Filho WS, Tirapelli ACN, Pereira TG, Wouters ATB, and Wouters F (2008). Intoxicação por *Sida carpinifolia* em ovinos. *Encontro Nacional de Diagnóstico Veterinário*, Campo Grande, MS.

Pedroso PMO, Von Hohendorf R, Oliveira LGS, Schmitz M, Cruz CEF, and Driemeier D (2009). *Sida carpinifolia* (Malvaceae) poisoning in fallow deer (*Dama dama*). *Journal of Zoo and Wildlife Medicine* 40:583-585.

Seitz AL, Colodel EM, Schmitz M, Gimeno EJ, and Driemeier D (2005). Use de lectin histochemistry to diagnose *Sida carpinifolia* (Malvaceae) poisoning in sheep. *Veterinary Record* 156:386-388.
Stegelmeier BL, Molyneux RJ, Elbein AD, and James LF (1995). The lesions of locoweed (*Astragalus mollissimus*), swainsonine, and castanospermine in rats. *Veterinary Pathology* 32:289-298.

# Chapter 50

# The Guinea Pig as an Animal Model for α-Mannosidosis

L.A. Cholich[1], E.J. Gimeno[2], P.G. Teibler[1], N. Jorge[1], and O.C. Acosta[1]

[1] *Cátedra de Farmacología, Facultad de Ciencias Veterinarias, Universidad Nacional del Nordeste, Sargento Cabral 2139, Corrientes 3400, Argentina;* [2] *Cátedra de Patología General, Facultad de Ciencias Veterinarias, Universidad Nacional de La Plata, Buenos Aires, Argentina*

## Introduction

α-Mannosidosis is a lysosomal storage disorder caused by deficient activity of the lysosomal α-mannosidase, commonly observed in Angus cattle and cats and occasionally in humans (Jolly and Walkley 1997; Rivero *et al.* 2001). The catabolism of glycoproteins by α-mannosidase is incomplete and the aberrant products accumulate in the cell, causing vacuolation depending upon the organs affected; the clinical signs are manifested in a variety of ways but neurological problems are commonplace and the most obvious (Molyneux *et al.* 1995; Hueza *et al.* 2005).

The ingestion of plants of the genera *Swainsona, Oxytropis, Astragalus, Sida,* and *Ipomoea* by herbivores induce an α-mannosidosis very similar to the genetic form although the disease is not identical. In the induced disease, the indolizidine alkaloid swainsonine inhibits lysosomal α-mannosidase but also Golgi mannosidase II, an enzyme associated with the processing of oligosaccharides during the glycosylation of proteins (James *et al.* 1970; Dorling *et al.* 1980; James and Panter 1989; Molyneux *et al.* 1995; de Balogh *et al.* 1999; Driemeier *et al.* 2000; Rodriguez-Armesto *et al.* 2004; Barbosa *et al.* 2006). The consequence is the accumulation of mannose-containing oligosaccharides in the lysosomes (Dorling *et al.* 1978). In addition, other isolated toxic components from *Ipomoea carnea* subsp *fistulosa* are calystegines $B_1$, $B_2$, and $C_1$ that show a potent lysosomal inhibitory activity, α and β galactosidases, and α-glucosidase (Molyneux *et al.* 1995).

Mice and rats have been used to study the effects of *Ipomoea* alkaloids (Haraguchi *et al.* 2003; Hueza *et al.* 2005; Stegelmeier e*t al.* 2008). Nevertheless, rodents are not good models because no neuronal lesions have been observed in rats (Hueza *et al.* 2005) or only with very high doses in mice (Stegelmeier *et al.* 2008). A colony of inherited α-mannosidosis guinea pigs has been established in order to evaluate different therapeutic strategies for storage disorders (Auclair and Hopwood 2007; Robinson *et al.* 2008).

Our aim was to evaluate the possible toxic effects promoted by *I. carnea* dry leaves in laboratory guinea pigs and propose a new experimental model for the reproduction of α-mannosidosis.

©CAB International 2011. *Poisoning by Plants, Mycotoxins, and Related Toxins* (eds F. Riet-Correa, J. Pfister, A.L. Schild, and T.L. Wierenga)

## Materials and Methods

Six male, 4-week-old, Hartley guinea pigs (80±19 g) were used. The food (small balls) was prepared by mixing commercial pellet material with *I. carnea* dry leaves 50:50. The treated animals (*n*=3) received the 'small balls' of 15 g each and fresh green vegetables *ad libitum*. The control group diet (*n*=3) consisted of fresh green vegetables and commercial pellets. Animals were euthanized after being anesthetized with i.p. injection of chloral hydrate (300 mg/kg) after 45 days of intoxication. Samples of brain, cerebellum, brain stem, liver, pancreas, and kidney were collected and fixed in 10% neutral buffered formalin, processed, and stained with hematoxylin and eosin (H&E) for histological examinations. Representative sections of the previously mentioned tissues were submitted to lectin histochemical procedures. Eight lectins were used, namely: Con-A (*Concanavalia ensiformis*), SBA (*Glycine max*), PNA (*Arachis hypogaea*), RCA - I (*Ricinus communis*-I), UEA-1 (*Ulex europaeus*-I), WGA (*Triticum vulgaris*), s-WGA (Succinyl-WGA), and LCA (*Lens culinary*) (Table 1).

**Table 1.** Lectins used in this study and their major specificities.

| Acronym | Lectin | Specificity | Concentration (μg/ml) |
|---|---|---|---|
| UEA-I | *Ulex europaeus-1* | α-L-Fuc | 30 |
| LCA | *Lens culinary* | α-Glc, α-D-Man | 30 |
| PNA | *Arachis hypogaea* | β-D-Gal (β 1-3) D-GalNAc | 10 |
| SBA | *Glycine maximus* | A-D-GalNAc; β-D-GalNAc | 30 |
| WGA | *Triticum vulgaris,* | ß-D-GlcNAc; NeuNAc | 30 |
| RCA-I | *Ricinus communis* | β-D-Gal and α-D-Gal | 30 |
| CON-A | *Concanavalina ensiformis* | α-D-Man; α-D-Glc | 30 |
| sWGA | Succinyl *Triticum vulgaris* | GlcNAc ß (1-4) Glc NAc | 30 |

Fuc: Fucose; Gal: Galactose;GalNAc: *N*-Acetyl galactosamine: Glc: Glucose; GlcNAc: *N*-Acetyl glucosamine; Man: mannose; NeuNAc: Acetyl neuraminic acid (sialic acid) (Goldstein and Hayes 1978).

## Results

Histopathological evaluation of H&E tissues from treated animals revealed the presence of small cytoplasmic vacuoles in hepatocytes, Kupffer cells, exocrine pancreas, and renal tubular epithelium cells. Neurons of all parts of the brain stem, mainly neurons of the pontine nuclei, were swollen and distended by vacuolation of the perikaryon.

In affected animals the exocrine pancreas and renal tubules were positive for WGA. The vacuolated neurons were stained with Con-A, WGA, s-WGA, and LCA. These cells did not stain in corresponding slides from control animals. The results of the lectin binding patterns for affected and control animals are summarized in Table 2.

## Discussion

This experimental study demonstrated that the intake of a feed containing 50% of dry leaves of *I. carnea* during 45 days induces a lysosomal storage disease in guinea pigs. Cytoplasmatic vacuolation was evident in cells of the liver, exocrine pancreas, kidney, medulla oblongata, and pons. The lesions are similar to the acquired α-mannosidosis secondary to the ingestion of plants of the genera *Swainsona, Oxytropis*, *Astragalus*, *Sida*, and *Ipomoea* by herbivores. The vacuolation of cortical neurons has also been observed in mice receiving very high doses of swainsonine (Stegelmeier *et al.* 2008). Huxtable and Dorling (1985) gave swainsonine in drinking water to rats for periods up to 200 days; they observed neuronal mannoside storage only in peripheral ganglia and in those areas of the brain not protected by the blood/brain barrier. On the other hand, a similar dose in guinea pigs leads to the development of extensive neuronal vacuolation in the central nervous system and peripheral ganglia in 4 weeks (Huxtable and Dorling 1982).

**Table 2.** Intensity of lectin binding on affected cells in organs evaluated from poisoned and normal guinea pigs.

| Cells | Lectins | | | | | | | |
|---|---|---|---|---|---|---|---|---|
| | Con A | sWGA | LCA | PNA | RCA-I | SBA | UEA-I | WGA |
| EP [a] | 3(0)[b] | 0 (0) | 3 (1) | 3 (1) | 1 (2) | 1 (2) | 1 (2) | 3 (1) |
| H | 2 (0) | 0 (0) | 2 (0) | 0 (0) | 1 (1) | 0 (0) | 1 (0) | 0 (0) |
| KC | 0 (0) | (0) | 3 (1) | 2 (1) | 1 (2) | 0 (0) | 1 (0) | 2 (0) |
| KTE | 2 (0) | 0 (0) | 2 (1) | 3 (1) | 1 (1) | 1 (1) | 0 (0) | 3 (1) |
| N | 3 (1) | 3 (0) | 3 (1) | 0 (0) | 0 (0) | 0 (0) | 0 (0) | 3 (1) |

[a] EP= Exocrine Pancreas, H=Hepatocytes, KC=Kupffer cells, KTE= Kidney tubular epithelium, N= Neurons of brain stem nuclei

[b]Numbers indicate staining intensity on a subjective estimated scale from 0 – unreactive to 3 – most reactive. Control results from normal guinea pigs are provided in parentheses.

Other studies also indicate that the central nervous system is protected against swainsonine by an extensive barrier of astrocytes and endothelial cells in mice (Bowen *et al.* 1993). Our results using dry leaves of *I. carnea* seem to indicate that the guinea pig is highly susceptible to the mixture of alkaloids present in the plant. At the present time, there is no explanation for the resistance of rats, mice, and hamsters. Additional work will be required to explain why the guinea pig, also a rodent, is susceptible.

Our study has also revealed the nature of stored material in lysosomal vacuoles using lectin histochemistry. The pattern of lectin staining observed in neurons partially agrees with the results reported for locoweed and swainsonine toxicosis and for mannosidosis in humans, cats, and calves (Alroy *et al.* 1985). In feline mannosidosis, WGA and Con-A recognized the undegraded glycoproteins and oligosaccharides stored in lysosomes of affected cells as was the case in *I. carnea* intoxication (Castagnaro 1990). The reaction was clear to sWGA, WGA which indicates the accumulation of β-D-*N*-acetyl-glucosamine and *N*-acetyl-neuraminic acid, and Con-A specific for α-D-mannose and α-D-glucose (Goldstein and Hayes 1978). The neuronal reaction to LCA is in agreement with the findings of Armién *et al.* (2007) and is a clear indication of the presence of α-glucose and α-D-mannose. This result is coincident with the staining pattern of the vacuoles found in pancreas, liver, and kidney except for sWGA. Also in *Ipomoea*-intoxicated goats, LCA reacted strongly in nervous tissues and negatively to very mild to liver, kidney, and pancreas (Armién *et al.* 2007).

The partial reaction of pancreatic, liver, and kidney cells with UEA-I, RCA-I, and SBA was observed in affected and control guinea pigs.

The exocrine acinar cells in pancreas of affected guinea pigs were positive for WGA, Con-A, PNA, and LCA which is partially in agreement with previous reports (Driemeier *et al.* 2000; Armién *et al.* 2007). The adult exocrine acinar cells of normal mice, rats, guinea pigs, and rabbits were expressed at a more or less high rate of binding of PNA, Con-A, UEA-I, RCA-I, and WGA (Muresan *et al.* 1982). The lectin binding properties revealed extensive and specific heterogeneity of the acinar cell population; this population appears homogeneous by classical light and electron microscopic preparations (Jeraldo *et al.* 1996). The partial staining obtained in our study with UEA-I, RCA-I, and SBA correspond to the lectin binding pattern of normal acinar cells in goats; this was corroborated by the lectin reactivity of control guinea pigs.

We conclude that *Ipomoea carnea* subsp. *fistulosa* induces a glycoprotein storage disease in guinea pigs, which makes it a valuable animal model. The discovery of a novel model for plant induced α-mannosidosis opens a wide array of possibilities for further studies. For instance, striking differences in susceptibility were observed when comparing the guinea pig with other rodents. This guinea pig model may be superior to using rats and mice for toxicology evaluation of individual alkaloids.

## Acknowledgements

EJG and OCA are research career members of CONICET (Consejo Nacional de Investigaciones Científicas y Técnicas). LACh is a research fellow of CONICET.

## References

Alroy J, Organ U, Ucci AA, and Grevris VE (1985). Swainsonine toxicosis mimics lectin histochemistry of mannosidosis. *Veterinary Pathology* 22:311-316.

Armién AG, Tokarnia CH, Peixoto PV, and Frese K (2007). Spontaneous and experimental glycoprotein storage disease of goats induced by *Ipomoea carnea* subsp *fistulosa* (Convolvulaceae). *Veterinary Pathology* 44:170-184.

Auclair D and Hopwood J (2007). Morphopathological features in tissues of α-mannosidosis guinea pigs at different gestational ages. *Neuropathological Applied Neurobiology* 33:572-85.

Barbosa RC, Riet-Correa F, Medeiros RMT, Lima EF, Gimeno EJ, Barros SS, Molyneux RJ, and Gardner DR (2006). Intoxication by *Ipomoea sericophylla* and *Ipomoea riedelii* in goats in the state of Paraíba, Northeastern Brazil. *Toxicon* 47:371-379.

Bowen D, Adir J, White SL, Bowen CD, Matsumoto K, and Olden K (1993). A preliminary pharmacokinetic evaluation of the antimetastatic immunomodulator swainsonine: clinical and toxic implications. *Anticancer Research* 13:841-844.

Castagnaro M (1990). Lectin histochemistry of the central nervous system in a case of feline α-mannosidosis. *Research Veterinary Science* 49:375-377.

de Balogh K, Dimande AP, Van Der Lugt JJ, Molyneux RJ, Naudé TW, and Welman WG (1999). A lysosomal storage disease induced by *Ipomoea carnea* in goats in Mozambique. *Journal Veterinary Diagnostic Investigation* 11:266-273.

Dorling PR, Huxtable CR, and Vogel P (1978). Lysosomal storage in *Swainsona* spp. toxicosis: an induced mannosidosis. *Neuropathological Applied Neurobiology* 4:285-291.

Dorling PR, Huxtable CR, and Colegate SM (1980). Inhibition of lysosomal α-mannosidase by swainsonine, an indolizidine alkaloid isolated from *Swainsona canescens*. *Biochemical Journal* 191:649-651.

Driemeier D, Colodel EM, Gimeno EJ, and Barros SS (2000). Lysosomal storage disease caused by *Sida carpinifolia* poisoning in goats. *Veterinary Pathology* 37:153-159.

Goldstein IJ and Hayes CE (1978). The lectins: carbohydrate binding proteins of plants and animals. *Advances in Carbohydrate Chemistry and Biochemistry* 35:127-340.

Haraguchi M, Gorniak SL, Ikeda K, Minami Y, Kato A, Watson AA, Nash RJ, Molyneux RJ, and Asano N (2003). Alkaloidal components in the poisonous plants, *Ipomoea carnea* (Convolvulacea). *Journal of Agricultural Food Chemistry* 51:4995-5000.

Hueza IM, Guerra JL, Haraguchi M, Naoki A, and Górniak SL (2005). The role of alkaloids in *Ipomoea carnea* toxicosis: a study in rats. *Experimental Toxicological Pathology* 57:53-58.

Huxtable CR and Dorling PR (1982). Swainsonine-induced mannosidosis. Animal model of human disease. *American Journal of Pathology* 107:124-126.

Huxtable CR and Dorling PR (1985). Mannoside storage and axonal dystrophy in sensory neurones of swainsonine-treated rats: morphogenesis of lesions. *Acta Neuropathologica* 68:65-73.

James LF and Panter KE (1989). Locoweed poisoning in livestock. In *Swainsonine and Related Glycosidase Inhibitors* (LF James, AD Elbein, RJ Molyneux, and CD Warren, eds), pp. 23-38. Iowa State University Press, Ames, Iowa.

James LF, van Kampen KR, and Hartley WJ (1970). Comparative pathology of *Astragalus* (locoweed) and *Swainsona* poisoning in sheep. *Veterinary Pathology* 7:116-125.

Jeraldo TL, Coutu JA, Verdier PA, McMillan PN, and Adelson JW (1996). Fundamental cellular heterogeneity of the exocrine pancreas. *Journal of Histochemistry and Cytochemistry* 44:215-220.

Jolly RD and Walkley SU (1997). Lysosomal storage disease of animals: an essay in comparative pathology. *Veterinary Pathology* 34:527-548.

Molyneux R, McKenzie R, and O'Sullivan B (1995). Identification of the glycosidase inhibitors swainsonine and calystegine B2 in Weir vine (*Ipomoea* sp Q6 [*aff. calobra*]) and correlation with toxicity. *Journal of Natural Products* 58:878-886.

Muresan V, Sarras MP, and Jamieson JD (1982). Distribution of acinar cells of the mammalian pancreas. *Journal of Histochemistry and Cytochemistry* 30:947-955.

Rivero R, Kautz S, Gomar MS, Barros SS, and Gimeno EJ (2001). Enfermedad de almacenamiento lisosomal en terneros del norte del Uruguay. *Veterinaria* (Montevideo) 36:5-9.

Robinson AJ, Crawley AC, Auclair D, Weston PF, Hirte C, Hemsley KM, and Hopwood JJ (2008). Behavioural characterisation of the a-mannosidosis guinea pig. *Behaviour Brain Research* 186:176–184.

Rodriguez-Armesto R, Repetto AE, Ortega HH, Peralta CJ, Pensiero JF, and Salvetti NR (2004). Intoxicación en cabras por ingestión de *Ipomoea hieronymi* var. *calchaquina* en la Provincia de Catamarca, Argentina. *Veterinaria Argentina* 21:332-341.

Stegelmeier BL, Molyneux RJ, Asano N, Watson AA, and Nash RJ (2008). The comparative pathology of the glycosidase inhibitors swainsonine, castanospermine, and calystegines A3, B2, and C1 in mice. *Toxicological Pathology* 36:651-659.

# Chapter 51

# Poisoning by *Solanum paniculatum* of Cattle in the State of Pernambuco, Northeastern Brazil

E.L.S. Guaraná[1], F. Riet-Correa[2], C.L. de Mendonça[1], R.M.T. Medeiros[2], N.A. Costa[1], and J.A.B. Afonso[1]

[1]*Clínica de Bovinos, Campus Garanhuns, Universidade Federal Rural de Pernambuco, PO Box 152, 55292-901, Garanhuns, PE, Brazil;* [2] *Hospital Veterinário, Centro de Saúde e Tecnologia Rural, Universidade Federal de Campina Grande, 58700-900, Patos, PB, Brazil*

## Introduction

Storage diseases are characterized by accumulation of indigestible metabolic products in the cells due to the deficient activity of one of a diverse number of catabolic lysosomal enzymes. They are classified as either genetic or acquired. Genetic storage diseases are denominated based on the metabolic byproduct that accumulates in the lysosomes, such as α-mannosidosis, whereas acquired genetic storage diseases result from the ingestion of plants that contain specific inhibitors of one or more catabolic lysosomal enzymes. *Astragalus*, *Oxytropis*, and *Swainsona* spp. are plants that cause storage diseases and are known as 'locoweed' – a term designated for the intoxication caused by the ingestion of these plants, which are found in western Canada, USA, northern Mexico, and Australia and affect horses, cattle, sheep, and goats (James *et al.* 1981; Smith 2006). Many lysosomal storage diseases of humans and animals affect the central nervous system with consequent neurological disorders (Ralph 1990).

In Brazil, storage diseases induced by the ingestion of plants include acquired glycoprotein lysosomal storage diseases caused by *Ipomoea carnea* subsp. *fistulosa* in goats (Armién *et al.* 2007; Oliveira *et al.* 2009), sheep (Armién *et al.* 2007), and cattle (Antoniassi *et al.* 2007), *I. sericophylla* and *I. riedelii* in goats (Barbosa *et al.* 2006), *Turbina cordata* in goats, horses, and cattle (Dantas *et al.* 2007), and *Sida carpinifolia* in goats, sheep, cattle, and horses (Driemeier *et al.* 2000), lipofuscinosis caused by *Phalaris angusta* in cattle (Gava *et al.* 1999), and neurolipidosis caused by *Solanum fastigiatum* var. *fastigiatum* in cattle (Riet Correa *et al.* 1983; Rech *et al.* 2006). Intoxication by *S. fastigiatum* occurs in Rio Grande do Sul, southern Brazil. This plant is a bush that reaches 1 m in height with broad leaves and white blossoms. It is a weed found mainly in pastures and abandoned lands and is locally known as 'joá-preto' or 'jurubeba'. Cattle have to consume large amounts of the plant to become intoxicated. Other *Solanum* species causing similar diseases are *S. kwebense* in South Africa (Pienaar *et al.* 1976), *S. dimidiatum* in the USA (Menzies *et al.* 1979), and *S. bonariense* in Uruguay (Verdes *et al.* 2006) which affect

©CAB International 2011. *Poisoning by Plants, Mycotoxins, and Related Toxins* (eds F. Riet-Correa, J. Pfister, A.L. Schild, and T.L. Wierenga)

bovines, and *S. cinereum* in Australia (Bourke 1997) and *S. viarium* in the USA (Porter *et al.* 2003) which affect goats.

The intoxication is characterized by periodic episodes of cerebellar origin, including loss of equilibrium, extension of the neck and fore limbs, hypermetria, nystagmus, opisthotonos, wide-based stance, falling to the side or backwards, and muscular tremors. The attacks occur mainly when the animals are disturbed or frightened. Between episodes, most cattle show no clinical signs but some may show permanent hypermetria, extension of the head, head tilting, or other abnormal positions of the head. The attacks can be induced by the head raising test (raising the head of the animal for about 60 s and then suddenly releasing it). The disease is chronic and most animals continue indefinitely with periodic attacks. Some cattle die as a consequence of misadventure or drowning during attacks (Riet-Correa *et al.* 1983; Rech *et al.* 2006).

Most cases show no gross alterations, but traumatic lesions can be observed (Riet-Correa *et al.* 1983; Rech *et al.* 2006). Occasionally the cerebellum is reduced in size (Rech *et al.* 2006). Histologic lesions, localized in the cerebellum, are degeneration and loss of Purkinje neurons which appeared enlarged with a clear homogeneous perikaryon, loss of Nissl substance, and fine diffuse vacuolation. Some nuclei appear with a globular aspect or pyknotic. Later these neurons disappear and are substituted by proliferation of the Bergmann glia. Numerous axonal spheroids are observed in the granular layer and white matter of the cerebellum and cerebellar peduncles. Wallerian degeneration with macrophages and vacuolation of the white matter is observed associated with the axonal spheroids (Riet-Correa *et al.* 1983; Rech *et al.* 2006).

In the first report of the disease in Rio Grande do Sul, lesions similar to those observed in human and animal gangliosidosis were observed with electron microscopy (Riet-Correa *et al.* 1983). In another report, lipidic inclusions similar to those observed in hereditary or induced lipidosis in human and animals were observed under electron microscopy in the perikaryon, axons, and dendrites of the Purkinje cells. It appears that these inclusions originate in the endoplasmic reticulum and are probably due to an interaction between the toxic compound of the plant and lipids from the affected cells forming complexes, which are not degraded (Barros *et al.* 1989).

The objective of this chapter is to report spontaneous outbreaks of poisoning by *S. paniculatum* in cattle in the state of Pernambuco and experimental reproduction of the intoxication in cattle in the states of Pernambuco, Paraíba, and Rio Grande do Sul.

## Spontaneous Poisoning

The outbreaks occurred during a 3-year period between September 2005 and December 2008 in three counties (Brejão, Pesqueira, and Poção) in the Agreste region (transition between the lush, rainy coast and the semiarid region of the deep interior) of the state of Pernambuco, affecting 2- to 5-year-old Holstein and Brown-Swiss cattle and their crosses used for milk production. Morbidity, mortality, and case fatality rates were 3 to 25%, 0 to 20%, and 0 to 60%, respectively. A sample from a plant known as *jurubeba* present in large amounts in the pastures on the farms was collected and identified as *S. paniculatum*. On inspection of the pastures the plant was growing mixed with grasses which probably favored its ingestion by cattle.

Clinical signs were periodical attacks with incoordination, neck and head extension, ataxia, hypermetria, intention tremors, nystagmus, loss of balance, and falling to the side or backwards. These episodes occurred when the animals were disturbed or frightened or were induced by the head raising test. Some animals showed permanent signs including abnormal posture, intention tremors, staggering gait with limbs in abduction, and progressive loss of weight. Biochemical and hematologic parameters were within normal ranges. The affected animals experienced progressive weight loss and some of them became prostrate and died within approximately 15 days. In other animals, according to the owners, there was a reduction in signs during the rainy season but periodic attacks still occurred.

Two cows were necropsied, one with periodic signs (Cow 1) and another with permanent signs (Cow 2). Macroscopic lesions were not observed in Cow 1, but Cow 2 with permanent cerebellar signs showed a small cerebellum with atrophy of the cerebellar folia. Histological lesions in Cow 1 consisted of fine vacuolization in the perikaryon of Purkinje cells which appeared pale and occasionally with a marginal nucleus. Loss of these neurons with substitution by Bergmann glia was also observed. Numerous axonal spheroids were observed in the granular layer of the cerebellum and in the cerebellar medulla. Axonal spheroids and vacuolated neurons were also observed in the Gracilis nucleus. Cow 2 showed severe atrophy of the molecular layer and severe loss of most Purkinje cells.

## Experimental Poisoning

The first experimental poisoning was with a *Solanum* sp. collected in Rio Grande do Sul (Barros *et al.* 1987), later identified as *S. paniculatum* (Franklin Riet-Correa, unpublished data). One calf with an initial weight of 86 kg ingested the fresh aerial parts of the green plant *ad libitum* 5 days a week. First signs were observed 260 days after ingestion and the animal was euthanized 421 days after first signs. Another calf weighing 105 kg received daily 400 g of ground dry leaves of *S. paniculatum* via a surgically implanted cannula, 5 days a week. First signs were observed 106 days after the start of plant administration and the animal was euthanized 36 days after first signs (Barros *et al.* 1987).

In another experiment with *S. paniculatum* collected in the state of Paraíba, dry and ground leaves, flowers, and fruits were dosed (5 g/kg BW daily) to two calves (#1 and 2) weighing 200 and 250 kg, respectively. Clinical signs appeared 102 (Calf 1) and 96 (Calf 2) days after the start of dosing, and they were euthanized 6 days and 1 day after first signs, respectively (Medeiros *et al.* 2004).

In a third experiment, dried and ground leaves, fruits, and flowers of *S. paniculatum* collected in the state of Pernambuco were administered daily mixed with the ration at the dose of 3 g/kg BW for a period of 150 days. After the first 3 months the dose was increased to 4 g/kg BW. First signs were observed 120 days after the start of dosing and the animal was euthanized 30 days after first signs (Guaraná *et al.* unpublished data).

Clinical signs of periodic attacks with cerebellar signs were similar to those previously reported in natural and experimental poisoning by *Solanum* spp. Gross lesions were not observed and histologic lesions of the cerebellum were vacuolization and loss of Purkinje cells, proliferation of Bergmann glia, and presence of axonal spheroids in the cerebellar granular layer and white matter (Barros *et al.* 1987; Medeiros *et al.* 2004; Guaraná *et al.* unpublished data). In the semi-thin toluidin blue-stained sections, the Purkinje cells exhibited 5-20 μm cytoplasmic vacuoles and a number of round dense bodies diffusely distributed within the perikaryon. Upon ultrastructural examination the perikarya of Purkinje cells contained numerous lipid inclusions similar to those found in the inherited neurolipidosis. These inclusions seemed to derive from the endoplasmic reticulum from which the lamellar bodies, vesiculo-membranous bodies, cytoplasmic membranous bodies, and dense bodies take their origin. Similar changes occurred in the axon cylinders and dendrites of these cells. The morphologic evidence suggests that the lipidic inclusions are the result of the formation of lipid complexes rather than a lysosomal defect such as occurs in inherited lipidosis (Barros *et al.* 1987).

## Discussion

Based on the epidemiological data, clinical signs, and histological lesions observed in spontaneous cases and in the experimental reproduction of similar clinical signs and lesions we conclude that the disease observed in the state of Pernambuco is caused by the ingestion of *S. paniculatum.* Cerebellar signs and lesions induced by *S. paniculatum* are very similar to those caused by other *Solanum* species around the world, including *S. kwebense* (Pienaar *et al.* 1976), *S. dimidiatum* (Menzies *et al.* 1979), *S. bonariense* (Verdes *et al.* 2006), *S. cinereum* (Bourke 1997), and *S. viarium* (Porter *et al.* 2003).

The toxic compound of *Solanum* spp. responsible for nervous signs is unknown but it is possible that its main toxin is an enzyme inhibitor or substance that favors the formation of lipid complexes that are resistant to metabolism (Barros *et al.* 1987). Studies of the roots of *Solanum* spp. have revealed a multiplicity of alkaloids and the presence of compounds that represent new structural types. There are reports of the isolation of a new steroidal alkaloid (paniculidine) and its nitrogenous glycoside (paniculine) from the roots of *S. paniculatum*. The roots of plants cultivated in Europe have yielded a glycoside (juribin) that through either enzymatic or acidic hydrolysis yielded 1 mole of D-glucose and an alkamine (jurubidine) which is a steroidal alkaloid with a new structural type (Schreiber 1968).

*S. paniculatum* and *S. fastigiatum*, both known as *jurubeba*, are largely used in northern Brazil and other Brazilian regions by humans as a medicinal plant for many purposes (Medeiros *et al.* 2004). The fruits are also used as a condiment and to produce a kind of wine. Due to the toxicity of these species to cattle and probably to goats, their use by humans for any food or medicinal purpose is not recommended.

## Acknowledgements

This work has been financially supported by the National Institute for Science and Technology for the Control of Toxic Plants (CNPq), grant number 573534/2008-0.

## References

Antoniassi NAB, Ferreira EV, Santos CEP, Campos JLE, Nakazato L, and Colodel EM (2007). Intoxicação espontânea por *Ipomoea carnea* subsp. *fistulosa* (Convolvulaceae) em bovinos no Pantanal Matogrossense. *Pesquisa Veterinária Brasileira* 27(10):415-418.

Armién AG, Tokarnia CH, Peixoto PV, and Frees K (2007). Spontaneous and experimental glycoprotein storage disease of goats induced by *Ipomoea carnea* subsp *fistulosa* (Convolvulaceae). *Veterinary Pathology* 44:170-184.

Barbosa RC, Riet-Correa F, Medeiros RMT, Lima EF, Barros SS, Gimeno JE, Molyneux RJ, and Gardner DR (2006). Intoxication by *Ipomoea sericophylla* and *Ipomoea riedelii* in goats in the state of Paraíba, Northeastern Brazil. *Toxicon* 47:371-379.

Barros SS, Riet-Correa F, Andujar MB, Barros CSL, Méndez MC, and Schild AL (1987). *Solanum fastigiatum* and *Solanum* sp poisoning in cattle: ultrastructural changes in the cerebellum. *Pesquisa Veterinária Brasileira* 7:1-5.

Barros CSL, Metzdorf LL, Santos MN, Barros SS, and Peixoto PV (1989). Intoxicação experimental por *Senecio brasiliensis* (Compositae) em ovinos. *Pesquisa Veterinária Brasileira* 9:55-67.

Bourke DA (1997). Cerebellar degeneration in goats grazing *Solanum cinereum* (Narrawa burr). *Australian Veterinary Journal* 75:363-365.

Dantas AFM, Riet-Correa F, Gardner DR, Medeiros RMT, Barros SS, Anjos BL, and Lucena RB (2007). Swainsonine-induced lysosomal storage disease in goats caused by the ingestion of *Turbina cordata* in Northeastern Brazil. *Toxicon* 49:11-116.

Driemeier D, Colodel EM, Gimeno EJ, and Barros SS (2000). Lysosomal storage disease caused by *Sida carpinifolia* poisoning in goats. *Veterinary Pathology* 37:153-159.

Gava A, Sousa RS, de Deus MS, Pilati C, Cristani J, Mori A, and Neves DS (1999). *Phalaris angusta* (Gramineae) como causa de enfermidade neurológica em bovinos no Estado de Santa Catarina. *Pesquisa Veterinária Brasileira* 19:35-38.

James LF, Hartley WJ, and Van Kampen KR (1981). Syndromes of *Astragalus poisoning* in livestock. *Journal American Veterinary Medical Association* 178:146-150.

Medeiros RMT, Guilherme RF, Riet-Correa F, Barbosa RC, and Lima EF (2004). Experimental poisoning by *Solanum paniculatum* (jurubeba) in cattle. *Pesquisa Veterinária Brasileira* 24 (Supl.):41.

Menzies JS, Bridges CH, and Bailey EM (1979). A neurological disease of cattle associated with *Solanum dimidiatum. Southwest Veterinary* 32:45-49.

Oliveira CA, Barbosa JD, Duarte MD, Cerqueira VD, Riet-Correa F, and Riet-Correa G. (2009). Intoxicação por *Ipomoea carnea* subsp. *fistulosa* (Convolvulaceae) em caprinos na Ilha do Marajó, Pará. *Pesquisa Veterinária Brasileira* 29(7):583-588

Pienaar JG, Kellerman TS, Basson PA, Jenkins WL, and Vahrmeijer J (1976). Maldronksiekte in cattle: a neuronopathy caused by *Solanum kwebense. Onderstepoort Journal Veterinary Research* 43:67-74.

Porter MB, Mac Kay RJ, Uhl E, Platt SR, and de Lahunta A (2003). Neurologic disease putatively associated with ingestion of *Solanum viarium* in goats. *Journal of the American Veterinary Medical Association* 223:501-504.

Ralph WS (1990). Sistema Nervoso Central. In *Patologia Veterinária Especial* (RG Thomson, ed.), pp. 579-643. Manole, São Paulo.

Rech RR, Rissi DR, Rodrigues A, Pierezan F, Piazer JVM, Kommers GD, and Barros CSL (2006). Intoxicação por *Solanum fastigiatum* (Solanaceae) em bovinos: epidemiologia sinais clínicos e morfometria das lesões cerebelares. *Pesquisa Veterinária Brasileira* 26: 183-189.

Riet-Correa F, Méndez MC, Schild AL, Summers BA, and Oliveira JA (1983). Intoxication by *Solanum fastigiatum* var. *fastigiatum* as a cause of cerebellar degeneration of cattle. *Cornell Veterinarian* 73:240-256.

Schreiber K (1968). Steroid Alkaloids: The *Solanum* group. In *The Alkaloids: Chemistry and Pharmacology: v. 10* (RHF Manske, ed.), pp. 1-82. Academic Press, New York.

Smith MO (2006). Doenças do Sistema Nervoso. In *Tratado de Medicina Interna de Grandes Animais* (BP Smith, ed.), pp. 872-1018. Manole, São Paulo.

Verdes JM, Moraña Gutiérrez F, Battes D, Fidalgo LE, and Guerrero F (2006). Cerebellar degeneration in cattle grazing *Solanum bonariense* ('naranjillo') in Western Uruguay. *Journal of Veterinary Diagnostic Investigation* 18:299-303.

# Chapter 52

# The Diagnostic Significance of Detecting *Rathayibacter toxicus* in the Rumen Contents and Feces of Sheep that may be Affected by Annual Ryegrass Toxicity

J.G. Allen and A.R. Gregory

*Animal Health Laboratories, Department of Agriculture and Food, Locked Bag No. 4, Bentley Delivery Centre WA 6983, Australia*

## Introduction

Animals develop annual ryegrass toxicity (ARGT), an often fatal neurological disease of livestock, if they eat annual ryegrass seedheads containing corynetoxins (CT) produced by the bacterium *Rathayibacter toxicus*. This is a significant disease of livestock in Western and South Australia but also occurs in South Africa (Finnie 2006).

ARGT is not a simple disease to diagnose. The clinical signs are distinctive but may be confused with those caused by some other diseases, and the pathology is not significant, consistent, or specific (Bourke 1994; Finnie 2006). A probable diagnosis of ARGT is made when there is evidence that the animals had access to toxic ryegrass and they displayed the expected clinical signs. The diagnosis is strengthened if there are supporting pathological changes and the potential toxicity of the feed is known. An ELISA for the CT has been developed (Than *et al.* 2004) but is not available for routine diagnostic testing. However, an indirect measure of the potential toxicity of feeds is provided by an ELISA for *R. toxicus* which was developed to support the export of oaten hay from Australia (Masters *et al.* 2006) and is available for diagnostic purposes in Western and South Australia. The toxicity risk associated with different levels of *R. toxicus* in pasture has been established (McKay and Riley 1993) and there is good correlation between the ELISAs for the CT and *R. toxicus* (Masters AM, unpublished data).

The certainty of diagnosis of ARGT would be improved if there was a test available to determine diagnostically significant levels of the CT in rumen contents or tissues. Since there are no such CT assays available but good associations have been demonstrated between clinical disease and levels of *R. toxicus* present in pastures (McKay and Riley 1993), it was decided to investigate if there might be diagnostically significant levels of *R. toxicus* present in the rumen contents or feces of animals that die from ARGT.

©CAB International 2011. *Poisoning by Plants, Mycotoxins, and Related Toxins* (eds F. Riet-Correa, J. Pfister, A.L. Schild, and T.L. Wierenga)

## Materials and Methods

Sheep used in two experiments evaluating the administration of hydroxypropyl β-cyclodextrin via controlled release devices (Allen *et al.*, Chapter 53, this volume) were used in the current investigation of changes in *R. toxicus* concentrations in rumen fluid and feces. The CT were administered by the inclusion of weighed amounts of toxic annual ryegrass in the ration; acute, subacute, and chronic fatal intoxications were produced.

Experiment 1: acute/subacute toxicity produced by the administration of an estimated daily dose rate of 0.04-0.25 mg CT/kg BW. Three sheep with no controlled release devices and three with one controlled release device were sampled at intervals during the experiment and at death if they died or required euthanasia because of severe clinical signs. Another five sheep (2 with no controlled release devices, 1 with one controlled release device, and 2 with two controlled release devices) were sampled at death during the experiment.

Experiment 2: chronic toxicity produced by the administration of an estimated daily dose rate of 0.04-0.08 mg CT/kg BW. The five sheep with no controlled release devices were sampled at intervals during the experiment. One sheep in this group was euthanized during the experiment but was not sampled at death.

Rumen samples were collected from live sheep by passing a 9 mm diameter plastic rumen tube with a heavy brass collection plunger attached to the end into the rumen. The collection plunger was cylindrical with a rounded and enclosed end, 15 mm diameter × 60 mm long. It had 12 rows of 7 perforations each 1 mm in diameter arranged around the surface of the cylinder. A 50 ml syringe was attached to the free end of the rumen tube to suck out rumen fluid; 40-50 ml were collected from each sheep. Fecal samples, about 40-70 ml in volume, were collected by hand from the rectum. In sheep found dead or euthanized because of severe clinical signs, rumen fluid was collected directly from the opened ventral sac of the rumen and fecal samples collected directly from the opened rectum. All rumen and fecal samples were transferred to clean 75 ml plastic containers immediately after collection. These were labeled and stored at 4°C until tested within 7 days.

All sampling of sheep was conducted approximately 24 h after they were last fed. Live sheep were always sampled in the morning before they were fed, and animals that died invariably did so overnight and were found at the first morning inspection. Similarly, sheep that required euthanasia were identified at the first morning inspection so were also sampled 24 h after last being fed.

An ELISA for *R. toxicus* in pasture and hay (Masters *et al.* 2006) was modified for use on rumen fluids and feces. The assay has a capture format and is specific for a soluble polysaccharide produced by the bacterium. It provided results in ELISA units (EU) per ml for rumen fluid or per g for feces. It was considered suitable to test this procedure in sheep receiving hydroxypropyl β-cyclodextrin because this compound is a binding agent for the CT and other hydrophobic compounds so was not expected to affect the water soluble polysaccharide measured by the *R. toxicus* assay.

## Results

### Experiment 1, acute/subacute toxicity

Rumen and fecal samples were collected on 32 occasions and rumen samples on a further three occasions. Three sheep were sampled on the day prior to the start of feeding the toxic ryegrass, 23 collections were made from sheep that were showing no clinical signs and were apparently healthy, one was made from a sheep that had convulsed on the day but survived, and eight collections were made at death from sheep that died or required euthanasia because of severe clinical signs. The relationship between the rumen and fecal ELISA results just failed to be significant ($P = 0.055$).

*Rumen fluid*

The three samples collected prior to the toxic ryegrass being fed returned results of 0 EU/ml. The ELISA results for the nine samples collected from sheep at death and the one sheep convulsing ranged from 52-2,000,000 EU/ml with a mean of 544,784 EU/ml while the results for the 23 samples collected from healthy sheep ranged from 8,400-2,940,000 EU/ml with a mean of 491,017 EU/ml. These two groups of results were not significantly different. The ELISA results for individual sheep varied greatly throughout the experiment (e.g. 43,000-2,940,000 and 52-340,000 EU/ml).

Daily intakes of toxic ryegrass in this experiment varied considerably (15-235 g/sheep) and it was not fed on 2 days. Not only did the dose rate vary during the experiment but it varied considerably between sheep on the same day (15-114 g/sheep). Since the *R. toxicus* is ingested via the toxic ryegrass seed, relationships between the amount of ryegrass seed consumed and the ELISA results were examined. There were positive significant linear relationships between the ELISA results for the rumen fluid and the amount of ryegrass ingested over the 3 days ($P = 0.032$, $R^2 = 0.144$) and 5 days ($P = 0.007$, $R^2 = 0.221$) prior to sampling. However, there was no significant relationship between the ELISA results for the rumen fluid and the amount of ryegrass ingested 1 day, and over the 10 days, prior to sampling, or the total amount of ryegrass ingested up to the time of sampling.

*Feces*

The three samples collected prior to the toxic ryegrass being fed returned results of 0 EU/g. The ELISA results for the nine samples collected from sheep at death and the one sheep convulsing ranged from 29-123,000 EU/g with a mean of 35,837 EU/g while the results for the 20 samples collected from healthy sheep ranged from 160-19,200 EU/g with a mean of 3062 EU/g. These two groups of results were significantly different ($P = 0.01$).

The ELISA results for individual sheep varied greatly throughout the experiment (e.g. 29-9,300 and 610-19,200 EU/g).

As with the rumen fluid samples relationships existed between the amounts of ryegrass seed consumed and the fecal ELISA results. There were positive significant linear relationships between the ELISA results for the fecal samples and the amount of ryegrass ingested over the 3 days ($P = 0.011$, $R^2 = 0.217$) and 5 days ($P = 0.001$, $R^2 = 0.333$) prior to sampling. However, there was no significant relationship between the ELISA results for the feces and the amount of ryegrass ingested 1 day, and over the 10 days, prior to sampling, or the total amount of ryegrass ingested up to the time of sampling.

**Experiment 2, chronic toxicity**

Rumen and fecal samples were collected on 117 occasions and rumen samples on a further 5 occasions. Five sheep were sampled on the day prior to the start of feeding the toxic ryegrass and 116 collections were made from sheep that were showing no clinical signs and were apparently healthy. One sheep was sampled on a day (day 120) that it convulsed. It continued to display clinical signs over the next three days and was then euthanized due to the severity of the clinical signs. No collection was made after it was euthanized. As in Experiment 1, the relationship between the rumen and fecal ELISA results just failed to be significant ($P = 0.056$).

*Rumen fluid*

The five samples collected prior to the toxic ryegrass being fed returned results of 0 EU/ml. The ELISA result from the single animal that was convulsing and then euthanized 3 days later was 410 EU/ml. The other 116 rumen fluid ELISA results from apparently healthy animals ranged from 8-3,444,000 EU/ml with a mean of 162,418 EU/ml. The ELISA results in individual sheep varied greatly throughout the experiment (e.g. 220-420,000, 307-2,358,000, 8-1,420,000, 410-1,590,000, and 54-3,444,000 EU/ml).

The intake of toxic ryegrass was varied during this experiment (31-100 g/sheep/day) but not to the extent it was in Experiment 1, and it was varied uniformly for the whole group so the daily doses to individual sheep on the same day did not vary greatly (4-12 g). In this experiment there were positive significant linear relationships between the ELISA results for the rumen fluid samples and the amount of ryegrass ingested on the day before sampling ($P = 0.0003$, $R^2 = 0.107$), and over the 5 days ($P = 0.0002$, $R^2 = 0.115$) and 10 days ($P = 0.0002$, $R^2 = 0.121$) prior to sampling.

*Feces*

The five samples collected prior to the toxic ryegrass being fed returned results of 0 EU/g. The ELISA result for the single animal that was convulsing and then euthanized 3 days later was 240 EU/g. The other 111 fecal ELISA results from apparently healthy animals ranged from 0-80,000 EU/g with a mean of 1218 EU/g. The ELISA results in individual sheep varied greatly throughout the experiment (e.g. 0-80,000, 23-1500, 3-2300, 22-3000, and 17-790 EU/g).

Unlike with the rumen fluid samples there were no significant relationships between the ELISA results for the fecal samples and the amount of ryegrass ingested over various periods prior to sampling.

## Discussion

This study established that the ELISA for detection of *R. toxicus* in hay and pasture (Masters *et al.* 2006) can be effectively used to detect the bacterium in the rumen contents or feces of sheep. Furthermore, detection of the bacterium provides good evidence that the animal has been consuming toxic ryegrass. This knowledge can be useful when ARGT is suspected in an animal that is believed not to have had access to toxic ryegrass. This situation may occur in a feedlot or if the disease is diagnosed in an area outside those in which ARGT is endemic. However, in Western Australia the causative agents of ARGT are so widely distributed through the main agricultural area (Roberts *et al.* 1994) that a significant proportion of sheep and cattle will return a positive test for *R. toxicus* in their rumen contents and feces.

Unfortunately, this study did not demonstrate the existence of a diagnostically significant level of *R. toxicus* in the rumen contents that might be used to make a definitive diagnosis of ARGT. The rumen fluid ELISA results for affected animals in the two experiments ranged from 52-2,000,000 EU/ml, a range that was no different to that found in apparently healthy animals of 8-3,444,000 EU/ml. In fact, the three greatest ELISA results were recorded in healthy sheep. One (2,940,000 EU/ml) was in a sheep in Experiment 1 that remained healthy for another 22 days before it developed severe clinical signs and required euthanasia. The other two (2,358,000 and 3,444,000 EU/ml) were in sheep in Experiment 2 that were still healthy after consuming toxic ryegrass for 120 days and remained healthy to the end of the experiment 23 days later. The CT are cumulative toxins (Jago and Culvenor 1987) and clinical signs do not manifest until almost a lethal dose has been ingested. Therefore, it is not surprising that large rumen fluid ELISA results may be found in animals that have shown no clinical signs of ARGT.

In contrast to the rumen fluid ELISA results, it is possible that a diagnostically significant fecal ELISA result exists. In Experiment 1 the mean fecal ELISA result for affected sheep was significantly ($P$ = 0.01) greater than the mean for unaffected sheep. Similarly, when all the results from both experiments were combined the mean fecal ELISA result for affected sheep was still significantly ($P$ = 0.007) greater than that for the unaffected sheep. There was overlap of the ELISA results, the range for affected sheep in both experiments being 29-123,000 EU/g and for unaffected sheep 0-80,000, but closer examination of the data reveals some important differences. In Experiment 1 the greatest ELISA result in the unaffected sheep was 19,200 EU/g while in Experiment 2 the greatest result was 80,000 EU/g, but the other 110 results from unaffected sheep in Experiment 2 were 12,200 EU/g or less. If this one result of 80,000 EU/g is considered an anomaly, then three of the dead sheep had ELISA results greater than the highest result for the healthy sheep (67,000, 121,000, and 123,000 EU/g). Adopting a two-fold safety factor it may be concluded on the basis of these results that a fecal ELISA result of ≥40,000 EU/g is highly indicative that ARGT was the cause of death. The results obtained show that many sheep that die from ARGT will have fecal ELISA results less than 40,000 EU/g, but if it is greater this will provide considerable support for a diagnosis of ARGT.

In both experiments some rather loose associations were demonstrated between the intake of toxic ryegrass and the ELISA results obtained in the rumen contents and feces. This is not surprising because the *R. toxicus* is in the toxic ryegrass but the associations were not consistent. In Experiment 1 associations existed between both the rumen fluid and fecal ELISA results and the ryegrass intakes over 3 and 5 days before sampling. However, there were no associations between rumen fluid and fecal ELISA results and the ryegrass intakes 1 day and over 10 days before sampling. In contrast to this, in Experiment 2 the rumen fluid ELISA results were associated with the ryegrass intakes 1 day and over the 5 and 10 days before sampling, and there were no associations between the fecal ELISA results and the ryegrass intakes. These differences could not be explained simply by the different ryegrass intakes in the two experiments, and is yet another example of the complexity of this disease.

## References

Bourke CA (1994). Tunicaminyluracil toxicity, an emerging problem in livestock fed grass or cereal products. In *Plant-Associated Toxins: Agricultural, Phytochemical and Ecological Aspects* (SM Colegate and PR Dorling, eds), pp. 399-404. CAB International, Wallingford, UK.

Finnie JW (2006). Review of corynetoxins poisoning of livestock, a neurological disorder produced by a nematode-bacterium complex. *Australian Veterinary Journal* 84:271-277.

Jago MV and Culvenor CC (1987). Tunicamycin and corynetoxin poisoning in sheep. *Australian Veterinary Journal* 64:232-235.

Masters AM, Gregory AR, Evans RJ, Speijers JE, and Sutherland SS (2006). An enzyme-linked immunosorbent assay for the detection of *Rathayibacter toxicus*, the bacterium involved in annual ryegrass toxicity, in hay. *Australian Journal of Agricultural Research* 57:731-742.

McKay AC and Riley IT (1993). Sampling ryegrass to assess the risk of annual ryegrass toxicity. *Australian Veterinary Journal* 70:241-243.

Roberts WD, Mlodawski G, Macdonagh A, Gibson R, and Bucat J (1994). The distribution of annual ryegrass toxicity in Western Australia. In *Plant-associated Toxins - Agricultural, Phytochemical and Ecological Aspects* (SM Colegate and PR Dorling, eds), pp. 51-56. CAB International, Wallingford, UK.

Than KA, Cao Y, Michalewicz A, Olsen V, Anderton N, Cockrum P, Colegate SM, and Edgar JA (2004). Analysis of corynetoxins: a comparative study of an indirect competitive ELISA and HPLC. In *Poisonous Plants and Related Toxins* (T Acamovic, CS Stewart, and TW Pennycott, eds), pp. 402-407. CAB International, Wallingford, UK.

# Chapter 53

# Annual Ryegrass Toxicity in Sheep is Not Prevented by Administration of Cyclodextrin via Controlled Release Devices

J.G. Allen[1], P.J. Martin[2], and A. Shiraishi[3]

*[1]Animal Health Laboratories, Department of Agriculture and Food, Locked Bag No. 4, Bentley Delivery Centre WA 6983, Australia; [2] Virbac (Australia) Pty Ltd, 361 Horsley Road, Milperra NSW 2214, Australia, currently PJM Scientific Pty Ltd, PO Box 723, Five Dock NSW 2046, Australia; [3] Argenta Manufacturing Ltd, 2 Sterling Avenue, Manurewa, Auckland, New Zealand*

## Introduction

Annual ryegrass toxicity (ARGT) is a major disease of livestock in Western Australia. It is caused when livestock eat ryegrass infected with the toxigenic bacterium *Rathayibacter toxicus*. The bacterium is introduced into the ryegrass by the nematode *Anguina funesta* that reproduces in galls formed within the seedhead of the ryegrass. If the bacterium is introduced into the nematode gall it may multiply to kill the nematodes and form a toxic bacterial gall. The toxins produced by *R. toxicus* are called corynetoxins (CT) (Allen 2004).

There is no satisfactory treatment for ARGT (Stewart *et al.* 1998; Allen 2004). However, Stewart *et al.* (1998) provided some hope when they reported the successful treatment of affected animals with intraperitoneal injections of hydroxypropyl β-cyclodextrin (HPβ-CD). Field trials with this treatment increased the survival rate in 7 out of 9 outbreaks of ARGT when it was administered soon after clinical signs appeared.

The cyclodextrins are water soluble cyclic oligosaccharides that form host-guest complexes with hydrophobic molecules, changing their physical and biochemical properties and often increasing their water solubility. Stewart *et al.* (1998) reported this phenomenon in laboratory studies with cyclodextrin and tunicamycin, a closely related compound to the CT (Edgar *et al.* 1982). They proposed that in poisoned sheep cyclodextrin circulating in blood formed strong complexes with the CT, reducing their toxic effects and increasing their water solubility to facilitate excretion.

Subsequent evaluations under practical field conditions of the HPβ-CD treatment of sheep with clinical ARGT did not result in similar successes to those reported by Stewart *et al.* (1998) (Allen and Bywater, unpublished). Unavoidable delays in the commencement of treatment and the need for multiple administrations of HPβ-CD, often over several days, no doubt contributed to the lack of success with the treatment.

©CAB International 2011. *Poisoning by Plants, Mycotoxins, and Related Toxins* (eds F. Riet-Correa, J. Pfister, A.L. Schild, and T.L. Wierenga)

It was proposed that if cyclodextrin could be made available before and during ingestion of the CT, it may complex with the toxins in the rumen before they are absorbed and prove to be an effective prophylactic treatment against ARGT. Administration of the cyclodextrin by way of a controlled release device (CRD) would enable continual release of the binding agent in the rumen and achieve the desired prophylactic treatment regime. We describe here experiments conducted to evaluate the efficacy of cyclodextrin in preventing ARGT when administered by way of a CRD. Different dose rates of cyclodextrin were achieved by using either one or two CRDs in a sheep.

## Materials and Methods

Two experiments using Merino wether weaners (32.6-36.7 kg) individually penned in an animal house were conducted in 2000 with the approval of the Experimentation Ethics Committee of the Department of Agriculture and Food. Each experiment had the same three treatment groups of five animals: (i) no CRD (controls); (ii) one CRD; and (iii) two CRDs. Experiment 1 evaluated acute/subacute toxicity over 29 days of exposure to CT and Experiment 2 chronic toxicity over 143 days.

The CT were administered by the inclusion of toxic ryegrass seed in the daily ration of the sheep. The toxic ryegrass was harvested on a property near Wongan Hills in Western Australia in December 1998 and stored dry in large bales until used. The concentration of CT in the harvested material in each bale was determined by HPLC (Cockrum and Edgar 1985). A single bale of toxic ryegrass seed was used for Experiment 1 and the first 108 days of Experiment 2. A second bale was used for the remainder of Experiment 2. Experiment 1 commenced with an estimated daily CT dose rate of 0.25 mg/kg BW and varied between 0.04 and 0.25 mg/kg during the experiment, except that on days 19 and 26 no toxic ryegrass was administered. Experiment 2 commenced with an estimated daily CT dose rate of 0.06 mg/kg BW and varied between 0.04 and 0.08 mg/kg during the experiment.

The sheep were fed 300-400 g of commercial sheep cubes, 200-350 g of oaten chaff, and 0-50 g of lupins together with the required amount of toxic ryegrass each day. Daily feed intake and the daily intake of toxic ryegrass were measured and when appetites declined the toxic ryegrass was milled and dosed by stomach tube in an aqueous slurry. At regular intervals the sheep were weighed and blood samples collected. The plasma activity of glutamate dehydrogenase (GLDH) was determined in all blood samples. In addition to causing pathological changes in the brain (Berry *et al.* 1980), the CT cause damage in the liver (Berry *et al.* 1982). Change in the plasma activity of GLDH has been used successfully to monitor the development of toxicity in ARGT (Davies *et al.* 1995).

Sheep were observed regularly and the endpoint for individual sheep was when they died suddenly or developed severe clinical signs and were euthanized. It was decided to end Experiment 1 when 80% of sheep in all treatments had died or been euthanized and Experiment 2 when 40% of the sheep in the two CRD treatment required euthanasia.

### Controlled release device and cyclodextrin specifications

The technology for the CRD was patented in 1974 (Laby 1974) and subsequently developed into commercial products by Captec (NZ) Ltd, now Argenta Manufacturing Ltd. The HPβ-CD was prepared in tablets by a dry manufacturing process and formulated to be 85% w/w with excipients that controlled its dissolution in the rumen. The tablets were inserted into the CRD and release was controlled by a spring and plunger designed to deliver a mean of 162 mg HPβ-CD/day (range 153-177 mg/day) over 62 days (range 56-65 days). The CRDs were initially administered 3 days before toxic ryegrass was added to the ration and in the chronic toxicity experiment additional CRDs were administered at 61 and 63 day intervals. All CRDs were weighed before administration, then retrieved, dried, and weighed again at the end of the experiments.

## Results

### Experiment 1: acute/subacute toxicity

Deaths during the course of Experiment 1 made comparisons of feed intakes and live weights difficult. In general, feed intake was variable over the course of the experiment although a noticeable decline was noted in individual sheep 1 or 2 days before they died. All sheep became selective in their intake towards day 14, so from day 15 the toxic ryegrass was administered by stomach tube. In general, all sheep lost some weight during the experiment. The treatments did not appear to influence the feed intakes or the live weight changes.

There were no significant differences between the average total CT dose rates and daily dose rates of HPβ-CD for each of the treatments (Table 1). The range in the total CT dose rates for each of the sheep that died was 1.26-3.84 mg/kg BW and for those sheep that survived 3.77-4.73 mg/kg BW. The average rate of release of HPβ-CD from the CRDs was 127 mg/day (range 97-187 mg/day). The rate of release measured was significantly ($P$ = 0.011) associated with how long the CRD remained in the sheep, with the rate being greater the longer the CRD was in place.

**Table 1.** Average total CT dose rates and daily HPβ-CD dose rates (in Experiment 2, only for period after day 121) in each of the treatments in Experiments 1 and 2.

| Experiment | Treatment | Total CT dose rate* (mg/kg live weight) | Daily HPβ-CD dose rate (mg/day) |
|---|---|---|---|
| 1 | 0 CRD | 2.60 (1.31-4.73) | |
| | 1 CRD | 3.21 (1.58-4.60) | 130 (113-146) |
| | 2 CRDs | 2.91 (1.26-4.37) | 252 (195-328) |
| 2 | 0 CRD | 8.17 (7.16-8.43) | |
| | 1 CRD | 8.43 | 172 152-180) |
| | 2 CRDs | 8.43 | 345 (326-357) |

* Figures in parentheses depict the range.

The average plasma GLDH activities were elevated above the normal reference range in all treatments by day 7 and increased over the remainder of the experiment. There were no significant differences between the treatments until the last day of the experiment when sheep with no CRD that were still alive had a significantly ($P < 0.05$) lower average plasma GLDH activity than the sheep with one or two CRDs that were still alive.

Table 2 shows the clinical signs observed and when they occurred together with when animals were found dead or euthanized.

**Table 2.** Clinical signs observed, when they occurred, and when sheep died or were euthanized in Experiment 1.

| Treat. | Sheep | Day of experiment | | | | | | | | |
|---|---|---|---|---|---|---|---|---|---|---|
| | | 6 | 7 | 9 | 10 | 11 | 21 | 27 | 28 | 29 |
| 0 | 1 | De | ------ | ------ | ------ | ------ | ------ | ------ | ------ | ------ |
| CRD | 2 | C | C, K | ------ | ------ | ------ | ------ | ------ | ------ | ------ |
| | 3 | | | C | De | ------ | ------ | ------ | ------ | ------ |
| | 4 | | | | | | | | | G, K |
| | 5 | | | | | | | | | S |
| 1 | 6 | | De | ------ | ------ | ------ | ------ | ------ | ------ | ------ |
| CRD | 7 | | C | | | De | ------ | ------ | ------ | ------ |
| | 8 | | | | | | C | De | ------ | ------ |
| | 9 | | | | | | | C | | C, K |
| | 10 | | | | | | | | | S |
| 2 | 11 | De | ------ | ------ | ------ | ------ | ------ | ------ | ------ | ------ |
| CRD | 12 | C | C, K | ------ | ------ | ------ | ------ | ------ | ------ | ------ |
| | 13 | | | | | | | | De | ------ |
| | 14 | | | | | | | C | | G, K |
| | 15 | | | | | | | | | S |

De = found dead; C = characteristic episodic convulsions; G = ataxia and/or characteristic 'rocking horse' gait; K = euthanized; S = survived and did not show clinical signs.

**Experiment 2: chronic toxicity**

The feed intakes were generally maintained until day 116 after which time they declined in all treatments. The average daily intakes for each treatment were not significantly different at any time during the experiment. The average live weights for the treatments changed in a similar manner to the feed intakes. All sheep generally gained weight until day 114 after which time they lost weight. At no time during the experiment were the average live weights for the treatments significantly different.

There were no significant differences between the average total CT dose rates and daily dose rates of HPβ-CD for each of the treatments (Table 1). The total CT dose rate for the sheep euthanized on day 123 was 7.16 mg/kg BW and for the remaining sheep that survived to day 143 it was 8.43 mg/kg BW. The CRDs administered 3 days before the start of feeding toxic ryegrass and on day 58 of the experiment were completely empty when retrieved on day 143. Release rates of the HPβ-CD could only be determined for those CRDs administered on day 121.

The average plasma GLDH activities were elevated above the normal reference range in all treatments by day 29 and remained at similar activities until day 109. They then increased four to six fold by day 133 before decreasing by about a third by day 143. At no time during the experiment were the average plasma activities of GLDH for the treatments significantly different.

The decline in feed intakes after day 116, the loss of weight after day 114, and the increase in plasma GLDH activities after day 109 all followed the start of feeding toxic ryegrass from the second bale on day 109.

Clinical signs consistent with ARGT were observed in only three sheep. One required euthanasia on day 123 and the other two on day 143. These observations are summarized in Table 3.

**Table 3.** Clinical signs observed, when they occurred, and when sheep were euthanized in Experiment 2.

| Treat. | Sheep | Day of experiment | | | | | | | | |
|---|---|---|---|---|---|---|---|---|---|---|
| | | 120 | 121 | 122 | 123 | 133 | 138 | 139 | 140 | 143 |
| 0 | 16 | C | T | T | C, K | ------ | ------ | ------ | ------ | ------ |
| CRD | 17 | | | | | | | | | S |
| | 18 | | | | | | | | | S |
| | 19 | | | | | | | | | S |
| | 20 | | | | | | | | | S |
| 1 | 21 | | | | | | | | | S |
| CRD | 22 | | | | | | | | | S |
| | 23 | | | | | | | | | S |
| | 24 | | | | | | | | | S |
| | 25 | | | | | | | | | S |
| 2 | 26 | | | | | | Do,Dp | Do,Dp | G,Dp | G, K |
| CRD | 27 | | | | | C | | | C | G, K |
| | 28 | | | | | | | | | S |
| | 29 | | | | | | | | | S |
| | 30 | | | | | | | | | S |

De = found dead; C = characteristic episodic convulsions; T = general body tremors, but particularly affecting the face; Do = down and reluctant to stand; Dp = depressed; G = ataxia and/or characteristic 'rocking horse' gait; K = euthanized; S = survived and did not show clinical signs.

## Discussion

The results of these experiments offer no evidence that HPβ-CD delivered via CRDs provided any protection against ARGT under the particular CT challenges investigated.

In Experiment 1, 80% of the sheep in each treatment died or required euthanasia. Furthermore, the plasma GLDH activities in sheep still alive on day 29 indicated that the sheep with no CRD were the least affected.

In Experiment 2, 20% of the sheep with no CRD required euthanasia during the experiment and 40% of the sheep with two CRDs developed severe clinical signs of ARGT and required euthanasia on day 143. Plasma activities of GLDH throughout the experiment indicated that sheep in all treatments responded to the toxicity in a similar manner.

In Experiment 1, seven sheep died or required euthanasia when they had received an estimated total CT dose rate of less than 2 mg/kg BW (range 1.26-1.68 mg/kg). This is well below the reported lethal dose of 3.2-5.6 mg CT/kg BW for CT delivered in the manner used in this experiment (Jago and Culvenor 1987; Davies *et al.* 1995). Subsequently in Experiment 2, one sheep required to be euthanized after receiving 7.16 mg CT/kg and 12 of the sheep consumed 8.43 mg CT/kg without exhibiting any clinical signs.

The reason for these divergent results was probably caused by heterogeneity in the concentration of the CT within the bales of toxic ryegrass seed and this may have related to the harvesting technique. Nevertheless, the CT intakes of the sheep in these experiments did result in clinical ARGT that was acute (6-7 days), subacute (11-29 days), and chronic (120-143 days) in nature thus providing a range of rates of CT exposure under which to evaluate the HPβ-CD preventative treatment.

The lack of efficacy of HPβ-CD delivered via a CRD is perhaps not surprising because only 5% of ingested CT are absorbed from the gut (Stuart *et al.* 1994). For an oral delivery of HPβ-CD to be effective, almost all of the toxins ingested would need to be bound to reduce the risk of toxicity.

While these studies failed to demonstrate that HPβ-CD administered via CRDs was an effective prophylactic treatment for ARGT, they did once again highlight the complexity of this disease and the difficulty in protecting sheep grazing toxic ryegrass pastures.

## Acknowledgements

Dr Steve Colegate arranged for the CSIRO Plant Toxins Research Group to conduct the CT assays on the harvested toxic ryegrass and Dr John Edgar from the same Group generously provided his time to discuss both the planning and progress of the project. Captec (NZ) Ltd generously manufactured the controlled release devices. This research was funded by Virbac (Australia) Pty. Ltd.

## References

Allen JG (2004). Annual ryegrass toxicity. In *Clinical Veterinary Toxicology* (KH Plumlee, ed.), pp. 422-424. Mosby, St Louis, Missouri.

Berry PH, Howell JMcC, and Cook RD (1980). Morphological changes in the central nervous system of sheep affected with experimental annual ryegrass (*Lolium rigidum*) toxicity. *Journal of Comparative Pathology* 90:603-617.

Berry PH, Richards RB, Howell JMcC, and Cook RD (1982). Hepatic damage in sheep fed annual ryegrass, *Lolium rigidum*, parasitized by *Anguina funesta* and *Corynebacterium rathayi*. *Research in Veterinary Science* 32:148-156.

Cockrum PA and Edgar JA (1985). Rapid estimation of corynetoxins in bacterial galls from annual ryegrass (*Lolium rigidum* Gaudin) by high-performance liquid chromatography. *Australian Journal of Agricultural Research* 36:35-41.

Davies SC, White CL, and Williams IH (1995). Increased tolerance to annual ryegrass toxicity in sheep given a supplement of cobalt. *Australian Veterinary Journal* 72:221-224.

Edgar JA, Frahn JL, Cockrum PA, Anderton N, Jago MV, Culvenor CCJ, Jones AJ, Murray K, and Shaw KJ (1982). Corynetoxins, causative agents of annual ryegrass toxicity; their identification as tunicamycin group antibodies. *Journal of the Chemistry Society Chemical Communication* 4:222-224.

Jago MV and Culvenor CC (1987). Tunicamycin and corynetoxin poisoning in sheep. *Australian Veterinary Journal* 64:232-235.

Laby RH (1974). Device for administration to ruminants. U.S. Patent No. 3,844,285.

Stewart PL, May C, and Edgar JA (1998). Protective effects of cyclodextrins on tunicaminyluracil toxicity. In *Plant Toxins and Other Natural Toxicants*. (T Garland and AC Barr, eds), pp. 179-184. CAB International, Wallingford, UK.

Stuart PL, Than KA, and Edgar JA (1994). New approaches to studying the fate of corynetoxins in whole animals and their products. In *Plant-Associated Toxins: Agricultural, Phytochemical and Ecological Aspect* (SM Colegate and PR Dorling, eds), pp. 143-148. CAB International, Wallingford, UK.

# Chapter 54

# Secondary Toxicity from the Ingestion of Meat, Offal, or Milk from Animals Consuming Corynetoxins is Unlikely

J.G. Allen[1] and B.P. Mullan[2]

[1]*Animal Health Laboratories, Department of Agriculture and Food, Locked Bag No. 4, Bentley Delivery Centre WA 6983, Australia;* [2]*WA Pork Research, Development and Technology Transfer, Department of Agriculture and Food, Locked Bag No. 4, Bentley Delivery Centre, WA 6983, Australia*

## Introduction

Annual ryegrass toxicity (ARGT) is an often fatal disease of livestock caused by the ingestion of pasture, fodder, or grain contaminated with toxic annual ryegrass (*Lolium rigidum*). The ryegrass seedheads are rendered toxic by a bacterium, *Rathayibacter toxicus*, which is carried into galls within the seedhead by the nematode *Anguina funesta*. The bacterium produces very toxic compounds called corynetoxins (CT) which belong to a group of compounds called tunicaminyluracil (TMU) antibiotics (Allen 2004).

ARGT occurs over large areas in Western Australia (Roberts *et al.* 1994) and South Australia (McKay *et al.* 1985), and contamination by toxic ryegrass of fodder and grain produced in these areas occurs. There is real potential for the CT to enter the food chain of other animals via meat, offal, and milk derived from livestock consuming the CT.

One previous study has investigated the potential for residual toxicity to be present in the meat of livestock that died from ARGT (Bourke and Carrigan 1993). They reported no effects in pigs fed potentially contaminated meat. However, in that study, the closely related TMU antibiotic tunicamycin (Edgar *et al.* 1982) was used as the toxin source instead of CT, the potentially contaminated meat was fed to pigs for only 40 days, and liveweight change and the non-appearance of any clinical signs were the only assessments of toxicity that were made. A more thorough investigation of residual toxicity in the products of livestock with ARGT would use CT as the toxin source and, since these compounds are cumulative toxins (Jago and Culvenor 1987), a feeding period of greater than 40 days. Also, since the clinical signs of ARGT only become apparent when close to the lethal dose has been consumed (Jago and Culvenor 1987), parameters that provide a far more sensitive indication of an effect of CT intake would be measured. The most sensitive indicator of exposure to CT available is inhibition of the microsomal enzyme uridine diphospho *N*-acetylglucosamine:dolichyl phosphate *N*-acetylglucosamine-1-phosphate transferase (GlcNAc-1-P transferase) (Stewart and May 1994).

©CAB International 2011. *Poisoning by Plants, Mycotoxins, and Related Toxins* (eds F. Riet-Correa, J. Pfister, A.L. Schild, and T.L. Wierenga)

The work reported here was designed to maximize the chances of determining whether meat, offal, or milk from animals consuming CT presented a risk of causing secondary toxicity. Clinical ARGT was produced in cattle by feeding them toxic ryegrass, and meat, offal, and milk were collected from these animals. These products were then fed to pigs and rats for 117 and 196 days, respectively, before these animals were euthanized and liver microsomal GlcNAc-1-P transferase activities determined. Pigs are considered to have a similar susceptibility to the CT as sheep and cattle (Bourke and Carrigan 1993) and adult female rats are about 15-fold more resistant to the CT than sheep (Peterson *et al.* 1996).

## Materials and Methods

These experiments were conducted in 1999 and 2000 with the approval of the Experimentation Ethics Committee of the Department of Agriculture and Food.

Fourteen Angus cross yearling steers (387-452 kg liveweight) were placed into individual pens in a feedlot and fed *ad lib* a typical ration based on grain and hay. The grain and hay was sourced from outside the area in which ARGT is endemic so was considered to present minimal risk of being contaminated with CT. Eight of the steers were exposed to CT by adding toxic ryegrass to their ration for 11 consecutive days. The first steer became clinically affected on the 11th day and all eight were affected by the 13th day. As soon as each steer exhibited clinical signs it was removed to an abattoir where it was slaughtered and all the meat was removed from the carcass and packed into cardboard meat cartons. Livers and kidneys were also removed and packed into separate cartons. All cartons were clearly labeled either 'contaminated meat' or 'contaminated offal' and frozen until required.

The six steers not fed toxic ryegrass were then slaughtered and meat, livers, and kidneys collected and packed into cartons labeled 'uncontaminated meat' and 'uncontaminated offal' and frozen until required.

A mature Friesian cow (539 kg liveweight) being milked in a dairy herd was kept in a small paddock near the dairy and fed toxic ryegrass in its post-milking supplement for each day for 32 days. It then developed clinical signs of ARGT, was immediately removed to the abattoir, and slaughtered. As with the poisoned steers all the meat, liver, and kidneys were collected, packed in cardboard meat cartons labeled 'contaminated meat' and 'contaminated offal', and frozen until required. During the period of poisoning the cow was milked separately and its milk was placed into 2 l plastic bottles that were dated and labeled 'contaminated milk' and then frozen. Only the milk collected in the 48 h prior to slaughter was kept for use in the subsequent experiment. A similar volume of milk was collected at a single milking of the remainder of the herd, labeled 'uncontaminated milk', and frozen.

### Pig experiment – feeding meat

Sixteen newly weaned (4.3-5.3 kg liveweight) Large White × Landrace pigs were individually penned and randomly allocated in equal numbers to one of four treatments: (i) 50% uncontaminated meat; (ii) 50% contaminated meat; (iii) 25% contaminated meat; and (iv) 12% contaminated meat.

The diets fed were based on normal commercial rations and were fed as a mash. From weaning to about 20 kg liveweight a weaner ration was fed, from 20 to 50 kg liveweight a grower ration was fed, and from 50 kg until slaughter at approximately 105 kg a finisher ration was fed. All pigs were fed only the commercial weaner ration for the first 4 days, then over the next 4 days the meat was gradually added to the ration so that by day 8 all pigs were on the full designed meat ration. The amount fed each day was slightly more than had been consumed the previous day so that there was usually a small residue.

Sufficient meat for 2 weeks of feeding was thawed at a time and passed twice through a mincer with a 0.6 cm die. The minced meat was kept refrigerated until added to the commercial ration on the day of feeding in the required proportion by actual weight. So for the 50% meat treatments, equal weights of minced meat and ration mash were mixed for feeding. There was no adjustment for dry weights. This resulted in the two 50% meat treatments being isonitrogenous and isoenergetic, but the other two meat treatments had different nitrogen and energy concentrations.

Daily feed intakes (all residues discarded) were measured and the pigs weighed weekly. All pigs reached the finish weight at the same time and all were slaughtered on day 117. In the abattoir a sample of the right central lobe of the liver was collected from each pig as soon after slaughter as possible. It was immediately placed in a screw-topped 40 ml collection vial and frozen in dry ice. It was then stored at -80°C until assayed for liver microsomal GlcNAc-1-P transferase activity (Stewart 1998).

**Rat experiment – feeding meat and offal and providing milk instead of water**

Forty-two, 5-week-old female Wistar rats (134-174 g) were randomly allocated to one of eight treatments: (i) 50% uncontaminated meat; (ii) 50% contaminated meat; (iii) 25% contaminated meat; (iv) 50% uncontaminated offal; (v) 50% contaminated offal; (vi) 25% contaminated offal; (vi) 100% uncontaminated milk; and (viii) 100% contaminated milk. There were five rats in each treatment except the 50% uncontaminated and 50% contaminated offal treatments in which there were six rats. All rats in a treatment were kept together in a single box and the experiment was conducted in an air-conditioned animal house.

As with the pig experiment the diets fed were based on a commercial ration and the components of each ration were mixed in the required proportions by actual weights. Sufficient meat and offal for 3 months of feeding was thawed at a time, minced, combined in the correct proportions, and then prepared as pellets. To avoid any possible loss of CT from the contaminated meat and offal, the maximum temperature used in the pelleting process was 50°C and forced airflow was used to dry the pellets. The resultant pellets were not completely dry so were stored frozen until a week before feeding when they were stored at 4°C until fed. Sufficient pellets were provided twice a day to ensure that there was always feed available.

Rats provided milk instead of water were fed a commercial diet that was adjusted to contain less protein and fat and more fiber than normally fed. Sufficient milk for 4 days was thawed at a time, homogenized, and then kept at 4°C until dispensed to the rats. The milk was provided twice a day with the total provided being 10-20 ml more than the rats drank the previous day. This minimized spoilage of the milk.

Weekly feed intakes for the treatment group were measured and the rats were weighed at the start and end of the experiment and twice in between. After 196 days the experiment was ended and all rats euthanized. Immediately after euthanasia a sample of liver was collected from each rat and dealt with as in the pig experiment.

# Results

## Pig experiment

All pigs appeared healthy throughout the experiment. Those fed the ration with uncontaminated meat ate less of the total ration than the pigs fed rations containing contaminated meat during the experiment, and pigs in the 50% uncontaminated meat treatment ate less meat than pigs in the 50% contaminated meat treatment ($P < 0.05$), but there were no significant differences between the treatments for average total liveweight change and average daily gain (Table 1). There were no significant differences between the average liver microsomal GlcNAc-1-P transferase activities of the treatments (Table 1).

**Table 1.** Average total feed and meat consumed, average total liveweight changes and average daily gains, and average liver microsomal GlcNAc-1-P transferase activities in pigs fed rations containing either uncontaminated meat or meat that is potentially contaminated with CT.

| Treatment[1] | Total feed consumed (kg) | Total meat consumed (kg) | Total liveweight change (kg) | Daily weight gain (g) | Transferase activity (cpm/mg protein) |
|---|---|---|---|---|---|
| 50% UM | 208.6 [a] | 104.3 [a] | 102.4 [a] | 826 [a] | 35,785 [a] |
| 50% CM | 211.3 [b] | 105.7 [b] | 99.6 [a] | 803 [a] | 38,615 [a] |
| 25% CM | 212.2 [b] | 53.0 [c] | 98.4 [a] | 793 [a] | 42,130 [a] |
| 12% CM | 212.1 [b] | 25.4 [d] | 104.1 [a] | 839 [a] | 37,153 [a] |

Within a column, values with different superscripts are significantly different ($P < 0.05$).
[1] UM=uncontaminated meat; CM=contaminated meat

## Rat experiment

All rats appeared healthy throughout the experiment. The rats weighed between 280 and 421 g at the end of the experiment. Total feed intakes and total intakes of meat, offal, and milk for the treatments (Table 2) could not be analyzed statistically because there were no individual animal data. There were no significant differences between treatments for average total liveweight change and average daily gain (Table 2). There were no significant differences between the average liver microsomal GlcNAc-1-P transferase activities of the treatments (Table 2) whether the analysis was done over all the treatments or over groupings of the treatments (i.e. meat, or offal, or milk treatments).

# Discussion

The results of these experiments indicate that it is highly unlikely that secondary toxicity will ever result from the consumption of meat, offal, or milk derived from livestock that have ingested CT. In this study the potentially contaminated meat, offal, and milk was collected from cattle demonstrating clinical signs of ARGT thus indicating that they had ingested either a near lethal or a lethal dose rate of CT. Such animals should have had the maximum possible residue concentrations of CT or biologically active metabolic products of the CT in their tissues. In spite of this, no inhibition of the activity of liver microsomal

GlcNAc-1-P transferase could be demonstrated in pigs and rats fed the potentially contaminated meat for 117 and 196 days, respectively, or rats fed potentially contaminated offal or provided potentially contaminated milk instead of water for 196 days.

**Table 2.** Average total feed and meat/offal/milk consumed, average total liveweight changes, average daily gains, and average liver microsomal GlcNAc-1-P transferase activities in rats fed rations containing either uncontaminated meat or offal, or meat or offal that is potentially contaminated with CT, or in rats provided uncontaminated milk or milk potentially contaminated with CT instead of water.

| Treatment[1] | Total feed consumed (g) | Total meat/offal/milk consumed (g or ml) | Total liveweight change (g) | Daily weight gain (g) | Transferase activity (cpm/mg protein) |
|---|---|---|---|---|---|
| 50% UM | 3,213 | 1,607 | 210 [a] | 1.1 [a] | 25,094 [a] |
| 50% CM | 3,403 | 1,702 | 178 [a] | 0.9 [a] | 24,720 [a] |
| 25% CM | 4,007 | 1,002 | 208 [a] | 1.1 [a] | 21,536 [a] |
| 50% UO | 3,678 | 1,839 | 197 [a] | 1.0 [a] | 21,787 [a] |
| 50% CO | 3,378 | 1,689 | 183 [a] | 0.9 [a] | 20,525 [a] |
| 25% CO | 3,763 | 941 | 185 [a] | 0.9 [a] | 17,914 [a] |
| UMk | 3,296 | 7,213 | 198 [a] | 1.0 [a] | 24,470 [a] |
| CMk | 3,259 | 7,248 | 180 [a] | 0.9 [a] | 24,214 [a] |

Within a column and treatment groupings, values with different superscripts are significantly different ($P < 0.05$).

[1] UM = uncontaminated meat; CM = contaminated meat, UO = uncontaminated offal; CO = contaminated offal; UMk = uncontaminated milk; CMk = contaminated milk

These results occurred in spite of the pigs consuming up to an equal quantity of potentially contaminated meat as their gain in liveweight (104 kg meat versus 102 kg liveweight gain, Table 1), the rats consuming up to 9.6 and 9.2 times the quantity of potentially contaminated meat or offal, respectively, as their gains in liveweight (1702 g meat versus 178 g liveweight gain, 1689 g offal versus 183 g liveweight gain, Table 2), and the rats drinking 40.3 times the quantity of potentially contaminated milk as their gain in liveweight (7248 ml milk versus 180 g liveweight gain, Table 2).

Inhibition of liver microsomal GlcNAc-1-P transferase is the most sensitive indicator of CT intake that is available (Stewart and May 1994). In sheep inhibition of activity of this enzyme occurs after the consumption of only 3-5% of the lethal dose (Stewart and May 1994). The results of this study indicate that if any CT or biologically active metabolic products of them were present in the meat, offal, or milk, they were present in concentrations that provided only the no observable effect level or less, even with the very high consumption rates used in the experiments.

## Acknowledgements

Staff on the Vasse and Medina Research Stations of the Department of Agriculture and Food provided considerable assistance in the conduct of these experiments, and Mr R Nicholls and Mr G Doncon provided valuable technical assistance. Mr P Stewart of the CSIRO Plant Toxins Research Group conducted all the liver microsomal GlcNAc-1-P transferase assays. The work was funded by Meat and Livestock Australia.

## References

Allen JG (2004). Annual ryegrass toxicity. In *Clinical Veterinary Toxicology* (KH Plumlee, ed.), pp. 422-424. Mosby, St Louis, Missouri.

Bourke CA and Carrigan MJ (1993). Experimental tunicamycin toxicity in cattle, sheep and pigs. *Australian Veterinary Journal* 70:188-189.

Edgar JA, Frahn JL, Cockrum PA, Anderton N, Jago MV, Culvenor CCJ, Jones AJ, Murray K, and Shaw KJ (1982). Corynetoxins, causative agents of annual ryegrass toxicity; their identification as tunicamycin group antibodies. *Journal of the Chemistry Society Chemical Communication* 4:222-224.

Jago MV and Culvenor CC (1987). Tunicamycin and corynetoxin poisoning in sheep. *Australian Veterinary Journal* 64:232-235.

McKay AC, Fisher JM, Giesecke R, and Crosby J (1985). Are farmers controlling annual ryegrass toxicity in South Australia? In *Plant Toxicology, Proceedings of the Australia-USA Poisonous Plants Symposium, Brisbane 1984* (AA Seawright, MP Hegarty, LF James, and RF Keeler, eds), pp. 559-568. Queensland Poisonous Plants Committee, Yeerongpilly.

Peterson JE, Jago MV, and Stewart PL (1996). Permanent testicular damage induced in rats by a single dose of tunicamycin. *Reproductive Toxicology* 10:61-69.

Roberts WD, Mlodawski G, Macdonagh A, Gibson R, and Bucat J (1994). The distribution of annual ryegrass toxicity in Western Australia. In *Plant-associated Toxins – Agricultural, Phytochemical and Ecological Aspects* (SM Colegate and PR Dorling, eds), pp. 51-56. CAB International, Wallingford, UK.

Stewart PL (1998). Activity of N-acetylglucosamine-1-phosphate transferase in sheep liver microsomes: *in vivo* and *in vitro* inhibition by tunicamycin. *Research in Veterinary Science* 64:31-35.

Stewart PL and May C (1994). Liver UDP-GlcNAc:dolichol phosphate GlcNAc-1-phosphate transferase activity as an indicator of ARGT. In *Plant-associated Toxins – Agricultural, Phytochemical and Ecological Aspects* (SM Colegate and PR Dorling, eds), pp. 149-154. CAB International, Wallingford, UK.

# Chapter 55

# Metabolism of the Endophyte Toxin Lolitrem B in Cattle Liver Microsomes

J.M. Duringer[1] and A.M. Craig[2]

*[1]Department of Environmental and Molecular Toxicology, Oregon State University, Corvallis, OR, 97331, USA; [2]Department of Biomedical Sciences, College of Veterinary Medicine, Oregon State University, Corvallis, OR, 97331, USA*

## Introduction

Perennial ryegrass is a perennial cool-season grass which has been deliberately infected with the endophytic fungus *Neotyphodium lolii* as it confers benefits such as insect resistance, growth enhancement, and drought tolerance to the plant, decreasing the use of pesticides, fertilizers, and irrigation (Joost 1995). Unfortunately, *N. lolii* exerts some of these benefits through the production of loline and ergot alkaloids which cause deleterious effects in cattle and other herbivores when endophyte-infected grasses are grazed or fed as hay (Oliver 2005). Over the last decade, plant breeders have intensified the infection of endophytes in order to increase the benefits to the plant which has consequently increased the concentration of alkaloids in these forages over time.

The Japanese Ministry of Health have highlighted their concern for the public health safety of the endophyte toxin lolitrem B to humans particularly in reference to instituting 'safe feed' standards. The loline alkaloid lolitrem B is responsible for the neurological syndrome known as ryegrass staggers which involves a tremoring response in the smooth musculature of affected animals due to inhibition of large conductance calcium-activated potassium channels (Dalziel *et al.* 2005). To date, the pharmacokinetics and metabolic pathway of this mycotoxin in livestock or humans is largely unknown.

Japanese researchers conducted a preliminary feeding trial where six Japanese Wagyu cattle were fed straw containing the ryegrass toxin lolitrem B (Miyazaki *et al.* 2004). They removed several tissues and found that lolitrem B was detected in the perirenal fat at 210 ppb in cattle displaying ryegrass staggers and at ~150 ppb in animals with no clinical signs. From this research, the Japanese concluded that 'until the metabolism of lolitrems and the toxicity of metabolites are investigated, it will not be possible to guarantee the safety of the products from cattle fed grass containing lolitrems.' Given the lipophilic nature of lolitrem B and the fact that the Kobe beef which the Japanese consume (as well as other animal by-products) may contain a high percentage of fat, humans may be unintentionally exposed to this tremorgenic toxin. Before a risk assessment analysis can be extended to humans, however, a thorough understanding of the fate and metabolism of lolitrem B in cattle and thus the compounds available for human consumption must be achieved.

©CAB International 2011. *Poisoning by Plants, Mycotoxins, and Related Toxins* (eds F. Riet-Correa, J. Pfister, A.L. Schild, and T.L. Wierenga)

Pilot studies in our lab indicated that new peaks corresponding to possible lolitrem B metabolites are detected via HPLC in the urine of cattle after they are fed lolitrem B-containing forage. This makes metabolic conversion of lolitrem B highly probable in cattle. Unfortunately, no data currently exist on the hepatic metabolism of lolitrem B in any animal species. Thus, the fate and metabolism of lolitrem B in cattle liver microsomes was studied. Kinetics and identification of metabolites are reported using *in vitro* assays in combination with liquid chromatography-mass spectrometry (LC-MS). The overall aim is to establish the specific compounds likely to be present in animal by-products that are available for human consumption which will allow for further research as to the possible human toxicity of this tremorgenic mycotoxin.

## Materials and Methods

### Bovine liver incubation

From multiple locations in the liver, 20 g of tissue were collected immediately after slaughter from ten Jersey steers processed at the Clark Meat Science Center, Oregon State University, USA. Tissues were homogenized and microsomes prepared by centrifugation separation (Duringer *et al.* 2004). Microsomal protein content was determined with Coomassie reagent using bovine serum albumin as the standard (Lowry *et al.* 1951). Liver microsomes (0.25 mg) were incubated with 2.9 μM lolitrem B and 50 mM potassium phosphate buffer (pH 7.5), 1.0 mM EDTA, 5 mM $MgCl_2$, and 2.0 mM NADPH for 1 h at 37°C. The incubation was stopped by submersion in ice cold water and proteins were separated by centrifugation.

### HPLC separation

A Zorbax RX-SIL, 4.6 × 250mm, 5μ column was used to separate lolitrem B and its metabolites at a flow rate of 0.5 ml/min over 20 min using a mobile phase of 4:1:0.02 dichloromethane:acetonitrile:water. Detection was by fluorescence with excitation and emission wavelengths being 268 and 440 nm, respectively.

### Mass spectrometry

A 3200 QTRAP (Applied Biosystems) was used for detection of lolitrem B and metabolites by positive electrospray ionization (ESI(+)). Settings were as follows: declustering potential = 81, entrance potential = 9, collision cell exit potential = 4, collision energy = 63, ion spray voltage = 5200, temperature = 600°C, gas 1 = 50, gas 2 = 50, curtain gas = 20. Multiple reaction monitoring (MRM) or enhanced mass spectrometry (EMS) survey scans were used. A *m/z* of 686.4 was determined for the parent mass (Figure 1). For MRM, 686.4 → 238.2 was the strongest transition and was selected for quantitation. 686.4 → 196.2 was the second strongest transition and was used as the confirmatory transition. For EMS IDA EPI, 70 of the most common metabolite transformation masses from 100-1100 amu were selected for EMS using Analyst software (Applied Biosystems). If one of these masses passed through the first quadrupole, an enhanced product ion (EPI) scan was triggered which scanned from 80-1100 amu to generate an MS/MS spectra for those metabolites. LightSight software (Applied Biosystems) was utilized to mine samples for predicted and new metabolites.

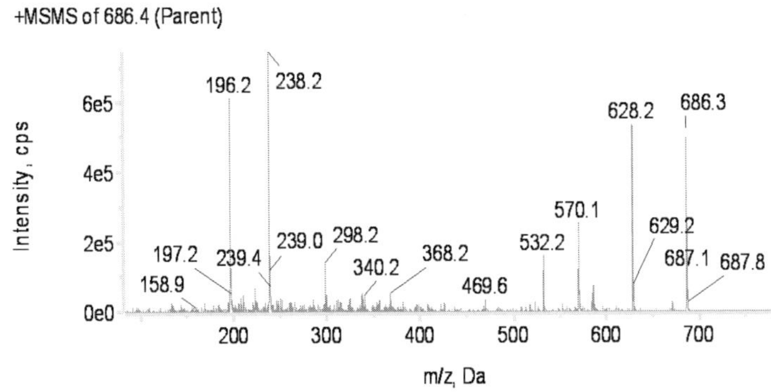


**Figure 1.** Mass spectral analysis of lolitrem B. Parent mass is 686.4 *m/z* with significant fragment ions at 238.2 and 196.2 *m/z*.

## Results and Discussion

After 1 h incubation with 2.9 μM lolitrem B, the average conversion rate was 67 nM lolitrem B metabolized/min/mg protein. This left about 65.7% (or 1.9 μM) lolitrem B remaining after 1 h. In the ten cattle that were tested, the range of metabolism was 40.4-85.3% lolitrem B metabolized in 1 h, showing the range of ability that poor to efficient metabolizers have in clearing lolitrem B from the liver (Figure 2).

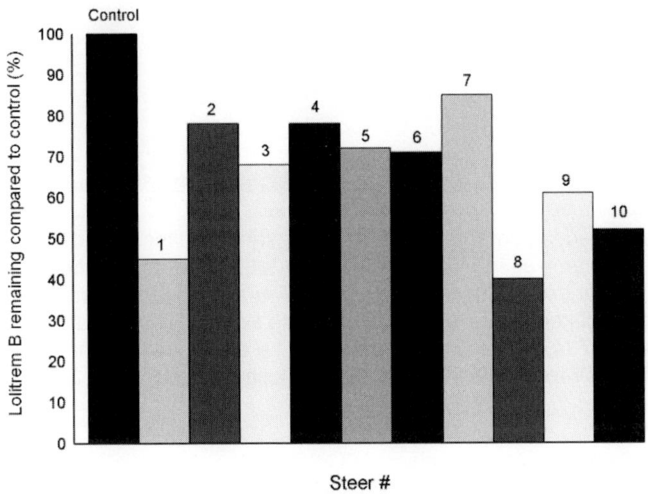


**Figure 2.** Percent of lolitrem B remaining in ten Jersey steer liver microsomes after a 1 h incubation with 2.9 μM lolitrem B as compared to a control.

The toxic threshold established in our laboratory through feeding trials and case study reports for ryegrass staggers is 0.44 nm lolitrem B/g plant material (or 2000 ppb) (Blythe *et al.* 2007). If a cow consumes ~12 kg toxic lolitrem B feed/day (as DM forage), it would get a total dose of 5.28 μmol lolitrem B/day. As is often the case in toxicology, however, the product formed from metabolism of the parent compound could be an equal or even stronger toxicant. Which compound(s) leaves the liver and elicits effects on the nervous system at large conductance calcium-activated potassium channels as well as the relative potency of such compounds are questions that remain to be answered.

When bovine liver microsome incubations were performed and run by LC-MS/MS, a control sample (no xenobiotic-metabolizing enzyme activation) was compared against one with an activated enzyme system to look for metabolites formed. Scans were run in MRM and enhanced product ionization (EPI) modes. When comparing data from both modes, a trioxidation metabolite was detected. In MRM, this consisted of a mass gain of 48.0 for a *m/z* of 734.4 (Figure 3). By EPI, this appeared as part of an additional transformation, that of trioxidation + oxidation + demethylation for a mass gain of 50.0 to an *m/z* of 736.4 found using the LightSight metabolite identification software available from Applied Biosystems (Figure 4). While these results appear to be promising as the first identification of lolitrem B metabolites, they are preliminary and need to be confirmed by additional incubations.

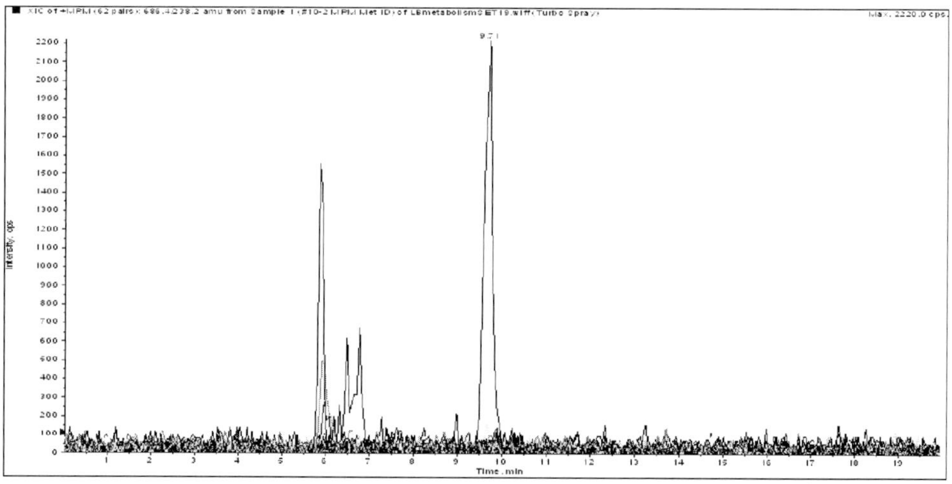


**Figure 3.** Extracted ion chromatogram (XIC) of steer #10 incubated for 1 h with 2.9 μM lolitrem B. Survey scan was in MRM mode using MetID to formulate the most common MRM biotransformation transitions. Peak at 780.6/169.2 *m/z* represents a mass gain of 94.2 which cannot be associated with any predicted Phase I or II metabolite. Double peak at 6.5 minutes represents 734.4/169.2 *m/z*, which could be a tridoxidation metabolite. Peak at 9.71 minutes represents the parent peak (lolitrem B) at 686.4/238.2 *m/z*.

## Conclusions

Lolitrem B is metabolized by bovine liver microsomes at an average rate of 67 nm/min/mg protein. However, a question as to which compounds leave the liver, enter systemic circulation and affect the large-conductance calcium-activated potassium channels responsible for causing ryegrass staggers, and to what potency, remain to be answered.

A trioxidation metabolite appears to be a common product of preliminary lolitrem B incubations. While this study represents the first identification of such lolitrem B metabolites, it needs to be confirmed by further incubations. In addition, Michaelis-Menton kinetics need to be performed in order to characterize the biochemistry of the enzymes involved in production of these metabolites.

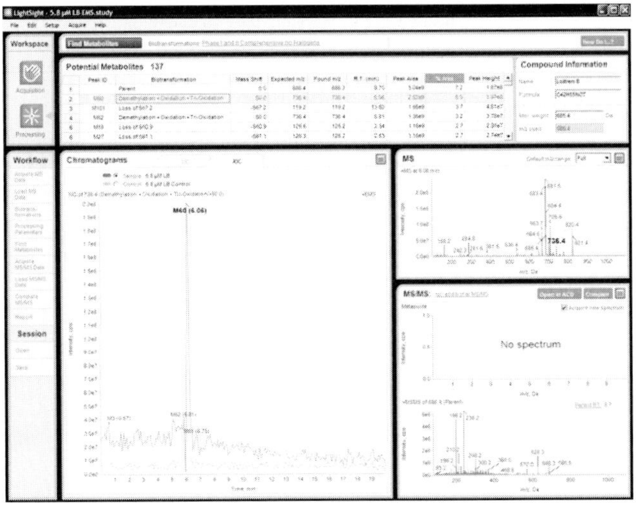


**Figure 4.** LightSight metabolite identification software window showing a trioxidation metabolite from a lolitrem B bovine liver microsome incubation. A comparison was run between a control and an activated sample incubated with 5.8 μM lolitrem B for 1 h.

## References

Blythe LL, Estill C, Males J, and Craig AM (2007). Determination of the toxic threshold of lolitrem B in cattle eating endophyte-infected perennial ryegrass. In *Proceedings of the 6th International Symposium on Fungal Endophytes of Grasses* (AJ Popay and ER Thom, eds), pp. 399-402. New Zealand Grassland Association, Christchurch, New Zealand.

Dalziel JE, Finch SC, and Dunlop J (2005). The fungal neurotoxin lolitrem B inhibits the function of human large conductance calcium-activated potassium channels. *Toxicology Letters* 155:421-426.

Duringer JM, Buhler DR, and Craig AM (2004). Comparison of hepatic *in vitro* metabolism of the pyrrolizidine alkaloid senecionine in sheep and cattle. *American Journal of Veterinary Research* 65:1563-1572.

Joost, R (1995). *Acremonium* in fescue and ryegrass: Boon or bane? A review. *Journal of Animal Science* 73:881-888.

Lowry OH, Rosebrough NJ, Farr AL, and Randall RJ (1951). Protein measurement with the folin phenol reagent. *Journal of Biological Chemistry* 193:265-275.

Miyazaki S, Ishizaki I, Ishizaka M, Kanbara T, and Ishiguro-Takeda Y (2004). Lolitrem B residue in fat tissues of cattle consuming endophyte-infected perennial ryegrass straw. *Journal of Veterinary Diagnostic Investigation* 16:340-342.

Oliver JW (2005). Pathophysiologic response to endophyte toxins. In *Neotyphodium in cool-season grasses*. (CA Roberts, CP West, and DE Spiers, eds), pp. 291-304. Blackwell Publishing, Ames, Iowa.

# TOXIC PLANTS AFFECTING OTHER SYSTEMS

# Chapter 56

# Further Investigations of *Xanthoparmelia* Toxicity in Ruminants

M. Raisbeck[1], R. Dailey[1], R. Siemion[1], D. Montgomery[1], J. Ingram[2], C. Jesse[3], and M. Vasquez[1]

[1]*Wyoming State Veterinary Laboratory, University of Wyoming, Laramie, WY 82070, USA;* [2]*Department of Clinical Sciences, College of Veterinary Medicine and Biomedical Sciences, Colorado State University, Fort Collins, CO 80523, USA;* [3]*Wyoming Dept Analytical Services, Laramie, WY 82070, USA*

## Introduction

During a 7 week period in the winter of 2004 approximately 400-500 elk (*Cervus canadensis*) in south-central Wyoming were poisoned by a lichen (*Xanthoparmelia chlorochroa*) which was previously described as good winter feed for wildlife (Thomas and Rosentreter 1992; Cook *et al.* 2007). Clinically, the condition was characterized by red-stained urine, ataxia, muscular weakness, recumbency, and death. Affected animals remained alert and exhibited normal mentation until they died from starvation, predation, or euthanasia. Several elk had multifocal to coalescing pale white/tan streaks in locomotor muscles of the limbs consistent with exertional myopathy. Muscle lesions were not observed in any elk recumbent less than 2 days. No significant histological lesions were observed in sections of brain, spinal cord, or peripheral nerves from any elk. The following experiments were undertaken in an attempt to define the natural history of this disease and to isolate the proximate toxin responsible for *X. chlorochroa* toxicity.

## Experimental Reproduction of Lichen Poisoning in Domestic Sheep

Lichen was collected from three widely separated sites throughout Wyoming (south-central, southeastern, and northwestern) and from two different seasons (spring 2004 and fall 2005) at the site of the original elk mortality and fed to 12 yearling ewe lambs in an attempt to reproduce the syndrome. The lichen was air-dried, chopped to about 1 cm, and mixed with ground lucerne hay before feeding at 2% body weight (BW) daily. After a 2-week acclimation period, the initial lichen ration consisted of 10% lichen and 90% lucerne. The percentage of lichen was increased by 10% per day until the diet consisted entirely of lichen. Blood was collected for serum chemistries and complete blood counts (CBC) on day 0, at the onset of clinical signs, or every 3 days if no clinical signs were observed. Each ewe was examined visually for signs of intoxication several times daily and neurologic

©CAB International 2011. *Poisoning by Plants, Mycotoxins, and Related Toxins* (eds F. Riet-Correa, J. Pfister, A.L. Schild, and T.L. Wierenga)

examinations were conducted every second day. Complete postmortem examinations were conducted when an animal became moribund or at the end of the study (21 days). In addition to routine histological specimens, multiple samples of skeletal muscle and nerve were extended and fixed to wooden tongue depressors for approximately 1 h prior to being fixed in buffered formalin.

The sheep readily consumed the lichen diet. Ewes had to remain on a 100% lichen diet for at least 2 days and in most cases longer to produce any clinical signs. Red urine containing neither myoglobin nor hemoglobin was the only sign seen consistently in all individuals and occurred only after graduating to a 100% lichen diet. Other signs were primarily locomotor. An initial subtle ataxia rapidly progressed to stiff-legged bunny-hopping, crossed front or hind limbs, and leaning on fences. Neurological examinations did not reveal evidence of central nervous dysfunction and in fact suggested a non-neurological etiology for the condition.

The onset and severity of clinical signs varied appreciably between lichen groups and between individuals within the same group (Table 1; Dailey *et al.* 2008a). Lichen collected immediately after the 2004 episode produced subjectively more severe clinical signs and in a shorter period of time than did lichen collected at other sites or collected at the original site 15 months later, suggesting that potency varies with season and site.

**Table 1.** Results of lichen-feeding trial. Lichen was collected from three sites: the site of the elk mortality (Red Rim, immediately after the mortality in 2004 and again in autumn 2005), southeast Wyoming (Monolith), and northwest Wyoming (McCulloch Peaks).

| Red Rim 2005 | Monolith 2006 | McCulloch Peaks 2006 | Red Rim 2004 |
|---|---|---|---|
| Red urine | Red urine | Red urine | Red urine |
| Subtle ataxia | Spastic | Spastic | Spastic |
| | Weakness | Weakness | Weakness |
| | | Bunny-hopping | Bunny-hopping |
| | | Recumbence | Recumbence |
| | | Onset in 5 days of 100% lichen | Onset in 2 days of 100% lichen |

Severity of toxicity between collections increases from left to right.

## Usnic Acid as the Putative Toxic Agent

A single unsubstantiated report (Beath 1939) attributed the toxicity of *X. chlorochroa* to usnic acid (UA). Since UA is readily available commercially and because we were able to measure percent-level concentrations of (+)-UA in lichen from the original outbreak, it seemed reasonable to test the hypothesis that UA is the active toxin in *X. chlorochroa* before examining other possibilities. Domestic ewes were utilized in a sequential up-down study design (Brownlee *et al.* 1953). A given ewe was fed half of its daily dose of UA on a small amount of hay then after the UA was consumed was fed sufficient additional lucerne hay to meet nutritional requirements. Usnic acid was fed for 7 days and the ewes were observed for an additional 3 days or until the animal showed clinical signs or clinicopathological evidence of poisoning. If an animal completed the 10-day experimental period with no effects, the dose was increased by 50% for the subsequent animal. Conversely, if a dose did cause signs, the dose for the subsequent animal was reduced by 50% of the difference between the preceding two doses. The initial dose (102 mg UA/kg BW) was calculated to be slightly less than the amount of UA received from eating the most toxic group of lichen in the previous study. As the doses increased the ewes became reluctant to eat the treated hay and after ewe #5, the daily dose was split in half and administered b.i.d. in a slurry of ground lucerne.

Ewes were examined several times a day for signs of toxicity and received detailed neurological examinations during the final 3 days of each iteration of the experiment. Blood was collected for clinicopathological examination and at the end of the final observation period each ewe was humanely euthanized with pentobarbital and subjected to intensive postmortem examination.

Results of the up-down toxicity study are summarized in Table 2. Only the animals receiving the highest two doses (776 and 667 mg/kg BW/day) exhibited any clinical signs. During the week of dosing both ewes displayed signs of abdominal discomfort and lethargy. Ewe #6 (776 mg/kg) became anorexic and stiff in her hind limbs after the sixth dose of UA and was found dead the morning of day 7. Ewe #7 (667 mg/kg BW/day) died the morning of day 8. The remaining ewes remained asymptomatic throughout the trial and red urine was never observed in any of the UA-dosed animals and all of the rest of the animals in the study remained asymptomatic. The subacute oral $LD_{50}$ of (+)-UA was estimated to be between 485 and 647 mg/kg/day for 7 days in domestic sheep.

**Table 2.** Results of usnic acid experiment.

| Dose & route | Clinical signs | Histopathology | Clinical pathology |
|---|---|---|---|
| 102 mg/kg *ad lib* | No effect | No lesions | |
| 153 mg/kg *ad lib* | No effect | No lesions | |
| 230 mg/kg *ad lib* | No effect | No lesions | |
| 345 mg/kg *ad lib* | No effect | No lesions | |
| 776 mg/kg *ad lib* | Died Day 6 | Severe myopathy | Elevated CK, AST, LDH |
| 647 mg/kg by gavage | Died Day 7 | Severe myopathy | Elevated CK, AST, LDH |
| 323 mg/kg by gavage | No effect | Severe myopathy | |
| 485 mg/kg by gavage | No effect | Severe myopathy | |

Elevated serum creatine kinase (CK) activity was observed in one asymptomatic and both symptomatic ewes. The former was merely a transient spike (1782 u/l) and may have resulted from sampling trauma (bruising). Creatine kinase in the symptomatic ewes was dramatically (9,950-101,500 u/l) and continuously elevated from day 4 onward and was accompanied by elevated (4,610-36,900 u/l) lactate dehydrogenase (LDH) and aspartate aminotransferase (AST, 6,900-38,000 u/l).

The symptomatic ewes were also the only animals with gross postmortem lesions attributable to UA. Both extensor and flexor muscles of the appendicular skeleton were very pale and edematous, with white chalky areas suggestive of mineralization (Dailey *et al.* 2008b). Axial skeletal muscles, myocardium, and diaphragm were not affected. In asymptomatic ewes changes were minimal and included widely scattered foci of rounded or swollen eosinophilic myocytes and variable blurring or loss of striations. In symptomatic ewes histologic lesions were confined to muscle groups with gross lesions and represented a spectrum of acute and subacute damage. Acute lesions included swollen hypereosinophilic and/or hypercontracted myofibers with central vacuolation. Many myofibers were fragmented. Macrophages infiltrated the degenerate myofibers and cytoplasmic mineralization was prominent in some areas of myodegeneration. Early myocyte regeneration, evidenced by centralization of nuclei, multinucleation and/or cytoplasmic basophilia, was common in the more acutely degenerative areas, suggesting an ongoing process.

Previous reports suggest that UA is primarily hepatotoxic in monogastrics such as rodents and people (Abo-Khatwa *et al.* 1996; Favreaux *et al.* 2002; Durazo *et al.* 2004; Han *et al.* 2004; Neff *et al.* 2004). However, our results indicate that it is primarily myotoxic in ruminants. The very minimal lesions that occurred in asymptomatic ewes had the appearance of artifacts but similar changes are reported to occur in 'irritable' muscle such as seen in Duschene's muscular dystrophy (Gaschen and Burgunder 2001). The lesions in the symptomatic animals suggest an active ongoing process occurring over several days. This interpretation is supported by continuously elevated CK, LDH, and AST activities.

## Chemical Comparison of Lichen from Different Sources

Lichen was collected from nine locations in Wyoming including the four used in the sheep experiment above. Thin layer chromatography and GC-MS analysis identified three major known components in lichen from the original die-off. An UPLC-MS/MS (ultra-high performance liquid chromatography-mass spectrometry/mass spectrometry) method was developed to compare usnic, salazinic, and norstictic acid concentrations between locations (Dailey 2008; Dailey *et al.* 2011) Briefly, 100 mg portions of ground air-dried lichen were extracted with acetone, an aliquot of the supernatant dried and reconstituted 7:3 water:acetonitrile, filtered, and analyzed by UPLC-MS.

There was no significant difference in UA or norstictic acid concentrations between the nine lichen sources by ANOVA. When comparing just the four lichen sources used in the sheep experiment above, UA concentrations were noticeably but not significantly lowest in the most potent of the lichen sources. Salazinic acid concentrations varied significantly between sources; however, most of this variation occurred in samples collected after the sheep bioassay. The variation in salazinic acid concentration did not parallel the potency of the lichen in bioassays.

## Clinical Cases

After publication of the original elk mortality we received inquiries and/or diagnostic samples from potential lichen poisonings in domestic animals. For diagnostic purposes and until better criteria can be developed diagnosis of lichen poisoning is based upon (i) clinical signs of ataxia, muscular weakness, red urine, and recumbency; (ii) significant (30-40%) amounts of *X. chlorochroa* in rumen contents; and (iii) elimination of differentials such as Pb and *Conium*. Eight cases, seven in cattle and one in elk, met these criteria during 2005-2008 and a summary of these cases is presented below.

There was no obvious geographic predilection as field cases occurred at sites separated by several hundred miles. Most occurred in late winter or early spring on arid sandy upland soils with sparse vegetation. Although short grass seemed to be a common factor, overall nutrition was not as body condition scores in cattle ranged from very thin to good condition. Rather, it appeared that short forage permitted grazing animals access to the lichen which is normally too small to be easily picked up in large quantities. Alternatively, the lichen toxin(s) content might be higher in winter. Concentrations of lichen substances such as UA have been reported to vary in other lichen species in response to drought and UV radiation (Fernández *et al.* 2006).

Morbidity was usually less than 10% although the 2008 elk episode was estimated to be much higher. Cattle usually recover with nursing care; however, there seems to be a 'window' of approximately 72 h after which they did not recover. Elk did not recover once recumbent. Red urine was a common finding but only one sample contained detectable myoglobin or hemoglobin. In two cattle cases red-stained feces were noted as well. There were no consistent clinical chemistry findings although slightly elevated CK and LDH were associated with animals that had been recumbent for more than a day. Urine was collected (where available) from diagnostic cases and from experimental ewes and was compared to control urine from the same species by TLC (Dailey, 2008). Lichen-intoxicated animals' urine contained a unique orange spot that turned dark red when sprayed with p-anisaldehyde. This compound was purified by preparative TLC and subjected to UPLC-MS and determined to have a molecular weight of 256.1.

## Conclusions

*X. chlorochroa* is potentially toxic to elk, cattle and sheep, with sensitivity being elk >> cattle > sheep. Other species (antelope, deer, horses, etc.) have not been examined experimentally nor have field cases been reported. Poisoning seems to require ingestion of a significant fraction of the total diet as lichen over a period of 1 to a few days. Clinical signs are suggestive of appendicular skeletal muscular weakness although further study will be required to completely rule out nervous involvement. Usnic acid previously reported to be the putative toxic agent is probably not the whole story as: (i) clinical signs produced by UA are different than those produced by lichen; (ii) the amount of UA in the most toxic lichen tested was several fold less than required to produce signs when fed as a pure substance; and (iii) the UA content of lichen of varying potencies did not parallel the toxicity of the lichen. It is probable that there is another as yet unidentified lichen substance in *X. chlorochroa* which acts synergistically with UA or somehow potentiates UA.

## References

Abo-Khatwa AN, Al-Robai AA, and Al-Jawhari DA (1996). Lichen acids as uncouplers of oxidative phosphorylation of mouse-liver mitochondria. *Natural Toxins* 4:96-102.

Beath OA (1939). Poisonous plants and livestock poisoning. *University of Wyoming Agricultural Experiment Station bulletin* 231:50-53.

Brownlee KA, Hodges JL, and Rosenblatt M (1953). The up-and-down method with small samples. *Journal of the American Statistical Association* 48:262-277.

Cook WE, Cornish TE, Williams ES, Brown B, Hiatt G, Kreeger TJ, Dailey RN, and Raisbeck MF (2007). *Xanthoparmelia chlorochroa* intoxication in Wapiti (*Cervus canadensis*). In *Poisonous Plants. Global Research and Solutions.* (KE Panter, TL Wierenga, and JA Pfister, eds) pp. 40-45. CAB International, Cambridge, MA.

Dailey RN (2008). Toxicity of *Xanthoparmelia chlorochroa* and the lichen substance (+)-usnic acid in ruminants, 138 pp. PhD thesis, University of Wyoming.

Dailey RN, Montgomery DL, Ingram JT, Siemion R, and Raisbeck MF (2008a). Experimental reproduction of tumbleweed shield lichen (*Xanthoparmelia chlorochroa*) poisoning in a domestic sheep model. *Journal of Veterinary Diagnostic Investigation* 20:760-5.

Dailey RN, Montgomery DL, Ingram JT, Siemion R, Vasquez M, and Raisbeck MF (2008b). Toxicity of the lichen secondary metabolite (+) usnic acid in domestic sheep. *Veterinary Pathology* 45:19-25.

Dailey RN, Siemion R, Jesse C, and Raisbeck MF (2011). UPLC-MS analysis of lichen substances from *Xanthoparmelia chlorochroa. Journal of AOAC, Int'l.* (in press)

Durazo FA, Lassman C, Han S, Saab S, Lee NP, Kawano M, Saggi B, Gordon S, Farmer DG, Yersiz H, Goldstein LI, Ghobrial M, and Busuttil RW (2004). Fulminant liver failure due to usnic acid for weight loss. *American Journal of Gastroenterology* 99:950-952.

Favreau JT, Ryu ML, Braunstein G, Orshansky G, Park SS, Coody GL, Love LA, and Fong T (2002). Severe hepatotoxicity associated with the dietary supplement LipoKinetix. *Annals of Internal Medicine* 136:590-595.

Fernández E, Quilhot W, Rubio C, Hidalgo ME, Diaz R, and Ojeda J (2006). Effects of UV radiation on usnic acid in *Xanthoparmelia microspora* (Müll. Arg. Hale). *Photochemistry and Photobiology* 82:1065-8.

Gaschen F and Burgunder JM (2001). Changes of skeletal muscle in young dystrophin-deficient cats: a morphological and morphometric study. *Acta Neuropathological* 101:591-600.

Han D, Matsumaru K, Rettori D, and Kaplowitz N (2004). Usnic acid-induced necrosis of cultured mouse hepatocytes: inhibition of mitochondrial function and oxidative stress. *Biochemical Pharmacology* 67:439-451.

Neff GW, Reddy KR, Durazo FA, Meyer D, Marrero R, and Kaplowitz N (2004). Severe hepatotoxicity associated with the use of weight loss diet supplements containing ma huang or usnic acid. *Journal of Hepatology* 41:1062-1063.

Thomas AE and Rosentreter R (1992). Utilization of lichens by pronghorn antelope in three valleys in east-central Idaho. *Idaho Bureau of Land Management Technical Bulletin* 92-93.

# Chapter 57

# Administration of *Senna occidentalis* Seeds to Juvenile Rats: Effects on Hematological Parameters and Immune Lymphoid Organs

D.P. Mariano-Souza[1], M.L. Pinheiro[2], C.A. Paulino[3], and S.L. Górniak[1]

*[1]Research Center of Veterinary Toxicology (CEPTOX), Department of Pathology, School of Veterinary Medicine and Animal Science, University of São Paulo, SP, 05508-900, Brazil; [2]Laboratory of Pharmacology, Department of Pathology, School of Veterinary Medicine and Animal Science, University of São Paulo, SP, 05508-900, Brazil; [3]University Bandeirante of São Paulo, SP, 02071-013, Brazil*

## Introduction

*Senna occidentalis* (*So*) (= *Cassia occidentalis*) from the family Caesalpinoideae is native to tropical South America but can be found throughout many tropical and subtropical regions of the world (Tokarnia *et al.* 2000). This plant is a major contaminant of maize, soybean, sorghum, wheat, and other cereal crops (Lal and Gupta 1973). Although most harvested cereals are mechanically cleaned and screened before being processed, *S. occidentalis* seeds can contaminate the final product because of their similarity in size and density to some grains, mainly sorghum. Natural and experimental intoxication with this plant has been described in many animal species including bovines (Barros *et al.* 1990). The most important lesion caused by *So* is the degeneration and necrosis of striated and cardiac muscles described in chicks (Haraguchi *et al.* 1998), rabbits (Tasaka *et al.* 2000), rats (Barbosa-Ferreira *et al.* 2005), and other animals. Moreover, hepatotoxic (Soyuncu *et al.* 2008) and neurotoxic (Barbosa-Ferreira *et al.* 2005) effects have been found in studies with *So* seeds. Previous work with chickens (Silva *et al.* 2003; Hueza *et al.* 2007) receiving low concentrations of *So* seeds suggest that this plant induced alterations in lymphoid organs. Similar results were obtained in adult rats treated with 4% *So* seeds in their food. This study seeks to verify the effects of *So* on hematological, inflammatory, and immunological responses in young rats as their immune system may be more sensitive to toxic insult than that of the adult (De Jong and Van Loveren 2007).

## Material and Methods

*So* seeds used in the experiments were obtained from the Research Center for Veterinary Toxicology-CEPTOX at Pirassununga, São Paulo, Brazil. Thirty male Wistar

©CAB International 2011. *Poisoning by Plants, Mycotoxins, and Related Toxins* (eds F. Riet-Correa, J. Pfister, A.L. Schild, and T.L. Wierenga)

rats, all 21 days old, were divided into groups: *So*4 ($n$=10) was fed a ration containing 4% *So* seeds; a control group ($n$=10) received commercial ration; and a peer-fed (PF) group ($n$=10) received the same amount of ration as consumed by the *So*4 group but with no *So*. Food consumption and weight gains were evaluated for 28 consecutive days. All animals were euthanized, and hematological and immunological parameters were examined.

## Results

The rats of *So*4 group showed decreases in food consumption, weight gains (Figures 1, 2), and weight of the thymus (Figure 3) and an increase in spleen weight (Figure 4) compared to controls. Peer-fed rats also had a decrease in thymus weight (Figure 3) compared to controls.

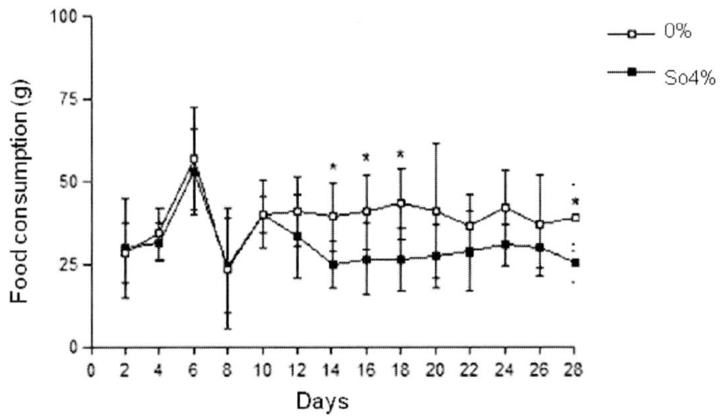


**Figure 1.** Food consumption (g, mean ± SD) of control and *So*4 rats fed for 28 consecutive days. Data were analyzed using the Kruskal-Wallis non-parametric test followed by Dunn´s multiple comparisons test. * $P < 0.05$ different from the adult control group. ** $P < 0.05$ different from the juvenile group that received the same treatment.

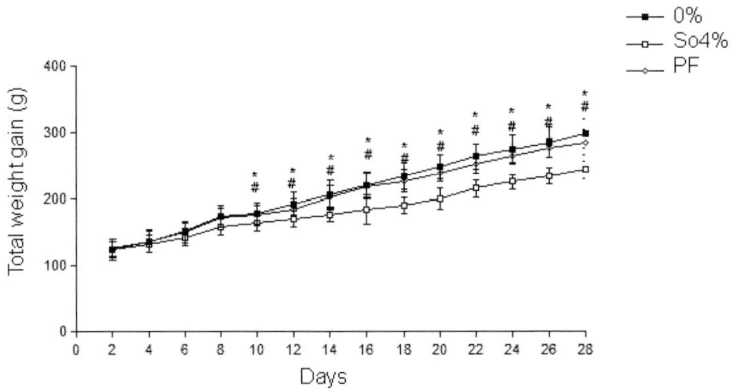


**Figure 2.** Total weight gain (g, mean ± SD) of control, *So*4, and PF rats fed for 28 days. Data were analyzed using ANOVA followed by Dunnett's test. * $P < 0.05$ different from control group. ** $P < 0.05$ different from the PF group (Student *t* test).

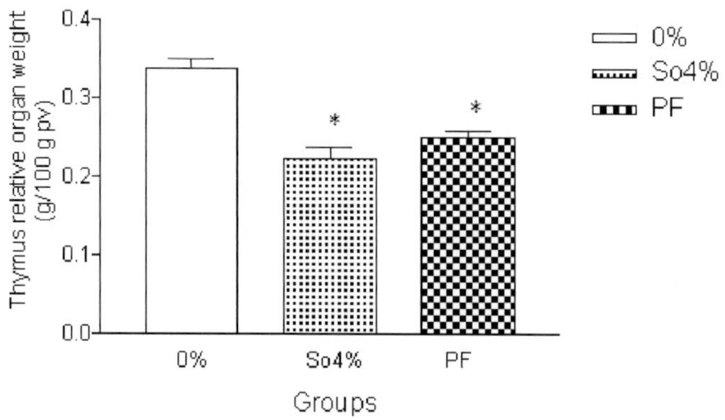


**Figure 3.** Thymus weight (g/100 g pv, mean ± SD) of control, *So*4, and PF rats fed for 28 days. Data were analyzed using ANOVA followed by Dunnett's test. * $P < 0.05$ different from control group.

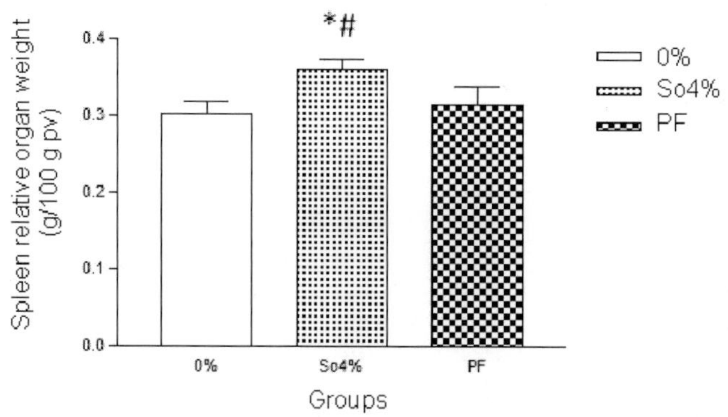


**Figure 4.** Spleen weight (g/100 g pv, mean ± SD) of control, *So*4, and PF rats fed for 28 days. Data were analyzed using ANOVA followed by Dunnett's test. * $P < 0.05$ different from control group. **$P < 0.05$ different from the PF group (Student *t* test).

The *So*4 rats had microcytic and hypochromic anemia (Table 1) characterized by a reduction in mean corpuscular volume (MCV), mean corpuscular hemoglobin (MCH), and mean corpuscular hemoglobin concentration (MCHC). The immune system analysis revealed that the *So*4 animals had a decrease in percentage of phagocytic neutrophils (Figure 5).

## Discussion

In this study we observed a decrease in food consumption by rats in the *So*4 group. Since the animals showed a reduction in food consumption only during the second week of *So* administration, this suggests that such a reduction was not associated with the low palatability of this plant but mainly with its anorexic effects. In fact, some data confirm that anorexia is associated with abusive human consumption of *Senna* when people take this plant as a laxative agent for weight loss (Soyuncu *et al.* 2008). Moreover, spontaneous intoxication with *So* in domestic animals (Barros *et al.* 1990) and several experimental studies performed in laboratory animals (Barbosa-Ferreira *et al.* 2005) showed that anorexia is a common feature in *So* toxicosis.

**Table 1.** Hematological parameters (means ± SD) of control, *So*4, and PF rats fed for 28 days.

| | Groups | | |
|---|---|---|---|
| Hematological Parameters | 0 | So4 | PF |
| RBC ($x10^6/mm^3$) | 6.4 ± 0.9 | 6.5 ± 0.1 | 6.8 ± 0.5 |
| WBC ($x10^6/mm^3$) | 6.1 ± 0.6 | 5.7 ± 0.3 | 5.8 ± 0.7 |
| HGB (g/d IL) | 15.3 ± 0.2 | 15.2 ± 0.7 | 16.5 ± 0.7 |
| HCT (%) | 42.2 ± 1.5 | $39.5 \pm 2.5^{ab}$ | 44.5 ± 2.8 |
| MCV (fl) | 75.3 ± 0.5 | $67.5 \pm 1.5^{ab}$ | 77.4 ± 1.0 |
| MCH (pg) | 27.5 ± 0.9 | 24.6 ± 0.9 | 25.7 ± 1.3 |
| MCHC (%) | 36.6 ± 1.0 | $23.8 \pm 1.5^{a}$ | 35.7 ± 1.6 |

[a] Significantly different from the adult control group at $P < 0.05$ (Kruskal-Wallis non-parametric test followed by Dunn´s multiple comparison test).
[b] Significantly different from the PF group at $P < 0.05$ (Student *t* test).

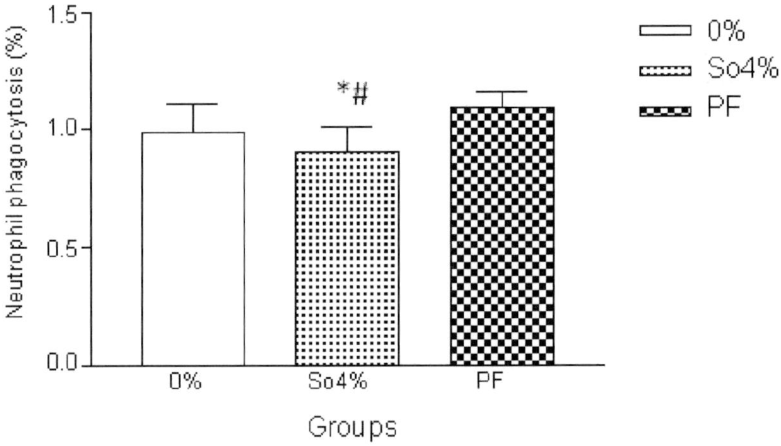


**Figure 5.** Mean percentage (%) of phagocytotic neutrophils from control, *So*4, and PF rats fed for 28 days. Data were analyzed using ANOVA followed by Dunnett's test. * $P < 0.05$ different from control group. **$P < 0.05$ different from the PF group (Student *t* test).

In the present study we observed a decrease in body weight gain in the *So*4 group. Hypothetically, this effect might be related only to the anorexia produced by the plant. However, it should be considered that PF rats did not show any alteration in weight gains hence other factors probably contributed to this effect. The plant contains anthranoids that are widely used as laxative agents (Fugh-Berman 2000). Similarly, we observed that *So*-treated rats had soft feces with increased fecal volume.

*So* often causes hepatotoxicity according to reports on its use for phytotherapeutic purposes in humans (Soyuncu *et al.* 2008). Experimental intoxication in different animal species such as rabbits (Tasaka *et al.* 2000), broiler chickens (Haraguchi *et al.* 1998), and rats (Barbosa-Ferreira *et al.* 2005) has suggested that hepatotoxicity is one of the main toxic effects of *So*. According to Beuers *et al.* (1991), the induction of the hepatotoxicity produced by *So* is probably due to the effect of anthraquinone, a compound in the plant.

Data from several studies with humans (Stickel *et al.* 2000) and experimental animals (Cui *et al.* 2008) show a direct relation between hepatotoxicity and weight loss after *So* consumption. Therefore, we suggest that another factor that possibly contributed to the loss of weight in the *So*4 group was hepatotoxicity.

The complexity of the immune system results in multiple potential target sites for the pathological effects of immunotoxic xenobiotics (De Jong and Van Loveren 2007). Our experiments showed that *So* seeds produced alterations in rat lymphoid organs and hematologic parameters as well as in percentage of phagocytic neutrophils. The analyses of the thymus from the *So*4 group revealed a decrease in size suggesting an immunotoxic effect like the one that occurred in chickens (Silva *et al.* 2003). These results provide the first evidence that *S. occidentalis* has a direct toxic effect on thymus as a target organ in mammals and suggest that alterations in lymphoid organs are probably associated with the direct toxic effects of this plant. However, while we observed a clear immunotoxic effect of *So* in rats, Bin-Hafeez *et al.* (2001), studying mice treated with aqueous extract of *So* for 2 weeks, demonstrated the potent immunoprotective effect of this plant. We should emphasize that in the aqueous extracts of *So* as used in the Bin-Hafeez experiment the liposoluble components such as anthraquinone are not present (de Witte 1993). When whole seeds are administered, as was the case in our study, the animals are exposed to these liposoluble substances. Thus, this finding supports the hypothesis that anthraquinone and other lipophilic substances promoted the immunosuppressive effect.

We observed an increase in the spleen weight in juvenile rats from the *So*4 group. In this context, general parameters like organ weight which may indicate target organ specific toxicity play an important role as a first indicator for the presence of direct immunotoxicity (De Jong and Van Loveren 2007). However, we presently do not have a way to clarify the toxic mechanism of *So* in the spleen. More work will be required in our laboratory in order to clarify this question.

It is well known that malnutrition has a great impact on the size of lymphoid tissues particularly the thymus (Savino 2002). Since our work and that of Silva *et al.* (2003) showed that *So* produces a significant decrease in food intake, we argue that alterations in the bursa of Fabricius and in the thymus could be due to the nutritional deficiency and not to a toxic effect of *So* itself. In fact, animals from the PF group also showed the same thymus changes as did *So*4 treated rats, further supporting this theory.

## Conclusion

Overall, the present study showed that *S. occidentalis* seeds lead to injury to both the lymphoid organs and the hematopoietic system. The PF group allowed us to verify that the observed effects are related to the direct toxic effect of *Senna* seeds and not due to a possible nutritional alteration caused by reduced feed ingestion. Our findings suggest that the evaluation of both systems should be an integral part of investigations on the chronic effects of *S. occidentalis* in different animal species.

## Acknowledgements

This research was supported by grants from CAPES and is part of Dr Souza's doctoral thesis, which will be presented to the Experimental and Comparative Pathology Program, School of Veterinary Medicine and Animal Science, University of São Paulo, Brazil.

## References

Barbosa-Ferreira M, Dagli ML, Maiorka PC, and Górniak SL (2005). Sub-acute intoxication, by *Senna occidentalis* seeds in rats. *Food Chemical Toxicology* 43:497-503.

Barros CSL, Pilati C, Andujar MB, Graça DL, Irigoyen LF, Lopes ST, and Santos CF (1990). Intoxicação por *Cassia occidentalis* (Leg Caesalpinoideae) em bovinos. *Pesquisa Veterinária Brasileira* 10:47-58.

Beuers U, Spengler U, and Pape GR (1991). Hepatitis after chronic abuse of senna. *Lancet* 337:372-373.

Bin-Hafeez B, Ahmad I, Haque R, and Raisuddin S (2001). Protective effect of *Cassia occidentalis* L. on cyclophosphamide-induced suppression of humoral immunity in mice. *Journal Ethnopharmacology* 75:13-18.

Cui L, Zhou QF, Liao CY, Fu JJ, and Jiang GB (2008). Studies on the toxicological effects of PFOA and PFOS on rats using histological observation and chemical analysis. *Archives of Environmental Contamination and Toxicology* 56:338-349.

De Jong WH and Van Loveren H (2007). Screening of xenobiotics for direct immunotoxicity in an animal study. *Methods* 41:3-8.

de Witte P (1993). Metabolism and pharmacokinetics of anthranoids. *Pharmacology* 1:86-97.

Fugh-Berman A (2000). Herb-drug interactions. *Lancet* 355:134-138.

Haraguchi M, Górniak SL, Calore EE, Cavaliere MJ, Raspantini PCF, Calore NMP, and Dagli MLZ (1998). Muscle degeneration in chickens caused by *Senna occidentalis* seeds. *Avian Pathology* 27:346-351.

Hueza IM, Latorre AO, Raspantini PC, Raspantini LE, Mariano-Souza DP, Guerra JL, and Górniak SL (2007). Effect of *Senna occidentalis* seeds on immunity in broiler chickens. *Journal of Veterinary Medicine. A, Physiology, Pathology, Clinical Medicine* 54:179-185.

Lal J and Gupta PC (1973). Anthraquinone glycoside from seeds of *Cassia occidentalis* Linn. *Experientia* 29:142-143.

Savino W (2002). The thymus gland is a target in malnutrition. *European Journal of Clinical Nutrition* 56:S46–S49.

Silva TC, Gorniak SL, Oloris SC, Raspantini PC, Haraguchi M, and Dagli ML (2003). Effects of *Senna occidentalis* on chick bursa of Fabricius. *Avian Pathology* 32:633-637.

Soyuncu S, Cete Y, and Nokay AE (2008). Portal vein thrombosis related to *Cassia angustifolia*. *Clinical Toxicology* 27:1-4.

Stickel F, Egerer G, and Seitz HK (2000). Hepatotoxicity of botanicals. *Public Health Nutrition* 3:113-124.

Tasaka AC, Calore EE, Cavaliere MJ, Dagli MLZ, Haraguchi M, and Górniak SL (2000). Toxicity testing of *Senna occidentalis* seed in rabbits. *Veterinary Research Communications* 24:573-582.

Tokarnia CH, Dobereiner J, and Peixoto PV (2000). *Plantas Tóxicas do Brasil*, 310 pp. Helianthus, Rio de Janeiro.

# Chapter 58

# *Mascagnia exotropica* Poisoning in Ruminants

D.L. Raymundo[1], E.M. Colodel[2], P.M. Bandarra[1], P.M.O. Pedroso[1], L. Sonne[1], K.L. Takeuti[1], C.E.F. Cruz[1], and D. Driemeier[1]

[1]*Setor de Patologia Veterinária, Universidade Federal do Rio Grande do Sul, Porto Alegre, RS, 91540-000, Brazil;* [2]*Laboratório de Patologia Veterinária, Universidade Federal de Mato Grosso, Cuiabá, MT, 78068-900, Brazil*

## Introduction

The consumption of plants that may cause sudden death has been associated with 60% of all the deaths caused by poisonous plants in Brazil (Tokarnia *et al.* 1990). The onset of clinical signs is very acute and at necropsy there are no significant lesions; however, hydropic degeneration within tubular epithelium of kidneys may be seen microscopically (Gava *et al.* 1998). The condition in Brazil may be caused after animal ingestion of plants from three families: Bignoniaceae, Malpighiaceae, and Rubiaceae (Tokarnia *et al.* 2000). *Palicourea marcgravii* (Rubiaceae) is the most important and is the primary poisonous plant causing cattle losses in Brazil (Tokarnia and Döbereiner 1986). Significant losses have also been linked to the consumption of *P. juruana* (Tokarnia and Döbereiner 1982), *P. grandiflora* (Tokarnia and Döbereiner 1981), and *P. aenofusca* (Tokarnia *et al.* 1983) from the same family, and *Pseudocalymma elegans* (Tokarnia *et al.* 1969), *Arrabidaea bilabiata* (Döbereiner *et al.* 1983), and *A. japurensis* (Tokarnia *et al.* 1981) from the Bignoniaceae family. Except for the South region and the state of Mato Grosso do Sul, plants in the Rubiaceae and Malpighiaceae families are distributed in most areas of the country (Tokarnia *et al.* 1990). There are also four *Mascagnia* species (Malpighiaceae) that have been linked to sudden death in ruminants, three of which are distributed from the midwestern to the northeastern regions in Brazil: *M. pubiflora* (Fernandes and Macruz 1964), *M. elegans* (Couceiro *et al.* 1976), and *M. rigida* (Tokarnia *et al.* 1961).

Only *M. exotropica* has been found in southern Brazil (Riet-Correa and Méndez 2007). It is a climbing shrub whose branches may grow up to and cover the top of medium-sized trees in woodland habitats, and the sprouts of the plant growing in the forest understory may easily be accessed and eaten by animals. Poisoning may also occur when animals ingest its leaves from fallen branches (Tokarnia *et al.* 2000). Deaths due to the consumption of this plant may reach 40% of small herds (Gava *et al.* 1998). This report concerns the clinical and pathological findings recorded in retrospective cases of *M. exotropica* poisoning in ruminants.

©CAB International 2011. *Poisoning by Plants, Mycotoxins, and Related Toxins* (eds F. Riet-Correa, J. Pfister, A.L. Schild, and T.L. Wierenga)

## Poisoning in Ruminants in Rio Grande do Sul

Retrospective cases of spontaneous and experimental *M. exotropica* poisoning were retrieved from the records of the Veterinary Pathology Sector of the Federal University of Rio Grande do Sul (SPV-UFRGS) in the period of 1997-2008. Clinical and epidemiological data were recorded during the farm visits. Samples obtained at necropsies were fixed in buffered 10% formalin, processed by standard histological methods, and stained by hematoxylin and eosin.

During the indicated period 5.12% (319) of the total recorded cases (6235) were caused by poisonous plants, 6.9% (17) of which were attributed to *M. exotropica* poisoning. Fourteen cattle, two sheep, and one goat died after spontaneous poisoning by the plant. Three cattle and two goats were experimentally poisoned with green leaves of the plant. The toxic dose to cattle was 10g/kg BW. Clinical signs were similar in the spontaneous and experimental cases and were triggered or enhanced by moving the animals. Affected animals were reluctant to move and showed tachycardia, jugular pulse or jugular engorgement even at resting, muscular tremors, sudden falls, lateral recumbence, paddling, and death, which occurred between 3 and 10 min after initiation of clinical signs. There were no changes at necropsy. Histopathologically, there were multifocal tumefaction and vacuolation in the epithelium of the distal convoluted tubules of kidneys (Figure 1).

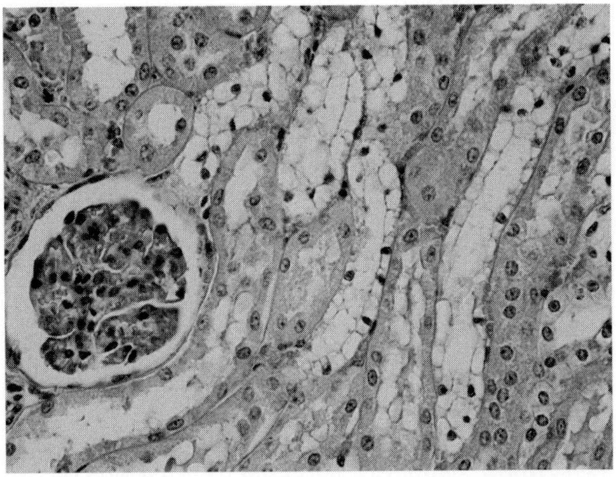

**Figure 1.** *Mascagnia exotropica* poisoning. Histological section of kidney; multifocal tumefaction and vacuolation within epithelium of the distal convoluted tubules.

## Conclusions

Diagnosis was based on epidemiological, clinical, and pathological findings. The presence of the plant in the area where animals were grazed is fundamental to confirm the condition. *M. exotropica* (green leaves) was toxic to cattle at 10 g/kg BW and clinical signs were triggered or enhanced by movement. *M. exotropica* poisoning is an important cause of death in cattle, goats, and sheep in southern Brazil.

## References


Couceiro JEM, Silva ACC, and Silva JA (1976). Observações e ensaios sobre a alegada intoxicação de bovinos por plantas, no Estado de Pernambuco. *Anais XV Congresso Brasileiro de Medicina Veterinária*, pp. 45-46. Rio de Janeiro.

Döbereiner J, Tokarnia CH, and Silva MF (1983). Intoxicação por *Arrabidaea bilabiata* (Bignoniaceae) em bovinos na Região Amazônica do Brasil. *Pesquisa Veterinária Brasileira* 3(1):17-24.

Fernandes NS and Macruz R (1964). Toxicidade da corona – *Mascagnia pubiflora* (Juss.) Griseb. (Malpighiaceae). *Arquivos do Instituto Biológico*, São Paulo, 31(1):1-4.

Gava A, Cristani J, Branco JV, Neves DS, Mondadori AJ, and Sousa RS (1998). Mortes súbitas em bovinos causadas pela ingestão de *Mascagnia* sp. (Malpighiaceae), no Estado de Santa Catarina. *Pesquisa Veterinaria Brasileira* 18(1):16-20.

Riet-Correa F and Méndez MC (2007). Intoxicações por plantas e micotoxinas. In *Doenças de Ruminantes e Eqüinos* (F Riet-Correa, AL Schild, RAA Lemos, and JRJ Borges, eds), 3rd edn, Vol. 2, pp. 191-194. Pallotti, Santa Maria.

Tokarnia CH and Döbereiner J (1981). Intoxicação por *Arrabidaea japurensis* (Bignoniaceae) em bovinos em Roraima. *Pesquisa Veterinária Brasileira* 1:7-17.

Tokarnia CH and Döbereiner J (1982). Intoxicação experimental por *Palicourea juruana* (Rubiaceae) em bovinos e coelhos. *Pesquisa Veterinária Brasileira* 2(1):17-26.

Tokarnia CH and Döbereiner J (1986). Intoxicação por *Palicourea marcgravii* (Rubiaceae) em bovinos no Brasil. *Pesquisa Veterinária Brasileira* 6(3):73-92.

Tokarnia CH, Canella CFC, and Döbereiner J (1961). Intoxicação por um 'tingui' (*Mascagnia rigida* Griseb.) em bovinos no Nordeste do Brasil. *Arquivos do Instituto Biológico Animal*, Rio de Janeiro, 4:203-215.

Tokarnia CH, Döbereiner J, Canella CFC, and Guimarães DJ (1969). Intoxicação experimental por *Pseudocalymma elegans* (Vell.) Kuhlm. em bovinos. *Pesquisa Agropecuária Brasileira* 4:195-204.

Tokarnia CH, Döbereiner J, and Silva MF (1981). Intoxicação por *Palicourea grandiflora* (Rubiaceae) em bovinos no Território de Rondônia. *Pesquisa Veterinária Brasileira* 1(3):85-94.

Tokarnia CH, Döbereiner J, Couceiro JEM, and Silva ACC (1983). Intoxicação por *Palicourea aeneofusca* (Rubiaceae), a causa de mortes súbitas em bovinos na Zona da Mata de Pernambuco. *Pesquisa Veterinária Brasileira* 3(3):75-79.

Tokarnia CH, Peixoto PV, and Döbereiner J (1990). Poisonous plants affecting heart function of cattle in Brazil. *Pesquisa Veterinária Brasileira* 10(1/2):1-10

Tokarnia CH, Döbereiner J, and Peixoto PV (2000). *Plantas Tóxicas do Brasil*, pp. 19-48. Helianthus, Rio de Janeiro.

# Chapter 59

# Relationship between a Peculiar Form of Hydropic-Vacuolar Degeneration of the Distal Convolute Tubules, Monofluoroacetate Poisoning, and Plants that Cause 'Sudden Death' in Brazil

P.V. Peixoto[1], V.A Nogueira[2], T.N. França[2], T.C Peixoto[3], J. Döbereiner[4], and C.H. Tokarnia[1]

[1]*Instituto de Zootecnia, Universidade Federal Rural do Rio de Janeiro (UFRRJ), Seropédica, RJ 23890-000, Brazil;* [2]*Instituto de Veterinária, UFRRJ, Seropédica, RJ 23890-000, Brazil;* [3]*Pós-graduação em Medicina Veterinária, UFRRJ, Seropédica, RJ 23890-000, Brazil;* [4]*Embrapa-CNPAB/Projeto Sanidade Animal, Seropédica, RJ 23890-000, Brazil*

## Introduction

Poisons containing sodium monofluoroacetate (MF), also known as 1080, have been banned in some countries including the USA and Brazil. However, the compound is still used in Australia and elsewhere for the control of rabbits, foxes, pigs, and wild dogs (McIlroy 1992).

MF competitively inhibits citrate aconitase resulting in blockade of the Krebs cycle and reduced production of ATP (Peters 1952). The cause of death is often dependent on the species and physiologic state. MF causes heart failure in cattle (Jubb *et al.* 1992), sheep (Schultz *et al.* 1982), horses, goats, rabbits, and monkeys (Chenoweth and Gilman 1946). It causes neurologic disease in humans (Gajdusek and Luther 1950), dogs, guinea pigs, mice, and hamsters. In cats and domestic pigs the effect is on both tissues (Chenoweth and Gilman 1946).

MF or potassium monofluoracetate are considered to be the poisonous principle of *Dichapetalum cymosum* in South Africa (Marais 1944; Kellerman *et al.* 1988) and *Gastrolobium* spp., *Oxylobium* spp., and *Acacia georginae* in Australia (Oelrichs and McEwan 1962). Poisoning by these plants has been described as sudden or with a peracute course.

In Brazil, 12 plants are known that cause sudden death and are responsible for great losses of cattle every year. The clinical course observed in animals poisoned by Brazilian sudden death-causing plants (BSDCP) is similar to that described in animals poisoned by MF-containing plants in Africa and Australia. To casual observers, poisoned Brazilian

©CAB International 2011. *Poisoning by Plants, Mycotoxins, and Related Toxins* (eds F. Riet-Correa, J. Pfister, A.L. Schild, and T.L. Wierenga)

animals generally do not manifest clinical signs. Suddenly they lie down or fall to the ground and die. Death is most likely due to cardiac arrest. With more careful clinical evaluation, subtle evidence of heart failure such as an engorged pulsing jugular vein is present (Tokarnia *et al.* 2000). Among such plants, *Palicourea marcgravii* stands out for its high toxicity (the lethal dose is 0.6 g fresh plant/kg body weight in cattle), wide distribution, good palatability, and cumulative effect (Tokarnia *et al.* 2000). It is responsible for about 500,000 deaths of adult cattle every year in Brazil (Tokarnia *et al.* 2002). By chromatography MF has been demonstrated in the leaves of *P. marcgravii* (Oliveira 1963; Krebs *et al.* 1994).

Döbereiner and Tokarnia (1959) identified in the kidney of cattle poisoned by *P. marcgravii* a lesion they called hydropic-vacuolar degeneration of the distal convoluted uriniferous tubules (HVDDT) (Figure 1). This is a consistent finding they consider typical for this poisoning. The lesion differs from the more common tubular epithelial hydropic degeneration by the severe cytoplasmic swelling/vacuolation and marked nuclear pkynosis of well delimitated groups of cells of the convoluted tubules. The lesions described by Döbereiner and Tokarnia affected almost exclusively cells of the distal tubules; only occasionally do collecting tubules show that lesion. Further studies confirmed that this lesion also developed in the kidney of cattle, sheep, goats, and rabbits naturally and experimentally poisoned by all the other BSDCP (Peixoto *et al.* 1987; Tokarnia *et al.* 2000). In MF poisoning in humans no specific references on the occurrence of HVDDT in the kidney could be found nor was that lesion described as characteristic. The role of MF in poisoning by BSDCP is uncertain. Tokarnia *et al.* (2000) concluded that *P. marcgravii* and likely the other 11 BSDCP contained sodium monofluoracetate and this is the poisonous principle responsible for the deaths. However, others have argued that there may be other toxins that contribute or synergize with MF (GH Habermehl, 1986, personal communication; Górniak 1988; Kemmerling 1996; González *et al.* 2000; Coelho *et al.* 2007).

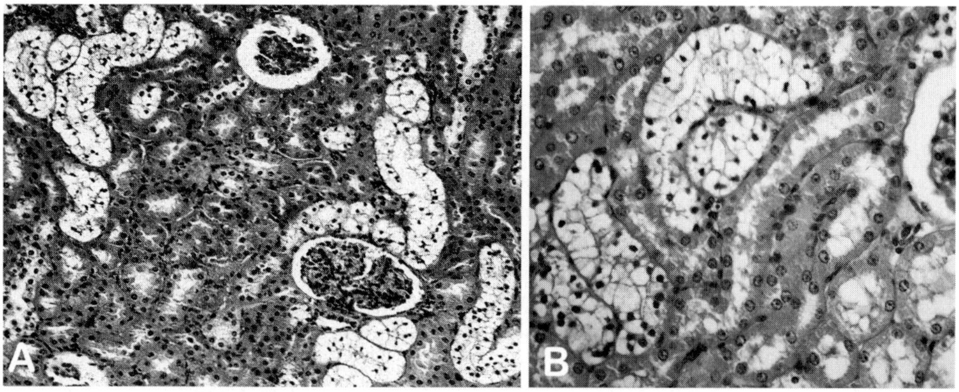


**Figure 1.** (A) HVD detected in the kidney of cattle poisoned by *P. marcgravii*, diagnosed by Döbereiner and Tokarnia (1959). (B) HVD in the kidney of a sheep (#31260; Table 1) poisoned by MF.

Besides the importance for diagnosis the eventual establishment of MF as the compound responsible for the deaths of animals that ingest BSDCP can have economic importance for the livestock industry. In Australia, genetic studies with the intention of rendering bacteria capable of metabolizing or destroying MF in the rumen of cattle have been developed as a mechanism that could be introduced as a prophylactic measure (Gregg *et al.* 1998). The use of antidotes such as glycerol monoacetate and acetamide (Kellerman *et al.* 1988) is not viable under most animal husbandry conditions.

The objectives of this study are to demonstrate that HVDDT is characteristic for the poisoning by MF at least in cattle and sheep. As similar lesions are found in animals that die from BSDCP, MF should be considered the toxic principle of these plants.

## Materials and Methods

These preliminary experiments were performed in the research animal housing of the Pathology Section of Projeto Sanidade Animal Embrapa/UFRRJ. Two adult half-bred Friesian cows (410 and 468 kg) and two crossbred sheep (one 6 months old weighing 19 kg and the second 3 years old weighing 31 kg) were dosed orally with 0.5 and 1.0 mg MF (Sigma Aldrich Co) per kg of body weight. The dose was diluted in 50 ml of distilled water for the cows and in 10 ml of water for the sheep.

This is a lethal dose and after the death of the animals postmortem examinations were done immediately. Samples of all the organs and of the central nervous system were collected; fixed in 10% formalin; dehydrated in ethanol; cleared in xylol; embedded in paraffin; sectioned with the microtome to the thickness of 5 μm; and stained with hematoxylin-eosin for microscopic examination.

## Results

Part of the experimental results presented in this chapter gave origin to the studies by Nogueira (2009) and Peixoto (2009) and are still in progress. The results of these experiments are presented in Table 1. Though the clinical course was brief, poisoned animals showed tachycardia, dyspnea, and slight loss of equilibrium as swaying and reluctance to stand. The cattle developed a prominent jugular pulse. In the final stages, cattle and sheep fell into lateral recumbency, with paddling movements of the legs.

**Table 1.** Experimental outline, results, and intensity of HVD in the kidneys of cattle and sheep poisoned by MF. All doses resulted in death.

| Animal ID | Weight (kg) | MF Dose (mg/kg) | Amount of MF (mg) | Symptom onset | Clinical course | HVD with nuclear picnosis |
|---|---|---|---|---|---|---|
| Cattle | | | | | | |
| 31209 | 468 | 0.5 | 234 | 1 h 55 min | 12 min[a] | +++[b] |
| 31210 | 410 | 1.0 | 410 | 3 h 32 min | 5 min[a] | +[c] |
| Sheep | | | | | | |
| 31260 | 31 | 0.5 | 15,5 | 14 h 6 min | 8 h | +++ |
| 31261 | 19 | 1.0 | 19 | 13 h 20 min | 10 min | + |

[a]Dramatic phase; [b]+++ severe, [c]+ light

At postmortem examination both cattle and sheep had dilated and blood-filled auricles and jugular veins. The lungs were congested and edematous. The cattle also developed slight to moderate subserosal edema at fixation sites of the gall bladder to the liver and slight edema of the mesentery between duodenum and pancreas.

The histological examination revealed, in all animals, HVDDT of the distal convoluted uriniferous tubules, associated with nuclear pkynosis. One cow also developed moderate hepatocellular vacuolation.

## Discussion

The two cattle and two sheep that received MF died quickly and developed HVDDT in the kidney. This is an identical lesion to that observed after the ingestion of *P. marcgravii* and the other sudden death-causing plants in Brazil.

Cellular swelling and hydropic degeneration are not common in the collecting duct tubular epithelium except the descriptions of kidney lesions associated with the ingestion of BSDCP. Although similar hydropic-vacuolar degeneration has been observed in cases of poisoning by dioxane (Jones and Hunt 1983) and sodium selenite (Khattab 2007), in these cases the alteration is not restricted to the distal convoluted tubules. Similarly other toxins such as cisplatin (Fillastre and Raguenez-Viotte 1989), potassium dichromate (Cristofori *et al.* 2007), glycerol 50% (Rodrigo *et al.* 2004), and solution of tartaric acid (Friedman and Kaplan 1943) cause specific hydropic-vacuolar degeneration of the proximal convoluted uriniferous tubules without affecting the distal tubules.

In many experimental studies and most of the reports of MF poisoning, HVDDT is not described. Occasionally, degeneration and tubular necrosis of the convoluted tubules is mentioned (Cho *et al.* 1982; Collicchio-Zuanaze 2006) in cattle and cats, respectively. It is difficult to determine why there was no specific description of these lesions that in our experience are consistent and highly specific. Part of the answer for this question could be due to the differences of the nomenclature. For instance, the analysis of the photomicrographs published by Cater and Peters (1961) who studied the renal lesions induced by the intraperitoneal inoculation of fluorocitrate (the poisonous metabolite of fluoroacetate) in mice shows the clear visualization of a lesion identical to HVDDT that we describe. However, that lesion was described as severe 'fatty degeneration' of the convoluted tubules. In the same way, careful analysis of the photomicrographs published by Collicchio-Zuanaze (2006) of MF-poisoned cats have similar HVDDT. In this work the authors described 'tubular degeneration and hyaline and tubular necrosis.' We consider that pyknotic nuclei indicate that the cells are unviable. Lim *et al.* (1975) reported in mice poisoned experimentally with sodium fluoride degeneration and necrosis of the convoluted tubules. Our evaluation of the published photomicrographs shows that the lesion is also similar to collecting duct HVDDT.

Our review of MF-associated lesions in humans found more severe coagulative necrosis of the renal tubules. This may be a species specific change as horses poisoned by *P. marcgravii* and *Pseudocalymma elegans* develop predominantly coagulative necrosis of uriniferous tubules and less often HVDDT (Tokarnia *et al.* 1993, 1995).

Another question is why HVDDT has not been described in poisoning by *Dichapetalum* spp. in Africa or *Gastrolobium* spp., *Oxylobium* spp., and *Acacia georginae* in Australia, plants that also contain MF (or potassium monofluoroacetate) as their toxic principle. It may be partially explained by the relatively small doses or longer duration that is likely to occur with these plants. Such poisoned animals develop chronic myocardial lesions that cause heart failure. In those conditions, the concentration of MF excreted in the kidney could be insufficient to cause the typical renal lesion. The administration of fractions (1/5 and 1/10) of the lethal dose of *P. marcgravii* and *P. elegans* to sheep reinforces that hypothesis: sheep that receive those small doses usually do not develop HVDDT unless they receive a new lethal dose of the plant (Tokarnia *et al.* 1986; Consorte *et al.* 1994).

There remains a last question: why all animals poisoned by BSDCP do not develop HVDDT. We believe that this is due to dose and duration. Larger doses result in a shorter clinical course and animals may die of cardiac arrest before the MF is eliminated at high concentrations in the urine and produces the renal lesion. Tokarnia and Döbereiner's (1986) findings support this as they found the longer the duration between the administration of *P. marcgravii* and the death of the animal, the greater the incidence and the intensity of HVDDT.

Similar trends have also been observed in the poisonings by other BSDCP (Tokarnia *et al.* 1981, 1983, 2004; Döbereiner *et al.* 1983 ). There seems to be a positive correlation between a long course and development of HVDDT. We found that HVDDT was less intense in the animals that ingested the largest dose of MF (1.0 mg/kg), which supports this hypothesis.

Another consideration is the individual and species variation. Other authors included in Eisler (1995) postulate that individual and species variation to MF may be attributed to the reduced ability to convert fluoroacetate to fluorocitrate.

In summary these preliminary findings suggest that HVDDT of the epithelium (swelling and vacuolar degeneration with nuclear pkynosis restricted to the distal convoluted tubules epithelium) is common and characteristic of MF poisoning in cattle and sheep. Though not conclusive the findings of HVDDT lesions in the four experimentally MF-poisoned animals and many of the experimental and clinical poisonings with BSDCP suggest that MF may be a toxic principle of these plants. At the present time, studies have shown the presence of MF in three of the sudden death-causing plants although it is not known if concentrations were sufficient to be lethal. Additional controlled research studies are needed to confirm the cause of death of animals that ingest BSDCP. Such studies will be invaluable in aiding our understanding of the toxicity of sudden death-causing plants in Brazil and elsewhere. This information will also be useful to livestock producers as they become aware of these toxic plants and manage their stock to minimize the risk of poisoning.

## References

Cater DB and Peters RA (1961). The occurrence of renal changes, resembling nephrosis, in rats poisoned with fluorocitrate. *British Journal of Experimental Pathology* 42:278-289.

Chenoweth MB and Gilman A (1946). Studies on the pharmacology of fluoroacetate. *Journal of Pharmacology and Experimental Therapeutics* 87:90-103.

Cho YJ, Lee CS, Kwak SD, and Park CK (1982). Pathological studies on the experimentally induced rodenticide poisoning in ruminant. *Korean Medical Database* 22:221-232.

Coelho EG, Amaral ACF, Ferreira JLP, Santos AG, Pinheiro MLB, and Silva JRA (2007). Calcium oxalate crystals and methyl salicylate as toxic principles of the fresh leaves from *Palicourea longiflora*, an endemic species in the Amazon state. *Toxicon* 49:407-409.

Collicchio-Zuanaze RC (2006). Perfil hematológico, bioquímico, histopatológico e toxicológico de gatos induzidos experimentalmente com monofluoroacetato de sódio,

167 pp. Tese de doutorado em Medicina Veterinária, Faculdade de Medicina Veterinária e Zootecnia, Universidade Estadual Paulista, Botucatu.

Consorte LB, Peixoto PV, and Tokarnia CH (1994). Intoxicação experimental por *Pseudocalymma elegans* (Bignoniaceae) em ovinos. *Pesquisa Veterinária Brasileira* 14:123-133.

Cristofori P, Zanetti E, Fregona D, Piaia A, and Trevisan A (2007). Renal proximal tubule segment-specific nephrotoxicity: An overview on biomarkers and histopathology. *Toxicologic Pathology* 35:270-275.

Döbereiner J and Tokarnia CH (1959). Intoxicação de bovinos pela 'erva-de-rato' (*Palicourea marcgravii* St. Hil.) no vale do Itapicuru, Maranhão. *Arquivos do Instituto de Biologia Animal*, Rio de Janeiro, 2:83-91.

Döbereiner J, Tokarnia CH, and Silva MF (1983). Intoxicação por *Arrabidaea bilabiata* (Bignoniaceae) em bovinos na Região Amazônica do Brasil. *Pesquisa Veterinária Brasileira* 3:17-24.

Eisler R (1995). Sodium monofluoroacetate (1080) hazards to fish, wildlife, and invertebrates: A synoptic review, 52 pp. *US National Biological Service, Biological Report 27, Patuxent Environmental Science Centre.*

Fillastre JP and Raguenez-Viotte G (1989). Cisplatin nephrotoxicity. *Toxicological Letters* 46:163-175.

Friedman M and Kaplan A (1943). Studies concerning the site of renin formation in the kidney. IV. The renin content of the mammalian kidney following specific necrosis of proximal convoluted tubular epithelium. *Journal of Experimental Medicine* 77:65-73.

Gajdusek DC and Luther G (1950). Fluoroacetate poisoning: A review and report of a case. *American Journal of Diseases of Children* 79:310-320.

González B, Suárez-Roca H, Bravo A, Salas-Auvert R, and Avila D (2000). Chemical composition and biological activity of extracts from *Arrabidaea bilabiata. Pharmaceutical Biology* 38:287-290.

Górniak SL (1988). Intoxicação por *Palicourea marcgravii:* Uma abordagem experimental, 99 pp. Tese de doutorado, Faculdade de Medicina Veterinária e Zootecnia, USP, São Paulo.

Gregg K, Hamdorf B, Henderson K, Kopecny J, and Wong C (1998). Genetically modified ruminal bacteria protect sheep from fluoroacetate poisoning. *Applied and Environmental Microbiology* 64: 3496-3498.

Jones TC and Hunt RD (1983). The urinary system, pp. 1443-1502. In *Veterinary Pathology*, 5th edn. Lea & Febiger, Philadelphia.

Jubb KVF, Kennedy PC, and Palmer N (1992). *Pathology of Domestic Animals*, vol. 3, 780 pp. 5th edn. Saunders Elsevier, Toronto.

Kellerman TS, Coetzer JAW, and Naudé TW (1988). *Plant Poisoning and Mycotoxicoses of Livestock in Southern Africa*. Oxford University Press, Cape Town.

Kemmerling W (1996). Toxicity of *Palicourea marcgravii*: Combined effect of fluoroacetate, N-methyltyramine and 2-methyltetrahydro-β-carboline. *Zeitschrift für Naturforschung* 51:59-64.

Khattab FKI (2007). Effects of sodium selenite on the ultrastructure of the kidney cortex in normal rats. *Journal of Applied Sciences Research* 3:803-810.

Krebs HC, Kemmerling W, and Habermehl G (1994). Qualitative and quantitative determination of fluoroacetic acid in *Arrabidaea bilabiata* and *Palicourea marcgravii* by $^{19}$F-NMR spectroscopy. *Toxicon* 32:909-913.

Lim JKJ, Jensen GK, and King Jr OH (1975). Some toxicological aspects of stannous fluoride after ingestion as a clear, precipitate-free solution compared to sodium fluoride. *Journal of Dental Research* 54:615-625.

Marais JSC (1944). Monofluoroacetic acid, the toxic principle of 'gifblaar *Dichapetalum cymosum* (Hokk) Engl. *Onderstepoort Journal of Veterinary Science Animal and Industry* 20:67-73.

McIlroy JC (1992). The effect on Australian animals of 1080-poisoning campaigns. *Proceedings of the 15th Vertebrate Pest Conference*, pp. 355-359. University of Nebraska, Lincoln.

Nogueira VA (2009). Lesões induzidas por monofluoroacetato de sódio em bovinos. Tese de Doutorado, Universidade Federal Rural do Rio de Janeiro, Seropédica, RJ (in progress).

Oelrichs PB and McEwan T (1962). The toxic principle of *Acacia georginae*. *Queensland Journal of Agricultural Sciences* 19:1-16.

Oliveira MM (1963). Chromatographic isolation of monofluoroacetic acid from *Palicourea marcgravii* St. Hil. *Experientia*, Basel, 19:586-587.

Peixoto TC (2009). Aspectos clínico-patológicos e laboratoriais do envenenamento por monofluoroacetato de sódio em ovinos. Dissertação de mestrado, Universidade Federal Rural do Rio de Janeiro, Seropédica, RJ (in progress).

Peixoto PV, Tokarnia CH, Döbereiner J, and Peixoto CS (1987). Intoxicação experimental por *Palicourea marcgravii* (Rubiaceae) em coelhos. *Pesquisa Veterinária Brasileira* 7:117-129.

Peters RA (1952). Lethal synthesis. *Proceedings of the Royal Society of London. Series B, Biological Sciences* 139:143-170.

Rodrigo R, Bosco C, Herrera P, and Rivera G (2004). Amelioration of myoglobinuric renal damage in rats by chronic exposure to flavonol-rich red wine. *Nephrology Dialysis Transplantation* 19:2237-2244.

Schultz RA, Coetzer JAW, Kellerman TS, and Naudé TW (1982). Observations on the clinical, cardiac and histopathological effects of fluoracetate in sheep. *Onderstepoort Journal of Veterinary Research* 49:237-245.

Tokarnia CH and Döbereiner J (1986). Intoxicação por *Palicourea marcgravii* (Rubiaceae) em bovinos no Brasil. *Pesquisa Veterinária Brasileira* 6:73-92.

Tokarnia CH, Döbereiner J, and Silva MF (1981). Intoxicação por *Palicourea grandiflora* (Rubiaceae) em bovinos no Território de Rondônia. *Pesquisa Veterinária Brasileira* 1:89-94.

Tokarnia CH, Döbereiner J, Couceiro JEM, and Silva ACC (1983). Intoxicação por *Palicourea aeneofusca* (Rubiaceae), a causa de 'mortes súbitas' em bovinos na Zona-da-Mata de Pernambuco. *Pesquisa Veterinária Brasileira* 3:75-79.

Tokarnia CH, Peixoto PV, and Döbereiner J (1986). Intoxicação experimental por *Palicourea marcgravii* (Rubiaceae) em ovinos. *Pesquisa Veterinária Brasileira* 6:121-131.

Tokarnia CH, Costa ER, Barbosa JD, Armién AG, and Peixoto PV (1993). Intoxicação experimental por *Palicourea marcgravii* (Rubiaceae) em eqüinos. *Pesquisa Veterinária Brasileira* 13:67-72.

Tokarnia CH, Peixoto PV, Armién AG, Driemeier D, and Barbosa JD (1995). Intoxicação experimental por *Pseudocalymma elegans* (Bignoniaceae) em eqüinos. *Pesquisa Veterinária Brasileira* 15:35-39.

Tokarnia CH, Döbereiner J, and Peixoto PV (2000). *Plantas Tóxicas do Brasil*, 310 pp. Editora Helianthus, Rio de Janeiro.

Tokarnia CH, Döbereiner J, and Peixoto PV (2002). Poisonous plants affecting livestock in Brazil. *Toxicon* 40:1635-1660.

Tokarnia CH, Barbosa JD, Oliveira CMC, Brito MF, Oliveira RB, and Barbas LA (2004). Aspectos epidemiológicos e clínico-patológicos comparados da intoxicação por *Arrabidaea bilabiata* (Bignoniaceae) em búfalos e bovinos. *Pesquisa Veterinária Brasileira* 24:74-79.

# Chapter 60

# Poisoning by *Mascagnia rigida* in Goats and Sheep

G.J.N. Galiza, J.S. Vasconcelos, T.S. Assis, J.A.S. Araujo, A.F.M. Dantas, R.M.T. Medeiros, and F. Riet-Correa

*Veterinary Hospital, CSTR, Federal University of Campina Grande, 58700-000 Patos, Brazil*

## Introduction

In Brazil, the most important group of poisonous plants is the group that causes sudden death associated with exercise, composed of 12 species: *Palicourea marcgravii*, *P. aeneofusca*, *P. juruana*, *P. grandiflora*, *Arrabidaea bilabiata*, *A. japurensis*, *Pseudocalyma elegans*, *Mascagnia rigida*, *M. elegans*, *M. pubiflora*, *M.* aff. *rigida*, and *M. exotropica* (Tokarnia *et al.* 2000; Riet-Correa and Méndez 2007). These plants are responsible for 60% of all deaths in cattle caused by toxic plants (Tokarnia *et al.* 1990). The primary toxic plant causing sudden deaths is *P. marcgravii* which has a wide distribution and is responsible for most cattle deaths due to plant poisonings in Brazil (Tokarnia *et al.* 1990).

*M. rigida* (Juss.) Griseb. is the most important poisonous plant in northeastern Brazil. It is also responsible for causing sudden death of cattle in northeastern Minas Gerais and north of Espirito Santo (Tokarnia *et al.* 2000). Although mainly a plant of semiarid regions *M. rigida* is found in the more humid and fertile areas (Tokarnia *et al.* 2000) and also occurs in the tropical wet climate of the coast of Paraíba (Vasconcelos *et al.* 2008a). Studies by thin layer chromatography suggest that its toxic compound is fluoroacetic acid (Cunha *et al.* 2006). Most animals die suddenly when they exert themselves (Riet-Correa *et al.* 2006). The natural intoxication occurs in cattle (Tokarnia *et al.* 1985; Medeiros *et al.* 2002; Vasconcelos *et al.* 2008a), goats (Oliveira *et al.* 1978; Vasconcelos *et al.* 2008b), and sheep (Silva *et al.* 2008; Vasconcelos *et al.* 2008b). Experimentally, the intoxication was reproduced in cattle (Tokarnia *et al.* 1961, 1987), goats (Paraguassu 1983; Vasconcelos *et al.* 2008b), sheep (Silva *et al.* 2008; Vasconcelos *et al.* 2008b), and rabbits (Tokarnia *et al.* 1987, 1994; Medeiros *et al.* 2002). The objective of this chapter is to review recent reports of poisoning by *M. rigida* in goats and sheep.

## Epidemiology

Outbreaks in the state of Paraíba occurred at the beginning of the rainy season when the plant sprouts before other forages or after the end of the rainy season when *M. rigida*

©CAB International 2011. *Poisoning by Plants, Mycotoxins, and Related Toxins* (eds F. Riet-Correa, J. Pfister, A.L. Schild, and T.L. Wierenga)

stays green as other forages senesce (Vasconcelos *et al.* 2008b). However, in outbreaks observed in the state of Rio Grande do Norte sheep were intoxicated during the rainy season with good forage availability (Silva *et al.* 2008). *M. rigida* had great variation in toxicity in experimental studies with cattle (Tokarnia *et al.* 1961, 1994) and rabbits (Tokarnia *et al.* 1985; Medeiros *et al.* 2002). In studies with sheep and goats the intoxications occurred with single doses of 10-20 g/kg body weight (Vasconcelos *et al.* 2008b) and with repeated daily doses of 10-20 g/kg BW to achieve 60g/kg BW (Silva *et al.* 2008). In field conditions it is possible that poisoning occurs after repeated ingestions of small doses (Tokarnia *et al.* 1990). The marked variation in the toxicity of the plant explains the variation in the occurrence of the disease in different regions and also between different farms in the same region (Riet-Correa *et al.* 2006). On some farms only sheep recently introduced to the pastures are affected whereas animals raised in the same pastures were not affected (Silva *et al.* 2008), suggesting that native animals are resistant to the intoxication or do not ingest the plant (Vasconcelos *et al.* 2008b). This situation is also frequently encountered with cattle. Preliminary experiments demonstrated that there are resistant animals, but it is not known if this resistance is hereditary or acquired.

## Clinical Signs

In goats and sheep clinical signs characteristic of fluoroacetate poisoning are engorgement of the jugular veins, reluctance to move, incoordination and unsteady gait, sternal recumbence, dyspnea, respiratory distress, depression, instability, muscular tremors, and falls (Silva *et al.* 2008; Vasconcelos *et al.* 2008b). In experimental intoxications the death occurs in a period of 4 min to approximately 28 h after the first clinical signs (Vasconcelos *et al.* 2008b). In spontaneous cases clinical signs always appear when the animals are exercising. Some less affected animals recover 24-48 h after the first signs if they stay quiet without being forced to move (Paraguassu 1983; Vasconcelos *et al.* 2008b). Farmers in the state of Paraíba report that kids born from goats grazing in pastures with *M. rigida* die suddenly immediately after colostrum ingestion. To test if the toxic compound of *M. rigida* causes sudden deaths in newborn lambs and kids, 2g/kg BW of the plant were given daily to two goats and five sheep in the 15 days previous to parturition. One sheep aborted two lambs 5 days before parturition. The four lambs of the other four sheep ingested the colostrum without problems. The kid from one goat ingested the colostrum and died suddenly 5 min later. The kid from the other goat died immediately after parturition before ingestion of colostrum. These results suggest that in goats the active principle of *M. rigida* is eliminated through the milk at toxic doses for the kids (Vasconcelos *et al.* 2008b). Ongoing experiments at the University of Campina Grande are studying *M. rigida* as a cause of neonatal mortality in kids and abortion in sheep and goats.

## Pathology

Macroscopic lesions are not observed but in some experimentally intoxicated goats and sheep alterations associated with acute cardiac insufficiency (mainly lung edema) were observed (Paraguassu 1983; Silva *et al.* 2008; Vasconcelos *et al.* 2008b). Other non-specific changes are increased lobular pattern of the liver, hydropericardium, and petechiae in the pleural surface and epicardium (Vasconcelos *et al.* 2008b). The main histological alteration is hydropic vacuolar degeneration and necrosis of epithelial cells of the renal tubules mainly in the cortical region. Other lesions were diffuse vacuolar degeneration in hepatocytes and lung edema (Silva *et al.* 2008; Vasconcelos *et al.* 2008b). Sheep intoxicated experimentally with repeated doses of *M. rigida* showed lymphocytic infiltration of the myocardium associated with edema and degeneration of myocytes (Silva *et al.* 2008). Diffuse vacuolization of the Purkinje fibers are also reported in experimentally poisoned goats, but it is not clear if this is a lesion or an artifact (Vasconcelos *et al.* 2008b).

## Diagnosis

Clinical signs linked to exercise and the presence of *M. rigida* are suggestive of the diagnosis. The histologic lesion of the kidneys is characteristic but is not observed in all cases (Riet-Correa *et al.* 2007). The differential diagnosis includes poisoning by other plants that cause sudden death such as *Palicourea aeneofusca* which occurs in the coastal region of the northeastern states of Alagoas, Pernambuco, and Paraíba (Vasconcelos *et al.* 2008b).

## Control and Prophylaxis

Removal of the plant by grubbing is difficult because the plant has a persistent root crown which facilitates regrowth after removal. The use of fences to isolate areas with *M. rigida* is a good control measure on some farms. Farmers should minimize animal movement or leave animals in pastures without the plant for at least 1 week to allow recovery and prevent deaths by *M. rigida*. After this time the animals can be moved without apparent risk.

## Acknowledgements

This work has been financially supported by the National Institute for Science and Technology for the Control of Toxic Plants (CNPq), grant number 573534/2008-0.

## References

Cunha LC, Gorniak SL, Haraguchi M, Riet-Correa F, Xavier FG, and Florio JC (2006). *Palicourea marcgravi* e *Mascagnia rigida*: um estudo por cromatografia em camada delgada (CCD). *II Simpósio de Pós-Graduação e XV Semana Científica Prof. Dr. Benjamin Eurico Malucelli*, São Paulo, in CD-ROM (Abstract).

Medeiros RMT, Geraldo Neto SA, Barbosa RC, Lima EF, and Riet-Correa F (2002). Sudden death caused by *Mascagnia rigida* in cattle in Paraíba, Northeastern Brazil. *Veterinary and Human Toxicology* 44:286-288.

Oliveira AC, Oliveira GC, Paraguassu AA, and Freire LMGM (1978). Intoxicação por um 'tingui' (*Mascagnia rigida* Griseb.) em caprinos na Bahia. p. 172. *XVI Congresso Brasileiro de Medicina Veterinária*, Salvador, Bahia (Abstract).

Paraguassu AA (1983). Intoxicação experimental por *Mascagnia rigida* Grisebach (Malpighiaceae) em caprinos no Nordeste do Brasil. 65 pp. Dissertação de Mestrado, Universidade Federal Rural do Rio de Janeiro, Itaguaí, RJ.

Riet-Correa F and Méndez MC (2007). Intoxicações por Plantas e Micotoxinas, pp. 99-219. In *Doenças de Ruminantes e Eqüídeos* (F Riet-Correa, AL Schild, RAA Lemos, and JRJ Borges, eds), pp. 99-219. Editora Pallotti, Santa Maria, RS.

Riet-Correa F, Medeiros RMT, and Dantas AFM (2006). *Plantas Tóxicas da Paraíba,* 158 pp. Centro de Saúde e Tecnologia Rural/SEBRAE/PB, Patos.

Silva IP, Lira RA, Barbosa RR, Batista JS, and Soto-Blanco B (2008). Intoxicação natural pelas folhas de *Mascagnia rigida* (Malpighiacea) em ovinos. *Arquivos do Instituto Biológico*, São Paulo, 75:229-233.

Tokarnia CH, Canella C, and Döbereiner J (1961). Intoxicação por um 'tinguí' (*Mascagnia rigida* Griseb.) em bovinos no Nordeste do Brasil. *Arquivos do Instituto de Biologia Animal*, Rio de Janeiro, 4:203-215.

Tokarnia CH, Döbereiner J, and Peixoto PV (1985). Intoxicação por *Mascagnia* aff. *rigida* em bovinos no Norte do Espírito Santo. *Pesquisa Veterinária Brasileira* 5:77-91.

Tokarnia CH, Döbereiner J, and Canella C (1987). Intoxicação experimental por *Mascagnia rigida* (Malpighiaceae) em coelhos. *Pesquisa Veterinária Brasileira* 7:11-16.

Tokarnia CH, Peixoto PV, and Döbereiner J (1990). Poisonous plants affecting heart function of cattle in Brazil. *Pesquisa Veterinária Brasileira* 10:1-10.

Tokarnia CH, Döbereiner J, and Peixoto PV (1994). Aspectos clínico-patológicos complementares da intoxicação por algumas plantas tóxicas brasileiras. *Pesquisa Veterinária Brasileira* 14:111-121.

Tokarnia CH, Döbereiner J, and Peixoto PV (2000). *Plantas Tóxicas do Brasil*, pp. 26-27. Editora Helianthus, Rio de Janeiro, RJ.

Vasconcelos JS, Riet-Correa F, Dantas AFM, Medeiros RMT, and Dantas AJA (2008a). Mortes súbitas causadas por *Palicourea aeneofusca* e *Mascagnia rigida* na Zona da Mata Paraibana. *Pesquisa Veterinária Brasileira* 28:457-460.

Vasconcelos JS, Riet-Correa F, Dantas AFM, Medeiros RMT, Galiza GJN, Oliveira DM, and Pessoa AFA (2008b). Intoxicação por *Mascagnia rigida* (Malpighiaceae) em ovinos e caprinos. *Pesquisa Veterinária Brasileira* 28:521-526.

Chapter 61

# Hematological, Biochemical, and Urinary Alterations of Enzootic Bovine Hematuria in Dairy Cows in the Caparaó Microregion, Espírito Santo State, Brazil

B.C. Favarato[1], G.B. Bof[1], E.V. de Oliveira[1], L.O. Trivilin[2], L.C. Porfírio[3], and L.C. Nunes[3]

[1]*Veterinary Medicine graduate student, Universidade Federal do Espírito Santo;* [2]*Veterinary Medicine postgraduate student, Universidade Federal do Espírito Santo;* [3]*Department of Veterinary Medicine, Universidade Federal do Espírito Santo, Alto Universitário, PO box 16, Alegre, Espírito Santo, Brazil, 29500-000*

## Introduction

*Pteridium arachnoideum* (bracken fern) is considered a poisonous plant globally, important not only for its cosmopolitan distribution and poisoning of livestock in various parts of the world but also for its high carcinogenic potential observed in animals and humans (Santos 2001). In cattle, bracken fern poisoning causes a chronic non-infectious disease called enzootic bovine hematuria (EBH). The main features of EBH include the development of hemangiomatous lesions on the wall of the urinary bladder causing intermittent hematuria and death by anemia (Radostitis *et al.* 2007).

Singh *et al.* (1973) studied the changes of bovine blood with EBH and found anemia characterized by reduction of hematocrit and hemoglobin due to progressive loss of blood in the urine. Falbo *et al.* (2005) observed serum biochemical changes of hypocalcemia and normophosphatemia, although they found increased fractional urinary excretion of both calcium and phosphorus. Other urinary findings include are proteinuria, high concentrations of calcium and magnesium, and normal excretion of phosphorus (Ghergariu *et al.* 1990).

Durão *et al.* (1995) in their studies found that macro and microhematuria are found in animals with EBH. The duration of these changes varies from one animal to another and intermittent hematuria can be separated by long periods of weeks to years.

In the southern region of the state of Espírito Santo, EBH is frequently noted; however, there is a paucity of information on the subject. Understanding the hematological, serum biochemical, and urinary changes can provide valuable information on pathophysiology of EBH. This study evaluates hematological, biochemical, and urinary changes of dairy cattle with clinical signs of EBH in the microregion of Caparaó in the southern region of Espírito Santo.

©CAB International 2011. *Poisoning by Plants, Mycotoxins, and Related Toxins* (eds F. Riet-Correa, J. Pfister, A.L. Schild, and T.L. Wierenga)

## Material and Methods

From August 2008 to January 2009, 18 crossbred Dutch dairy cows of various ages were selected as they showed clinical signs of EBH, especially dark urine. All properties that were included in this study had a heavy infestation of *P. arachnoideum* and a history of consumption of this plant by cattle.

For each animal body score, heart and respiratory rates, capillary reperfusion times, mucous membrane coloration, and body temperature were recorded. Venous blood was collected from the tail vein for hemogram and serum biochemical analysis. A sample of 5 ml of blood was treated with the anticoagulant ethylenediaminetetraacetate (EDTA) to determine the hematocrit, serum protein levels, and leukogram analysis. Another 5 ml of blood were placed in tubes without anticoagulant and the serum was separated and collected for determination of creatinine, urea, calcium, phosphorus, magnesium, and fibrinogen concentrations.

Samples of 10 ml of urine were collected by spontaneous micturition for urinalysis. All samples for hemogram and urinalysis were properly packed and sent for processing at the Laboratory of Clinical Pathology of the Veterinary Hospital of the Federal University of Espírito Santo. Samples for biochemical analysis were sent to a private laboratory.

The hemogram was performed according to Schalm and Jain (1986). For the biochemical analysis the samples were subjected to centrifugation of 3000 g for 5 min to obtain the serum. Commercial kits were used for the measurements of creatinine, urea, calcium, phosphorus, and magnesium. Dosage of plasma fibrinogen was performed by the method of capillary and precipitation at 56°C by refractometry.

Samples of urine were subjected to physical examination of volume, color, appearance, and odor. Density was measured by refractometry. Semiquantitative determinations of protein, acetone, glucose, bile pigments, bile salts, urobilinogen, and hemoglobin were carried out through strip reagents (Uritest®). For the analysis of urine sediment, samples were centrifuged at 1500 g for 5 min then the supernatant was discarded and the precipitate homogenized and evaluated in a Neubauer chamber.

Statistical analysis included descriptive analysis of the clinical, hematological, and biochemical changes observed.

## Results and Discussion

Data for body score, cardiac and respiratory frequency, capillary refill time, mucus coloration, and body temperature are presented in Tables 1 and 2.

Cattle with EBH had low body scores, high capillary refill times, and pale mucous membranes. Moreira-Souto *et al.* (2006) also reported progressive weight loss, pallor of mucous membranes, and severe intermittent hematuria for months in cattle with EBH. They attributed the anemia and other changes to bracken fern-induced neoplasms in the urinary bladder.

We found that EBH cattle had decreased blood hematocrits probably due to continuous loss of blood in the urine. Similar findings and conclusions of severe anemia with significant reduction of the hematocrit were previously reported as Singh *et al.* (1973) found considerable reduction in hematocrit, hemoglobin, and erythrocyte counts in animals with EBH.

We found no significant changes in the leukogram, total plasma protein, and plasma fibrinogen when compared to values from unaffected animals (Schalm and Jain 1986).

Occasionally, however, some animals developed leukocytosis that may be associated with secondary infections, or systemic stress. Eight animals (44.44%) showed hypoproteinemia.

**Table 1.** Values for body score[1], heart rate (HR), respiratory rate (RR), capillary perfusion time (CP), mucus coloration (MC), and body temperature (BT, $^0$C) of dairy cattle with EBH in the Caparaó microregion, Espírito Santo, Brazil, between August 2008 and January 2009.

| Animal | Body score | HR (bpm) | RR(mov/min) | CP (s) | MC | BT |
|---|---|---|---|---|---|---|
| 1 | 3 | 84 | 29 | 4 | pale | 37.3 |
| 2 | 2.5 | 120 | 28 | 2 | normal | 38.3 |
| 3 | 3 | 80 | 26 | 2 | normal | 38.2 |
| 4 | 3.5 | 100 | 18 | 3 | pale | 38.2 |
| 5 | 3 | 65 | 19 | 2 | normal | 38.5 |
| 6 | 3 | 54 | 22 | 1 | normal | 39.2 |
| 7 | 3 | 64 | 22 | 1 | normal | 38.2 |
| 8 | 4 | 72 | 32 | 3 | pale | 38.7 |
| 9 | 3.5 | 86 | 40 | 3 | pale | 39.1 |
| 10 | 3.5 | 64 | 24 | 3 | pale | 38.9 |
| 11 | 4 | 83 | 40 | 2 | normal | 38.6 |
| 12 | 4 | - | 32 | 2 | normal | 39.2 |
| 13 | 4 | 68 | 32 | 3 | normal | 38.4 |
| 14 | 4 | 100 | 16 | 3 | pale | 38.5 |
| 15 | 4 | - | 16 | 2 | pale | 38.8 |
| 16 | 3 | 120 | 32 | 3 | pale | 37.7 |
| 17 | 2.5 | 104 | 44 | 2 | normal | 38.7 |
| 18 | 3 | 80 | 32 | 3 | pale | 38.8 |

[1]The range for body condition score is 1-5 with 1 being thin.

**Table 2.** Values for hematocrit (Ht), total plasma protein (PPT), fibrinogen (Fb), and total leukocyte counts of dairy cattle with EBH in the Caparaó microregion, Espírito Santo, Brazil, between August 2008 and January 2009*.

| Animal | Ht (%) | PPT (g/dl) | Fb (g/dl) | Total leucometry |
|---|---|---|---|---|
| 1 | 9 | 3.4 | 200 | 8,650 |
| 2 | 36 | 7.8 | 800 | 7,050 |
| 3 | 24 | 7.6 | 200 | 10,900 |
| 4 | 27 | 8.4 | 1,600 | 7,400 |
| 5 | 28 | 8.5 | 200 | 7,900 |
| 6 | 33 | 7.2 | 600 | 9,900 |
| 7 | 28 | 8.2 | 1,000 | 15,700 |
| 8 | 21 | 6.6 | 1,000 | 9,600 |
| 9 | 19 | 6.6 | 400 | 6,550 |
| 10 | 19 | 6.6 | 600 | 7,900 |
| 11 | 28 | 6.6 | 600 | 8,700 |
| 12 | 38 | 7.4 | 600 | 19,050 |
| 13 | 29 | 8.5 | 400 | 16,600 |
| 14 | 19 | 5.2 | 600 | 7,650 |
| 15 | 18 | 5.8 | 400 | 6,200 |
| 16 | 24 | 7.6 | 200 | 16,300 |
| 17 | 26 | 8.2 | 800 | 23,500 |
| 18 | 22 | 6.6 | 200 | 18,450 |

* performed according to Schalm and Jain (1986).

Other serum biochemical values are reported in Table 3, and included decreased serum creatinine concentrations. This may be an inconsistent alteration as Singh *et al.* (1973) found that serum creatinine concentrations were increased significantly in EBH cattle.

**Table 3.** Concentrations of calcium, phosphorus, urea, magnesium, and serum creatinine of dairy cattle with EBH in the Caparaó microregion, Espírito Santo, Brazil, between August 2008 and January 2009*.

| Animal | Calcium (mg/dl) | Phosphorus (mg/dl) | Urea (mg/dl) | Magnesium (mg/dl) | Creatinine (mg/dl) |
|---|---|---|---|---|---|
| 1 | 5.73 | 10.74 | 34.63 | 2.08 | 1.02 |
| 2 | 11.3 | 11.56 | 49.34 | 2.56 | 0.94 |
| 3 | 12.56 | 8.02 | 23.94 | 2.18 | 0.75 |
| 4 | 10.43 | 10.75 | 23.19 | 2.35 | 0.63 |
| 5 | 10.57 | 9.96 | 18.42 | 1.8 | 0.18 |
| 6 | 7.82 | 9.21 | 24.06 | 2.55 | - |
| 7 | 17.15 | 9.62 | 14.21 | 2.42 | 0.83 |
| 8 | 11.85 | 13.24 | 9.72 | 1.92 | 0.52 |
| 9 | 15.47 | 13.13 | 19.07 | 1.96 | 0.19 |
| 10 | 9.41 | 15.33 | 15.67 | 2 | 0.7 |
| 11 | 16.87 | 13.09 | 20.29 | 2.24 | 1.16 |
| 12 | 15.58 | 12.98 | 23.81 | 3.04 | 1.16 |
| 13 | 17.43 | 11.96 | 19.2 | 2.15 | 1.05 |
| 14 | 8.06 | 7.5 | 11.59 | 1.98 | 0.22 |
| 15 | 7.5 | 6.9 | 11.32 | 1.76 | 0.29 |
| 16 | 2.26 | 5.93 | 28.55 | 1.98 | 0.7 |
| 17 | 8.16 | 9.62 | 23.21 | 2.23 | 0.6 |
| 18 | 14.35 | 9.21 | 34.51 | 2.69 | 0.29 |

* The biochemical tests were obtained using commercial kits.

We also found variable changes in serum urea concentrations when compared to reference values (Kaneko *et al.* 1997). Three animals (16.66%) had increased, eight animals (44.44%) had decreased, and seven (38.88%) had normal BUN (blood urea nitrogen) concentrations. Luz (2007) reported that high concentrations of urea can result from renal dysfunction, malignant tumors, water depletion, decreased blood flow to the kidneys, and shock. However, in these conditions there is a usually a concomitant increase in creatinine. It may be that urea changes are related to hypoproteinemia subsequent to liver damage and altered protein metabolism.

Seven animals (38.88%) had hypercalcemia, seven (38.88%) hypocalcemia, and four animals (22.22%) had normal calcium levels. It is prudent in cases of hypercalcemia in bracken poisoning to investigate other possible causes since the change in plasma calcium may be related to different pathological processes or even from deficient or excessive intake. Rajendran *et al.* (1983) found normal plasma levels of calcium in animals experimentally intoxicated by bracken fern. Moreover, Singh *et al.* (1973) observed macrohematuria in animals with hypocalcemia whereas Ghergariu *et al.* (1990) observed microhematuria. Thus, it is suggested that changes in calcium levels may occur along with bleeding.

Seventeen animals (94.44%) had hyperphosphatemia and five (27.77%) hypermagnesemia. These results may imply a dysfunction in the renal excretion of phosphate ions but also changes in the parathyroid gland function and metabolism of vitamin $D_3$. Junior (2004a) mentions that in the absence of renal failure the main cause of hyperphosphatemy is hypoparathyroidism. Therefore, these results cannot be overlooked and where possible thyroid function should be evaluated by hormonal dosages for the differential diagnosis. Junior (2004b) also reports that alteration in renal function and organic overload of magnesium can cause hypermagnesemia.

Results of urinalysis are presented in Table 4. Macrohematuria was observed in 14 animals (77.77%) and in four (2.22%) microhematuria was confirmed by centrifugation. This finding is extremely important for the diagnosis of EBH, mainly for the differential diagnosis with diseases causing hemoglobinuria.

**Table 4.** Concentrations of urinary hemoglobin (Hb), urobilinogen (Ub), bilirubin (Bb), protein (Pt), nitrites (Nt), ketone bodies (Cc), ascorbic acid (Asc), glucose (Gl), pH, specific gravity (Sg), and urinary hematuria (HRA) measured in the urinalysis of dairy cattle with EBH in the Caparaó microregion, Espírito Santo, Brazil, between August 2008 and January 2009*.

| Animal | Hb | Ub | Bb | Pt | Nt | Cc | Asc | Gl | pH | Sg | HRA |
|---|---|---|---|---|---|---|---|---|---|---|---|
| 1 | 250 | normal | + | 500 | neg | + | neg | neg | 7 | 1.015 | Macro |
| 2 | 50 | normal | + | 100 | neg | + | + | neg | 6 | 1.020 | Macro |
| 3 | 250 | normal | ++ | 100 | neg | + | neg | neg | 7 | 1.005 | Micro |
| 4 | 50 | normal | + | 100 | neg | + | neg | neg | 8 | 1.005 | Macro |
| 5 | neg | normal | ++ | 100 | neg | neg | neg | neg | 8 | 1.000 | Micro |
| 6 | neg | normal | + | 100 | neg | + | neg | neg | 8 | 1.000 | Micro |
| 7 | neg | normal | + | 100 | neg | neg | neg | neg | 8 | 1.000 | Micro |
| 8 | 250 | 4 | + | 500 | + | + | neg | - | 9 | 1.005 | Macro |
| 9 | 50 | 2 | ++ | 500 | + | + | neg | neg | 8 | 1.000 | Macro |
| 10 | 250 | 2 | + | 100 | + | + | neg | neg | 8 | 1.000 | Macro |
| 11 | 250 | neg | neg | 30 | neg | + | neg | neg | 8 | 1.000 | Macro |
| 12 | 250 | neg | + | 30 | neg | neg | neg | neg | 8 | 1.000 | Macro |
| 13 | 250 | neg | + | 100 | neg | + | neg | neg | 7 | 1.005 | Macro |
| 14 | 250 | neg | + | 100 | neg | + | neg | neg | 8 | 1.005 | Macro |
| 15 | 250 | neg | neg | 30 | neg | + | neg | neg | 8 | 1.015 | Macro |
| 16 | 250 | 2 | ++ | 500 | + | ++ | neg | neg | 9 | 1.005 | Macro |
| 17 | 250 | 2 | + | 500 | + | + | neg | neg | 8 | 1.005 | Macro |
| 18 | neg | normal | neg | 30 | neg | + | neg | neg | 7 | 1.005 | Macro |

* The urinary parameters were obtained by semi-quantitative biochemical test using Uritest®.

In this study the examination of urinary sediment revealed an increased number of red cells which is abnormal for healthy animals. Also, Falbo *et al.* (2005) found an average of 351 cells/ml in animals with microhematuria and 3661 cells/ml in samples with macrohematuria, indicating abnormality when compared with normal values of 305 cells/ml standardized by the same authors. Both in samples with micro or macrohematuria, the values of urinary hemoglobin were high; however, three samples with microhematuria were negative for hemoglobin in the reagent strip.

Chemical examination of urine by reagent strip analysis showed proteinuria with protein concentrations from 30 to 500 mg/dl and an average of 195.55 mg/dl (++). According to Kaneko *et al.* (1997), it is normal to see trace amounts of protein in urine. As Ghergariu *et al.* (1990) also observed moderate proteinuria both in animals with micro and macrohematuria, it is likely that proteinuria is directly related to urinary tract hemorrhage.

The values of urobilinogen were high in only five (27.77%) of the 18 samples and hyposthenuria was observed in all samples evaluated. Urine pH remained within normal limits, i.e. neutral to alkaline for herbivores.

## Conclusions

Hematological findings for EBH are anemia, hypoproteinemia, and leukocytosis. There was hyperphosphatemia and variable serum levels of calcium and urea. For the urinalysis, macro and microhematuria, proteinuria, and hyposthenuria were observed. In this study the clinical, biochemical, and hematological disorders in animals with EBH were not specific for the disease, therefore there must be corroborating epidemiological data to make a specific diagnosis.

## Acknowledgements

This work was financially supported by Fundação de Apoio à Ciência e Tecnologia do Espírito Santo and from the National Council of Technological and Scientific Development and by the Centro de Ciências Agrárias of the Universidade Federal do Espírito Santo.

## References

Durão JFC, Ferreira ML, and Cabral A (1995). Aspectos anatomopatológico e clínicos da hematúria enzoótica dos bovinos. *Revista Portuguesa de Ciências Veterinárias* 90:132-137.

Falbo MK, Reis ACF, Balarin MRS, Bracarense APFRL, Araújo JRJP, Okano W, and Sandini IE (2005). Alterações hematológicas, bioquímicas, urinárias e histopatológicas na intoxicação natural pela samambaia *Pteridium aquilinum* (L.) Kühn. *Semina* 16:547-558.

Ghergariu S, Bale G, and Oros NA (1990). Unele modificari hematologice, biochimice sanguine si urinare la taurine intr-o-zona de hematurie enzootica. *Revista de Zootehnie Si Medicina Veterinara* 5:15-23.

Junior JF (2004a). Fósforo na medicina de urgência. In http://www.medicinacomplementar.com.br/tema280205. Accessed 20/11/2008.

Junior JF (2004b). Magnésio na medicina de urgência. In http://www.medicina complementar.com.br/tema280205. Accessed 20/11/2008.

Kaneko JR, Harvey JW, and Bruss ML (1997). *Clinical Biochemistry of Domestic Animals*, Academic Press, San Diego.

Luz LM (2007). Nitrogênio Uréico Sangüíneo (BUN). In http://www.mundovestibular.com.br/articles/964/1/NITROGENIO-UREICO-SANGUINEO-BUN/Paacutegina1.htm. Accessed 10/11/2008.

Moreira-Souto MA, Kommers GD, Barros CSL, Rech RR, and Piazer JVM (2006). Neoplasmas da bexiga associados à hematúria enzoótica. *Ciência Rural* 36:1647-1650.

Radostits OM, Gay CC, Hinchcliff KW, and Constable PD (2007). *Veterinary medicine. A textbook of the diseases of cattle, horses, sheep, pigs and goats*, 10th edn. Saunders, London.

Rajendran MP, Chennakesavalu M, Narayana Rao CV, Viraghavan K, and Damodaram S (1983). Experimental production of enzootic bovine haematuria with bracken fern. *Indian Veterinary Journal* 60:173-178.

Santos RC (2001). Avanços na pesquisa com o broto de samambaia usado como alimento em Minas Gerais. *Revista de Pesquisa e Pós-graduação* 1:1-6.

Schalm OM and Jain NC (1986). *Veterinary Hematology*, 4th edn. Lea & Fabiger, Philadelphia, USA.

Singh AK, Joshi HC, and Ray SN (1973). Studies in bovine haematuria. I. Haematological and biochemical observations on the blood of cattle suffering from haematuria. *Indian Journal of Animal Science* 43:296-299.

# Chapter 62

# Upper Urinary Tract Lesions Associated with Enzootic Bovine Hematuria

L.C. Nunes[1], D.M. Donatele[1], C.M. Scardua[2], M.D. Dórea[3], L.N. Monteiro[4], C.C. Bernardo[5], and A. Calais Jr[5]

*[1]Department of Veterinary Medicine, Universidade Federal do Espírito Santo, Alto Universitário, box 16, Alegre, Espírito Santo, Brazil, 29500-000; [2]Veterinary Practitioner, João Neiva, Espírito Santo, Brazil; [3]Medicine Veterinary postgraduate student of the Universidade Federal do Espírito Santo; [4]Veterinary Medical Residency in Animal Pathology of the Universidade Estadual Paulista Julio de Mesquita Filho, Botucatu, Sao Paulo, Brazil; [5]Veterinary Medicine graduate student, Universidade Federal do Espírito Santo, Brazil*

## Introduction

Enzootic bovine hematuria (EBH) is a chronic non-infectious disease caused by bracken fern (*Pteridium arachnoideum*). It is characterized by intermittent hematuria and death by anemia due to the development of hemangiomatous lesions on the walls of the urinary bladder (Radostits *et al.* 2007). Blood losses occur without bone marrow replacement (aplastic anemia) and the disease may affect pregnant cows causing abortion (Marçal *et al.* 2001). The diagnosis of EBH is based on epidemiology, clinical signs, and macroscopic and microscopic lesions in the urinary bladder (Moreira-Souto *et al.* 2006).

In the early stages of the disease microhematuria occurs which may be imperceptible (subclinical stage); gradually the urine becomes dark, characterized by macrohematuria (clinical stage) (Jubb *et al.* 1991). In this phase the animal shows weakness, pale mucus, anemia, edema, drop in milk production, dysuria, tenesmus, arched back, urethral obstructions, uremia, and death (Marçal and Gaste 1991; Radostits *et al.* 2007).

The nature of the bladder tumors associated with the ingestion of *P. arachnoideum* is quite peculiar: epithelial and mesenchymal tumors can be observed in the same animal (Tokarnia *et al.* 2000).

In Brazil EBH is responsible for significant economic losses in many areas where *P. arachnoideum* is abundant (Tokarnia *et al.* 2000; Moreira-Souto *et al.* 2006). Data from the southern region of Espírito Santo, specifically in the Caparaó microregion, confirmed that this plant is present in all ten counties of the microregion and 56.4% of bovines with any disease present clinical signs of hematuria (Silva *et al.* 2009).

Although many aspects of neoplastic lesions caused by *P. arachnoideum* are already known, little is known about lesions in the upper urinary tract associated with EBH. This study describes the macro and microscopic aspects of the lesions found in the upper urinary tract of dairy cattle with EBH in the counties of Iuna and Ibatiba in southern Espírito Santo, Brazil.

## Material and Methods

Only cattle with signs of EBH (bloody urine) and lesions in the upper urinary tract (renal pelvis and ureters) were included in this study. Between the years 2007-2008 three cattle with those lesions were necropsied in the Animal Pathology Service of the Universidade Federal do Espírito Santo, Alegre, Espírito Santo, Brazil. All animals were from the microregion of Caparaó where EBH is endemic and *P. arachnoideum* infestation in pastures is heavy.

The animals were necropsied and samples of different tissues were fixed in 10% formalin, embedded in paraffin, cut at 5 μm, stained with hematoxylin-eosin, and examined by optical microscopy.

## Results and Discussion

Bovine 1, an 8-year-old female, showed bilateral hydronephrosis with purulent exudate in the ureter. The right kidney had occlusive thrombosis in the renal artery and multiple infarcts. Abscesses were observed in the left kidney and confirmed by microscopic examination. In the urinary bladder red lesions of hemangious origin were observed and diagnosed histologically as cavernous hemangiomas.

Bovine 2, a 10-year-old female, presented with bilateral hydronephrosis and a white mass with areas of calcification in the renal pelvis. The mass was characterized histologically as transitional cell carcinoma which could be a metastasis of the urinary bladder. Because rarely do transitional cell carcinomas metastasize retrograde to the renal pelvis it is more likely this was a primary transitional cell carcinoma of the renal pelvis. Several neoplasms were also present in the urinary bladder that were identified as transitional cell carcinomas and hemangiomas. A viscous exudate with a gray coloration was observed in the ureter.

Bovine 3, an 8-year-old steer, presented with accentuated abdominal distention and uroperitoneum, bilateral hydronephrosis, rupture of left ureter, and fibrinous peritonitis. The left kidney was encased with extensive fibrosis. Abscesses and diffuse interstitial fibrosis were observed in both kidneys. *Corynebacterium* spp. was isolated from the urethral exudate. In this animal a large urethral clot or occlusion was also observed. The neoplasms of the urinary bladder were identified as adenocarcinomas, transitional cell carcinomas, and hemangiosarcomas.

Since 1967 it has been reported that the different clinical forms of chronic bracken fern poisoning do not occur with equal frequency in different Brazilian regions (Tokarnia *et al.* 1969). Epidemiological data obtained on 27 farms in the region of Jaguari, Rio Grande do Sul State, revealed that tumors of the upper digestive tract (UDTT) including papillomas, transforming papillomas, and squamous cell carcinomas were more frequent than EBH (Moreira-Souto *et al.* 2006). A similar situation occurs in the state of Santa Catarina (Gava *et al.* 2002). In contrast, in Espírito Santo in the Caparaó microregion there is a higher prevalence of EBH than tumors of the upper digestive tract (Silva *et al.* 2009).

The diversity of the neoplasms observed in cattle with EBH is surprising especially in comparison with the small variation in the occurrence of bladder tumors in other species of domestic animals (Peixoto *et al.* 2003). These authors found 22% of bovines with EBH had non-neoplastic urinary tract alterations that included vascular proliferation, vascular ectasia, hemorrhage, lymphocytic nodules, diffuse lymphocytic infiltration, fibrosis, proliferation of myxoid stroma, and proliferative inflammatory granulomas or pseudotumors.

Wosiaki (2000) reported lesions of hyperplasia, hypoplasia, desquamation, and degeneration of the urinary epithelium and diffuse mononuclear infiltrates in cattle with EBH. They suggested that these findings may be associated with papillomavirus infection type 2 (BPV-2).

In this study less common lesions in the upper urinary tract were found associated with EBH. These lesions included mononuclear inflammatory reaction (acute nephritis) in 63.4% of the cases and tubular renal degeneration in 34.2% (Christian *et al.* 2004).

The urinary bladder was found to have more frequent lesions including extensive chronic inflammation. This may be because the mucosa of the bladder is constantly infected by external agents present in the urine. In the microbiological culture of Bovine 3, *Corynebacterium* spp. was isolated from the urethral exudates. Bacteria of the genus *Corynebacterium* are common in the mucosa and skin of mammals (Gomes 2009). *C. renale* is frequently isolated from cases of pyelonephritis. *C. pilosum* also occurs in the urine and vagina of approximately 4% of healthy cows and rarely in cases of cystitis and pyelonephritis. *C. cystitidis* is widely distributed and is a common cause of cystitis and pyelonephritis. This microorganism was never isolated from healthy cows but is a commensal isolate of the prepuce of 90% of the bulls.

In this study, lesions of the upper urinary tract included ureter and renal pelvis hydronephrosis, abscesses, and presence of grayish exudates. The most severe lesions were observed in the steer. This fact may be associated with the length of the urethra that can be more easily obstructed by the presence of clots or neoplasms. According to Gomes (2009), the affected urethra and subsequently the ureters become distended and the mucosa has areas of necrosis. The kidneys very often are enlarged, the pelvis is dilated, and the papillary region presented necrosis and abscess formation. The renal pelvis can contain liquid which is gray and slimy with an odorless exudate mixed with fibrin, clots, necrotic debris, and calcareous material. Large number of bacilli with characteristics of diphteroids are found free or attached to fragments of necrotic tissue.

## Conclusions

We conclude that bovines affected with EBH develop serious lesions in the upper urinary tract probably caused by ascending infection and pressure from the lower urinary tract lesions such as urethral obstruction. It is believed that these lesions are increased by secondary bacterial infections that can be responsible for the death of some affected cattle.

## Acknowledgements

This work had financial support from the Fundação de Apoio à Ciência e Tecnologia do Espírito Santo and support from the Centro de Ciências Agrárias of the Universidade Federal do Espírito Santo.

## References


Christian GE, Alfonso CC, Rosa PC, Néstor FP, and Roberto ER (2004). Caracterización de las lesiones encontradas en Bovinos con hematuria vesical enzoótica en la Zona de Oxapampa, Pasco. *Revista de Investigaciones Veterinarias del Peru* 15:25-36.

Gava A, Neves DS, Gava D, Moura ST, Schild AL, and Riet-Correa F (2002). Bracken fern (*Pteridium aquilinum*) poisoning in cattle in southern Brazil. *Veterinary and Human Toxicology* 44:362-365.

Gomes MJP (2009). *Gênero* Corynebacterium. *Laboratório de Bacteriologia da Faculdade de Veterinária, Universidade Federal do Rio Grande do Sul.* available at: http://www.ufrgs.br/labacvet/pdf/coryne200901.pdf. Accessed 01/05/2009.

Jubb K, Kennedy P, and Palmer N (1991). *Patología de los animales domésticos*, pp. 453-458. Hemisferio Sur, Uruguay.

Marcal WS and Gaste L (1991). Perspectiva terapéutica para la hematuria enzoótica de los bovinos – estudio clínico preliminar. In *Proceedings 46th Conferencia Anual da Sociedade Paulista de Medicina Veterinaria*, 48 pp.

Marçal WS, Gaste L, Reitchert Netto NC, Gargantini M, Fernandes RP, and Monteiro AA (2001). Ocorrência de intoxicação aguda em bovinos pela samambaia (*Pteridium aquilinum* L. Kuhn) no norte do Paraná- Brasil. *Semina* 22:139-144.

Moreira-Souto MA, Kommers GD, Barros CSL, Rech RR, and Piazer JVM (2006). Neoplasmas da bexiga associados a hematúria enzoótica. *Ciência Rural* 36:1647-1650.

Peixoto PV, França TN, Barros CSL, and Tokarnia CH (2003). Histopathological aspects of bovine enzootic hematuria in Brazil. *Pesquisa Veterinária Brasileira* 23:65-81.

Radostits OM, Gay CC, Blood DC, and Hinchcliff KW (2007). *Veterinary medicine. A textbook of the diseases of cattle, horses, sheep, pigs and goats*, 2065 pp. W. B. Saunders, London.

Silva MA, Scárdua CM, Dórea MD, Nunes LC, Martins IVF, and Donatele DM (2009). Prevalência de hematúria enzoótica bovina em rebanhos leiteiros na microrregião do Caparaó, Sul do Espírito Santo, entre 2007 e 2008. *Ciência Rural* 39:1847-1850.

Tokarnia CH, Döbereiner J, and Canella CFC (1969). Ocorrência da hematúria enzoótica e de carcinomas epidermóides no trato digestivo superior em bovinos no Brasil. II. Estudos complementares. *Pesquisa Agropecuária Brasileira* 4:209-224.

Tokarnia CH, Döbereiner J, and Peixoto PV (2000). *Plantas tóxicas do Brasil*, 320 pp. Helianthus, Rio de Janeiro.

Wosiacki SR (2000). Papiloma vírus bovino tipo 2 em bexiga de bovinos na hematúria enzoótica; detecção utilizando a reação em cadeia pela polimerase e estudo histopatológico. Dissertação de Mestrado, CCA, Universidade Estadual de Londrina.

# Chapter 63

# Similarities between Non-Neoplastic Urinary Bladder Lesions in Bovine Enzootic Hematuria and those Induced by Radiotherapy in Humans

L.G. Oliveira[1], T.N. França[2], L.I. Oliveira[4], P.V. Peixoto[3], and M.F. Brito[2]

[1]*Curso de Pós-graduação em Medicina Veterinária, UFRRJ, Seropédica, RJ 23890-000, Brazil;* [2]*Instituto de Veterinária, UFRRJ, Seropédica, RJ 23890-000, Brazil;* [3]*Instituto de Zootecnica, Universidade Federal Rural do Rio de Janeiro (UFRRJ), Seropédica, RJ 23890-000, Brazil;* [4]*Universidade Estácio de Sá (UNESA), Rio de Janeiro, RJ 22783-320, Brazil*

## Introduction

*Pteridium* spp. are very important toxic plants not only because they are cosmopolitan, but also due to the different types of poisoning they cause in various animal species (Evans 1987; Tokarnia *et al.* 2000), and their carcinogenic potential to humans (Alonso-Amelot and Avendano 2002). In Brazil, there are two species that belong to this genus and they are responsible for significant economic losses: *P. arachnoideum* which occurs mainly in the south and southeast regions and *P. caudatum* which occurs mainly in deforested areas of the Amazonian region (Peixoto and Tokarnia, unpublished). In Brazil, hemorrhagic diathesis, upper digestive tract carcinomas, and especially in the southeast region, enzootic hematuria had been reported in cattle. Bovine enzootic hematuria (BEH) occurs due to the radiomimetic action of a norsesquiterpene called ptaquiloside which is present in the plant (Hirono *et al.* 1984). Microscopically, both non-neoplastic and neoplastic alterations that occur in the bladders of cattle suffering from BEH exhibit an almost perfect match with those that occur in the corresponding condition in humans (Peixoto *et al.* 2003). Although greater emphasis has been given to neoplasms as the cause of enzootic hematuria, this disease can also be associated with non-neoplastic processes.

In humans, exposure to radiotherapy for the treatment of uterine or prostatic neoplasms can cause diverse non-neoplastic alterations such as hemorrhage, cystic cystitis (Suresh *et al.* 1993), and frequently squamous metaplasia in addition to vesical neoplasms (Suresh *et al.* 1993; Baker and Young 2000). Radiation-associated urothelial dysplasias can hardly be distinguished from *in situ* carcinomas (ISCs) (Baker and Young 2000) even though many authors classify them as grade IV dysplasias (Peixoto *et al.* 2003; Murphy *et al.* 2004). Cells bearing abundant generally lamellar or acidophilic cytoplasm as well as presence of cytoplasmic eosinophilic granules, nuclear hyperchromasia, and intraepithelial cysts tend to be considered evidence of the influence of radioactivity in the urothelium (Suresh *et al.* 1993). In addition, there is a close relationship between radiation and direct

©CAB International 2011. *Poisoning by Plants, Mycotoxins, and Related Toxins* (eds F. Riet-Correa, J. Pfister, A.L. Schild, and T.L. Wierenga)

damage to blood vessels in the bladder in humans (Suresh *et al.* 1993; Baker and Young 2000; Murphy *et al.* 2004). Endothelial degeneration and necrosis can be found after exposure to X-rays whereas conspicuous thickening and hyalinization of the blood vessel walls and reduction of their lumen are later events (Murphy *et al.* 2004). Other vesicular lesions associated with radiotherapy consist of inflammation, ulceration of the urothelium, and edema and fibrosis of the lamina propria (Baker and Young 2000; Murphy *et al.* 2004). In the early stages the inflammatory process in the lamina propria can be mixed; however, as time elapses there is a tendency for it to become lymphocytic or lymphoplasmocytic, perivascular, and in some cases perineural (Suresh *et al.* 1993).

After the 1986 nuclear accident near Chernobyl in the Ukraine, Romanenko *et al.* (2003) evaluated 164 patients with chronic cystitis that had been exposed to ionizing radiation for a long period. The histological examination of bladder biopsies revealed that 97% of the cases exhibited various degrees of cellular pleomorphism and nuclear hyperchromasia which were generally associated with thickening of the urothelium. Brunn's nests and ISCs occurred in 73% of the cases. All the patients exhibited proliferative cystitis, cystic cystitis, and glandular and/or squamous metaplasia which were accompanied by inflammation. Large necrotic and hyaline areas with few but prominent lymphocytic foci of infiltrating macrophages, histiocytes, and plasma cells were observed in the connective tissue of the lamina propria. Neovascularization, frequently with ectasic erythrocyte-laden vessels and hemorrhage, was detected in 62% of the cases. Proliferation of microvessels in the lamina propria was found in 46% of the cases.

It is important to highlight that concomitant epithelial and stromal neoplasms which have been described in cases of exposure to radiation in humans (Murphy *et al.* 2004) were also observed in bovines with BEH even though these alterations were outside of the scope of this study. Radiation has been reported to increase cellular pleomorphism and degree of malignancy in transitional carcinomas even at small doses (Murphy *et al.* 2004). Part of these non-neoplastic lesions are formally mentioned to occur in bladders of cattle affected by BEH (Peixoto *et al.* 2003).

The aim of this study is to establish a parallel between unreported non-neoplastic vesicular lesions observed in cattle affected by BEH with those that occur in the bladders of humans subjected to radiotherapy or accidentally exposed to radiation.

## Material and Methods

Ninety bladders were collected from cattle killed in slaughterhouses located in southern Rio de Janeiro state or subjected to necropsy in areas of southeastern Brazil invaded by *P. arachnoideum*. The findings used as criteria for the selection of the specimens included macroscopic hematuria, petechiae, ecchymoses, suffusions, hematomas, nodules, cysts, plaques, erosions, and ulcerations in the bladder. Bladders from animals with a known history of BEH were also collected even if no macroscopic lesions were present.

The bladders were longitudinally cut open through the anterior (ventral) side from the pelvic urethra to the urachal remnant area in the bladder floor in order to preserve the trigone region. Macroscopic lesions were scored, measured with calipers, and described. The collected bladders were then preserved with 10% buffered formalin. Fragments of the selected bladders were collected from areas with prominent lesions as well as areas with minimal or no macroscopic lesions, especially from the trigone, and individually processed for histopathological exams.

The histological evaluation was based mainly on the classification used by Murphy *et al.* (2004), Ordónez and Rosai (1996), Rammany (2008), and Meuten (2004).

## Results

### Macroscopic lesions

Focal or diffuse macroscopic alterations were generally observed in the mucous membrane; however, in the most severe cases the other layers of the bladder wall were also compromised. The mucous membrane exhibited variable degrees of thickening and irregular, wrinkled surface (Figure 1). Neoplastic and non-neoplastic lesions of diverse forms (polypoid, papilliform, pedunculated, ulcerated or nonulcerated nodules, and plaques or depressions with smooth or irregular surface) and sizes (between 1 mm and 2 cm in diameter) were present in a considerable number of bladders and caused distortion of the anatomic aspect and occasional deformation of the trigone with partial occlusion of the ureteral ostia. These alterations were generally accompanied by hemorrhages (multiple petechiae, ecchymoses, hematomas). Hematuria was observable in 17.7% of the cases and varied in degree from a moderately reddish liquid to a great number of blood clots with marked bladder distension.

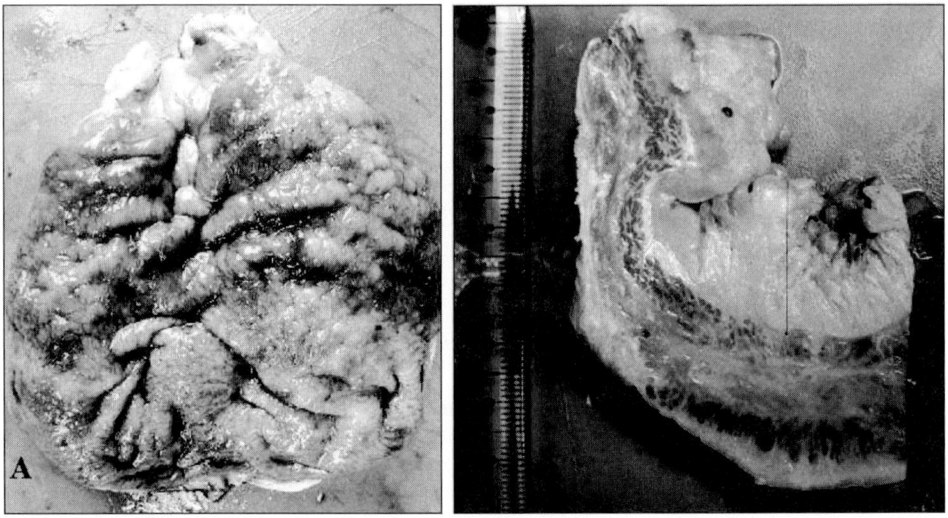


**Figure 1.** (A) Urinary bladder 39. Diffuse wrinkled appearance of the mucous membrane. (B) Transversal section of bladder 24 (fixed material) with marked thickening of the wall.

### Microscopic alterations

Non-neoplastic alterations were observed in all the bladders studied and were accompanied or not by neoplastic processes. Moderate and high-grade dysplasia was observed in the urothelium (94.4%); in addition there were hyperplasic processes (95.5%) (Figure 2) generally in the form of Brunn's nests, high grade dysplasia, and carcinoma *in situ* (Figure 3). Hyaline, mucinoid vacuoles, and cytoplasmic acidophilic granules were observed and were especially associated with high-grade dysplasia (Figure 4A). Clear cell (Figure 4B) or sometimes squamous glandular (Figure 5A), intestinal (Figure 5B), and rarely Paneth cells metaplasia were found in 76.6% of the bladders. There was formation of Brunn's nests (Figure 6 A, B) accompanied by cystic cystitis (48.8%) (Figure 6 C, D). Urothelial microcysts probably formed by rupture of cells which contained cytoplasmic vacuoles were also present.

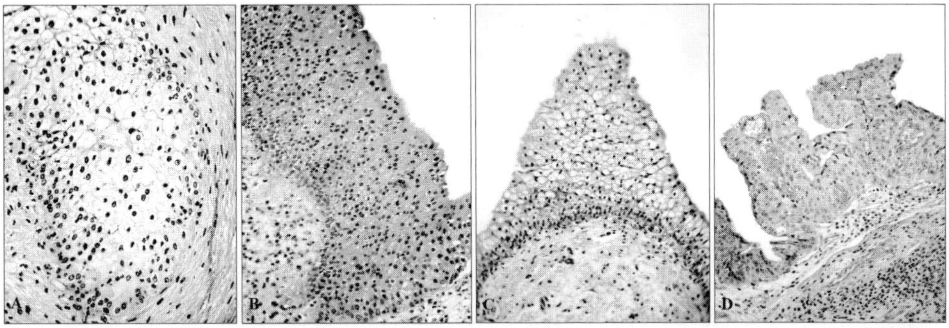

**Figure 2.** (A) Urinary bladder 125. Hyperplasic projection of the urothelium into the lamina propria in the form of Brunn's nest; notice mild dysplasia and metaplasia in clear cells. (HE, obj. 25X). (B) Bladder 6 and (C) Bladder 18. Urothelial hyperplasia with 15 or more cell layers. (D) Bladder 2. Micropapillary hyperplasic proliferation.

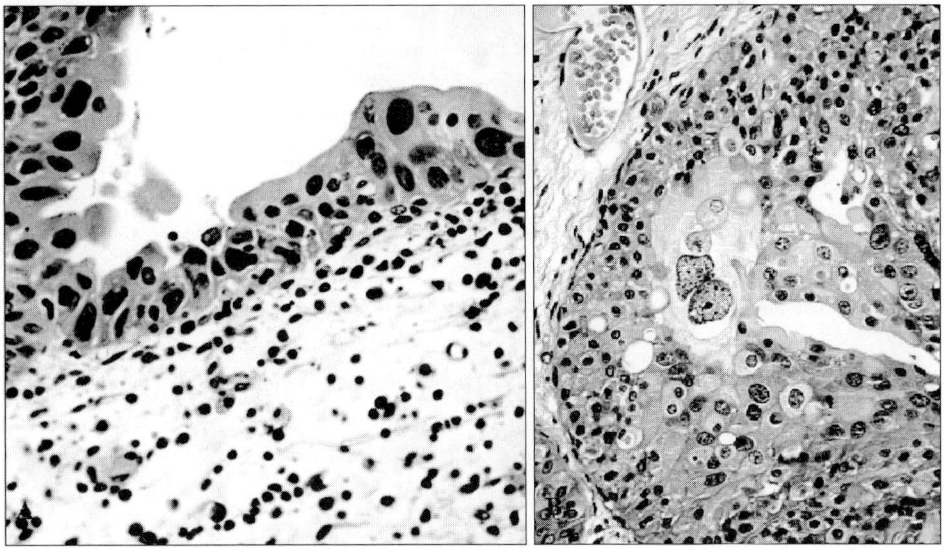

**Figure 3.** Urothelial carcinomas *in situ* (or grade IV dysplasia). (A) Loss of structural arrangement, bizarre multinucleated cells (B) and marked pleomorphism.

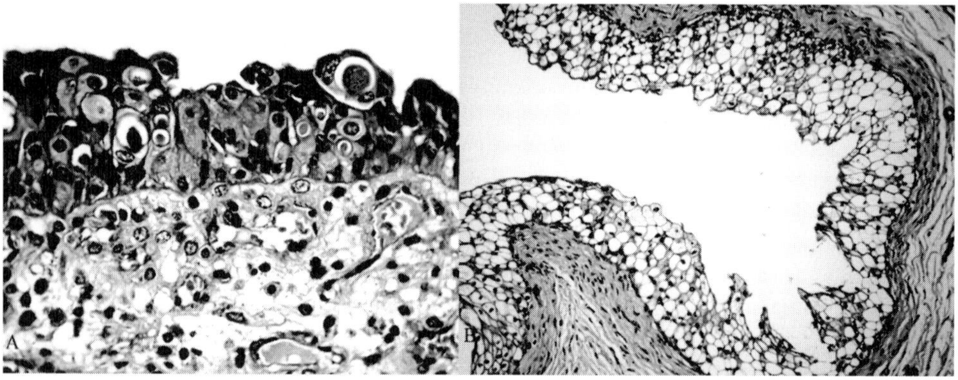

**Figure 4.** (A) Bladder 3. Grade III dysplastic urothelium; note the presence of prominent cytoplasmic vacuoles with hyaline substance. (B) Bladder 25. Clear cell metaplasia in the urothelium.

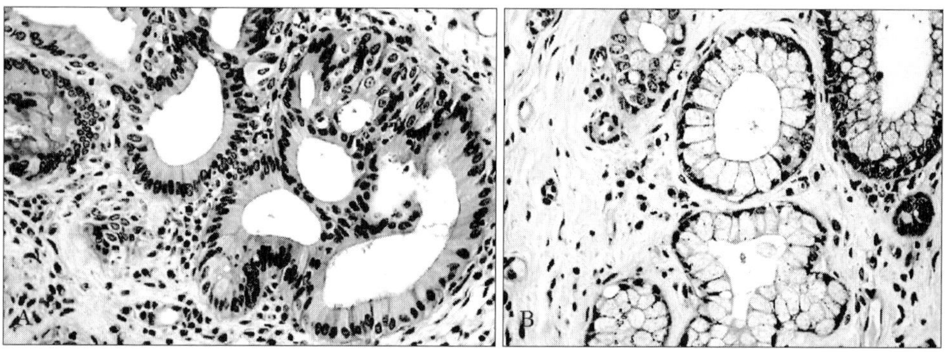

**Figure 5.** Urothelial metaplasia. (A) Bladder 93. Glandular cystitis. (B) Bladder 126. Intestinal metaplasia.

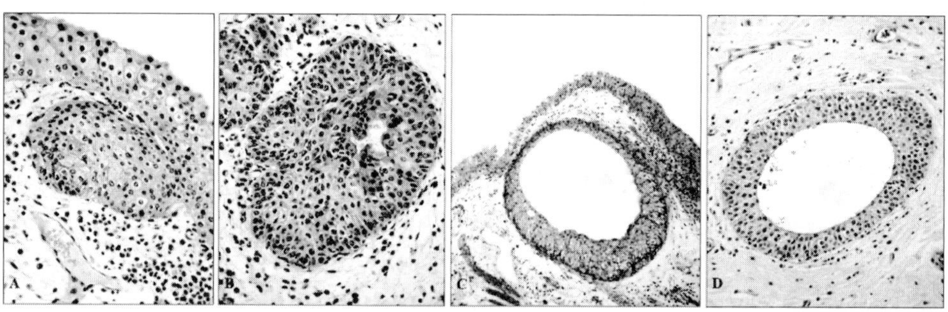

**Figure 6.** (A) Hyperplasic projection of the urothelium into the lamina propria in the form of Brunn's nest. (B) Brunn's nest without urotelial correlation; notice central degeneration and necrosis, the first step to cystic cystitis. (C and D) Bladder 90. Cystic cystitis.

Among the interstitial alterations, fibrosis of the lamina propria was present in 81.1% of the bladders examined and was generally underneath the urothelium. This type of lesion was more evident when high-grade dysplastic processes, Brunn's nests, and more frequently ISCs coincided or when present in the surroundings of neoplastic projections that invaded the lamina propria. Myxoid stromal metaplasia (26.6%), whether diffuse or with multiple foci in the lamina propria or in the stroma of polyps, was associated with benign or malignant vascular proliferation in most of the cases.

Alterations in the blood vessels occurred in 72.2% of the cases; blood vessel proliferation was prominent and usual in the stroma of polyps and also adjacent to the basal lamina of dysplastic urothelium (Figure 7A). A pattern was characterized by diffuse or multifocal proliferation of immature blood vessels in the lamina propria (Figure 7B). Capillary proliferation perpendicularly to the orientation of fibrosis was common (Figure 7C). However, in general vascular proliferation did not exhibit any organized pattern (Figure 7D). Blood vessels exhibited, in decreasing order of frequency, marked congestion and dilation, perivascular fibrosis, degeneration of the muscle layer and thickening of the wall with fibrosis, endothelial activation and degeneration, and occasionally necrosis of the wall. Alterations in lymph vessels such as marked ectasia and cell proliferation in the lamina propria were infrequent. Hemorrhages in the lamina propria were generally not associated with any previous lesions or were associated with vascular proliferation and neoplasia or even urothelial neoplasia.

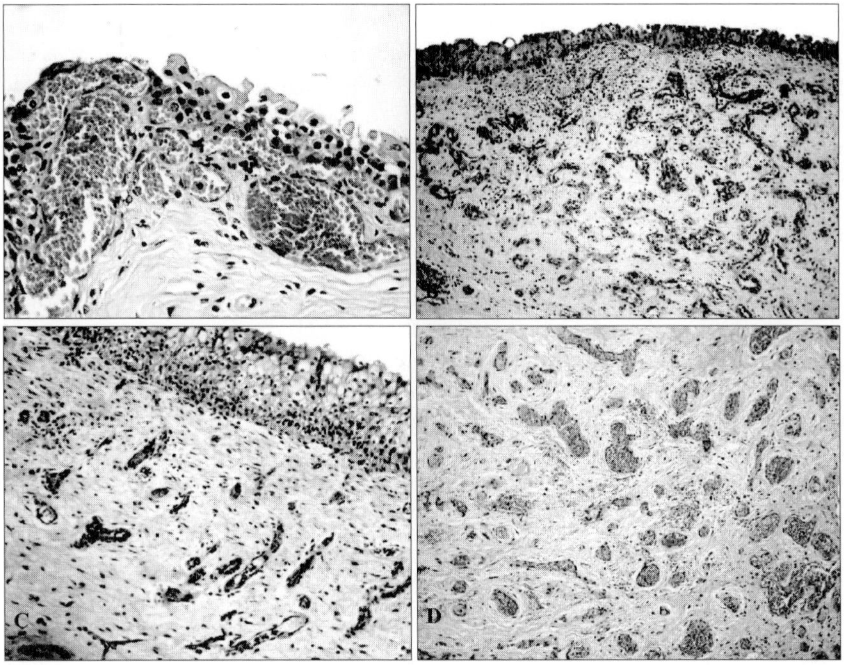


**Figure 7.** Non-neoplastic vascular and fibroblastic proliferative foci. (A) Bladder 63. Vascular proliferation adjacent to the urothelium (HE, obj. 40). (B) Bladder 2. Immature angioblastic proliferation in the lamina propria, with fibrosis (HE, obj. 10X). (C) Bladder 95. Angioblastic proliferation perpendicular to the fibroblastic proliferation. (D) Bladder 3. Vascular dilation and proliferation and diffuse perivascular fibroblastic proliferation (HE, obj. 10X).

Inflammatory processes were observed in all cases and varied from mild diffuse lymphoplasmocytic infiltration to multiple prominent lymphocytic foci around blood vessels or in the vicinity of neoplastic proliferative areas in the lamina propria. A more prominent inflammatory process was frequently observed adjacent to dysplastic or carcinomatous urothelium. Leukostasis in dilated lymph vessels and edema of the lamina propria were more prominent when associated with severe inflammation. Lymphocytic perivasculitis in the lamina propria was found in many cases. Degeneration of the detrusor musculature and small muscle bundles of the lamina propria, lymphoplasmocytic perineuritis, fibrosis, and perineural edema were less frequent findings.

## Discussion

There is a near perfect morphologic match between non-neoplastic vesical lesions described in cases of BEH and those caused by exposure to radiation in the bladders of humans. Hyperplasic, dysplastic, and metaplastic processes associated with cystic cystitis, Brunn's nests, and ISCs similar to those observed here have been described in humans exposed to radiation (Suresh *et al.* 1993).

In considering urothelial alterations, cattle with BEH developed dysplasia with or without accumulation of hyaline mucinoid substance and intraepithelial cysts, which are lesions described in the bladders of humans subjected to radiotherapeutic treatment. It is also important to mention that radiation intensifies dysplastic processes and pleomorphism in urothelial dysplasias (Baker and Young 2000; Murphy *et al.* 2004), a fact that may explain the higher frequency of high-grade urothelial dysplasias and carcinomas and also corroborate the hypothesis of benign urothelial or mesenchymal neoplasm malignization as postulated by Peixoto *et al.* (2003). In our study, some urothelial and urothelial neoplasms exhibited areas with clear transition to malignancy in spite of generally displaying benign characteristics.

Likewise, interstitial fibrosis in the lamina propria, proliferation with vascular dilation and thickening of the tunica media are findings considered to be strong evidence of the influence of radiation on the bladder in humans (Suresh *et al.* 1993; Baker and Young 2000; Murphy *et al.* 2004). Some of these alterations have been previously observed by Peixoto *et al.* (2003) in bovines which suffered from BEH.

In our cases, the inflammatory infiltrate, which was mainly lymphoplasmocytic, would be more intense when accompanied by dysplastic and neoplastic processes. This indicates the presence of long-term insult by ptaquiloside similar to the injury observed by Suresh *et al.* (1993) in the bladders of humans exposed to radiation.

The high frequency of mesenchymal neoplasms, especially vascular, observed almost exclusively in the bladders of humans exposed to radiation (Baker and Young 2000; Murphy *et al.* 2004) as well as the multifocal characteristic of the neoplasms, in particular of the ISCs, coincides with what we observed in BEH.

Numerous epithelial and mesenchymal neoplastic lesions, both malignant and benign, were also observed, however, those were not the focus of this work.

## Conclusion

We found great similarity between the histological pattern of both neoplastic and non-neoplastic lesions in the bladders of bovines poisoned by *Pteridium* spp. and the reported alterations observed in the bladders of humans exposed to radiotherapy, especially subjects with delayed lesions associated with previous periods of radiotherapy (Suresh *et al.* 1993; Baker and Young 2000; Murphy *et al.* 2004; Rammany *et al.* 2008) or after chronic exposure to environmental radiation (Romanenko *et al.* 2003).

## References

Alonso-Amelot ME and Avendano M (2002). Human carcinogenesis and bracken fern: review of the evidence. *Current Medicinal Chemistry* 9(6):675-86.

Baker PM and Young RH (2000). Radiation-induced pseudocarcinomatous proliferations of the urinary bladder: a report of 4 cases. *Human Pathology* 31:678-683.

Evans IA (1987). Bracken carcinogenicity. In *Reviews on Environmental Health* (GV James, ed.), vol. 7, pp. 161-199. International Quarterly Scientific. Reviews Freund Publishing House, Tel Aviv.

Hirono I, Aiso S, Yamaji T, Mori H, Yamada K, Niwa H, Ojika M, Wakamatsu K, Kigoshi I, Niiyama K, and Ousaki Y (1984). Carcinogenicity in rats of ptaquiloside isolated from bracken. *Gan* 75:833-836

Meuten DJ (2004). Tumours of the urinary system. In *Tumours in Domestic Animals* (DJ Meuten), pp. 524-525. Iowa State Press, Iowa.

Murphy WM, Grignon DJ, and Perlman EJ (2004). Tumors of the urinary bladder. In *Tumors of the kidney bladder and related structures* (WM Murphy, DJ Grignon, and EJ Perlman, eds), pp. 241-351. American Registry of Pathology, Washington, DC.

Ordónez NG and Rosai J (1996). Urinary tract: Kidney, renal pelvis and ureter: Bladder and male urethra. In *Surgical Pathology* (J Rosai and S Ackerman, eds), vol. 1, cap. 17, pp. 1059-1220. Mosby, St. Louis.

Peixoto PV, França TN, Barros CSL, and Tokarnia HC (2003). Histopathological aspects of Bovine Enzootic Hematuria in Brazil. *Pesquisa Veteterinária Brasileira* 23(2):65-81.

Rammany DM (2008). Genital urinary tract. In *Web Pathology* (DM Rammany). http://webpathology.com/index.asp.

Romanenko A, Morimura K, Wanibuchi H, Wei M, Zaparin W, Vozianov A, and Fukushima S (2003). Urinary bladder lesions induced by persistent chronic low-dose ionizing radiation. *Cancer* 94:328-333.

Suresh UR, Smith VJ, Lupton EW, and Haboubi NY (1993). Radiation disease of the urinary tract: histological features of 18 cases. *Journal of Clinical Pathology* 46:228-231.

Tokarnia CH, Döbereiner J, and Canella CFC (1969). Ocorrência de hematúria enzoótica e de carcinomas epidermóides no trato digestivo superior em bovinos no Brasil II. Estudos complementares. *Pesquisa Agropecuária Brasileira Seção Veteterinária* 4:209-224.

Tokarnia CH, Döbereiner J, and Peixoto PV (2000). *Plantas tóxicas do Brasil*, 310 pp. Editora Helianthus, Rio de Janeiro.

# Chapter 64

# Immunosuppression Induced by *Pteridium aquilinum* facilitates the Development of Lung Carcinogenesis

B.D. Caniceiro[1], A.O. Latorre[1], H. Fukumasu[1], M. Haraguchi[2], and S.L. Górniak[1]

*[1]Department of Pathology, Faculty of Veterinary Medicine and Animal Sciences, University of São Paulo, Ave. Prof. Dr Orlando Marques de Paiva, 87, 05508-270, São Paulo, SP, Brazil; [2]Biological Institute, Ave. Conselheiro Rodrigues Alves, 1252, 04014-002, São Paulo, SP, Brazil*

## Introduction

*Pteridium aquilinum*, popularly known as 'bracken fern', is one of most important toxic plants in the world. Its ingestion is associated not only with poisoning of livestock in various parts of the world but also with the development of cancers in humans and animals (Alonso-Amelot 1999; Sugimura 2000). Moreover, it is hypothesized that this plant also induces immunosuppression because in cattle infected with bovine papilloma virus (BPV) and fed on *P. aquilinum*, papillomas become malignant carcinomas which in healthy cattle is self-limiting (Borzacchiello *et al.* 2003).

Our earlier studies in mice showed that *P. aquilinum* reduces natural killer (NK) cell cytotoxicity (Latorre *et al.* 2009). These cells play important roles in cancer immunosurveillance (Yang *et al.* 2006) and suppression of NK activity may cause increased cancer metastasis and decreased host survival (Whiteside 2006). Further, mice with lung cancer and high NK activity had reduced lung metastasis compared to immunosuppressed mice (Trinchieri 1989).

The objective of the present study is to evaluate the relationship between immunosuppression caused by *P. aquilinum* and lung carcinogenesis induced by ethyl carbamate (EC).

## Material and Methods

### Mice

Subjects were 60-day-old female C57Bl/6 mice bred in the Department of Pathology at the School of Veterinary Medicine and Animal Sciences. The mice were maintained

©CAB International 2011. *Poisoning by Plants, Mycotoxins, and Related Toxins* (eds F. Riet-Correa, J. Pfister, A.L. Schild, and T.L. Wierenga)

under controlled conditions of temperature (22-25°C), relative humidity (50-65%), and lighting (12 h/12 h light/dark cycle). Drinking water and standard diet (Nuvilab-CR1®, Nuvital Nutrientes LTDA) were provided *ad libitum*. All procedures using the mice followed the 'Guide for the Care and Use of Laboratory Animals' (NIH publication No. 85-23) and were reviewed and approved by the Bioethics Committee of the FMVZ-USP (process #1511/2008).

### Reagents

Ethyl carbamate (Urethane) was obtained from Sigma Chemical Co (Saint Louis, USA). Thiamine (vitamin B1), PA methanol, chloroform, and acetic acid were from Labsynth (São Paulo, Brazil). EDTA was from Merck & Co (Darmstadt, Germany). Dopalen (ketamine) and Anasedan (xylazine) were from Vetbrands ™ (São Paulo, Brazil).

### *Pteridium aquilinum*

*P. aquilinum* buds were collected at Pirassununga, São Paulo, Brazil, in February 2008. The buds were maintained frozen at -80°C until extract preparation. For extract preparation, the frozen buds (1 kg) were ground and put under pressure to yield a viscous residue that was weighed to calculate the dose. The extract was maintained frozen at -80°C and immediately before being given to mice was suspended in distilled water and used at a dose equivalent to 30 g of plant/kg body weight (BW).

The mice were treated for 13 weeks. The administration of the *P. aquilinum* extract was by gavage once daily for 14 consecutive days and thereafter for 5 days/week for 11 weeks at the same time of day. In addition, all mice were supplemented with Vitamin $B_1$ in water (10 mg/ml) as proposed by Schacham *et al.* (1970) throughout the treatment period to avoid the effects of thiaminase 1, a component of *P. aquilinum* (Fenwick 1988). The body weight of all mice was measured every 3 days during the first 14 days for dose adjustment and thereafter twice/week for the duration of the study.

### Induction and evaluation of lung carcinogenesis

The protocol proposed by Miller *et al.* (2003) for induction of lung carcinogenesis was used with some modifications. The mice were treated weekly by intraperitoneal (i.p.) injection with EC at 1mg/g BW for 11 weeks and euthanized 30 weeks after starting the treatment. The lungs were collected and carefully inflated with metacarn (60:30:10 methanol, chloroform, and acetic acid, v/v) to count the number of macroscopic lesions. After macroscopic analyses, the lungs were fixed for 8 h and placed in 95% alcohol until processing. Throughout the treatment the mice were weighed weekly for dose adjustment.

The pulmonary lesions were evaluated macro- and microscopically. The macroscopic parameters evaluated were: incidence (% animals with injury), number of lesions/animal, and multiplicity (number of lesions/animal with injury). For the microscopic evaluation, the lesions observed were classified as pre-neoplastic and neoplastic.

### Experimental design

Mice were separated into four groups: negative control (Co); immunosuppression (Pt–30 g *P. aquilinum*/kg BW); carcinogenesis (E–1 mg EC/g BW), and carcinogenesis and immunosuppression (PE–30 g *P. aquilinum*/kg BW and 1 mg EC/g BW). The mice from

Co and E groups received water by gavage and mice from Pt and PE groups were treated with *P. aquilinum* by the same route. The treatment with EC started on day 15 of the experiment; the mice from E and PE groups were i.p. injections with EC and mice from the Co and Pt groups were treated with PBS.

### Statistical analysis

The data were analyzed using GraphPad Prism 4.00® software (GraphPad Software, Inc., San Diego, CA) by one-way analysis of variance (ANOVA) followed by Dunnett's test for multiple comparisons. Percentage data from three or more groups were compared by the Kruskal-Wallis test followed by Dunn's test for multiple comparisons, and percentage data from two groups were compared by the Mann-Whitney test. All data were expressed as the mean ± SEM and differences were considered to be statistically significant at $P < 0.05$.

## Results

The macroscopic evaluation showed a higher number of lesions/mouse in the PE group compared with E group that approached significance ($P = 0.08$ Mann Whitney Test) (Figure 1); the other parameters were not significantly altered. The macroscopic evaluation showed a higher number of lesions/mouse in the PE group compared with E group that approached significance ($P = 0.08$ Mann Whitney Test) (Figure 1); the other parameters were not significantly altered. Finally, observed decrease in weight gain of mice in the groups treated with ethyl carbamate (Ur and PU) during the treatment period when compared to the groups Co and Pt ($P < 0.0001$ Kruskal-Wallis Test, $P < 0.001$ post-test Dunn's Multiple Comparison Test). However, there was the same weight gain among all groups when assessed during the period of the experiment (data not shown).

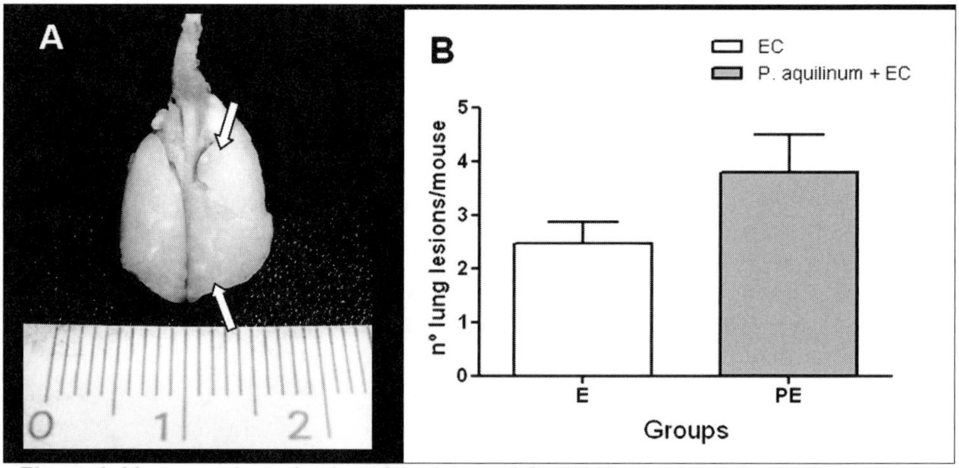


**Figure 1.** Macroscopic evaluation of the lungs of C57BL/6 mice treated with *P. aquilinum* and/or ethyl carbamate (EC). A – Lung insufflated with metacarn (arrows indicate tumoral lesions of ≈1 mm). B – Number of lesions per animal.

Furthermore, we carried out a histopathologic analysis of lungs after staining with hematoxylin and eosin (H&E), examining the lungs for the presence of pre-neoplastic and neoplastic lesions. The pre-neoplastic lesions observed here were classified according to the latest classification of lung proliferative lesions in mice (Nikitin *et al.* 2004):

- Alveolar hyperplasia: solitary or multiple foci of increased cellularity distal to terminal bronchioles. The background of bronchoalveolar architecture remains detectable and epithelial cells are usually single layered. Round to oval hypertrophic type II pneumocytes with abundant eosinophilic cytoplasm line alveolar wall.
- Epithelial hyperplasia: characterized by an increase in number of cuboidal, columnar, ciliated, or mucous cells without atypia. Cells maintain normal architecture of bronchioles and alveoli.
- Lymphocytic hyperplasia: characterized by proliferation of lymphoid tissue associated with the bronchioles (BALT).
- Alveolar hyperplasia in appearance on bronchiolar epithelium (bronchiolization): the alveolar walls are lined by cuboidal to columnar cells with features of bronchiolar differentiation, such as formation of cilia, Clara cell resemblance, and presence of mucous granules. Foci of consolidation may indicate early stages of adenoma formation. Macrophages may be present in the alveolar lumens.

The neoplastic lesions observed here were also classified according to the classification above. These lesions consist of adenomas that are characterized by areas composed of cuboid to columnar cells lining the alveoli. The size is usually less than 5 mm in diameter and retains preexisting alveolar structure. The different types are classified as follows:

- Solid adenoma: round to oval cells fill alveolar spaces. Cells usually have abundant eosinophilic cytoplasm with fine granularity and/or vacuoles.
- Papillary adenoma: consists primarily of papillary structures lined by cuboidal to columnar cells. Cells forming papillary structures are frequently more hyperchromatic and atypical, which is regarded as an indication of potential progression toward malignancy.
- Mixed adenoma: in this type of adenoma both papillary and solid structures are present.

Thus, there was a higher percentage of neoplastic lesions in the PE group compared with E group (E = 33.33%; PE = 60%) (Figure 2); however, there were no significant alterations in the percentage of pre-neoplastic lesions.

## Discussion and Conclusion

The NK cells play several important roles in defense against intracellular microbes and in cancer immunosurveillance (Abbas *et al.* 2007). Therefore, it was very important to determine if the reduced NK cytotoxicity observed in mice treated with *P. aquilinum* (Latorre *et al.* 2009) facilitates cancer development. The results obtained here showed that *P. aquilinum* increased the number of lesions/mouse and the percentage of neoplastic lesions.

We conclude that the immunosuppression caused by *P. aquilinum* facilitates the development of lung carcinogenesis in mice. Moreover, we hypothesize that this effect may be directly involved in the development of cancer in humans and animals that feed on bracken fern.

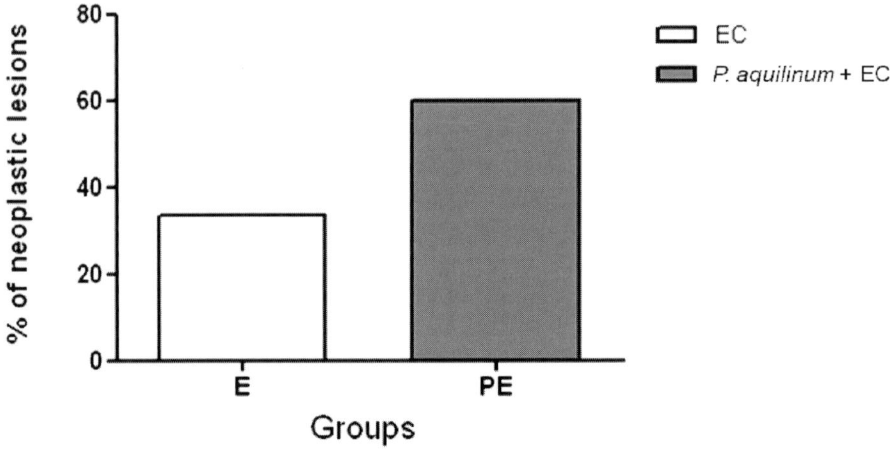


**Figure 2.** Incidence of lung neoplastic lesions of C57BL/6 mice treated with *P. aquilinum* and/or ethyl carbamate (EC).

## Acknowledgements

Beatriz D. Caniceiro was supported by a fellowship from FAPESP (Proc. 2008/50073-6), Brazil.

## References

Abbas AK, Litchman AH, and Pober JS (2007). *Cellular and Molecular Immunology*, 6th edn. Elsevier Saunders, London.

Alonso-Amelot ME (1999). Helecho macho, salud animal y salud humana. *Revista Faculdade de Agronomia (*LUZ*)* 16:528-547.

Borzacchiello G, Ambrosio V, Roperto S, Poggiali F, Tsirimonakis E, Venuti A, Campo MS, and Roperto F (2003). Bovine Papillomavirus Type 4 in Oesophageal Papillomas of Cattle from the South of Italy. *Journal of Comparative Pathology* 128(2-3):203-206.

Fenwick GR (1988). Bracken (*Pteridium aquilinum*) – toxic effects and toxic constituents. *Journal of the Science of Food and Agriculture* 46:147-173.

Latorre AO, Furlan MS, Sakai M, Fukumasu H, Hueza IM, Haraguchi M, and Górniak SL (2009). Immunomodulatory effects of *Pteridium aquilinum* on natural killer cell activity and select parts of the cellular immune response of mice. *Journal of Immunotoxicology* 6(2):104-114.

Miller YE, Dwyer-Nield LD, Keith RL, Le M, Franklin WA, and Malkinson AM (2003). Induction of a high incidence of lung tumors in C57BL/6 mice with multiple ethyl carbamate injections. *Cancer Letters* 198(2):139-144.

Nikitin AY, Alcaraz A, Anver MR, Bronson RT, Cardiff RD, Dixon D, Fraire AE, Gabrielson EW, Gunning WT, Haines DC, Kaufman M H, Linnoila RI, Maronpot RR, Rabson AS, Reddick RL, Rehm S, Rozengurt N, Schuller HM, Shmidt EN, Travis WD, Ward MJ, and Jacks T (2004). Classification of proliferative pulmonary lesions of the mouse: recommendations of the mouse models of Human cancers consortium. *Cancer Research* 64(7):2307-2316.

Schacham P, Philp RB, and Gowdey CW (1970). Antihematopoietic and carcinogenic effects of bracken fern (*Pteridium aquilinum*) in rats. *American Journal of Veterinary Research* 31(1):191-197.

Sugimura T (2000). Nutrition and dietary carcinogens. *Carcinogenesis* 21(3):387-395.

Trinchieri G (1989). Biology of natural killer cells. *Advances in Immunology* 47:187-376.

Whiteside TL (2006). Immune suppression in cancer: Effects on immune cells, mechanisms and future therapeutic intervention. *Seminars in Cancer Biology* 16(1):3-15.

Yang Q, Goding SR, Hokland ME, and Basse PH (2006). Antitumor Activity of NK Cells. *Immunologic Research* 36(1-3):13-25.

# Chapter 65

# Outbreak of Acute Poisoning by Bracken Fern (*Pteridium aquilinum*) in Cattle

L. Sonne, P.M. Bandarra, D.L. Raymundo, P.M.O. Pedroso, A.G.C. Dalto, J.S. Leal, C.E.F. Cruz, and D. Driemeier

*Setor de Patologia Veterinária, Universidade Federal do Rio Grande do Sul, Porto Alegre, RS, Brazil*

## Introduction

The consumption of *Pteridium aquilinum* (Polypodiaceae) by cattle has been associated with poisoning on all continents except Antarctica. In Brazil, the variety *arachnoideum* has been identified (Tokarnia *et al.* 2000; França *et al.* 2002). Ingestion of bracken fern may cause three distinct forms of poisoning: an acute form characterized by hemorrhagic diathesis (Anjos *et al.* 2008) and two chronic forms characterized by tumors in the urinary bladder (known as bovine enzootic hematuria; Özkul and Aydin 1996; Sardon *et al.* 2005; Carvalho *et al.* 2006) and in the upper digestive tract (Tokarnia *et al.* 1969; Souto *et al.* 2006). Animals may ingest the plant especially when they are hungry, during shortages of suitable forage, and when bracken fern is sprouting. All parts of the plant are toxic (Riet-Correa and Méndez 2007) and the most important toxic component in the plant is ptaquiloside, which is a glycoside sesquiterpenoid that may induce carcinogenicity beyond a radiomimetic effect that causes hemorrhagic diathesis (Carvalho *et al.* 2006).

The acute form may affect cattle at any age and has a high mortality (close to 100%). Depression, epistaxis, lacrimation, dyspnea, blood clots in the feces, petechiae in mucous membranes, and fever may be seen in acute cases (Osebold 1951; Sippel 1952) which may fatally terminate in 2 or 3 days, while subacute cases often can survive longer, from 4 to 10 days (Osebold 1951). Hematologic exams show thrombocytopenia, leucopenia, and non-regenerative normocytic normochromic anemia. The hemorrhages are attributed to the severe reduction in the thrombocyte population (Osebold 1951; Anjos *et al.* 2008). At necropsy varying hemorrhages (from petechiae to large ecchymoses) may affect skin, subcutaneous tissue, heart, lung, urinary bladder, skeletal muscles, mucous membranes of abomasum and intestine, and serous surfaces (Craig and Davies 1940; Sippel 1952). Liver infarcts (Sippel 1952) and necrotic lesions (Osebold 1951), epistaxis, blood in feces, and ulcers in the abomasum and intestine are also observed (Marçal *et al.* 2002). Microscopic lesions in acute poisoning include bone marrow aplasia with accentuated decrease of the granulocytic and megakaryocytic series (Marçal *et al.* 2002). Scattered hemorrhages, bacteria, and thrombosed vessels associated with infarcts mainly in liver and kidneys have also been described (Anjos *et al.* 2008). There is no treatment for bracken fern poisoning

©CAB International 2011. *Poisoning by Plants, Mycotoxins, and Related Toxins* (eds F. Riet-Correa, J. Pfister, A.L. Schild, and T.L. Wierenga)

(Sippel 1952; Riet-Correa and Méndez 2007). This paper concerns an outbreak of bracken fern poisoning in cattle and discusses data on the platelet counts of the affected animals.

## Materials and Methods

In total, 47 out of 203 finishing open cows were affected and died in an outbreak of bracken fern poisoning in a herd from a farm located in the Vacaria municipality, northeastern region of Rio Grande do Sul, Brazil. Animals were kept in a poor pasture infested with *P. aquilinum*. Three animals were necropsied and samples from organs were collected, fixed in buffered 10% formalin, routinely processed for histology, and stained by hematoxylin and eosin. Blood samples were collected from 39 of the 168 remaining cows; however, 16 samples were not suitable for analysis.

## Results and Discussion

Clinical signs of the affected group of cows included apathy, depression, emaciation, irregular and roughened coat, weakness, skin hemorrhages, bloody feces, bloody nasal discharge, gait instability, recumbence, and death. Cows that were necropsied presented epistaxis (1/3), blood clots in trachea and lung (1/3), abomasal hemorrhages and ulcers (2/3), and hepatic pale areas (1/3). Decreased population of cells in bone marrow whose space was replaced by fat and dilatation of the medullar sinusoids were the main microscopic findings. Hemorrhages in lung (1/3) and abomasum (2/3) and ulcers in abomasum were also observed histologically. Thrombosed vessels with bacteria and infarct areas were seen in kidney (1/3) and liver (1/3), and hemosiderosis was present in mesenteric lymph nodes (2/3). Blood analysis from 11 animals demonstrated decreased platelet counts (Table 1); all animals died between 1 and 2 weeks after the visit to the farm. Death due to thrombocytopenia has been linked to low platelet counts such as $10/mm^3$ (Anjos *et al.* 2008). In these cases platelet counts of animals that died were equal to or lower than $62/mm^3$; however, the deaths occurred in 1 to 2 weeks after blood sample collection therefore there was sufficient time to develop severe thrombocytopenia.

## Conclusions

Diagnosis was based on epidemiological, clinical, pathological and hematological findings. It is highly probable that the main underlying causes for this outbreak were the lack of suitable forage and the introduction of animals to a new paddock infested with bracken fern. Platelet counts allowed detection of apparently normal, but affected animals that were not showing clinical signs. Therefore, platelet counts may be useful for detecting animals sub-clinically poisoned by *P. aquilinum*, and may allow a more precise evaluation of the magnitude of the problem.

**Table 1.** Platelet counts in 23 cows from an outbreak of bracken fern poisoning in southern Brazil, and outcome of the poisoning 1-2 weeks after the initial visit to the farm.

| Cow | Platelet counts[1] | Outcome | Cow | Platelet counts | Outcome |
|---|---|---|---|---|---|
| 1 | 2,000/mm$^3$ | died | 13 | 173,000/mm$^3$ | survived |
| 2 | 4,000/mm$^3$ | died | 14 | 186,000/mm$^3$ | survived |
| 3 | 6,000/mm$^3$ | died | 15 | 249,000/mm$^3$ | survived |
| 4 | 8,000/mm$^3$ | died | 16 | 265,000/mm$^3$ | survived |
| 5 | 12,000/mm$^3$ | died | 17 | 286,000/mm$^3$ | survived |
| 6 | 14,000/mm$^3$ | died | 18 | 322,000/mm$^3$ | survived |
| 7 | 16,000/mm$^3$ | died | 19 | 367,000/mm$^3$ | survived |
| 8 | 22,000/mm$^3$ | died | 20 | 455,000/mm$^3$ | survived |
| 9 | 34,000/mm$^3$ | euthanized | 21 | 498,000/mm$^3$ | survived |
| 10 | 36,000/mm$^3$ | died | 22 | 499,000/mm$^3$ | survived |
| 11 | 62,000/mm$^3$ | died | 23 | 508,000/mm$^3$ | survived |
| 12 | 152,000/mm$^3$ | survived* | | | |

[1]Platelets – Reference value for cattle: 100,000-800,000.
* All the animals from No. 12 downward on the list were alive when the last contact was made with the farm personnel (4 weeks after the initial visit).

## References


Anjos BL, Irigoyen LF, Fighera RA, Gomes AD, Kommers GD, and Barros CSL (2008). Intoxicação aguda por samambaia (*Pteridium aquilinum*) em bovinos da região Central do Rio Grande do Sul. *Pesquisa Veterinária Brasileira* 28:501-507.

Carvalho T, Pinto C, and Peleteiro MC (2006). Urinary bladder lesions in bovine enzootic haematuria. *Journal of Comparative Pathology* 134:336-346.

Craig JF and Davies GO (1940). Some observations on bracken poisoning. *The Veterinary Record* 52:499.

França TN, Tokarnia CH, and Peixoto PV (2002). Enfermidades determinadas pelo princípio radiomimético de *Pteridium aquilinum* (Polypodiacea). *Pesquisa Veterinária Brasileira* 22:85-96.

Marçal WS, Gaste L, Reichert Netto NC, and Monteriro FA (2002). Intoxicação aguda pela samambaia (*Pteridium aquilinum* L. Kuhn), em bovinos da raça Aberdeen Angus. *Archives of Veterinary Science* 7:77-81.

Osebold JW (1951). An approach to the pathogenesis of fern poisoning in the bovine species. *Journal of American Veterinary Medical Association* 121:440-441.

Özkul IA and Aydin Y (1996). Tumors of the urinary bladder in cattle and water buffalo in the Black Sea region of Tukey. *British Veterinary Journal* 152:473-475.

Riet-Correa F and Méndez MDC (2007). Intoxicações por plantas e micotoxinas: Plantas que causam fibrose hepática. In *Doenças de Ruminantes e Eqüideos* (F Riet-Correa, AL Schild, RAA Lemos, and JRJ Borges, eds), vol. 2, pp. 99-219. Pallotti, Santa Maria.

Sardon D, De La Fuente I, Calomge E, Perez-Alenza MD, Castaño M, Dunner S, and Peña L (2005). H-ras immunohistochemical expression and molecular analysis of urinary lesions in grazing adult cattle exposed to bracken fern. *Journal of Comparative Pathology* 132:195-201.

Sippel L (1952). Bracken fern poisoning. *Journal of Veterinary Medical Association* 121:9-13.

Souto MAM, Kommers GD, Barros CSL, Piazer JVM, Rech RR, Riet-Correa F, and Schild AL (2006). Neoplasias do trato alimentar superior de bovinos associadas ao consumo espontâneo de samambaia (*Pteridium aquilinum*). *Pesquisa Veterinária Brasileira* 26:112-122.

Tokarnia CH, Döbereiner J, and Canella CFC (1969). Ocorrência da hematúria enzoótica e de carcinomas epidermóides no trato digestivo superior em bovinos no Brasil. II. Estudos complementares. *Pesquisa Agropecuária Brasileira* 4:209-224.

Tokarnia CH, Döbereiner J, and Peixoto PV (2000). *Plantas Tóxicas do Brasil,* 310 pp. Helianthus, Rio de Janeiro.

# Chapter 66

# Immunosuppressive Effects of *Pteridium aquilinum* on Natural Killer Cells of Mice and its Prevention with Selenium

A.O. Latorre[1], B.D. Caniceiro[1], M. Haraguchi[2], and S.L. Górniak[1]

*[1]Department of Pathology, Faculty of Veterinary Medicine and Animal Sciences, University of São Paulo, Prof. Dr. Orlando Marques de Paiva 87, 05508-270, São Paulo, Brazil; [2]Biological Institute, Conselheiro Rodrigues Alves 1252, 04014-002, São Paulo, Brazil*

## Introduction

Bracken fern (*Pteridium aquilinum*) is one of the most common plants in the world (Taylor 1990) and its ingestion has been associated with induction of cancer in humans and animals (Alonso-Amelot 1999; Shahin *et al.* 1999; Alonso-Amelot and Avendano 2002). Several compounds–tannins and flavonols–have been isolated from bracken fern; however, only a few compounds such as quercetin and ptaquiloside have been reported to possess carcinogenic potential in experimental animals. Quercetin was reported to cause urinary bladder and intestinal tumors in rats (Pamukcu *et al.* 1980) but genotoxicity studies in mice did not show this effect (Ngomuo and Jones 1996). In addition, a critical evaluation of the available literature on the biological effects of quercetin concluded that dietary intake levels do not produce adverse health effects (Harwood *et al.* 2007). On the other hand, a body of evidence suggests that ptaquiloside, a norsesquiterpene glucoside, is the main carcinogenic compound found in bracken fern (Hirono *et al.* 1987). Ptaquiloside is soluble in water and in an aqueous alkaline solution (pH range 8 to 11) this substance is readily converted to an unstable dienone that contains a highly reactive cyclopropyl group that reacts rapidly with amino acids, nucleosides, nucleotides, and DNA (Ojika *et al.* 1987), eventually leading to cancer.

In humans, epidemiological studies revealed an increased risk of esophageal and gastric cancer in people who consume bracken fern or its toxins either directly as crosiers or rhizomes (Sugimura 2000; Abnet 2007) or indirectly through the consumption of milk from cows feeding on bracken fern (Alonso-Amelot *et al.* 1996; Alonso-Amelot 1997).

In cattle, bracken fern causes an acute hemorrhagic syndrome associated with depression of the bone marrow characterized by leucopenia and thrombocytopenia (Riet-Correa *et al.* 2001; Do Nascimento França *et al.* 2002). Chronic exposure induces carcinomas of the urinary bladder known as bovine enzootic hematuria (BEH) (Peixoto *et al.* 2003; Carvalho *et al.* 2006) and carcinomas of the upper alimentary tract (Tokarnia *et al.* 2000; Do Nascimento França *et al.* 2002).

©CAB International 2011. *Poisoning by Plants, Mycotoxins, and Related Toxins* (eds F. Riet-Correa, J. Pfister, A.L. Schild, and T.L. Wierenga)

An increased risk for the development of these bracken fern-induced carcinomas has been associated with the bovine papilloma virus (BPV). This association is supported by BPV-2 induced urinary bladder lesions and clinical signs of BEH in adult cattle (Campo *et al.* 1992) and BPV-4 with squamous cell carcinomas of the upper alimentary tract (Borzacchiello *et al.* 2003; Moreira Souto *et al.* 2006). Consequently, it is possible that the observed increases in cancer could be related to induction of an overall immunosuppression by the plant or its various constituents. This study was conducted to evaluate the immunosuppressive effects of bracken fern in mice.

## Immunosuppresive Effects of *Pteridium aquilinum* in Mice

Earlier histological studies performed in C57BL/6 mice administered an extract of *P. aquilinum* by gavage over a period of 14-30 days revealed a significant reduction in splenic white pulp area. A delayed-type hypersensitivity (DTH) and reduced IFNγ production by NK cells during $T_H1$ priming were observed. Finally, NK cell numbers in the spleen were reduced and the innate response in these hosts assessed by analysis of NK cell cytotoxic functionality was also diminished (Latorre *et al.* 2009). In this study we determined if NK cell cytotoxicity was a function of reduced activity or a decrease in viability. Furthermore, we evaluated the effect of selenium (Se) treatment in prevention of or reduction in NK cell cytotoxicity (Latorre *et al.* 2007).

## Material and Methods

### Aqueous extract of *Pteridium aquilinum*

A voucher specimen (No. MH 513) was deposited at the Laboratory of Chemistry of Natural Products, Centre of Animal Health, Biological Institute of Sao Paulo. *P. aquilinum* var. *arachnoideum* was identified by Dr Jefferson Prado, a specialist in Pteridophyta identification from the Botanical Institute of Sao Paulo. *P. aquilinum* buds were collected in Pirassununga, Sao Paulo, Brazil, in February 2008. The buds were frozen at -80°C until extract preparation. The aqueous extract of the frozen ground buds was prepared as described by Burkhalter *et al.* (1996) with modifications. Briefly, the frozen ground buds (20 g) were extracted with cold RPMI medium (20 ml) under agitation for 1 h in the dark. After filtration, the aqueous extract (AEP) was maintained frozen at -20ºC until use.

### Preparation of spleen cell suspensions and treatment *in vitro*

The spleen was collected from each of six male C57BL/6 mice to prepare cell suspensions. For each mouse, the tissue was gently fragmented by squeezing the distal end of the syringe into the plate in cold RPMI medium. The erythrocytes present in the suspension were then lysed with 0.4% ammonium chloride sterile solution. Splenocytes were centrifuged at 1200 rpm (4°C, 8 min) and the pelleted cells were re-suspended in RPMI complete medium supplemented with 10% FBS. The samples were incubated in 6-well plates for 2 h at 37ºC in a humidified atmosphere containing 5% $CO_2$ to separate non-adherent from adherent cells. Cultures of non-adherent cells were performed for each treatment as follows: untreated, Se 0.1 mM, AEP 4 mg/ml, and AEP 4 mg/ml + Se 0.1 mM, and incubated for 1 h at 37°C in a humidified atmosphere with 5% $CO_2$. After treatment, the cells were washed and adjusted to $5 \times 10^6$ cells/ml in complete RPMI medium.

### Natural killer (NK) cell activity assay

To assay for NK cell cytotoxicity the non-adherent cells isolated from the spleen were used as effector cells and the Ehrlich ascites tumor cells as the target cells. Quadruplicate cell cultures from each treatment were done with $5 \times 10^5$ effector cells and $5 \times 10^3$ target cells stained with CFSE (ratio 100:1) for 4 h at 37°C in a humidified atmosphere containing 5% $CO_2$. Spontaneous death was determined by incubating Ehrlich ascites tumor cells only in complete RPMI medium. Propidium iodide (PI) was then added and the samples acquired by flow cytometry. Data were analyzed using FlowJo 7.2.2® software. The level of NK cell cytotoxicity was expressed as:

$$\text{Cytotoxicity (\%)} = \frac{\text{dead targets in samples (\%)} - \text{spontaneously dead targets}}{100 - \text{spontaneously-dead targets (\%)}} \times 100$$

### MTT assay

The yellow tetrazolium MTT (3-(4,5-dimethylthiazolyl-2)-2,5-diphenyltetrazolium bromide) is reduced by metabolically active cells in part by the action of dehydrogenase enzymes to generate reducing equivalents such as NADH and NADPH. The resulting intracellular purple formazan can be solubilized and quantified by spectrophotometric means.

For measurement of non-adherent cell viability the cells were then incubated in a 96-well microtiter plate for 24 h at 37°C in a humidified atmosphere containing 5% $CO_2$. Quadruplicate cultures from each treatment were done with $5 \times 10^5$ cells/well and after 21 h of incubation 10 μl of MTT solution (5 mg/ml MTT in PBS) were added. At the end of the incubation period 100 μl of acidic isopropanol (0.04 M HCl in absolute isopropanol) were added to dissolve purple formazan and the plate was maintained in the dark for 1 h at room temperature. The absorbance was measured at 570 nm in a microtiter plate reader.

### Statistical analysis

The data were analyzed using GraphPad Prism 5.00® software (GraphPad Software, Inc., San Diego, CA). All data were expressed as the mean ± SD and differences were considered to be statistically significant at $P < 0.05$.

## Results

The treatment with aqueous extract of *P. aquilinum* reduced NK cell cytotoxicity and did not alter cell viability (Table 1). Selenium co-treatment further prevented the reduction of NK cell cytotoxicity but reduced cell viability (Table 1).

**Table 1.** Percentage of NK cell cytotoxicity and cell viability of splenocytes treated *in vitro* with aqueous extract of *Pteridium aquilinum* (AEP) 4 mg/ml and/ or selenium (Se) 0.1 mM.

| | Untreated | AEP | AEP+Se | Se |
|---|---|---|---|---|
| Cytotoxicity (%) | 27.67 ± 3.84 | 21.27 ± 5.96[a] | 28.15 ± 5.52 | 31.24 ± 5.35[b] |
| Cell Viability (abs) | 0.048 ± 0.011 | 0.045 ± 0.009 | 0.027 ± 0.006* | 0.024 ± 0.003* |

Data are expressed as mean ± SD ($n$=6). [a] $P$ = 0.0379, [b] $P$ = 0.0240 Two-way ANOVA.
*$P$ < 0.001 Repeated Measures ANOVA followed by Dunn's Multiple Comparisons post test

## Discussion and Conclusions

Our earlier studies showed that *P. aquilinum* caused a reduction of NK cell cytotoxicity and reduced the number of these cells in the spleen of the mice (Latorre *et al.* 2009). Because these cells play several important roles in defense against intracellular microbes and in cancer immunosurveillance (Abbas *et al.* 2007), it was important to verify if the reduced NK cytotoxicity was from a reduction in the number of NK cells or whether it was due to reduced cytotoxic activity. In addition we evaluated if selenium could prevent the decline of activity of these cells. These results demonstrate that the aqueous extract of *P. aquilinum* reduced the cytotoxicity of NK cells but did not alter cell viability. Furthermore, it was observed that selenium co-treatment prevented the decrease in NK cytotoxicity. Among the actions of selenium in the immune system was described improvement in cytotoxic activity of NK cells (Petrie *et al.* 1989; McKenzie *et al.* 1998) and this action likely related to the results shown in this study.

We conclude that *P. aquilinum* reduces NK cell cytotoxicity due to a diminution of cell activity and that selenium supplementation can prevent this immunotoxic effect. Additionally, we suggest that this reduction of NK cell cytotoxicity affects the ability of these cells to lyse tumor cells prior to stimulation and may be involved in the development of cancer in humans and animals that ingest bracken fern.

## Acknowledgements

Andreia O. Latorre was supported by a fellowship from FAPESP (Proc. 07/50313-4), Brazil.

## References

Abbas AK, Litchman AH, and Pillai S (2007). Cellular and Molecular Immunology 6th edn. In *Immunity to Tumors,* (AK Abbas, AH Litchman, and S Pillai, eds), pp. 397-418. Saunders Elsevier, Philadelphia.

Abnet CC (2007). Carcinogenic food contaminants. *Cancer Investigation* 25:189-196.

Alonso-Amelot ME (1997). The link between bracken fern and stomach cancer: Milk. *Nutrition* 13:694-696.

Alonso-Amelot ME (1999). Bracken fern, animal health and human health. *Revista Faculdade de Agronomia (LUZ)* 16:528-547.

Alonso-Amelot ME and Avendano M (2002). Human carcinogenesis and bracken fern: A review of the evidence. *Current Medicinal Chemistry* 9:675-686.

Alonso-Amelot ME, Castillo U, Smith BL, and Lauren DR (1996). Bracken ptaquiloside in milk. *Nature* 382:587.

Borzacchiello G, Ambrosio V, Roperto S, Poggiali F, Tsirimonakis E, Venuti A, Campo MS, and Roperto F (2003). Bovine papillomavirus type 4 in oesophageal papillomas of cattle from the south of Italy. *Journal of Comparative Pathology* 128:203-206.

Burkhalter PW, Groux PMJ, Candrian U, Hübner P, and Lüthy J (1996). Isolation, determination and degradation of ptaquiloside – A bracken fern (*Pteridium aquilinum*) carcinogen. *Journal of Natural Toxins* 5:141-160.

Campo MS, Jarrett WFH, Barron R, Oneil BW, and Smith KT (1992). Association of Bovine Papillomavirus Type-2 and Bracken Fern with Bladder-Cancer in Cattle. *Cancer Research* 52:6898-6904.

Carvalho T, Pinto C, and Peleteiro MC (2006). Urinary bladder lesions in bovine enzootic haematuria. *Journal of Comparative Pathology* 134:336-346.

Do Nascimento França T, Tokarnia CH, and Peixoto PV (2002). Diseases caused by the radiomimetic principle of *Pteridium aquilinum* (Polypodiaceae). *Pesquisa Veterinária Brasileira* 22:85-96.

Harwood M, Danielewska-Nikiel B, Borzelleca JF, Flamm GW, Williams GM, and Lines TC (2007). A critical review of the data related to the safety of quercetin and lack of evidence of *in vivo* toxicity, including lack of genotoxic/carcinogenic properties. *Food and Chemical Toxicology* 45:2179-2205.

Hirono I, Ogino H, Fujimoto M, Yamada K, Yoshida Y, Ikagawa M, and Okumura M (1987). Induction of Tumors in ACI Rats Given a Diet Containing Ptaquiloside, a Bracken Carcinogen. *Journal of the National Cancer Institute* 79:1143-1149.

Latorre AO, Akinaga RM, Caniceiro BD, Haraguchi M, and Górniak SL (2007). Selenium supplementation abrogated *Pteridium aquilinum* spleen damage in mice. *Brazilian Journal of Toxicology* 20(suppl. 3):13.

Latorre AO, Furlan MS, Sakai M, Fukumasu H, Hueza IM, Haraguchi M, and Górniak SL (2009). Immunomodulatory Effects of *Pteridium aquilinum* on Natural Killer Cell Activity and Select Aspects of the Cellular Immune Response of Mice. *Journal of Immunotoxicology* 6:104-114.

McKenzie RC, Rafferty TS, and Beckett GJ (1998). Selenium: An essential element for immune function. *Immunology Today* 19:342-345.

Moreira Souto MA, Kommers GD, Barros CSL, Piazer JVM, Rech RR, Riet-Correa F, and Schild AL (2006). Neoplasms of the upper digestive tract of cattle associated with spontaneous ingestion of bracken fern (*Pteridium aquilinum*). *Pesquisa Veterinária Brasileira* 26:112-122.

Ngomuo AJ and Jones RS (1996). Genotoxicity studies of quercetin and shikimate *in vivo* in the bone marrow of mice and gastric mucosal cells of rats. *Veterinary and Human Toxicology* 38:176-180.

Ojika M, Wakamatsu K, Niwa H, and Yamada K (1987). Ptaquiloside, a potent carcinogen isolated from bracken fern *Pteridium-aquilinum* var *latiusculum* – structure elucidation based on chemical and spectral evidence, and reactions with amino-acids, nucleosides, and nucleotides. *Tetrahedron* 43:5261-5274.

Pamukcu AM, Yalciner S, Hatcher JF, and Bryan GT (1980). Quercetin, a rat intestinal and bladder carcinogen present in bracken fern (*Pteridium-aquilinum*). *Cancer Research* 40:3468-3472.

Peixoto PV, Franca TD, Barros CSL, and Tokarnia CH (2003). Histopathological aspects of bovine enzootic hematuria in Brazil. *Pesquisa Veterinaria Brasileira* 23:65-81.

Petrie HT, Klassen LW, Klassen PS, O'Dell JR, and Kay HD (1989). Selenium and the immune response: 2. Enhancement of murine cytotoxic T-lymphocyte and natural killer cell cytotoxicity *in vivo*. *Journal of Leukocyte Biology* 45:215-220.

Riet-Correa F, Schild AL, Méndez MDC, and Lemos RAA (2001). Plantas de ação mutagênica e anti-hematopoiética. In *Doenças de Ruminantes e Eqüinos* (F Riet-Correa, AL Schild, MDC Méndez, and RAA Lemos eds), pp. 265-267. Varela Editora e Livraria LTDA., São Paulo.

Shahin M, Smith BL, and Prakash AS (1999). Bracken carcinogens in the human diet. *Mutation Research-Genetic Toxicology and Environmental Mutagenesis* 443:69-79.

Sugimura T (2000). Nutrition and dietary carcinogens. *Carcinogenesis* 21:387-395.

Taylor JA (1990). The bracken problem: A global perspective. *AIAS Occasional Publication* 40:3-19.

Tokarnia CH, Döbereiner J, and Peixoto PV (2000). Plantas de ação radiomimética. In *Plantas Tóxicas do Brasil,* (CH Tokarnia, J Döbereiner, and PV Peixoto, eds), pp. 178-187. Editora Helianthus, Rio de Janeiro.

# Chapter 67

# Toxic Nephrosis in Cattle from Pernambuco State, Northeastern Brazil Associated with the Ingestion of *Thiloa glaucocarpa*

E.G. de Miranda Neto[1], A.L.L. Pereira[2], J.C.A. Souza[3], C.L. Mendonça, F. Riet-Correa[2], A.F.M. Dantas[2], N.A. Costa[2], R.O. Rego[2], A.P. Silva Filho[2], and J.A.B. Afonso[2]

[1]*Hospital Veterinário, Centro de Saúde e Tecnologia Rural, Universidade Federal de Campina Grande, 58700-900, Patos, PB, Brazil;* [2] *Clínica de Bovinos, Campus Garanhuns, Universidade Federal Rural de Pernambuco, PO Box 152, 55292-901, Garanhuns, PE, Brazil*

## Introduction

*Thiloa glaucocarpa* is a nephrotoxic shrubby tree belonging to the family Combrataceae known by the names 'sipaúba' and 'vaqueta'. It is one of the most characteristic plants of the semiarid vegetation (caatinga) in northeastern Brazil. *T. glaucocarpa* occurs mainly in the state of Piauí, coastal Ceará, western Bahia, and northeastern Minas Gerais. Outbreaks of poisoning by this plant occur annually at the beginning of the rainy season. Case fatality rate is over 75%. Estimated annual losses exceed 1000 primarily adult cattle in Piauí and 500 cattle in Ceará (Tokarnia *et al.* 1981). Under natural conditions, intoxication from *T. glaucocarpa* is only reported in cattle (Tokarnia *et al.* 2000) and the poisoning was reproduced experimentally in this species (Tokarnia *et al.* 1981; Silva 1987). The poisoning was also reproduced in rabbits but in this species liver lesions are more marked that kidney lesions (Tokarnia *et al.* 1988).

Outbreaks of the poisoning occur at the start of the rainy season approximately 10-25 days after the first rain and cattle became intoxicated only during a short period lasting 5-8 days. There are two hypotheses to explain why outbreaks occur in such a short period of time. The first is that cattle only eat *T. glaucocarpa* – one of the first plants to green up after the rains begin – when it is newly sprouted and stop eating this plant (or eat it in an insufficient amount to became intoxicated) when other vegetation is available. The second hypothesis is that the plant loses toxicity as it matures, but experiments with cattle demonstrated that plants collected after the occurrence of the disease and also mature leaves were toxic (Tokarnia *et al.* 1981). The incidence of the disease varies from year to year; if the rainy season starts with heavy and continuous rains the incidence is low but if rains are scarce the incidence is higher. After intermittent rains, *T. glaucocarpa* sprouts more readily than other plants thus is more available for grazing cattle. With continuous rains there is

©CAB International 2011. *Poisoning by Plants, Mycotoxins, and Related Toxins* (eds F. Riet-Correa, J. Pfister, A.L. Schild, and T.L. Wierenga)

simultaneous blooming of nearly all the plants and *T. glaucocarpa* is not ingested or is ingested in lower quantities (Tokarnia *et al.* 2000).

In cattle experimentally intoxicated by *T. glaucocarpa*, first clinical signs are observed 1 to several days after the start of ingestion depending on the dose ingested. However, under natural conditions it is likely that clinical signs appear after several days of ingesting the plant. The clinical manifestation period is sub-acute, in general 5-20 days. However, there are cases where the clinical course was shorter (Tokarnia *et al.* 1981). In experiments carried out by Silva (1987), the clinical manifestation period was 2-37 days.

The characteristic clinical signs of the disease are subcutaneous edema mainly in the buttocks but also in the perineal, supra-mammary, abdominal, scrotal, and preputial regions (Tokarnia *et al.* 1981). Ascites with increased volume of the abdomen is also observed. In some cases subcutaneous edema is not seen. Cases with subcutaneous edema are named as *popa inchada* (swollen buttocks) by the farmers and cases without edema are called *venta seca* (dry nostrils). Other signs observed in both forms of the intoxication are depression, anorexia, decreased rumination, ruminal stasis, serous-hemorrhagic nasal discharge, and dry feces covered with mucus followed by bloody diarrhea. The animals lie down for long periods and show incoordination when moved. Progressive weight loss, coarse hair, polydipsia, restlessness, and recumbence are also observed (Tokarnia *et al.* 1981). In experimental cases serum levels of urea, creatinine, and total bilirubin were increased and albumin and bile salts were present in the urine (Silva 1987).

Necropsy findings consist of subcutaneous edema, ascites, hydrothorax, hydropericardium, and edema in the mesentery, perirenal tissue, and the folds of the abomasum. The kidneys are nearly always pale with red spots on the surface and in the parenchyma. Petechiae, ecchymosis, and suffusion are observed in the serous membranes, epicardium, endocardium, mucosa of the abomasum, large and small intestines, nostrils, pharynx, larynx, trachea, and esophagus. Occasionally there are large areas of necrosis covered with fibrin and ulcers in the pharynx, larynx, trachea, and esophagus. The liver may have a lighter color and increased lobular pattern. Congestion and hemorrhages appear in the mucosa of the abomasum and small intestine. The small intestine at times has reddish contents. The colon contents are dried and covered with mucus and blood. Occasionally the content of the colon and cecum appears liquid or pasty with the presence of liquid or coagulated blood (Tokarnia *et al.* 1981, 2000). The most significant and constant histological alteration is a toxic tubular nephrosis affecting part of the tubules of the renal cortex. Necrotic epithelial cells are transformed into amorphous eosinophil masses filling up the entire tubule and externally delimited by the basal membrane. Occasionally the epithelium is finely vacuolated or with hyaline drops degeneration. Hyaline cylinders and less frequently cell detritus are observed the lumen of the tubules in both the cortex and medulla. In nearly all cases, there is tubular dilatation with flattened epithelial cells. Interstitial edema is occasionally observed in the cortex. Paracentral necrosis is frequently observed in the liver (Tokarnia *et al.* 2000). The main toxins responsible for the nephrotoxicity of *T. glaucocarpa* are the tannins vescalagin, castalagin, stachyurin, and casuarinin (Itakura *et al.* 1987).

There is no known treatment; prophylaxis consists of removing cattle from the areas invaded by *T. glaucocarpa* at least 5 days following the first rain at the beginning of the rainy season and for a period of approximately 1 month (Tokarnia *et al.* 2000).

The objective of this chapter is to report outbreaks of nephrosis that occurred in the state of Pernambuco in areas invaded by *Thiloa glaucocarpa* but in a different period of the year than previously reported.

## Toxic Nephrosis in Cattle in Pernambuco

Thirty-two cases of toxic nephrosis were reported in the files of the Bovine Clinics of the Federal Rural University of Pernambuco, in the City of Garanhuns, state of Pernambuco. The cases occurred on seven farms in 2000-2002, 2007, and 2008 in the municipalities of Saloá, Buíque, Capoeiras, Paranatama, and São João, all located in the Agreste region (transition area between the lush, rainy coast and the semiarid region of the deep interior). Twenty-three of the 32 affected bovines died. Outbreaks occurred between August and October at the start of the dry season. All animals affected were females of Holstein/Zebu mixed breed. The animals had been raised in extensive systems, fed on native pastures, and received mineral salt. *T. glaucocarpa* was present in all paddocks where the disease occurred.

The owners reported that the animals appeared weak and depressed with anorexia, swelling on the posterior portion of the thigh, vulva, and anus, and small blackish feces. Eight of these cases were submitted to a clinical exam. Clinical signs were apathy, congested conjunctiva, dehydration, absence of rumination, dried nostrils, diminished or absent appetite, and a reduction in ruminal and intestinal movements. Four individuals had subcutaneous edema in the perineal region–one extending to the mammary gland, two in the vulva, and two affecting the caudal face of the thighs. Blackish hardened feces covered with mucus were observed in two animals. There was enlargement and increased sensitivity in the left kidney of two animals.

From the eight animals sent to the clinic, three recovered after 7 to 35 days; three died; one was euthanized; and another was dead on arrival. On blood analysis of seven animals, mean erythrocyte values remained within normal ranges for the species with the exception of two animals that exhibited alterations in total plasma proteins: one exhibited hypoproteinemia and the other exhibited hyperproteinemia due to dehydration. One bovine exhibited hypoalbuminemia with hyperglobulinemia and low albumin/globulin ratio indicating the chronic nature of the disease. Another showed hyperfibrinogenemia, probably due to the intensity of tissue damage caused by the disease. Leukocytosis by neutrophilia was found in five animals. Renal function tests were performed on three individuals; urea and creatinine values ranged from 19.14 to 524 mg/dl and 6.99 to 19.12 mg/dl, respectively.

Macroscopic findings were edema of the subcutaneous tissue of the perineum, abdomen and submandibular region, and edema of the semi-tendonous and semi-membranous muscles. The kidneys were swollen and pallid with a yellow to brown color and the kidney surface and parenchyma showed petechial hemorrhages and cortical-medullar and medullar congestion. Petechial hemorrhages and suffusions were observed in the perirenal fat tissue. Petechiae and ulcers were present in the mucosa of the jejunum and ileum which had dried contents and considerable amounts of mucus. The esophageal mucosa had multifocal ulcers. The wall of the abomasum was edematous and five animals had ascites and hydrothorax with a considerable amount of cloudy liquid.

On histopathological examination of three bovines, the kidney had severe degeneration and necrosis of the tubular epithelial cells mainly in the proximal tubules with formation of granular cylinders in the cortical and cortical-medullar region. There were also hyaline cylinders in the interior of the tubules mainly in the medullar region and less frequently in the cortical region. In some areas, there were hyaline drops in the interior of the epithelial cells. Dilatation of some tubules in the external cortical region, mild mononuclear lymphoplasmocytic interstitial infiltrate, and congestion and hemorrhages of

the cortical-medullar were also observed. Interstitial edema and tubular regeneration in the external cortical region were observed in one case.

## Discussion

Clinical signs and lesions observed in the affected animals were similar to those reported in *Thiloa glaucocarpa* poisoning. Some of the lesions and signs observed in these outbreaks including edema in the perineal region and buttocks (Tokarnia *et al.* 1981) have not been reported in other nephrotoxic Brazilian plants (*Amaranthus* spp. and *Setaria* spp.) (Riet-Correa and Méndez 2007). *T. glaucocarpa* was present on the farms where the disease occurred, suggesting that the disease was caused by the ingestion of this plant. However, the time of the year in which the disease occurred is not in accordance with the reported epidemiology of the disease. In previous reports the disease only occurred for a few days after the first rains when *T. glaucocarpa* was newly sprouting (Tokarnia *et al.* 1981). The outbreaks reported in this chapter occurred from August to September at the start of the dry season. There are no reports of outbreaks in other periods of the year but mature leaves of *T. glaucocarpa* are still toxic under experimental conditions (Tokarnia *et al.* 1981), which suggests that the poisoning can occur if there are regional conditions that induce the animals to consume the plant. Those conditions need to be investigated. Another possibility is that the nephrosis was caused by another unknown nephrotoxic plant.

## Acknowledgements

This work has been financially supported by the National Institute for Science and Technology for the Control of Toxic Plants (CNPq), grant number 573534/2008-0.

## References

Itakura Y, Habermehl G, and Mebs D (1987). Tannins occurring in the toxic Brazilian plant *Thiloa glaucocarpa*. *Toxicon* 25:1292-1300.

Riet-Correa F and Méndez MC (2007). Intoxicações por Plantas e Micotoxinas. In *Doença de Ruminantes e Eqüídeos* (F Riet-Correa, AL Schild, RAA Lemos, and JRJ Borges, eds), pp. 99-221. Pallotti, Santa Maria.

Silva SV (1987). Aspectos clínicos, laboratoriais e anátomo-histopatológicos na intoxicação experimental por Sipaúba (*Thiloa glaucocarpa* Eichl.) em bovinos no Estado do Piauí, 89 pp. MS Dissertation, UFRPE, Recife.

Tokarnia CH, Döbereiner J, Canella CFC, Couceiro JEM, Silva ACC, and Araujo FV (1981). Intoxicação de bovinos por *Thiloa glaucocarpa* (Combretaceae) no nordeste do Brasil. *Pesquisa Veterinária Brasileira* 1:111-132.

Tokarnia CH, Peixoto PV, and Döbereiner J (1988). Intoxicação experimental pelas folhas e extratos de *Thiloa glaucocarpa* (Combretaceae) em coelhos. *Pesquisa Veterinária Brasileira* 8:61-74.

Tokarnia CH, Döbereiner J, and Peixoto PV (2000). *Plantas tóxicas do Brasil*, 320 pp. Helianthus, Rio de Janeiro.

# Chapter 68

# Osteolathyrism in Calves in Uruguay

C. García y Santos, S. Sosa, A. Capelli, W. Pérez, R. Domínguez, C. Aldecoa, C. Franco, and G. Moreira

*Laboratorio de Toxicología, Facultad de Veterinaria, Universidad de la República, Av. Lasplaces 1550, CP 1600, Montevideo, Uruguay*

## Introduction

Plant poisoning is a major problem in grazing cattle and sheep throughout the world and in Uruguay. Economic losses due to plant poisoning include animal deaths, lower yield, decreased reproductive performance, and the occurrence of other associated diseases. Diagnostic testing, treatment of sick animals, management measures, plant control, and land depreciation also contribute to economic losses (Riet-Correa and Medeiros 2000). In our country there are more than 30 toxic plant species of veterinary importance (Rivero *et al.* 2000; Riet-Correa and Medeiros 2001). Until now there have been no reports of intoxication by *Lathyrus* species.

Plants of the genus *Lathyrus* are responsible for causing the syndrome called lathyrism which is diagnosed both in human and animal populations in different regions of the world. Lathyrism was reported by Hippocrates in the 4th century BC (Bruneton 2001). The genus *Lathyrus* includes more than 160 species commonly known as 'arvejillas'. They belong to the Fabaceae family and Leguminosae-Papilionoideae subfamily. They are herbaceous or sub-herbaceous, annual or perennial, climbing by means of tendrils and angular stems. In Uruguay there are seven species of *Lathyrus* including *L. hirsutus*, an annual biannual plant, 20 to 100 cm tall, leaves with two pinnae, with petioles, and branched upper tendrils. Flowers are violaceous or bluish, grouped in clusters of 1 to 4 flowers each. The fruit is a legume pod covered by long hairs (Lombardo and Muñoz 1980; Lombardo 1982).

Because of its high protein content, ranging from 25% to 75%, numerous species of the genus *Lathyrus* are used for human and animal diets (White *et al.* 2002). Toxic compounds are concentrated in seeds (Barceloux 2008).

Some species such as *L. sativus* produce a clinical condition known as neurolathyrism, more frequently seen in humans but also seen in animals. This species contains β-N-oxalylamino-L-alanine (BOAA) that produces an irreversible non-progressive degeneration of the spinal cord (Steyn 1933; Spencer and Schaumburg 1983; Spencer *et al.* 1985, 1988, 1991; Spencer 1995; Haque *et al.* 1996; Chowdhury and Davis 1998).

Other *Lathyrus* spp. contain 4-glutamyl-3-aminopropionitrile, a non-protein amino acid that acts by inhibiting the enzyme lysyl oxidase. This decreases the amount of cross-linked elastin and collagen causing weakness of the locomotor system (osteolathyrism) and increased fragility of blood capillaries (angiolathyrism). These clinical pictures are seen

©CAB International 2011. *Poisoning by Plants, Mycotoxins, and Related Toxins* (eds F. Riet-Correa, J. Pfister, A.L. Schild, and T.L. Wierenga)

mainly in animals (Shore *et al.* 1984). One case of osteolathyrism caused by *L. hirsutus* was reported in cattle in Alabama, USA (Sugg *et al.* 1944).

Clinical signs of osteolathyrism are salivation, stiffness in either forelimbs, hindlimbs, or both, decreased responses to external stimuli, reluctance to move, incoordination, and kyphosis. Differential diagnosis includes other neurotoxic plants and mycotoxins, calcinogenic plants, hypervitaminosis D, hypervitaminosis A, copper deficiency, and traumatic reticuloperitonitis, among others (Radostits *et al.* 2002).

This chapter reports spontaneous outbreaks of intoxication by *Lathyrus hirsutus* in calves in Uruguay and its experimental reproduction in the same species.

## Spontaneous Outbreaks of Osteolathyrism in Uruguay

Seven outbreaks during the years 2004-2008 were studied. Cases occurred in November in the Departments of San José and Canelones. Twenty-seven animals, mostly calves (Holstein, Normand, Hereford, Limousin, and crosses) were affected. Morbidity was 60% and mortality 0%. The outbreaks occurred in previously planted pastures or natural grasslands that had been invaded by the plant. Clinical signs similar to those reported in osteolathyrism were observed 15 days after grazing on these pastures. Signs worsened when the animals were excited. Sick animals remained recumbent or had difficulty moving, therefore they were unable to feed properly leading to a significant reduction in weight gain. Intoxicated animals recovered within days after being removed from the pastures but showed substantial weight loss (Pereira *et al.* 2007).

Farms where outbreaks occurred were visited and plants collected for botanic identification. The most prevalent weed was identified as *L. hirsutus*. Seeds of the plant were found in feces and microhistological analysis of leaf epidermis showed evidence of the weed. Diagnosis was based on epidemiological data, presence of the plant, and clinical signs.

## Experimental Reproduction of the Disease

For the experimental reproduction, suspected plants were collected from one of the affected farms. Three Holstein calves weighing an average of 90 kg received 3%, 2%, and 1%, respectively, of their body weight of green plant (aerial parts) for 30 days. A fourth calf received 0.9% of his body weight of dried pods and seeds for 10 days. On day 10 the last calf was euthanized with thiopentone and exsanguinated. Fragments of liver, heart, kidney, adrenal glands, lungs, spleen, esophagus, duodenum, jejunum, mesenteric lymphatic nodules, aorta, articular cartilage, muscle, tendon, bone, and brain were fixed with 10% buffered formalin for histological analysis at the Laboratorio de Patologia de Facultad de Veterinaria, UdelaR, Montevideo, Uruguay, and at the DILAVE Miguel C. Rubino, Laboratorio Regional Noroeste, Paysandú, Uruguay. Aorta, tendon, bone, and muscle samples were fixed in glutaraldehyde and sent to the Laboratório Regional de Diagnóstico, Faculdade de Veterinária, UFPel, Pelotas, RS, Brazil, for ultrastructural analysis.

In calves fed with the aerial parts of the fresh plant no clinical signs of intoxication were observed. After 7 days the calf that received the dried pods and seeds showed clinical signs similar to those observed in field cases. No lesions were observed either macroscopically or histologically. Ultrastructural analysis is under way.

Clinical signs observed in the experiment are similar to those of the natural outbreak of *L. hirsutus* intoxication (Sugg *et al.* 1944; Pereira *et al.* 2007). The calf that showed clinical signs was the one that received the dried pods and seeds of the plant. This is attributed to the fact that 4-glutamyl-3-aminopropionitrile is more concentrated in the dried pods and seeds (Bianchi and Gualtieri 1964).

## Conclusions

The outbreaks studied and the experimental reproduction lead us to conclude that *Lathyrus hirsutus* is a new toxic vegetal species in Uruguay. This poisoning is important in the area studied, where there are mainly small farms and any reduction in weight gain can result in a significant economic loss for these producers.

## Acknowledgements

We thank the farmers that allowed us to carry out epidemiological studies and experimental reproductions. We also thank the veterinarians that reported the clinical cases and the Pharmaceutical Chemists for performing chemical studies. We thank Dr Rodolfo Rivero and Dr Antonio Moraña for histopathologic studies and especially Dr Jorge Moraes for revising this paper. Project CSIC I+D (UdelaR).

## References

Barceloux DG (2008). *Medical Toxicology of Natural Substances: Foods, Fungi, Medicinal Herbs, Plants and Venomous Animals*, 1158 pp. John Wiley and Sons, New York, USA.

Bianchi M and GualtieriG (1964). Osteo-cartilaginous alterations due to lathyrism. *Archivio di Ortopedia* 77:576-572.

Bruneton J (2001). *Plantas tóxicas-Vegetales peligrosos para el Hombre y los animales,* 527 pp. Editorial Acribia S.A., Zaragoza, Spain.

Chowdhury SD and Davis RH (1998). Influence of dietary osteolathyrogenson the eggshell quality of laying hens. *British Poultry Science* 39:497-499.

Haque A, Hossain M, Wouters G, and Lamboin F (1996). Epidermiological Study of Lathyrism in Northwestern Districts of Banglandesh. *Neuroepidemiology* 15(2):83-91.

Lombardo A (1982). *Flora Montevidensis*. (Intendencia Municipal de Montevideo – Servicio de Publicaciones y Prensa, Tomo I, ed.), 316 pp. Montevideo, Uruguay.

Lombardo A and Muñoz J (1980). *Plantas Trepadoras* (Intendencia Municipal de Montevideo, Servicio de Publicaciones y Prensa, ed.), 111 pp. Montevideo, Uruguay.

Pereira R, Capelli A, Dominguez R, Arago S, Pérez W, Alonso E, and García y Santos C (2007). Intoxicación espontánea por ingestión de *Lathyrus hirsutus* en terneros del Uruguay. *V Jornadas Técnicas,* 74 pp. Facultad de Veterinaria, Montevideo, Uruguay, 21-23 November 2007.

Radostits OM, Gay CC, Blood DC, and Hinchcliff KW (2002). *Medicina Veterinaria*: *Tratado de enfermedades del ganado bovino, ovino, porcino, caprino y equino,* vol. II, 9th edn, 1008 pp. Editorial Mc Graw-Hill interamericana, Madrid.

## Materials and Methods

Forty-seven female quail (*Coturnix coturnix*) were used in the study. Three days prior to each experiment the quail were housed individually in a temperature-controlled environment on a 12 h light:12 h dark cycle with free access to water and food. They were fed a standard quail ration (Purina, São Lourenço da Mata, PE, Brazil). Two experiments were conducted: a toxicological study of cyanide and evaluation of the possible interference of cyanide exposure with diet selection.

The toxicological study was performed with 27 female quail that were assigned to three groups that received by gavage 0, 1, or 3 mg of potassium cyanide (KCN)/kg body weight (BW)/day for 7 consecutive days. The food consumption of each animal was measured daily and dosed animals were closely monitored for signs of poisoning. The body weight of each animal was recorded on the first and last days of the experiment and the gains in body weight were calculated. Blood samples were collected for hematology analysis and a biochemical panel was conducted 24 h after the last dose of cyanide. The hematology analysis included counts of red blood cells (RBC) and white blood cells (WBC) and measurement of the hematocrit (PCV) and mean corpuscular volume (MCV). Serum biochemistries were measured using a CELM SBA-200 automated chemistry analyzer using Laborlab reagents (Guarulhos, SP, Brazil). Determinations included glucose, cholesterol, total proteins, alanine aminotransferase (ALT), and aspartate aminotransferase (AST). Tissues from the pancreas, thyroid gland, liver, kidney, and the whole central nervous system were collected for histopathological study. Paraffin-embedded sections were stained with hematoxylin and eosin (H&E).

The trial of diet selection was conducted with 20 female quail. The birds had access to two separate rations: a conventional quail ration and the same ration supplemented with 1% $NaSO_4$. The food consumption of each animal was measured daily for each ration and the dosed animals were closely monitored for signs of poisoning. The body weight of each animal was recorded on the first and last days of the experiment and gains in body weight were calculated. After 2 days the quail were randomly assigned to two groups, dosed with 0 (control) or 3 mg of KCN/kg BW/day for 5 consecutive days. After this period two quail per treatment group were used for histopathological study and the others were observed for an additional 4 days without dosing. Individual consumption of each ration was measured daily during all periods of the study.

The statistical design for the study was a completely randomized design with two treatments. Statistical analysis was done using analysis of variance (ANOVA) with the Dunnett test as a post-test using the software GraphPad Prism v.4 for Mac. Statistical significance was set at $P < 0.05$. The results were presented as the means with their standard errors.

## Results

Throughout the toxicological study most animals remained clinically normal; however, one quail developed moderate trembling and vocalization after the first day of cyanide dosing. On the second day this quail developed convulsions and died. No changes were found in the hematological and blood chemical panels. At postmortem examination, no macroscopic lesions were found in any tissue of this bird. Histological changes were found only in the other animals dosed with cyanide and consisted of increased number of vacuoles in the colloid of the thyroid glands, mild hepatic periportal vacuolation, and evidence of spongiosis in the mesencephalon.

Throughout the diet selection trial, no clinical signs were found in any quail. There were no significant differences in food consumption or in ration preference (Table 1).

**Table 1.** Daily food consumption (g) of quails treated with 0 (control) or 3 mg/kg of KCN fed conventional ration and 1% sodium sulfate ($NaSO_4$) supplemented ration.

| | Control (*n*=10) | | KCN (*n*=10) | |
|---|---|---|---|---|
| | Conventional ration | Sulfur supplemented | Conventional ration | Sulfur supplemented |
| Before treatment | | | | |
| day 1 | 13.1±1.96 | 16.3±1.66 | 13.4±2.14 | 14.7±1.72 |
| day 2 | 13.0±1.78[a] | 18.0±1.28[a] | 13.9±2.26 | 14.6±2.21 |
| Cyanide treatment | | | | |
| day 1 | 12.8±1.78 | 14.3±1.29 | 13.3±2.33 | 11.5±2.02 |
| day 2 | 15.0±1.68 | 16.8±1.50 | 14.2±2.05 | 12.4±1.86 |
| day 3 | 15.9±1.02 | 14.5±1.43 | 13.3±2.08 | 15.1±1.69 |
| day 4 | 16.2±1.91 | 16.4±1.65 | 14.6±1.51 | 17.8±0.82 |
| day 5 | 15.3±1.74 | 16.9±1.93 | 14.0±1.70 | 16.8±1.12 |
| After treatment | | | | |
| day 1 | 10.6±1.33 | 10.5±1.45 | 7.73±1.01 | 10.5±0.75 |
| day 2 | 21.0±0.74 | 21.0±1.55 | 18.6±1.15 | 20.3±0.79 |
| day 3 | 12.03±2.01 | 17.4±1.65 | 13.1±1.41 | 12.9±1.44 |
| day 4 | 22.8±0.96 | 22.5±1.78 | 20.8±1.13 | 19.0±1.99 |

[a] Means present significant difference ($P < 0.05$)

## Discussion

Degenerative damage to the liver promoted by KCN has been found in rabbits (Okolie and Osagie 1999), rats (Sousa *et al.* 2002), and goats (Soto-Blanco *et al.* 2005, 2008). In the present work, quail treated with KCN presented mild hepatic periportal vacuolation but did not show alterations in their biochemical panel. Thus it is feasible that the liver is a target tissue for cyanide toxicity in this species but its relative importance is undetermined at present.

Disturbances of glucose metabolism and fibrocalculous pancreatic diabetes or malnutrition-related diabetes have been associated with chronic exposure to cyanide through consumption of cassava in humans (McMillan and Geevarghese 1979; Kamalu 1995; Petersen 2002). However, no disturbance in glucose levels or pancreatic morphology was detected in the quails dosed with cyanide in the current study. These data and those from previous studies (Soto-Blanco *et al.* 2001b, 2005, 2008) suggest that cyanide does not induce a diabetogenic effect in quail.

Development of hypothyroidism and goiter has been linked to long-term consumption of cyanogenic plants by both humans (Adewusi and Akindahunsi 1994) and animals (Kamalu and Agharanya 1991). Thiocyanate, the main product of the transformation of cyanide in the organism, is probably the factor responsible because this ion competes with iodide for uptake by the thyroid gland resulting in hypothyroidism (Delange and Ermans 1996). Quail treated with cyanide in our work revealed an increased number of resorption vacuoles in the follicles of the thyroid; this is similar to the findings of previous work in several animal species (Soto-Blanco *et al.* 2001a, 2002a, 2008).

The central nervous system is an important target of prolonged cyanide toxicity (Tylleskär *et al.* 1995; Spencer 1999; Soto-Blanco *et al.* 2002b, 2005, 2008). Cyanide may promote neurotoxic effects through inhibition of cellular respiration because neurons have relatively high metabolic rates with little ability for anaerobic metabolism. Furthermore, cyanide interferes with several neurotransmitters including γ-aminobutyric acid (GABA; Cassel *et al.* 1991), glutamic acid (Cassel *et al.* 1991), acetylcholine (Owasoyo and Iramain 1980), dopamine (Cassel *et al.* 1995), excitatory amino acids (McCaslin and Yu 1992; Gunasekar *et al.* 1996), and nitric oxide (Gunasekar *et al.* 1996). In the present study, there was evidence of spongiosis in the mesencephalon of quail dosed with cyanide.

The purpose of the second part of this work was to determine whether cyanide interferes with diet selection in quail. Several studies have reported interference with the amount of energy and/or protein in a diet after ingestion of a toxin (Guglielmo *et al.* 1996; Villalba *et al.* 2002, 2004; Villalba and Provenza 2005). Given that the main route of detoxification of cyanide is via enzymatic combination with sulfur (Sousa *et al.* 2003), the quail were offered a conventional diet enriched with this compound to test whether they chose it in preference to the conventional ration. However, the ingestion of sulfur was not affected by exposure to cyanide. Cyanide inhibits several cellular enzymes including cytochrome oxidase which is a key enzyme in the cellular respiratory chain, inhibition of which results in cellular hypoxia and cytotoxic anoxia (Ballantyne 1987). Lambs infused with amygdalin, a cyanogenic glycoside, preferred foods with higher ratios of energy:protein than those chosen by controls (Villalba *et al.* 2002), which could be attributed to a modification of ingestive behavior to compensate for the effect of cyanide. A possible hypothesis is that the post-ingestive effects of toxins may modify ingestive behavior in relation to macronutrients but not to micronutrients. Another possibility is that cyanide dose was not high enough to cause sufficient toxicity that additional sulfur was required. It is less likely that post-ingestive feedback mechanisms are different in avian species from those in mammals. Future studies are needed to test these hypotheses.

In conclusion, exposure to cyanide promotes damage to the liver and central nervous system in quail. On the other hand, the ingestion of sulfur by quail was not affected by exposure to cyanide.

## References

Adewusi SRA and Akindahunsi AA (1994). Cassava processing, consumption, and cyanide toxicity. *Journal of Toxicology and Environmental Health* 43:13-23.

Ballantyne B (1987). Toxicology of cyanides. In *Clinical and Experimental Toxicology of Cyanides* (B Ballantyne and TC Marrs, eds), pp. 41-126. Wright, Bristol.

Cassel G, Karlsson L, and Sellstrom A (1991). On the inhibition of glutamic acid decarboxylase and gamma-aminobutyric acid transaminase by sodium cyanide. *Pharmacology and Toxicology* 69:238-241.

Cassel G, Koch M, and Tiger G (1995). The effects of cyanide on the extracellular levels of dopamine, 3,4-dihydroxyphenylacetic acid, homovanillic acid, 5-hydroxyindoleacetic acid and inositol phospholipid breakdown in the brain. *Neurotoxicology* 16:73-82.

DeGabriel JL, Moore BD, Foley WF, and Johnson CN (2009). The effects of plant defensive chemistry on nutrient availability predict reproductive success in a mammal. *Ecology* 90:711-719.

Delange FM and Ermans AM (1996). Iodine deficiency. In *Werner and Ingbar's The Thyroid: a Fundamental and Clinical Text* (LE Braverman and RD Utiger, eds), 7th edn, pp. 296-316. Lippincott-Raven, Philadelphia.

Foley WJ, McLean S, and Cork SJ (1995). Consequences of biotransformation of plant secondary metabolites on acid-base metabolism in mammals – a final common pathway? *Journal of Chemical Ecology* 21:721-743.

Guglielmo CG, Karasov WH, and Jakubas WJ (1996). Nutritional costs of a plant secondary metabolite explain selective foraging by ruffed grouse. *Ecology* 77: 1103-1115.

Gunasekar PG, Sun PW, Kanthasamy AG, Borowitz JL, and Isom GE (1996). Cyanide-induced neurotoxicity involves nitric oxide and reactive oxygen species generation after N-methyl-D-aspartate receptor activation. *Journal of Pharmacology and Experimental Therapeutics* 277:150-155.

Illius AW and Jessop NS (1995). Modeling metabolic costs of allelochemical ingestion by foraging herbivores. *Journal of Chemical Ecology* 21:693-719.

Jones DA (1998). Why are so many food plants cyanogenic? *Phytochemistry* 47:155-162.

Kamalu BP (1995). The adverse effects of long-term cassava (*Manihot esculenta* Crantz) consumption. *International Journal of Food Science and Nutrition* 46:65-93.

Kamalu BP and Agharanya JC (1991). The effect of a nutritionally-balanced cassava (*Manihot esculenta* Crantz) diet on endocrine function using the dog as a model. 2-Thyroid. *British Journal of Nutrition* 65:373-379.

McCaslin PP and Yu XZ (1992). Cyanide selectively augments kainate- but not NMDA-induced release of glutamate and taurine. *European Journal of Pharmacology and Environmental Toxicology, Pharmacology Section* 228:73-75.

McMillan DE and Geevarghese PJ (1979). Dietary cyanide and tropical malnutrition diabetes. *Diabetes Care* 2:202-208.

Okolie NP and Osagie AU (1999). Liver and kidney lesions and associated enzyme changes induced in rabbits by chronic cyanide exposure. *Food and Chemical Toxicology* 37:745-750.

Owasoyo JO and Iramain CA (1980). Acetylcholinesterase activity in rat brain: effect of acute cyanide intoxication. *Toxicology Letters* 6:1-3.

Petersen JM (2002). Tropical pancreatitis. *Journal of Clinical Gastroenterology* 35:61-66.

Provenza FD, Villalba JJ, Dziba LE, Atwood SB, and Banner RE (2003). Linking herbivore experience, varied diets, and plant biochemistry diversity. *Small Ruminant Research* 49:257-274.

Soto-Blanco B and Górniak SL (2003). Milk transfer of cyanide and thiocyanate: cyanide exposure by lactation in goats. *Veterinary Research* 34:213-220.

Soto-Blanco B, Górniak SL, and Kimura ET (2001a). Physiopathological effects of chronic cyanide administration to growing goats – a model for cyanogenic plants ingestion. *Veterinary Research Communications* 25:379-389.

Soto-Blanco B, Sousa AB, Manzano H, Guerra JL, and Górniak SL (2001b). Does prolonged cyanide exposure have a diabetogenic effect? *Veterinary and Human Toxicology* 43:106-108.

Soto-Blanco B, Maiorka PC, and Górniak SL (2002a). Effects of long-term low-dose cyanide administration to rats. *Ecotoxicology and Environmental Safety* 53:37-41.

Soto-Blanco B, Maiorka PC, and Górniak SL (2002b). Neuropathologic study of long term cyanide administration to goats. *Food and Chemical Toxicology* 40:1693-1698.

Soto-Blanco B, Stegelmeier BL, and Górniak SL (2005). Clinical and pathological effects of short-term cyanide repeated dosing to goats. *Journal of Applied Toxicology* 25:445-450.

Soto-Blanco B, Stegelmeier BL, Pfister JA, Gardner DR, and Panter KE (2008). Comparative effects of prolonged administration of cyanide, thiocyanate and chokecherry (*Prunus virginiana*) to goats. *Journal of Applied Toxicology* 28:356-363.

Sousa AB, Soto-Blanco B, Guerra JL, Kimura ET, and Górniak SL (2002). Does prolonged oral exposure to cyanide promote hepatotoxicity and nephrotoxicity? *Toxicology* 174:87-95.

Sousa AB, Manzano H, Soto-Blanco B, and Górniak SL (2003). Toxicokinetics of cyanide in rats, pigs and goats after oral dosing with potassium cyanide. *Archives of Toxicology* 77:330-334.

Spencer PS (1999). Food toxins, AMPA receptors, and motor neuron diseases. *Drug Metabolism Review* 31:561-587.

Tylleskär T, Banea M, Bikangi N, Nahimana G, Persson LÅ, and Rosling H (1995). Dietary determinants of a non-progressive spastic paraparesis (Konzo): a case-referent study in a high incidence area of Zaire. *International Journal of Epidemiology* 24:949-956.

Villalba JJ and Provenza FD (2005). Foraging in chemically diverse environments: energy, protein, and alternative foods influence ingestion of plant secondary metabolites by plants. *Journal of Chemical Ecology* 31:123-138.

Villalba JJ, Provenza FD, and Bryant JP (2002). Consequences of the interaction between nutrients and plant secondary metabolites on herbivore selectivity: benefits or detriments for plants? *Oikos* 97:282-292.

Villalba JJ, Provenza FD, and Han G (2004). Experience influences diet mixing by herbivores: implications for plant biochemistry diversity. *Oikos* 107:100-109.

# Chapter 70

# Toxicity to Honey Bees from Pollen from Several Plants in Northeastern Brazil

L.X. Mesquita[1], P.B. Maracajá[1], S.M. Sakamoto[1], O. Malaspina[2], and B. Soto-Blanco[1]

*[1]Department of Animal Sciences, Universidade Federal Rural do Semi-Árido (UFERSA), BR 110 Km 47, Mossoró, RN, 59625-900, Brazil; [2]Department of Biology, Universidade Estadual Paulista Júlio de Mesquita Filho, Rio Claro, SP, Brazil*

## Introduction

Secondary plant compounds are well known to affect the nutritional quality of plants and to limit growth and reproduction in herbivores. The optimal defense hypothesis postulates that a plant should defend its most valuable parts if resources are limited (Rhoades 1979). Higher concentration of secondary compounds should be found in tissues that are more valuable to the plant. Thus younger and reproductive tissues are better protected than other parts of the plant because removal of these parts is detrimental to the plant (Karban and Baldwin 1997).

Several plant species contain secondary compounds in nectar and pollen that could be toxic to pollinators including bees (Adler 2000; Adler and Irwin 2005; Praz *et al.* 2008). For example, almond (*Amygdalus communis* L., Rosaceae) contains the cyanogenic glycoside amygdalin that releases cyanide. Amygdalin is found in the nectar and pollen of almond trees and the consumption of this pollen can be toxic to honey bees (Kevan and Ebert 2005). Experimental approaches are necessary to identify plant species that produce pollen that is toxic to bees. In the present study the toxic potential of the pollen of *Azadirachta indica*, *Mimosa tenuiflora*, *Piptadenia stipulacea*, and *Ricinus communis* to honey bees (*Apis mellifera*) was tested.

## Materials and Methods

### Plant material

Plant species used in this study were *Azadirachta indica* A. Juss. (Meliaceae), *Mimosa tenuiflora* (Willd.) Poir. (Mimosaceae), *Piptadenia stipulacea* (Benth.) Duche (Mimosaceae), and *Ricinus communis* L. (Euphorbiaceae). Pollen samples were collected near Mossoró city, RN, in northeastern Brazil (5°11'15"S and 37°20'39"W) at an altitude of 16 m above sea level. The climate is characterized as semiarid with mean annual

©CAB International 2011. *Poisoning by Plants, Mycotoxins, and Related Toxins* (eds F. Riet-Correa, J. Pfister, A.L. Schild, and T.L. Wierenga)

temperature of 27.4°C while the mean annual rainfall and mean relative humidity are 674 mm and 68.9%, respectively. Voucher specimens (*A indica*: #7162; *M. tenuiflora*: #9591; *P. stipulacea*: #9599; *R. communis*: #4520) were deposited at the Dárdano de Andrade-Lima (MOSS) Herbarium, Universidade Federal Rural do Semi-Árido (UFERSA), Mossoró, RN, Brazil. Pollen material was dried at 40°C for 48 h and powdered.

**Animals and experimental design**

Honeycombs that contained pupae of Africanized honey bees (*Apis mellifera*) were collected from the apiary of UFERSA. Newly emerged forager bees identified on the basis of their body size and coloration were used for the experiment. All the bees used were of the same age. Twenty bees were put into a wooden box (11 × 11 × 7 cm). The boxes were kept in an acclimatized chamber (BOD) at 32°C and 70% humidity. The basal control food was sugar and honey (5:1). Pollen samples from *A. indica*, *M. tenuiflora*, *P. stipulacea*, and *R. communis* were added to the basal food at levels of 2.5%, 5.0%, and 10.0% (w/w). Each prepared food was offered to 180 bees and the bees were observed daily until the last one died. Dead bees were opened to confirm presence of food on digestive apparatus.

**Statistical analysis**

Statistical analyses were performed using GraphPad Prism (v.4 for Mac). Median survival times with 95% confidence intervals (CI) were estimated using Kaplan–Meier survival analysis. Differences in the time distributions between groups were tested for statistical significance using the log-rank test.

## Results

The median survival times of the bees fed *P. stipulacea* were 5 days for the control group, 3 days for the 2.5% group, 4 days for the 5% group, and 3 days for the 10% group. There was a significant difference ($P < 0.0001$) at log-rank test between the survivals and the control group differed ($P < 0.0001$) from all treatment groups.

The median survival times of the bees fed *A. indica* were 5.5 days for the control group and 5 days for the groups fed 2.5%, 5%, and 10%. There was a significant difference ($P < 0.05$) between the survival curves and the control group differed ($P < 0.05$) from the group fed 5%.

The median survival times of the bees fed *M. tenuiflora* were 8 days for the control group, 9 days for the 2.5% group, 7 days for the 5% group, and 8 days for the 10% group. The median survival times of the bees fed *R. communis* were 7.5 days for the control group, 7 days for the 2.5% group, 6 days for the 5% group, and 9 days for the 10% group. There was no significant difference ($P > 0.05$) between the survival curves of the control groups and the groups of bees fed with different doses of *M. tenuiflora* and *R. communis*.

## Discussion

*Azadirachta indica* is a plant that is largely grown for use as a pesticide and insect repellent (Karunamoorthi *et al.* 2009). Chemical evaluation of the flowers identified 5 sesquiterpenes, 3 aromatics, 17 fatty acids, 5 fatty acid esters, 3 steroids, and 8 hydrocarbons (Siddiqui *et al.* 2009). A skin prick test on human volunteers revealed that the pollen of *A. indica* is allergenic (Chakraborty *et al.* 1998); the allergenicity is attributable to two proteins (Karmakar and Chatterjee 1994). In this work, only slight toxicity of *A. indica* pollen to honey bees was observed.

*Mimosa tenuiflora* is a xerophilous plant which is very common in degraded areas in the semiarid region of the study. The ingestion of leaves from *M. tenuiflora* is responsible for congenital malformations (cleft lip, unilateral corneal opacity, ocular bilateral dermoids, buphthalmos with a cloudy brownish appearance of the anterior chamber due to an iridal cyst, and segmental stenosis of the colon) in ruminants (Pimentel *et al.* 2007). In Mexico the cortex of *M. tenuiflora* is a popular medicine for the treatment of skin burns and wounds (Rivera-Arce *et al.* 2007). However, the results of the present study indicate that the pollen of *M. tenuiflora* is not toxic to bees.

*Piptadenia stipulacea* is a woody plant species that grows at a high density in the semiarid area of northeastern Brazil especially on abandoned agricultural sites (Pereira *et al.* 2003). The uses of this plant include construction, fuel (Lucena *et al.* 2007), and animal forage (Almeida *et al.* 2006). Field observations by Brazilian beekeepers have revealed that *P. stipulacea* is toxic to bees during the flowering season. In the present work, it was verified that ingestion of the pollen of this plant significantly reduced the survival of honey bees, which confirmed the toxic effect suggested by the beekeepers' observations.

*Ricinus communis* is a plant that is distributed widely across the world. In the semiarid region of northeastern Brazil, *R. communis* is cultivated in large numbers mainly for the production of biodiesel. However, it is well known as a poisonous plant in which the principal toxin is ricin, a large water soluble glycoprotein. All parts of the plant are toxic but the seeds contain the highest concentration of ricin (Soto-Blanco *et al.* 2002; Garland and Bailey 2006). However, data from the present study indicate that the pollen of *R. communis* is not toxic to bees.

The great majority of flowering plants rely on insects or other animals for pollination and bees are the most important pollinating insects (Westerkamp 1996). The significance of the toxicity of pollen is poorly understood. The function of secondary compounds in nectar has been explained by several hypotheses including encouragement of specialist pollinators, avoidance of nectar robbers (insects, birds, or other flower visitors that remove nectar without pollinating), prevention of microbial degradation of nectar, disturbance of the behavior of pollinators, and the result of previous evolutionary forces that are no longer acting on the plant (Adler 2000). With regard to pollen, the relationship between bees and flowers was considered to be one of balanced mutual exploitation. In fact, bees store pollen and nectar to feed their larvae. They collect large amounts of pollen grains very efficiently, making these grains generally unavailable for pollination (Westerkamp 1996). Thus, the presence of secondary compounds in pollen grains could be a strategy designed to restrict the loss of pollen to bees.

In conclusion, only one of the plants we tested had a toxic effect and the magnitude of this effect was small. Our results confirm what beekeepers had observed in field situations.

## References

Adler LS (2000). The ecological significance of toxic nectar. *Oikos* 91:409-420.

Adler LS and Irwin RE (2005). Ecological costs and benefits of defenses in nectar. *Ecology* 86:2968-2978.

Almeida ACS, Ferreira RLC, Santos MVF, Silva JAA, Lira MA, and Guim A (2006). Avaliação bromatológica de espécies arbóreas e arbustivas de pastagens em três municípios do Estado de Pernambuco. *Acta Scientiarum, Animal Sciences* 28:1-9.

Chakraborty P, Gupta BS, Chakraborty C, Lacey J, and Chanda S (1998). Airborne allergenic pollen grains on a farm in West Bengal, India. *Grana* 37:53-57.

Garland T and Bailey EM (2006). Toxins of concern to animals and people. *Revue Scientifique et Technique de l'Office Internacional des Epizooties* 25:341-351.

Karban R and Baldwin IT (1997). *Induced responses to herbivory*. The University of Chicago Press, Chicago, Wisconsin.

Karmakar PR and Chatterjee BP (1994). Isolation and characterization of two IgE-reactive proteins from *Azadirachta indica* pollen. *Molecular and Cellular Biochemistry* 131:87-96.

Karunamoorthi K, Mulelam A, and Wassie F (2009). Assessment of knowledge and usage custom of traditional insect/mosquito repellent plants in Addis Zemen Town, South Gonder, NorthWestern Ethiopia. *Journal of Ethnopharmacology* 121:49-53.

Kevan PG and Ebert T (2005). Can almond nectar & pollen poison honey bees? *American Bee Journal* 145:507-509.

Lucena RFP, Albuquerque UP, Monteiro JM, Almeida CFCBR, Florentino ATN, and Ferraz JSF (2007). Useful plants of the semiarid Northeastern region of Brazil – a look at their conservation and sustainable use. *Environmental Monitoring and Assessment* 125:281-290.

Pereira IM, Andrade LA, Sampaio EVSB, and Barbosa MRV (2003). Use-history effects on structure and flora of caatinga. *Biotropica* 35:154-165.

Pimentel LA, Riet-Correa F, Gardner D, Panter KE, Dantas AFM, Medeiros RMT, Mota RA, and Araújo JAS (2007). *Mimosa tenuiflora* as a cause of malformations in ruminants in the Northeastern Brazilian semiarid rangelands. *Veterinary Pathology* 44:928-931.

Praz CJ, Müller A, and Dorn S (2008). Specialized bees fail to develop on non-host pollen: do plants chemically protect their pollen? *Ecology* 89:795-804.

Rhoades DF (1979). Evolution of plant chemical defence against herbivores. In *Herbivores: their interaction with secondary plant metabolites* (GA Rosenthal and DH Janzen, eds), pp.3-54. Academic Press, New York.

Rivera-Arce E, Chávez-Soto AA, Herrera-Arellano A, Arzate S, Agüero J, Feria-Romero IA, Cruz-Guzmán A, and Lozoya X (2007). Therapeutic effectiveness of a *Mimosa tenuiflora* cortex extract in venous leg ulceration treatment. *Journal of Ethnopharmacology* 109:523-528.

Siddiqui BS, Ali ST, Rajput MT, Gulzar T, Rasheed M, and Mehmood R (2009). GC-based analysis of insecticidal constituents of the flowers of *Azadirachta indica* A. Juss. *Natural Products Research* 23:271-283.

Soto-Blanco B, Sinhorini IL, Górniak SL, and Schumaher-Henrique B (2002). *Ricinus communis* cake poisoning in a dog. *Veterinary and Human Toxicology* 44: 155-156.

Westerkamp C (1996). Pollen in bee–flower relations. Some considerations on melittophily. *Botanica Acta* 109:325-332.

# Chapter 71

# Vetch (*Vicia villosa*) Poisoning in Cattle in the State of Santa Catarina

A. Gava, F.H. Furlan, S.D. Traverso, L.O. Veronezi, and F. Jönck

*Laboratory of Animal Pathology, University of Santa Catarina State, Av. Luiz de Camões, 2090, Conta Dinheiro, 88520-000, Lages – Santa Catarina/Brazil*

## Introduction

*Vicia* spp. is an annual or perennial legume popularly known as vetch or vicas. Due to their high nutritional value they are used in regions with a temperate or subtropical climate. The vetch species with highest economical interest are *V. villosa* and *V. sativa* (Bastos and Miotto 1996). In Santa Catarina State, vetch is mainly sown in winter, generally associated with oat (*Avena sativa*) or ryegrass (*Lolium multiflorum* and *L. perenne*), and grazing occurs more intensely during spring.

Different species of vetch are responsible for causing granulomatous disease in cattle (Panciera *et al.* 1966; Peet and Gardner 1986; Harper *et al.* 1993; Barros *et al.* 2001). Three distinct clinical syndromes are attributed to vetch intake although the most common and best studied occurs due to a systematic granulomatous disease that affects mainly Holstein and Aberdeen Angus cattle. This disease is clinically characterized by dermatitis, pruritus, fever, conjunctivitis, diarrhea, and weight loss (Panciera *et al.* 1992; Barros *et al.* 2001; Fighera and Barros 2004). At necropsy, the main lesion is the formation of grey nodules in several tissues. On histologic examination these nodules consist of granulomatous inflammation with multinucleated giant cells (Panciera *et al.* 1992; Barros *et al.* 2001; Fighera and Barros 2004).

## Vetch (*Vicia villosa*) Poisoning

The background reports, epidemiological data, and clinical signs were collected from owners and veterinarians during visits to the locations where the disease occurred. Six outbreaks were seen and five cows were necropsied. Sections of all tissues were collected, fixed in 10% buffered neutral formalin for 48 h, embedded in paraffin, sectioned at 4 µm, stained with hematoxylin and eosin, and examined by light microscopy.

The disease was diagnosed only in adult animals in locations where the pastures were formed by oat or ryegrass associated with *V. villosa*. Morbidity ranged from 5% to 35% and lethality from 28% to 100%. The illness occurred in October and November and clinical signs began after 50 to 120 days of grazing in pastures contaminated by *V. villosa*.

©CAB International 2011. *Poisoning by Plants, Mycotoxins, and Related Toxins* (eds F. Riet-Correa, J. Pfister, A.L. Schild, and T.L. Wierenga)

The sick animals initially manifested a rough hair coat with formation of elevated areas on the skin. With the occurrence of pruritus, lesions evolved into the formation of alopecic plaques that spread out from the head and neck to the rest of the body. Pustules formed on the udder that broke releasing a yellow exudate. The animals also presented anorexia, hyperthermia, and a sudden fall in milk production. In some animals, conjunctival hyperemia with tearing, nasal hemorrhage, and dark-colored urine were also observed.

At necropsy, multifocal to coalescent grey nodules were observed mainly in the lymph nodes, kidneys, spleen, liver, and heart. Furthermore, in one cow there were grey grooves along the skeletal muscle fibers. Upon histologic examination multifocal infiltration by lymphocytes, plasma cells, macrophages, and multinucleated giant cells was observed in the lymph nodes, kidneys, spleen, liver, and heart. Differential diagnosis includes other systemic granulomatous diseases with similar clinical signs and lesions. These diseases are named 'disease similar to vetch poisoning' (Panciera *et al.* 1992) or 'pyrexia, pruritis, and hemorrhagic syndrome' and are associated with the consumption by cattle of citrus pulp (Saunders *et al.* 2000; Gava and Barros 2001; Iizuka *et al.* 2005), silage containing the Sylade preservative (Matthews and Shreeve 1978), hay containing additives (Andrews *et al.* 1983), and feed containing the industrial byproduct di-ureido isobutane (DUIB, Breukink *et al.* 1978). In Brazil, diseases similar to vetch poisoning are only reported in cattle consuming citrus pulp (Gava and Barros 2001), thus this is the main differential diagnosis considered in this country.

Most of the cases of vetch poisoning are associated with the consumption of *V. villosa* and/or its hybrids and subspecies (Burroughs *et al.* 1983; Peet and Gardner 1986; Odriozola *et al.* 1991; Johnson *et al.* 1992; Harper *et al.* 1993; Barros *et al.* 2001; Fighera and Barros 2004). Some cases are related to the consumption of *V. benghalensis* (Green and Kleynhans 1989; Harper *et al.* 1993). In Brazil, particularly in the southern region, vetch poisoning was unknown until 2001. Previously, the vetch used for cattle feed was *V. sativa*, a species not related to the granulomatous disease. Afterwards, there was an increase in the planting of both species but because *V. villosa* is more aggressive it took over the locations where vetch is cultivated.

## Conclusion

The bovine disease characterized by alopecia mainly on the head and neck, sudden drop in milk production, fever, and hemorrhage may be linked to the intake of *V. villosa*. The main differential diagnosis is with citrus pulp poisoning as pulp poisoning has also been diagnosed in the State of Santa Catarina.

## References

Andrews AH, Longstaffe JA, Newton AC, and Musa I (1983). Acute fatal haemorrhagic syndrome in dairy cows. *The Veterinary Record* 112:614.

Barros CSL, Fighera RA, Rozza DB, Rech RR, Sallis SV, and Langohr IM (2001). Systemic granulomatous disease in cattle in Rio Grande do Sul, Brazil, associated with grazing vetch (*Vicia* spp). *Pesquisa Veterinária Brasileira* 21(4):162-171.

Bastos NR and Miotto STS (1996). O gênero *Vicia* (Leguminosae – Faboideae) no Brasil. *Pesquisas Botânicas* 46:85-180.

Breukink HH, Holzhauer C, and Westenbrock ACJM (1978). Pyrexia with dermatitis in dairy cows. *The Veterinary Record* 103:221.

Burroughs GW, Neser JA, Kellerman TS, and Van Niekerk FA (1983). Suspected hybrid vetch (*Vicia villosa* crossed with *Vicia dasycarpa*) poisoning of cattle in the Republic of South Africa. *Journal of the South African Veterinary Association* 54:75-79.

Fighera RA and Barros CSL (2004). Systemic granulomatous disease in Brazilian cattle grazing pasture containing vetch (*Vicia* spp.). *Veterinary and Human Toxicology* 46(2):62-66.

Gava A and Barros CSL (2001). Intoxicação por polpa cítrica. In *Doenças de Ruminantes e Eqüinos* (F Riet-Correa, AL Schild, MC Méndez, and RAA Lemos, eds) vol. 2, pp. 212-215. Editora Varella, São Paulo.

Green JR and Kleynhans JE (1989). Suspected vetch (*Vicia benghalensis*) poisoning in a Friesland cow in the Republic of South Africa. *Journal of the South African Veterinary Association* 60(2):109-10.

Harper PA, Cook RW, Gill PA, Fraser GC, Badcoe LM, and Power JM (1993). Vetch toxicosis in cattle grazing *Vicia villosa* ssp *dasycarpa* and *V. benghalensis*. *Australian Veterinary Jounal* 70(4):140-144.

Iizuka A, Haritani M, Shiono M, Sato M, Fukuda O, Hagiwara A, Miyazaki S, Tanimura N, Kimura K, Nakazawa K, Kobayashi M, Takahashi T, Saito T, and Fukai K (2005). An outbreak of systemic granulomatous disease in cows with high milk yields. *The Journal of Veterinary Medical Science* 67(7):693-9.

Johnson B, Moore J, Woods LW, and Galey FD (1992). Systemic granulomatous disease in cattle in California associated with grazing hairy vetch (*Vicia villosa*). *Journal of Veterinary Diagnostic Investigation* 4:360-362.

Matthews JG and Shreeve BJ (1978). Pyrexia/pruritis/haemorrhagic syndrome in dairy cows. *The Veterinary Record* 103:408-409.

Odriozola E, Paloma E, Lopez T, and Campero C (1991). An outbreak of *Vicia villosa* (hairy vetch) poisoning in grazing Aberdeen Angus bulls in Argentina. *Veterinary and Human Toxicology* 33:278-280.

Panciera RJ, Johnson L, and Osburn BI (1966). A disease of cattle grazing hairy vetch pasture. *Journal of the American Veterinary Medical Association* 148:804-808.

Panciera RJ, Mosier DA, and Ritchey JW (1992). Hairy vetch (*Vicia villosa* Roth) poisoning in cattle: update and experimental induction of disease. *Journal of Veterinary Diagnostic Investigation* 4:318-325.

Peet RL and Gardner JJ (1986). Poisoning of cattle by hairy or wooly-pod vetch, *Vicia villosa* subspecies *dasycarpa*. *Australian Veterinary Journal* 63:381-382.

Saunders GK, Blodgett DJ, Hutchins TA, Prater RM, Robertson JL, Friday PA, and Scarrat WK (2000). Suspected citrus pulp toxicosis in dairy cattle. *Journal of Veterinary Diagnostic Investigation* 12:269-271.

# Chapter 72

# *Baccharis pteronioides* Toxicity

B.L. Stegelmeier[1], Y. Sani[2], and J.A. Pfister[1]

[1]*USDA-ARS Poisonous Plant Research Laboratory, Logan, Utah 84341, USA;* [2]*Research Institute for Veterinary Science, Bogor, Jawa Barat*

## Introduction

*Baccharis* spp. (Asteraceae) are native American plants with over 400 species and varieties in both North and South America (Abad *et al.* 2006). A variety of toxins including diterpenic lactones, sesquiterpenes, flavonoids, saponins, tannins, phenolic compounds, and essential oils have been isolated and described from *Baccharis* species (Grance *et al.* 2008). In South America *B. coridifolia* commonly poisons livestock in southern Brazil, Uruguay, Argentina, and Paraguay (Tokarnia *et al.* 1992; Barros 1993, 1998). Experimental poisoning has also been described in cattle, sheep, horses, rabbits, and mice (Tokarnia and Dobereiner 1975, 1976; Dobereiner *et al.* 1976; Costa *et al.* 1995; Rodrigues and Tokarina 1995; Varaschin and Alessi 2003; Rissi *et al.* 2005). Signs of poisoning produced by these South American *Baccharis* spp. are variable and include anorexia, diarrhea, constipation, muscular tremors, tachypnea, tachycardia, recumbency, and death. The associated lesions include reddening, edema, and erosions of the stomach with hemorrhagic gastroenteritis and widespread lymphoid necrosis. (Tokarina and Dobereiner 1975; Barros 1998; Varaschin and Alessi 2003; Rissi *et al.* 2005). All parts of the South American plant can be toxic with higher toxin concentrations in the flowers and seeds (Jarvis *et al.* 1988). Preliminary studies suggest there are multiple toxins including roridin, miotoxin, miophitocen, and verrucarin (Busam and Habermehl 1982; Habermehl *et al.* 1985; Jarvis *et al.* 1996).

As early as 1920 in North America *B. pteronioides* was associated with livestock poisoning. Marsh *et al.* (1920) reported that the lethal dose for sheep was 'near one pound'. Marsh speculated that poisoning was related to lack of alternative forage and that stockmen should learn to recognize the plant and avoid exposure when forage is limited. Other than describing the intestinal lesions as 'burned with potash', there are no good descriptions of the gross or histologic lesions. Since Marsh's report poisonings have been sporadic (Manley *et al.* 1982). The *B. pteronioides* toxin is not known nor are there any comparisons of this species with other toxic *Baccharis* species.

In North America *B. pteronioides* poisoning is characterized by anorexia, lethargy, weakness, and sudden death. Postmortem findings are variable with most having hemorrhagic enteritis. The purpose of this study is to develop a small animal model of *B. pteronioides* toxicity, describe the clinical, gross, and histologic lesions of poisoning, and initiate chemical studies to identify the toxin and characterize and describe its toxicity.

©CAB International 2011. *Poisoning by Plants, Mycotoxins, and Related Toxins*
(eds F. Riet-Correa, J. Pfister, A.L. Schild, and T.L. Wierenga)

## Materials and Method

Leaves of *B. pteronioides* (Utah State University, Logan Utah, Intermountain Herbarium voucher #2828) were stripped from the branches, freeze dried, finely ground (to pass through a 1 mm screen), thoroughly mixed, and stored at −4°C until used. Forty-eight Syrian hamsters were divided into four groups and dosed twice daily by oral gavage with 0, 50, 100, and 200 mg of *B. pteronioides* mixed with 2 ml of peanut oil. The 0 mg group was dosed with 200 mg of finely ground lucerne as a volume and carrier negative control. After 10 days of treatment the hamsters were euthanized, serum was collected by cardiac puncture, and tissues were collected for histologic evaluation. More details of this study and its findings have been previously published (Stegelmeier *et al.* 2009).

## Results and Discussion

All but one hamster dosed with 200 mg developed diarrhea, became reluctant to move, and stopped eating. Because of this, three of the 200 mg group were necropsied early after 8 days of dosing. At necropsy all but one of the high dose animals had multiple, variably sized hemorrhagic infarctions in the liver. The hepatocytes were swollen with open, enlarged nuclei and abundant nuclear pseudoinclusions. Many hepatic vessels were dilated with multifocal vasculitis, fibrinous vascular degeneration, and fibrin thrombi. Scattered randomly in multiple hepatic lobules there were large zones of hemorrhagic and coagulative necrosis (infarctions). These infarcts were surrounded by fibrin and infiltrates of neutrophils with fewer numbers of macrophages, multinucleated giant cells, and lymphocytes. The hamsters dosed with 100 mg *Baccharis* exhibited prominent centrilobular hepatocellular swelling. The swollen hepatocytes had vacuolated cytoplasm with prominent nuclear pseudoinclusions. The gastric mucosa often contained numerous yeast organisms. No significant histologic lesions were identified in the controls or hamsters treated with 50 mg.

Treated animals developed extensive hepatocellular swelling and degeneration that progressed to vasculitis and hemorrhagic infarction in the high dose group. The mucosa of the large and small intestine was expanded with lymphocytes and plasma cells. The mucosal surface was covered with numerous bacteria and smaller numbers of yeast organisms. Many of the submucosal lymphoid tissues were edematous and necrotic. As the low and medium dose animals were similar to controls, the biochemical changes were dependent on the development of the extensive vasculitis and infarctions that characterized the 200 mg/day group. These hepatic and vascular changes are similar to those reported in livestock.

Brazilian *B. cordiofolia* and *B. megapotamica* poisoning in livestock produces necrosis of gastrointestinal mucosa and lymphoid tissues (Barros 1998; Driemeier *et al.* 2000; Tokarnia *et al.* 2002; Varaschin and Alessi 2003). Similar necrotic lesions involving secondary lymphoid germinal centers, the lymph nodes, spleen, intestine, and thymus have been described in rodents and rabbits gavaged with ground plant (Habermehl *et al.* 1985; Varaschin and Alessi 2003). Though less severe, we observed similar lesions as many of the high dose hamsters developed hemorrhagic gastroenteritis and mild submucosal lymph node necrosis. Similar lymphoid necrosis is associated with stress and endogenous corticosteroid release and recent work has shown that certain trichothecenes exacerbate lymphoid cytotoxicity or apoptosis induced by a variety of toxins and cellular messengers (Uzarski *et al.* 2003). In this study the lymphoid necrosis was only seen in animals with extensive hepatic necrosis and hemorrhagic enteritis. Further work is needed to determine if *B. pteronioides* toxins are directly involved in producing lymphoid necrosis.

More work is needed to isolate and identify the *B. pteronioides* toxins. If the toxins are similar to many of the trichothecenes isolated from other *Baccharis* plants they are likely to be similar to the roridins, miotoxins, miophytocens, and verrucarol that are produced by the soil fungi *Myrothecium* spp., which are absorbed in the root and translocated to the vegetative plant parts (Habermehl *et al.* 1985; Barros 1998). Although effects and toxicity of mycotoxins are fairly well understood, such preharvest fungal-plant interactions are relatively unexplored and present new research challenges to better explain the conditions and interactions that result in poisonous plant problems.

## Conclusion

We were able to reproduce clinical and histologic lesions of *B. pteronioides* poisoning in hamsters that were similar to those reported in previous field cases and feeding trials. These findings indicate that at high doses *B. pteronioides* is toxic and produces lesions that may be similar to bacterial endotoxemia-produced vasculitis and infarction. Research to purify and identify the toxin, the toxic dose, and mechanism of toxicity is ongoing.

## References

Abad MJ, Bessa AL, Ballarin B, Aragon O, Gonzales E, and Bermejo P (2006). Anti-inflammatory activity of four Bolivian *Baccharis* species (Compositae). *Journal of Ethnopharmacology* 103:338-344.

Barros CSL (1993). Intoxicações por plantas que afetam o tubo digestivo. Intoxicação por *Baccharis coridifolia*. In *Intoxicações Por Plantas e Micotoxicoses em Animais Domésticos* (F Riet-Correa, MC Mendez, and AL Schild, eds), pp. 159-169. Editorial Hemisfero Sur, Montevideo.

Barros CSL (1998). Livestock poisoning by *Baccharis coridifolia*. In *Toxic Plants and Other Natural Toxicants* (T Garland and A Barr, eds), pp. 569-572. CAB International, New York.

Busam L and Habermehl GG (1982). Accumulation of mycotoxins by *Baccharis coridifolia*: a reason for livestock poisoning. *Naturwissenschaften* 69:392-393.

Costa E, Costa J, Armien AG, Barbosa G, and Peixoto PV (1995). Intoxicação experimental por *Baccharis coridifolia* (Compositae) em equinos. *Pesquisa Veterinária Brasileira* 15:19-26.

Dobereiner J, Resende A, and Tokarnia CH (1976). Intoxicação experimental por *Baccharis coridifolia* em coelhos. *Pesquisa Agropecuária Brasileira, Serie Veterinária* 11:27-35.

Driemeier D, Cruz C, and Loretti AP (2000). *Baccharis megapotamica* var Weirii poisoning in Brazilian cattle. *Veterinary and Human Toxicology* 42:220-221.

Grance SR, Teixeira MA, Leite RS, Guimaraes EB, de Siqueira JM, de Oliveira WF, Filiu SV, and Vieira MD (2008). *Baccharis trimera*: Effect on hematological and biochemical parameters and hepatorenal evaluation in pregnant rats. *Journal of Ethnopharmacology* 117:28-33.

Habermehl GG, Busam L, Heydel P, Mebs D. Tokarnia CH, Dobereiner J, and Spraul M (1985). Macrocyclic trichothecenes: cause of livestock poisoning by the Brazilian plant *Baccharis coridifolia*. *Toxicon* 23:731-745.

Jarvis BB, Midiwo JO, Bean GA, Boul-Nasr MB, Barros CS, and Bassam-Aboul-Nasr M (1988). The mystery of trichothecene antibiotics in *Baccharis* species. *Journal of Natural Products* 51:736-744.

Jarvis BB, Wang C, and Cox MS (1996). Brazilian *Baccharis* toxins: livestock poisoning and isolation of macrocyclic trichothecenes glucosides. *Natural Toxins* 4:58-61.

Manley GD, Edds GT, and Sundlof SF (1982). Cattle deaths from poisonous plant. *Folia Morphologica – Praha* 11:20.

Marsh CD, Clawson AB, and Eggleston WW (1920). *Baccharis pteronioides* as a poisonous plant of the southwest. *Journal of the American Veterinary Medical Association* 57:430-434.

Rissi DR, Rech RR, Fighera RA, Cagnini DQ, Commers GD, and Barros CSL (2005). Intoxicaçã experimental por *Baccharis coridifolia* em bovinos. *Pesquisa Veterinária Brasileira* 25:111-114.

Rodrigues RL and Tokarnia CH (1995). Fatores que influenciam a toxidez de *Baccharis coridifolia* (Compositae): um estudo experimental em coelhos. *Pesquisa Veterinária Brasileira* 15:51-69.

Stegelmeier BL, Sani Y, and Pfister JA (2009). *Baccharis pteronioides* toxicity in livestock and hamsters. *Journal Veterinary Diagnostic Investigation* 21:208-213.

Tokarnia CH and Dobereiner J (1975). Intoxicação experimental em bovinos por 'mio-mio', *Baccharis coridifolia. Pesquisa Veterinária Brasileira* 10:79-97.

Tokarnia CH and Dobereiner J (1976). Intoxicação experimental em ovinos por 'mio-mio', *Baccharis coridifolia. Pesquisa Veterinaria Brasileira* 11:19-26.

Tokarnia CH, Peixoto PV, and Gava A (1992). Intoxicação experimental por *Baccharis megapotamica* var. *megapotamica* e var. *Weirii* (Compositae) em bovinos. *Pesquisa Veterinária Brasileira* 2:19-31.

Tokarnia CH, Dobereiner J, and Peixoto PV (2002). Poisonous plants affecting livestock in Brazil. *Toxicon* 40:1635-1660.

Uzarski RL, Islam Z, and Pestka J (2003). Potentiation of trichothecene-induced leukocyte cytotoxicity and apoptosis by TNF-alpha and Fas activation. *Chemical and Biological Interactions* 146:105-119.

Varaschin MS and Alessi AC (2003). Poisoning of mice by *Baccharis coridifolia*: an experimental model. *Veterinary and Human Toxicology* 45:42-44.

# Chapter 73

# Toxicity of *Dieffenbachia* spp. with a Focus on Livestock Poisoning

A.C. Dantas[1], J.A. Guimarães[1], A.C.L. Câmara[2], J.A.B. Afonso[1], and C.L. Mendonça[1]

[1]*Clínica de Bovinos, Campus Garanhuns, Universidade Federal Rural de Pernambuco, PO Box 152, 55292-901, Garanhuns, Pernambuco, Brazil;* [2]*Hospital Escola de Grandes Animais, Universidade de Brasília, Galpão 4, Granja do Torto, 70636-200, Brasília, Distrito Federal, Brazil*

## Introduction

Since 1807 *Dieffenbachia* species have been used for their toxic effects. In the Caribbean and West Indies this tropical shrub with thick waxy leaves and fleshy stems was used for torturing slaves and to sabotage crime witnesses (Kissman 1961; Arditti and Rodriguez 1982). The plant causes local irritation with mucosal swelling, pain, and the inability to talk thus the plant is known as 'dumb cane' and 'mother-in-law's tongue' (Arditti and Rodriguez 1982). In Brazil the plant is called 'comigo-ninguém-pode' (Tokarnia *et al.* 2000).

*Dieffenbachia* spp. are frequently used as an ornamental office or house plant. For this reason there are a large number of poisoning cases involving children, victims of practical jokes, horticulturists (Gardner 1994), and small animals (pets) (Plumlee 2002; Lightfoot and Yeager 2008). This also explains the rare reports of poisoning in livestock because of their restricted access to the plant (Osweiler 1998; Radostits *et al.* 2002). The objective of this paper is to review the clinical signs and experimental and spontaneous poisoning by the houseplant *Dieffenbachia* spp.

## Mechanism of Toxicity

Many houseplants of the Araceae family contain insoluble calcium oxalate crystals and have a different form of toxicity than the plants that contain soluble oxalates which cause renal toxicosis (nephrolithiasis) (Froberg *et al.* 2007). These plants are popular and include elephant's ear, schefflera, caladium, dumbcane (*Dieffenbachia* spp.), pothos, many ivy varieties, philodendron, peace lilies (*Spathiphyllum* spp.), and calla lilies (*Zantedeschia* spp.) (Burrows and Tyrl 2001).

In certain plants such as *Dieffenbachia* spp., the toxic properties are caused by both mechanical and chemical effects. Insoluble oxalate is in the form of calcium oxalate needle-

©CAB International 2011. *Poisoning by Plants, Mycotoxins, and Related Toxins* (eds F. Riet-Correa, J. Pfister, A.L. Schild, and T.L. Wierenga)

shaped crystals or raphides contained in oval-shaped cells called idioblasts. The idioblast cells have an opening on both ends of the cell. When mechanical force is applied to the idioblasts, the raphides fire out of the sharp crystals and are propelled a distance of 2 to 3 cell lengths and are embedded into the mucous membranes, tongue, and throat. Some authors reports no clinical signs if the contact with the skin or mucosa occurs with the intact plant (Rauber 1985; Gardner 1994; Osweiler 1998; Burrows and Tyrl 2001; Radostits *et al.* 2002; Cumpston *et al.* 2003). Others believe that chemical effect is probably due to the presence of saponins, cyanogenic glycosides, proteolytic enzymes, and alkaloids in the leaves that could produce kinins and act as chemical mediators of inflammation (Kuballa *et al.* 1980; Gardner 1994; Corazza *et al.* 1998; Osweiler 1998). Although the exact nature of this putative toxic substance has not been determined, some candidates include a proteinaceous substance, a substance with a proteolytic property, or a substance that affects bradykinin activity (Rauber 1985; Cumpston *et al.* 2003).

The onset of pain can be immediate or can occur up to 2 h after chewing on the plant. The animal can have increased salivation, vocalization, anorexia, and depression. Swelling inside the mouth can occur but is usually not sufficient to cause airway obstruction (Pedaci *et al.* 1999). Mild vomiting or diarrhea is possible if the animal swallows a large amount of the plant (Plumlee 2002).

## Clinical Signs

Every year in Brazil many cases of spontaneous poisoning by toxic ornamental plants involving children and pets are reported; special concern is given to the Araceae family (Dip *et al.* 2004). The best known and most toxic member of this family is *Dieffenbachia seguine*, a taxonomical synonym to *D. picta* Schott and *D. maculata.* This plant is widespread in all parts of Brazil, occurring from subtropical areas to the equatorial rainforest in northern Brazil (Gardner 1994; Tokarnia *et al.* 2000).

Some authors report minimal or no effects after exposure of oral mucosa in humans or animals; more typically chewing of the leaves, petioles, and stems results in painful oropharyngeal edema, inability to talk, and profuse salivation (Gardner 1994; Osweiler 1998; Radostits *et al.* 2002; Cumpston *et al.* 2003; Froberg *et al.* 2007). Depending on the plant part consumed and on type of exposure many clinical symptoms are described (Corazza *et al.* 1998). Reports in humans include: ocular injuries (Chiou *et al.* 1997; Hsueh *et al.* 2004); contact dermatitis causing vesicular burns, pruritus, erythema, and/or edematous lesions (Sanchez-Morrilas 2005); acute airway compromise (Cumpston *et al.* 2003); vesicles and ulceration of the oral mucosa, esophagitis, and aortoesophageal fistula (Gardner 1994; Snajdauf *et al.* 2005).

Experimental intoxication in livestock is limited. In their experiments with ornamental plants Tokarnia *et al.* (2000) achieved mild signs of poisoning in cattle and sheep after ingestion of *D. picta.* Initial signs were observed immediately or within 40 min after chewing of the plant. Animals were totally recovered in 10 days. All experimental animals showed clinical signs associated with the plant's local irritative effects; these included edema of the tongue, lips, and face, profuse salivation (ptyalism), difficulty in food apprehension, and necrosis of the mouth epithelium. Pathological findings consisted of accentuated necrotic-degenerative alterations of the tongue epithelium, gaps between basal and spinous layers, polymorphonuclear infiltrates, and coagulative necrosis of muscular fibers adjacent to the tongue (Tokarnia *et al.* 2000).

Some reviews of spontaneous poisoning focus on humans (Froberg *et al.* 2007) and pets including dogs, cats, and birds (Plumlee 2002; Lightfoot and Yeager 2008). One paper of spontaneous poisoning in livestock reports clinical and biochemical findings in a female crossbred goat raised extensively in northeastern Brazil. Clinical examination revealed hyperthermia (39.6ºC), dehydration of about 10%, subcutaneous edema from the submandibular to the xiphoid area also involving the esophagus, protruded and edematous tongue with focal laceration areas on the dorsal side, ptyalism, and rumen and gastrointestinal hypomotility. The goat also showed severe bloat as a consequence of a possible edema in the esophageal mucosa and/or increased size of the mediastinal lymph nodes. Biochemical alterations consists of a severe rise in creatinine phosphokinase (364.3 U/l; reference values: 0.8-8.9 U/l) caused by coagulative necrosis of the muscular fibers adjacent to the tongue and/or esophageal mucosa. Aspartate aminotransferase, gamma glutamyltransferase, and creatinine levels were within the reference values for the species (Dantas *et al.* 2007).

## Treatment

Treatment involves symptomatic care. In pets the mouth should be rinsed and the animal can be offered small amounts of milk or soft food to decrease the pain. If the amount of plant ingestion is unusually large, the animal can be given gastrointestinal protectants and anti-inflammatory drugs. Most animals recover uneventfully within 24 h (Plumlee 2002). Livestock can be successfully treated with systemic steroids, diuretics, and transfaunation (Dantas *et al.* 2007).

Contact dermatitis from raphide-containing plants may respond to topical or systemic steroids. Symptomatic care with an H2-antagonist may also be beneficial (Froberg *et al.* 2007). A recent study also shows the ability of eugenol to reduce tongue edema induced by *D. picta* in mice (Dip *et al.* 2004). Corneal irritation may be treated with a cycloplegic or steroidal eye drop (Chiou *et al.* 1997).

## Conclusions

Although *Dieffenbachia* spp. is used frequently as an ornamental plant, the plant is dangerous because it contains needle-shaped crystals of calcium oxalate that cause severe irritation; after ingestion these crystals may cause skin and mucosa damage in any animal species. Pets and humans are most often affected because of exposure to household or garden plants. Livestock are not often affected because of their lack of exposure to these plants.

## References

Arditti J and Rodriguez E (1982). *Dieffenbachia*: uses, abuses and toxic constituents: a review. *Journal of Ethnopharmacology* 5(3):293-302.

Burrows GE and Tyrl RJ (2001). *Araceae juss*. In *Toxic plants of North America.* (GE Burrows, RJ Tyrl, eds), pp. 105-119. Ames, Iowa State University Press, Iowa.

Chiou AG, Cadez R, and Bohnke M (1997). Diagnosis of *Dieffenbachia* induced corneal injury by confocal microscopy. *British Journal of Ophthalmology* 81(2):168-169.

Corazza M, Romania I, Polib F, and Virgili A (1998). Irritant contact dermatitis due to *Dieffenbachia* spp. *Journal of European Academy of Dermatology and Venereology* 10(1):87-89.

Cumpston KL, Vogel SN, Leikin JB, and Erickson TB (2003). Acute airway compromise after brief exposure to a Dieffenbachia plant. *Journal of Emergency Medicine* 25(4):391-397.

Dantas AC, Guimarães JA, Câmara ACL, Afonso JAB, Mendonça CL, Costa NA, and Souza MI (2007). Intoxicação natural por comigo-ninguém-pode (*Dieffenbachia* sp.) em caprino. *Ciência Veterinária nos Trópicos* 10(2-3):119-123.

Dip EC, Pereira NA, and Fernandes PD (2004). Ability of eugenol to reduce tongue edema induced by *Dieffenbachia picta* Schott in mice. *Toxicon* 43(6):729-735.

Froberg B, Ibrahim D, and Furbee RB (2007). Plant poisoning. *Emergency Clinics of North America* 25(2):375-344.

Gardner DG (1994). Injury to the oral mucous membranes caused by the common houseplant, *Dieffenbachia*: a review. *Oral Surgery, Oral Medicine, Oral Pathology* 78(5):631-633.

Hsueh KF, Lin PY, Lee SM, and Hsieh CF (2004). Ocular injuries from plant sap of genera *Euphorbia* and *Dieffenbachia*. *Journal of the Chinese Medical Association* 67:93-98.

Kissman KG (1961). Knowledge of poisonous plants in the United States – brief history and conclusions. *Economic Botanic* 38(1):119-130.

Kuballa B, Lugnier AA, and Anton R (1980). Study of *Dieffenbachia* induced edema in mouse and rat hindpaw: respective role of oxalate needles and trypsin-like protease. *Toxicology and Applied Pharmacology* 58:444-451.

Lightfoot TL and Yeager JM (2008). Pet bird toxicity and related environmental concerns. *Veterinary Clinics Exotic Animal Practice* 11(2):229-259.

Osweiler GD (1998). Toxicoses relacionadas com plantas. In *Toxicologia Veterinária* (GD Osweiler, ed.), pp. 386-439. Artes Médicas, Porto Alegre, Rio Grande do Sul.

Pedaci L, Krenzelok EP, Jacobsen TD, and Aronis J (1999). *Dieffenbachia* species exposures: an evidence-based assessment of symptom presentation. *Veterinary and Human Toxicology* 41(5):335-338.

Plumlee KH (2002). Plant hazards. *Veterinary Clinics Small Animal Practice* 32(2):383-395.

Radostits OM, Gay CC, Blood DC, and Hinchcliff KW (2002). Doenças causadas por toxinas de plantas, fungos, cianofitas, clavibactérias e por venenos de carrapatos e animais vertebrados. In *Clínica veterinária – um tratado de doenças dos bovinos, ovinos, suínos, caprinos e eqüinos* (OM Radostits, CC Gay, DC Blood, and KW Hinchcliff, eds), pp.1432-1543. Guanabara Koogan, Rio de Janeiro.

Rauber A (1985). Observations on the idioblasts of *Dieffenbachia*. *Journal of Toxicology and Clinical Toxicology* 23(2-3):79-90.

Sanchez-Morillas L (2005). Contact dermatitis due to *Dieffenbachia*. *Contact Points* 53:172-173.

Snajdauf J, Mixa V, Rygl M, Vyhnánek M, Morávek J, and Kabelka Z (2005). Aortoesophageal fistula – an unusual complication of esophagitis caused by *Dieffenbachia* ingestion. *Journal of Pediatric Surgery* 40:29-31.

Tokarnia CH, Döbereiner J, and Peixoto PV (2000). *Plantas tóxicas do Brasil*, 310 pp. Helianthus, Rio de Janeiro.

# Chapter 74

# Morphological, Morphometric, and Histochemical Analysis of the Large Intestine of Rabbits Intoxicated with *Solanum glaucophyllum* (duraznillo blanco)

C.N. Zanuzzi[1,2,3], C.G. Barbeito[1,2,3], M.L. Ortiz[1], P.A. Fontana[1], E.L. Portiansky[1,3], and E.J. Gimeno[1,3]

[1]*Institute of Pathology;* [2]*Department of Histology and Embryology, School of Veterinary Sciences, National University of La Plata, 60 y 118 (1900) La Plata, Buenos Aires, Argentina;* [3]*Members of CONICET (National Research Council)*

## Introduction

*Solanum glaucophyllum* (= *S. malacoxylon*) is a calcinogenic plant responsible for enzootic calcinosis of ruminants in South America, a disease that causes considerable economic losses (Worker and Carrillo 1967; Puche and Bingley 1995). This plant contains high levels of 1,25-dihydroxyvitamin $D_3$ as glycoside derivatives in its leaves. The chronic ingestion of this material generates a hypervitaminosis D-like state and soft tissue mineralization. The clinically intoxicated animals present stiffness, painful gait, xyphosis, anorexia, loss of body condition, and in the most severe cases advanced cachexia (Worker and Carrillo 1967).

Vitamin D receptors are present in multiple tissues and $1,25(OH)_2D_3$ has pleiotropic effects in its target organs. The activation of vitamin D receptors can increase or decrease specific gene transcription and consequently modify the synthesis of the products coded by them (Bikle 2007). In addition to the well known effects of vitamin D on mineral homeostasis it participates in immunomodulation and the regulation of cell proliferation and differentiation (Gimeno *et al.* 2000; Bikle 2007; Fontana *et al.* 2009).

In the intestine vitamin D enhances the absorption efficiency of dietary calcium and phosphate (Bikle 2007). In addition, vitamin D is a key regulator of gastrointestinal homeostasis as a participant in intestinal epithelium differentiation and proliferation (Suda *et al.* 1990; Holt *et al.* 2002), detoxification (Kutuzova and DeLuca 2007), and in the preservation of the mucosal barrier integrity (Kong *et al.* 2008).

The differential carbohydrate expression of cells is of great value as a differentiation indicator. The terminal glycosylation sequences expressed by cells reflect the expression of the corresponding glycosyltransferases and glycosidases (Biol-N'garagba and Luisot 2003), thus, carbohydrates are considered a secondary product of gene expression (Gimeno and

©CAB International 2011. *Poisoning by Plants, Mycotoxins, and Related Toxins*
(eds F. Riet-Correa, J. Pfister, A.L. Schild, and T.L. Wierenga)

Barbeito 2004). The different types of glycoconjugates in tissue sections can be shown using conventional histochemistry techniques such as periodic acid-Schiff (PAS) and Alcian Blue (AB) (Spicer and Schulte 1992). Additional information is obtained using lectin histochemistry. Lectins are a heterogeneous group of proteins or glycoproteins of plant and animal origin that bind to specific terminal carbohydrates (Goldstein and Hayes 1978) and they are useful to study the process of cell differentiation in the intestine (Gelbert *et al.* 1992; Falk *et al.* 1994).

Little is known about the possible effects of high doses of vitamin D on the gastrointestinal tract (Razzaque and Lanske 2006). In addition, there are few studies on cell differentiation changes in domestic animals under plant-induced hypervitaminosis D (Barros and Gimeno 2000; Gimeno *et al.* 2004; Zanuzzi *et al.* 2008; Fontana *et al.* 2009). Thus, we analyzed the morphological and morphometric changes as well as the histochemical and lectin histochemical carbohydrate pattern in the colon and rectum of rabbits intoxicated by *S. glaucophyllum* (Sg).

## Materials and Methods

Twenty-five 3-month-old New Zealand male rabbits were used. All animals were clinically healthy. They were fed with a standard diet free of calcinogenic substances and water *ad libitum.* Every animal was housed in an individual cage. All the procedures were carried out according to the 'Guide for the Care and Use of Laboratory Animals' of the National Research Council (National Academy Press, 1996, Washington, USA). Ten animals were experimentally intoxicated *per os* with 125 mg/animal of powdered Sg leaves twice a week until they were killed. Five of them were killed 15 days after the beginning of the intoxication (I1515 group), whereas the other five were left for another 15 days (I3030 group). Five more animals were intoxicated for 15 days but killed after 45 days (probably recovered group–PRG1545). Two nutritionally restricted groups (NRG) were used to determine the influence of an anorexia state. These animals received the same amount of food as the intoxicated animals. Two animals were nutritionally restricted for 15 days and then returned to the *ad libitum* diet until they were killed 15 days later (NRG1530); the other two were restricted for 30 days and then killed (NRG3030). Six rabbits were used as controls. The body weight of each animal was recorded once a week. Clinical signs were observed and recorded every day during the entire study. Samples of colon and rectum of each animal were rinsed in PBS, fixed in 10% neutral buffered formalin, and embedded in paraffin. Sections of 3 μm were stained with hematoxylin and eosin for qualitative examination. Some slides were used for the conventional histochemical techniques of PAS and Alcian Blue (pH 2.5, 1.0, and 0.5) to analyze the carbohydrate composition of the mucin produce by the cells of the surface and glandular epithelium (Cook 1990).

For the lectin histochemistry study the slides were dewaxed and rehydrated and then incubated with 0.03% $H_2O_2$ in methanol for 30 min at room temperature to inhibit endogenous peroxidase activity. Slides were then treated with 1% bovine serum albumin (BSA) in phosphate buffer solution (PBS) for 30 min and incubated overnight with biotinylated lectins. The seven lectins (Lectin Kit BK 1000, Vector Laboratories, Inc., Burlingame, CA, USA) with different carbohydrate specificity used were the following: Con-A (*Canavalia ensiformis*, specifically binding α-D-Man and α-D-Glc); DBA (*Dolichos biflorus*, with binding specificity to α-D-GalNAc); SBA (*Glycine maximus*, binding specificity to α-D-GalNAc, β-D-GalNAc and α and β-Gal); PNA (*Arachis hypogea*, that specifically binds β-D-Gal and (β1-3) GalNAc); RCA-1 (*Ricinus communis*-

1, binding specificity β-D-Gal and α-D-Gal); UEA-1 (*Ulex europaeus*-1, binding specificity α–L–Fuc); and WGA (*Triticum vulgaris*, binding specificity α-D GlcNAc and NeuNAc) (Goldstein and Hayes 1978). The optimal lectin concentration was 30 μg/ml in PBS for all lectins except for PNA (10 μg/ml). The horseradish peroxidase streptavidin SA-5704 (Vector Laboratories, Inc., Burlingame, CA, USA), used as a detection system, was incubated for 30 min. Slides were rinsed three-fold in PBS for 5 min each time. Liquid 3,3'-diaminobenzidine tetrahydrochloride (DAB) was used as chromogen (DakoCytomation, Carpinteria, CA, USA). Negative controls for lectin staining included exposure to horseradish-peroxidase and substrate medium without lectin. The dark golden-brown DAB hydrogen peroxide reaction product showed the positively stained structures. Mayer's hematoxylin was used for counterstaining. The lectin binding pattern of goblet cells and enterocytes (glycocalix and apical cytoplasm) was evaluated. The intensity of lectin binding was subjectively scored from 0 to 3 with 0 = negative, 1 = weak, 2 = moderate, and 3 = strong. Lectin controls were performed by the addition of inhibitory sugars at a final concentration of 0.01 M.

For morphometric analysis images of each sample section were captured from a microscope (Olympus BX61 system microscope, Tokyo, Japan) with an objective magnification of 40× through an attached digital video camera (EvolutionVF, QImaging, USA) and digitized with a 24 bits RGB TIFF format. The images were processed and analyzed using the ImagePro Plus v6.2 program (Media Cybernetics, Silver Spring, MA, USA). The following parameters were evaluated: area, length, width, perimeter of the crypts, and the thickness of the intestinal wall and muscular layer. The ANOVA test was used to evaluate differences among groups. The Bonferroni test was used as a post hoc index. Significant differences were defined as those with $P < 0.05$.

## Results

### Morphological study

Colon and rectum lamina propria and submucosa of both intoxicated groups appeared moderately hypercellular and edematous with lymphangiectasia, which in more severe cases extended up to the muscular layer. The infiltration consisted of mononuclear cells such as macrophages, lymphocytes, and plasmocytes. The colonic wall appeared thinner in both intoxicated groups in comparison with control animals. The colon and rectum sections of some animals from PRG and both NRG groups showed a morphological pattern similar to that of controls whereas others resembled that described for the intoxicated animals.

### Morphometric study

In colon the most remarkable changes were observed in the parameters of the crypts of the PRG since the area, perimeter, length, and width significantly increased. Neither the thickness of the intestinal wall nor muscular layer was significantly affected between experimental groups.

In the rectum the area, perimeter and length of crypts were slightly reduced in I1515 group and both NRG but the change was not statistically significant. Values for the thickness of mucosa-submucosa layer considered as a whole were significantly decreased in I1515 animals whereas intermediate values between control and I1515 group were present in the PRG and both NRG.

### Histochemical and lectin histochemical studies

Neither PAS nor AB technique results showed differences between the studied groups. The surface epithelium of the colon and rectum showed a weak to moderate reactivity with PAS and AB solutions. The epithelium of superficial, middle, and deep crypts was more weakly labeled. The mucin of goblet cells of both intestinal sections positively reacted with PAS and AB solutions.

The lectin histochemical study revealed differences in the carbohydrate composition of both intestinal sections. There was a reduction in DBA binding to the glycocalix of the surface and crypt epithelium of the colon and rectum of both intoxicated groups. Similar changes with SBA were observed in the colon of both intoxicated groups and also in NRG1530. UEA-1 reactivity varied from null to strong between groups in the colon whereas in the rectum the binding pattern was null in control group and variable in other groups. Goblet cells were heterogeneously labeled.

## Discussion

Our results showed different morphological and histochemical changes in the colon and rectum in response to *S. glaucophyllum* intoxication. In the intestine proliferation, differentiation, and death are naturally occurring processes that are controlled by multiple nutritional and hormonal factors at all stages of life from prenatal to adulthood (Dauca *et al.* 1990; Biol-N'garagba and Louisot 2003; Chaudhry *et al.* 2008). The mechanisms involved in this intestinal adaptation, also known as enteroplasticity, include morphological, physiological, and functional aspects. The intestine exerts adaptive responses under different situations such as resection, metabolic alterations, fasting, and malnutrition (Drozdowski and Thomson 2006, 2009).

In the colon of the intoxicated group no significant differences in the parameters were found. Despite the partial nutritional restriction of the NRG no significant change was observed either. This is in disagreement with several studies that describe atrophic modification in the intestine after fasting (Dunel-Erb *et al.* 2001). On the other hand, there was a significant increase in the area and length of the crypts of the PRG. The longer crypts found in most of the animals from PRG may reflect not only an adaptive response during the recovery period but also a hypercompensatory effect. Deregulation of the homeostatic mechanisms of the proliferative crypt cells and individual metabolic differences during the time of intoxication and at the onset of the recovery period may explain that result. It is known that after prolonged fasting morphological and functional rehabilitation during the refeeding period depends on differences in the lipid and protein mobilization at the end of the fasting time. The metabolic state finally defines the nutritional needs for the proper adaptation as well as the magnitude of the morphological and functional changes (Thouzeau *et al.* 1995; Dunel-Erb *et al.* 2001; Habold et *al.* 2007).

The morphometric parameters studied in the rectum were not statistically significant between groups. However, the lowest values were found in the I1515 animals suggesting that longer intoxication time may provide an adaptive capacity. With respect to the thickness of the mucosa-submucosa layer there was a significant reduction in the I1515 group whereas intermediate values between control and I1515 group were found in both NRG. Reduction of the mucosa thickness was reported in the small intestine of young rats under protein privation but not in adults (Rodrigues *et al.* 1985). Neither crypt width nor muscular layer thickness was affected during intoxication, recovery, or nutritional restricted time.

Several studies have shown that diet components such as carbohydrates, lipids, and proteins stimulate the expression of genes involved in the process of intestinal adaptation (Drozdowski and Thomson 2006, 2009). In addition, glucocorticosteroids, growth hormone, and growth factors such as IGF-1, keratinocyte growth factor, epidermal growth factor, and glucagon-like peptide 2 participate as well (Biol-N' garagba and Louisot 2003; Drozdowski and Thomson 2006, 2009). The changes present in the rectum of I1515 animals could be the result of a complex multifactorial process in which the high doses of vitamin D might have a key participation. A combination of hypervitaminosis D state and nutritional restriction may be simultaneously acting to produce these results.

Changes in cell renewal might be responsible for the morphological alteration observed during the intoxication and nutritional restricted time in both intestinal sections. Modifications in the intestinal kinetics have been previously described under different conditions (Aldewachi *et al.* 1975; Xiao *et al.* 2001; Razzaque and Lanske 2006; Habold *et al.* 2007). However, there is no information on changes in tissue renewal in animals intoxicated with *S. glaucophyllum*. Further, little is known about the role of vitamin D in intestinal cell proliferation and differentiation (Suda *et al.* 1990; Biol-N'garagba and Louisot 2003).

In addition to the morphological changes we also found differences in the lectin binding pattern of the intoxicated animals especially with DBA. Since modifications in the binding pattern of SBA and UEA-1 were also found in the NRG it seems that vitamin D is not the only factor responsible for that modification. It may be important to also consider changes as a result of the anorexia state that characterize the clinical disease. Changes in diet composition can alter intestinal flora homeostasis, its interaction with the intestinal epithelium, and consequently produce modifications in the expression of carbohydrate at the brush border enterocytes (Sharma and Schumacher 1995). In addition, several hormones and growth factors also participate (Mahmood and Torres-Pinedo 1985; Biol-N'garagba and Louisot 2003). Thyroid hormones and glucocorticoids have been involved in the process of intestinal glycosylation. They have been implicated in the modulation of gene transcription of glycosyltransferases, enzymes involved in the glycosylation process (Biol-N'garagba *et al.* 2002; Biol-N'garagba and Louisot 2003). Thus, we suggest that vitamin D as a steroidal hormone-like glucocorticoid may regulate the transcription of glycosyltransferases and glycosidases. In addition, it may also induce the synthesis of enzymes such as ornithine decarboxylase and spermidine N-acetyltransferase, required for the metabolism of polyamines. Polycationic components are involved in enterocyte proliferation and differentiation (Suda *et al.* 1990; Biol-N'garagba *et al.* 2002).

In this work we described histopathological and morphometrical changes in the large bowel from *S. glaucophyllum* intoxication. The modifications in the glycosylation pattern may indicate a new role for vitamin D as a regulator of intestinal glycosylation. Future studies on cell proliferation and death will help to understand the morphological changes described.

## Acknowledgements

Partially supported by grants from the Agencia Nacional de Promoción Científica y Tecnológica (ANPCyT) and the Academia Nacional de Agronomía y Veterinaria.

## References


Aldewachi HS, Wright N, Appleton D, and Watson A (1975). The effect of starvation and refeeding on cell population kinetics in the rat small bowel mucosa. *Journal of Anatomy* 119:105-121.

Barros SS and Gimeno EJ (2000). Cell differentiation and bone protein synthesis in the lungs of sheep with spontaneous calcinosis. *Journal of Comparative Pathology* 123:270-277.

Bikle DD (2007). What is new in vitamin D? *Current Opinion in Rheumatology* 19:383-388.

Biol-N'garagba MC and Louisot P (2003). Regulation of the intestinal glycoprotein glycosilation during postnatal development: role of hormonal and nutritional factors. *Biochimie* 85:331-352.

Biol-N'garagba MC, Greco S, George P, Hugueny I, and Louisot P (2002). Polyamine participation in the maturation of glycoprotein fucosylation, but not sialylation, in rat small intestine. *Pediatric Research* 51:625-634.

Chaudhry KK, Mahmood S, and Mahmood A (2008). Hormone induced expression of brush border lactase in suckling rat intestine. *Molecular and Cellular Biochemistry* 312:11-16.

Cook H (1990). Carbohydrates. In *Theory and Practice of Histological Techniques.* 3rd edn (JD Bancroft and A Stevens, eds), pp. 191-195. Churchill Livingstone, Edinburgh London, Melbourne and New York.

Dauca M, Bouziges F, Colin S, Kedinger M, Keller MK, Schilt J, Simon-Assmann P, and Haffen K (1990). Development of the vertebrate small intestine and mechanisms of cell differentiation. *International Journal of Development Biology* 34: 205-218.

Drozdowski L and Thomson AB (2006). Intestinal mucosa adaptation. *World Journal of Gastroenterology* 12:385-406.

Drozdowski L and Thomson AB (2009). Intestinal hormones and growth factors: effects on the small intestine. *World Journal of Gastroenterology* 15:385-406.

Dunel-Erb S, Chevalier C, Laurent P, Bach A, Decrock F, and Le Maho Y (2001). Restoration of the jejunal mucosa in rats refed after prolonged fasting. *Comparative Biochemistry and Physiology Part A: Molecular Integrative Physiology* 129:933-947.

Falk P, Roth KA, and Gordon J (1994). Lectins are sensitive tools for defining the differentiation programme of mouse gut epithelial cell lineages. *Gastrointestinal and Liver Physiology* 29:987-1003.

Fontana PA, Zanuzzi CN, Barbeito CG, Gimeno EJ, and Portiansky EL (2009). Evaluation of immunotoxic effects of plant induced-hypervitaminosis D on cattle thymus. *Pesquisa Veterinária Brasileira* 27:267-275.

Gelbert H, Whiteley H, Ballard G, Scott J, and Kuhlenschmidt M (1992). Temporal lectin histochemical characterization of porcine small intestine. *American Journal of Veterinary Research* 53:1873-1880.

Gimeno EJ and Barbeito CG (2004). Glicobiología, una nueva dimensión para el estudio de la biología y de la patología. In *Anales de la Academia Nacional de Agronomia y Veterinaria* 58:1-34. Buenos Aires.

Gimeno EJ, Costa EF, Gomar MS, Massone AR, and Portiansky EL (2000). Effects of plant-induced hypervitaminosis D on cutaneous structure, cell differentiation and cell proliferation in cattle. *Journal of Veterinary Medicine A* 47:201-21.

Gimeno EJ, Portiansky EL, Gomar MS, Costa EF, Masone AR, Alonso CR, Dallorso ME, and Barros SS (2004). Calcinosis in ruminants due to plant poisoning. Contributions on the pathogenesis. In *Poisonous plants and related toxins* (T Acamovic, CS Stewart, and TW Pennycott, eds), pp. 84-89. CAB International, Cambridge, Massachusetts, USA.

Goldstein IJ and Hayes CE (1978). The lectins: carbohydrate-binding proteins of plants and animals. *Advances in Carbohydrate Chemistry & Biochemistry* 35:127-340.

Habold C, Reichardt F, Foltzer-Jourdainne C, and Lignot JH (2007). Morphological changes of the rat intestinal lining in relation to body stores depletion during fasting and refeeding. *Pflűgers Archiv* 455:323-332.

Holt P, Arber N, Halmos B, Forde K, Kissileff H, McGlynn KA, Moss SF, Fan K, Yang K, and Lipkin M (2002). Colonic epithelial cell proliferation decreases with increasing levels of serum 25-hydroxyvitamin D. *Cancer Epidemiology Biomarkers and Prevention* 11:113-119.

Kong J, Zhang Z, Musch MW, Ning G, Sun J, Hart J, Bissonnette M, and Li YC (2008). Novel role of the vitamin D receptor in maintaining the integrity of the intestinal mucosal barrier. *American Journal of Physiology–Gastrointestinal and Liver Physiology* 294:208-216.

Kutuzova GD and DeLuca HF (2007) 1,25-Dihydroxyvitamin D3 regulates genes responsible for detoxification in intestine. *Toxicology and Applied Pharmacology* l218:37-44.

Mahmood A and Torres-Pinedo R (1985). Effect of hormone administration on the sialylation and fucosylation of intestinal microvillus membranes of suckling rats. *Pediatric Research* 19:899-902

Puche RC and Bingley JB (1995). *Calcinosis of Cattle in Argentina.* 1st English edn. Universidad Nacional de Rosario Editora, Rosario.

Razzaque MS and Lanske B (2006). Hypervitaminosis D and premature aging: lessons learned from Fgf23 and Klotho mutant mice. *Trends in Molecular Medicine* 12:298-305.

Rodrigues MA, de Camargo JL, Coelho KI, Montenegro MR, Angelini AY, and Burini RC. (1985). Morphometric study of the small intestinal mucosa in young, adult, and old rats submitted to protein deficiency and rehabilitation. *Gut* 26:816-821.

Sharma R and Schumacher U (1995). The influence of diets and gut microflora on lectin binding patterns of intestinal mucins in rats. *Laboratory Investigation* 73:558-564.

Spicer SS and Schulte BA (1992). Diversity of cell glycoconjugates shown histochemically: a perspective. *Journal of Histochemistry and Cytochemistry* 40:1-48.

Suda T, Shinki T, and Takahashi N (1990). The role of vitamin D in bone and intestinal cell differentiation. *Annual Review of Nutrition* 10:195-211.

Thouzeau C, Le Maho Y, and Larue-Achagiotis C (1995). Refeeding in fasted rats: dietry self-selection according to metabolic status. *Physiology and Behaviour* 58:1051-1058.

Worker NA and Carrillo BJ (1967). 'Enteque seco'. Calcification and wasting in grazing animals in Argentina. *Nature* 215:72-74.

Xiao ZQ, Moragoda L, Jaszewski R, Hatfield JA, Fligiel SEG, and Majumdar APN (2001). Aging is associated with increased proliferation and decreased apoptosis in the colonic mucosa. *Mechanisms of Ageing and Development* 122:1849-1864.

Zanuzzi CN, Fontana PA, Barbeito CG, Portiansky EL, and Gimeno EJ (2008). Paneth cells: histochemical and morphometric study in control and *Solanum glaucophyllum* intoxicated rabbits. *European Journal of Histochem*istry 52:93-100.

# Chapter 75

# Enzootic Calcinosis of Sheep in Uruguay

C. García y Santos[1], A. Capelli[1], S. Sosa[1], W. Pérez[1], R. Domínguez[1], R. Pereira[1], F. Bonino[1], J.M. Goyen[1], and E. Alonso[2]

[1]*Laboratorio de Toxicología, Facultad de Veterinaria, Universidad de la República, Av. Lasplaces 1550, CP 1600, Montevideo, Uruguay;* [2]*Laboratorio de Botánica, Facultad de Química, Universidad de la República, Montevideo, Uruguay*

## Introduction

Enzootic calcinosis is a chronic intoxication of ruminants and horses caused by the ingestion of calcinogenic plants. These plants contain glycosides conjugated to vitamin $D_3$ or its derivatives and cause hypercalcemia, hyperphosphatemia, and mineralization of soft tissues. This produces a severe physical deterioration and depreciation in animals causing significant economic losses (Gimeno 2000). Several calcinogenic plants have been studied to date: *Solanum glaucophyllum* (Carrillo and Worker 1967; Camberos and Davis 1969; Riet-Correa *et al.* 1975; Gimeno 1977), *Nierembergia veitchii* (Barros *et al.* 1970; Riet-Correa *et al.* 1987, 1993), *Cestrum diurnum* (Krook *et al.* 1975a, b), *Trisetum flavescens* (Braun *et al.* 2000; Dirksen *et al.* 2003), *Solanum torvum* (Morris *et al.* 1979), *Stenotaphrum secundatum* (Arnold and Fincham 1997), *Solanum esuriale* (O'Sullivan 1976), and *Solanum verbascifolium* (Tustin *et al.* 1973). This work describes two outbreaks of enzootic calcinosis diagnosed in Uruguay between 2005 and 2007, one caused by *Solanum glaucophyllum* (Garcia y Santos *et al.* 2007) and the other by *Nierembergia rivularis* (Etcheverry *et al.* 2008).

## Poisoning by *Solanum glaucophyllum* in Sheep

In the outbreak of poisoning by *S. glaucophyllum* which occurred during 2006 in Soca, 8th Police District of Canelones, Corriedale, Hampshire Down and crossbred adult sheep were affected. The area involved was natural grassland with streams. Deaths occurred throughout the year and mortality was estimated at 15% of 100 animals involved. Necropsies were performed on three female adult sheep that died without any clinical signs at the Department of Pathology of the Veterinary Faculty. Fragments of liver, kidney, heart, aorta, medium sized arteries, lung, lymphatic nodules, intestine, and brain were fixed in 10% formalin, routinely processed for histopathology, and stained by hematoxylin and eosin (H&E). Diagnosis of intoxication was based on epidemiology, presence of the calcinogenic plant, clinical signs, and gross and histological lesions. The toxicity of *S. glaucophyllum* was tested in rabbits. Plants were collected in the problem pasture; leaves

©CAB International 2011. *Poisoning by Plants, Mycotoxins, and Related Toxins* (eds F. Riet-Correa, J. Pfister, A.L. Schild, and T.L. Wierenga)

were dried at room temperature for 48 hours, then in a heater at 60°C for 48 h, crushed, and pelleted. Six New Zealand rabbits weighing 2-3 kg received a total dose of 100 mg/kg for 2 days. Two other rabbits were used as controls and were fed a commercial feed. Rabbits were examined clinically and later euthanized with thiopentone and necropsied. Viscera fragments were collected in 10% buffered formalin, routinely processed for histopathological examination, and stained by H&E and von Kossa. The rabbits fed with the plant showed anorexia, apathy, and marked depression 4 days after the beginning of the experiment. Two rabbits had diarrhea and one had convulsions before death. Some animals died during the experiment and the remaining were euthanized and necropsied on day 10. Control rabbits were also euthanized and necropsied on day 10. Histology of the arteries revealed calcified areas and stained positive with von Kossa for calcium. Clinical signs observed in the rabbits were similar to those reported by Moraña *et al.* (1994). Histological findings were similar to those observed in the spontaneously affected sheep and to other natural and experimental intoxications by calcinogenic plants (Eckell *et al.* 1960; Carrillo and Worker 1967; Dobereiner *et al.* 1971; Riet-Correa *et al.* 1975, 1987, 1993; Gill *et al.* 1976; Neumann *et al.* 1977; Mello 2003; Barros *et al.* 2006; García y Santos *et al.* 2006, 2007; Rissi *et al.* 2007).

## Poisoning by *Nierembergia rivularis* in Sheep

*Nierembergia rivularis* (Solanaceae) is a prostrate plant that grows intermingled with native vegetation and has creeping stems with obovated or spatulate leaves and white flowers (Burkart 1979). In Uruguay, this species is found in the departments of Colonia, Soriano, Tacuarembó, Río Negro, Rivera, and Rocha (Alonso 2008, personal communication). The outbreak from *N. rivularis* was in Rivera, 6th Police District; affected animals were grazing in a native pasture in which forage was scarce due to a serious drought in that area (Garcia y Santos *et al.* 2006). The deaths occurred only in summer and autumn between December 2005 and February 2006. From a total of 200 Corriedale and crossbred sheep from different age groups, morbidity was 10% and mortality 6%. Clinical signs of intoxication were anorexia, cachexia, stiffness, and kyphosis. Gross and histological lesions were characterized by calcium salt deposition on the medial layer of the arteries. Four Corriedale sheep were used for the experimental reproduction of the disease. Three were forced to graze in an area where *N. rivularis* was present. The fourth, used as a control, grazed in an area in the same paddock free of *N. rivularis*. Macroscopic alterations of enzootic calcinosis were observed in the three experimental sheep. No lesions were observed in the control sheep. Histological findings were mineralization and fragmentation of the tunica intima and media of various arteries, with the presence of giant cells. Thin layer chromatography of a chloroform extract from *N. rivularis* revealed the existence of vitamin $D_3$ or one of its forms.

## Conclusions

*Solanum glaucophyllum* and *Nierembergia rivularis* cause enzootic calcinosis spontaneously in sheep in Uruguay. Chromatographic studies revealed that *N. rivularis* contains vitamin $D_3$ or one of its forms.

## References


Arnold RM and Fincham IH (1997). Manchester wasting disease: a calcinosis caused by a pasture grass (*Stenotaphrum secundatum*) in Jamaica. *Tropical Animal Health and Production* 29:174-176.

Barros SS, Pohlenz J, and Santiago C (1970). Zur Kalzinose beim Schaf. *Deutsche Tieraerztliche Wochenschrift* (abstract) 77:321-356.

Barros SS, Soraes MP, and Gimeno EJ (2006). Macrophages and giant cell proliferation associated with bone protein synthesis and calcification in the thrachea and bronchi of rabbits intoxicated with *Solanum glaucophyllum*. *Veterinary Pathology* 43:494-499.

Braun U, Diener M, Camenzind D, Flückiger M, and Thoma R (2000). Enzootic calcinosis in goats caused by golden oat grass (*Trisetum flavescens*). *Veterinary Record* 146:161-162.

Burkart A (1979). Parte V: Dicotiledóneas metaclamídeas. In *Flora ilustrada de Entre Ríos* (INTA), pp. 434-443. 1ª Colección Científica del INTA, Argentina.

Camberos HR and Davis GK (1969). Acción de *Solanum malacoxylon* sobre balance mineral en ovinos. *Gaceta Veterinaria* 3:466-474.

Carrillo BJ and Worker NA (1967). Enteque seco: arteriosclerosis y calcificación metastásica de origen tóxico en animales a pastoreo. *Revista Investigaciones Agropecuarias INTA* Argentina 4(2):9-30.

Dirksen G, Sterr K, and Hermanns W (2003). Enzootic calcinosis in sheep after consumption of golden oat grass (*Trisetum flavescens* L., P.B.). *Deutsche Tieraerztliche Wochenschrift* 110(12):475-483.

Dobereiner J, Tokarnia CH, Costa JBD, Campos JLE, and Dayrell MS (1971). 'Espichamento', intoxicação de bovinos por *Solanum malacoxylon* no Pantanal de Mato Grosso. *Pesquisa Agropecuária Brasileira* 6:91-117.

Eckell OA, Gallo GG, Mertin AA, and Portella RA (1960). Observation sobre el Enteque Seco de los bovinos. *Revista de la Facultad de Agronomía y Veterínaria de La Plata* 6:193-211.

Etcheverry GP, Goyen JM, and Pereira R (2008). Intoxicación por *Nierembergia rivularis* en ovinos de Uruguay. Tesis Facultad de Veterinaria, Universidad de la República, 60 pp.

García y Santos C, Pérez W, Mosca V, Pereira R, Seoane A, Rodríguez M, Moraes J, and Rivero R (2006). Calcinosis Enzoótica en ovinos de Uruguay. In *XXXIV Jornadas Uruguayas de Buiatría*, pp. 195-196. Paysandú, Uruguay.

Garcia y Santos MC, Pereira R, Capelli A, Domínguez R, Bonino F, Goyen JM, and Arago S (2007). Intoxicación espontánea en ovinos por ingestión de *Solanum glaucophyllum* (*malacoxylon*) en Uruguay. In *XXV Jornadas Uruguayas de Buiatría*, pp. 284-285. Paysandú, Uruguay.

Gill BS, Singh M, and Chopra AK (1976). Enzootic calcinosis in sheep: clinical signs and pathology. *American Journal Veterinary Research* 37(5):545-552.

Gimeno EJ (1977). Estudios sobre 'Enteque seco'. Algunas consideraciones históricas. *Veterinaria Argentina* 39(322):382-388.

Gimeno EJ (2000). Calcinosis enzoótica en rumiantes: un problema vigente de la ganadería nacional. *Academia Nacional de Agronomía y Veterinaria.* Sesión pública extraordinaria. Tomo LIV, p. 202-234. Buenos Aires, Argentina.

Krook L, Wasserman RH, Shivley JN, Tashjian AH Jr, Brokken TD, and Morton JF (1975a). Hypercalcemia and calcinosis in Florida horses: implication of the shrub, *Cestrum diurnum*, as the causative agent. *Cornell Veterinarian* 65:26-56.

Krook L, Wasserman RH, McEntee K, Brokken TD, and Melbourne TB (1975b). *Cestrum diurnum* poisoning in Florida cattle. *Cornell Veterinarian* 65(10):557-575.

Mello JRB (2003). Calcinosis–calcinogenic plants. *Toxicon* 41:1-12.

Moraña JA, Barros SS, Driemeier D, and Flores YE (1994). Gastropatia em coelhos experimentalmente induzida por planta calcinogênica (*Solanum malacoxylon*). *Pesquisa Agropecuária Brasileira* 14:35-42.

Morris KM, Simonite JP, Pullen L, and Simpson JA (1979). *Solanum torvum* as a causative agent of enzootic calcinosis in Papua, New Guinea. *Research Veterinary Science* 27(2):264-266.

Neumann F, Nobel TA, and Bogin E (1977). Enzootic calcinosis in sheep and C-cells hyperplasia of the thyroid. *Veterinary Record* 101(18):364-366.

O'Sullivan BM (1976). Humpy back of sheep. Clinical and pathological observations. *Australian Veterinary Journal* 52:414-418.

Riet-Correa F, Riet-Correa I, and Bellagamba C (1975). Calcificación metastásica enzoótica (enteque seco) en bovinos del Uruguay. *Veterinaria* 12(60):15-23.

Riet-Correa F, Schild AL, Mendez MC, Wasserman R, and Krook L (1987). Enzootic calcinosis in sheep caused by the ingestion of *Nierembergia veitchii* (Solanaceae). *Pesquisa Veterinária Brasileira* 7(3):85-95.

Riet-Correa F, Mendez MC, Schild AL, and Petiz CA (1993). Enzootic calcinosis in sheep. Experimental reproduction with *Nierembergia veitchii* (Solanaceae). *Pesquisa Veterinária Brasileira* 13(1/2):21-24.

Rissi D, Rubia R, Pierezan F, Kommers GD, and Barros CSL (2007). Intoxicação em ovinos por *Nierembergia veitchii*: observações em quatro surtos. *Ciência Rural* 37(5):1393-1398.

Tustin RC, Pienaar CH, Schmidt JM, Faul A, van der Walt K, Boyazoglu PA, and de Boom HP (1973). Enzootic calcinosis of sheep in South Africa. *Journal South Africa Veterinary Association* 44(4):383-395.

# Chapter 76

# Enzootic Calcinosis in Ruminants from Central Brazil

K.M.R. Guedes[1], E.M. Colodel[2], M.B. Castro[1], V.S. Mustafa[1], D.D. Moraes[1], J.L. Reis Jr[1], J.R.J. Borges[3], F.M. Boabaid[2], D.G. Ubiali[2], L.P. de Arruda[2], and F. Riet-Correa[4]

*[1]Laboratory of Veterinary Pathology, University of Brasília, Brasília, DF, 70910-970, Brazil; [2]Laboratory of Veterinary Pathology, Federal University of Mato Grosso, Cuiabá, MT, 78068-900, Brazil; [3] Large Animal Hospital, University of Brasília, Brasília, DF, 70636-100, Brazil; [4] Laboratory of Animal Pathology, Federal University of Campina Grande, Patos, PB, 58700-970, Brazil*

## Introduction

Enzootic calcinosis occurs frequently in livestock in the central-western region of Brazil which is composed of the states of Goiás, Tocantins, Mato Grosso, Mato Grosso do Sul, and the Federal District where Brasília, the nation's capital, is located. This region represents 18.86% of the Brazilian territory (Michels *et al.* 2006) and has nearly 71 million cattle (34.8% of the Brazilian cattle population) (IBGE 2005). There are also in the region 937,000 sheep (6.24% of the Brazilian sheep population). Enzootic calcinosis is also reported in sheep in Minas Gerais (southeastern Brazil) with a sheep population of 188,000 animals.

Since 2004 a disease characterized clinically and pathologically by soft tissue mineralization has been investigated by the laboratories of veterinary pathology from the University of Brasília (UnB) and Federal University of Mato Grosso (UFMT). It affects ruminants including cattle, sheep, and goats from different farms located in the central-western region and Minas Gerais. The clinical and pathological features of the disease are comparable to the previously reported enzootic calcinosis (EC) caused by calcinogenic plants (Döbereiner *et al.* 1971; Okada *et al.* 1977; Riet-Correa *et al.* 1987) which is characterized by hypercalcemia, hypercalcitoninism, hypoparathyroidism, osteopetrosis, and systemic mineralization of soft tissues. Cardiovascular and respiratory systems are most commonly affected. These lesions are frequently associated with decreased body condition, milk and meat production, and increased mortality. Enzootic calcinosis is caused by the ingestion of *Trisetum flavescens* in Germany (Dirksen *et al.* 1973) and Austria (Libiseller *et al.* 1976), *Cestrum diurnum* in the USA (Krook *et al.* 1975) and Cuba (Durand *et al.* 1999), *Solanum torvum* in New Guinea (Morris *et al.* 1979), and *S. malacoxylon* (=*glaucophylon*) in Argentina (Carrillo and Worker 1967) and Uruguay (Riet-Correa *et al.* 1975). In Brazil EC is caused by *S. malacoxylon* in the Pantanal of Mato

©CAB International 2011. *Poisoning by Plants, Mycotoxins, and Related Toxins* (eds F. Riet-Correa, J. Pfister, A.L. Schild, and T.L. Wierenga)

Grosso, affecting cattle (Döbereiener *et al.* 1971), and by *Nierembergia veitchi* in Rio Grande do Sul State, southern Brazil, affecting mainly sheep (Riet-Correa e*t al.* 1987).

Numerous EC outbreaks with important economic losses have been reported in the central-western region of Brazil. However, the economic impact remains unknown. The objective of this paper is to report some epidemiologic data, clinical signs, and pathology of 42 outbreaks of EC in central-western Brazil and Minas Gerais. These data are important for the determination of the etiology of the disease which is still unknown.

## Material and Methods

### Spontaneous outbreaks

Animals affected with EC were identified during necropsies performed routinely at the Large Animal Hospital at the University of Brasília and the Laboratory of Veterinary Pathology at the Federal University of Mato Grosso. Lesions of enzootic calcinosis were observed in animals from 42 farms and most of these farms were visited to collect data about epidemiology and clinical signs of the disease. Additionally, pastures were evaluated for the presence of invasive weeds and potentially poisonous plants. During visits at least one affected animal from each farm was euthanized to assess the gross changes and to collect samples for histopathology. The majority of animals submitted for necropsy were adult except for a 4-month-old sheep. Multiple tissues were sampled including heart, thoracic and abdominal aorta, lungs, liver, kidneys, small and large intestines, thyroid, bones, and central nervous system. Tissues were placed in 10% buffered formalin for at least 48 h followed by routine processing and paraffin embedding. Sections 4-5 μm thick were stained with hematoxylin and eosin. Monthly serum samples were collected from 10% of the herd from two selected farms. Farm 1 is located in the municipality of Edilândia, Goiás State, and Farm 2 is located in the administrative region of São Sebastião, Distrito Federal. Calcium and phosphorus levels were determined from serum samples to verify seasonal variations of these minerals and to determine the relationship of serum Ca and P levels with the presence of certain weeds and the distribution of rainfall.

### Experimental intoxication

Invasive weeds were selected for experimental administration to rabbits and sheep in order to identify plants with calcinogenic properties. Weed selection was based on seasonality, evidence of animal consumption in farms where outbreaks were reported, and frequency of those plants in the paddocks. The following plants were administered experimentally to rabbits or sheep: *Ageratum canyzoides*, *Chamaecrista desvauxii*, *Cissus erosa*, *Elephantopus mollis*, *Eragrostis* sp., *Eupatorium odoratum*, *Hymenaea courbaril*, *Jatropha* sp., *Ludwigia octavali*, *Mimosa hirsutissima*, *Murdania nudiflora*, *Rhynchanthera novemnervia*, *Scoparia dulcis*, *Sida santaremensis*, *Tibouchina* sp., *Vernonia brasiliana*, *Waltheria* sp., *Arachis* sp., *Cnidoscolies cnicadendron*, *Eeleusine tristachya*, *Elephantopus mollis*, *Eleusine tristacya*, *Ludwigia leptocarpa*, *Ludwigia* sp., *Mimosa pellita*, *Sida cerradoensis*, *Sida ciliaris*, *Sida spinosa*, and *Vernonia ferruginea*.

In experiments with rabbits, 2-months-old or older male and female New Zealand rabbits were used. Serum levels of calcium and phosphorus were measured prior to and at the end of plant ingestion. The fresh recently collected aerial part of each tested plant species was given as feed to two animals for at least 30 days. The plants were administered immediately after collections or were kept at 3-5°C for no more than 10 days. Water and commercial rabbit feed were additionally provided *ad libitum*. A control animal had access to water and commercial rabbit feed only.

In experiments with sheep 14 1- to 3-year-old sheep were used. Each animal was kept in an individual pen for 40-60 days. During this time the aerial part of selected weed was given as feed at the minimal dose of 20 g/kg mixed with 200 g of silage. Commercial feed (200 g) concentrate and water were provided daily for each animal for voluntary ingestion. Serum samples were collected weekly for calcium and phosphorus measurement.

At the end of the experiments animals were euthanized with an overdose of barbiturate. Necropsies were performed and tissues were collected for histopathological examination. Collected samples and tissues were processed for histologic studies according to the methodology previously described in this study.

## Results

### Spontaneous disease

Forty-two farms with cases of enzootic calcinosis were identified (Table 1). Nineteen farms were in Mato Grosso State, 19 in Distrito Federal, six in Goiás State, two in Minas Gerais State, and one in Tocantins State. Sheep were affected in 20 farms, cattle in 18, and goats in four.

On the farms visited the herds were raised extensively on *Brachiaria* spp. pastures, but in most paddocks the pastures were markedly degraded with a variety of invasive weeds. However, in some affected farms animals were kept on native pasture free of *Brachiaria* spp. In some herds mineral, energy, or protein supplementation was provided. In most outbreaks animals became ill after the beginning of the rainy season with progressive wasting, weakness, and death. Some animals showed clinical improvement from June-May to September-October after the end of the rainy period. However, in general those animals showed early clinical signs at the start of the following rainy season (October to March). The only apparent successful control method was to move herds from a degraded pasture to a newly planted pasture with no invasive weeds. The frequency of the disease was variable depending on each farm but sheep were more severely affected than cattle and goats with higher morbidity and mortality rates (Table 1).

All affected species presented similar clinical signs characterized by decreased body condition, kyphosis, and lameness with some animals walking on their knees. Lethargy and dyspnea were observed in the most severely affected animals. Heart failure, lateral recumbence, and death were observed in more advanced clinical cases. Animals from the same areas that died of causes other than EC frequently showed variable degrees of soft tissue mineralization. Calcium and phosphorus serum concentrations were increased in most animals. The biochemical analysis from the two investigated farms showed increased serum levels of Ca and P from October to December at both farms, which coincided with the period of heaviest rains during the period.

**Table 1.** Some epidemiologic data of 41 farms where enzootic calcinosis was diagnosed.

| Species | Breed | Herd size | Deaths[a] | Location | First diagnosis |
|---|---|---|---|---|---|
| Bovine | Holstein | 100 | 70 | Flores de Goiás GO | May 2004 |
| Ovine | Santa Inês | 220 | 45 | Poconé MT | Mar 2005 |
| Ovine | Santa Inês | 270 | 30 | Alto Paraguai MT | May 2005 |
| Bovine | Nelore | 780 | 2 | Rondonópolis MT | Dec 2005 |
| Ovine | Santa Inês | 300 | 70 | Nova Brazilandia Mt | Jan 2006 |
| Bovine | Nelore | 70 | 2 | Poconé MT | Jan 2006 |
| Bovine | Mixed | 30 | 12 | São José do Povo MT | Jan 2006 |
| Bovine | Nelore | 50 | 1 | São José do Povo MT | Jan 2006 |
| Bovine | Mixed | 120 | 4 | São José do Povo MT | Jan 2006 |
| Bovine | Girolanda | 45 | 4 | Rondonópolis MT | Jan 2006 |
| Ovine | Santa Inês | 450 | 70 | Santo Antônio de Leverger MT | Feb 2006 |
| Ovine | Santa Inês | 80 | 12 | Cuibá MT | Feb 2006 |
| Ovine | Santa Inês | 240 | 10 | Paracatu MG | Mar 2006 |
| Bovine | NI | 40 | 1 | Paranoá DF | Aug 2006 |
| Caprine | NI | 300 | 1 | Edilândia GO | Aug 2006 |
| Ovine | Santa Inês | 110 | 10 | São Miguel do Araguaia GO | Aug 2006 |
| Bovine | Holstein | 200 | 1 | Brasília DF | Sep 2006 |
| Ovine | Santa Inês | 148 | 20 | Santo Antônio do Descoberto GO | Sep 2006 |
| Bovine | Nelore | 472 | 1 | Mimoso GO | Nov 2006 |
| Bovine | Mixed | 40 | 2 | São José do Povo MT | Jan 2007 |
| Ovine | Dorper | 230 | 2 | Cuiabá MT | Jan 2007 |
| Ovine | Sana Inês | 180 | 17 | Nobres MT | Jan 2007 |
| Ovine | Santa Inês | 55 | 1 | Brazlândia DF | Feb 2007 |
| Bovine | Girolando | NI | 1 | Ni | Feb 2007 |
| Ovine | Santa Inês | 35 | 1 | Paracatu MG | Feb 2007 |
| Ovine | Mixed | 215 | 85 | Araguaçu TO | Mar 2007 |
| Ovine | Bergamasca | -[b] | 2 | Brasília DF | Mar 2007 |
| Ovine | Santa Inês | NI | NI | NI | May 2007 |
| Ovine | Santa Inês | 47 | 8 | São Sebastião DF | Jun 2007 |
| Bovine | Holstein | NI | 1 | NI | Jul 2007 |
| Ovine | Santa Inês | NI | 1 | NI | Jul 2007 |
| Bovine | Girolando | NI | 1 | Brasília DF | Aug 2007 |
| Bovine | Girolando | NI | 1 | Brazlândia DF | Jan 2008 |
| Caprine | Saanen | 60 | 1 | Santo Antônio de Leverger MT | Jan 2008 |
| Caprine | Saanen | 46 | 5 | Santo Antônio de Leverger MT | Jan 2008 |
| Caprine | Saanen | 70 | 12 | Nova Olimpia MT | Jan 2008 |
| Ovine | Santa Inês | 101 | 3 | Sobradinho DF | Feb 2008 |
| Bovine | Girolando | NI | 1 | Paranoá DF | Mar 2008 |
| Bovine | Nelore | 4380 | 1 | Povoado JK GO | Apr 2008 |
| Bovine | Mixed | NI | 1 | NI | Jun 2008 |
| Ovine | Mixed | 140 | 23 | Paranatinga MT | Feb 2009 |
| Ovine | Santa Inês | 700 | 90 | Paranatinga MT | Feb 2009 |

[a] Deaths due to EC
[b] Herd size was variable
NI = Not informed

Gross and histological findings observed mainly in cardiovascular, respiratory, and musculoskeletal systems were similar in all affected species. There was mineral deposition in variable degrees in arteries especially in the aorta, carotid, and iliac arteries. Mineralized arteries were characterized by an irregular and chalky endothelial surface with an inelastic, firm, and thickened surface. Other frequent areas of mineralization were in the heart, particularly in the left ventricle papillary muscles and atrioventricular valves. Lungs also showed lesions associated with mineralization such as failure to collapse and multifocal elevated white and gritty areas on the pleural surface. Mineralization of the renal corticomedullary junction and skeletal muscles were more commonly seen in sheep. The main histologic findings were medial and intimal mineralization of the aorta, carotid, and other arteries and mineralization in alveolar septa of the lungs, tendons, ligaments, endocardium, renal tubules, and renal arteries. Osseous metaplasia was frequently observed in the animals with more severe lesions.

### Experimental intoxication

None of the 33 rabbits and 14 sheep that ingested selected weeds to detect calcinogenic effect of the plants developed clinical changes and mineralization of soft tissues was not observed during necropsies or histologic examination. Serum Ca and P concentrations were within normal values.

## Discussion

The epidemiological, clinical, and pathological findings in this study are similar to those reported in enzootic calcinosis caused by the ingestion of the calcinogenic plants mentioned earlier. The majority of affected herds were kept on pastures without supplementation. However, the disease was also present in some herds where mineral and/or energy and protein supplementation was provided. The absence of known calcinogenic plants, the seasonal occurrence of the disease in paddocks with degraded pastures invaded by weeds, and the increased serum values of Ca and P during the rainy season strongly suggest that the disease in central-western Brazil is caused by an unknown calcinogenic plant. The seasonal incidence of the disease reported in outbreaks of EC caused by other plants (Döbereiner *et al.* 1971; Gimeno 1977; Riet-Correa *et al.* 1987) suggests that the occurrence during the rainy period of the year is associated with the vegetative stage of some invasive weed. Despite the failure to reproduce the disease by the administration of different plants to rabbits and sheep, new attempts with other plants or at different dosages should be done. Previous papers revealed a marked variation between different calcinogenic plants; *Solanum malacoxylon* induces calcinoses in cattle at doses of 3-5 g/kg BW. In contrast, doses of 388 g/kg of *Nierembergia veitchii* are necessary to induce clinical signs in sheep (Riet-Correa *et al.* 1993).

Different from other poisonings by calcinogenic plants that predominately affects adult animals of mainly one species, the EC reported in this paper affects different species including cattle, sheep, and goats and also young animals as observed in a 4-month-old sheep that also developed the disease. EC causes important economic losses to the livestock industry in the central-western region of Brazil. The disease is an important limiting factor for the expansion of ovine production in this region and various institutions are investigating the etiology of this condition.

## Acknowledgements

The authors would like to acknowledge the researchers from the Plant Science team from UFMT for the botanic identification of the tested weeds. The authors are grateful for the financial support provided by INCT for the Control of Plant Poisonings, MCT, CNPq (Grant 573534/2008-0).

## References

Carrillo BJ and Worker NA (1967). Enteque seco: arteriosclerosis y calcificación metastásica de origem tóxico em animales a pastoreo. *Revista Investigaciones Agropecuarias*, B. Aires, *Série Patologia Animal* 4:9-30.

Dirksen G, Plank P, Hanichen T, and Spiess A (1973). Enzootic calcinosis in cattle. VI. Experimental calcinosis in the rabbit due to selective feeding of *Trisetum flavescens*. *Deutsche Tierärztliche Wochenschrift* 80:148-151.

Döbereiner J, Tokarnia CH, Costa JBD, Campos JLE, and Dayrell MS (1971). 'Espichamento', intoxicação de bovinos por *Solanum malacoxylon* no Pantanal de Mato Grosso. *Pesquisa Agropecuária Brasileira* 6:91-117.

Durand R, Figueredo JM, and Mendoza E (1999). Intoxication in cattle from *Cestrum diurnum*. *Veterinary and Human Toxicology* 41:26-27.

Gimeno EJ (1977). Estudio histopatologico del enteque seco experimental en ratas y revision bibliografica de las calcinosis. Tesis Universidad Nacional de La Plata, Argentina, 150 pp.

IBGE (2005). http://www.ibge.gov.br/home/presidencia/noticias/noticia_visualiza. php?id_ noticia=499&id_pagina=1

Krook L, Wasserman RH, McEntee K, Brokken TD, and Teigland MB (1975). *Cestrum diurnum* poisoning in Florida cattle. *The Cornell Veterinarian* 65:557-575.

Libiseller R, Glawischnig E, Kohler H, and Swoboda R (1976). Calcinosis in cattle in Austria. III. Experimental production of calcinosis in sheep and rabbits with green oats (*Trisetum flavescens*) from the pannonic climatic zone. *Zentralbl Veterinary Medicine* 23:1-30.

Michels I, Rodrigues JD, and Lucena LP (2006). Proposta de Elaboração de Estudo de Cadeia Produtiva da Ovinocultura em Mato Grosso do Sul. *Relatório Final SEBRAE*, pp. 32-33. Campo Grande – MS.

Morris KM, Simonite JP, Pullen L, and Simpson JA (1979). *Solanum torvum* as a causative agent of enzootic calcinosis in Papua, New Guinea. *Research in Veterinary Science* 27:264-6.

Okada KA, Carrillo BJ, and Tilley M (1977). *Solanum malacoxylon* Sendtner: A toxic plant in Argentina. *Economic Botany* 31:225-236.

Riet-Correa F, Riet-Correa I, and Bellagamba C (1975). Calcificación metastática enzoótica (enteque seco) en bovinos del Uruguay. *Veterinaria*, Uruguay, 12:15-23.

Riet-Correa F, Schild AL, Mendez MC, Wasserman R, and Krook L (1987). Enzootic calcinosis in sheep caused by the ingestion of *Nierembergia veitchii* (Solanaceae). *Pesquisa Veterinária Brasileira* 7:85-95.

Riet-Correa F, Mendez MC, Schild AL, and Petiz CA (1993). Enzootic calcinosis in sheep. Experimental reproduction with *Nierembergia veitchii*. *Pesquisa Veterinária Brasileira* 13: 21-24.

Chapter 77

# Radiographic Monitoring of Lesions Induced by *Solanum malacoxylon* (Solanaceae) Poisoning in Rabbits

D.G. Ubiali[1], P.B. Néspoli[1], F.M. Boabaid[2], M.I.V. Silva[1], C.A. Pescador[1], M.A. Souza[1], L. Nakazato[1], and E.M. Colodel[1]

[1]*Veterinary Pathology Laboratory, Federal University of Mato Grosso, Cuiabá, MT, 78068-900, Brazil;* [2] *Veterinary Pathology Sector, Federal University of Rio Grande do Sul, Porto Alegre, RS, Brazil*

## Introduction

Calcinogenic plants are an important cause of economic losses in livestock production in different countries. Calcinosis is a term used when chronic diseases lead to soft tissue calcification mainly in the cardiovascular and respiratory system (Carrillo and Worker 1967; Tokarnia *et al.* 2000). The steroidal glycoside $1,25(OH)_2D_3$ was identified in *S. malacoxylon* and other calcinogenic plants and acts as an active metabolite of vitamin $D_3$. It is associated with mineralization of tissues, acting directly on the hormonal metabolism of calcium (Tokarnia *et al.* 2000; Mello 2003).

Most plants with calcinogenic action belong to the Solanaceae family (*Solanum malacoxylon* (=*glaucophyllum*), *S. torvum*, *S. esuriale*, *S. verbascifolium*, *Cestrum diurnum*, *Nierembergia veitchii*, and *N. rivularis*) while *Triseum flavescens* and *Stenotaphrum secundatum* are representatives of the Graminae family. Bovines, buffalo, horses, sheep, goats, and swine can develop calcinosis with different regional manifestations (Mello 2003). The disease in cattle exhibits a seasonal pattern and is known as 'espichamento' in the Pantanal wetlands of Mato Grosso state, Brazil, and has been related to ingestion of *S. malacoxylon* (Döbereiner *et al.* 1971). In Brazil this plant inhabits flooded areas of the Pantanal as well as some municipalities in Rio Grande do Sul state (Schild 1991; Tokarnia *et al.* 2000; Riet-Correa and Méndez 2007).

The diagnosis is made based on epidemiological data, clinical signs, blood biochemistry, and necropsy findings (Tokarnia *et al.* 2000; Riet-Correa and Méndez 2007). In animals with clinical signs of calcinosis it is possible to observe improvement in body condition when they are removed from pastures containing calcinogenic plants and a rapid clinical manifestation in subsequent years after re-exposure in problematic areas (Barros *et al.* 1992). The objective of this study was to evaluate by radiographic monitoring the lesions caused by experimental poisoning with *S. malacoxylon* in rabbits.

©CAB International 2011. *Poisoning by Plants, Mycotoxins, and Related Toxins* (eds F. Riet-Correa, J. Pfister, A.L. Schild, and T.L. Wierenga)

## Material and Methods

The experiment was conducted at the Laboratory of Veterinary Pathology, Federal University of Mato Grosso (LPV-UFMT). Ten adult rabbits (*Oryctolagus cuniculus*) (four males and six females) with an average weight of 3 kg were used. The animals were kept in cages and given commercial diets with levels of 0.38% calcium and 0.07% phosphorus *ad libitum* with free access to water. Six rabbits of similar age and weight were used as controls.

*S. malacoxylon* was collected at Fazenda Campo Largo, North Pantanal, municipality of Poconé, Mato Grosso, Brazil, in July 2008. The leaves were separated from the stems and dried in an oven at 65C . Dry matter of the leaves was 24.5%. The dry leaves were ground and stored under refrigeration. The plant was administered orally by forced feeding. Nine rabbits received a daily dose of 0.01 g/kg and one rabbit received a weekly dose of 0.05 g/kg. The dose, frequency of administration, and radiographic results are shown in Table 1.

**Table 1.** The experiment design and results of radiographic observations of lesions induced by *Solanum malacoxylon* poisoning in rabbits.

| Rabbit | Dose g/kg | Frequency | Number of doses | Number of doses at first positive image | Lesion intensity at last image |
|---|---|---|---|---|---|
| 1 | 0.01 | Daily | 17 | 14 | ++ |
| 2 | 0.01 | Daily | 25 | 21 | +++ |
| 3 | 0.01 | Daily | 25 | 21 | + |
| 4 | 0.01 | Daily | 15 | 21 | ++ |
| 5 | 0.01 | Daily | 24 | 21 | + |
| 6 | 0.01 | Daily | 15 | 14 | + |
| 7 | 0.01 | Daily | 21 | 14 | +++ |
| 8 | 0.01 | Daily | 59 | 19 | +++ |
| 9 | 0.05 | Weekly | 11 | 7 | ++ |
| 10 | 0.01 | Daily | 25 | Negative | Negative |
| Control | --- | --- | --- | Negative | Negative |

Intensity classification of radiographic alterations: + Mild, ++ Moderate, +++ Severe.

Seven experimental rabbits were euthanized and necropsied at the end of the administration. In three rabbits there were severe radiographic changes after they consumed 21 and 25 daily doses of 0.01 g/kg and 11 weekly doses of 0.05 g/kg, respectively, and at this point administration of the plant was discontinued. These animals were radiographically monitored once a month for 1 year.

The radiographic assessments were performed at the Radiodiagnosis Sector at the Veterinary Hospital, UFMT, at weekly intervals using an X-ray machine operating at 15 mA, 55Kv per 0.05 s. The animals were positioned in right lateral and ventrodorsal recumbency. In the first position the animals were radiographed with extension of the neck, cranial traction of the forelimbs with forearms parallel to the neck, and the elbow flexed at an angle of 90°. The images of the aorta were classified as to intensity of calcification as negative, mild, moderate, and severe. The rabbits were clinically monitored by weekly measurement of weight and daily ingestion of diet. The doses of *S. malacoxylon* were recalculated after each weighing of the rabbits.

During necropsy fragments of the aorta, lung, thyroid gland, heart, kidney, humerus, trachea, tendon, liver, esophagus, brain, cerebellum, spinal cord, skeletal muscle, stomach, intestine, and spleen were collected, fixed in 10% formalin solution, processed by standard histological methods, and stained by hematoxylin and eosin and by the Von Kossa method for mineral deposition. At the end of the experiment control rabbits were euthanized and underwent necropsy as previously described.

## Results and Discussion

The clinical signs observed in all rabbits during the ingestion of *S. malacoxylon* were characterized by reduction in feed intake, severe weight loss, tachycardia, and pasty yellow feces. The clinical course ranged from 15 to 60 days. All the experimental rabbits became ill, six died spontaneously, one was euthanized, and three were monitored for 1 year after discontinuation of administration of the plant. During this period the three rabbits increased body weight and feed intake was similar to the ingestion of the control rabbits.

The main finding observed was aorta calcification. The contours of the vessel were well defined with density similar to bone tissue which ranged from a mild form to a severe form. The mild images show the aorta with contours defined in thoracic portion and severe images manifested as increased radiopacity extending from the cardiac, thoracic, and cranial abdominal portions. The intensity of radiographic lesions was proportional to the number of doses of *S. malacoxylon* administered shown as an increase in radiopacity of the aorta and increase in distinctness of its contours. In rabbits 8 and 9, which ingested 59 0.01 g/kg daily doses and 11 0.05 g/kg weekly doses of *S. malacoxylon*, respectively, there was mineralization of tracheal rings and a diffuse increase in lung density. At the end of the period of ingestion of *S. malacoxylon*, the radiographic findings in the ten rabbits were: 1 (10%) negative image, 3 (30%) mild, 3 (30%) moderate, and 3 (30%) severe lesions. On average, mild and severe lesions were observed respectively 18 and 35 days throughout the experimental poisoning of rabbits with 0.01 g/kg daily doses of *S. malacoxylon.* Control animals had no images of calcification at the beginning and at the end of the experiment.

Macroscopic findings showed correlation with both positive and negative radiographic findings. However, in the rabbit with a negative image there was a mild microscopic multifocal calcification of the aorta. The intensity of the findings coincided with the degree of calcification observed in necropsy in ten rabbits of this experiment. This indicates that the radiographic method showed high sensitivity in the detection of mineralization of the aorta in rabbits ingesting *S. malacoxylon.* Barros *et al.* (1992) performed a radiographic monitoring of four sheep spontaneously poisoned by *N. veitchii* and observed a slight decrease in the radiopacity of carotid arteries.

At necropsy the cardiovascular and respiratory findings were constant and consisted of diffuse mineralization of soft tissues, with greater intensity in the aorta and lungs. The deposition of minerals in arteries was constant, and the artery walls were thickened, inelastic, wrinkled in appearance and brittle. The lungs of all necropsied animals exhibited an expanded appearance with multifocal white plaques in the pleural surface, prominent especially in the borders of the diaphragmatic lobes. In the heart, white streaks were identified in valves and in the epicardium. The thyroid gland increased in volume. Mild mineralization was noted in the kidneys as white streaks on the cortex and medulla. The liver exhibited increased lobular pattern on the capsular surface. Two animals exhibited gastric ulcers. The intestinal contents were pasty and yellow in the colon and rectum.

Microscopically the main alterations were observed in the cardiovascular, respiratory, endocrine, renal, and skeletal systems. The aorta showed severe mineralization of the media layer. The mineralization is seen as fine granular deposits that are heavily marked by hematoxylin. Arteries and arterioles of the heart, kidneys, and lung showed different degrees of mineralization of the media layer and proliferation of the intima layer with tumefaction, fragmentation, and eosinophilia of elastic fibers. In the lungs there was thickening of the alveolar septa with deposition of mineral granules. In the heart there was variable intensity of multifocal mineralization of cardiac fibers and fiber necrosis. The kidney exhibited calcification of the basal layer of the epithelium of the renal pelvis. There was calcification of the renal tubular membrane and glomerular capsule in the rabbit poisoned by 0.05 g/kg weekly doses of *S. malacoxylon*. Hyperplasia of thyroid C cells and thickening of the cortical layer of the humerus and its trabeculae were frequently observed as well as calcification of the tracheal epithelium and dystrophic calcification of the cartilage of the tracheal rings. Von Kossa staining for mineral deposition revealed the presence of mineral in septa lungs, kidney tubules, tracheal cartilage, circular muscle layer of the intestine, muscle fibers of the diaphragm, and in arteries of the heart, lungs, and kidneys. In the three rabbits in which *S. malacoxylon* was discontinued after 21 and 25 daily doses and 11 weekly doses an increase in feed intake and weight gain was noticed but there was no regression in radiographic alterations after 1 year of monitoring. One control rabbit was euthanized at the end and the aorta showed normal morphology.

## Conclusions

The lesions induced by ingestion of *S. malacoxylon* in rabbits were radiographically observed for a period of 1 year after the cessation of administration of the plant. These findings suggest that radiographic monitoring can be an important support for *in vivo* diagnosis of enzootic calcinosis.

## References


Barros SS, Driemeier D, Santos MN, and Gerreiro JAM (1992). Evolução clínica e reversibilidade das lesões da calcinose enzoótica dos ovinos induzida por *Nierembergia veitchii*. *Pesquisa Veterinária Brasileira* 12:5-10.

Carrillo BJ and Worker NA (1967). 'Enteque seco', calcification and wasting in grazing animals in the Argentine. *Nature* 215:72-74.

Döbereiner J, Tokarnia CH, Costa JBD, Campos JLE, and Dayrell MS (1971). 'Espichamento', intoxicação de bovinos por *S. malacoxylon*, no Pantanal de Mato Grosso. *Pesquisa Agropecuária Brasileira* 6:91-117.

Mello JRB (2003). Calcinosis–calcinogenic plants. *Toxicon* 41(1):1-12.

Riet-Correa F and Méndez MDC (2007). Intoxicações por plantas e micotoxinas. In *Doenças de ruminantes e eqüídeos* (F Riet-Correa, AL Schild, RAA Lemos, and JRJ Borges, eds), pp. 182-184. Palotti, Santa Maria.

Schild AL (1991). Intoxicações por plantas calcinogênicas. In *Intoxicações por plantas e micotoxinas em animais domésticos* (F Riet-Correa F, MDC Méndez, and AL Schild, eds), pp. 259-269. Hemisfério Sul do Brasil, Pelotas.

Tokarnia CH, Döbereiner J, and Vargas PV (2000). *Plantas tóxicas do Brasil*, 310 pp. Helianthus, Rio de Janeiro.

# Chapter 78

# Spontaneous Intoxication by *Solanum malacoxylon* in *Bubalus bubalis* in Northern Pantanal of Mato Grosso, Brazil

C.E.P Santos[1], L.C. Marques[2], J.C. Canola[2], and J.A. Silva[3]

*[1]Department of Veterinary Medical Clinic, Federal University of Mato Grosso; [2] School of Animal and Veterinary Sciences, São Paulo State University 'Júlio de Mesquita Filho', Jaboticabal, São Paulo, Brazil; [3]Matogrossense Enterprise of Research, Assistance and Rural Extension, Brazil*

## Introduction

*Solanum malacoxylon* (=*glaucophyllum*) is a plant of the Solanaceae family known as 'espichadeira' in Brazil and causes an enzootic calcinosis (espichamento) in cattle in the Pantanal Matogrossense (Döbereiner *et al.* 1971; Tokarnia and Döbereiner 1974). The disease is also common in Argentina (Gimeno 1977) and Uruguay (Riet-Correa *et al.* 1975) where it is known as 'enteque seco'. The active principle of *S. malacoxylon* is a glycosidic derivative of $1,25(OH)_2D_3$ (calcitriol). This compound is absorbed directly in the intestine and causes degeneration and calcification of elastic fibers, hypercalcemia, and hyperphosphatemia (Riet-Correa and Méndez 2007). It is a plant of marshy habitats and loamy soils (Tokarnia *et al.* 2000). In natural conditions, the poisoning has been reported in ovines, equines (Carrillo and Worker 1967), bovines (Döbereiner *et al.* 1971), and swine (Campero and Odriozola 1990). The highest incidence occurs during the dry season due to the ingestion of leaves on the ground as animals do not graze directly on the bushes. The objective of this work is to report a natural outbreak of enzootic calcinosis in buffalo in the North Pantanal area, Poconé County, state of Mato Grosso, Brazil. Buffalo were introduced into the Pantanal because of their importance for meat and milk production and ecotourism. No other reports of *S. malacoxylon* poisoning in buffalo have been found.

## Materials and Methods

Epidemiological, clinical, and pathological data were obtained by visiting the farm where the cases occurred. One seriously affected 3-year-old buffalo cow was euthanized with barbiturates due to its untreatable state and necropsied. Blood samples were collected for biochemical analysis and fragments of different tissues were fixed in 10% formalin, routinely processed for histology, and stained with hematoxylin and eosin. In addition, ultrasonographic examinations were performed using a portable scanner (Pie Medical). The

©CAB International 2011. *Poisoning by Plants, Mycotoxins, and Related Toxins* (eds F. Riet-Correa, J. Pfister, A.L. Schild, and T.L. Wierenga)

flexor tendon of the buffalo cow was examined percutaneously using an 8.0 Mhz linear probe.

## Results and Discussion

The farm is located at 16º21'53.5"S and 056º41'07"W with mean altitude of 160 m. A buffalo herd composed of 37 animals of different ages was moved to the area in October 2006 because of their adaptability to flooded areas. In January 2007 some animals began to lose weight. During local inspection a high infestation by *S. malacoxylon* in the area was verified. Three buffalo exhibited rigid gait with arched backs, retracted abdomen, and keeping their front limbs slightly flexed in order to support themselves on the tips of their hooves. Other clinical findings were difficulty in rising and kneeling after rising. The animal that was euthanized and necropsied was cachectic and exhibited hypotrichosis. There was calcification in the aorta and other arteries of smaller caliber as well as in the cardiac valves. Histologic examination revealed slight multifocal hemosiderosis in the spleen. In the lungs there were slight multifocal interstitial pneumonia with presence of macrophages. In the aorta, there was extensive chondroid metaplasia with discrete areas of calcification. Ultrasonographic findings showed calcifications of the flexor tendons in transverse and longitudinal sections. Serum analysis revealed increased levels of calcium (10.3 mg/dl) and phosphorus (8.6 mg/dl).

The clinical, pathological, and biochemical findings were consistent with previous reports of calcinosis in other species (Döbereiner *et al.* 1971; Tokarnia and Döbereiner 1974; Riet-Correa *et al.* 1975; Braun *et al.* 2000; Tokarnia *et al.* 2000).

In this study, an important clinical finding was greatly reduced milk production. This clinical sign was also observed in goats with enzootic calcinosis caused by *Trisetum flavescens* (Braun *et al.* 2000). Other findings involved serious hypotrichosis. Franz *et al.* (2007) reported abnormal hair growth in cows experimentally poisoned by *T. flavescens*.

Ultrasonography can be used as a diagnostic tool to detect calcified areas in tendons. Clinical trials using other calcinogenic plants have demonstrated the potential of ultrasonography as a diagnostic tool in living bovines and sheep exhibiting signs of calcinosis, which allows the identification of the problem at an early stage (Franz *et al.* 2007).

## Conclusions

These results are consistent with the diagnosis of natural poisoning by *S. malacoxylon*. The occurrence of the disease in the rainy season is probably due to the behaviour of buffalo that are used to grazing native plants in flooded areas. There is no information regarding the susceptibility of buffalo to *S. malacoxylon* poisoning compared to bovines.

## Acknowledgements

Financial support was obtained from FAPEMAT. Thanks to the Araras Eco Lodge for providing the animal and CA Pescador for histological examinations.

## References

Braun U, Diener M, Camenzind D, Flückiger M, and Thoma R (2000). Enzootic calcinosis in goats caused by golden oat grass (*Trisetum flavescens*). *Veterinary Record* 146:161-162.

Campero CM and Odriozola E (1990). A case of *Solanum malacoxylon* toxicity in pigs. *Veterinary and Human Toxicology* 32(3):238-239.

Carrillo BJ and Worker NA (1967). Enteque seco: arteriosclerosis y calcificación metastásica de origem tóxico en animales a pastoreo. *Revista de Investigaciones Agropecuarias*, Buenos Aires, *Sér. Patologia Animal* 4(2):9-30.

Döbereiner J, Tokarnia CH, Costa JBD, Campos JLE, and Dayrel MS (1971). 'Espichamento', intoxicação de bovinos por *Solanum malacoxylon* no Pantanal de Mato Grosso. *Pesquisa Agropecuária Brasileira Sér. Veterinária* 6:91-117.

Franz S, Gasteiner J, Schilcher F, and Baumgartner W (2007). Use of ultrasonography to detect calcifications in cattle and sheep fed *Trisetum flavescens* silage. *Veterinary Record* 161:751-754.

Gimeno EJ (1977). Estudio histopatologico del enteque seco experimental en ratas y revision bibliografica de las calcicosis, 150 pp. PhD Dissertation en Ciencias Veterinarias, Universidad Nacional de La Plata, Argentina.

Riet-Correa F and Méndez MC (2007). Intoxicações por plantas e micotoxinas. Plantas calcinogênicas. In *Doenças de Ruminantes e Eqüídeos* (F Riet-Correa, AL Schild , RAA Lemos, and JRJ Borges, eds), pp. 99-219, vol. 2. Pallotti, Santa Maria.

Riet-Correa F, Riet-Correa I, and Bellagamba C (1975). Calcificación metastática enzoótica (enteque seco) en bovinos del Uruguay. *Veterinaria*, Uruguay, 12:15-23.

Tokarnia CH and Döbereiner J (1974). 'Espichamento', intoxicação de bovinos por *Solanum malacoxylon*, no Pantanal de Mato Grosso. II. Estudos complementares. *Pesquisa Agropecuária Brasileira, Sér. Vet.* 9:53-62.

Tokarnia CH, Döbereiner J, and Peixoto PV (2000). *Plantas Tóxicas do Brasil* 310 pp. Helianthus, Rio de Janeiro.

# Chapter 79

# Experimental Poisoning by *Nierembergia rivularis* in Sheep of Uruguay

C. García y Santos[1], G. Etcheberry[1], J.M. Goyen[1], R. Pereira[1], W. Pérez[1], and A. Ruiz[2]

[1]*Facultad de Veterinaria, Universidad de la República, Av. Lasplaces 1550, CP 1600 Uruguay;* [2]*Facultad de Química, Universidad de la República, Uruguay*

## Introduction

Recently an enzootic calcinosis was diagnosed in sheep grazing in areas invaded by the plant *Nierembergia rivularis* (= *N. repens*) in a farm located in Bañado Grande, 6th Police District, Route 44, 30 km northeast from Ansina Town, coordinates 31°47'12.7"S, 55°10'26.4"W, Rivera, northern region of Uruguay. The objective of this paper is to report the experimental poisoning in sheep ingesting *N. rivularis*.

## Material and Methods

The experiment was done on the same farm where the natural intoxication occurred. Four Corriedale crossbred sheep were used, one as control. The animals, weighing an average of 20 kg, had been raised in a field without any known toxic plants.

Before the start of the experiment the sheep were identified, clinically examined, dewormed with doramectin and closantel, and immunized against clostridial diseases. Fecal analyses for gastrointestinal nematodes were performed at 30 day intervals. An area of 625 $m^2$ where the plant was dominant was demarcated using an electric fence and the three experimental sheep were put inside. The electric fence was rotated every 15 days over the course of 3 months so that the animals always had access to pasture containing *N. rivularis*. The control sheep grazed on the same farm in a paddock free of *N. rivularis*. All the animals had access to shade and water *ad libitum*. The animals were observed by farm workers in order to make sure that the aforementioned conditions were fulfilled. Every 15 days a clinical examination was performed on each animal and they were weighed and blood samples were collected. Serum calcium levels were determined by photometric colorimetric (Calcium liquicolor) analysis and serum phosphorus levels by UV photometric analysis (Phosphorus liquirapid), both from Human Gesellschaft für Biochemica und Diagnostica mbH®.

Ninety days after the beginning of the experiment the animals were sent to the Veterinary Faculty. Radiologic and ultrasonographic examinations were performed. The

©CAB International 2011. *Poisoning by Plants, Mycotoxins, and Related Toxins* (eds F. Riet-Correa, J. Pfister, A.L. Schild, and T.L. Wierenga)

animals were then euthanized with thiopental. Samples from lungs, trachea, heart, aorta, esophagus, intestine, peritoneum, kidneys, adrenal glands, liver, spleen, and central nervous system were obtained and fixed in buffered formalin 10%. They were routinely processed and stained with hematoxylin-eosin. Selected sections were stained by von Kossa for calcium.

Plant epidermis and feces from the animals were processed for microhistology. Extracts of *N. rivularis* were studied by thin-layer chromatography at the Pharmacognosis Laboratory of the Chemistry Faculty.

## Results and Discussion

Cardiac and respiratory frequencies, ruminal motility, and temperature remained within normal values throughout the experiment. Nematode egg counts never exceeded 800 eggs/g showing a mild nematode burden and Happich-Boray analysis for liver flukes was negative.

Clinically the most significant observation was that the animals were unable to follow the flock as they were left behind during herd movement and showed dyspnea. When subjected to physical efforts, the signs of fatigue were more obvious. Calcium serum levels increased when ingestion of *N. rivularis* increased. X-rays showed an increased radiopacity of the aortic arch as the only anomaly. Ultrasonography revealed an increase in ecogenicity at the corticomedullary junction of the kidney. The cardiac valves had a normal appearance.

The most remarkable necropsy findings were a slight increase in the amount of pericardial fluid, small whitish areas inside the heart atria, and a very hard and stiff aorta with a whitish, rough, and striated internal surface. White elevated areas were observed on the surface of the apical lung lobes and on the surface and borders of the diaphragmatic lobes. The surface of the liver had a whitish spotted appearance and marked congestion. A diffuse white mottling was present throughout the parenchyma. The kidneys had congestion and a slightly mottled area close to the corticomedullary junction. No alterations were seen in other organs. Histologically the elastic fibers of the intima and middle layer of the aorta and other medium sized arteries showed degeneration and mineralization. Within the wall of arteries there were giant cells and macrophages.

Trichomes and epidermal fragments of *N. rivularis* were found in feces of the experimental animals. Thin-layer chromatography of plant extracts revealed the presence of vitamin $D_3$ metabolites or its other forms.

Anomalies in skeletal conformation such as kyphosis, slight flexion of fore limbs, and rigid gait similar to those reported in *N. veitchii* poisoning (Riet-Correa *et al.* 1987) were not observed. Other studies show a constant hypercalcemia when there is constant access to the calcinogenic plant (Riet-Correa *et al.* 1987, 1993). According to other authors (Greco 2005; Guyton and May 2007), calcium homeostasis takes place within narrow time limits, no longer than 1 h. So, if the plant is not offered at a constant rate serum calcium will vary as apparently occurred during the course of our study. Throughout the experiment serum phosphorous levels did not show any significant variation. In contrast, the calcium/phosphorous ratio changed in the same way as calcium serum values. X-rays showed signs of calcification similar to those observed in goats (Braun *et al.* 2000; Soto 2005, personal communication). Ultrasonography revealed kidney anomalies similar to those reported in other calcinosis (Franz *et al.* 2007).

Macroscopic and histologic findings were characteristic of enzootic calcinosis (Eckell *et al.* 1960; Carrillo and Worker 1967; Dobereiner *et al.* 1971; Gill *et al.* 1976; Neumann *et al.* 1977; Riet-Correa *et al.* 1987; Gimeno 2000; Barros *et al.* 2006; García y Santos *et al.* 2006; Rissi *et al.* 2007). Lesions observed in experimental sheep were less intense than those found in the natural outbreak observed in the same farm (García y Santos *et al.* 2006). This difference can be attributed to the short period of time that the animals spent in the pasture, to the amount of plant present, or to individual susceptibility of the animals.

Trichomes and epidermal fragments found in feces of the animals confirmed the ingestion of *N. rivularis*. This technique was used in previous experiments with the same objective (Panter *et al.* 1987; Yagueduú *et al.* 1998, 2000). Thin-layer chromatography of plant extracts revealed the presence of vitamin $D_3$ metabolites or other forms, confirming that *N. rivularis* is a calcinogenic plant.

The design used in this study was selected based on the low height, low availability, and high palatability of the plant using sheep, the same species affected during spontaneous poisoning (García y Santos *et al.* 2006). The use of electric fencing ensured that the animals ingested the suspect plant allowing for the necessary conditions to characterize a toxic plant, namely the use of same animal species in which the natural intoxication occurred, an abundance of the suspected toxic plant, and grazing pressure sufficient to force consumption (Tokarnia *et al.* 2000). However, this design did not allow the determination of the toxic dose of the plant.

At the present time, a multidisciplinary team from the Veterinary and Chemistry Faculties is working on the quantification of the active principles, the determination of the toxic dose of the plant, and the study of other potentially calcinogenic plants in Uruguay.

## Conclusion

The calcinogenic plant *N. rivularis*, which contains vitamin $D_3$ metabolites, was the cause of an outbreak of enzootic calcinosis in sheep in 2005.

# Acknowledgements

We thank the farmers that received us and allowed us to carry out epidemiological studies and experimental reproductions, the veterinarians that reported the clinical cases, and the Pharmaceutical Chemists for performing chemical studies. We also thank Dr Rodolfo Rivero and Dr Antonio Moraña for histopathologic studies and especially Dr Jorge Moraes for revising this manuscript. Project funded CSIC I+D (UdelaR).

## References

Barros SS, Soraes MP, and Gimeno EJ (2006). Macrophages and giant cell proliferation associated with bone protein synthesis and calcification in the trachea and bronchi of rabbits intoxicated with *Solanum glaucophyllum*. *Veterinary Pathology* 43:494-499.

Braun U, Diener M, Camenzind D, Flückiger M, and Thoma R (2000). Enzootic calcinosis in goats caused by golden oat grass (*Trisetum flavescens*). *Veterinary Record* 146:161-162.

Carrillo BJ and Worker NA (1967). Enteque seco: arteriosclerosis y calcificación metastásica de origen tóxico en animales a pastoreo. *Revista Investigaciones Agropecuarias INTA Argentina* 4(2):9-30.

Dobereiner J, Tokarnia CH, Costa JBD, Campos JLE, and Dayrell MS (1971). 'Espichamento', intoxicação de bovinos por *Solanum malacoxylon*, no Pantanal de Mato Grosso. *Pesquisa Agropecuaria Brasileira* 6:91-117.

Eckell OA, Gallo GG, Martín AA, and Portela RA (1960). Observaciones sobre el 'Enteque Seco' de los bovinos. *Revista de la Facultad de Ciencias Veterinarias*, La Plata, 6:5-91.

Franz S, Gasteiner J, Schilcher F, and Baumgartner W (2007). Use of ultrasonography to detect calcifications in cattle and sheep fed *Trisetum flavescens* silage. *Veterinary Record* 161(22):751-754.

García y Santos C, Pérez W, Mosca V, Pereira R, Seoane A, Rodríguez M, Moraes J, and Rivero R (2006). Calcinosis Enzoótica en ovinos de Uruguay. In *XXXIV Jornadas Uruguayas de Buiatría,* pp. 195-196. Paysandú, Uruguay.

Gill BS, Singh M, and Chopra AK (1976). Enzootic calcinosis in sheep: clinical signs and pathology. *American Journal Veterinary Research* 37(5):545-552.

Gimeno EJ (2000). Calcinosis enzoótica en rumiantes: un problema vigente de la ganadería nacional. *Academia Nacional de Agronomía y Veterinaria.* Sesión pública extraordinaria. Tomo LIV, pp. 202-234. Buenos Aires, Argentina.

Greco D and Stabenfeldt GH (2005). In *Fisiología Veterinaria* (JG Cunningham, ed.), 3rd edn. Elsevier, Madrid, pp. 341-372.

Guyton AC and May JE (2007). Hormona paratiroidea, calcitonina, metabolismo del calcio y fósforo, vitamina D, huesos y dientes. In *Tratado de fisiología médica* (AC Guyton and JE Hall, eds), pp. 978 -995. 10th edn. Elsevier, Madrid.

Neumann F, Nobel TA, and Bogin E (1977). Enzootic calcinosis in sheep and C-cell hyperplasia of the thyroid. *Veterinary Record* 101(18):364-366.

Panter KE, Ralphs MH, Smart RA, and Duelke B (1987). Death camas poisoning in sheep: A case report. *Veterinary and Human Toxicology* 29(1):45-48.

Riet-Correa F, Schild AL, Méndez MC, Wasserman R, and Krook L (1987). Enzootic calcinosis in sheep caused by the ingestion of *Nierembergia veitchii* (Solanaceae). *Pesquisa Veterinária Brasileira* 7(3):85-95.

Riet-Correa F, Mendez MC, Schild AL, and Petiz CA (1993). Enzootic calcinosis in sheep. Experimental reproduction with *Nierembergia veitchii* (Solanaceae). *Pesquisa Veterinária Brasileira* 13(1/2):21-24.

Rissi D, Rubia R, Pierezan F, Kommers GD, and Lombardo de Barros CS (2007). Intoxicação em ovinos por *Nierembergia veitchii*: observações em quatro surtos. *Ciência Rural* 37(5):1393-1398.

Tokarnia CH, Döbereiner J, and Peixoto PV (2000). *Plantas Tóxicas do Brasil.* 310 pp. Ed Helianthus, Río de Janeiro.

Yagueduú C, Cid MS, and Lopez T (1998). Microhistological analysis of sheep gastrointestinal content to confirm poisonous plant ingestion. *Journal Range Management* 51:655-660.

Yagueduú C, Cid MS, Lopez T, and Brizuela MA (2000). Exactitud y precisión en la cuantificación por microanálisis de *Cestrum Parqui* L'Herit en el contenido digestivo de ovinos en pastoreo. *Veterinaria Argentina* 17(170):757-767.

# Chapter 80

# Spontaneous Nitrate/Nitrite Poisoning in Cattle Fed with Oats (*Avena sativa*) and Ryegrass (*Lolium multiflorum*) in the State of Santa Catarina, Brazil

F. Jönck[1], A. Gava[1], F.H. Furlan[1], S.D. Traverso[1], L.O. Veronezi[1], V. Borelli[1], E. Gheller[2], and R.A. Casagrande[1]

[1]*Animal Pathology Laboratory, University of Santa Catarina State, Av. Luiz de Camões, 2090, Conta Dinheiro, 88520-000/Lages – Santa Catarina/Brazil;* [2]*Laticínios Tirol Ltda., Treze Tílias – Santa Catarina/Brazil*

## Introduction

Nitrate and nitrite poisoning is a common problem in veterinary medicine. The primary source of exposure to nitrate by animals is ingestion of plants and water high in nitrates (Van Dijk *et al.* 1983; Boermans 1990; Choon *et al.* 1990; Riet-Alvariza 1993; Cheeke 1998; Tokarnia *et al.* 2000; Medeiros *et al.* 2003; Ozmen 2003; McKenzie *et al.* 2004). When ruminants ingest plants with high nitrate levels, rumen bacteria reduce these compounds to nitrite. Once absorbed, the nitrites oxide the hemoglobin iron ion, turning it into methemoglobin. As methemoglobin is unable to react with oxygen, cellular anoxia occurs. When methemoglobin levels reach 30-40% clinical signs occur and levels from 80-90% cause death. The susceptibility of different species to poisoning depends on the capacity to transform nitrates into nitrites; swine are most susceptible followed by cattle, sheep, and horses (Van Dijk *et al.* 1983; Boermans 1990; Cheeke 1998; Radostits *et al.* 2000). During poisoning there is a clear decrease in blood pressure due to the nitrate ion having a relaxing effect in the smooth muscle of the small blood vessels (Van Dijk *et al.* 1983), which can lead to hypoxia due to peripheral circulatory insufficiency. However, this effect is less significant when compared to methemoglobin formation (Radostits *et al.* 2000).

Nitrates absorbed from the soil by plant roots are usually incorporated into plant tissues as amino acids, proteins, and other nitrogenous compounds. The primary location for nitrate conversion is in green leaves during the growing stage (Ozmen 2003). Abnormal nitrate accumulation in plants is influenced by many factors such as fertilization with nitrogen fertilizers or organic matter from animals, rapid plant growth when rain occurs after drought, treatment with herbicides, and soil characteristics including temperature, acidity, and deficiency of phosphorus, sulfur, or molybdenum (Riet-Alvariza 1993; Cheeke 1998; Radostits *et al.* 2000).

©CAB International 2011. *Poisoning by Plants, Mycotoxins, and Related Toxins*
(eds F. Riet-Correa, J. Pfister, A.L. Schild, and T.L. Wierenga)

Mortality of cattle grazing on grasses cultivated in super-fertilized soils has been reported (Rosenberger 1975; Riet-Alvariza 1993; Jones *et al.* 2000; Radostits *et al.* 2000). In Brazil, only three outbreaks of nitrate poisoning have been reported in cattle, all of them in the State of Paraíba (Medeiros *et al.* 2003). In the Western plateau and high Itajaí valley of the State of Santa Catarina the use of cultivated grazing areas has significantly increased in the past few years. Reports of deaths and abortions in cattle in these locations have become common, mostly in cattle kept in areas farmed exclusively with oats (*Avena sativa*) and/or ryegrass (*Lolium multiflorum*). The development of technologies applied to agriculture to increase feed production for cattle has contributed to the appearance of feed-related illnesses not previously observed.

## Nitrate/Nitrite Spontaneous Poisoning

Nitrate/nitrite poisoning in cattle in the State of Santa Catarina was mainly observed between June and October, which corresponds to the highest peak in the production of oats and ryegrass. In general, the outbreaks were observed in the first weeks after rain preceded by dry weather in locations where there was super-fertilized soils with swine or poultry organic matter or by excessive chemical fertilization. The onset of clinical signs appeared suddenly and consisted of brown mucous membranes, breathing difficulty, muscular trembling, staggering, constant urination, weakness, tachycardia, and bloat. Death occurred after a clinical manifestation period of 15-30 min.

In locations where the illness occurred, abortions were frequently observed even in cows without other clinical signs of the disease. Morbidity ranged from 5-25% and lethality from 0-85%. The main necropsy findings were dark-colored blood with more fluidity and slightly brown-colored lungs and encephalon. In skeletal muscles and left myocardium an intense red color was observed. In one cow, an accentuated red color was observed in the mucosa of the abomasum. Microscopically, no significant lesions were observed.

Fourteen animals that manifested the initial clinical signs of the disease were treated intravenously with 4 mg/kg of 1% methylene blue. All animals recovered after an average of 30 min. The farm outbreaks of nitrate/nitrite poisoning ceased after the replacement of oats and/or ryegrass by other types of pastures. However, in one location cases of nitrate/nitrite poisoning continued even after the ryegrass was cut and resprouted three times and only ceased after the animals' removal from the location.

Clinical signs were similar to those reported in other outbreaks of nitrate poisoning (Van Dijk *et al.* 1983; Boermans 1990; Medeiros *et al.* 2003; Ozmen 2003). In cattle, the methemoglobin peak occurred around 5 h after nitrate intake. Deaths occurred within 12-24 h after the intake of the feed although in severe poisonings the clinical course was shorter, generally 1.5 to 4 h after the intake of feed with high nitrate levels (Boermans 1990; Ozmen 2003). Abortions were reported frequently as a consequence of the poisoning and damage to fetuses probably occurred due to serious anoxia (Radostits *et al.* 2000).

The concentration of nitrate considered sufficient to cause intoxication is above 1.5% (dry matter basis (DMB); Boermans 1990; Olson *et al.* 2002; Ozmen 2003). Samples of oats and ryegrass collected from two locations of the present study showed concentrations of nitrate that ranged from 1.89% to 3.36% DMB. This demonstrates that the oat and ryegrass leaves in this location had sufficient nitrate to cause intoxication in cattle.

## Conclusions

The sudden appearance of clinical signs in cattle kept in oat and/or ryegrass pastures with excessive nitrogen fertilization in the State of Santa Catarina are caused by the high concentration of nitrate in these grasses. The diagnosis of nitrate/nitrite poisoning in cattle was made after epidemiological evaluation and observations of brown mucous membranes, breathing difficulty, bloat, dark-colored blood, accentuated red color in the left myocardium and skeletal muscles, and evaluation of nitrate concentration in suspicious plants. The diagnosis was confirmed by the clinical recovery of affected animals after treatment with 1% methylene blue and by the cessation of outbreaks after the replacement of oats and ryegrass by other pastures with low levels of nitrate.

## References

Boermans HJ (1990). Diagnosis of nitrate toxicosis in cattle, using biological fluids and a rapid chromatographic method. *American Journal of Veterinary Research* 51(3):491-495.

Cheeke PR (1998). *Natural toxicants in feeds, forages and poisonous plants*, 479 pp. Interstate Publishers, Danville.

Choon Y, Brandow RA, and Howlett P (1990). An unusual cause of nitrate poisoning in cattle. *Canadian Journal of Veterinary Research* 31(2):118.

Jones CJ, Hunt RD, and King NW (2000). *Patologia Veterinária*, 1415 pp. Manole, Barueri.

McKenzie RA, Rayner, AC, Thompson GK, Pidgeon GF, and Burren BR (2004). Nitrate-nitrite toxicity in cattle and sheep grazing *Dactylotenium radulans* (button grass) in stockyards. *Australian Veterinary Journal* 82(10):630-634.

Medeiros RTM, Riet-Correa F, Tabosa IM, De Souza AV, Da Silva ZA, Junior GS, and Barbosa RC (2003). Nitrate and nitrite poisoning in cattle caused by the ingestion of *Echinochloa polystachya* and *Pennisetum purpureum* in the semiarid region of the state of Paraíba. *Pesquisa Veterinária Brasileira* 23(1):17-20.

Olson OE, Emerick RJ, and Whitehead EI (2002). Forage Nitrate Poisoning – A summary. *Cooperative Extension Service South Dakota State University – US Department of Agriculture.* www.agbioplus.sdstate.edu/articles/FS420.pdf.

Ozmen O (2003). Nitrate poisoning in cattle fed *Chenopodium album* hay. *Veterinary and Human Toxicology* 45(2):83-84.

Radostits OM, Gay CC, Blood DC, and Hinchcliff KW (2000). *Veterinary Medicine*, 881pp. W.B. Saunders, London.

Riet-Alvariza F (1993). Intoxicación por nitratos y nitritos. In *Intoxicações por plantas e micotoxicoses em animais domésticos* (F Riet-Correa, MC Mendez, and AL Schild, eds), pp. 291-297. Editorial Agropecuária Hemisféric Sur, Montevideo.

Rosenberger G (1975). Avvelenamenti. In *Malatie Del Bovino* (G Rosemberg, ed.), pp. 1120-1184. Editrice Essegivi, Piacenza.

Van Dijk A, Lobsteiyn AJH, Wensing T, and Breunkink HJ (1983). Treatment of nitrate intoxication in a cow. *The Veterinary Record* 112(12):272-279.

Tokarnia CH, Döbereiner J, and Peixoto PV (2000). *Plantas Tóxicas do Brasil*, 310 pp. Helianthus, Rio de Janeiro.

# Chapter 81

# Poisoning of Sheep by Shells of *Jatropha curcas* Seeds

O.R. Ferreira[1], S.S. Brito[1], F.C.S Chagas[1], A.T. Ramos[1], F.Y.M. Hosomi[2], F.G. de Lima[3], P.C. Maiorka[2], V.L. Araújo[1], J.N.M. Neiva[1], M.C.S. Fioravante[3], and V.M. Maruo[1]

[1]*Escola de Medicina Veterinária e Zootecnia, Universidade Federal do Tocantins, Campus de Araguaína, BR 153, km 112, Zona rural, CEP 77.804-970, Brazil;* [2]*Faculdade de Medicina Veterinária e Zootecnia, Universidade de São Paulo, Av. Prof. Dr. Orlando Marques de Paiva, 87, CEP 05508-270, Brazil;* [3]*Escola de Veterinária, Campus Samambaia, Universidade Federal de Goiás, Caixa postal 131, CEP 74 001 970, Brazil*

## Introduction

*Jatropha curcas* known as physic nut, purging nut, piñoncillo, Habb-El-Meluk, black vomit nut, coral plants, Ratanjyot, and Barbados nut (depending on the region) is a member of the Euphorbiaceae family (Makkar *et al.* 1998; Froberg 2007). This species is native to the South American tropics, especially Brazil; it is commonly found and utilized throughout most of the tropical and subtropical regions of the world (Kumar and Sharma 2008). The physic nut is one of the most important oleaginous plants because its seeds can be used as key raw material for making industrial products such as paints, varnish, and cosmetics (Ratree 2004). Currently, *J. curcas* is being harvested for biofuel production (Gübitz *et al.* 1999).

*J. curcas* is a species that has received much attention recently for the production of plant oils. Current estimates suggest that there are now 2.5 million ha of *J. curcas* planted in India and China alone with plans for an additional 9.3 million ha by 2010 (King *et al.* 2009). In addition to being a source of oil *J. curcas* provides a meal that may serve as a highly nutritious and economic protein supplement in animal feed (Kumar and Sharma 2008). The seeds weigh about 0.75 g and contain 30-32% protein, indicating good nutritional value (Makkar *et al.* 1997).

According to Froberg *et al.* (2007), all parts of the plant are toxic, especially the seeds, which are generally toxic to humans and animals. Indirect exposure takes place through consumption of animal products contaminated with toxic phorbol esters such as honey produced by bees (Goel *et al.* 2007) and meat and milk produced from animals that feed on diets contaminated with these toxic components (Zayed *et al.* 1998). The seeds have been reported as highly toxic in toxicity studies (Adam and Magzoub 1975; Ahmed and Adam 1979; Gadir *et al.* 2003). The toxicity is attributed to saponins, tannins, lectins (curcin), protease inhibitors, phytate, and phorbol esters (Haas *et al.* 2002; Gadir *et al.* 2003).

©CAB International 2011. *Poisoning by Plants, Mycotoxins, and Related Toxins* (eds F. Riet-Correa, J. Pfister, A.L. Schild, and T.L. Wierenga)

Curcin, a toxic protein isolated from the seeds, was found to inhibit protein synthesis during *in vitro* studies (Stirpe *et al.* 1976). The high concentration of phorbol esters present in *J. curcas* seeds has been identified as the main toxic agent responsible for their toxicity (Makkar *et al.* 2007). The term phorbol is used to describe the family of naturally occurring compounds that can be referred to as tigliane diterpene (Goel *et al.* 2007). The phorbol esters are analogues of diacylglycerol, an activator of many isoforms of protein kinase C (PKC). PKC acts as a regulator of many cellular processes. Because diacylglycerol has a short biological half-life in the cell, the activation of PKC is usually only transient. Activation of PKC by phorbol-esters, however, is much more prolonged (King *et al.* 2009).

The toxicity of *J. curcas* has been reported when feeding plants containing these esters to various animal models such as goats (Adam and Magzoub 1975; Ahmed and Adam 1979; Gadir *et al.* 2003), sheep (Ahmed and Adam 1979), calves (Ahmed and Adam 1979), mice (Stirpe *et al.* 1976; Panigrahi *et al.* 1984), rats (Panigrahi *et al.* 1984; Gandhi *et al.* 1995), and fish (Becker and Makkar 1998). Poisoning caused by ingestion of seeds or oil of *J. curcas* is characterized by changes mainly in cardiovascular, respiratory, and digestive systems. The main clinical signs of toxicosis include loss of appetite, abdominal pain, general weakness, and weight loss with watery diarrhea (Adam and Magzoub 1975; Gadir *et al.* 2003; Froberg *et al.* 2007; Oliveira 2008).

However, interest in these constituents of *J. curcas* seeds is not restricted to the toxicity of the seeds. Molluscicidal activity of the seed extracts as well as widespread use of the seeds in traditional medicine may also be associated with the presence of these substances (Haas *et al.* 2002).

The ability to use *J. curcas* meal as animal feed not only improves the economics of production but also means the crop would produce both fuel and feed (King *et al.* 2009). Makkar *et al.* (1997) demonstrated that there is variation among different provenances of *J. curcas* in secondary plant metabolites and toxic components, which may be caused by genetic differences or by different environments.

Despite the numerous reports about the toxicity of seeds and the press cake, a byproduct of oil extraction, little is known about the toxicity of the shells from *J. curcas* seeds. The objective of this research is to evaluate the toxicity of shells of *J. curcas* seeds with focus on the histopathological effects.

## Material and Methods

Shells of *J. curcas* seeds were donated by a biofuel producing company located in Paraíso, state of Tocantins. Twenty 8-month-old clinically healthy male sheep were assigned to four groups of five animals each. Treatment groups received 15% (T1), 30% (T2), and 40% (T3) of their diet as shells of *J. curcas* seeds with the remainder of the diet as *Panicum maximum* cv. Mombaça. A control group received only grass. The animals were kept for 19 days in individual metabolic cages receiving water *ad libitum*.

Necropsies were performed on all sheep after death or euthanasia. Specimens of liver, kidneys, intestine, rumen, reticulum, omasum, abomasum, spleen, lungs, and heart were fixed in 10% formalin, embedded in paraffin wax, sectioned at 3-5 μm and stained with hematoxylin and eosin (HE) for histopathological examination.

## Results

All animals that received diets with shells of *J. curcas* seeds had impaired appetite, loss of body condition, and depression. Generally, poorly formed wet feces were produced in small quantities. On the 10th day of treatment one T3 animal died. It had dark and fetid diarrhea, dehydration, nasal discharge, and severe breathing difficulty.

At necropsy animals of the T1 group showed more remarkable changes than the animals from T2 and T3. Cardiovascular and digestive systems were the most affected. The changes in the respiratory tract included trachea and lungs full of froth, and firm, congested, and crackled lungs. Gastrointestinal tract changes were hyperemic mucosa in abomasum, congestion of intestinal serosa and mesenteric vessels, enlarged mesenteric lymph nodes, and congested bowel. In some animals mucous exudate and bloody content were observed in the bowel. Light yellow transparent liquid containing fibrin was observed in the thoracic and abdominal cavities and pericardial sac. Less frequent findings included enlarged pre-scapular lymph nodes (1/15), suffusions in the serosa of the rumen (2/15), flaccid heart (3/15), kidney congestion (1/15), and ulcers in the abomasum (6/15).

Histopathological analysis revealed degenerative changes in the kidneys and liver as well as inflammatory changes in segments of the gastrointestinal tract. These findings were observed in all treated groups; however, T1 presented more intense changes when compared to other groups. Mild to moderate tubular degeneration was seen in kidneys. Slightly eosinophilic amorphous material was deposited in the glomerulus. Other findings include Bowman's capsule thickening and calcium crystals in the renal pelvis. Liver showed mild cholestasis (3/15), midzonal and/or centrilobular vacuolar degeneration, diffuse and severe congestion, sinusoid dilatation, and periportal infiltrate with macrophages, lymphocytes, plasma cells, and neutrophils. In the gastrointestinal tract the inflammatory exudate was composed of macrophages, lymphocytes, and plasma cells. In mesenteric lymph nodes, the changes were mild to severe edema with some atypical lymphocytes and plasma cells, macrophages containing hemosiderin, and disorganized lymphoid follicles.

In T1 animals, mild to moderate congestion and mild infiltration of lymphocytes, neutrophils, and macrophages were observed in the lungs. In other treatments the lungs were congested. In the spleen of T1 animals mild or moderate swelling and lymphoid tissue disorganization were observed and as in other groups some macrophages containing hemosiderin were also found.

## Discussion and Conclusions

The results of the experiment reported here showed that shells from *J. curcas* seeds are toxic to sheep with fatal consequences and that the degree of poisoning is dose related. The important signs of toxicity included dehydration and loss of condition. Diarrhea was probably due to intestinal lesions.

Phytochemical studies demonstrated the presence of purging oil, tannins, sterols, terpenes, and the toxalbumins crotin and curcin in the seeds of *Croton macrostachys* and *J. curcas* seeds (Ahmed and Adam 1979). *Croton* species which contain phorbol esters produce similar effects when ingested by farm animals, suggesting that the toxic effect of physic nut is due to the presence of phorbol esters (Gadir *et al.* 2003). The phorbol esters in addition to tumor promotion induce a remarkable diversity of other biological effects at exceptionally low concentrations. These are responsible for skin irritant effects and tumor promotion because they stimulate protein kinase C, which is involved in signal transduction and developmental processes of most cells and tissues, producing a variety of biological effects in a wide range of organisms (Goel *et al.* 2007).

Lack of appetite and probably the impaired digestive efficiency resulting from severe damage could explain the loss of condition of *J. curcas* poisoned sheep. It seems likely that dehydration resulted from reduced water intake combined with fluid loss from the alimentary tract. According to Gadir *et al.* (2003), this as well as the hemorrhage and congestion could be due to altered permeability of the capillaries by the active constituents present in shells from the seeds. Depletion of hepatic glycogen caused by the interference of toxic factors in *J. curcas* with carbohydrate metabolism in the liver could explain the depression observed in sheep that was described by Gadir *et al.* (2003).

Experiments using *J. curcas* seed with rats showed inflammatory changes in the alimentary tract and degeneration and necrosis of the liver as well as hemorrhage in the kidney and lung (Gandhi *et al.* 1995; Makkar and Becker 1999). Similar lesions were seen in goats, sheep, and calves, with the additional findings of hemorrhage in the rumen, reticulum, and spleen, hemorrhagic and/or catarrhal abomasitis, severe fatty change in liver, and accumulation of straw-coloured fluid in serous cavities (Ahmed and Adam 1979; Gadir *et al.* 2003).

We conclude that the shells of *J. curcas* seeds obtained as a byproduct of the oil extraction cannot be used as a meal in animal feed due to its toxicity. The alterations found in the animals that received shells of *J. curcas* seeds were very similar to those found in experiments conducted in animals fed *J. curcas* seeds. Since the seeds contain large amount of phorbol esters it is possible that shells produce toxicity due the presence of this chemical.

Further studies should be undertaken to characterize the phorbol esters contained in the shells of *J. curcas* seeds and to evaluate methods of detoxification of this byproduct from the biofuel industry.

## References

Adam SEI and Magzoub M (1975). Toxicity of *Jatropha curcas* for goats. *Toxicology* 4:347-354.

Ahmed OMM and Adam SEI (1979). Toxicity of *Jatropha curcas* in calves. *Veterinary Pathology* 16:476-482.

Becker K and Makkar HPS (1998). Toxic effects of phorbolesters in carp (*Cyprinus carpio* L.). *Veterinary and Human Toxicology* 40:82-86.

Froberg GBMD, Ibrahim D, and Furbee MDRB (2007). Plant poisoning. *Emergency Medicine Clinics of North America* 25:475-433.

Gadir WSA, Onsa TO, Ali WEM, El Badwin SMA, and Adam SEI (2003). Comparative toxicity of *Croton macrostachys*, *Jatropha curcas* and *Piper abusynica* seeds in Nubian goats. *Small Ruminant Research* 48:61-67.

Gandhi VM, Sherian KM, and Mulky MJ (1995). Toxicological studies on ratanjyot oil. *Food and Chemical Toxicology* 33:39-42.

Goel G, Makkar HPS, Francis G, and Becker K (2007). Phorbol esters: structure, biological activity, and toxicity in animals. *International Journal of Toxicology* 26:279-288.

Gübitz GM, Mittelbach M, and Trabi M (1999). Exploitation of the tropical oil seed plant *Jatropha curcas* L. *Bioresource Technology* 67:73-82.

Haas W, Sterk H, and Milttelbach M (2002). Novel 12-Deoxy-16-hydroxyphorbol diesters isolated from the seed oil of *Jatropha curcas*. *Journal of Natural Products* 65:1434-1440.

King AJ, Jesus AC, Freudenberg M, Ramiaramanana D, and Graham A (2009). Potential of *Jatropha curcas* as source of renewable oil and animal feed. *Journal of Experimental Botany* 30:351-357.

Kumar A and Sharma S (2008). An evaluation of multipurpose oil seed crop for industrial uses (*Jatropha curcas* L.): A review. *Industrial Crops and Products* 28:1-10.

Makkar HPS and Becker K (1999). Nutritional studies on rats and fish (carp *Cyprinus carpio*) fed diets containing unheated and heated *Jatropha curcas* meal of a non-toxic provenance. *Plant Foods and Human Nutrition* 53:182-292.

Makkar HPS, Becker K, Sporer F, and Wink M (1997). Studies on nutritive potential and toxic constituents of different provenances of *Jatropha curcas*. *Journal of Agricultural and Food Chemistry* 45:3152-3157.

Makkar HPS, Aderibigbe AO, and Becker K (1998). Comparative evaluation of non-toxic and toxic varieties of *Jatropha curcas* for chemical composition, digestibility, protein degradability and toxic factors. *Food Chemistry* 2:207-215.

Makkar HPS, Francis G, and Becker K (2007). Bioactivity of phytochemicals in some lesser-known plants and their effects and potential applications in livestock and aquaculture production systems. *Animal* 1:137-1391.

Oliveira LI, Jabour FF, Nogueira VA, and Yamasaki EM (2008). Intoxicação experimental com as folhas de *Jatropha gossipifolia* (Euphorbiaceae) em ovinos. *Pesquisa Veterinária Brasileira* 28:275-278.

Panigrahi IS, Francis BJ, Cano LA, and Surbage S (1984). Toxicity of *Jatropha curcas* seeds from México to rats and mice. *Nutrition Reports International* 29:1089-1099.

Ratree S (2004). A preliminary study on physic nut (*Jatropha curcas* L.) in Thailand. *Pakistan Journal of Biological Sciences* 7:1620-1623.

Stirpe F, Pession-Brizzi A, Lorenzoni E, Strocchi P, Montanaro L, and Sperti S (1976). Studies on the proteins from the seeds of *Croton tiglium* and of *Jatropha curcas*. Toxic properties and inhibition of protein synthesis *in vitro*. *Biochemistry Journal* 156:1-6.

Zayed SMAD, Farghaly M, Gminski HTR, and Hecker E (1998). Dietary cancer risk from conditional cancerogens in produce of livestock fed on species of spurge (Euphorbiaceae) III. Milk of lactating goats fed on the skin irritant herb *Euphorbia peplus* is polluted by tumor promoters of the ingenane diterpene ester type. *Journal of Cancer Research and Clinical Oncology* 124:301-306.

# Chapter 82

# Toxicology Study of Ethanolic Extract from Aerial Parts of *Jatropha gossypiifolia* L. in Rats

S.R. Mariz[1], G.S. Cerqueira[2], W.C. Araújo[2], J.G. Dantas[2], J.A. Ramalho[2], T.V. Palomaro[2], N.L. Barbosa Jr[2], J.C. Duarte[2], H.B. dos Santos[2], K. Olveira[2], M.S.T. de Araújo[2], M.F.F.M. Diniz[2], and I.A. de Medeiros[2]

[1]*Unidade Acadêmica de Medicina, Centro de Ciências Biológicas e da Saúde, Universidade Federal de Campina Grande, Av. Juvêncio Arruda, n. 795, Bodocongó, CEP: 58.430-800, Campina Grande (PB);* [2]*Universidade Federal da Paraíba, Cidade Universitária, CEP: 58.059-900, João Pessoa (PB)*

## Introduction

*Jatropha gossypiifolia* L. (Euphorbiaceae), informally called 'pião-roxo' in Brazil, is a highly toxic plant. This species grows around the world especially in some tropical countries. The most important chemical compounds found in this plant are organic acids, alkaloids, terpenes, steroids, flavonoids, lignanes, and tannins (Mariz *et al.* 2006). Apart from its toxic component which causes digestive and metabolic disturbances (Mariz *et al.* 2006, 2008; Oliveira *et al.* 2008), the plant has therapeutic uses as can be observed by experiments demonstrating its hypotensive effect (Abreu *et al.* 2003). In order to contribute to the development of a herbal drug from this species, the objective of this research was to evaluate the toxicity of the ethanolic extract (EE) from aerial parts of *J. gossypiifolia* in rats. The experiments were carried out according to Brazilian regulations for ethical phytotherapy studies (Brasil 2004).

## Methodology

The aerial parts (leaves and stems) of *J. gossypiifolia* were collected in the municipality of Santa Rita, state of Paraíba, Brazil, from June to August 2004 and identified at 'Lauro Pires' Herbarium at Paraíba Federal University (UFPB) where a representative sample of the species was registered under the code Agra & Góis 4192 (JPB). The ethanolic extract (EE) of *J. gossypiifolia*, the product evaluated in this study, was prepared according to common methods (Mariz *et al.* 2006). The yield of dried extract was 7.9% (w/w) (Mariz *et al.* 2006).

For the acute toxicity test the EE was administered *per os* (po) to male and female Wistar rats (*Rattus norvegicus albinus*) at doses of 1.2, 1.8, 2.7, 4.0, and 5.0 g/kg body weight (BW). Each group (*n*=12) was observed for 14 days after the treatment and

©CAB International 2011. *Poisoning by Plants, Mycotoxins, and Related Toxins* (eds F. Riet-Correa, J. Pfister, A.L. Schild, and T.L. Wierenga)

monitored for clinical signs (Almeida *et al.* 1999), lethality, body weight gain, and food and water consumption. At the end of the experiment the surviving animals were euthanized for biochemical, hematological, and histologic analysis.

In the chronic toxicity trial the rats were divided into three experimental groups and administered daily doses of 45, 135, and 405 mg/kg. Each group ($n$=10 per gender and dose) was dosed daily by gavage for a period of 13 weeks. In addition to the parameters cited above, body temperature, tail glucose level, and changes in behavior were analyzed by the Open Field and Rotarod methods. At the end of the experiment 40% of the animals from each group were euthanized and examined by the methods mentioned in the acute toxicity experiment.

## Results and Discussion

The acute dosing study showed that only doses equal to or higher than 1.8 g/kg BW (po) cause important disturbances. Motor incoordination with hind limb paralysis was observed in the Open Field and Rotarod methods. Other signs were neurological depression, digestive disorders, and weight loss. In male rats treated with 5 g/kg po some alterations suggested signs of acute toxicity including increased serum activities of creatinine and aspartate amino transaminase, increased serum concentrations of sodium and potassium, decreased serum levels of urea, albumin, and amylase, leucopenia, and mild inflammation of liver and lungs on histologic examination. The acute dosing study shows that $LD_{50}$ was estimated between 4 and 5 g/kg po in male rats but higher than 5 g/kg po in female rats. Such results suggest that acute toxicity from EE is not very important because changes to the $LD_{50}$ were observed only when higher doses were used (Mariz *et al.* 2006, 2008).

The most important results in the chronic dosing study were the damage on locomotor activity (Figure 1) and digestive disorders. Also, lethality rate was 46.6% (405 mg/kg) among male rats and 13.3% among females (doses of 135 mg/kg and 405 mg/kg). Furthermore, histologic lesions were observed in liver, kidney, and lungs. Mild chronic pericholangiolitis, foci of necrosis, mild lobular fibrosis in zone 3, mild venular congestion, and hyperplasia of Kupffer cells were observed in the liver. The lungs mainly in male rats presented chronic interstitial pneumonitis with extensive areas with congested capillaries and lymphocytic exudation causing septum thickening with restrictions of the corresponding alveolar spaces and BALT hyperplasia. In the kidneys the peripheric adipose tissue showed focal adiponecrosis (hystiocitic exudadation with xanthomatous standard). These histopathological alterations indicated an inflammatory response with stimulation of the immune system.

## Conclusions

The data presented here, especially the animal mortality during prolonged treatment, indicate the high chronic toxicity of the ethanol extract from *J. gossypiifolia*. Further work will be necessary to refine the chemistry by conducting more detailed fractionations of the plant extract for future testing as well as examining the chemical composition of these more refined fractions to determine the active compounds.

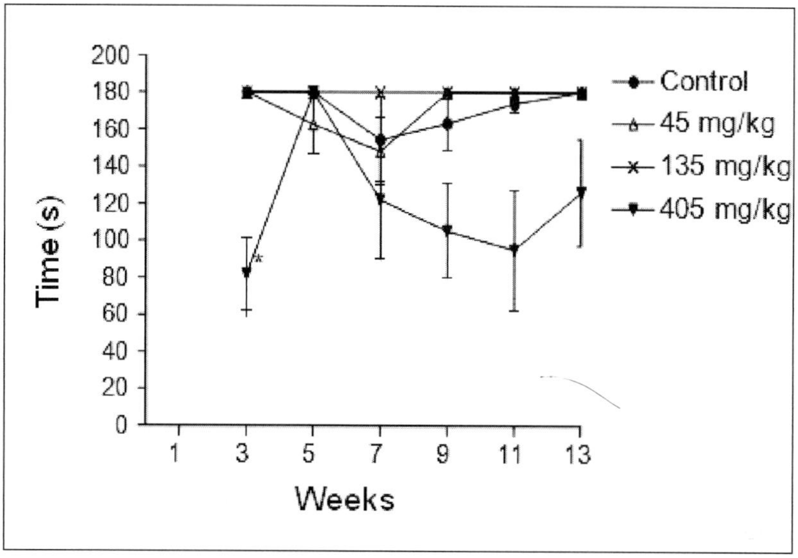


**Figure 1.** Time of permanence (in seconds) in the Rota Rod of female rats under prolonged treatment (13 weeks) with different doses of EE of *J. gossypiifolia*. The values express the mean (± s.e.m.) of each group (*n*=5). *Values statistically different from the control group (ANOVA followed by Tukey, $P < 0.05$).


## Acknowledgements

The authors would like to thank the *Comissão de Aperfeiçoamento de Ensino Superior* (CAPES) for having sponsored this research through the Program of Interinstitutional Qualification (PQI).


## References


Abreu IC, Marinho ASS, Paes AMA, Freire SMF, Olea RSG, Borges MOR, and Borges ACR (2003). Hypotensive and vasorelaxant effects of the ethanolic extract from *Jatropha gossypiifolia* L. in rats. *Fitoterapia* 74:651-657.

Almeida RN, Falcão ACGM, Diniz RST, Quintans-Júnior LJ, Polari RN, Barbosa-Filho JM, Agra MF, Duarte JC, Ferreira CD, Antoniolli AR, and Araújo CC (1999). Metodologia para avaliação de plantas com atividade no Sistema Nervoso Central e alguns dados experimentais. *Revista Brasileira de Farmacognosia* 80:72-76.

Brasil (2004). Ministério da Saúde. Agência Nacional de Vigilância Sanitária (ANVISA). Resolução – RE n°90/2004. Normas para estudos toxicológicos de produtos fitoterápicos. *Diário Oficial da República Federativa do Brasil,* Poder Executivo, Brasília, DF, 12 March 2004.

Mariz SR, Cerqueira GS, Araújo WC, Duarte JC, Melo AFM, Santos HB, Oliveira K, Diniz MFFM, and Medeiros IA (2006). Estudo toxicológico agudo do extrato etanólico de partes aéreas de *Jatropha gossypiifolia* L. em ratos. *Revista Brasileira de Farmacognosia* 16:372-378.

Mariz SR, Araújo MST, Cerqueira GS, Araújo WC, Duarte JC, Diniz MFFM, and Medeiros IA (2008). Avaliação histopatológica em ratos após tratamento agudo com o extrato etanólico de partes aéreas de *Jatropha gossypiifolia* L. *Revista Brasileira de Farmacognosia* 18(2):213-216.

Oliveira LI, Jabour FF, Nogueira VA, and Yamasaki EM (2008). Intoxicação experimental com as folhas de *Jatropha gossypifolia* (Euphrobiaceae) em ovinos. *Pesquisa Veterinária Brasileira* 28(6):275-278.

# MYCOTOXINS AND OTHER TOXINS

# Chapter 83

# Changes in Carbohydrate Expression in the Cervical Spinal Cord of Mice Intoxicated with Perivitellin PV2 from *Pomacea canaliculata*

P.E. Fernández[1], V. Frassa[2], E.J. Gimeno[1], M.S. Dreon[2], and H. Heras[2]

*[1]Instituto de Patología B. Epstein, Cátedra de Patología General, Facultad de Ciencias Veterinarias, Universidad Nacional de La Plata (UNLP), 60 y 118, 1900 La Plata, Argentina; [2]Instituto de Investigaciones Bioquímicas de La Plata (INIBIOLP), CONICET CCT La Plata, Facultad de Medicina(UNLP), Calles 60 y 120, 1900, La Plata, Argentina*

## Introduction

*Pomacea canaliculata* (Gastropoda: Ampullariidae) is a freshwater South American snail species, well adapted both to temperate regions where thermal changes during the year are large (Albrecht *et al.* 1999, 2005) and to subtropical and tropical regions where periods of drought may alternate with periods of excessive rainfall (Pizani *et al.* 2005). These edible snails are extremely invasive. They deposit their eggs above the waterline in a calcareous, coloured clutch and have developed a protective and nourishing perivitelline fluid surrounding the embryos. *P. canaliculata* eggs are provided with a multifunctional perivitellin called ovorubin which plays critical roles for embryo development such as photoprotection, antioxidant, trypsin inhibitor, and nutrient provision and also provides the eggs with a pink-reddish colour (Heras *et al.* 1998, 2007; Dreon *et al.* 2004b, 2006). Ovorubin functions are complemented by perivitellin PV2 which has been described as a source of nutrients during embryogenesis (Heras *et al.* 1998) and for potent neurotoxic activity (Heras *et al.* 2008).

In previous work we identified perivitellin PV2 as the first mollusc neurotoxin genetically encoded outside the cone-snail family Conidae (Heras *et al.* 2008). Upon intraperitoneal injection in mice the toxin provoked death with signs suggesting neurological damage. We examined the possible effects of the purified toxin on the central nervous system. Clinical signs, histopathology, and immunocytochemical studies revealed damage mostly in the spinal cord of mice. Experiments showed chromatolysis and a decreased response to the Ca-binding protein calbindin D-28K associated with a significant increase of TUNEL-positive cells of the dorsal horn neurons. This was particularly evident in laminae II and III following the laminar model proposed by Paxinos and Watson (1986). These results suggest that calcium buffering and apoptosis may play a role in the neurological disorders induced by the toxin in mammalian central nervous system.

©CAB International 2011. *Poisoning by Plants, Mycotoxins, and Related Toxins* (eds F. Riet-Correa, J. Pfister, A.L. Schild, and T.L. Wierenga)

The next aim of our work was focused on the characterization of changes in carbohydrate moieties in the dorsal horn of the spinal cord of control and PV2-intoxicated mice using lectin-histochemical techniques.

Cell surface carbohydrates are likely to play important roles in the function of the nervous system in which the use of lectins, plant-derived proteins which bind specifically but non-immunologically to saccharides, has proven to be an effective approach to study the functional relevance of glycoconjugates.

## Materials and Methods

Female BALBcAnN mice, 4 weeks old, weighing 19–21 g from a colony started with a stock provided by NIH USA and bred in a specific pathogen free (SPF) environment, were employed in the experiments. All the studies performed with animals were carried out in accordance with the Guide for the Care and Use of Laboratory Animals (National Research Council 1996).

Groups of six mice were injected i.p. with a single dose of PBS (control group) or with 1 mg/kg PV2 (19 µg per mouse). After 40 h animals were anesthetized by an injection of ketamine hydrochloride (40 mg/kg i.p.) plus xylazine (8 mg/kg i.m.) and then perfused transcardially with 50 ml (PBS) followed by 50 ml of 4% paraformaldehyde in PBS.

Samples of spinal cord were carefully removed, kept in the same fixative for 3 h, and then in Bouin for 3 days. Representative 5-7 µm sections were stained with hematoxylin and eosin for histological examination of general morphology. Nine lectins were used: Con A, DBA, SBA, PNA, RCA-I, UEA-1, JAC, WGA, and sWGA (Table 1).

**Table 1.** Lectins used in this study and their major specificities.

| Acronym | Lectin | Specificity (µl/ml) | Concentration |
|---|---|---|---|
| UEA-1 | *Ulex europaeus-1* | α -L-Fuc* | 30 |
| DBA | *Dolichos biflorus,* | α-D-GalNAc | 30 |
| PNA | *Arachis hypogaea* | β-D-Gal (β 1-3) D-GalNAc | 10 |
| SBA | *Glycine maximus* | α-D-GalNAc; β –D-GalNAc | 30 |
| WGA | *Triticum vulgaris,* | ß-D-GlcNAc; NeuNAc | 30 |
| RCA-1 | *Ricinus communis* | ß-Gal | 30 |
| CON-A | *Concanavalina ensiformis* | α-D-Man; α-D-Glc | 30 |
| JAC | *Artocarpus integrifolia* | Gal ß-1-3 GalNAc | 30 |
| sWGA | *Succinyl Triticum vulgaris* | GlcNAc ß (1-4) Glc NAc | 30 |

*Fruc: Fucose; Gal: Galactose; GalNAc: N-Acetyl galactosamine; Glc: Glucose; GlcNAc: N-Acetyl glucosamine; Man: mannose; NeuNAc: Acetyl neuraminic acid (sialic acid)

We followed this protocol for the lectin histochemistry: paraffin sections were deparaffinized with xylene and endogenous peroxidase was quenched. They were then hydrated, washed in phosphate-buffered saline, and incubated with biotinylated lectins for 1 h followed by incubation for 10 min with horseradish peroxidase-streptavidin (Vector Laboratories Inc, USA) which binds to biotin with high affinity. The horseradish peroxidase was activated by incubation for 4-10 min with a buffered Tris-HCL solution (0.05M, pH 6) containing 0.02% 3',3'-diaminobenzidine tetrahydrochloride (DAB) and 0.05% $H_2O_2$. Positively-stained cells were demonstrated by the dark golden brown DAB-$H_2O_2$ reaction product.

## Results

The labeling of positively-stained lectins involved different areas of the spinal cord, nevertheless, comparisons between control and intoxicated animals showed that the most significant changes were detected in the grey matter with SBA lectin. This lectin is a galactose-binding lectin that binds to oligosaccharide structures in which the terminal residue is derived from galactose or N-acetylgalactosamine. Thus, SBA-lectin neuronal staining which were moderately positive in laminae II and III of the grey matter of spinal cord in control animals was strongly positive in the intoxicated animals. The rest of the lectins, despite showing different staining patterns in the grey matter, did not show any remarkable differences between control and intoxicated mice. UEA-1 lectin was the only one which displayed negative results for both control and inoculated animals.

## Discussion

Recently we have identified a novel toxin isolated from the eggs of the freshwater mollusc *P. canaliculata* which, upon intraperitoneal injection, produces a strong neurotoxicity in mice (Heras *et al.* 2008). After assaying egg extracts, total lipids, and several protein fractions, we found that the toxicity was associated with only one of the proteins. The biochemical, biophysical, and immunological characterization of this toxic protein indicated that it was indistinguishable from PV2, one of the major proteins of the egg perivitellin fluid. This protein was first described more than 10 years ago and a considerable body of information is available. PV2 is moderately lipidated and glycosylated and composed of two subunits of 67 and 31 kDa. (Garín *et al.* 1996; Heras *et al.* 1998; Dreon *et al.* 2004a). Interestingly, the N-terminal amino acid sequences of PV2 subunits showed no homology with other related or non-related proteins (Dreon *et al.* 2002). Moreover, immunological studies have shown that PV2 does not cross react with perivitellins from *P. scalaris*, a related snail species (Ituarte *et al.* 2008), further highlighting the singularity of this protein.

Neuronal calcium plays a key role in neuron survival as well as programed cell death (apoptosis) and pathological neuronal degeneration and it has been shown that disturbances in calcium regulation can have potentially lethal effects (Bastianelli 2003). The molecular basis of neurodegeneration in the central nervous system is related to an increased concentration of cytosolic free calcium (Ramirez-Exposito and Martinez-Martos 1998). Thus, the regulation of calcium binding protein expression is a critical factor for agents that disrupt calcium homeostasis in neurons (Lema Tome *et al.* 2006). The absence of calcium buffering proteins would lead to intracellular calcium accumulation rapidly reaching toxic amounts (Iacopino *et al.* 1990; Iacopino and Christakos 1990; Mattson *et al.* 1991; Bastianelli 2003). In fact, neurons rich in calcium-binding proteins, especially calbindin and parvalbumin, are supposed to be relatively resistant to degeneration in a variety of acute and chronic disorders (McMahon *et al.* 1998). In particular, calbinding D28k is normally expressed in restricted neuronal populations of the mammalian brain where it plays a crucial role in protecting neurons against excitotoxic insults (Lee *et al.* 2004). Neurons containing calbindin were immunostained in the superficial dorsal horn of the spinal cord in laminae II and III in control rodents, similarly to what has previously been described for rats. In these rodents it has been suggested they may be important in primary afferent processing (Yamamoto *et al.* 1989; Yoshida *et al.* 1990). These regions displayed a significantly decreased immunostaining for calbindin. Moreover, our data showed that apoptotic cell death was restricted to the same localization in selected portions of the dorsal horn as the neuronal population which depicted the reduction in calbindin expression. This is congruent with many investigations reporting that the participation of calbindin as a calcium buffer can avoid apoptosis in susceptible cells from the central nervous system (Wernyj *et al.* 1999). We therefore suggest that reduced calbindin expression may proceed to marked degeneration of neurons at the laminae II and III or may accelerate the mechanism of neuronal degeneration by further facilitating the calcium-mediated cytotoxic events. The decrease in calbindin levels may be considered an early indicator of neurological disorders induced by acute PV2 toxicity. Moreover, a lower expression of calbindin has also been reported in the aged brain and in several neurodegenerative disorders such as Parkinson's disease, Alzheimer's disease, Huntington's corea, brain ischemia, and epilepsy (Iacopino and Christakos 1990; Krzywkowski *et al.* 1995).

We have shown that PV2 toxin administration to mice may induce neuronal apoptosis within the dorsal horn of the spinal cord and that a depletion of calbindin expression is involved (Heras *et al.* 2008). However, the mechanism of toxin-induced apoptosis in these areas needs further investigation.

The fact that PV2 is a large protein excludes any direct action on the spinal cord due to low permeability of the blood-nerve barrier. However, its two subunits are small enough to be transported across the blood-nerve or blood-brain barriers and it is therefore reasonable to hypothesize that they could penetrate by any of the mechanisms postulated for proteins such as fluid-phase endocytosis (Poduslo *et al.* 1994). In addition, it has been reported that protein glycosylation as in the case of PV2 increases its permeability across the barrier (Poduslo and Curran 1994). At present we do not know whether one or both subunits or their fragments reach the CNS, but the effect is manifest in the spinal cord.

The present work enhances current knowledge on the action of PV2 in the spinal cord of mice providing data on the lectin histochemical staining patterns. The most significant changes were detected in the grey matter with SBA lectin. This lectin is a galactose-binding lectin that binds to oligosaccharide structures in which the terminal residue is derived from galactose or N-acetylgalactosamine.

Glycoconjugates with terminal galactose residues were already localized in rat sensory small neurons of the dorsal horn of the spinal cord (Streit *et al.* 1986; Silverman and Kruger 1990). These studies illustrate significant lectin-reactive cell surface carbohydrate expression in sensory cell populations, probably including a substantial proportion of nociceptive axons which transmit specific sensory information from cutaneous and muscle receptors to second-order neurons in the spinal cord. Anatomical studies have revealed that dorsal root ganglion neurons branch in a directed manner in the spinal cord (Smith 1983). These observations suggest that there may be specific cues that enable sensory axons to find their appropriate spinal targets (Vaughn and Grieshaber 1973). There is, however, increasing evidence in support of the idea that axon-target interactions are mediated by cell-surface recognition molecules (Thanos *et al.* 1984) probably involving the recognition of cell-surface oligosaccharides by complementary carbohydrate-binding proteins. The mechanisms of cell adhesion may also contribute to sensory axon guidance and synaptic connectivity and interaction (Dodd and Jessell 1986). Interestingly, the SBA positive region is specifically located in laminae II and III, the same exact region where apoptosis and calbindin depleted neurons are affected by this neurotoxin.

Further studies about the relationship between immunohistochemical and lectinhistochemical data in the dorsal grey horn in the spinal cord of mice are needed to elucidate the way in which the toxin affects the neuronal synapses and transmission of nociceptive stimuli. In addition, biophysical, molecular biology, and pharmacological analyses of the toxin are currently under way.

## Acknowledgements

HH and EJG are research career members of CONICET. MSD is a research career member of CIC and VF is a research fellow of CONICET.

## References

Albrecht EA, Carreño NB, and Castro-Vazquez A (1999). A quantitative study of environmental factors influencing the seasonal onset of reproductive behaviour in the South American apple-snail *Pomacea canaliculata* (Gastropoda: Ampullariidae). *Journal of Molluscian Studies* 65:241-250.

Albrecht EA, Koch E, Carreño NB, and Castro-Vazquez A (2005). Control of the seasonal arrest of copulation and spawning in the apple snail *Pomacea canaliculata* (Prosobranchia: Ampullariidae): Differential effects of food availability, water temperature, and day length. *Veliger* 47:169-174.

Bastianelli E (2003). Distribution of calcium-binding proteins in the cerebellum. *Cerebellum* 2:242-262.

Dodd J and Jessell TM (1986). Cell surface glycoconjugates and carbohydrate-binding proteins: possible recognition signals in sensory neurone development. *Journal Experimental Biology* 124:225-238.

Dreon MS, Lavarías S, Garín CF, Heras H, and Pollero RJ (2002). Synthesis, distribution, and levels of an egg lipoprotein from the apple snail *Pomacea canaliculata* (Mollusca: Gastropoda). *Journal Experimental Zoology* 292:323-330.

Dreon MS, Heras H, and Pollero RJ (2004a). Characterization of the major egg glycolipoproteins from the perivitellin fluid of the apple snail *Pomacea canaliculata. Molecular Reproduction and Development* 68:359-364.

Dreon MS, Schinella G, Heras H, and Pollero RJ (2004b). Antioxidant defense system in the apple snail eggs, the role of ovorubin. *Archives of Biochemistry and Biophysics* 422: 1-8.

Dreon MS, Heras H, and Pollero RJ (2006). Biochemical composition, tissue origin and functional properties of egg perivitellins from *Pomacea canaliculata. Biocell* 30:359-365.

Garín CF, Heras H, and Pollero RJ (1996). Lipoprotein of the egg perivitellin fluid of *Pomacea canaliculata* snails (Mollusca: Gastropoda). *Journal Experimental Zoology* 276:307-314.

Heras H, Garín CF, and Pollero RJ (1998). Biochemical composition and energy sources during embryo development and in early juveniles of the snail *Pomacea canaliculata* (Mollusca: Gastropoda). *Journal Experimental Zoology* 280:375-383.

Heras H, Dreon MS, Ituarte S, and Pollero RJ (2007). Egg carotenoproteins in neotropical Ampullariidae (Gastropoda: Arquitaenioglossa). *Comparative Biochemistry and Physiology* C 146:158-167.

Heras H, Frassa MV, Fernández PE, Galosi CM, Gimeno EJ, and Dreon MS (2008). First egg protein with a neurotoxic effect on mice. *Toxicon* 52:1-8.

Iacopino AM and Christakos S (1990). Specific reduction of calcium-binding protein (28-kilodalton calbindin-D) gene expression in aging and neurodegenerative diseases. *Proceedings of the National Academy of Sciences USA* 87:4078-4082.

Iacopino AM, Rhoten WB, and Christakos S (1990). Calcium binding protein (calbindin-D28k) gene expression in the developing and aging mouse cerebellum. *Molecular Brain Research* 8:283-290.

Ituarte S, Dreon MS, Pollero RJ, and Heras H (2008). Isolation and partial characterization of a new lipo-glyco-carotenoprotein from *Pomacea scalaris* (Gastropoda: Ampullariidae). *Molecular Reproduction and Development* 75:1441-1448.

Krzywkowski P, De Bilbao F, Senut MC, and Lamour Y (1995). Age-related changes in parvalbumin- and GABA-immunoreactive cells in the rat septum. *Neurobiology of Aging* 16: 9-40.

Lee BH, Lee KH, Kim UJ, Yoon DH, Sohn JH, Choi SS, Yi IG, and Park YG (2004). Injury in the spinal cord may produce cell death in the brain. *Brain Research* 1020:37-44.

Lema Tome CM, Bauer C, Nottingham C, Smith C, Blackstone K, Brown L, Hlavaty C, Nelson C, Daker R, Sola R, Miller R, Bryan R, and Turner CP (2006). MK801-induced caspase-3 in the postnatal brain: inverse relationship with calcium binding proteins. *Neuroscience* 141:1351-1363.

Mattson MP, Rychlik B, Chu C, and Christakos S (1991). Evidence for calcium-reducing and excito-protective roles for the calcium-binding protein calbindin-D28k in cultured hippocampal neurons. *Neuron* 6:41-51.

McMahon A, Wong BS, Iacopino AM, Ng MC, Chi S, and German DC (1998). Calbindin-D28k buffers intracellular calcium and promotes resistance to degeneration in PC12 cells. *Molecular Brain Research* 54:56-63.

National Research Council (1996). *Guide for the Care and Use of Laboratory Animals,* pp. 64-66. Academic Press, Washington, USA.

Paxinos G and Watson C (1986). *The Rat Brain in Stereotaxic Coordinates,* pp. 25-31, Academic Press, Sydney.

Pizani NV, Estebenet AL, and Martin PR (2005). Effects of submersion and aerial exposure on clutches and hatchlings of *Pomacea canaliculata* (Gastropoda: Ampullariidae). *American Malacological Bulletin* 20:55-63.

Poduslo JF and Curran GL (1994). Glycation increases the permeability of proteins across the blood-nerve and blood-brain barriers. *Molecular Brain Research* 23: 157-162.

Poduslo JF, Curran GL, and Berg CT (1994). Macromolecular permeability across the blood-nerve and blood-brain barriers. *Proceedings of the National Academy of Sciences USA* 91:5705-5709.

Ramirez-Exposito MJ and Martinez-Martos JM (1998). The molecular basis of neurodegenerative processes in the central nervous system. *Revista de Neurología* 26:91-100.

Silverman JD and Kruger L (1990). Selective neuronal glycoconjugate expression in sensory and autonomic ganglia: relation of lectin reactivity to peptide and enzyme markers. *Journal of Neurocytology* 19:789-801.

Smith CS (1983). The development and postnatal organization of primary afferent projections to the rat thoracic spinal cord. *Journal of Comparative Neurology* 22:29-43.

Streit WJ, Schulte BA, Balentine JD, and Spicer SS (1986). Evidence for glycoconjugates in nociceptive primary sensory neurons and its origin from the Golgi complex. *Brain Research* 377:1-17.

Thanos S, Bonhoeffer F, and Rutishauser U (1984). Fiber-fiber interaction and tectal cues influence the development of the chicken retino tectal projection. *Proceedings of the National Academy of Sciences USA* 81:1906-1910.

Vaughn JE and Grieshaber JA (1973). A morphological investigation of an early reflex patterning in developing rat spinal cord. *Journal of Comparative Neurology* 148:177-210.

Wernyj RP, Mattson MP, and Christakos S (1999). Expression of calbindin-D28k in C6 glial cells stabilizes intracellular calcium levels and protects against apoptosis induced by calcium ionophore and amyloid beta-peptide. *Molecular Brain Research* 64:69-79.

Yamamoto T, Carr PA, Baimbridge KG, and Nagy JI (1989). Parvalbumin- and calbindin D28k-immunoreactive neurons in the superficial layers of the spinal cord dorsal horn of rat. *Brain Research Bulletin* 23:493-508.

Yoshida S, Senba E, Kubota Y, Hagihira S, Yoshiya I, Emson PC, and Tohyama M (1990). Calcium-binding proteins calbindin and parvalbumin in the superficial dorsal horn of the rat spinal cord. *Neuroscience* 37:839-848.

# Chapter 84

# Zearalenone: An Estrogenic Mycotoxin with Immunotoxic Effects

I.M. Hueza, V.K. Tanabe, J.C. Benassi, L.E.R. Raspantini, and S.L. Górniak

*Research Center of Veterinary Toxicology – Department of Pathology, School of Veterinary Medicine and Animal Science, University of São Paulo, São Paulo, SP 13635-900, Brazil*

## Introduction

Zearalenone (ZEA) is a non-steroidal estrogenic mycotoxin produced by a number of *Fusarium* species of fungi, including *F. culmorum*, *F. graminearum*, and others (Caldwell *et al.* 1970). These *Fusarium* species are common soil fungi present on almost all continents and are known to infest wheat, barley, rice, maize, and other crops, contaminating foodstuffs and animal feeds worldwide (Zinedine *et al.* 2007).

The concentration of ZEA in food and feed material for animals varies from a few micrograms up to 250 mg/kg of feed material and in cereals for human consumption from 1.0 ng to 170 mg/kg of grains (Binder *et al.* 2007). ZEA is a stable compound both during storage/milling and processing/cooking food as indicated by its presence in some grain products such as bread, brewed beers, and processed feeds (Zinedine *et al.* 2007).

Natural exposure to ZEA has been implicated as a cause of hyperestrogenism in farm animals especially in pigs, the most susceptible animal species to ZEA toxicosis. It has been also found that human exposure to ZEA in Puerto Rico led to precocious pubertal changes in thousands of young children between 1978 and 1981 (Sàenz de Rodriguez 1984). Thus, both human and animal feed contamination and animal exposure to ZEA could be considered a public health concern.

ZEA and its metabolites exert their estrogenic activity when they compete with natural estrogens in order to bind to estrogenic receptors. After binding, the receptor-ZEA complex migrates to the nucleus where it binds to estrogen-responsive elements thereby activating gene transcription (Riley 1998).

Estrogen has a widespread role in animal physiology, including immune system responses. This interest grew from the recognition that autoimmune disorders such as systemic lupus erythematosus (SLE), Sjogren's syndrome, and rheumatoid arthritis affect proportionally more females than males. Thus, the aim of the present study was to verify if ZEA, like estrogen, can induce alterations in some immune parameters.

©CAB International 2011. *Poisoning by Plants, Mycotoxins, and Related Toxins*
(eds F. Riet-Correa, J. Pfister, A.L. Schild, and T.L. Wierenga)

## Material and Methods

Thirty-six ovariectomized female Wistar rats (10 weeks old) were divided into three equal groups: one control (Co), one treated daily by gavage for 28 days with 3.0 mg/kg of ZEA diluted in DMSO, and one *pairfed* (PF) group that received an amount of diet equivalent to that consumed by their ZEA-treated partners in order to determine if abnormal immune responses are due to the direct effect of ZEA or if they result from dietary deficiencies. Rats from the Co group received only vehicle in the same way over the same period. At the end of the experimental period the thymus and spleen were removed from euthanized rats and weighed. The spleen was macerated to obtain a single splenocyte suspension. Bone marrow cell suspensions were achieved by flushing the marrow cavity of the left femur of each rat. Cells were counted in a Neubauer chamber. Each thymus was fixed in 10% buffered formalin, embedded in paraffin, sectioned at 5 μm thickness, and stained with H&E. Image Pro Software (Media Cybernetic, Bethesda, MD) was used for section analysis and the ratio of cortical to medullar areas (C/M) was determined and calculated by the formula C/M = Total cortex area/[(Total area) - (Total cortex)].

Before rats were euthanized, they were anesthetized for blood collection and blood cell counts and after euthanasia peritoneal macrophages (MO) were harvested to evaluate phagocytosis, hydrogen peroxide ($H_2O_2$), and nitric oxide (NO) production.

The methods used to study MO phagocytosis were based on those described by Rabinovitch and DeStefano (1973). Briefly, a total of 200 cells were counted on each slide from each rat and the MO phagocytosis index (PI) was calculated as follows: PI = number of phagocytic activity × 100÷200 adherent cells counted, i.e. as PI = % of MO with phagocytized zymosan particles. The mean of four counts obtained from two slides of each rat was used to express PI index.

$H_2O_2$ concentration was calculated from absorbance measurements as described by Pick and Mizel (1981). Spontaneous and PMA-induced $H_2O_2$ production experiments were repeated four times for each rat in each group and the mean value of the four counts was used to determine $H_2O_2$ concentration.

Finally, NO concentration in the supernatant of cultures of MO incubated with LPS (100 ng/ml) or vehicle for 24 h were measured using the Griess reagent. In brief, 100 μl of Griess reagent (freshly prepared) was mixed with an equal amount of cell culture supernatant and then incubated at room temperature for 10 min. The absorbance of the samples was then measured at 540 nm in the Multiskan EX reader. All experiments for NO measurements were performed in triplicate.

The Student *t* test was used to compare two groups and a significant F test ($P < 0.05$) in the analysis of variance (ANOVA) was followed by Dunnett's *post hoc* test to compare controls with ≥ 2 treatment groups. The data are expressed as mean ± S.E.M.

## Results

Figure 1 shows the total food consumption and total body weight gain of ovariectomized rats. Female rats treated with ZEA for 28 days showed a decreased food intake and a significantly lower body weight gain when compared with control rats; however, rats from the PF group which received the same amount of food as those rats treated with the mycotoxin did not show any alteration in body weight gain when compared with control female rats.

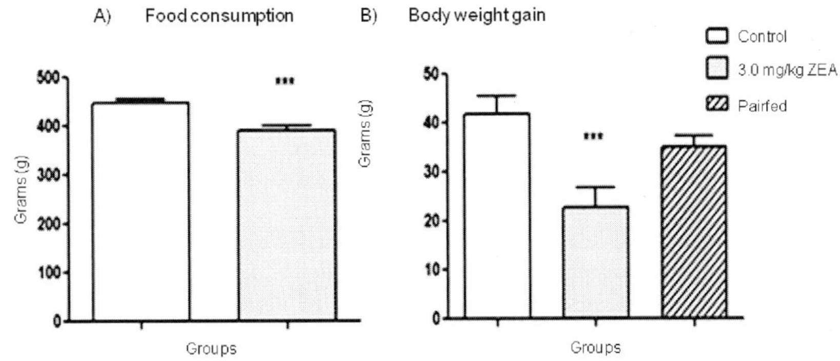


**Figure 1.** Food consumption (A) and body weight gain (B) of ovariectomized female Wistar rats exposed to 3.0 mg/kg BW of zearalenone (ZEA) for 28 days. Results represent means ± SEM. *** Significantly different from control ($P < 0.001$).

The immunopathology studies revealed a statistical reduction in thymus relative weight (Figure 2A) and an enlargement of the spleen of rats treated with 3.0 mg/kg of ZEA when compared to those lymphoid organs from control animals (Figure 2B). No other immune parameter analyzed, i.e. splenic and bone marrow cellularity or ratio of thymic cortical to medullar areas (C/M), was statistically affected by ZEA treatment. In addition, ZEA did not promote differences in macrophage activity between the groups, i.e. phagocytosis, $H_2O_2$ and NO production/release (data not shown).

Upon histologic examination the thymus from rats treated with ZEA showed reduced number of cells in the medullar region compared with thymus from untreated female rats.

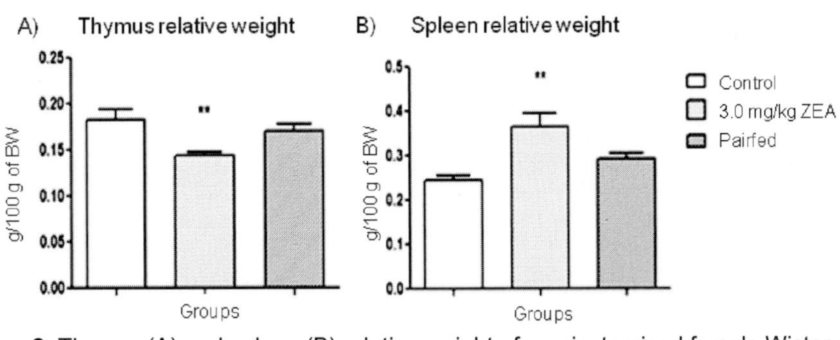


**Figure 2.** Thymus (A) and spleen (B) relative weight of ovariectomized female Wistar rats exposed to 3.0 mg/kg BW of zearalenone (ZEA) for 28 days. Results represent means ± SEM. ** Significantly different from control ($P < 0.01$).

## Discussion

Estrogens influence many physiological processes in mammals including reproduction, cardiovascular health, bone integrity, cognition, and behavior. Given this widespread role, it is not surprising that estrogen is also implicated in the development or progression of numerous diseases including autoimmune diseases (Silman and Oliver

2010). The potential role of estrogen as a contributor to this differential in disease incidence was also supported by anecdotal reports that SLE could frequently flare during pregnancy (Petri *et al.* 1991), a time when levels of estrogen are elevated.

Because ZEA has estrogenic activity we hypothesized that this compound may also induce alterations in some immune parameters. The present study clearly showed that this mycotoxin can interfere in thymus homeostasis leading to a reduction in its relative weight and a disruption in thymus follicular morphology.

It is well known that T cells mature into thymus and the interaction of thymocytes and thymus epithelial cells is essential to identify self-reactive T cells that will be eliminated by apoptosis (Nossal 1994). Considering the reduction in thymus relative weight and the disruption of its normal morphology that we found, this suggests that alterations in finely-tuned molecular interactions between thymocytes and the thymic environment can interfere in the process of central tolerance, predisposing self-reactive T lymphocytes to escape to colonize secondary organs there to be activated by self-antigens and resulting in auto-immune diseases.

The data obtained from spleen (splenomegaly) must be a result of enhanced concentration of red blood cells in the red pulp, which is a direct toxic effect of ZEA since there were no alterations in white blood cell counts.

It is important to emphasize that these results were not a consequence of reductions in body weight gain in the ZEA-treated female rats since animals from the PF group did not have the same alterations in their lymphoid organs.

The results from these studies are encouraging us to conduct further work to verify the role of ZEA in T lymphocyte development and function. Future work will require proper protocols and models to determine if ZEA can act as does estrogen to induce immunological disorders that can lead to serious diseases like autoimmune reactions in both domestic animals and human beings.

## Acknowledgements

This study was supported by grants from Fundação de Amparo à Pesquisa do Estado de São Paulo – FAPESP, Brazil (Proc. No. 06/57174-7 and 06/60397-8).

## References

Binder EM, Tan LM, Chin LJ, Handl J, and Richard J (2007). Worldwide occurrence of mycotoxins in commodities, animal feed and feed ingredients. *Animal Feed Science and Technology* 137:265-282.

Caldwell RW, Tuite J, Stob M, and Baldwin R (1970). Zearalenone production by *Fusarium* species. *Applied Microbiology* 20:31-34.

Nossal GJV (1994). Negative selection of lymphocytes. *Cell* 76:229-239.

Petri MD, Howard D, and Repke J (1991). Frequency of lupus flare in pregnancy. The Hopkins lupus pregnancy center experience. *Arthritis and Rheumatism* 34:1538-1545.

Pick E and Mizel D (1981). Rapid microassays for the measurement of superoxide and hydrogen peroxide production by macrophages in culture sing an automatic enzyme immunoassay. *Journal of Immunological Methods* 46:211-226.

Rabinovitch M and DeStefano MJ (1973). Macrophage spreading *in vitro*: I. Inducers of spreading. *Experimental Cell Research* 77:323-334.

Riley RT (1998). Mechanistic interaction of mycotoxins: theoretical considerations. In *Mycotoxins in Agriculture and Food Safety* (KK Sinha and D Bhatnagar, eds), pp. 227-253. Marcel Dekker, New York.

Sàenz de Rodriguez CA (1984). Environmental hormone contamination in Puerto Rico. *New England Journal of Medicine* 310:1741-1742.

Silman AJ and Oliver J (2010). Epidemiology of rheumatic diseases. In *ABC of Rheumatology* (A Adebajo, ed.), pp. 167-170. Wiley-Blackwell, West Sussex, UK.

Zinedine A, Soriano JM, Moltó JC, and Mañes J (2007). Review on the toxicity, occurrence, metabolism, detoxification, regulations and intake of zearalenone: An oestrogenic mycotoxin. *Food and Chemical Toxicology* 45:1-18.

# Chapter 85

# Ethanol Poisoning in Cattle by Ingestion of Waste Beer Yeast in Brazil

P.V. Peixoto[1], L.A.C. Brust[2], M.F. Brito[3], T.N. França[3], P. Malafaia[1], and C.H. Tokarnia[1]

*[1]Instituto de Zootecnica, Universidade Federal Rural do Rio de Janeiro (UFRRJ), Seropédica, RJ 23890-000, Brazil; [2]Curso de Pós-graduação em Ciências Veterinárias, UFRRJ, Seropédica, RJ 23890-000, Brazil; [3]Instituto de Veterinária, UFRRJ, Seropédica, RJ 23890-000, Brazil*

## Introduction

Recently the consumption of alcohol in Brazil and worldwide has increased. To meet this increased demand, the brewery industry has increased alcohol production which has generated large quantities of byproducts especially from beer and barley for animal feed.

In humans, the symptoms of ethanol poisoning vary directly with alcohol intake and the clinical signs are well known (Diamond 1997). However, ethanol poisoning in animals has rarely been reported. Its occurrence is associated with the ingestion of brewery or distillery residues (Humphreys 1988; Stöber 2002) or with the direct consumption of alcoholic beverages (Ratcliffe and Zuber 1977; Stöber 2002). Cases of alcohol intoxication have been described in pigs (Bell *et al.* 1950; Becker *et al.* 1954; Bruning and Yokoyama 1988; Stöber 2002), dogs (Ratcliffe and Zuber 1977; Thrall *et al.* 1984; Kammerer *et al.* 2001), and birds (Allen *et al.* 1981; Fitzgerald *et al.* 1990).

The intoxication can also occur during the digestion of certain fermented foods such as fruits (Kammerer *et al.* 2001; Stöber 2002) and even by fermentation of food during digestion (Abe *et al.* 1971). In cattle there are two citations with little details on occurrence of the intoxication from a slurry containing alcohol (Rubarth 1967; Abe *et al.* 1971). When animals ingest large amounts of ethanol, initially the nervous system is affected and later the heart and respiration. Acute intoxication is characterized by periods of excitement followed quickly by collapse, coma, and death by respiratory paralysis (Humphreys 1988).

In cattle, alcohol poisoning has been reproduced in calves through the direct administration of ethanol orally or through intestinal fistula and also by feeding of milk substitutes which ferment in the gastrointestinal tract to produce toxic levels of ethanol (Abe *et al.* 1971). Another experiment consisted in giving live beer yeast through a ruminal fistula to adult cattle (Bruning and Yokoyama 1988); clinical signs were characterized by loss of appetite, reluctance to maintain posture, lethargy, stupor (Bruning and Yokoyama 1988), incoordination (Abe *et al.* 1971; Bruning and Yokoyama 1988), spastic movements,

©CAB International 2011. *Poisoning by Plants, Mycotoxins, and Related Toxins* (eds F. Riet-Correa, J. Pfister, A.L. Schild, and T.L. Wierenga)

depression, sleepiness, slow respiratory frequency, and alcoholic odor in the exhaled air (Abe *et al.* 1971). Symptoms of severe alcohol intoxication seem similar to those observed in humans. Other observed signs were frequent tympanism and difficulty in coordinating movement and mastication. Ethanol is easily absorbed in the gastrointestinal tract and can be detected in the plasma 30 min after ingestion (Abe *et al.* 1971).

Ethanol intoxication of cattle is characterized by an excitement state followed by a depressive phase (Hibbs *et al.* 1984, 1986; Stöber 2002). Cattle may demonstrate avidity for liquids or food that contains ethanol. In the first phase rumination is absent and there is ruminal paralysis, tympanism, oscillation and staggers or abrupt falls and recumbency, reduction of lactation (milk production), reddened mucous membranes, engorged episcleral vessels, tachycardia, heart arrhythmia, and dyspneic breathing with alcoholic odor at expiration. In severe cases, colic (kicks to the abdomen) and very aggressive behavior may occur. The depressive stage is characterized by recumbency. The snout is dry and the body surfaces cold and insensitive which may be accompanied by temporary convulsive contractions of the neck or members, grinding of the teeth, wailing, and stertors.

In cases of severe alcohol intoxication there is coma and death due to respiratory paralysis; those animals that survive recover slowly within 2 days. Animals that received doses of 4.5 kg of waste beer yeast showed regression of the clinical signs within 6 h and had normal appetite 12 h after administration of the product (Bruning and Yokoyama 1988); those that received higher doses recovered more slowly.

Reports on macro- and microscopic alterations in animals intoxicated by ethanol are scarce. In cattle a characteristic ethyl odor in the ruminal content is noticed and a rosy coloration of the mucous membrane in the ventral area of the rumen; sometimes hemorrhagic abomasitis can be observed with hemorrhages in the subserosa and congestion in the encephalon and parenchymatous organs (Stöber 2002).

The waste beer yeast, a byproduct in beer production, has been used to feed cattle in confinement on several farms in the State of Rio de Janeiro and occasionally has caused fatal poisoning. The objective of this paper is to report the occurrence of several cases of intoxication by alcohol contained in waste beer yeast in cattle in the municipality of Piraí, State of Rio de Janeiro, as well as to describe the experimental reproduction of that condition.

## Material and Methods

The observation of a natural case of ethanol poisoning occurred during a visit to a farm in Piraí that used waste beer yeast as part of the ration in the confinement feeding of cattle for slaughter. Information about the occurrence of poisoning by beer yeast was collected through epidemiologic questionnaires on farms in the municipalities of Volta Redonda and Barra Mansa where yeast is routinely used for animal feeding.

The preliminary study was done in the Sector of Pathology, 'Projeto Sanidade Animal Embrapa/UFRRJ', in Seropédica, Rio de Janeiro (Brust 2009, personal communication). Two cattle weighing 242.5 kg (Bovine 1) and 124.0 kg (Bovine 2) were used. The waste beer yeast was collected directly from the stockpiling tanks of the product on the farm or from the truck-tank, deposited in 50 l plastic containers, and transported to the site of the experiment where it was stocked for 2 days before its administration. Chemical analyses performed on some 'yeast' samples in the Analytical Laboratory of Foods and Beverages/UFRRJ revealed an alcohol content of about 5%. The 'yeast' was given to each animal through a ruminal probe in doses of 20 l (82.47 ml/kg, Bovine 1) and 17 l (137.09

ml/kg, Bovine 2). The animals were examined before and after the administration of the yeast with special attention to possible neurological alterations, posture, attitude, appetite, heart and respiratory frequency, odor of the exhaled air, mobility, degree of ruminal distention, rectal temperature, and aspect of the mucous membranes, feces, and urine. Urine and blood samples were collected before and during the experiment to evaluate the ethanol concentrations by gas chromatography, performed at the Instituto Hermes Pardini/Minas Gerais.

## Results and Discussion

We found no information on the use of waste beer yeast in ruminant rations. However, brewery residue as a cause of ethanol poisoning in cattle was reported by Stöber (2002) and Rubarth (1967). In the State of Rio de Janeiro, moreover, several farms have used this byproduct as cattle feed and all have had occasional cases of intoxication, sometimes fatal.

We verified that cattle are intoxicated by waste beer yeast mainly when they consume more than they are habituated to. The lack of waste beer yeast for 3 or 4 days is described by farmers as the main factor responsible for intoxication. The lack of alcohol results in animals consuming the product avidly. This avid consumption may increase the risk of intoxication; however the risk may decrease over time as the alcohol concentration decreases from evaporation.

The clinical signs we observed began about 30 min after the ingestion of large quantities of the product and are characterized by obnubilation, staggering, falls, atypical postures, tympanism, and death in minutes or a few hours. According to one farmer, the animals die quickly after showing severe tympanism. We observed a natural case of intoxication by waste beer yeast during a visit to one farm where at least 19 cattle died from ethanol poisoning during the last 8 years. The milk cow was accustomed to ingesting 2 l of waste beer yeast daily then had unlimited access. The cow developed ataxia, laid down suddenly, could not get up, and became obnubilated with muscular tremors, deep breathing, and tachypnea. Later, the cow got up, showed mild ataxia, incoordination in the front quarter, and elevated tail and recovered during the day.

Under experimental conditions, clinical signs were seen a few minutes after the administration of the product; at the beginning the animals flexioned slightly the digits and had moderate instability which deteriorated to stumbling and falling into sternal recumbency with an inability to stand up. Animals were unstable after being assisted to stand.

The clinical examination revealed apathy, obnubilation, closed eyelids, increased heart rate (104 bpm), reduced rumen movements, light tympanism, and pasty feces. When in sternal recumbency, pendular movements of the head were observed with difficulty in maintaining the head upright. When in lateral recumbency, there was horizontal nystagmus that disappeared when the animal was put into a sternal position. The exhaled air and the eructed gas had an alcohol odor. The signs persisted for almost 18 h then gradually subsided with recovery the following day.

The clinical signs observed in cattle intoxicated naturally and experimentally by waste beer yeast are similar to those mentioned in cattle intoxicated experimentally by ethanol (Abe *et al.* 1971; Bruning and Yokoyama 1988; Stöber 2002). Death in natural cases of ethanol poisoning have been attributed to severe bloat. Although mentioned in the literature, bloat has not been thought to be primarily responsible for cattle death (Abe *et al.* 1971; Bruning and Yokoyama 1988; Stöber 2002). We observed only slight to moderate bloat in the two experimental cattle. Severe bloat may be more problematic if exacerbated by recumbence (Radostits *et al.* 2002); anecdotal accounts suggest bloat is a greater problem when waste yeast contains a higher alcohol concentration.

Even though the use of waste beer yeast for cattle feed is increasing in the State of Rio de Janeiro, cases of intoxication are not common. Losses can be avoided if livestock producers limit animal access especially if animals are without this material for several days. In addition, dilution of beer yeast with 50% water prevents intoxication.

Analyses of blood from the two animals revealed that ethanol levels in plasma rose from 0.8 mg/dl and 0.3 mg/dl (before administration of the beer yeast) to 253.4 mg/dl and 503.3 mg/dl, respectively, and in the urine from 11.4 mg/dl and 3.3 mg/dl (before administration of the product) to 104.2 mg/dl and 137.6 mg/dl, respectively, after administration. In calves intoxicated by ethanol the levels detected in plasma varied from 7.1 mg/dl to 21.2 mg/dl (Bruning and Yokoyama 1988). The higher levels in our study correlate with the severity of clinical signs exhibited by the animals (Bruning and Yokoyama 1988).

Many authors consider that spontaneous alcoholism does not exist among animals as proven by the rare occurrence and the resistance to ethanol ingestion (Woods and Winger 1974). We observed, however, that cattle seem to like alcohol-containing food as in the case of the waste beer yeast (with up to 5% of ethanol); their behavior when given access to the product left no doubt that the cattle had developed a special affinity for the waste beer yeast.

## Conclusion

Although waste beer yeast is being used with success in feeding cattle, the use of the yeast has the potential to intoxicate this species. Preventive measures such as avoiding interruption of the beer yeast supply, controlling animal access after periods of deprivation, and diluting the product with water can avoid or minimize that risk.

## References

Abe RK, Morrill JL, Bassett R, and Oehme FW (1971). Ethanol intoxication in calves fed certain milk replacers. *Journal Dairy Science* 54(2):252-257.

Allen NK, Aakhus-Allen SRA, and Walser MM (1981). Toxic effects of repeated ethanol intubations to chicks. *Poultry Science* 60(5):941-943.

Becker DE, Nesheim MC, Terrill SW, and Jensen AH (1954). Factors in the formulation of a semi-synthetic diet for amino acid studies with the pig. *Abstract Journal Animal Science* 13:975.

Bell JM, Williams HH, Loosli JK, and Maynard LA (1950). The effect of methionine supplementation of a soybean oil meal-purified ration for growing pigs. *Journal Nutrition* 40:551.

Bruning CL and Yokoyama MT (1988). Characteristics of live and killed brewer's yeast slurries and intoxication by intraruminal administration to cattle. *Journal Animal Science* 66:585-591.

Diamond I (1997). Alcoolismo e uso abusivo de etanol. In *Cecil/Tratado de Medicina Interna,* 20th edn (JC Benett and FC Plum, eds), vol. 1, pp. 55-58. Guanabara Koogan, Rio de Janeiro.

Fitzgerald SD, Sullivan JM, and Everson RJ (1990). Suspected ethanol toxicosis in two wild cedar waxwings. *Avian Diseases* 34:488-490.

Hibbs CM, Thilsted JP, Robb J, and Anspaugh V (1984). Ethanol toxicosis in cattle. *Proceedings of Annual Meeting of American Association of Veterinary Laboratory Diagnosticians* 27:413-416.

Hibbs CM, Smith GS, Hallford DM, Thilsted JP, Robb J, Trujillo P, and Anspaugh V (1986). Accidental and experimental ethanol toxicosis in cattle. *Proceedings of the 14th World Buiatrics Congress*, pp. 733-737. Dublin.

Humphreys DJ (1988). Organic compounds, III: Miscellaneous. In V*eterinary Toxicology*. 3th edn (DJ Humphreys, ed.), 183 pp. Baillière Tindall, London, UK

Kammerer M, Sachot E, and Blanchot D (2001). Ethanol toxicosis from the ingestion of rotten apples by a dog. *Veterinary and Human Toxicology* 43(6):349-350.

Radostits OM, Gay CC, Blood DC, and Hinchcliff KW (2002). Doenças do trato alimentar II. In *Clínica Veterinária,* 9th edn (OM Radostits, CC Gay, DC Blood, and KW Hinchcliff, eds), pp. 235-310. Guanabara Koogan, Rio de Janeiro.

Ratcliffe RC and Zuber RM (1977). Acute ethyl alcohol poisoning in dogs. *Australian Veterinary Journal* 53:48-49.

Rubarth S (1967). Leber und gallenwege. In *Joest's Handbuch der Speziellen Pathologischen Anatomie der Heustiere,* 3rd edn (J Dobberstein, G Pallaske, and H Stünzi, eds), p. 128. Paul Parey Verlag, Berlin.

Stöber M (2002). Äthylalkoholvergiftung. In *Innere Medizin und Chirurgie des Rindes,* 4th revd edn (G Dirksen, HD Grunder, and M Stöber, eds), 1132 pp. Blackwell Verlag GmbH, Berlin.

Thrall MA, Freemyer FG, Hamar DW, and Jones RL (1984). Ethanol toxicosis secondary to sourdough ingestion in a dog. *Journal American Veterinary Medical Association* 184:1513-1514.

Woods JH and Winger GD (1974). Alcoholism and animals. *Preventive Medicine* 3(1):49-60.

# Chapter 86

# Immunotoxic and Toxic Evaluation of Subchronic Exposure to Saxitoxin in Rats

F. Pípole[1], A.O. Latorre[1], L.R. Carvalho[2], and I.M. Hueza[1]

[1]*Department of Pathology, School of Veterinary Medicine and Animal Sciences, University of São Paulo, São Paulo, SP 13635-900, Brazil;* [2]*Phycology Section, Botany Institute of São Paulo, São Paulo, SP 04301-902, Brazil*

## Introduction

Cases of saxitoxin (STX) poisoning have been reported since the 19th century (Combe 1828 in Permewan 1888; Gessner *et al.* 1997). This toxin belongs to the group of paralytic shellfish poisoning (PSP) and can be synthesized by different organisms: in the oceans by dinoflagellates, in freshwater by several cyanobacteria genus, and on the land by some frogs (Llewellyn 2006; Van Apeldoorn *et al.* 2007). Because exposure to these toxins and cases of toxicosis occur in both animals and humans they are a public health concern.

There are approximate 30 variants of PSP toxins and all of them are toxic to mammals (Llewellyn 2006). The toxicity of individual PSP variants is often compared to pure STX and expressed as saxitoxin equivalents (Shimizu 1987). The differences in toxicity of each compound are related to the degree of sulfation of the molecule (Table 1) so they can be divided into three groups: a nonsulphated class which includes the most potent toxins such as STX; those that have only one sulfate group with moderate to high toxicity such as gonyautoxins; and C1-C4 toxin compounds which contain two sulfates and are considered the least toxic category (Oshima 1995). The analogs of STX have the ability to interconvert under certain conditions.

A measure of the lethality of STX and its derivatives in mice under standard conditions is expressed in mouse units (MU). A MU (approximately 0.18 μg of saxitoxin) is defined as the minimum amount of toxin that kills a 20 g mouse in 15 min when 1 ml of the extract is injected intraperitoneally (Schantz 1984, 1986).

The toxicity of STX is very severe. As little as 1 nmol/kg is enough to kill some animals when it is administered directly to their bloodstream. Oral administration is approximately 300 to 900-fold less toxic than intravenous injection depending upon the experimental animal model employed. Subcutaneous injection of STX is equipotent to intraperitoneal injection in mice (Table 2). After ingestion of food or water containing STX, the toxins are readily absorbed through the gastrointestinal mucosa. Once the toxin is absorbed, the severity and progression of toxicosis depends on the amount of toxin ingested and the rate of clearance from the body (Kao and Nishiyama 1993).

©CAB International 2011. *Poisoning by Plants, Mycotoxins, and Related Toxins* (eds F. Riet-Correa, J. Pfister, A.L. Schild, and T.L. Wierenga)

**Table 1.** Toxicity of some STX analogs (Rezanka and Dembitsky 2006). The relative toxicity of each compound is measured according to the degree of sulfation of the molecule. Saxitoxin (STX) is the reference compound.

| STX analogs | Net charge | Relative toxicity |
|---|---|---|
| Decarbamoylsaxitoxin deSTX | +2 | 0.513 |
| Decarbamoylneosaxitoxin deneoSTX | +2 | --- |
| Decarbamoylgonyautoxin deGTX2 | +1 | 0.651 |
| Decarbamoyl gonyautoxin deGTX1 | +1 | --- |
| Decarbamoyl gonyautoxin deGTX3 | +1 | 0.754 |
| Decarbamoyl gonyautoxin deGTX4 | +1 | --- |
| Saxitoxin STX | +2 | 1.000 |
| Neosaxitoxin neoSTX | +2 | 0.924 |
| Gonyautoxin GTX2 | +1 | 0.359 |
| Gonyautoxin GTX1 | +1 | 0.994 |
| Gonyautoxin GTX3 | +1 | 0.638 |
| Gonyautoxin GTX4 | +1 | 0.726 |
| Gonyautoxin GTX5 | +1 | 0.064 |
| Gonyautoxin GTX6 | +1 | --- |
| Epigonyautoxin GTX8 | 0 | 0.006 |
| C3 | 0 | 0.013 |
| Gonyautoxin GTX8 | 0 | 0.096 |
| C4 | 0 | 0.058 |

**Table 2.** $LD_{50}$ toxicity values for STX administered via different routes to vertebrates (Llewellyn 2006).

| Animal | Administration route | $LD_{50}$ (μg/kg) | $LD_{50}$ (nmol/kg) |
|---|---|---|---|
| Cat | Oral | 254 | 844 |
| Chicken | Intravenous | 3 | 1 |
| Dog | Oral | 181 | 601 |
| Guinea pig | Oral | 135 | 449 |
| Mouse | Intraperitoneal | 8-0 | 26-33 |
| | Intravenous | 3.4-8.5 | 11-28 |
| | Subcutaneous | 132 | 43 |
| | Oral | 263 | 874 |
| Pigeon | Oral | 91 | 302 |
| Rabbit | Intravenous | 4 | 1 |
| | Oral | 181 | 601 |
| Rat | Intraperitoneal | 10.5 | 35 |
| | Intramuscular | 7 | 23 |
| | Oral | 192 | 638 |

The mechanism by which STX promotes its toxic action, which has been known for a long time, consists of blocking voltage-gated sodium channels of mammalian neurons. This causes different symptoms such as paralysis, hypotension, dyspnea, and respiratory failure. The positive charge of the guanidine groups of STX binds to the negative charges of the carboxyl groups at the opening of the pore of the sodium channels on the extracellular side of the plasma membrane in nerve and muscle cells. Thus, the toxins block the influx of sodium through the channel (Llewellyn 2006).

STX is also reported to bind to calcium and potassium channels and interfere with neuronal nitric oxide synthase (Llewellyn 2006). In addition, it is suggested that STX can also inhibit calcium release-activated $Ca^{2+}$ channels (CRAC) present on excitable cells like neurons and muscle fibers (Su *et al.* 2004). Despite the fact that cells from the immune system do not have excitable properties as neuronal or muscle cells have, they require a great influx of calcium to be activated and proliferate. Both kinds of immune cells, T and B lymphocytes, employ CRAC channels to increase intracellular calcium levels. Thus, the purpose of the present study was to evaluate the toxic and immunotoxic effects of subchronic exposure of saxitoxins to rats.

## Material and Methods

Male adult Wistar rats (10 weeks old) were obtained from the USP colony in the Department of Pathology in the School of Veterinary Medicine and Animal Science. All rats received food and water *ad libitum* and were maintained under controlled conditions of temperature (22-25°C), relative humidity (50-65%), and lighting (12 h/12 h light/dark cycle). Food consumption and body weight were measured every other day. Experiments were carried out in accordance with the ethical principles for animal research adopted by the Bioethics Committee of the School of Veterinary Medicine and Animal Science, University of São Paulo.

Forty rats were randomly divided into four equal groups and treated with pure STX as follows: control (0.0), S10 (10 μg/kg BW), S30 (30 μg/kg BW), and S90 (90 μg/kg BW). They were treated by gavage for 28 days. On experimental day 29, rats were killed in order to collect spleen and thymus to evaluate their relative organ weight and cellularity and bone marrow tissue was harvested for cellularity analysis. The *in vitro* proliferation of the lymphocytes was also evaluated. In addition, sample tissues of CNS, thymus, spleen, mesenteric lymph nodes, liver, and kidneys were harvested for histopathological evaluation.

## Results

In the present study, no statistically significant differences ($P < 0.05$) were observed in body weight gain and food intake (Figure 1) among groups of animals. In addition, the analysis of lymphoid organs and lymphocyte proliferation from rats treated with STX did not show any alterations when compared with data obtained from the untreated group of animals (Figure 2). Moreover, histological analyses did not reveal any morphological alterations.

## Discussion

The results of this study revealed that with the doses used in this experiment STX did not promote any toxic effects on rats treated orally for 28 days. Even with the higher dose employed (90 μg/kg BW) which is approximately half the oral $LD_{50}$ in rats (Llewellyn 2006) we did not find any alterations in the parameters evaluated.

In order to verify that the STX used in this experiment was indeed toxic, three other rats were treated by gavage with the oral $LD_{50}$ dose of STX in rats. These animals died within a period of 20 min (data not shown).

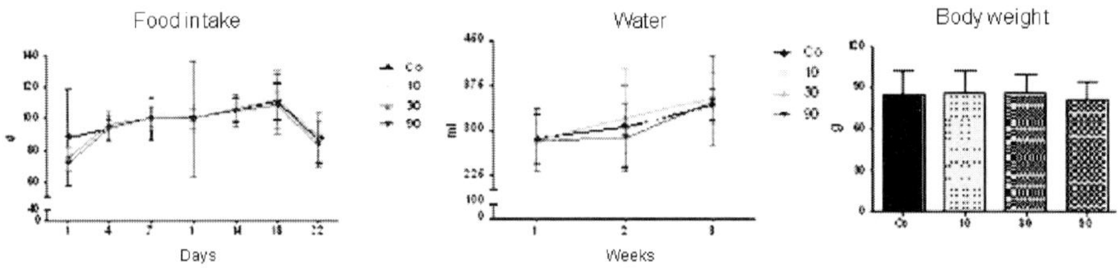


**Figure 1.** Analysis of food intake, water, and body weight gain of rats treated with 10, 30, and 90 µg/kg of saxitoxin by gavage for 28 days.

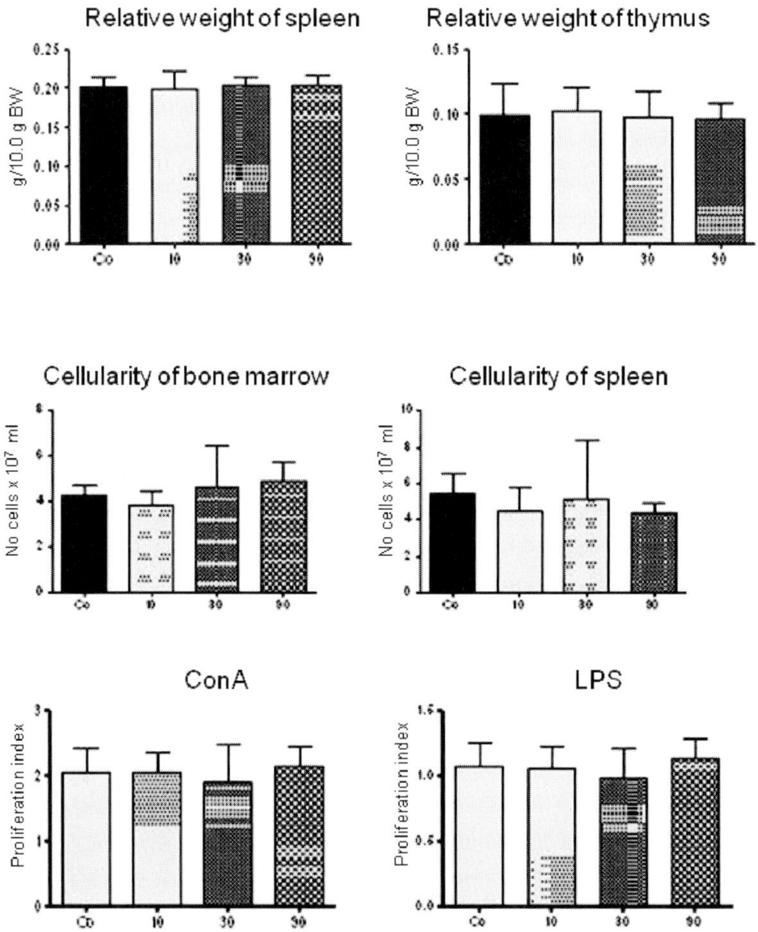


**Figure 2.** The graphics represent the relative weight of thymus and spleen, the cellularity of spleen and bone marrow, and proliferation of B and T lymphocytes to different stimulus (con-A and LPS) in rats treated with 10, 30, and 90 µg/kg of saxitoxin by gavage for 28 days.

Despite the fact that little is known about the toxicokinetics of STXs, it is known that STXs are metabolized by the oxidation of the tetrahydropurine nucleus in liver, kidney, and lungs to neosaxitoxins, a less toxic compound (García *et al.* 2009). However, there is no information about the rate of clearance of STX from the body which would be valuable information to assess the toxicity of STX in chronically exposed animals. Consequently, toxicokinetic studies need to be performed to determine the rate of clearance of STX from the body. Additionally, studies need to be performed to determine if STX, similar to phenobarbital (Maranhão 2005), can induce enzymatic activity which would promote the detoxification of STX.

## References


García C, Navarro AR, Diaz JC, Torres R, and Lagos N (2009). Evidence of *in vitro* glucuronidation and enzymatic transformation of paralytic shellfish toxins by healthy human liver microsomes fraction. *Toxicon* 53:206-213.

Gessner BD, Bell P, Doucette GJ, Moczydlowski E, Poli MA, Dolah PV, and Hall S (1997). Hypertension and identification of toxin in human urine serum following a cluster of mussel associated Paralytic Shellfish Poisoning Outbreaks. *Toxicon* 35(5):711-722.

Kao CY and Nishiyama A (1993). Paralytic shellfish poisoning. In *Algal toxins in seafood and drinking water* (IR Falconer, ed.), pp. 75-86. Academic Press, San Diego, California.

Llewellyn LE (2006). Saxitoxin, a toxic marine natural product that targets a multitude of receptors. *Natural Product Reports* 23:200-222.

Maranhão MVM (2005). Anesthesia and cerebral palsy. *Revista Brasileira de Anestesiologia* 55(6):680-702.

Oshima Y (1995). Post-collumn derivatization HPLC methods for Paralytic Shellfish Poisons. In *Manual Harmful Marine Algae* (GM Hallegraeff, DM Anderson, and AD Cembella, eds), pp. 81-94. UNESCO Publishing, Paris, France.

Parmewan GR (1888). Fatal case of poisoning by mussels, with remarks on the action of the poison. *Lancet* 2:568.

Rezanka T and Dembitsky VM (2006). Metabolites Produced by Cyanobacteria Belonging to several Species of the Family Nostocaceae. *Folia Microbiologica* 51(3):159-182.

Schantz EJ (1984). Histirical Perspective on Paralytic Shellfish Poison. In *Seafood Toxins* ACS Series (EP Ragelis, ed.), pp. 99-111. American Chemical Society.

Schantz EJ (1986). Chemistry and biology of saxitoxins and related toxins. *Annals of the New York Academy of Sciences* 479:15-23.

Su AI, Wiltshire R, Batalov S, Lapp H, ching KA, Block D, Zhang J, Soden R, Hayakawa M, Kreiman G, Cooke MP, Walker JR, and Hogenesch JB (2004). A gene atleas of the mouse and human protein-encoding transcriptomes. *Proceedings of the National Academy of Sciences USA* 101:6062-6067.

Shimizu SE (1987). Dinoflagelate toxins. In *Biology of Dinoflagelates* (JFR Taylor, ed.), pp. 282-315. Blackwell Scientific, Oxford.

Van Apeldoorn ME, Van Egmond HP, Speijers GJA, and Bakker GJI (2007). Toxins of Cyanobacteria. *Food & Nutrition Research* 51:7-60.

# Chapter 87

# *Geitlerinema unigranulatum* (Cyanobacteria) Extract Induces Alterations in Microcirculation and Ischemic Injury

C.R. Dogo[1,2], F.M. Bruni[3], C.L. Sant'Anna[2], M. Rangel[4], C. Lima[3], L.R. de Carvalho[2], and M. Lopes-Ferreira[3]

*[1]Post-Graduate Program in Plant Biodiversity and Environment, Botanic Institute of São Paulo, Brazil; [2] Phycology Section, Botanic Institute of São Paulo, Av. Miguel Estéfano, 3687 Água Funda 04301-902 – São Paulo/SP – Brazil; [3] Center for Applied Toxinology, Butantan Institute; [4] Laboratory of Immunopathology, Butantan Institute, Av. Vital Brasil, 1500 05503-900 – São Paulo/SP – Brazil*

## Introduction

Cyanobacteria are photosynthetic prokaryotic organisms able to inhabit any environment with traces of light and humidity, but they prefer to grow in freshwater. Cyanobacteria produce a wide range of special metabolites, especially cyanotoxins. These toxins are classified according to their effect on mammals: hepatotoxins; neurotoxins; cytotoxins; and dermatotoxins.

Hepatotoxins are typically microcystins which are cyclic heptapeptides formed by five amino acids that are common to all analogs and two variable amino acids that define each type. Other hepatotoxins are nodularins which are pentapeptides that have the same effect as the microcystins because they have the same toxic amino acid in their structures: ADDA [3-amino,2,6,8-trimethyl,9-methoxy,10-fenildeca-(4,6-dien)-dienoic acid] (Carvalho *et al.* 2008).

The neurotoxins consist of saxitoxins and anatoxins, all of which act quickly. The saxitoxins, known as paralytic shellfish poisons (PSP), act by blocking sodium channels and nerve transmissions leading to paralysis and death due to respiratory failure (Van Apeldoorn *et al.* 2007; Carvalho *et al.* 2008). Anatoxin-a and its variants imitate the effects of acetylcholine and anatoxin-a(S) inhibits the enzyme acetylcholinesterase causing convulsions, paralysis, salivation, and death by respiratory failure (Wiegand and Pflugmacher 2005; Carvalho *et al.* 2008).

Unlike the neurotoxic alkaloids, the amino acid β-methyl-L-amino-alanine (BMAA) produced by cyanobacteria accumulates in the superior frontal gyrus tissues and has a chronic and neurodegenerative effect causing symptoms similar to Parkinson's disease or amyotrophic lateral sclerosis (Cox *et al.* 2003; Murch *et al.* 2004).

©CAB International 2011. *Poisoning by Plants, Mycotoxins, and Related Toxins* (eds F. Riet-Correa, J. Pfister, A.L. Schild, and T.L. Wierenga)

The cytotoxin cylindrospermopsin and its variants are cytotoxic guanidine alkaloids that affect mainly the liver, heart, kidneys, and lungs. These cyanotoxins are peculiar because of their slow action. In mouse bioassays 5 to 7 days are necessary before observing the lethal effect (Van Apeldoorn *et al.* 2007; Carvalho *et al.* 2008).

The lipopolysaccharides (LPS) are dermatotoxins and are cell wall components common to all cyanobacteria. When they come into contact with the skin, these toxins cause allergies and skin irritation and, if ingested, cause neutropenia, thrombocytopenia, and changes in glucose levels (Wiegand and Pflugmacher 2005; Carvalho *et al.* 2008).

Cyanobacterial blooms are very common in freshwaters and are maintained by environmental factors such as light, temperature, and concentrations of certain nutrients (nitrogen and phosphorus). These blooms form dense layers of cells on the surface of the water and cause several ecological problems. Their capacity for producing toxins makes cyanobacteria a threat to animals and human health. This risk is recognized by specific legislation that established the monitoring of cyanobacteria and its toxic metabolites in water supplies, seafood, and fisheries (Falconer *et al.* 1999).

Due to the potential risk to human health a great deal of attention has been paid to the bloom-forming cyanobacterial genera such as *Microcystis*, *Cylindrospermopsis*, *Aphanizomenon*, *Anabaena*, and *Planktothrix*. Little is known about non-bloom-forming genera of cyanobacteria commonly present in freshwater such as *Geitlerinema*.

Zagatto *et al.* (1998) evaluated the toxicity of two strains of *Geitlerinema amphibium* isolated from drinking water supplies in São Paulo City, Brazil, and concluded that both were toxic ($LD_{50}$ 687 mg/kg BW). The toxin(s) produced by these strains were not identified. Based on a bioassay using mice Dogo and Carvalho (2006) showed that the SPC 920 *G. unigranulatum* strain isolated from the same drinking water supply (Guarapiranga Reservoir) is also toxic. However, the symptoms shown by the infected mice suggest that the cyanotoxin produced by this strain is different from all other known cyanotoxins, indicating the need for more studies on this organism and its toxins.

The mouse bioassay is a standard assay to evaluate the toxicity of strains or cyanobacterial blooms to mammals (Harada *et al.* 1999). However, even though it is helpful for classifying cyanotoxins this assay does not allow observation of reactions that occur during the organism intoxication (Carvalho 2006).

Intravital microscopy is a technique that allows one to see the microcirculation of mice *in vivo*. It is possible to observe the dynamics of physiological and pathological events occurring in microvessels including the figurative elements of blood, components of plasma, hemodynamic changes, and morphological changes of vascular walls (Raud and Lindborn 1994). Since the area studied using this technique is the cremaster muscle, it is possible to determine changes in muscle fibers (Clissa *et al.* 2006; Conceição *et al.* 2006; Magalhães *et al.* 2006; Junqueira *et al.* 2007). These data are valuable tools for determining the identity of the toxic agent and for clarifying the mechanism of toxin action.

In this study we observed the effects of toxic methanolic extract of *G. unigranulatum* (SPC 920) on the microcirculation and muscle fibers of mice using intravital microscopy.

## Material and Methods

### Cyanobacterium

The strain SPC 920-*G. unigranulatum* belonging to the Cyanobacteria Culture Collection of Institute of Botany was isolated from the Guarapiranga Reservoir in São

Paulo, Brazil, in August 2002. The strain was kept under controlled conditions: ASM-1 medium, temperature 22 ±1°C, irradiance 15-20 μmol/m$^2$/s and continuous illumination.

**Preparation of methanolic extract from SPC 920**

The biomass of *G. unigranulatum* (10 l) was dried and extracted with MeOH/$H_2O$ 75:25 v/v (5×) by exposure to ultrasound (40×30 s, 50 W) and the extract was centrifuged at 1045 g for 50 min. The supernatant was collected, concentrated under vacuum, and the methanolic extract (crude extract) was weighed (Fastner *et al.* 1998).

**Animals**

In the days preceding each experiment, the male mice (*Swiss* strain weighing 18-22 g) were kept in the vivarium of the Special Laboratory of Applied Toxinology of the Institute Butantan, São Paulo, Brazil. The animals were in groups of five per box with water and ration *ad libitum*. The boxes were kept under controlled conditions: temperature 23 ±1°C, 12/12 h light/dark cycles, and a ventilation system.

**Bioassay**

To confirm the toxicity of the strain aliquots of lyophilized methanolic extract, 20, 16.5, 10, and 5 mg) were suspended in saline and injected intraperitoneally (ip) in male mice. Symptoms were observed and upon death postmortem examination was performed (Harada *et al.* 1999).

**Analysis by intravital microscopy**

The mice were anesthetized with sodium pentobarbital (Hypnol® Cristália, 50 mg/kg) and kept on a plate with a constant temperature of 37˚C. The cremaster muscle was exposed by surgical manipulation of the testicles. After exposure the muscle was set on a transparent area of the plate positioned on the 'chariot' of the optical microscope. The muscle was kept moist by irrigation with 0.15 M PBS (phosphate buffer saline) (Lomonte *et al.* 1994). These observations required the use of an optical microscope, the Image A.1 Carl-Zeiss, attached to a camera, the AxioCam ICcI.

**Administration of the methanolic extract of strain SPC 920**

*Topical*

Three doses of methanolic extract (20, 40, and 120 μg) diluted in 20 μl of sterile saline solution were applied to the cremaster muscle of the mice. Each treatment was tested in triplicate (*n*=3). The observations were undertaken immediately after administration.

*Intraperitoneal route*

The doses used were 6, 125, 250, 500, and 1000 mg/kg BW and the analysis carried out after periods of 30 and 120 min after the inoculation. Each dose was tested in triplicate (*n*=3). Sterile saline solution was used as a negative control for all doses and times.

## Results and Discussion

Intravital microscopy is widely used in studies of toxins of poisonous and venomous animals (Lopes-Ferreira *et al.* 2002; Conceição *et al.* 2007) but this study represents the first time it has been used to assess the effects of cyanotoxins *in vivo*. These observations are important because responses to many infectious and toxic agents begins in the microcirculation.

It was observed that crude extract doses of 20 and 40 µg do not cause changes in the microcirculatory system. A dose of 120 µg caused venular stasis immediately after its application but this effect was only temporary. Based on these results, tests were developed to verify the dose- and time-dependence by intraperitoneal application.

The first dose injected (1000 mg/kg BW) caused death in the animals at times ranging from 50 min to 48 h. Intravital microscopy observation showed that this dose caused venular stasis and thrombus formation with subsequent involvement of arterioles; in topical administration these effects were not observed. These lesions were very clear at both periods of observation (30 and 120 min).

The dose of 500 mg/kg BW caused death in mice in up to 2 hours. This result is somewhat surprising considering that mice treated with doses of 1000 mg/kg BW sometimes lived as long as 48 h. This discrepancy might be explained by different susceptibilities of animals used in the tests or by variation in the amount of toxic substance present in different extracts of *G. unigranulatum*. This result occurs even with well known cyanotoxins (Rapala *et al.* 1997; Tonk *et al.* 2005). The doses of 250 and 125 mg/kg did not cause death in the animals. However, these doses caused the same changes in the microcirculatory system as the 500 mg/kg dose: thrombus formation and impairment of the arterioles. The dose of 6 mg/kg caused an increase in the number of leukocytes and venular stasis, but these changes were moderate compared to the higher doses.

With different doses and different periods of exposure (30 and 120 min) the pattern of clinical signs observed was the same as described before. These tests also showed that the intensity of effects, independent of the dose, increases in proportion to the time period; for the 120 min time period it was possible to observe partial venular stasis immediately after exposing the cremaster muscle in most trials. This result differed from that of the 30 min time period where this effect was only observed 5 to 10 min after exposing the muscle.

These tests using the intravital microscopy technique are unique in the study of new cyanotoxins. Therefore, it was necessary to establish protocols for these analyses which are well documented in cases of poisons and toxins of animals. The establishment of these protocols allows for consistency and continuity in studies using fractions from the pre-purification crude extract of *G. unigranulatum* to isolate and characterize toxin(s) responsible for the toxic effects.

## References

Carvalho LR (2006) Cianotoxinas. In *Manual ilustrado para identificação e contagem de cianobactérias planctônicas de águas continentais brasileiras* (CL Sant'anna, MTP Azevedo, LF Agujaro, MC Carvalho, LR Carvalho and RCR Souza, eds), pp. 9-19. Interciência, Rio de Janeiro.

Carvalho LR, Haraguchi M, and Górniak SL (2008). Intoxicação produzida por algas de água doce. In *Toxicologia aplicada à medicina veterinária* (HS Spinosa, SL Górniak, and J Palermo Neto, eds), pp. 621-640. Editora Manole, Barueri.

Clissa PB, Lopes-Ferreira M, Della-Casa MS, Farsky SHP, and Moura-Da-Silva AM (2006). Importance of jararhagin disintegrin-like and cysteine-rich domains in the early events of local inflammatory response. *Toxicon* 47:591-596.

Conceição K, Konno K, Melo RL, Marques EE, Hituma-Lima CA, Lima C, Richardson M, Pimenta DC, and Lopes-Ferreira M (2006). Orpotrin: a novel vasoconstrictor peptide from the venom of the brazilian stingray *Potomotrygom* gr. *orbignyi*. *Peptides* 27:3039-3046.

Conceição K, Bruni FM, Pareja-Santos A, Antoniazzi MM, Jared C, Lopes-Ferreira M, Lima C, and Pimenta DC (2007). Unusual profile of leukocyte recruitment in mice induced by a skin secretion of the tree frog *Phyllomedusa hypochondrialis*. *Toxicon* 49(5):625-633

Cox PA, Banack SA, and Murch SJ (2003). Biomagnification of cyanobacterial neurotoxins and neurodegenerative disease among the Chamorro people of Guam. *Proceedings of the Nacional Academy of Sciences of the United States of America* 110(23):13380-13383.

Dogo CR and Carvalho LC (2006). In *Congreso Basileiro de Ficologia 11*, Itajaí, p. 66 (abstract).

Falconer I, Bartram J, Chorus I, Kuiper Goodman, T, Utkilen H, Burch M, and Codd GA (1999). Safe Levels and Safe Practices. In *Toxic Cyanobacteria in Water. A Guide to their Public Health Consequences, Monitoring and Management* (I Chorus and J Bartram, eds), pp. 55-178. E & FN Spon on behalf of WHO, London.

Fastner J, Flieger I, and Neumann V (1998). Optimized extraction of microcystins from field samples: a comparison of different solvents and procedures. *Water Research* 32:3177-3181.

Harada K, Kondo F, and Lawton L (1999). Laboratory analysis of cyanotoxins. In. *Toxic Cyanobacteria in Water*. A guide to their public health consequences, monitoring and management (I Chorus and J Bartram, eds), pp. 369-405. E & FN SPON, New York.

Junqueira MEP, Grund LZ, Orii NM, Saraiva TC, Lopes CAM, Lima C, and Lopes-Ferreira M (2007). Analysis of the inflammatory reaction induced by the catfish (*Cathorops spixii*) venoms. *Toxicon* 49(7):909-919.

Lomonte B, Lungren J, Johansson B, and Bagge U (1994). The dynamics of local tissue damage induced by *Bothrops asper* venom and myotoxin II on the mouse cremaster muscle; an intravital. *Toxicon*, 32:41-55.

Lopes-Ferreira M, Moura-da-Silva AM, Piran-Soares AA, Angulo Y, Lomonte B, Gutíerrez JM, and Farsky SHP (2002). Hemostatic effects induced by *Thalassophryne nattereri* fish venom: a model of endothelium-mediated blood flow impairment. *Toxicon* 40:1141-1147.

Magalhães KW, Lima C, Piran-Soares AA, Marques EE, Hiruma-Lima CA, and Lopes-Ferreira M (2006). Biological and biochemical properties of the Brazilian *Potomotrygon* stingrays: *Potomotrygon* cf. *scobina* and *Potomotrygon* gr. *orbignyi*. *Toxicon* 47:575-583.

Murch SJ, Cox PA, Banack SA, Steele JC, and Sacks OW (2004). Occurrence of β-methilamino-L-alanine (BMAA) in ALS/PDC patients from Guam. *Acta Neurologica Scandinavica* 110:267-269.

Rapala J, Sivonen K, Lyra C, and Niemelä (1997). Variation of microcystins, cyanobacterial hepatotoxins, in *Anabaena* spp. as a function of growth stimuli. *Applied and Environmental Microbiology* 63(6):2206-2212.

Raud J and Lindborn L (1994). Studies by intravital microscopy of basic inflammatory mechanisms and acute allergic inflammation. In *The Handbook of*

*Immunopharmacology: Immunopharmacology of the Microcirculation* (SD Brain), pp. 127-170. Academic Press, London.

Tonk L, Visser PM, Christiansen G, Dittmann E, Snelder EO, Wieder C, Mur LR, and Huisman J (2005). The microcystin composition of the cyanobacterium *Planktothrix agaddhii* changes toward a more toxic variant with increasing light intensity. *Applied and Environmental Microbiology* 71(9):5177-5181.

Van Apeldoorn ME, Van Egmond HP, Speijers GJA, and Bakker GJI (2007). Toxins of Cyanobacteria. *Molecular Nutrition and Food Research* 51:7-60.

Wiegand C and Pflugmacher S (2005). Ecotoxycological effects of selected cyanobacterial secondary metabolites: a short review. *Toxicology and Applied Pharmacology* 203:201-218.

Zagatto PA, Aragão MA, Domingues DF, Buratini SV, and Araujo RPA (1998). Avaliação ecotoxicológica do reservatório do Guarapiranga, SP, com ênfase à problemática das algas tóxicas e algicidas. *Anais do IV Congresso Latino-Americano de Ficologia* 63-81.

# Chapter 88

# Production of a Saxitoxin Standard from Cyanobacteria

F. Pípole[1], I.M. Hueza[1], C.L. Sant'Anna[2], and L.R Carvalho[2]

*Department of Pathology, School of Veterinary Medicine and Animal Sciences, University of São Paulo São Paulo/SP – Brazil; [2] Phycology Section, Botany Institute of São Paulo, São Paulo/SP – Brazil*

## Introduction

Saxitoxins (STX) or paralytic shellfish poisons (PSP) are a group of alkaloid toxins that can sicken and even kill people. Saxitoxins have been recognized as a public health threat since 1793. It was first believed that these toxins were restricted to the marine environment and that they were specifically produced by dinoflagellates, which are food for filter-feeding bivalve shellfish such as oysters, mussels, scallops, and clams. It is now known that these toxins are also produced by several freshwater cyanobacteria and some frogs (Llewllyn 2006) and that they can accumulate in shrimp and fish (Strangetti 2007; Linares *et al.* 2009).

In fresh water saxitoxins are associated with cyanobacteria blooms and can be found in 20% of these blooms. Saxitoxins are produced by *Planktothrix*, *Aphanizomenon*, *Cylindrospermopsis*, *Anabaena*, and *Lyngbya* spp. (Falconer 2005). The most well known episode of cyanobacteria-related (*Anabaena circinalis*) saxitoxin contamination occurred in Australia in the Darling River in 1990. The reservoirs of coastal cities became contaminated and hundreds of animals died (Humpage and Rositano 1994). Additionally, many other blooms of saxitoxin-producing cyanobacteria have been described in recent years (Lagos *et al.* 1999; Yunes *et al.* 2003; Molica *et al.* 2005). Therefore, STX should be monitored in water supplies and also by the fishing industry.

Saxitoxins act by blocking the voltage-gated sodium channels of mammalian neurons which causes various symptoms including paralysis, hypotension, dyspnea, and respiratory failure (Llewellyn 2006). Saxitoxins are guanidine alkaloids with a carbamate group and their molecular weights range between 240 and 500 Da. They are colorless and hygroscopic solids that decompose in alkaline environments (Carvalho *et al.* 2008). There are about 30 saxitoxin analogs which are classified according to their ionic charges. The first group is known as the saxitoxins and these have neutral pH; the second group is known as the gonyautoxins and have a $1^+$ charge; and the third group is known as the sulfocarbamoil toxins, or C-toxins and have a $2^+$ charge (Lagos 2002).

Saxitoxin monitoring was initially carried out only with mouse bioassays using a technique established by the World Health Organization (WHO) to detect these toxins in

©CAB International 2011. *Poisoning by Plants, Mycotoxins, and Related Toxins* (eds F. Riet-Correa, J. Pfister, A.L. Schild, and T.L. Wierenga)

mollusks. This technique continues to be used and has been extended to include the identification of other cyanotoxins. However, the development of chromatographic and immunoassay methods as well as the introduction of mass spectrometry into the laboratory routine have provided monitors with a number of options for the detection of saxitoxins in water and food samples. However, with the exception of mass spectrometry these techniques for qualitative and quantitative analysis are all based on the use of standards.

Saxitoxin standards are currently very expensive because their production requires tonnes of mollusks which must be stored at low temperatures. The production of the saxitoxin standards occurs in batch mode following the shellfish seasonal collection and causes environmental degradation (Alfonso *et al.* 1993; Lagos 2006, personal communication). The difficulty in obtaining saxitoxins, their high toxicity, and the demand for them for non-scientific purposes have resulted in an expensive product sold by only a few international laboratories.

The production of these standards from strains of cyanobacteria has many advantages. This type of production can be continuous, it does not require crude biomass storage in refrigerators, eliminates the risks inherent in handling huge amounts of easily degradable material, is both cheaper and better than STX standards from mollusks, and also contributes to environment preservation. The purpose of this study was to establish a method for the production of a saxitoxin standard from cyanobacteria.

## The Organism

*Raphydiopsis brookii* Hill 1972, strain SPC 338 is kept in the Cyanobacteria Culture Collection, at the Institute of Botany, Brazil. *Raphydiopsis brookii* is a filamentous cyanobacterium belonging to the family Nostocaceae. Its trichomes are solitary, straight or slightly curved, and not constricted. The cylindrical cells present gas vesicles and the apical cell is long and acuminate. Heterocysts are absent and the akinetes are subapical (Sant'Anna *et al.* 2007).

## Cell Culture

This cyanobacterium was cultured in 5 l culture bottles in ASM-1 medium, pH 7.4 at 22±1°C, under continuous light at an irradiance of 45-50 mmol/m$^2$/s and a moderate aeration rate. The cells were grown until the late-exponential growth phase (about 3 weeks) at which point the culture was harvested.

## Purification Process

The cultured material was lyophilized, extracted with 0.1 M acetic acid and ultrasonication, and centrifuged. The supernatant was lyophilized and subjected to purification by chromatographic methods. The fractionation and purification were both bio-guided step by step to ensure the acquisition of the toxin. The saxitoxin purity was determined by HPLC-FLD analysis. This purification method is in the process of being patented.

## High Performance Liquid Chromatographic Analysis

Saxitoxins were measured using pre-column derivatization HPLC with a fluorescence on-line detection method (Lawrence *et al.* 1995). Each sample (20 µl) was injected into a reversed-phase Zorbax ODS column (250×4.6 mm, 5 µm), and the following mobile phases were used at pH 6: A=0.1 M ammonium formate and B=95:5 (v/v) 0.1 M ammonium formate/acetonitrile. The fluorimetric detector was set at an excitation wavelength of 340 nm and an emission wavelength of 390 nm.

## Bioassay

Male Swiss-Webster mice (18-22 g) were reared at the central vivarium of the Instituto Butantan. Five animals were housed per cage; the animals were divided into experimental and control groups and maintained under 12 h light/dark cycles in a well-ventilated room at 23±1°C. All animals received humane care and the studies were conducted in accordance with the Ethical Principles of the Committee on Ethics of Instituto Butantan.

Animals were injected intraperitoneally (i.p.) with crude extract and with fractions obtained from all purification steps, all diluted in sterile 0.9% NaCl solution (the bio-guided assay). Time to death, signs of poisoning, and other symptoms were observed up to 72 h after injection.

STX extracted (Figures 1A and 1B) from *Raphydiopsis brookii* strain SPC 338 showed 95% purity (Figure 1C) as determined by HPLC analysis, which is similar to STX standards obtained from mussels (Figure 1D).

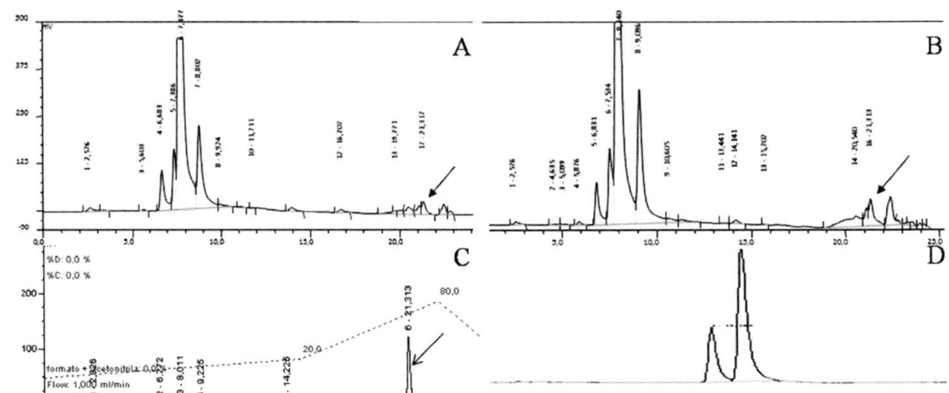


**Figure 1.** A: chromatogram of crude extract; B: chromatogram of a semi-purified sample; C: chromatogram of purified saxitoxin; D: chromatogram of a saxitoxin standard from mollusks. Conditions as described in the text. The arrow indicates the saxitoxin peak (Rt = 21.3 min).

The extract obtained from filter-feeding bivalve shellfish (oysters, mussels, scallops, clams) contained a number of saxitoxins that are very difficult to separate. The extract from cultured cyanobacteria contains few saxitoxin analogs and is easier to purify (Garcia *et al.* 2005; Vale 2006).

## Conclusions

The cyanobacterium standard contains only saxitoxin while the standard obtained from mollusks contains both saxitoxin and an analog. By selecting another cyanobacteria species it may be possible to obtain different analogs at a lower price. The biomass is easily obtained in culture, its production is not seasonal, and no special storage equipment is required. Thus, we conclude that this method represents an excellent option for obtaining STX standards.

## Aknowledgements

This research was financially supported by CAPES and CNPq.

## References

Alfonso A, Vieytes MR, Botana AM, Goenaga X, and Botana LM (1993). Preparation of mixtures of paralytic shellfish toxin (PSP) standards from mussels hepatopancreas. *Fresenius Journal of Analytical Chemistry* 345:212-216.

Carvalho LR, Haraguchi M, and Gorniak SL (2008). Intoxicação produzida por algas de água doce. In: *Toxicologia Aplicada à Veterinária* (HS Spinosa, SL Górniak, and J Palermo, eds), pp. 1-8. Editora Manole, São Paulo.

Falconer IR (2005). *Cyanobacterial toxins of drinking water supplies* 900 pp. CRC Press, Boca Ratón, Florida.

Garcia C, Bravo MC, Lagos M, and Lagos N (2005). Paralytic shellfish poisoning: post-mortem analysis of tissue and body fluid samples from human victims in the Patagônia fjords. *Toxicon* 43:149-158.

Humpage AR and Rositano J (1994). Paralytic shellfish poison from Australian cyanobacterial blooms. *Australian Journal of Marine and Freshwater Research* 45:761-771.

Lagos N (2002). Principales toxinas de origen fitoplanctónico: identificación y cuantificación mediante cromatografia líquida de alta resolucion (HPLC). In *Floraciones algales nocivas em el cono sur americano* (EA Sar, ME Ferrario, and B Reguera, eds), pp. 57 -76. Instituto Español de Oceanografia, Madrid.

Lagos N, Onodera H, Zagatto PA, Andrinolo D, Oshima Y, and Azevedo SMFO (1999). The first evidence of paralytic shellfish toxins in the freshwater cyanobacterium, *Cylindrospermopsis raciborskii*, isolated from Brazil. *Toxicon* 37: 1359-1373.

Lawrence JF, Ménard C, and Cleroux C (1995). Evaluation of pre-chromatographic oxidation for liquid chromatographic determination of paralytic shellfish poisons in shellfish. *Journal of AOAC International* 75(2):514-520.

Linares JP, Ochoa JL, and Martínez AG (2009). Retention and tissue damage of PSP and NSP toxins in shrimp: is cultured shrimp a potential vector of toxins to human population? *Toxicon* 53:185-195.

Llewellyn LE (2006). Saxitoxin, a toxic marine natural product that targets a multitude of receptors. *Natural Product Reports* 23:200-222.

Molica RJR, Oliveira EJA, Carvalho PVVC, Costa ANSF, Cunha MCC, Melo GL and Azevedo SMFO (2005). Occurrence of saxitoxins and na anatoxin-a(s)-like anticholinesterase in a Brazilian drinking water supply. *Harmful Algae* 4:743-753.

Sant´anna CL, Melcher SS, Carvalho MC, Gemelgo MP, and Azevedo MTP (2007). Planktic cyanobacteria from Upper Tietê Basin, SP, Brazil. *Revista Brasileira de Botânica* 30:1-17.

Strangetti BG (2007). Monitoração toxinológica do pescado comercializado nos municípios de São Sebastião e Caraguatatuba, SP, Brasil, 257 pp. Masters Dissertation, Universidade de São Paulo.

Vale P (2006). Implementacao de tecnicas de HPLC e LC-MS para estudo de perfis de biotoxinas marinhas em plancton e em bivalves. *Revista Portuguesa de Ciencias Veterinarias* 101:163-180.

Yunes JS, Cunha NT, Barros LP, Proença LAO, and Monserrat JM. (2003). Cyanobacterial neurotoxins from Southern Brazil. *Comments in Toxicology* 9:103-115.

# Chapter 89

# Differential Diagnosis between Plant Poisonings and Snakebites in Cattle in Brazil

P.V. Peixoto[1], F.S. Graça[2], S.A. Caldas[2], A.P. Aragão[2], T.N. França[3], and C.H. Tokarnia[1]

[1]*Departamento de Nutrição Animal e Pastagem, Universidade Federal Rural do Rio de Janeiro (UFRRJ), Seropédica, RJ 23890-000, Brazil;* [2]*Curso de Pós-graduação em Ciências Veterinárias, UFRRJ, Seropédica, RJ;* [3]*Departamento de Epidemiologia e Saúde Pública, UFRRJ, Seropédica, RJ*

## Introduction

In Brazil, poisonous plants are among the three main causes of death in adult cattle grazing on rangelands. Conservative estimates indicate that annually about 1 million cattle (0.5% of the total population) die from poisonous plants (Riet-Correa and Medeiros 2001). Many of these deaths are attributed to snakebites by farmers and veterinarians. In agreement with a recent study, however, the number of cattle deaths due to snakebites has been overestimated in Brazil (Tokarnia and Peixoto 2006). Although the main focus of this study is the importance of poisoning by plants, for informative reasons we outline the clinical-pathological aspects and situations in which snakebites occur in Brazil. Plants that could induce clinical signs or lesions that may be confused with incidents of snakebite are considered. The objective of this study is to provide information to facilitate the differential diagnosis between snakebites and plant poisonings by veterinarians in Brazil and other countries.

## General Considerations

Generally speaking throughout Brazil, poisoning of livestock by specific plants is often not diagnosed; on the other hand, diseases caused by other agents are often attributed to toxic plants. This situation is partly due to the fact that many field veterinarians do not perform enough postmortem examinations in part because of fear of contamination with infectious agents such as *Bacillus anthracis* and the rabies virus (the cause of diseases which are also confused with plant poisoning in this country) or otherwise because of a lack of favorable conditions. Often snakes are accused independently of the effect of the poisonous plant which caused the death. Practicing veterinarians also are not aware that there are differences between the clinical picture caused by the bites of snakes of the genera

©CAB International 2011. *Poisoning by Plants, Mycotoxins, and Related Toxins* (eds F. Riet-Correa, J. Pfister, A.L. Schild, and T.L. Wierenga)

*Crotalus* and *Bothrops* which are responsible for almost all snakebites in Brazil. More important still is that most veterinarians do not take into account that the clinical-pathological picture might significantly differ depending on the animal species envenomed by the snakebite. This is one of the causes for the many misunderstandings and incongruities found in the literature on the subject. The fact that most North American rattlesnakes are capable of inducing severe lesions at the site of the bite similar to that by snakes of the genus *Bothrops* also contributes to the confusion (Tokarnia and Peixoto 2006).

## Poisoning by Plants and Snakebites by *Crotalus* in Livestock

There is little information on the occurrence of snakebite from the genus *Crotalus* in livestock in Brazil. There have been, however, several experimental studies in cattle which provide information about the clinic-pathological and toxicological aspects of snakebites by the genus in Brazil (Araújo *et al.* 1963; Belluomini *et al.* 1982; Birgel *et al.* 1983; Lago 1996; Graça *et al.* 2008).

### Differentiation of snakebites by *Crotalus*

In cattle envenomed by *Crotalus* clinical signs are primarily in the nervous system and are characterized by progressive flaccid paralysis. This clinical picture is virtually indistinguishable from that observed in cases of botulism. It is also important to consider that cattle do not show myoglobinuria as occurs in about 40% of humans bitten by Brazilian rattlesnakes. This was demonstrated in 92 cases of cattle experimentally envenomed by *Crotalus durissus terrificus* (Belluomini *et al.* 1982; Saliba *et al.* 1983) and confirmed in the studies of Lago *et al.* (2004) and Graça *et al.* (2008). Hemorrhages are not prominent in postmortem examinations (many confuse postmortem hemoglobin imbibition with hemorrhages). Coagulation necrosis in groups of fibers of skeletal muscles are typical lesions (Graça *et al.* 2008). Despite that lesion there is no myoglobinuria in cattle envenomed by *Crotalus*. Therefore, if we consider only the signs *Crotalus* snakebites theoretically should be differentiated from poisoning by plants which cause disturbances of the central nervous system (CNS) or in skeletal muscles.

### Differentiation from plants that affect the CNS

We do not know of plants that affect primarily the central nervous system and produce clinical signs similar to those caused by Brazilian rattlesnakes. In Brazil plants that affect the CNS usually induce symptoms of cerebellum-vestibular or pontino-cerebellar disturbances. Plants that cause necrosis or hepatic cirrhosis sometimes induce apathy or somnolence but progressive flaccid paralysis is not observed and the macro- and microscopic lesions are very different.

### Differentiation from plants that cause muscular necrosis

The only Brazilian plant that could possibly show clinical signs similar to those from bites by *Crotalus* in cattle is *Senna occidentalis*. The paralysis caused by deficit/blockage of neurotransmission can be similar to secondary systemic muscular incapacity due to extensive areas of degeneration/necrosis of skeletal muscles caused by that plant. Normal awareness is not affected by either poisoning. Many animals poisoned by *S. occidentalis*, however, show evidence of myoglobinuria. The muscle lesions are often visible through macroscopic inspection. Sometimes the plant also causes regressive lesions in the heart (Tokarnia *et al.* 2000).

**Differentiation from plants that cause sudden death (SD)**

Toxic plants that cause sudden death should not be considered in a differential diagnosis with *Crotalus* bites. Although snakes of this genus can produce and store enough venom to kill up to six cattle of 500 kg, death does not occur until some time has elapsed (usually more than 6 h) after the bite. Moreover, the flaccid paralysis differs from the picture of sudden death. In cases of poisoning by these plants on farms with extensive herds (more than 50,000 head of cattle such as ranches that exist in the Amazon region), the animals are generally found dead without apparent lesions. If a postmortem examination is performed the cause of death may not be known because neither condition results in significant macroscopic lesions. Only if fragments of the kidneys and skeletal muscles are collected for histological examination can the differentiation be made. In the case of SD there is frequently the characteristic hydropic-vacuolar degeneration of the distal convoluted renal tubules (Tokarnia *et al.* 2000) and in the case of South American *Crotalus* bites there is coagulative muscle necrosis (Graça *et al.* 2008). On smaller farms the affected animal may be noticed and because the clinical signs are so different no doubt exists about the correct diagnosis. The differentiation can be further clarified by knowing the distribution and habitat of *Crotalus* snakes and of plants causing SD. There are 12 SD-causing plants in Brazil distributed throughout the five large regions of the country whereas the occurrence of *Crotalus* snakes is restricted. In the Amazon Region *Crotalus* only occurs in limited areas and reports of bites by this snake practically can be dismissed. Cattle in this area typically die only when they have access to the border of forests or 'capoeiras' (areas overgrown by young vegetation) which is the main habitat of *Palicourea marcgravii* or when moved (exercised). *P. marcgravii* is responsible for 80% of cattle deaths in such areas (Tokarnia *et al.* 2000). *P. marcgravii* requires shade to thrive but does not grow in full sun and does not grow well under closed canopies in mature forests. Knowledge of the specific distribution of poisonous snakes and plants causing sudden death may be helpful.

**Differentiation from plants that cause hemolytic anemia**

Unlike what happens in humans, these plants cause the elimination of red urine but they should not be considered in the differential diagnosis because *Crotalus durissus terrificus* does not induce myoglobinuria in cattle.

## Poisoning by Plants and Bites by *Bothrops/Lachesis* Snakes

There are few reports of fatal snakebites from the genus *Bothrops* in ruminants in Brazil (Méndez and Riet-Correa 2007; Tokarnia *et al.* 2008). Some experimental studies were performed in cattle (Araújo *et al.* 1963; Belluomini *et al.* 1982; Caldas *et al.* 2008; Aragão 2009). Although there are common characteristics of the clinical signs caused by snakes of this genus, there are several aspects of snakebites that are species specific. There are also differences in susceptibility of domestic animal species to the effects of the different fractions of venom from snakes. For instance, the experimental inoculation of the venom of *B. alternatus* caused death of cattle by hemorrhage especially around the inoculation site and adjacent areas (Caldas *et al.* 2008). Swelling and edema were essentially constituted of blood. The hemorrhagic tendency was such that one cow died due to hypovolemic shock while another animal had the hemorrhage stanched only by the application of a garrote at the base of the tail (site of the puncture to obtain blood samples for laboratory examination). In bites by *B. jararaca* the swelling at the site of inoculation was related to hemorrhage and edema whilst the poison of *B. jararacussu* induced mainly edema at the inoculation site and adjacent areas (Aragão 2009). All the animals that received the poison of *B. jararacussu* died with lung edema. We cannot define the pathogenic mechanism of that lesion but it is reasonable to think that shock phenomena could be involved. Thus, the main differential diagnosis should be made with plants that cause extensive or systemic hemorrhages or that produce localized subcutaneous edema.

### Plants that cause death due to significant hemorrhage

Hemorrhagic deaths are restricted to plants with radiomimetic effects that induce thrombocytopenia and spontaneous hemorrhages as in poisoning by *Pteridium arachnoideum* and *P. caudatum* in Brazil. The local increase of volume (with or without presence of holes from the fangs of the snake) is diagnostic for bites by *Bothrops* spp. Moreover, the hemorrhages caused by toxic plants tend to be more diffuse (systemic) while snake venom induces more severe hemorrhages at the site of the bite, the severity of which is directly proportional to the proximity to the inoculation site (Caldas *et al.* 2008). Macroscopically in the case of poisoning by *Pteridium* spp. in cattle, pale areas of coagulation necrosis (infarcts) in liver and heart are very frequent. Microscopically, destruction of the bone marrow is found. The knowledge of the distribution and habitat of the snakes is not useful as this genus occurs throughout Brazil. Although these snakes prefer more humid areas there is always overlap with the habitat of *Pteridium* spp. which also grows throughout the country. Snakebite must be differentiated from plants that produce subcutaneous edemas. Those can include: (i) cardiotoxic plants (chronic poisoning) that cause localized edema mainly on the sternum; (ii) nephrotoxic plants that cause subcutaneous edemas which begin in the rear parts of the legs and extend cranially; and (iii) photosensitizing plants that can produce subcutaneous localized edemas especially in the head of sheep. We consider it sufficient to mention only these plants since the other clinical-pathological characteristics are far different from those of snakebites. In the case of bites by snakes of the *Lachesis* genera, at least in humans, there is the local lesion and CNS disturbances but there is no information on bites from this genus in ruminants.

## References

Aragão AP (2009). Envenenamento experimental por *Bothrops jararaca* e *Bothrops jararacussu* em ovinos: aspectos clínico-patológicos e laboratoriais, 98 pp. Dissertação de Mestrado, Universidade Federal Rural do Rio de Janeiro, Instituto de Veterinária, Seropédica.

Araújo P, Rosenfeld G, and Belluomini HE (1963). Toxicidade de venenos ofídicos. II. Doses mortais para bovinos. *Arquivos do Instituto Biológico* 30:43-48.

Belluomini HE, Araújo P, Rosenfeld G, Leinz FF, and Birgel EH (1982). Symptomatologie der experimentellen Crotalustoxin-Vergiftung bei Rindern, die einer spezifischen

Serumtherapie unterworfen wurden. *Deutsche Tierärztliche Wochenschrift* 89(11):444-448.

Birgel EH, Belluomini HE, and Leinz FF (1983). Auswertung der Urinbefunde bei Rindern mit experimenteller *Crotalus*-Vergiftung. *Zentralblatt Veterinär Medizin* 30:283-289.

Caldas SA, Tokarnia CH, França TN, Brito MF, Graça AS, Coelho CD, and Peixoto PV (2008). Aspectos clínico-patológicos e laboratoriais do envenenamento experimental por *Bothrops alternatus* em bovinos. *Pesquisa Veterinária Brasileira* 28(6):303-312.

Graça FAS, Peixoto PV, Coelho CD, Caldas SA, and Tokarnia CH (2008). Aspectos clínicos e patológicos do envenenamento crotálico experimental em bovinos. *Pesquisa Veterinária Brasileira* 28(6):261-270.

Lago LA (1996). Avaliação clínica e laboratorial de bovinos submetidos ao envenenamento crotálico experimental – *Crotalus durissus terrificus* – Laurenti, 1768 – *Crotamina* positivo, 62 pp. Dissertação de Mestrado, Universidade Federal de Minas Gerais, Escola de Veterinária, Belo Horizonte.

Lago LA, Marques Júnior AP, Melo MM, Lago EP, Oliveira NJF, and Alzamora Filho F (2004). Perfil bioquímico sorológico de bovinos inoculados experimentalmente com veneno crotálico iodado livre e iodide incorporado em lipossomes. *Arquivo Brasileiro de Medicina Veterinária e Zootecnia* 56(5):653-657.

Méndez MC and Riet-Correa F (2007). Envenenamento botrópico. In *Doenças de Ruminantes e Eqüídeos* (F Riet-Correa, AL Schild, RAA Lemos, and JRJ Borges, eds), pp. 31-38. Pallotti, Santa Maria.

Riet-Correa F and Medeiros RMT (2001). Intoxicações por plantas em ruminantes no Brasil e no Uruguai: importância econômica, controle e riscos para a saúde pública. *Pesquisa Veterinária Brasileira* 21(1):38-42.

Saliba AM, Belluomini HE, and Leinz FF (1983). Experimentelle *Crotalus*-Vergiftung bei Rindern: Anatomisch-pathologische Studie. *Deutsche Tierärztliche Wochenschrift* 90:513-517.

Tokarnia CH and Peixoto PV (2006). A importância dos acidentes ofídicos como causa de mortes em bovinos no Brasil. *Pesquisa Veterinária Brasileira* 26(2):55-68.

Tokarnia CH, Döbereiner J, and Peixoto PV (2000). *Plantas Tóxicas do Brasil,* 310 pp. Ed. Helianthus, Rio de Janeiro.

Tokarnia CH, Brito MF, Malafaia P, and Peixoto PV (2008). Acidente ofídico em ovinos causado por *Bothrops jararaca. Pesquisa Veterinária Brasileira.* 28(12):643-648.

# Chapter 90

# The Use of a Guinea Pig Model in Detecting Diplodiosis, a Neuromycotoxicosis of Ruminants

R.A. Schultz, L.D. Snyman, K.M. Basson, and L. Labuschagne

*Toxicology Section, ARC-Onderstepoort Veterinary Institute, Private bag x05, Onderstepoort, 0110 South Africa*

## Introduction

Diplodiosis is a neuromycotoxicosis of cattle and sheep grazing on harvested maize fields in winter. Together with facial eczema in New Zealand and lupinosis in Australia it is rated as one of the most important mycotoxicoses of ruminants in the world (Kellerman *et al.* 2005).

Diplodiosis is one of the most commonly diagnosed nervous disorders of cattle and sheep in southern Africa; in South Africa alone it is regarded as being responsible for about 2% of all livestock mortalities from plant poisonings and mycotoxicoses (Kellerman *et al.* 1996). The disease, induced by the ingestion of maize (*Zea mays*) infected with *Stenocarpella* (=*Diplodia*) *maydis*, is characterized by ataxia, paresis, and paralysis. Poisoning is reversible as prompt removal of stock from the source together with good nursing usually results in complete recovery.

A complication of diplodiosis is that pregnant cows and ewes which have been exposed to infected maize can produce stillborn or non-viable offspring. Such perinatal losses have been noticed even in seemingly healthy herds or flocks in which, although being exposed to *S. maydis*, the pregnant animals never showed any overt signs of poisoning.

Since not all strains of *S. maydis* are toxigenic, a method to distinguish between toxin producing and non-toxin producing strains is urgently needed especially from the point of view of controlling the disease. In this respect isolation and characterization of the toxin(s) is a priority, *inter alia* because this will allow chemical monitoring of fields for toxicity in a grazing system where maize stover is an essential source of roughage for stock in winter. Apart from assessing the potential risk of pastures for stock, knowing the nature of the toxin(s) will be useful in setting legal limits for the level of toxin(s) in grain for human and animal consumption, and for studying the pathophysiology of the condition.

©CAB International 2011. *Poisoning by Plants, Mycotoxins, and Related Toxins*
(eds F. Riet-Correa, J. Pfister, A.L. Schild, and T.L. Wierenga)

## Materials and Methods

Crude extracts of *S. maydis* cultures, prepared according to a method developed in the Toxicology Section (Snyman 2004, personal communication), were used for trials in laboratory animals which were kept in cages and had free access to water and feed pellets. The initial trials were performed with cultures (Kellerman *et al.* 1991) inducing clinical signs reminiscent of diplodiosis. Crude extracts of the cultures were dosed to mice and guinea pigs to select the most appropriate laboratory animal model.

In two experiments female guinea pigs ($n$=10; $n$=9) weighing between 140 and 180 g were dosed with two different crude culture extracts (*c.* 3 ml equivalent to 75 g culture). The nature and degree of the clinical signs were recorded to evaluate the reproducibility of neurotoxicity in the animals.

Twenty culture samples of *S. maydis* were received from the ARC-Grain Crop Institute to test for the presence of the neurotoxic metabolite of *S. maydis*. These samples had been cultured from maize (Flett and McLaren 1994) obtained from the maize-producing area of South Africa which was infected with *S. maydis.* A crude extract of each sample (*c.* 3 ml equivalent to 75 g culture) was dosed to a guinea pig. The weights of the 20 guinea pigs varied between 101 and 165 g.

## Results and Discussion

The guinea pig was selected as the laboratory animal model of choice as the typical paretic signs followed on a latent period and its ability to recover with good nursing. These are consistent with those found in the disease in livestock. A small number of animals, however, died after they had received very high levels of the neurotoxin.

Clinical signs in the guinea pigs which lasted for 1 to 4 days included weakness, reluctance to move, and hind-limb paresis which progressed to lateral recumbency and paralysis. The signs observed within the first 24 h were evaluated on a scale of 0 to 5:

0–Normal;
1–Unwilling to move but with the head held high, sometimes hopping around;
2–Displays weakness or paretic signs with slight tremors of the head and ears;
3–Weak, lateral recumbency, and manifesting paddling movements with righting attempts virtually absent;
4–Paralysis with hind limbs stretched backwards;
5–Dead.

The clinical signs varied between 3 and 5 (extremely or fatally affected; Experiment 1) and 1 and 2 (mildly affected; Experiment 2) 24 hours after dosing the culture samples. Within the two trials the severity of the clinical signs was reproducible as depicted in the histograms (Figure 1). Using the clinical signs of the guinea pigs as a yardstick (unaffected, mildly or extremely affected) to evaluate the toxicity of the 20 samples received, the cultures could be classified as non-toxic, mildly toxic, or extremely toxic (Table 1).

**Table 1.** Distribution of toxicity (non-toxic to very toxic) in 20 crude extracts from samples of *Stenocarpella maydis* cultures using the guinea pig model.

| Unaffected (0 on the scale) | Mildly affected (1-2 on the scale) | Extremely affected (3-5 on the scale) |
|---|---|---|
| 5 | 10 | 5 |

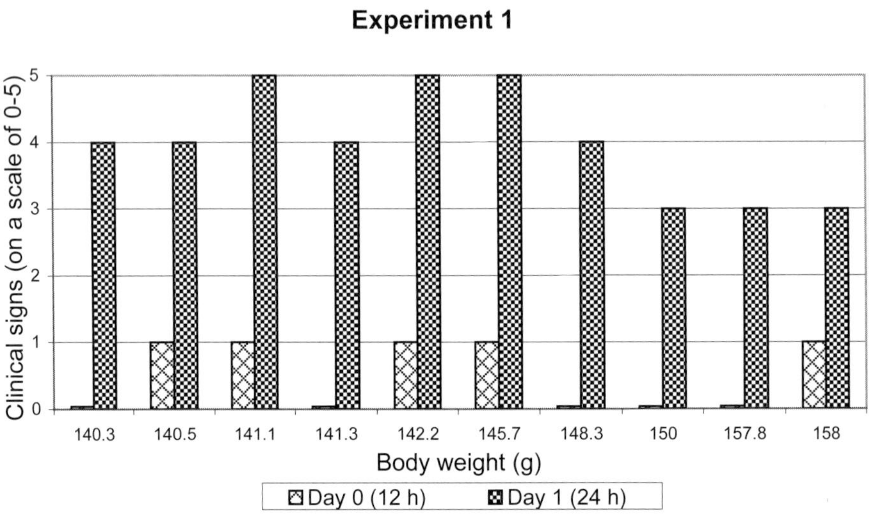


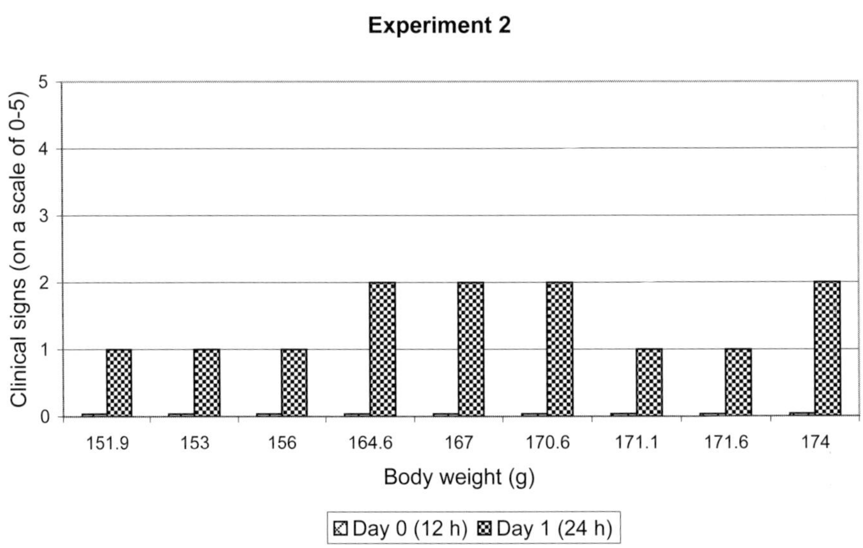


**Figure 1.** Clinical signs in guinea pigs dosed with crude extracts of *Stenocarpella maydis* cultures (*c.* 3 ml equivalent to 75 g).

## Conclusion

The guinea pig has been established as an appropriate bio-assay model for the identification of neurotoxin(s) in crude extracts of *S. maydis* cultures. The availability of a laboratory animal model is, in addition, an essential step towards the isolation of the active principle(s) of *S. maydis*. The long term aims of the project are to assess the potential risk posed to stock by affected pastures and the development of measures to control the disease.

## Acknowledgements

Funding was provided by the Gauteng Department of Agriculture Conservation and Environment (GDACE) through the Directorate: Technology Development and Support (TDS) and LASEC (Laboratory and Scientific Equipment Company SA (Pty) Ltd).

## References

Flett BC and McLaren NW (1994). Optimum disease potential for evaluating resistance to *Stenocarpella maydis* ear rot in corn hybrids. *Plant Disease*, June 587-589.

Kellerman TS, Prozesky L, Schultz RA, Rabie CJ, Van Ark H, Maartens BP, and Lübben A (1991). Perinatal mortality in lambs of ewes exposed to cultures of *Diplodia maydis* (=*Stenocarpella maydis*) during gestation. *Onderstepoort Journal of Veterinary Research* 58:297-308.

Kellerman TS, Naudé TW, and Fourie N (1996). The distribution, diagnoses and estimated economic impact of plant poisonings and mycotoxicoses in South Africa. *Onderstepoort Journal of Veterinary Research* 63:65-90.

Kellerman TS, Coetzer JAW, Naudé TW, and Botha CJ (2005). *Plant Poisonings and Mycotoxicoses of Livestock in Southern Africa*. 2nd edn, pp. 63-66. Oxford University Press, Cape Town.

# TOXIC COMPOUNDS AND CHEMICAL METHODS

# Chapter 91

# Acute Toxicity of Selenium Compounds Commonly Found in Selenium-accumulator Plants

T.Z. Davis[1], B.L. Stegelmeier[1], B.T. Green[1], K.D. Welch[1], K.E. Panter[1], and J.O. Hall[2]

[1]*USDA-ARS Poisonous Plant Research Laboratory, Logan, Utah 84341, USA*; [2]*Veterinary Diagnostic Laboratory, Utah State University, Logan, UT 84341, USA*

## Introduction

Selenium (Se) is an essential trace element required by mammals and poultry. Although essential, Se has a very narrow window between deficiency and toxicity. Selenium-accumulating plants such as *Astragalus* spp., *Stanleya pinnata*, and *Aster* spp. are commonly found in various regions of the western USA. Primary selenium accumulator plants can store up to 10,000 ppm Se as predominantly selenate and methylselenocysteine (MeSeCys) (Shrift and Virupaksha 1965; Pickering *et al.* 2000; Freeman *et al.* 2006) and be extremely toxic to livestock or wildlife that graze them. During a recent 4 year period over 500 sheep were poisoned on selenium accumulator plants growing on reclaimed mine sites in southeastern Idaho. In many of the deaths the selenium accumulator plant western aster (*Aster ascendens*) was determined as the cause of death. Ingestion of a few grams of western aster containing 4000 to 6000 ppm Se will result in death of sheep within approximately 24 h (Wilhelm *et al.* 2007). Preliminary studies indicate that selenate and MeSeCys are likely the predominant forms of selenium in western aster. The objective of this study was to compare the acute toxicosis and toxicokinetics of Se in lambs orally dosed with selenate, MeSeCys, selenomethionine (the most common form of Se in non-primary accumulator forages), and the selenium-accumulator plant western aster.

## Materials and Methods

### Animals and experimental setup

One day prior to initiation of the study seventeen 8- to 12-week-old sheep were weighed, bled, and randomly divided into five groups with four sheep in two groups and three sheep in three groups. Each group received one of the following doses: 6 mg of Se/kg

©CAB International 2011. *Poisoning by Plants, Mycotoxins, and Related Toxins* (eds F. Riet-Correa, J. Pfister, A.L. Schild, and T.L. Wierenga)

BW as sodium selenate ($n$=3), MeSeCys ($n$=4), selenomethionine ($n$=3), or western aster ($n$=3). The control group ($n$=4) did not receive any selenium. Respiratory samples were collected at 0.5, 1, 1.5, 2, 2.5, 3, 4, 6, and 8 h post-dosing. Whole blood and serum samples were collected 10, 20, 30 min, and 1, 2, 3, 4, 6, 8, 12, 18, 24, 48, 72, 120, and 168 h after dosing. Tissues were collected at the time of death of each animal or at 7 days after administration of the selenium.

Individual doses of sodium selenate and western aster were prepared and dissolved in approximately 10 ml of water and administered intraruminally via an intragastric tube. The MeSeCys and selenomethionine was prepared and dosed in the same manner as the sodium selenate except that it was dissolved in 5% ethanol. The control sheep were administered a 5% ethanol solution in the same manner.

**Collection and preparation of respiratory samples**

Expired air samples were collected from the sheep using the method described by Tiwary and co-workers (Tiwary *et al.* 2005). Briefly, a rebreathing apparatus was used to collect 2 l of expired air into Tedlar bags (SKC Inc, Eighty Four, PA, USA). Following collection of the expired air samples the air was passed over activated charcoal columns (8 mm outer diameter and 110 mm long) having 400 and 200 mg of sorbent in compartments I and II, respectively (Anasorb SCS conconut charcoal, SKC Inc, Eighty Four, PA, USA). The air was uniformly drawn over the charcoal column using a vacuum pump (Gilian 3500 pump, Sensidyne Inc, Clearwater, FL, USA) attached to polyvinyl tubing. The pump was set at a constant flow rate of 1 l/min and was allowed to run for exactly 2 min before it was turned off. The charcoal columns were capped immediately after disconnecting from the pump and were stored at room temperature (~22°C) in a dark room until analysis of the samples was performed.

Activated charcoal was removed from the column and added to a 15 ml metal-free tube. Selenium liberation from the column was performed as optimized by Tiwary *et al.* (2005). Three ml of solvent (50:50 ratio of absolute ethanol and water) was added to the tube and the tube was placed on a rotary shaker for 2 h. Tubes were then centrifuged at 500 g for 10 min. One ml of supernatant was added to 8.5 ml of 18.3 mega ohm water and 0.5 ml of trace metal grade nitric acid. Samples were analyzed within 48 h of extraction by inductively coupled plasma-mass spectrometry (ICP-MS) using an ELAN 6000 (Perkin Elmer, Shelton, CT, USA) at the atomic mass of 78 and 82. Selenium standards were prepared in the same solution with standard curves and quality control samples tested after every fifth sample. Sensitivity of the ICP-MS analysis was 1 ng/ml or 10 μg/extract.

**Tissue digestion and preparation**

Tissues were digested and Se concentration determined by inductively coupled plasma mass spectrometry analysis using the method of Tiwary *et al.* (2006). Briefly, 1 g (wet weight) of each tissue was put into a Teflon digestion tube with 2 ml of trace metal grade nitric acid. The tubes were heated at 90ºC for 2 h with intermittent unscrewing of caps to release the pressure. The tubes were then allowed to cool and total volume of the contents was brought to 3 ml by adding trace metal grade nitric acid. The contents were subsequently transferred to polypropylene trace metal-free centrifuge tubes and 0.5 ml of the digest was transferred into another trace metal-free tube containing 9.5 ml of ultrapure water (18.2 MΩ/cm). After vortexing the samples were analyzed by ICP-MS.

**Whole blood and serum preparation**

Whole blood and serum were digested and prepared for analysis using the following method. Seven hundred fifty microliters of the sample was introduced into a Teflon digestion tube. An equal amount (750 μl) of trace metal grade nitric acid was added to the digestion tubes and the caps were sealed. The tubes were then heated at 90°C for 2 h without unscrewing of the caps. After digestion tubes were allowed to cool and contents were transferred to another trace metal-free tube. One milliliter of the digest was transferred into another trace metal-free tube containing 9.0 ml of ultrapure water to make up a 5% nitric acid matrix. After vortexing the samples were analyzed by ICP-MS.

**Procedure for selenium analysis**

Samples prepared as per the aforementioned digestion methods were analyzed using the ELAN 6000 ICP-MS. Quantification of Se was performed by the standard addition method using a four-point standard curve. A quality control sample (in similar matrix) was analyzed after every five samples and analysis was considered acceptable if the Se concentration of the quality control sample fell within ± 5% of the standard/reference value for the quality control.

## Results

Sheep that were administered MeSeCys, selenomethionine, selenate, and western aster had signs of depression, reduced food intake, tachypnea, and labored breathing. When forced to move the sheep would walk a few steps and stand with their necks outstretched while taking short rapid breaths. The onset of clinical signs was observed 6 h after administration of MeSeCys and 8 h after administration of selenate, selenomethionine, and western aster. Seven hours after administration of the selenium one sheep in the 6 mg Se/kg MeSeCys group died; the remaining three sheep in the same group died at 8, 8.5, and 11.5 h post-dosing. All three sheep in the selenate group died between 18 and 36 h post-dosing. Two of the four sheep in the western aster group died at 18 and 22 h post-dosing. Two of the three sheep in the selenomethionine group died at 22 and 31 h post-dosing. However, the sheep in the selenomethionine and western aster groups that did not die appeared to be recovered by 72 h post-dosing.

Breath of sheep dosed with MeSeCys and selenomethionine had a noticeable garlic odor within 30 min of dosing although the odor was stronger in sheep dosed with MeSeCys. The garlic-like odor was much slower to appear on the breath of sheep dosed with selenate and western aster and it never did reach the same intensity as sheep dosed with MeSeCys. The concentration of Se in 2 l of expired air is shown in Figure 1. Selenium reached peak concentrations (μg/2 l of air) of 6.485 ± 2.730, 2.872 ± 2.408 and 1.624 ± 1.017 in sheep dosed with MeSeCys, selenomethionine, and selenate, respectively.

Serum selenium concentrations in the dosed lambs are shown in Figure 2. The Se concentrations in serum peaked 4 h post-dosing at 3.078 ± 0.444 ppm in lambs administered MeSeCys. Selenium concentrations in lambs administered selenomethionine and selenate peaked 8 h post-dosing at 3.193 ± 0.337 ppm and 2.847 ± 0.237 ppm, respectively. Selenium concentrations in lambs administered western aster peaked 12 h post-dosing at 2.970 ± 0.255 ppm.

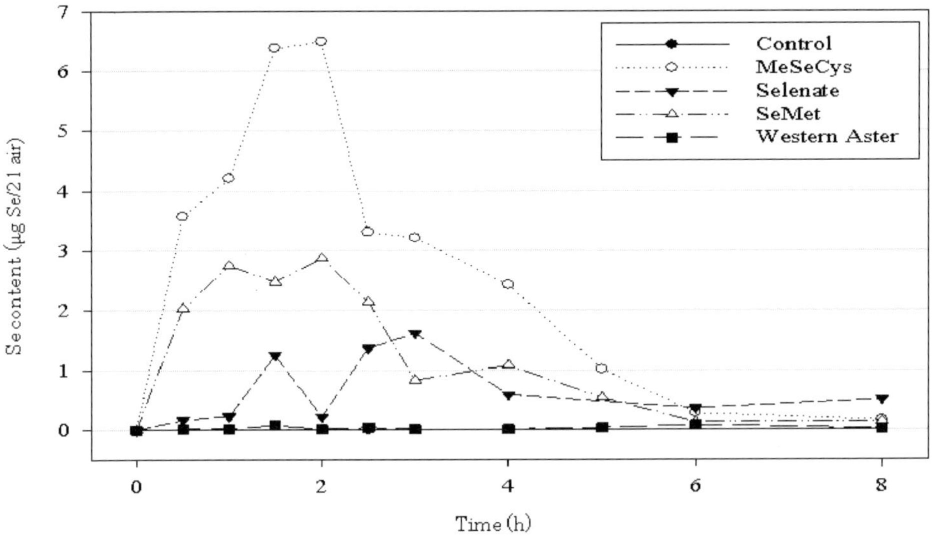


**Figure 1.** Respiratory elimination profile of Se in lambs administered 6 mg Se/kg as MeSeCys, selenomethionine, selenate, and western aster.

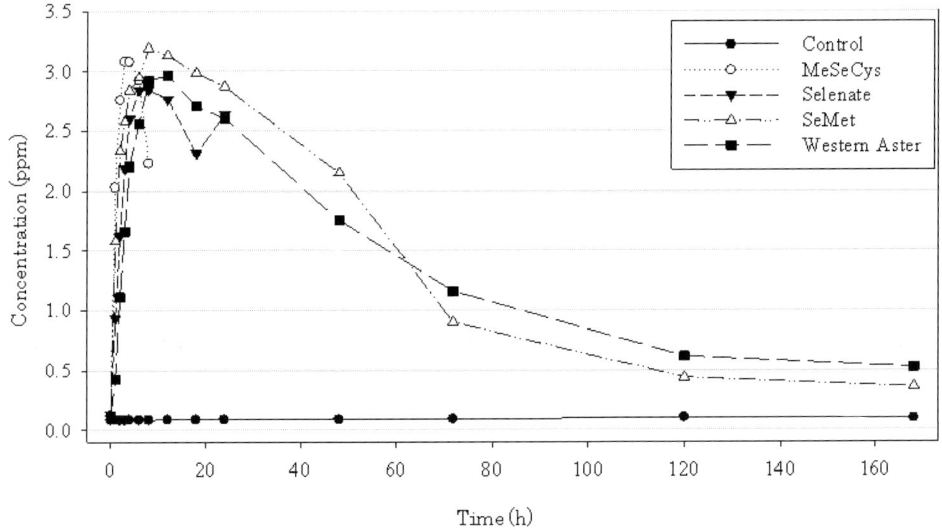


**Figure 2.** Serum selenium concentrations of lambs administered 6 mg Se/kg as MeSeCys, selenomethionine, selenate, and western aster.

Whole blood selenium concentrations in the dosed lambs are shown in Figure 3. Selenium concentrations in whole blood peaked sooner (6 h post dosing) and at higher concentrations (6.436 ± 1.521 ppm) in lambs administered MeSeCys than any other form of Se. Selenium concentrations in whole blood for sheep administered selenomethionine, selenate, and western aster peaked at 2.547 ± 0.098 ppm (6 h post-dosing), 1.892 ± 0.218 ppm (8 h post-dosing), and 1.876 ± 0.330 ppm (12 h post-dosing), respectively.

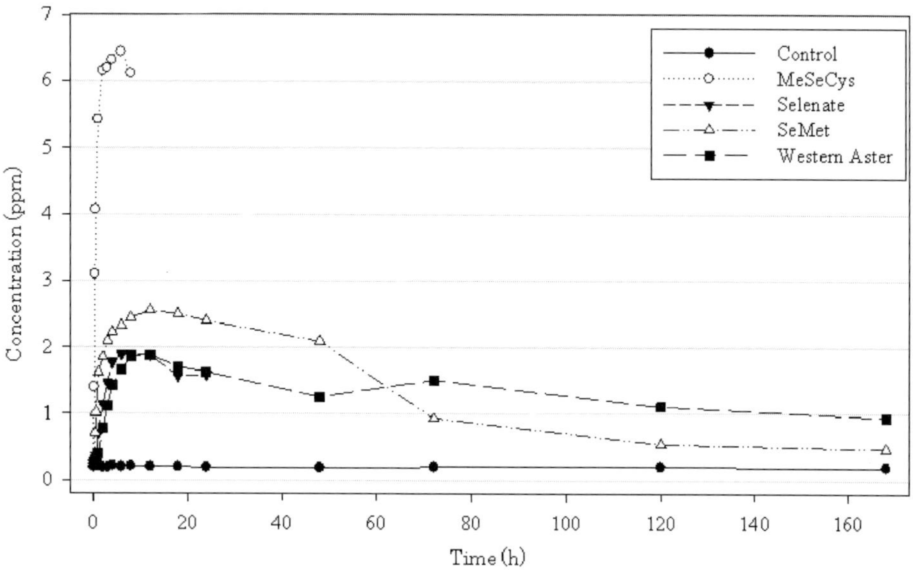


**Figure 3.** Whole blood selenium concentrations (ppm) of lambs administered 6 mg Se/kg as MeSeCys, selenomethionine, selenate, and western aster.

Mean Se concentrations in skeletal muscle, ventricle, lung, kidney, and liver of dosed lambs at the time of death or at 7 days post-dosing are reported in Table 1. Due to differences in time of deaths, lambs that died prior to the end of the study (7 days) would have had differing amounts of time to metabolize, distribute, and eliminate the Se. Thus, direct comparison at a single point in time was only possible at the termination of the study. It is of diagnostic importance to know concentrations that may occur in tissues at the time of death from Se poisonings. Selenium concentrations in all tissues of the control sheep were within the normal reference range.

## Discussion

The various chemical forms of selenium (MeSeCys, selenomethionine, selenate, and Se in western aster) that were dosed have very different rates of absorption and elimination. MeSeCys is much more rapidly absorbed than is selenomethionine and selenate indicating that MeSeCys may be absorbed in the rumen rather than later in the gastrointestinal tract where most forms of Se are absorbed. Sheep dosed with MeSeCys eliminate much more Se via respiration within 8 h after administration of Se than do lambs dosed with selenate or Se in western aster indicating a more efficient metabolic route to the volatile Se metabolites. The reason for the more rapid and efficient respiratory elimination of Se in lambs dosed with MeSeCys is most likely due to its more efficient conversion to methylselenol via the β-lyase pathway (Ohta *et al.* 2009). Methylselenol is an intermediate in the conversion of most Se forms to dimethylselenide and dimethyldiselenide, which are the most common forms of Se eliminated via respiration.

**Table 1.** Skeletal muscle, ventricle, lung, kidney, and liver Se concentrations (ppm, wet weight) at 7 days after dosing or at time of death in lambs administered 6 mg/kg Se.

| Se dosage mg Se/kg BW | | Muscle mean±SD | Ventricle mean±SD | Lung mean±SD | Kidney mean±SD | Liver mean±SD |
|---|---|---|---|---|---|---|
| *Control* | | | | | | |
| 0 mg | *n*=4 | 0.117±0.011 | 0.203±0.024 | 0.189±0.023 | 1.057±0.053 | 0.283±0.017 |
| *Se administered as MeSeCys* | | | | | | |
| (6 mg) | *n*=4 | (0.856±0.193) | (5.594±0.481) | 3.725±1.043 | 7.717±1.562 | 13.102±1.254 |
| *Se administered as Selenate* | | | | | | |
| (6 mg) | *n*=3 | (0.621±0.077) | (3.213±0.836) | (1.595±0.262) | (4.645±0.177) | (9.170±0.375) |
| *Se administered as Selenomethionine* | | | | | | |
| 6 mg | *n*=1 | 0.392 | 1.003 | 1.156 | 2.208 | 11.217 |
| (6 mg) | *n*=2 | (1.292±0.192) | (3.224±0.575) | (2.349±0.024) | (7.686±0.483) | (17.155±8.928) |
| *Se administered as Western Aster* | | | | | | |
| 6 mg | *n*=1 | 0.261 | 0.531 | 0.684 | 1.164 | 4.685 |
| (6 mg) | *n*=2 | (0.776±0.064) | (4.534±1.513) | (2.447±0.464) | (4.661±0.367) | (11.786±1.989) |

Peak Se concentrations in the serum were very similar for lambs in all groups dosed with Se but lambs dosed with MeSeCys reached peak serum Se concentrations in 4 h whereas peak Se concentrations were reached in 8, 8, and 12 h for lambs dosed with selenate, selenomethionine, and western aster, respectively. In contrast, peak Se concentrations in whole blood were very different when comparing lambs dosed with MeSeCys to the other three groups. The peak Se concentration in whole blood was reached sooner (6 h vs 8 or 12 h) and it was approximately 3.5 times greater than the Se concentrations from lambs dosed with selenate and western aster and 2.5 times greater than peak Se concentration in lambs dosed with selenomethionine. Additionally Se concentrations were 1.3, 1.7, 2.3, 1.7, and 1.4 times greater in muscle, ventricle, lung, kidney, and liver of lambs dosed with MeSeCys compared to lambs dosed with selenate. However, Se concentrations were higher in muscle, kidney, and liver in sheep dosed with selenomethionine when compared to sheep dosed with MeSeCys. The lower concentration in these tissues in sheep dosed with MeSeCys may be explained by their more rapid death (~8 h vs 24 h) thus having less time to distribute the Se to other tissues.

When diagnosing Se toxicity by measuring Se concentrations in respired air, whole blood, or tissues it is important to know the form of Se that was ingested because respiratory elimination, tissue accumulation, and whole blood kinetics are very different for different selenium forms.

## References

Freeman JL, Zhang LH, Marcus MA, Fakra S, McGrath SP, and Pilon-Smits EAH (2006). Spatial imaging, speciation, and quantification of selenium in the hyperaccumulator plants *Astragalus bisulcatus* and *Stanleya pinnata. Plant Physiology* 142:124-134.

Ohta Y, Kobayashi Y, Konishi S, and Hirano S (2009). Speciation analysis of selenium metabolites in urine and breath by HPLC- and GC-Inductively coupled plasma-MS after administration of selenomethionine and methylselenocysteine to rats. *Chemical Research Toxicology* 22:1795-1801.

Pickering IJ, Prince RC, Salt De, and George GN (2000). Quantitative, chemically specific imaging of selenium transformation in plants. *Proceedings National Academy Sciences USA* 97:10717-10722.

Shrift A and Virupaksha TK (1965). Seleno-amino acids in selenium-accumulating plants. *Biochim Biophys Acta* 100:65-75.

Tiwary AK, Panter KE, Stegelmeier BL, James LF, and Hall JO (2005). Evaluation of the respiratory elimination kinetics of selenium after oral administration in sheep. *American Journal Veterinary Research* 66:2142-2148.

Tiwary AK, Stegelmeier BL, Panter KE, James LF, and Hall JO (2006). Comparative toxicosis of sodium selenite and selenomethionine in lambs. *Journal of Veterinary Diagnosistic Investigation* 18:61-70.

Wilhelm A, Stegelmeier BL, Panter KE, and Hall JO (2007). Respiratory elimination of selenium in sheep given the accumulator plant *Symphotrichum spathulatum* (western mountain aster). *Proceedings, Western Section, American Society of Animal Science* 58:229-232.

# Chapter 92

# Agricultural and Pharmaceutical Applications of Chilean Soapbark Tree (*Quillaja saponaria*) Saponins

P.R. Cheeke

*Department of Animal Sciences, Oregon State University, Corvallis, OR 97331*

The Chilean soapbark tree (*Quillaja saponaria*) is native to semiarid regions of Chile. Its bark is a rich source of saponins. For hundreds of years, quillaja bark has been used in Chile by indigenous people to prepare shampoo because of the profuse foaming properties of quillaja saponins. During the 20th century to the present quillaja extract has had numerous applications such as a foaming agent in beverages, preparation of vaccine adjuvants, ore separation in mining, as a nematocidal agent in crop production, and as an animal feed additive. These applications and numerous others are a consequence of the surfactant activity of quillaja saponins. Quillaja saponins have a triterpenoid nucleus and two carbohydrate side chains. The nucleus (sapogenin) is lipid soluble while the side chains are water soluble, accounting for the surfactant properties. In addition to saponins quillaja contains polyphenolics and oligosaccharides. Traditionally the quillaja bark has been used as a source of saponins; a new process utilizes the entire woody biomass (San Martin and Briones 1999).

## Physiological Effects of Saponins

Saponins have diverse biological activities, many of which are a consequence of cholesterol binding. Saponins form irreversible complexes with cholesterol and other steroids such as bile acids. The hydrophobic portion of the molecules (the sapogenin) associates (lipophilic bonding) with the hydrophobic sterol nucleus in a stacked micellar aggregation (Oakenfull and Sidhu 1989). Saponins have hypocholesterolemic effects because they bind with cholesterol and bile acids in the gut, preventing enterohepatic recycling of cholesterol. Saponins are antiprotozoal agents because of their binding to cholesterol in protozoal cell membranes, causing microlesions and cell lysis (McAllister *et al.* 2001). Quillaja extract is used as an anticoccidial agent in cattle (Desert King International, unpublished research report). Quillaja extract is directly toxic to the protozoan *Histomonas meleagridis* that causes histomonosis (blackhead disease) in chickens and turkeys (Grabensteiner *et al.* 2007). Quillaja saponins have antiviral activity (Roner *et al.* 2007). Immersion of juvenile shrimp in a solution of sea water containing

©CAB International 2011. *Poisoning by Plants, Mycotoxins, and Related Toxins*
(eds F. Riet-Correa, J. Pfister, A.L. Schild, and T.L. Wierenga)

quillaja saponin significantly increased the resistance of the shrimp to a bacterial pathogen, *Vibrio alginolyticus* (Su and Chen 2007). In crop production a commercial quillaja extract preparation, QL-Agri, is used effectively as a nematocide. Thus quillaja saponins have anti-protozoal, antiviral, antibacterial, and nematocidal activity largely mediated via cholesterol binding.

## Applications of Quillaja in Animal Production

Dietary quillaja saponin administered to sows in late gestation reduces the incidence of stillborn piglets (Ilsley and Miller 2005): 13.25% in controls and 7.67% with quillaja. These authors also observed an immunostimulatory effect in the piglets with increased plasma IgG and IgA concentrations (Ilsley *et al.* 2005).

Numerous feeding trials with broiler chickens have demonstrated that quillaja powder as a feed additive has similar effects as antibiotic growth promotants. The results in the following trials were obtained with a combination of quillaja powder and *Yucca schidigera* whole plant powder. The product is referred to as Nutrafito Plus and contains approximately 85% quillaja powder. The trials were conducted with broiler chickens in Mexico (Trial 1) and Texas A&M University (Trial 2). The results (Table 1) have been published by Cheeke and Otero (2008). In both trials the Nutrafito Plus treatments showed growth promotant activity and improved feed/gain similar to or better than achieved with the positive controls containing antibiotic. The results suggest that a quillaja-yucca feed additive is a potential replacement for dietary growth promotant antibiotics. A possible mode of action of the saponin-containing Nutrafito Plus is an improvement in intestinal morphology. Schwarz *et al.* (2002) in Brazil observed that dietary quillaja powder resulted in beneficial changes in intestinal morphology, including increased villi length, decreased crypt of Lieberkuhn depth, and decreased mucosal thickness (Table 2).

**Table 1.** Growth and feed conversion (feed/gain) of broiler chickens fed quillaja-yucca powder (Nutrafito Plus).

| Treatment | Final body weight (kg) | Feed/Gain |
|---|---|---|
| Trial 1 (Mexico) | | |
| Negative Control | 2.411 $^{a}$ | 2.118 $^{a}$ |
| Positive control (PC) | 2.462 $^{b}$ | 2.077 $^{b}$ |
| PC + 100 ppm NF | 2.520 $^{c}$ | 2.029 $^{c}$ |
| PC + 150 ppm NF | 2.513 $^{c}$ | 2.029 $^{c}$ |
| PC + 100 ppm NF + F | 2.518 $^{c}$ | 2.035 $^{c}$ |
| PC + 150 ppm NF + F | 2.515 $^{,c}$ | 2.023 $^{c}$ |
| Trial 2 (Texas A&M) | | |
| Negative Control (NC) | 2.410 $^{a}$ | 1.79 $^{a}$ |
| NC + BMD50 | 2.570 $^{b}$ | 1.76 $^{a,b}$ |
| NC + 100 ppm NP | 2.620 $^{b}$ | 1.77 $^{a,b}$ |
| NC + 150 ppm NP | 2.610 $^{b}$ | 1.76 $^{b}$ |
| NC + 100 ppm NP + BMD50 | 2.620 $^{b}$ | 1.75 $^{b}$ |

$^{a,b,c}$ differ at $P < 0.05$

NF = Nutrafito Plus; F = Flavomycin; BMD50 = bacitracin

**Table 2.** Effects of dietary quillaja powder on broiler performance and intestinal mucosa (Schwarz *et al.* 2002).

| | Dietary treatment | | |
|---|---|---|---|
| | Negative control | Positive control | 300 ppm quillaja |
| Weight gain (g) | 2794 | 2793 | 2766 |
| Feed/gain | 1.90 | 2.05 | 1.95 |
| Villi height (μm) | 640.6 [b] | 631.5 [b] | 897.4 [a] |
| Crypt depth (μm) | 131.1 [b] | 134.9 [b] | 119.2 [a] |
| Mucosal thickness (μm) | 358.6 [b] | 268.9 [b] | 247.7 [a] |

## Conclusions

The saponins and other phytochemicals in the biomass of the Chilean soapbark tree (*Quillaja saponaria*) have numerous positive applications in animal production. The inclusion of low concentrations (100-150 ppm) of whole plant quillaja powder as a feed additive increases animal performance and improves gastrointestinal health.

## References

Cheeke PR and Otero R (2008). New alternative to replace antibiotic growth promoters. *World Poultry* 24(4):14-15.

Grabensteiner E, Arshad N, and Hess M (2007). Differences in the *in vitro* susceptibility of mono-eukaryotic cultures of *Histomonas meleagridis*, *Tetratrichomonas gallinarum* and *Blastocystis* sp. to natural organic compounds. *Parasitology Research* 101:193-199.

Ilsley SE and Miller HM (2005). Effect of dietary supplementation of sows with quillaja saponins during gestation on colostrum composition and performance of piglets suckled. *Animal Science* 80:179-184.

Ilsley SE, Miller HM, and Kamel C (2005). Effects of dietary quillaja saponin and curcumin on the performance and immune status of weaned piglets. *Journal of Animal Science* 83:82-88.

McAllister TA, Annett CB, Cockwill CL, Olson ME, Wang Y, and Cheeke PR (2001). Studies on the use of *Yucca schidigera* to control giardiosis. *Veterinary Parasitology* 97:85-99.

Oakenfull D and Sidhu GS (1989). Glycosides. In *Toxicants of Plant Origin* (PR Cheeke, ed.), vol. II, pp. 97-141. CRC Press, Boca Raton, Florida.

Roner MR, Sprayberry J, Spinks M, and Dhanji S (2007). Antiviral activity obtained from aqueous extracts of the Chilean soapbark tree (*Quillaja saponaria* Molina). *Journal of General Virology* 88:275-285.

San Martin R and Briones R (1999). Industrial uses and sustainable supply of *Quillaja saponaria* saponins. *Economic Botany* 53:302-311.

Schwarz KK, Franco SG, Fedalto LM, Borges SA, Fischer da Silva AV, and Pedroso AC (2002). Efeitos de antimicrobianos, probioticos, prebioticos e simbioticos sobre o desempenho e morfologia do jejuno de frangos. *Brazilian Journal of Poultry Science* (Suppl.) 4:75.

Su BK and Chen JC (2007). Effect of saponin immersion on enhancement of the immune response of white shrimp *Litopenaeus vannamei* and its resistance against *Vibrio alginolyticus*. *Fish and Shellfish Immunology* doi:10.1016/j.fsi.2007.09.002.

# Chapter 93

# Concentration and Effect in Mice of the Essential Oil Pulegone from *Mentha pulegium*, a Suspected Toxic Plant in Eastern Uruguay

J.M. Verdes[1], A. Moraña[1], V. Dehl[1], A. Ruiz-Díaz[1,2], E. Dellacasa[2], and F. Dutra[3]

[1]*Facultad de Veterinaria, Universidad de la República, Av. Alberto Lasplaces 1550, CP 11600, Montevideo, Uruguay;* [2]*Facultad de Química, Universidad de la República, Montevideo, Uruguay;* [3]*Dirección de Laboratorios Veterinarios 'Miguel C. Rubino', Treinta y Tres, Uruguay*

## Introduction

*Mentha pulegium* (L.) of the family Lamiaceae (Labiateae) is a weed native to Eurasia which occurs in rice farms in eastern Uruguay (Bonilla *et al.* 2002). *M. pulegium* has been suspected of being toxic to grazing cattle since 2001 during the summer drought, particularly in Uruguayan farms that alternate rice culture with beef cattle production. In one farm 20 out of 230 mainly Hereford and Hereford crossbred cattle exhibited respiratory distress, weight loss, and acute death in four cases (Machado and Dutra, personal communication).

Essential oil distilled from *M. pulegium* L. (pennyroyal oil) is an aromatic mint-like oil commonly used as a flavoring and fragrance agent (Gordon *et al.* 1982), in herbal medicine to induce menstruation and abortion in women among other effects (Sullivan *et al.* 1979; Chen *et al.* 2003; Ciganda and Laborde 2003; Soares *et al.* 2005), and as flea repellant in domestic animals (Sudekum *et al.* 1992).

The presence of pulegone in pennyroyal oil has also been associated with toxic effects, causing mainly acute death with centrilobular hepatic necrosis and diffuse pulmonary damage in rodents (Gordon *et al.* 1982), dogs (Sudekum *et al.* 1992), and humans (Sullivan *et al.* 1979). Pulegone can be oxidized by cytochrome P450 to reactive metabolites such as menthofuran, which is partly responsible for the toxicity observed in mice, rats, and humans (Gordon *et al.* 1987; Mizutani *et al.* 1987; Thomassen *et al.* 1990).

We report here the relative composition of the oil obtained from aerial parts of *M. pulegium* growing wild in farms where bovines were suspected to be affected by its ingestion. Additionally, the hepatotoxicity of the essential oil and its potential abortive effects in mice were evaluated under experimental conditions.

©CAB International 2011. *Poisoning by Plants, Mycotoxins, and Related Toxins* (eds F. Riet-Correa, J. Pfister, A.L. Schild, and T.L. Wierenga)

# Materials and Methods

### Botanical identification and chemical analysis

Considering possible environmental effects on the chemical characteristics of *M. pulegium*, samples of fresh leaves and stems representing the entire plant population studied were randomly collected at Los Ajos, Rocha Department (Uruguay, 54°10'W, 33°40'S) during the austral summer 2006 (January to February). Voucher specimens of *M. pulegium* were deposited at the Herbarium of the Facultad de Química, Universidad de la República, Montevideo (catalogue number MVFQ 4299). The essential oil was obtained from fresh leaves and stems by classical steam distillation for 2 h in a Clevenger-type apparatus (European Directorate 2002).

Samples of essential oil were analyzed by gas chromatography (GC) and gas chromatography-mass spectrometry (GC-MS) according to Lorenzo *et al.* (2002). Oil components were identified by comparison of their linear retention indices (LRIs) in the two columns (determined in relation to a homologous series of n-alkanes) with those of pure standards or literature reports. Comparison of fragmentation patterns in the MS with those stored on the GC-MS databases was also performed. The percentages of each component were reported as raw values without standardization.

### Biological activity and $LD_{50}$ in mice

In order to characterize the biological activity and to determine the minimum lethal dose of *M. pulegium* oil in mice (Gad and Chengelis 1992), five groups of six CD1 female pregnant mice, confirmed by presence of a vaginal plug after 2 days in contact with a male, were injected intraperitonially (i.p.) with a single dose of 1 ml of essential oil diluted at different levels with DMSO 0.5 % (v/v).

The groups were dosed as follows: 0 g/kg BW (control group), 0.5, 1, 2, and 4 g/kg BW (10-fold hepatic and lung damage dose used by Gordon *et al.* 1982).

The groups were closely observed for 48 h to record survival rates. Those mice that acutely died after injection or within 48 h post-treatment were necropsied and samples of blood, lung, heart, liver, kidney, and uterus were taken. Pregnant mice surviving at 48 h post-treatment were kept under daily observation until parturition in order to quantify litter size and viable newborns at birth.

Serological liver function tests were done including measurement of alanine aminotransferase (ALT) and aspartate aminotransferase (AST) activities. To evaluate the occurrence of necrosis and other histological hepatic lesions, liver samples were fixed in 10% buffered formalin and 5 μm paraffin-embedded sections were stained with hematoxylin and eosin for microscopic examination.

# Results and Discussion

The main constituents of the oil were pulegone (51.62%), isomenthone (20.97%), and menthone (14.33%) (Table 1).

The mice that received 4 g/kg BW showed severe respiratory distress, depression, coma, and death after 5-10 min, exhibiting cyanotic mucous membranes without other pathological or serological alterations.

**Table 1.** Percentage composition of the essential oil of *M. pulegium* and linear retention indices (LRI) of the components.

| Peak number | L.R.I.* | Constituent** | Percentage*** |
|---|---|---|---|
| 2 | 928 | α-pinene | 0.47 |
| 3 | 965 | sabinene | 0.09 |
| 4 | 968 | β-pinene | 0.41 |
| 5 | 978 | 3-octanone | 0.06 |
| 6 | 986 | β-myrcene | 0.36 |
| 7 | 982 | 1-octen-3-ol | 3.42 |
| 8 | 1021 | limonene | 0.93 |
| 9 | 1021 | 1,8-cineole | 0.08 |
| 10 | 1146 | menthone | 14.33 |
| 11 | 1159 | isomenthone | 20.97 |
| 12 | 1161 | neomenthol | 1.08 |
| 13 | 1178 | menthol | 0.79 |
| 14 | 1182 | isomenthol | 0.97 |
| 15 | 1241 | pulegone | 51.62 |
| 16 | 1247 | piperitone | 1.07 |
| 17 | 1410 | β-caryophyllene | 0.43 |
| 18 | 1445 | α-humulene | 0.63 |
| Monoterpene hydrocarbons | | | 2.26 |
| Oxygenated monoterpenes | | | 90.91 |
| Sesquiterpene hydrocarbons | | | 1.06 |
| Oxygenated sesquiterpenes | | | n.d.*** |
| Others | | | 3.48 |
| Total identified (%) | | | 97.71 |

*The components are reported according their elution order on SE-52. **Peak identifications are based on comparison of LRI values on two columns with those from pure standards or reported in the literature and on comparison of MS with file spectra. ***Relative proportions of the essential oil constituents were expressed as percentages obtained by peak-area normalization, all relative response factors being taken as one. Percentages were obtained on SE-52.

The group treated with 2 g/kg BW had similar clinical signs and death after a period from 2 to 24 h post treatment. Histological examination of liver samples showed evidence of mild and focal centrilobular necrosis of hepatocytes. There was also an increase in activities of ALT (1339 ± 138 U/l) and AST (2200 ± 658 U/l).

The mice injected with 1 g/kg BW presented respiratory distress with full recovery after 2 h and the group dosed with 0.5 g/kg BW showed mild respiratory signs and altered gait with complete recovery after 1 h.

An ANOVA test between groups of mice that recovered after treatment (controls, 0.5, and 1 g/kg BW) indicated no difference in litter size ($P \geq 0.05$) and a normal number of viable newborns were observed in these groups at birth.

## Conclusions

*Mentha pulegium* essential oil distilled from samples from Los Ajos (Rocha, Uruguay), when injected i.p. into mice was toxic, causing respiratory distress, depression, and death within 5 to 10 min (at 4 g/kg BW) or 2 to 24 h (at 2 g/kg BW). Focal centrilobular hepatic necrosis and altered enzyme activity of ALT and AST were also confirmed at 2 g/kg BW probably due to its major constituent, pulegone (Gordon *et al.* 1982). In our experimental conditions the abortive effect described in human folk medicine was not observed in pregnant mice that survived to 48 h post treatment at doses from 0.5 to 1 g/kg BW.

The respiratory distress, hepatic effects, and acute deaths described at high doses appear to be similar to those clinical features previously described in suspected field cases of *M. pulegium* toxicity in cattle.

## Acknowledgements

Financial support was provided by Comisión Sectorial de Investigación Científica (CSIC, UdelaR, Uruguay) and Comisión de Investigación y Desarrollo Científico (Facultad de Veterinaria, UdelaR, Uruguay). We thank Emilio Machado DVM (Rocha, Uruguay) and Prof Carmen García y Santos (Facultad de Veterinaria, UdelaR, Uruguay) for their valuable comments and Claudio Borteiro for manuscript proofreading.

## References

Bonilla O, Zorrilla G, Deambrosi E, and Deal E (2002). Unidad de Producción Arroz-Ganadera (UPAG) INIA 'Treinta y Tres'. *Revista del Plan Agropecuario (Vacunos de carne)*:36-42.

Chen LJ, Lebetkin EH, and Burka LT (2003). Comparative disposition of (R)-(+)-pulegone in B6C3F1 mice and F344 rats. *Drug Metabolism and Disposition* 31:892-899.

Ciganda C and Laborde A (2003). Herbal infusions used for induced abortion. *Journal of Toxicology. Clinical Toxicology* 41:235-239.

European Directorate for the Quality of Medicines (2002). *European Pharmacopoeia* 4th edn. European Directorate for the Quality of Medicines – Council of Europe. Maisonnneuve SA, Sainte Ruffine, France.

Gad SC and Chengelis CP (1992). *Animal Models in Toxicology*. Marcel Dekker, New York.

Gordon WP, Forte AJ, Mc Murtry RJ, Gal J, and Nelson SD (1982). Hepatotoxicity and pulmonary toxicity of pennyroyal oil and its constituent terpenes in the mouse. *Toxicology and Applied Pharmacology* 65:413-424.

Gordon WP, Huitric AC, Seth CL, Mc Clanahan RH, and Nelson SD (1987). The metabolism of the abortifacient terpene, (R)-(+)-pulegone, to a proximate toxin, menthofuran. *Drug Metabolism and Disposition* 15:589-594.

Lorenzo D, Paz D, Dellacassa, E Davies P, Vila R, and Cañigueral S, (2002) Essential Oils of *Mentha pulegium* and *Mentha rotundifolia* from Uruguay. *Brazilian Archives of Biology and Technology* 45:519-524.

Mizutani T, Nomura H, Nakanishi K, and Fujita S (1987). Effects of drug metabolism modifiers on pulegone-induced hepatotoxicity in mice. *Research Communications in Chemical Pathology and Pharmacology* 58:75-83.

Soares PMG, Assreuy AMS, Souza EP, Lima RF, Silva TO, Fontenele SR, and Criddle DN (2005). Inhibitory effects of the essential oil of *Mentha pulegium* on the isolated rat myometrium. *Planta Medica* 71:214-218.

Sudekum M, Poppenga RH, Raju N, and Braselton WE (1992). Pennyroyal oil toxicosis in a dog. *Journal of the American Veterinary Medical Association* 200:817-818.

Sullivan JB, Rumack BH, Thomas H, Peterson RG, and Bryson P (1979). Pennyroyal oil poisoning and hepatoxicity. *The Journal of the American Medical Association* 242:2873-2874.

Thomassen D, Slattery JT, and Nelson SD (1990). Menthofuran-dependent and independent aspects of pulegone hepatotoxicity: roles of glutathione. *The Journal of Pharmacology and Experimental Therapeutics* 253:567-572.

# Chapter 94

# Effect of MDL-Type Alkaloids on Tall Larkspur Toxicosis

K.D. Welch, D.R. Gardner, K.E. Panter, B.T. Green, D. Cook, J.A. Pfister, B.L. Stegelmeier, and T.Z. Davis

*USDA-ARS Poisonous Plant Research Laboratory, Logan, Utah 84341, USA*

## Introduction

Larkspurs (*Delphinium* spp.) are one of the most serious toxic plant problems on foothill and mountain rangelands in the western USA (Pfister *et al.* 1999). Total costs to the livestock industry have been estimated to be millions of dollars annually (Nielsen *et al.* 1994). The toxicity of larkspur plants is due to more than 18 norditerpenoid alkaloids which occur as one of two types: the 7, 8-methylenedioxylycoctonine (MDL)-type including deltaline and 14-O-acetyldictyocarpine (14-OAD) and the *N*- (methylsuccinimido) anthranoyllycoc-tonine (MSAL)-type including methyllycaconitine (MLA) (Figure 1) (Pfister *et al.* 1999). Although the MSAL-type alkaloids are much more toxic (Manners *et al.* 1991, 1993), the MDL-type alkaloids are generally more abundant (Pfister *et al.* 1999; Gardner *et al.* 2002).

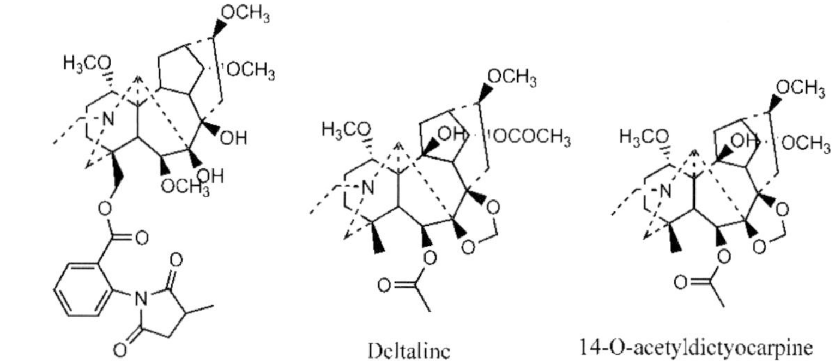

Deltaline

14-O-acetyldictyocarpine

Methyllycaconitine (MLA)

**Figure 1.** Structures of select norditerpenoid alkaloids in tall larkspur species.

Current management recommendations for grazing cattle on larkspur-containing ranges are based primarily on the concentration of MSAL-type alkaloids in the larkspur

©CAB International 2011. *Poisoning by Plants, Mycotoxins, and Related Toxins* (eds F. Riet-Correa, J. Pfister, A.L. Schild, and T.L. Wierenga)

(Pfister *et al.* 2002; Ralphs *et al.* 2002). *D. barbeyi* is one of the more problematic species of tall larkspur plants due to its high concentration of MLA. However, the most abundant norditerpenoid alkaloids in most *D. barbeyi* populations are the less toxic MDL-type alkaloids deltaline or 14-OAD (Manners *et al.* 1993; Pfister *et al.* 1999; Gardner *et al.* 2002). The relative concentration of these two alkaloids is location dependent with deltaline being more abundant in some populations while 14-OAD predominates in others (Gardner *et al.* 2002). Although the toxicities of MLA, 14-OAD, and deltaline have been determined individually (Manners *et al.* 1991; Panter *et al.* 2002) it is not known what effects the large concentration of deltaline or 14-OAD in these plants has on the toxicity of MLA. The contributions of MDL-type alkaloids on the overall toxicity of larkspurs were evaluated. First, the effects of deltaline and 14-OAD on the toxicity of MLA were assessed by comparing the lethality of i.v. administration of these alkaloids in mice. Second, the effective doses of tall larkspur collections that contain different ratios of MDL to MSAL-type alkaloids were determined in cattle.

## Materials and Methods

### Alkaloid preparation and analysis

Samples were quantitatively analyzed for total alkaloid content and MSAL-type alkaloid content using a Fourier transform infrared spectroscopy (FTIR) method previously described (Gardner *et al.* 1997). The purified larkspur alkaloids used in this study were extracted from *D. barbeyi* (Pelletier *et al.* 1981, 1989). Alkaloids, both individually purified alkaloids and a total alkaloid extract, were suspended in physiological buffered saline solution and the pH was lowered with HCl to achieve solubility. Ammonium hydroxide was then added to the solutions to raise the pH to as close to physiological pH (5.5 to 7.0) as possible while still retaining solubility. Solutions were stored in sterile injection vials at 4°C until use. No adverse effects were seen after injections of solutions (0.05 to 0.2 ml) with lower pH (5.5 to 7.0). The total alkaloid extract was analyzed by FTIR for measurement of MSAL-type and total alkaloids.

### Plant material

*D. barbeyi* was collected in the flowering stage during July 2003 near Manti, Utah (39°03.154'N 111°30.752'W, Poisonous Plant Research Laboratory (PPRL) collections number 03-12, at an elevation of approximately 3000 m). *D. glaucescens* was collected in the flowering stage during July 2008 near Dillon, Montana (45°25.888'N, 112°42.524'W, PPRL collections number 08-07, at an elevation of approximately 2500 m). The plant material was air-dried and ground to pass through a 2.4 mm mesh and mixed. After processing the plant was stored in plastic bags away from direct light at ambient temperature in an enclosed shed until use. Alkaloid analyses were performed prior to the start of the study.

### Animals

Median lethal dose ($LD_{50}$) was determined using male Swiss Webster mice (Simonson Laboratories Inc., Gilroy, CA) weighing 23±2 g. Between 0.05 and 0.2 ml of the purified alkaloid(s) in buffered saline were injected via the tail vein. Mice were observed for clinical effects and mortality and the $LD_{50}$ of the solutions was determined using a modified up and down method (Bruce 1987). This method is preferred because fewer animals are required, however, it results in unbalanced numbers in each group. The $LD_{50}$ values were calculated using SAS Proc Probit in a logistic regression (SAS V. 9, SAS Inst. Inc., Cary, NC).

Sixteen Angus steers (2 years old, 492±28 kg) were used for this study. The cattle were maintained on lucerne/grass hay with a mineral supplement. The cattle were fasted overnight prior to the day of the experiment. The steers were weighed and then restrained in a squeeze chute. Baseline physiological measurements of the cattle were recorded just prior to the administration of a single larkspur dose based on MSAL-type alkaloid content (mg MSAL-type alkaloids/kg BW). The dried finely ground larkspur was suspended in approximately 8 l of tap water and administered via oral gavage. After oral dosing the animals were monitored for 48 h for the development of clinical signs including muscle weakness and trembling, a decrease in G.I. motility, shuffling gait, and collapse. Twenty-four hours after oral dosing the animals were again restrained in a squeeze chute and physiological measurements obtained.

### Physiological monitoring of cattle

Heart rate in cattle was monitored as outlined previously (Green *et al.* 2009a). Briefly, data were recorded using an AD Instruments Powerlab and signals were amplified with an Octal Bioamp amplifier. Heart rate was monitored using 3M Red Dot model 2670 repositionable monitoring electrodes secured in place with a gel-based formulation of cyanoacrylate adhesive. The leads were placed as described by Chen *et al.* (2002) with the positive electrode placed on the right scapula and the negative electrode on the sternum adjacent to the heart. A ground electrode was attached to the perineum. The heart rate signal was amplified with a gain range of ±500 μV. The heart rate signal was filtered with a mains filter, 60 Hz notch filter, 120 Hz low-pass; 0.1 Hz high-pass filter and digital band-pass filter with a high cut-off frequency of 45 Hz and a low cut-off frequency of 0.1 Hz. The cyclic measurements feature of ADI Chart software package was used to calculate heart rate in beats/min. The heart rate in each animal was allowed to stabilize before analysis (typically 5 min). After stabilization a 5 min period of heart rate was sampled. Five minute periods of heart rate were measured prior to the dosing of cattle with larkspur and 24 h after dosing.

### Analysis and statistics

Data are expressed as the mean±SD. Confidence (fiducial) intervals (95%) were calculated for $LD_{50}$ values using logistic regression. Statistical analyses were performed using SigmaStat for Windows (version 3.1). Statistical comparisons of $LD_{50}$ values between groups were performed using ANOVA with a posthoc test of significance between individual groups. Statistical comparisons between two groups (0 and 24 h) were made using a standard Student's *t*-test. Differences were considered significant when $P < 0.05$.

## Results

The acute toxicity of MLA was compared to the toxicity of MDL-type alkaloids administered individually versus their co-administration as mixtures with MLA having the following composition: 1:1, 1:5, and 1:25 MLA to MDL-type alkaloid. The $LD_{50}$ for MLA alone was 4.4±0.7 mg/kg BW whereas the $LD_{50}$ for deltaline alone was 113.3±6.4 mg/kg BW. Even though deltaline was approximately 25 times less toxic than MLA the co-administration of deltaline with MLA affected lethality (Figure 2). There was a dose-dependent increase ($P < 0.05$) in toxicity as the ratio of deltaline to MLA was increased from 1:1 to 1:5 to 1:25 with their respective $LD_{50}$ values of 2.7±0.3, 2.5±0.2, and 1.9±0.1 mg/kg. Similar results ($P < 0.05$) were obtained when 14-OAD was co-administered with MLA (Figure 2). There were no differences in the clinical signs or the time to death among any of the treatment groups.

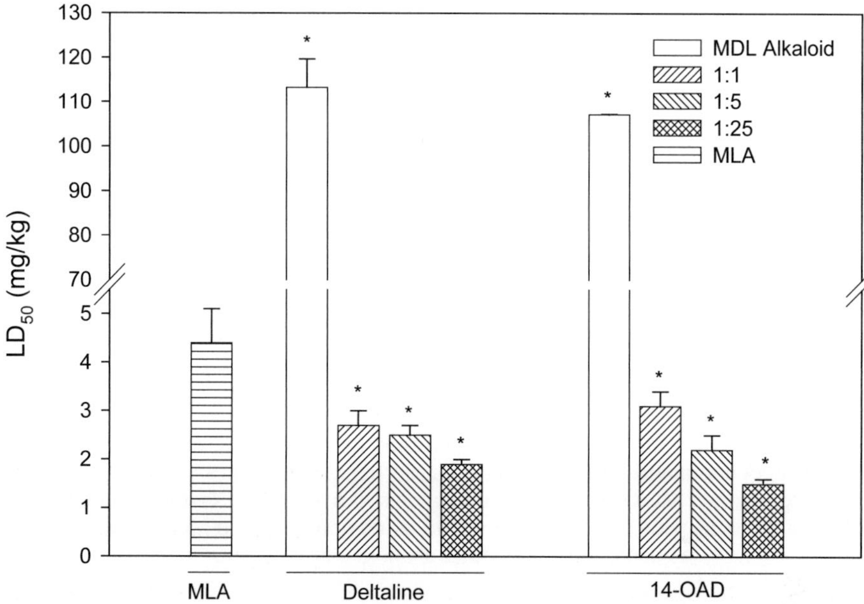


**Figure 2.** The effect of co-administration of various MDL-type alkaloids on the toxicity of MLA. The data represent the $LD_{50}$ of MLA alone, MDL-type alkaloids alone, and MLA plus MDL-type alkaloids at ratios of 1:1, 1:5, and 1:25 MLA to MDL-type alkaloids. Results represent the mean ± SD of 24 to 86 mice per group; *$P < 0.05$ as compared to the MLA group.

To assess the validity of the additive effect of MDL-type alkaloids on the toxicity of MLA, toxicity of a total alkaloid extract from *D. barbeyi* was tested. The total alkaloid extract contained approximately a 1:5 ratio of MLA to MDL-type alkaloids as determined by FTIR with deltaline and 14-OAD being the predominant MDL-type alkaloids in the extract. The $LD_{50}$ of the total alkaloid extract (2.0±0.2 mg/kg BW) was lower ($P < 0.05$) than that of pure MLA and was very similar ($P > 0.05$) to that of the mixtures of MLA and either deltaline or 14-OAD at a 1:5 ratio ($LD_{50}$: 2.5±0.2 and 2.0±0.2 mg/kg BW, respectively) (Figure 3).

For the experiments with cattle, two different populations of tall larkspur were collected, a *D. barbeyi* and a *D. glaucescens* collection. Samples from each population were analyzed for total alkaloid content and MSAL-type alkaloid content using the FTIR method. The *D. barbeyi* collection contained 16.0 mg/g of total alkaloids of which 3.9 mg/g were MSAL-type alkaloids (Table 1). Thus, the Manti larkspur had a 3.1 to 1 ratio of MDL- to MSAL-type alkaloids. The *D. glaucescens* collection contained 13.4 mg/g of total alkaloids of which 8.2 mg/g were MSAL-type alkaloids (Table 1). Thus, this population of larkspur had a 0.6 to 1 ratio of MDL- to MSAL-type alkaloids. The concentration of MSAL-type alkaloids in these collections were used as the basis for calculating the doses that were given to cattle.

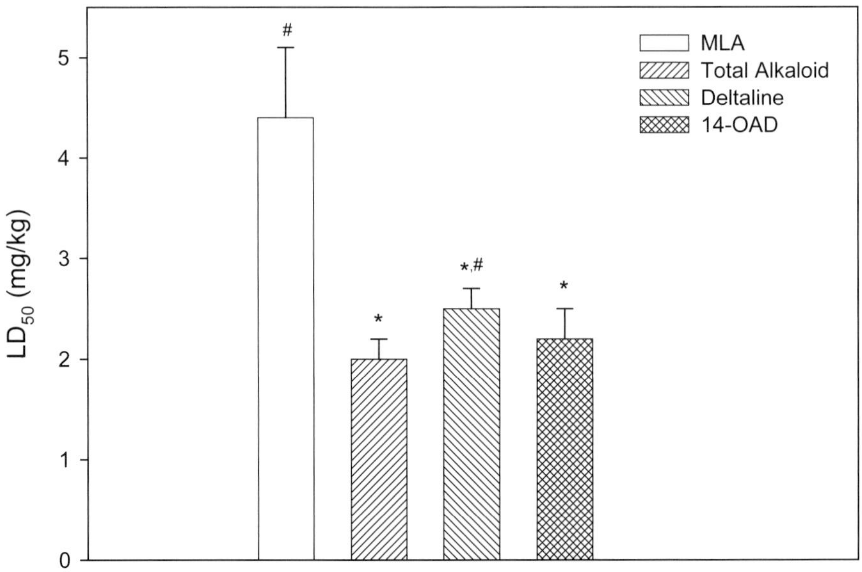


**Figure 3.** Comparison of the toxicity of a total alkaloid extract from *D. barbeyi* versus the co-administration of purified alkaloids. The data represent the $LD_{50}$ of MLA alone, a total alkaloid extract, and MLA plus MDL-type alkaloids at a 1:5 ratio. Results represent the mean±SD of 25 to 86 mice per group; $^{*}P < 0.05$ as compared to the MLA group; $^{\#}P < 0.05$ as compared to the total alkaloid extract group.

**Table 1.** The MSAL-type alkaloid and total alkaloid content of various tall larkspur populations.

| Larkspur | MSAL mg/g | MDL mg/g | Total Alkaloid mg/g | MDL : MSAL |
|---|---|---|---|---|
| *D. barbeyi*; Manti, UT | 3.9 | 12.1 | 16.0 | 3.1 |
| *D. glaucescens*; Dillon, MT | 8.2 | 5.1 | 13.4 | 0.6 |

For the cattle study we considered an effective dose: the amount of plant material that would significantly increase the heart rate and elicit clinical signs of poisoning. Our reference point for starting the experiment was the Manti collection with a dose of 8 mg MSAL/kg BW as previous research in our laboratory has shown that this dose causes an elevation in heart rate and muscle weakness but generally not to the extent that the animal becomes recumbent. Treatment of five steers with *D. barbeyi* at 8 mg MSAL/kg BW increased ($P = 0.014$) heart rate from a baseline of 75±9 beats/min (bpm) to 102±16 bpm 24 h after dosing (Table 2). A sixth steer was dosed with *D. barbeyi*, however, at 24 h it was sternally recumbent and consequently heart rate analysis was not performed. Treatment of four steers with *D. glaucescens* at 12 mg MSAL/kg BW did not change heart rate 24 h after treatment ($P = 0.353$). The heart rate was 62±10 bpm at baseline vs. 73±17 bpm at 24

h. Increasing the dose of the *D. glaucescens* collection to 14 and 15 mg MSAL/kg BW did not alter ($P > 0.20$) the heart rate. A dose of 18 mg MSAL/kg BW of the *D. glaucescens* collection was required to increase ($P = 0.041$) heart rate from a baseline of 65±14 bpm to 100±21 bpm at 24 h.

**Table 2.** Dose-response relationships of different tall larkspur populations for producing changes in heart rate in cattle[a].

| | Dose mg Alkaloid/kg BW | | | Heart Rate[b], bpm | |
|---|---|---|---|---|---|
| Larkspur | MSAL | MDL | Total | 0 h | 24 h |
| *D. barbeyi*; Manti, UT | 8 | 25 | 33 | 75± 9 | 102±16* |
| *D. glaucescens*; Dillon, MT | 12 | 8 | 20 | 62±10 | 73±17 |
| | 14 | 9 | 23 | 61±14 | 69± 5 |
| | 15 | 10 | 25 | 55± 3 | 74±18 |
| | 18 | 11 | 29 | 65±14 | 100±21* |

[a]Cattle were orally dosed with varying amounts of different tall larkspur populations and heart rate was monitored at 0 and 24 h.
[b] Data represent the mean±SD of heart rate from 3-6 animals. * $P < 0.05$ as compared with baseline (0 h).

In addition to changes in heart rate we also monitored cattle for overt clinical signs of poisoning including muscle weakness and trembling, a decrease in GI motility, shuffling gait, and collapse. In every dose for both plant populations there was an obvious visual change in fecal consistency with animals typically producing dry feces 24 h after treatment. However, the cattle had no obvious difficulty defecating until the dose administered also caused an increase in heart rate. The cattle dosed with the *D. barbeyi* collection at 8 mg MSAL/kg BW showed fine muscular tremors in the head and shoulder regions 7 h after dosing. Muscle weakness was even more pronounced 24 h after dosing as one of six animals was sternally recumbent. The cattle dosed with the *D. glaucescens* collection did not show any clinical signs of poisoning until a dose of 15 mg MSAL/kg BW was reached and then the signs were very minor. Cattle dosed with *D. glaucescens* at 18 mg MSAL/kg BW had very noticeable clinical signs with two of six animals sternally recumbent 24 h post dosing.

## Discussion

Previous research has demonstrated that the MSAL-type alkaloids are much more toxic than the MDL-type alkaloids (Manners *et al.* 1993, 1995). Consequently, current management recommendations for grazing cattle on larkspur-containing ranges are based primarily on the concentration of MSAL-type alkaloids in larkspur (Pfister *et al.* 2002; Ralphs *et al.* 2002). However, in many species of tall larkspur the MDL-type alkaloids are generally more abundant (Pfister *et al.* 1999; Gardner *et al.* 2002). Until now, it was not clear if a high concentration of MDL-type alkaloids in larkspur plants increases the toxicity or if the toxicity of larkspur plants is solely attributable to the MSAL-type alkaloids.

The results from the mouse experiments demonstrated that using the two most abundant MDL-type–alkaloids, deltaline and 14-OAD, were essentially the same in that they both caused a dose-dependent increase in the toxicity when co-administered with

MLA (Figure 2). The effect of these two alkaloids on the toxicity of MLA was additive as their co-administration with MLA at 1:1, 1:5, and 1:25 ratios resulted in decreases in the $LD_{50}$ by approximately 25%, 50%, and 60%, respectively, versus that of MLA alone. Solutions containing both MLA and a MDL-type alkaloid showed increased toxicity compared to MLA alone, suggesting that MDL-type alkaloids exacerbate MSAL toxicity and therefore play an important part in the toxicity of larkspur plants.

One key aspect of this study was demonstrating that the toxicity of a mixture of pure MLA and MDL-type alkaloids at a 1:5 ratio had similar toxicities as a solution of a total alkaloid extract from *D. barbeyi* that contained approximately 5 times as much MDL-type alkaloids (predominantly deltaline and 14-OAD) as MLA. The slightly lower $LD_{50}$ value for the total alkaloid extract could be explained by the small amount of 14-deactylnudicauline, another MSAL-type alkaloid found in the extract. These results suggest that use of purified compounds gives a close approximation to the overall toxicity of the plant itself.

We dosed cattle with ground plant material collected from populations of tall larkspur that have inherently different concentrations of MDL- and MSAL-type alkaloids. We used ground plant material for two main reasons. First, it would be difficult to isolate and purify sufficient amounts of pure norditerpenoid alkaloids to dose cattle. Second, the utilization of plant material more closely reflects the grazing situation associated with larkspur poisoning of cattle than giving alkaloid extracts. Two different populations of tall larkspurs known to contain a wide spectrum of MDL- to MSAL-type alkaloid ratios were chosen for the study. We hypothesized that a larkspur population with a high MDL-type alkaloid concentration will be more toxic than a larkspur population with a low MDL-type alkaloid concentration, given similar MSAL-type alkaloid content.

The results of this study clearly demonstrate that as the ratio of MDL- to MSAL-type alkaloids decreased, the amount of plant material required to raise heart rate in cattle increased. A decrease in the MDL- to MSAL-type alkaloid ratio from 3.1:1 to 0.6:1 required the dose to be increased from 8 to 18 mg MSAL/kg BW in order to achieve an elevated heart rate. Coincidentally the dose that was observed to elevate heart rate was above 26 mg total alkaloid/kg BW for each population. Consequently it could be argued that surpassing a threshold of total alkaloid is all that is required to elevate heart rate. However, we recently found a correlation between the increase in heart rate associated with larkspur intoxication and serum MLA concentrations ($P$ = 0.0001) but not deltaline ($P$ = 0.2), an MDL-type alkaloid (Green *et al.* 2009b). Additionally, we observed that cattle dosed at 37.6 mg total alkaloid/kg BW with a tall larkspur population that contains almost exclusively MDL-type alkaloids showed no elevation in heart rate (unpublished data). These results are in agreement with our mouse portion of the study and demonstrate that MDL-type alkaloids increase the toxicity of larkspur plants by potentiating the toxicity of the MSAL-type alkaloids.

Even though higher concentrations of MSAL-type alkaloids are required to elicit clinical signs in animals dosed with larkspur containing reduced concentrations of MDL-type alkaloids, the difference in the total amount of plant material dosed was small and well within the quantity that a cow could eat in a rangeland setting. The amount of dried plant material required for an effective dose of the *D. barbeyi* collection was 960±41 g and 1066±49 g for the *D. glaucescens* collection. Even though the concentration of the MSAL-type alkaloids is the most important factor, the results from this study suggest that the MDL-type alkaloids play an important role in the toxicity of larkspur plants by potentiating the toxicity of the MSAL-type alkaloids.

It is noteworthy that the effect of the MDL-type alkaloids appears to be more pronounced in cattle than mice. The results from the mouse study indicate that a change in the MDL:MSAL from 1:1 to 5:1 resulted in a 7% change in the $LD_{50}$. However, in the cattle study a change in the MDL:MSAL from 1:1 to 3:1 resulted in 63% difference in the dose based on MSAL-type alkaloid content required for an effective dose. There are a number of potential reasons for the differences between these two studies including: (i) a difference in the endpoints used in the two studies, a lethal dose versus an effective dose; (ii) a difference in the nicotinic acetylcholine receptors between cattle and mice which could result in the MDL-type alkaloids being more toxic in cattle than in mice; (iii) a difference in dosing purified compounds i.v. versus dosing ground plant material orally; and (iv) differences in the metabolism and subsequent toxicokinetic profile of norditerpenoid alkaloids in cattle and mice. The data from the mouse study suggested that there was no effect of the MDL-type alkaloids on the elimination of MLA from the mice (data not shown). However, it is possible that in cattle large quantities of MDL-type alkaloids hinder the elimination of the MSAL-type alkaloids thus increasing the bioavailability of the more toxic MSAL-type alkaloids and effectively increasing the toxic potential of the plant material. A recent study by Green *et al.* (2009b) demonstrated that the increase in heart rate in cattle poisoned with larkspur is directly correlated with the serum MLA concentrations and not the serum deltaline concentrations. Additional studies will be performed in the future to determine if the elimination of MSAL-type alkaloids differs between the two populations of larkspur used in this study.

One note of caution for making management recommendations based on the results of this study is that the animals in this study were dosed with a single bolus dose using ground plant material. A single bolus dose of ground plant material does not accurately represent the conditions under which animals are poisoned on the range. It has been demonstrated that there are three distinct thresholds involved in tall larkspur toxicosis (Pfister *et al.* 2002). First, a subclinical toxicosis that results in reduced tall larkspur consumption for 1 to 3 days but no overt signs nor overall reductions in consumption of other forage. Second, a short-acting toxicosis with overt clinical signs results in reduced food intake for several days but no long term effects. Third, a potentially fatal toxicosis with severe clinical signs that may result in death. It has been postulated that cyclic consumption enables cattle to generally regulate larkspur consumption below the second threshold in a typical range setting, which allows most cattle the opportunity to use an otherwise nutritious plant (Pfister *et al.* 2002). Consequently, future studies need to be conducted using various dosing regimens and forms of larkspur that more realistically mimic grazing situations before final recommendations are made.

In conclusion, the MSAL-type alkaloids such as MLA cause greater toxicity than MDL-type alkaloids and are the primary factors responsible for the toxicity of larkspur plants. Consequently, for a larkspur plant to be toxic to livestock a sufficient quantity of MSAL-type alkaloids is required. However, MDL-type alkaloids appear to potentiate the overall toxicity of the MSAL-type alkaloids and should be considered when predicting potential toxicity of larkspur populations. Therefore, when chemical analyses are performed on larkspur plants to assess their toxic potential the concentration of both the MSAL-type and total alkaloids should be determined with more weight given to the MSAL-type alkaloids. Finally, the results from this study indicate that larkspur plants containing large amounts of MDL-type alkaloids in addition to high MSAL-type alkaloid content should be considered potentially more dangerous to cattle than plants with only high MSAL-type alkaloids.

## Acknowledgements

The authors wish to thank Kendra Dewey and Scott Larsen for their expert technical support; Al Maciulis, Rex Probst, and Danny Hansen for assistance with animal care and handling; and Jessie Roper and Anita McCollum for making the plant collections.

## References

Bruce RD (1987). A confirmatory study of the up-and-down method for acute oral toxicity testing. *Fundamental and Applied Toxicology* 8:97-100.

Chen W, Nemoto T, Kobayashi T, Saito T, Kasuya E, and Honda Y (2002). ECG and heart rate determination in fetal cattle using a digital signal processing method. *Animal Science Journal* 73:545-551.

Gardner DR, Manners GD, Ralphs MH, and Pfister JA (1997). Quantitative analysis of norditerpenoid alkaloids in larkspur (*Delphinium* spp.) by Fourier transform infrared spectroscopy. *Phytochemical Analysis* 8:55-62.

Gardner DR, Ralphs MH, Turner DL, and Welsh SL (2002). Taxonomic implications of diterpene alkaloids in three toxic tall larkspur species (*Delphinium* spp.). *Biochemical Systematics and Ecology* 30:77-90.

Green BT, Pfister JA, Cook D, Welch KD, Stegelmeier BL, Lee ST, Gardner DR, Knoppel EL, and Panter KE (2009a). Effects of larkspur (*Delphinium barbeyi*) on heart rate and electrically evoked electromyographic response of the external anal sphincter in cattle. *American Journal of Veterinary Research* 70:539-546.

Green BT, Welch KD, Gardner DR, Stegelmeier BL, Davis TZ, Cook D, Lee ST, Pfister JA, and Panter KE (2009b). Serum elimination profiles of methyllycaconitine and deltaline in cattle following oral administration of larkspur (*Delphinium barbeyi*). *American Journal of Veterinary Research* 70:926-931.

Manners GD, Pfister JA, Ralphs MH, Panter KE, and Olsen JD (1991). Larkspur chemistry: Toxic alkaloids in tall larkspurs. *Journal of Range Management* 45:63-67.

Manners GD, Panter KE, Ralphs MH, Pfister JA, Olsen JD, and James LF (1993). Toxicity and chemical phenology of norditerpenoid alkaloids in the tall larkspurs (*Delphinium* species). *Journal of Agricultural and Food Chemistry* 41:96-100.

Manners GD, Panter KE, and Pelletier SW (1995). Structure-activity relationships of norditerpenoid alkaloids occurring in toxic larkspur (*Delphinium*) species. *Journal of Natural Products* 58:863-869.

Nielsen DB, Ralphs MH, Evans JS, and Call CA (1994). Economic feasibility of controlling tall larkspur on rangelands. *Journal of Range Management* 47:369-372.

Panter KE, Manners GD, Stegelmeier BL, Lee S, Gardner DR, Ralphs MH, Pfister JA, and James LF (2002). Larkspur poisoning: toxicology and alkaloid structure-activity relationships. *Biochemical Systematics and Ecology* 30:113-128.

Pelletier SW, Kulanthaivel P, and Olsen JD (1989). Alkaloids of *Delphinium barbeyi*. *Phytochemistry* 28:1521-1525.

Pelletier SW, Daily Jr OD, Moody NV, and Olsen JD (1981). Isolation and structure elucidation of alkaloids of *Delphinium glaucescens* Ryb. *The Journal of Organic Chemistry* 46:3284-3293.

Pfister JA, Gardner DR, Panter KE, Manners GD, Ralphs MH, Stegelmeier BL, and Schoch TK (1999). Larkspur (*Delphinium* spp.) poisoning in livestock. *Journal of Natural Toxins* 8:81-94.

Pfister JA, Ralphs MH, Gardner DR, Stegelmeier BL, Manners GD, Panter KE, and Lee ST (2002). Management of three toxic *Delphinium* species based on alkaloid concentrations. *Biochemical Systematics and Ecology* 30:129-138.

Ralphs MH, Gardner DR, Turner DL, Pfister JA, and Thacker E (2002). Predicting toxicity of tall larkspur (*Delphinium barbeyi*): measurement of the variation in alkaloid concentration among plants and among years. *Journal of Chemical Ecology* 28:2327-2341.

# Chapter 95

# LC/MS/MS Analysis of the Daphnane Orthoester Simplexin in Poisonous *Pimelea* Species of Australian Rangelands

M.T. Fletcher, K.Y.S. Chow, R.G. Silcock, and J.A. Milson

*Department of Employment, Economic Development and Innovation, Health and Food Sciences Precinct, PO Box 156, Archerfield Qld 4108, Australia*

## Introduction

*Pimelea* species (also known as riceflowers) are ephemeral native plants found throughout inland regions of Queensland (Qld), New South Wales (NSW), South Australia (SA), and the Northern Territory (NT), extending over about one-quarter of Australia's pastoral lands. Three species of *Pimelea* (*P. simplex*, *P. elongata*, and *P. trichostachya*) are poisonous to livestock and potentially fatal to cattle with serious economic consequences through loss of production, stock deaths, and the costs of agistment. The associated poisoning syndrome in cattle is unique to Australia and characterized by pulmonary venule constriction leading to right ventricular dilation and subcutaneous edema of brisket and head. Consumption of plant material can also lead to acute diarrhea in cattle and sheep.

Feeding trials in the early1970s established *Pimelea* spp. as the cause of this syndrome (Clark 1971a, b, 1973; McClure and Farrow 1971) and the primary toxin was identified as the novel daphnane orthoester simplexin **1** (Roberts *et al.* 1975; Freeman *et al.* 1979). A number of compounds of related structure have also been isolated from these *Pimelea* species including huratoxin **2** and12β-acetoxyhuratoxin **3** (Zayed *et al.* 1977; Freeman *et al.* 1979; Hafez *et al.* 1983). However the incidence of poisoning remains difficult to predict and there is a lack of clear understanding of why some properties or animals are affected by *Pimelea* poisoning when others are not. In this study, liquid chromatography/mass spectrometry/mass spectrometry (LC/MS/MS) analysis of more than 700 plant samples enabled toxin levels to be related to plant species, stage of growth, and other environmental factors to provide a sound basis for further epidemiological studies.

©The State of Queensland (through the Department of Employment, Economic Development and Innovation) 2011. *Poisoning by Plants, Mycotoxins, and Related Toxins* (eds F. Riet-Correa, J. Pfister, A.L. Schild, and T.L. Wierenga)

## Materials and Methods

### Plant collections

*Pimelea* plant specimens were collected in affected regions of central Australia and the stage of growth, nature of the site, and location coordinates recorded together with a separate pressed sample for identification by the Queensland Herbarium. Air-dried samples were separated into aerial portion, main stem, and root. Aerial portion included flower heads, seeds, leaves, and branches and represents the portion of the plant most likely to be consumed by grazing cattle. Each portion was milled and stored frozen prior to analysis.

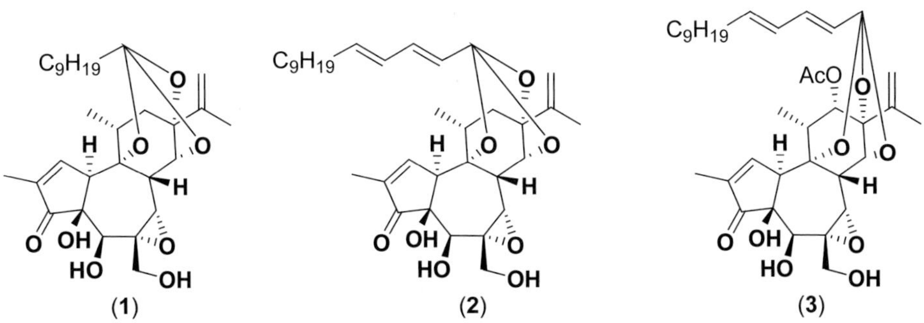


**Figure 1.** Chemical components identified in *P. trichostachya*, *P. simplex*, and *P. elongata* including the major toxin simplexin (1) and related compounds huratoxin (2) and and 12β-acetoxyhuratoxin (3).

### Field Weathering Studies

A known amount of coarsely chopped aerial shoot material (litter) and ripe seed samples for each of the three species was put into individual mesh bags and placed in fenced plots at four different locations in random grid arrangements in early 2007. The mesh bag weathering trials were located near Longreach (Qld), Mitchell (Qld), Marree (SA), and Broken Hill (NSW). The majority of the mesh bags were pressed onto the soil surface by wire mesh and a small number of additional bags from *P. simplex* and *P. trichostachya* were buried at shallow depth at the Longreach site. Bags were retrieved for testing at regular intervals over the next 2 years (three replicates at each collection).

### Plant extraction

Milled plant material (0.5 g) was shaken overnight with 80% methanol in water (20 ml). A portion of the extract (2 ml) was transferred to a glass tube and solvent evaporated under nitrogen. The residue was taken up in dichloromethane (4 ml) and washed with sodium chloride solution (5 ml). The dichloromethane extract was dried, solvent evaporated, and the residue partitioned between acetonitrile (4 ml) and hexane (10 ml). The hexane layer was washed with acetonitrile (2 ml). Solvent was evaporated from the combined acetonitrile extract and the residue taken up in methanol for LC/MS/MS analysis. All analyses were calculated on a dry weight basis.

### Simplexin standard

Simplexin standard was obtained by extraction of a milled bulk stem and root sample of *P. trichostachya* (AQ751555) followed by solvent partitioning and chromatography. The identity of the isolated simplexin was confirmed by NMR comparisons with literature values (Freeman *et al.* 1979) and shown to be ≥ 95% pure by HPLC-ELSD.

### LC/MS/MS analysis

LC separations were performed on a 2.1×100 mm SunFire $C_{18}$ 3.5 μm column (Waters) with an initial eluent of 90% methanol/water containing 0.1% formic in each solvent, increased to 100% methanol in 15 min, and held at this concentration for a further 10 min gradient elution. MS detection was by atmospheric pressure chemical ionization in positive mode (APCI+) on a Quattro Premier triple quadrupole mass spectrometer (Micromass).

Simplexin was quantified in plant samples by multiple reaction monitoring (MRM) in the MS/MS mode, monitoring 533>253 where 533 is the protonated molecular ion of simplexin ($[M+H]^+$) and 253 is one of its predominant daughter ions. A secondary MRM 533>267 was used as a confirmatory transition. Levels of simplexin in plant extracts were quantitated by comparison with external simplexin standard solutions prepared in methanol (0.5-5.5 mg/l). Related orthoesters could be detected by analogous transitions (e.g. huratoxin MRMs of 585>253 and 585>267).

## Results and Discussion

More than 700 *Pimelea* plant samples have been analyzed in phytochemical studies conducted across *P. elongata*, *P. simplex* subsp. *continua*, *P. simplex* subsp. *simplex*, and *P. trichostachya* samples collected in this project from various locations in Qld, SA, and NSW. There is a somewhat surprising uniformity of toxin composition across these taxa albeit with significant variations in level dependent on stage of growth and species.

### Simplexin levels in *Pimelea* species and plant parts

Simplexin was the major analyte in all taxa with varying minor levels of related components including huratoxin which was consistently present, particularly in green young plant material. Simplexin levels in both *P. trichostachya* and *P. elongata* were higher (580 and 540 mg/kg in flowering foliage, respectively) compared with *P. simplex*, which had maximum simplexin levels of only 255 mg/kg (Figure 2). Levels of huratoxin were somewhat higher in *P. simplex* (relative to simplexin) than in *P. trichostachya* or *P. elongata* and this is seemingly consistent with the report by Freeman *et al.* (1979) that one sample of *P. simplex* contained almost equal amounts of huratoxin and simplexin.

Whilst the toxin profile in each of the three species is similar there were distinct differences in the location of toxins in plant parts of each species. Representative analyses of flowering/post-flowering samples of each species are shown in Table 1. In *P. elongata*, highest simplexin levels were seen in root and flower heads, but with significant levels also in branches, stem, and leaves. In *P. simplex* flower heads and roots contained similar simplexin levels with very little toxin detected in branches, stem, and leaves. Flower heads of *P. trichostachya* contained high simplexin levels with much lower levels seen in other plant parts, including roots.

**Simplexin levels at different growth stages**

The concentration of different toxins in a plant is often linked to plant growth stage and health (or vigor). Toxin levels measured in *Pimelea* plants of different growth stages showed that the simplexin level is generally higher in pre-flowering to flowering plants and decreases through flowering to post-flowering stages (Figure 2). However, these results showed considerable variation in simplexin concentrations between different populations of the same species, even at the same growth stage, as indicated by wide ranges of results in Figure 2.

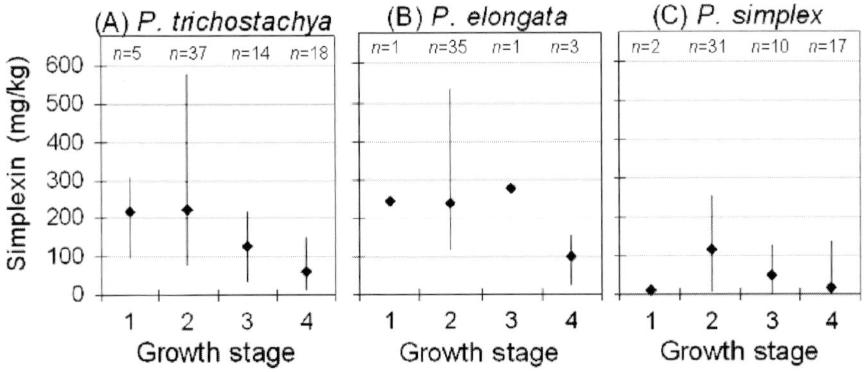


**Figure 2.** Simplexin content of (A) *P. trichostachya*, (B) *P. elongata*, and (C) *P. simplex* aerial plant material showing range of concentration (vertical bar) and mean (♦) at four growth stages: 1 = pre-flowering; 2 = flowering; 3 = post flowering/late seeding; 4 = dead/dry stalks.

**Table 1.** Simplexin distribution in plant parts of representative flowering/post-flowering specimens of each *Pimelea* species.

| | Simplexin concentration (mg/kg) | | |
|---|---|---|---|
| Sub-sample | *P. elongata* | *P. simplex* subsp. *simplex* | *P. trichostachya* |
| Flowers & seeds | 341 | 253 | 709 |
| Branches | 161 | <LOQ | 70 |
| Main stem | 195 | <LOQ | 48 |
| Leaves | 244 | 22 | 49 |
| Root | 409 | 281 | 66 |

LOQ = limit of quantification

To investigate the impact of stage of growth alone it is necessary to exclude the impact of environmental factors such as soil type, moisture, and temperature and this can best be achieved by examining multiple samples collected from the same population at the one site. To this end multiple *P. elongata* samples at various growing stages were collected from a property near Bollon, Qld. These plants had grown in close proximity to each other and were collected simultaneously for analyses. The mean levels of simplexin measured in the combined aerial (above ground) portion of the plant and the root from each stage of growth are shown in Table 2. This demonstrates the decline in simplexin level in the post-flowering stage to about one-fifth that in the pre-flowering sample. Simplexin levels in the roots are less variable.

Similarly, *P. trichostachya* samples at different development stages collected from a single site near Jericho (Qld) at various time intervals also illustrate the reduction of the toxin level in the aerial sub-sample as the plants grew and aged (Table 3). Flowers and seeds have a much higher simplexin level than other foliage (Table 1) and loss of flowers/seeds within the seed dispersal phase correlates with the observed decrease of simplexin level in bulk aerial foliage of post-flowering plants.

**Table 2.** Simplexin concentration in *P. elongata* at different stages of growth collected from a single location near Bollon, Qld on the same date.

| Stage of growth | Plant height above ground (cm) | Simplexin concentration (mg/kg) | |
|---|---|---|---|
| | | aerial | root |
| pre-flowering/flowering | 7 | 589 | not available |
| flowering | 11 | 404 | 661 |
| flowering | 17 | 410 | 743 |
| post-flowering | 23 | 134 | 559 |

**Table 3.** Simplexin concentration in aerial sub-samples of *P. trichostachya* at different stages of growth, collected from a single location near Jericho, Qld over several weeks.

| Stage of growth | Plant height above ground (cm) | Simplexin concentration (mg/kg) |
|---|---|---|
| pre-flowering | 15 | 309 |
| flowering, seeding | 21 | 219 |
| post-flowering, late seeding | 33 | 191 |

## Weathering Effects on the Degradation of Simplexin in Seeds and Litter

Weathering trials investigating the degradation of simplexin in plant material have demonstrated more rapid breakdown of the toxin in litter compared to seeds under the same weathering conditions. The simplexin concentration of the three replicates at each time interval were averaged and plotted against length of exposure at each site (Figure 3).

Litter of *P. trichostachya* had the highest concentration of simplexin in the three species. In this species, the simplexin content of the litter showed a gradual decline with only low levels remaining at 18 months. The simplexin level was approximately 20 mg/kg at all sites by 9-18 months. The *P. simplex* plant material collected for the weathering trial was dry aged material and contained comparatively low levels of simplexin with initial simplexin levels of less than 15 mg/kg, making weathering results less conclusive.

Unlike the results from the litter samples no significant decrease has occurred in the seed samples after 18 months of exposure. The simplexin content of the mesh bag seed samples remains fairly constant in *P. trichostachya* (Figure 3) at all three sites. As the toxin is concentrated in the interior of the seed the seed coat is probably serving its purpose to protect the embryo and slowing the rate of the sample decomposition and so the toxin as well. With *P. simplex* seed there was some variation in simplexin levels during the trial (Figure 3) and this may represent some inconsistency in the sample makeup for this species which contained a large proportion of unfilled seeds and extraneous plant debris. *P. elongata* seed bags were only exposed at one site, Longreach (Qld). Unfortunately insects (presumably ants) have removed virtually all seed from these bags hence there are no chemical analysis results for seeds of this species. *P. elongata* litter (also at only one site) contained only low levels of simplexin and is not plotted.

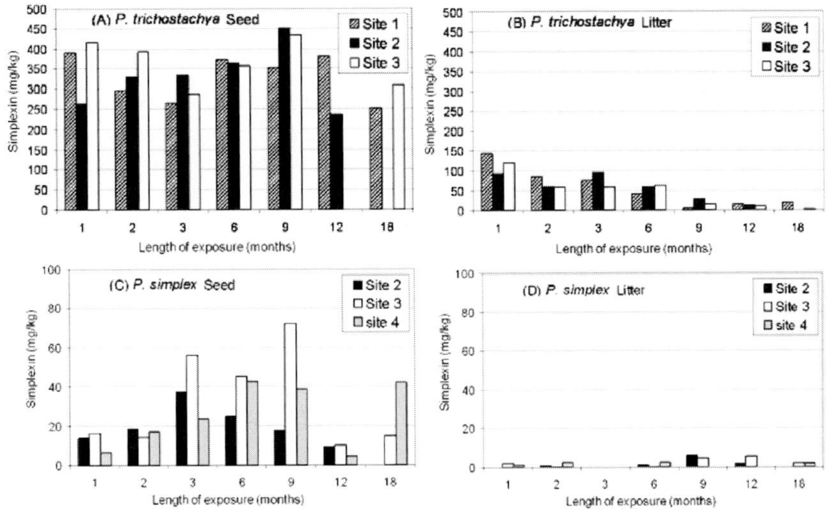


**Figure 3.** Simplexin content of *P. trichostachya* and *P. simplex* subsp. *continua* plant material weathered in mesh bags at three sites for up to 18 months. (Site 1 = Mitchell Qld, Site 2 = Broken Hill NSW, Site 3 = Longreach Qld, and Site 4 = Maree, SA). All Site 2 samples at 18 months and Site 3 *P. trichostachya* 12 month seed samples were lost.

## Conclusion

Young green pimelea plants are generally avoided by stock, which is fortunate as plants at this stage contain high levels of toxin. Older dry *Pimelea* plants growing among palatable herbage and grasses are more likely to be inadvertently consumed by stock. Dry and dead stalks have reduced simplexin levels but these levels can still be significant, particularly where adhering seeds are still present. Toxin levels in weathered seeds persist for many months, while levels in other plant parts are diminished.

## Acknowledgements

This study was partly funded by the Australian Government Natural Heritage Trust through AgForce Queensland.

## References

Clark IA (1971a). St George disease of cattle. *Australian Veterinary Journal* 47:123.
Clark IA (1971b). A note on the pathogenesis of St George disease of cattle. *Australian Veterinary Journal* 47:285-286.

Clark IA (1973). The pathogenesis of St George disease of cattle. *Research in Veterinary Science* 14:341-349.

Freeman PW, Ritchie E, and Taylor WC (1979). The constituents of Australian *Pimelea* spp. I. The isolation and structure of the toxin of *Pimelea simplex* and *P. trichostachya* Form B responsible for St. George disease of cattle. *Australian Journal of Chemistry* 32:2495-2506.

Hafez A, Adolf W, and Hecker E (1983). Active principles of the thymelaeaceae. III. Skin irritant and cocarcinogenic factors from *Pimelea simplex*. *Planta Medica* 49:3-8.

McClure TJ and Farrow BR (1971). Chronic poisoning of cattle by desert rice flower (*Pimelea simplex*) and its resemblance to St. George disease as seen in north-western New South Wales. *Australian Veterinary Journal* 47:100-102.

Roberts HB, McClure TJ, Ritchie E, Taylor JD, and Freeman PW (1975). The isolation and structure of the toxin of *Pimelea simplex* responsible for St George disease of cattle. *Australian Veterinary Journal* 51:325-326.

Zayed S, Hafez A, Adolf W, and Hecker E (1977). New tigliane and daphnane derivatives from *Pimelea prostrata* and *Pimelea simplex*. *Experientia* 33:1554-1555.

# Chapter 96

# The Physiological Effects and Toxicokinetics of Tall Larkspur (*Delphinium barbeyi*) Alkaloids in Cattle

B.T. Green, K.D. Welch, J.A. Pfister, D. Cook, B.L. Stegelmeier, S.T. Lee, D.R. Gardner, and K.E. Panter.

*USDA-ARS Poisonous Plant Research Laboratory, Logan, Utah 84341, USA*

## Introduction

Toxic larkspur (*Delphinium*) species have been responsible for large losses to the cattle industry in western North America since the beginning of the 20th century (Marsh *et al.* 1916; Pfister *et al.* 1999, 2003). The initial symptoms of larkspur poisoning in cattle include lack of appetite, general uneasiness, nausea, rapid pulse and respiration, and a stiff staggering gait (Marsh *et al.* 1934; Nation *et al.* 1982). As the poisoning proceeds bloating, respiratory depression, tremors in locomotor muscles leading to more generalized tremors, failure of voluntary muscular coordination, and finally collapse to sternal or lateral recumbency occur (Olsen *et al.* 1990). The poisoning of cattle from the consumption of larkspur has been attributed to diterpenoid alkaloids produced by the plant and found in high concentration in plant tissues. Total diterpenoid alkaloid content can represent 3% of plant dry weight and is a mixture of 10-15 alkaloids belonging to three structural classes– the norditerpenoid alkaloids; $C_{20}$-diterpenoid alkaloids; and *bis*-diterpenoid alkaloids–that vary in relative abundance between species of larkspur (Olsen *et al.* 1990). The norditerpenoids are the most toxic and can be further subdivided into two main structural groups, the *N*-(methylsuccinimido) anthranoyllycoctonine type (MSAL-type) and 7,8-methylenedioxylycoconine type (MDL-type) norditerpenoid alkaloids of which the MSAL-type alkaloids are the most toxic (Olsen *et al.* 1990; Manners *et al.* 1993, 1995). The most abundant members of the MSAL-type alkaloids group include methyllycaconitine (MLA), nudicauline, and 14-deacetylnudicauline (Gardner *et al.* 1997). Plants high in MSAL-type alkaloids are thought to be the most toxic to cattle and the concentrations of these alkaloids have been used for the prediction of plant toxicity (Pfister *et al.* 2002; Ralphs *et al.* 2002).

Of the MSAL-type alkaloids found in larkspur species MLA is the most thoroughly investigated. MLA has been described as possessing curariform activity and is a potent competitive blocker of nicotinic acetylcholine receptors in autonomic neurons and voluntary striated muscle (Benn and Jacyno 1983). MLA at nanomolar concentrations is a potent and selective competitive antagonist of $\alpha_7$-nicotinic acetylcholine receptors and has an affinity value in the nanomolar range at $\alpha_7$-nicotinic acetylcholine receptors and in the

©CAB International 2011. *Poisoning by Plants, Mycotoxins, and Related Toxins* (eds F. Riet-Correa, J. Pfister, A.L. Schild, and T.L. Wierenga)

micromolar range at muscle-type $\alpha_4\beta_2$ and $\alpha_3\beta_4$-nicotinic acetylcholine receptors (Ward *et al.* 1990; Alkondon *et al.* 1992; Lopez *et al.* 1998; Sharples and Wonnacott 2001). Less is known about the toxicological properties of nudicauline and 14-deacetylnudicauline. Nudicauline has nanomolar affinity at $\alpha_7$-nicotinic acetylcholine receptors and functional studies have shown that nudicauline and 14-deacetylnudicauline have $IC_{50}$ values in the micromolar range in the lizard sciatic nerve extensor digitorum longus preparation (Hardick *et al.* 1996; Dobelis *et al.* 1999).

Larkspur has received significant research attention at higher alkaloid concentrations which elicit acute clinical signs of poisoning (Pfister *et al.* 1999). However, there is little information available on doses of larkspur alkaloids which cause subacute physiological effects in cattle. The present study was designed to test the hypothesis that toxic larkspur produces subacute alterations in heart rate and external anal sphincter tone after its ingestion in cattle and that these effects can be mitigated by inhibitors of acetylcholinesterase.

## Materials and Methods

### Animals

Mixed-breed beef cows were used in the dose-response study and black Angus steers were used in the toxicokinetic study. The cattle were maintained on lucerne-grass hay with a mineral supplement for at least three weeks before and between dosing trials. Animals were handled frequently so that they were very tractable, and they were habituated to the experimental protocol as previously described (Green *et al.* 2009a). Baseline physiological measurements of the cattle were recorded just prior to the administration of a single larkspur dose. The dried finely ground larkspur or dried finely ground pasture grass mixture (used as a control) was administered via oral gavage in approximately 11 l of tap water. For the dose-response studies, after oral dosing the animals were monitored as previously described and released to an individual pen and monitored as previously described (Green *et al.* 2009a). For the toxicokinetic studies, 18 h prior to the start of the study, a 16 ga indwelling catheter was placed in the jugular vein of each steer as previously described (Green *et al.* 2009a). All animal work was done under veterinary supervision with the approval and supervision of the Utah State University Institutional Animal Care and Use Committee.

### Plant material

Tall larkspur (*Delphinium barbeyi*) in the early flowering stage was collected during July of 2003 above Manti, Utah, USA (lat 39°03.154'N, long 111°30.752'W, PPRL collections number 03-12) at an elevation of approximately 3000 m above sea level. A voucher specimen was deposited at the Utah State University Herbarium (#237494). The plant material was air-dried, ground to pass through a 2.38 mm mesh, and mixed using a Gehl Mix-All model 55 (Gehl Company, West Bend, WI, USA). After processing, the plant was stored in plastic bags away from direct light at room temperature until use. An improved mixed pasture grass hay was similarly ground and used as a control for the dosing of dried ground plant material.

### Chemical analysis

Replicate samples ($n$=5) were analyzed for alkaloid content using methods previously described (Gardner *et al.* 1997, 1999).

### Physiological monitoring

All data were simultaneously recorded using an AD Instruments Powerlab, and signals were amplified with an Octal Bioamp amplifier (AD Instruments Inc, Colorado Springs, CO, USA). Heart rate was monitored using 3M Red Dot model 2670 repositionable monitoring electrodes (3M Corporation, St Paul, MN, USA) cemented in place with a gel-based formulation of cyanoacrylate adhesive (Henkel Consumer Adhesive, Inc, Avon, OH, USA). A large human anal canal electromyography (EMG) probe was obtained from SRS Medical (Redmond, WA, USA) and used to measure electrically evoked responses of the external anal sphincter (i.e. the EMG response). The leads were placed as described by Chen *et al.* (2002). A ground electrode was attached to the perineum. Signals were filtered as previously described (Green *et al.* 2009a).

### Data analysis

Data are expressed as the mean±SE of each physiological response. The kinetic profiles of MLA and deltaline were analyzed using standard pharmacokinetic software. A curve-stripping procedure was used to determine the basic pharmacokinetic parameters of rate for the elimination phase of the MLA and deltaline concentration curves. The following parameters were determined: $t_{1/2}=0.693/k_{elimination}$, $C_{max}$, $T_{max}$, and area under the curve (AUC). The $t_{1/2}$ is the elimination half life, and $C_{max}$ and $T_{max}$ describe the maximum serum alkaloid concentration and time of maximal serum alkaloid concentrations, respectively. A trapezoidal method was used to determine the AUC of a concentration vs time graph. Determination of alkaloid dose-response relationships through nonlinear regression methods and statistical analysis of neostigmine and physostigmine data were performed using GraphPad Prism version 4.03 for Windows (GraphPad Software, San Diego, CA, USA). Comparisons of the EMG response to baseline values were made after normalization of the EMG response after larkspur treatment as a percentage of the baseline response measured in each animal using a one-sample $t$-test. Comparisons between a control mean to a single treatment mean were made by paired two-tailed $t$-test. Comparisons between control and treatment means were made by ANOVA followed by Dunnett's test, and comparisons between multiple means post-ANOVA were made with a Tukey-Kramer test. Dose-response data was analyzed and curves statistically compared as described (Miller 2003). In all cases, the limit for statistical significance was set at $P < 0.05$.

## Results

Changes in heart rate and EMG response were detectable at 24 h after the administration of dried ground larkspur containing norditerpenoid alkaloids to cattle. The administration of this plant extract in equivalent MSAL-type alkaloid doses ranging between 0.5-11.8 mg/kg increased heart rate and decreased the tone of the external anal sphincter in a dose-dependent manner (Figures 1 and 2).

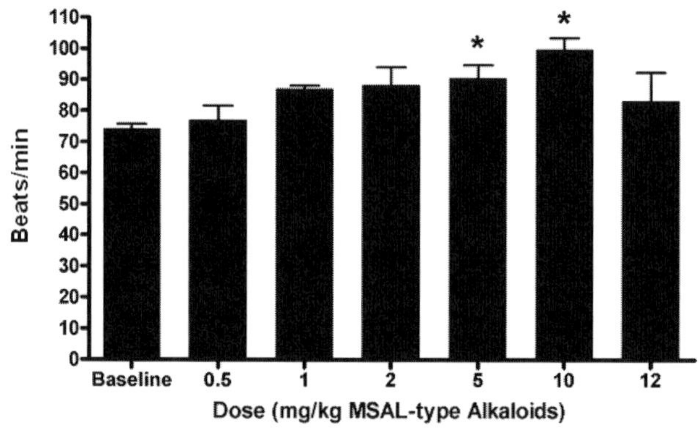


**Figure 1.** Dose-response relationship of norditerpenoid alkaloids for producing changes in heart rate in cattle. Cattle were orally dosed with varying amounts of larkspur containing 5.8 mg/g dry weight of MSAL-type alkaloids and monitored for heart rate (beats/minute; bpm). Each data point represents three animals for all doses except for the equivalent 10.4 mg/kg alkaloid dose which represents responses in nine animals. The $ED_{50}$s for the heart rate was 1.7 mg/kg (95% confidence intervals; 0.3 to 9.0). *= $P < 0.05$ compared to baseline, ANOVA, Dunnett's test. Data obtained from Green *et al.* (2009a).

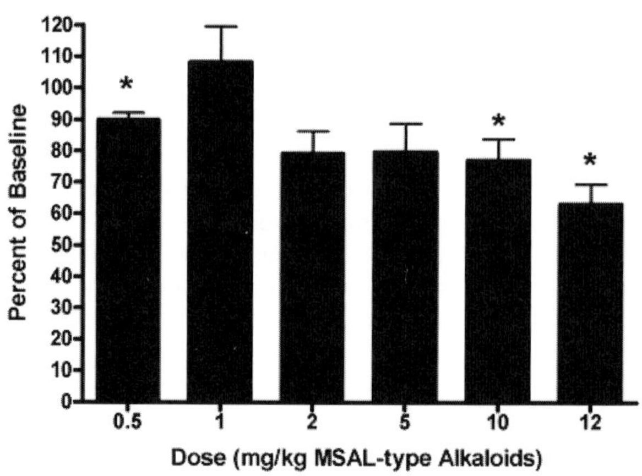


**Figure 2.** Dose-response relationships of norditerpenoid alkaloids for producing changes in EMG responses in cattle. Cattle were orally dosed with varying amounts of larkspur containing 5.8 mg/g dry weight of MSAL-type alkaloids and monitored for EMG response (percent of control) at time = 0 and 24 h after dosing. Each data point represents the mean±SE of responses at 24 h from three animals for all doses except for the equivalent 10.4 mg/kg alkaloid dose which represents responses in nine animals. The $ED_{50}$s for the EMG response was 47.6 mg/kg (95% confidence intervals 3.5 to 655.7 mg/kg). *$P < 0.05$, response versus 100%, one-sample *t*-test. Data obtained from Green *et al.* ( 2009a).

The baseline heart rate and baseline RMS value of the EMG response evoked by electrical stimulation were 74±2 bpm and 187±10 μV respectively from 27 samples of the 12 cattle used in this study. Of the six larkspur doses given only equivalent doses of 4.9 and

10.4 mg/kg of toxic alkaloids increased the heart rate when compared to baseline values 24 h after oral dosing ($P < 0.0001$ one-way ANOVA, $P < 0.05$, and 0.01, respectively, Dunnett's test, $n$=3 animals and 9 animals for 4.9 and 10.4 mg/kg, respectively). The maximum increase in heart rate (100±4 bpm, $n$=9 animals) was observed at an equivalent alkaloid dose of 10.4 mg/kg. The EMG response to larkspur was different from baseline at 0.5, 10.4, and 11.8 mg/kg equivalent doses of toxic alkaloids ($P$ = 0.045, 0.009, 0.027, respectively, one-sample $t$-test, $n$=3, 3, and 9 animals at 0.5, 10.4 and 11.8 mg/kg, respectively).

The highest equivalent dose of toxic larkspur alkaloids used in this study was 14.5 mg/kg in three cows. Physiological responses from these animals were not recorded at 24 h after larkspur administration because severe muscle weakness and postural instability led them into sternal or lateral recumbency. Cattle were given a 0.02 mg/kg intramuscular dose of neostigmine if they were unable to rise after being approached by a researcher and were so weak that they could not ambulate and would be likely to easily fatigued during physiologic testing (Table 1).

For the toxicokinetic component of the study, five steers were orally dosed with dried ground larkspur at an equivalent dose of 10.4 mg/kg MLA and 11.0 mg/kg deltaline, heart rate was recorded continuously, and venous blood was sampled periodically over 96 h. The five steers showed no overt clinical signs of poisoning during the entire 96 h of the experiment. The heart rate peaked 17 h after dosing with a mean of 79±5 bpm ($n$=4; one steer had removed its monitoring leads at 17 h and they were subsequently reattached). This was significantly different from both the mean baseline heart rate of 59±2.0 bpm ($P$ = 0.0043, two-tailed $t$-test, $n$=5) and the mean rate of the negative controls at 17 h (52±4 bpm, $P$ = 0.0049, $t$-test, $n$=4). After reaching the maximum heart rate at 17 h it declined to a value of 54±6 bpm at 96 h ($n$=5) (Figure 3). The results of the toxicokinetic analysis are described elsewhere (Green *et al.* 2009b).

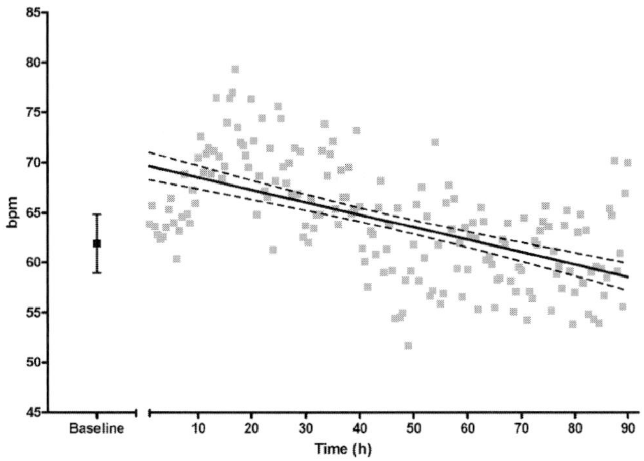


**Figure 3.** The mean heart rate over 90 h of five steers gavaged with dried ground larkspur at an equivalent dose of 10.4 mg/kg MLA and 11.0 mg/kg deltaline. Baseline data were recorded 30 min prior to oral dosing. Baseline data are presented as the mean heart rate in bpm±SE. The best fit line with corresponding 95% confidence intervals is displayed from t = 0 to t = 90 h.

**Table 1.** Clinical observations of animals dosed with 14.54 mg/kg MSAL-type alkaloids in the form of dried ground larkspur.

| Animal number | Predominant breed | Neostigmine (mg/kg i.m.) | Time | Clinical signs[a] |
|---|---|---|---|---|
| 5 | Hereford | | 0930 | Dosed with larkspur |
| | | | 1200 | none |
| | | | 1300 | none |
| | | | 1730 | Sternal recumbency, tried to stand upon approach |
| | | 0.02 | 1849 | Sternal recumbency, labored breathing, very weak |
| | | | 1900 | Standing, no signs |
| | | | 1000 | No signs of intoxication |
| 7 | Angus | | 0915 | Dosed with larkspur |
| | | | 1200 | none |
| | | | 1300 | none |
| | | | 1400 | Collapse to lateral recumbency, labored breathing |
| | | 0.02 | 1403 | Dosed with neostigmine |
| | | | 1408 | Standing |
| | | | 1413 | Periodic collapse |
| | | | 1650 | Sternal recumbency |
| | | 0.02 | 1030 | Very weak, sternal recumbency, labored breathing |
| 8 | Angus | | 0900 | Dosed with larkspur |
| | | | 1200 | none |
| | | | 1300 | none |
| | | 0.02 | 1530 | Sternal recumbency, labored breathing |
| | | | 1540 | Standing, periodic collapse |
| | | | 1730 | Sternal recumbency |
| | | | 1408 | Standing |
| | | | 1413 | Standing, periodic collapse |
| | | | 1650 | Sternal recumbency |
| | | 0.02 | 1000 | Very weak, sternal recumbency, labored breathing |

[a]Typical sequence of clinical signs at equivalent doses of 10.5 mg/kg toxic alkaloids which are similar to that described by Pfister *et al.* (1994a, b).

## Discussion

The larkspur collection used in this study contained norditerpenoid alkaloids of which deltaline and MLA are in the highest abundance (Green *et al.* 2009a). When the maximum equivalent MSAL alkaloid dose (14.54 mg/kg) used in this study was administered to cows the animals exhibited signs of neuromuscular blockade between 5 and 8 h and two of the three animals remained intoxicated for over 24 h (Table 1). There was a large amount of variation in the response of animals to the maximum larkspur dose administered with one animal standing that was apparently unaffected and two animals that were eventually recumbent. When lower doses of larkspur alkaloids were given to cattle the effects appeared to be due to the inhibition of ganglionic neurotransmission and resulting changes in heart rate. However, additional investigations are needed to more fully describe the mechanisms by which MSAL-type alkaloids act in cattle.

Neostigmine at a dose of 0.02 mg/kg i.m. was used to 'rescue' intoxicated animals in recumbency. This dose of neostigmine has been commonly used in cattle to stimulate rumenoreticular motility (Kahn 2005). Larkspur mortality in cattle is the result of recumbence due to the curare-like effects of norditerpenoid alkaloids and the inhibition of eructation (Pfister *et al.* 1999). Reversal of recumbence and most importantly lateral recumbence by the use of anticholinesterase agents may provide a means to alter the medical outcome of larkspur poisoning. Further research is needed to investigate the utility of neostigmine in field applications and to reverse the effect of MSAL-type alkaloids ingested by cattle in lethal amounts.

In summary, norditerpenoid alkaloids affect multiple nicotinic cholinergic receptors and alter neurotransmission at autonomic ganglia and the neuromuscular junction. Furthermore, neostigmine appears to be effective in reversing the acute neuromuscular and cardiac effects and the longer-term cardiac actions of toxic larkspur alkaloids. MLA and deltaline reach maximum serum concentrations by 10 hours after dosing and MLA has a slower $t_{1/2}$ of 20.5 hours when compared to deltaline at 8.2 hours. The longer MLA clearance suggests that a withdrawal time of 7 days be used to allow poisoned animals to clear these toxins. More work is needed to better determine the biologic effect and mechanisms of MSAL and MDL interactions in combined intoxication and to determine why some animals or species are more susceptible to poisoning.

## Acknowledgements

The authors wish to thank Ed Knoppel, Kendra Dewey, Scott Larsen, and Anita McCollum for their expert technical support and Jessie Roper, Al Maciulis, Rex Probst, and Danny Hansen for assistance with animal care and handling.

## References

Alkondon M, Pereira EF, Wonnacott S, and Albuquerque EX (1992). Blockade of nicotinic currents in hippocampal neurons defines methyllycaconitine as a potent and specific receptor antagonist. *Molecular Pharmacology* 41:802-808.

Benn MH and Jacyno JM (1983). The toxicology and pharmacology of diterpenoid alkaloids. In *Alkaloids: Chemical and Biological Perspective* (SW Pelletier, ed.), pp. 153-210. John Wiley and Sons, New York.

Chen W, Nemoto T, Kobayashi T, Saito T, Kasuya E, and Honda Y (2002). ECG and heart rate determination in fetal cattle using a digital signal processing method. *Animal Science Journal* 73:545-551.

Dobelis P, Madl JE, Pfister JA, Manners GD, and Walrond JP (1999). Effects of *Delphinium* alkaloids on neuromuscular transmission. *Journal of Pharmacology and Experimental Therapeutics* 291:538-546.

Gardner DR, Manners GD, Ralphs MH, and Pfister JA (1997). Quantitative analysis of norditerpenoid alkaloids in larkspur (*Delphinium* spp.) by fourier transform infrared spectroscopy. *Phytochemical Analysis* 8:55-62.

Gardner DR, Panter KE, Pfister JA, and Knight AP (1999). Analysis of toxic norditerpenoid alkaloids in *Delphinium* species by electrospray, atmospheric pressure chemical ionization, and sequential tandem mass spectrometry. *Journal of Agriculture and Food Chemistry* 47:5049-5058.

Green BT, Pfister JA, Cook D, Welch KD, Stegelmeier BL, Lee ST, Gardner DR, Knoppel EL, and Panter KE (2009a). Effects of larkspur (*Delphinium barbeyi*) on heart rate and electrically evoked electromyographic response of the external anal sphincter in cattle. *American Journal of Veterinary Research* 70:539-546.

Green BT, Welch KD, Gardner DR, Stegelmeier BL, Davis TZ, Cook D, Lee ST, Pfister JA, and Panter KE (2009b). The serum elimination profiles of MLA and deltaline in cattle orally dosed with *Delphinium barbeyi*. *American Journal of Veterinary Research* 70(7):926-931.

Hardick DJ, Blagbrough IS, Cooper G, Potter BV, Critchley T, and Wonnacott S (1996). Nudicauline and elatine as potent norditerpenoid ligands at rat neuronal *alpha*-bungarotoxin binding sites: importance of the 2-(methylsuccinimido)benzoyl moiety for neuronal nicotinic acetylcholine receptor binding. *Journal of Medicinal Chemistry* 39:4860-4866.

Kahn C (2005). The ruminant digestive system. In *Merck Veterinary Manual*, 9th edn, pp. 1993-1995. Merck and Co, Whitehouse Station, New Jersey.

López MG, Montiel C, Herrero CJ, García-Palomero E, Mayorgas I, Hernández-Guijo JM, Villarroya M, Olivares R, Gandía L, McIntosh JM, Olivera BM, and García AG (1998). Unmasking the functions of the chromaffin cell *alpha*7 nicotinic receptor by using short pulses of acetylcholine and selective blockers. *Proceedings of the National Academy of Sciences of the United States of America* 24:14184-14189.

Manners GD, Panter KE, Ralphs MH, Pfister JA, Olsen JD, and James LF (1993). Toxicity and chemical phenology of norditerpenoid alkaloids in the tall larkspurs (*Delphinium* species). *Journal of Agriculture and Food Chemistry* 41:96-100.

Manners GD, Panter KE, and Pelletier SW (1995). Structure-activity relationships of norditerpenoid alkaloids occurring in toxic larkspur (*Delphinium*) species. *Journal of Natural Products* 58:863-869.

Marsh CD, Clawson AB, and Marsh H (1916). Larkspur Poisoning of Livestock. *United States Department of Agriculture Bulletin* 365.

Marsh CD, Clawson AB, and Marsh H (1934). Larkspur or Poison Weed. *United States Department of Agriculture Bulletin* 988.

Miller JR (2003). GraphPad Prism Version 4.0 Step-by-Step Examples. GraphPad Software Inc., San Diego, California.

Nation PN, Benn MH, Roth SH, and Wilkens JL (1982). Clinical signs and studies of the site of action of purified larkspur alkaloid, methyllycaconitine, administered parenterally to calves. *Canadian Veterinary Journal* 23:264-266.

Olsen JD, Manners GD, and Pelletier SW (1990). Poisonous properties of Larkspur (*Delphinium* spp.). *Collectanea Botanica (Barcelona)* 19:141-151.

Pfister JA, Panter KE, and Manners GD (1994a). Effective dose in cattle of toxic alkaloids from tall larkspur (*Delphinium barbeyi*). *Veterinary and Human Toxicology* 36:10-11.

Pfister JA, Panter KE, Manners GD, and Cheney CD (1994b). Reversal of tall larkspur (*Delphinium barbeyi*) poisoning in cattle with physostigmine. *Veterinary and Human Toxicology* 36:511-514.

Pfister JA, Gardner DR, Panter KE, Manners GD, Ralphs MH, Stegelmeier BL, and Schoch TK (1999). Larkspur (*Delphinium* spp.) poisoning in livestock. *Journal of Natural Toxins* 8:81-94.

Pfister JA, Ralphs MH, Gardner DR, Stegelmeier BL, Manners GD, Panter KE, and Lee ST (2002). Management of three toxic *Delphinium* species based on alkaloid concentrations. *Biochemical Systematics and Ecology* 30:129-138.

Pfister JA, Gardner DR, Stegelmeier BL, Hackett K, and Secrist G (2003). Catastrophic cattle loss to low larkspur (*Delphinium nuttallianum*) in Idaho. *Veterinary and Human Toxicology* 45:137-139.

Ralphs MH, Gardner DR, Turner DL, Pfister JA, and Thacker E (2002). Predicting toxicity of tall larkspur (*Delphinium barbeyi*): measurement of the variation in alkaloid concentration among plants and among years. *Journal of Chemical Ecology* 28:2327-2341.

Sharples CGV and Wonnacott S (2001). Neuronal nicotinic receptors. *Tocris Reviews* 19.

Ward JM, Cockcroft VB, Lunt GG, Smillie FS, and Wonnacott S (1990). Methyllycaconitine: a selective probe for neuronal *alpha*-bungarotoxin binding sites. *FEBS Letters* 270:45-48.

# Chapter 97

# Lupine-Induced 'Crooked Calf Disease' in Washington and Oregon: Identification of the Alkaloid Profiles of *Lupinus sericeus*, *Lupinus sulphureus*, and *Lupinus leucophyllus*

D. Cook, S.T. Lee, D.R. Gardner, J.A. Pfister, K.D. Welch, B.T. Green, T.Z. Davis, and K.E. Panter

*USDA-ARS Poisonous Plant Research Laboratory, Logan, Utah 84341, USA*

## Introduction

Lupines (*Lupinus* spp.) are common plant species found on western US rangelands (Kingsbury 1964). Lupines are highly adaptable occurring in desert to alpine ecosystems. There are approximately 150 lupine species in the Intermountain West and Great Basin and these species may contain a variety of piperidine and/or quinolizidine alkaloids (Wink 1992). These alkaloids have been implicated in plant-herbivore interactions and possibly plant-microbe interactions (Wink 1992). Furthermore, many of these alkaloids can be toxic and/or teratogenic to livestock thus causing losses to livestock producers (Kingsbury 1964).

Historically lupines have caused large losses in sheep due to acute intoxication (Kingsbury 1964). In the latter part of the 19th century thousands of sheep died from lupine poisoning and isolated cases of smaller losses of sheep continue today (Kingsbury 1964). In addition ingestion of lupine by cattle can cause congenital birth defects in calves termed 'crooked calf disease' (Palotay 1959; Wagnon 1960). Crooked calf disease is the result of reduced fetal movement during days 40-100 of gestation that causes the limbs and spine to develop in misaligned or contracted positions (Shupe *et al.* 1967, 1968). The quinolizidine alkaloid anagyrine (Keeler 1976) and some piperidine alkaloids (Keeler and Panter 1989) can reduce fetal movement during this critical period of gestation (Panter *et al.* 1990).

Lupine-induced crooked calf disease continues to pose a problem in several western states. For example, lupine-induced crooked calf disease cases are documented in northeastern Oregon (OR) and the Channel Scablands of east-central Washington (WA) on a yearly basis. Three major lupine species are found on these rangelands *Lupinus sulphureus* (sulphur lupine), *L. leucophyllus* (velvet lupine), and/or *L. sericeus* (silky lupine). The objective of this chapter is to highlight the characteristic alkaloid profiles of *L. sulphureus*, *L. leucophyllus*, and *L. sericeus* in northeastern Oregon and the Channel Scablands of east-central Washington. For further details concerning this research one is referred to two recent publications (Lee *et al.* 2007; Cook *et al.* 2009).

©CAB International 2011. *Poisoning by Plants, Mycotoxins, and Related Toxins* (eds F. Riet-Correa, J. Pfister, A.L. Schild, and T.L. Wierenga)

## Materials and Methods

### Plant material

Field collections of *L. sulphureus*, *L. leucophyllus*, and *L. sericeus* were made from mid-May to mid-July in 2006-2008 and immediately frozen on dry ice. Voucher specimens were pressed at each location and were subsequently classified by staff at the Intermountain Herbarium at Utah State University, Logan, Utah. In addition, herbarium specimens of *L. sulphureus* from the Marion Owneby Herbarium at Washington State University, the Oregon State University Herbarium, the University of Washington Herbarium, the University of British Columbia Herbarium, and the Intermountain Herbarium at Utah State University were sampled for subsequent alkaloid extraction. Vegetative and floral tissues were sampled. Specimens of question were verified to be authentic *L. sulphureus* specimens by staff at the Intermountain Herbarium at Utah State University.

### Sample extraction and alkaloid determination

Plant material was extracted according to a procedure reported by Lee *et al.* (2007). GC/FID analysis was performed to identify unique chemotypes according to a procedure reported by Cook *et al.* (2009). GC/MS analysis was performed to confirm chemotypes and identify the indivividual alkaloids according to a procedure reported by Lee *et al.* (2007). Individual alkaloids were identified using procedures and protocols reported by Kinghorn and Balandrin (1984), Wink *et al.* (1995), Lee *et al.* (2007), and Cook *et al.* (2009).

## Results and Discussion

### *Lupinus sulphureus*

Seven chemotypes were identified from the collections of *L. sulphureus*. Each chemotype was unique in its chemical composition and location.

Chemotype A collected near Ritzville, WA contained a single alkaloid that was confirmed to be ammodendrine. Plants with this chemotype grow in south-central Washington extending up the middle of the state of Washington into southern British Columbia.

Chemotype B collected near Ukiah, OR contained two alkaloids that were confirmed to be *N*-methyl ammodendrine and ammodendrine. Plants with this chemotype grow primarily in the southwest corner of Umatilla County and the eastern part of Morrow County in Oregon. In addition a small number of plants of this chemotype are dispersed throughout the region that chemotype A covers. All plants of this chemotype contain *N*-methyl ammodendrine and ammodendrine.

Chemotype C collected near Pendleton, OR contained four alkaloids that were confirmed to be gramine, 5,6-dehydrolupanine, lupanine, and anagyrine. Plants with this chemotype grow only in Umatilla and Union Counties in Oregon. All plants of this chemotype contain lupanine and anagyrine.

Chemotype D collected near Anatone, WA contained seven alkaloids that were determined to be sparteine, a potential 11,12-dehydrosparteine isomer, 11,12 dehydrosparteine, epiaphylline, 5,6-dehydrolupanine, lupanine, and anagyrine. Plants with this chemotype grow in the southeastern corner of Asotin County in Washington. All plants of this chemotype contain lupanine, anagyrine, and sparteine.

Chemotype E collected near Pomeroy, WA contained 12 alkaloids that were determined to be α-isosparteine, sparteine, a potential 11,12-dehydrosparteine isomer, 11,12 dehydrosparteine, 5,6-dehydro-α-isolupanine, α-isolupanine, lupanine, 11,12-dehydrolupanine, 7-hydroxylupanine, thermopsine, 10,17-dioxo-β-sparteine, and 17-oxolupanine. Plants with this chemotype grow in Asotin, Garfield, Columbia, and Walla Walla Counties in Washington. In addition, plants of this chemotype grow in Umatilla, Union, and Wallowa Counties in Oregon. α-Isolupanine and thermopsine were found in all plants of this chemotype.

Chemotype F collected near Coppei, WA contained nine alkaloids that were determined to be α-isosparteine, sparteine, 5,6-dehydro-α-isolupanine, α-isolupanine, lupanine, aphylline, thermopsine, 17-oxolupanine, and a potential 17 oxo-lupanine isomer. Plants with this chemotype grow in Walla Walla and Columbia Counties in Washington. Aphylline, α-isolupanine, thermopsine, and a tentative assignment of a 17-oxolupanine isomer were detected in greater than 90% of the samples representing this chemotype.

Chemotype G collected near Tollgate, OR contained nine alkaloids that were determined to be sparteine, 11,12 dehydrosparteine, dehydrolupanine, lupanine, 11,12-dehydrolupanine, thermopsine, 7-hydroxylupanine, 17-oxolupanine, and anagyrine. Plants with this chemotype grow in northern Umatilla County and are located between chemotype C, E, and F. Lupanine and sparteine were detected in all samples of this chemotype.

### *Lupinus leucophyllus*

Two chemotypes of *L. leucophyllus* were identified from the plant collections. Chemotype A collected near Ritzville, WA contained seven alkaloids that were determined to be 5,6-dehydrolupanine, lupanine, (2*R*)-hydroxyaphyllidine (also known as (-) argyrolobine), (2*S*)-hydroxyaphyllidine, (2*R*,9*R*)-dihydroxyaphyllidine, (2*S*,9*R*)-dihydroxyaphyllidine, and anagyrine. Preliminary data suggest that plants with this chemotype are found throughout the channel scablands of Washington and into the Blue Mountains of Oregon. Chemotype B collected near Pendleton, OR contained nine alkaloids that were identified to be tetrahydrorhombofoline, angustifoline, α-isolupanine, lupanine, hydroxytetrahydrorhombofoline, 13α-hydroxylupine, 13α-tigloyloxylupanine, and 13β-tigloyloxylupanine. One major peak remained unknown; however, it is likely to be tigloyloxylupanine with a hydroxyl moiety. Preliminary data suggest that plants with this chemotype are found only in the Blue Mountains of Oregon. Further collections are needed to define the characteristic alkaloids of each chemotype and the distribution of both these chemotypes.

### *Lupinus sericeus*

A single chemotype was identified from the collections of *L. sericeus*. Four major alkaloids were identified in this chemotype: sparteine, 11,12 dehydrosparteine, lupanine, and 7-hydroxylupanine. None of these alkaloids are teratogenic. Further collections are needed to determine if this is the only chemotype and the extent of its distribution.

## Conclusions

Three important conclusions can be drawn from these data.

First, each chemotype identified poses a different risk to livestock due to its alkaloid composition (Panter *et al.* 1997; Lee *et al.* 2007). For example, chemotypes C and D of *L. sulphureus* and chemotype A of *L. leucophyllus* contain the teratogen anagyrine while chemotype A and B of *L. sulphureus* contain the suspected teratogen ammodendrine. In addition chemotypes E, F, and G contain thermopsine which induces myopathy in livestock (Keeler and Baker 1990). This clearly demonstrates that taxonomic identification of a lupine species is not sufficient to determine risk and that alkaloid analysis must be performed on each lupine population to determine risk.

Second, in considering potential risk to livestock, distribution and density of the poisonous plant must be considered (Kingsbury 1964; Panter *et al.* 1997). Each chemotype of *L. sulphureus* for the most part has a distinct distribution with defined boundaries. Interestingly, none of the chemotypes appear to follow notable geographical features such as watersheds. It is interesting to note the broad geographical range of chemotypes A and E in contrast to the narrower geographical range of chemotypes F and G. Also notable is the fact that all field collections of *L. sulphureus* at a particular location in this survey have the same chemical phenotype.

Third, this work suggests that the qualitative nature of the alkaloid profile in *L. sulphureus* remains constant and is not significantly modified by the environment. This conclusion is supported by the fact that the field collections have the same chemical phenotypes as the herbarium specimens from identical locations that were collected over 100 years ago. Furthermore, this suggests the alkaloid composition of herbarium specimens is not modified as a result of long term storage at room temperature. This does not establish, however, that the quantitative amounts of these alkaloids do not vary between years. Quantitative assessment of the alkaloids over time merits further investigation. Further research is needed to show that this observation holds true for *L. leucophyllus* and *L. sericeus*.

We are currently not able to explain why there is such a large diversity in the alkaloid composition between populations of *L. sulphureus* and *L. leucophyllus* although some possibilities merit consideration and discussion:

1. These lupines may represent distinct varieties or species. For example, the same lupine species may have similar alkaloid profiles as is the case for *L. polyphyllus* from North America or the *L. linearis-L. gibertianus* complex from South America (Planchuelo-Ravelo *et al.* 1993; Wink *et al.* 1995). Alternatively, the same lupine species may have multiple alkaloid profiles as is the case for *L. argenteus*, *L. formosus*, *L. leucophyllus*, and *L. sulphureus* (Wink and Carey 1994; Lee *et al.* 2005, 2007).
2. These alkaloid profiles may be a result of chemical warfare between the plant and herbivores. In certain instances one chemotype is more susceptible to herbivores than another chemotoype (Berenbaum and Zangerl 1988). Furthermore, isomers of the same compound can have differential toxicity to herbivores (Berenbaum *et al.* 1989).
3. The individual populations may be a result of hybridization between another population and/or another lupine species. For example, *L. polyphyllus* var. *polyphyllus* and *L. arcticus* var. *subalpinus* intergrade in terms of their alkaloid profiles where the two species overlap (Majak *et al.* 1994).

To address these possibilities we plan to pursue phylogenetic analysis of the field collections representing collections of each of the distinct alkaloid profiles. In addition, we are pursuing taxonomic studies to identify morphological characters that may separate some of these groups based upon alkaloid composition. In conclusion, this study clearly demonstrates that taxonomic identification of a lupine species is not sufficient to determine risk and that alkaloid analysis must be performed on each population to determine risk.

## References

Berenbaum MR and Zangerl AR (1988). Stalemates in the coevolutionary arms race: syntheses, synergisms, and sundry other sins. In *Chemical Mediation of Coevolution* (KC Spenser, ed.), pp. 113-132. Academic Press, San Diego, California.

Berenbaum MR, Zangerl AR, and Lee K (1989). Chemical barriers to adaptation by a specialist herbivore. *Oecologia* (Berlin) 80:501-506.

Cook D, Lee ST, Gardner DR, Pfister JA, Welch KD, Green BT, Zavis TZ, and Panter KE (2009). The alkaloid profiles of *Lupinus sulphureus*. *Journal of Agriculture and Food Chemistry* 57:1646-1653.

Keeler RF (1976). Lupin alkaloids from teratogenic and nonteratogenic lupins. III. Identification of anagyrine as the probable teratogen by feeding trials. *Journal of Toxicology and Environmental Health* 1:878-889.

Keeler RF and Baker DC (1990). Myopathy in cattle induced by alkaloid extracts from *Thermopsis montana*, *Laburnum anagyroides*, and a *Lupinus* sp. *Journal of Comparative Pathology* 103:169-182.

Keeler RF and Panter KE (1989) Piperidine alkaloid composition and relation to crooked calf disease-inducing potential of *Lupinus formosus*. *Teratology* 40:423-432.

Kinghorn AD and Balandrin MF (1984) Quinolizidine alkaloids of the leguminosae: Structural types, analysis, chemotaxonomy, and biological activities. In *Alkaloids: Chemical and Biological Perspectives* (SW Pelletier, ed.), vol. 4, pp. 105-148. John Wiley & Sons, Inc., New York.

Kingsbury JM (1964). *Poisonous Plants of the United States and Canada.* pp. 333-341. Prentice-Hall, Englewood Cliffs, New Jersey.

Lee ST, Molyneux RJ, Panter KE, Chang CWT, Gardner DR, Pfister JA, and Garrossian M (2005). Ammodendrine and *N*-methylammodendrine enantiomers: isolation, optical rotation, and toxicity. *Journal of Natural Products* 68:681-685.

Lee ST, Cook D, Panter KE, Gardner DR, Ralphs MH, Motteram, ES, Pfister JA, and Gay CC (2007). Lupine induced 'Crooked Calf Disease' in Washington and Oregon: identification of the alkaloid profile in *Lupinus sulfureus*, *Lupinus leucophyllus*, and *Lupinus sericeus*. *Journal of Agriculture and Food Chemistry* 55:10649-10655.

Majak W, Keller WJ, Duan Z, Munro D, Smith RA, Davis AM, and Ogilvie RT (1994). Alkaloid distribution in two species of *Lupinus* in Central British Columbia. *Phytochemistry* 36:883-885.

Palotay JL (1959) 'Crooked Calves.' *Western Veterinarian.* 6:16-20.

Panter KE, Bunch TD, Keeler RF, Sisson DV, and Callan RJ (1990). Multiple congenital contractures (MCC) and cleft palate induced in goats by ingestion of piperidine alkaloid-containing plants: Reduction in fetal movement as the probable cause. *Clinical Toxicology* 28:69-83.

Panter KE, Gardner DR, Gay CC, James LF, Mills R, Gay JM, and Baldwin TJ (1997). Observations of *Lupinus sulphureus*-induced 'crooked calf disease'. *Journal of Range Management* 50:587-592.

Planchuelo-Ravelo AM, Witte L, and Wink M (1993). Quinolizidine alkaloid profiles of South American Lupins: *Lupinus linearis* and the *Lupinus gibertianus* complex. *Zeitschrift für Naturforschung* 48c:702-706.

Shupe J, Binns W, James L, and Keeler R (1967). Lupine, a cause of crooked calf disease. *Journal American Veterinary Medical Association* 151:198-203.

Shupe J, Binns W, James L, and Keeler R (1968). A congenital deformity in calves induced by the maternal consumption of lupin. *Australian Journal of Agricultural Research* 19:335-340.

Wagnon KA (1960). Lupine poisoning as a possible factor in congenital deformities in cattle. *Journal of Range Management* 13:89-91.

Wink M (1992). The role of quinolizidine alkaloids in plant-insect interactions. In *Insect-Plant Interactions* (EA Bernays ed.), pp. 131-166. CRC Press, Boca Raton, Florida.

Wink M and Carey DB (1994). Variability of quinolizidine alkaloid profiles of *Lupinus argenteus* (Fabaceae) from North America. *Biochemical Systematics and Ecology* 22:663-669.

Wink M, Meißner C, and Witte L (1995). Patterns of quinolizidine alkaloids in 56 species of the genus *Lupinus*. *Phytochemistry* 38:139-153.

# Chapter 98

# Comparative Study of Monocrotaline Toxicity on Peritoneal Macrophage Activity When Dosed for 14 or 28 Days

J.C. Benassi[1], M. Haraguchi[2], S.L. Górniak[1], and I.M. Hueza[1]

*[1]Research Center for Veterinary Toxicology (CEPTOX), Department of Pathology, School of Veterinary Medicine and Animal Science, University of São Paulo, CEP 05508-900, Brazil; [2] Biological Institute of São Paulo, SP 04014-002, Brazil*

## Introduction

Monocrotaline (MCT) is a pyrrolizidine alkaloid present in poisonous plants of the *Crotalaria* genus. Species of this genus can be found worldwide, mainly in tropical and subtropical areas (Williams and Molyneux 1987). In Brazil *C. spectabilis* is cultivated for use as a natural fertilizer inasmuch as it fixes nitrogen in the soil (Bokhtiar *et al.* 2003). Livestock are often at risk of poisoning by this plant when they have access to *Crotalaria* spp.-treated areas, especially during the dry season.

The main active principle in *Crotalaria* spp. is the pyrrolizidine alkaloid MCT. However, this alkaloid must be transformed by microsomal liver enzymes into a pyrrole compound, dehydromonocrotaline (MCTP), in order to activate the molecule so that it possesses a strong bonding affinity for nucleophilic groups (Wang *et al.* 2005).

The main toxic effects observed in livestock by MCT are related to hepato- and nephrotoxicity. Hepatic periacinar necrosis, megalocytosis of hepatocytes with neutrophilic infiltration, glomerulo-nephritis, and alterations in renal tubular epithelium cells are observed histologically (Figueredo *et al.* 1987; Wilson *et al.* 1992). In addition to these many alterations rodents and humans intoxicated with MCT display progressive pulmonary hypertension (PH) that progresses to *cor pulmonale* (Pan *et al.* 1993). The PH pathogenic mechanism is still unknown.

In a study performed by Sasaki *et al.* (2007), pulmonary endothelial cells of rats exposed to MCT were unable to produce nitric oxide (NO). The proposed PH mechanism for the effect of MCT was a decrease in L-arginine availability which is a substrate for nitric oxide synthases (NOS), a family of enzymes involved in NO synthesis. In addition, in a previous study performed in our lab (Hueza *et al.* 2009) peritoneal macrophages (MO) of rats treated for 14 days with MCT were unable to release NO even though other MO inflammatory activities were not affected.

Nevertheless, most immunotoxic protocols proposed by regulatory agencies such as the US Environmental Protection Agency (US-EPA) and others have suggested 28 days of

©CAB International 2011. *Poisoning by Plants, Mycotoxins, and Related Toxins* (eds F. Riet-Correa, J. Pfister, A.L. Schild, and T.L. Wierenga)

treatment. Thus, the aim of the present study was to evaluate if MCT administration to rats for 28 days induces the same modulatory effect on NO production by peritoneal MO.

## Materials and Methods

Powdered *C. spectabilis* seeds (1 kg), defatted with hexane, were exhaustively extracted with 92% ethanol. The combined extracts were filtered and concentrated under reduced pressure to remove the solvent. The residue was dissolved in a mixture of ethanol-water (1:1), applied to a column of Amberlite IR-120B (H+ form), and extracted with 0.5 M ammonium hydroxide ($NH_4OH$). The resulting eluate was concentrated under reduced pressure, resulting in a white residue that was in turn re-crystallized from methanol (three times) to yield colorless crystals (2.3 g). These crystals were then compared with standard MCT (Sigma, St Louis, MO) using thin-layer chromatography (Molyneux and Roitman 1980) as well as elemental and spectral ($^1H$ and $^{13}C$ NMR) analyses. The purity of the extracted product was determined to be 99.26%. All analyses were performed at Analytical Laboratory of the Institute of Chemistry, University of São Paulo.

To prepare the doses used in the present studies, MCT was first dissolved and mixed for 15 min in 1% phosphoric acid solution. The pH was adjusted to 7 with NaOH and distilled water was then added to achieve the final concentrations of 0.3, 1.0, and 3.0 mg MCT/ml for use herein.

### Innate immunity: macrophage activity

Forty male Wistar rats (10 weeks old) were divided into four equal groups (one control and three experimental groups) that received 0.0, 0.3, 1.0, or 3.0 mg MCT/kg by gavage once a day for 28 days. On the 29th day, the animals were killed and peritoneal MO were collected to evaluate MO activity.

Phagocytosis: the methods used to study MO phagocytosis were based on those described by Rabinovith and De Stefano (1973a, b) with modifications introduced by Passeti (1993). Briefly, 200 cells were counted for each slide for each rat and the MO phagocytosis index (PI) was calculated as follows: PI=number of phagocytic activity × 100÷200 adherent cells counted, i.e. PI = % of MO with phagocytized zymosan particles. The mean of four counts obtained from two slides from each rat was used to express PI index.

$H_2O_2$ release: spontaneous and phorbol myristate-acetate solution (PMA)-induced $H_2O_2$ release by MO was measured by the method of Russo *et al.* (1989). $H_2O_2$ concentration was calculated from absorbance measurements as described by Pick and Mizel (1981). Spontaneous and PMA-induced $H_2O_2$ production experiments were repeated four times for each rat in each group and the mean value of the four counts was used to determine $H_2O_2$ concentration.

Finally, NO concentration in the supernatant of cultures of MO incubated with LPS (100 ng/ml) or vehicle for 24 h were measured using the Griess reagent (Green *et al.* 1982). In brief, 100 µl of Griess reagent (freshly prepared) was mixed with an equal amount of cell culture supernatant and then incubated at room temperature for 10 min. The absorbance of the samples was then measured at 540 nm in the Multiskan EX reader. All experiments for NO measurements were performed in triplicate.

Analysis of variance (ANOVA) was used to compare the means of various groups followed by Dunnet's *post hoc* test for detection of significant differences among the groups, with the level of significance set at $P < 0.05$. Data are expressed as mean ± SEM.

## Results and Discussion

There were no differences ($P > 0.05$) among various treatment groups in phagocytosis or in NO production by peritoneal MO. Conversely, rats treated with 3.0 mg/kg of MCT had enhanced $H_2O_2$ production both spontaneously and PMA-induced when compared to untreated rats (Table 1 and Figure 1).

**Table 1.** Spontaneous and PMA-induced $H_2O_2$ production (Abs 620 nm) by peritoneal macrophages of rats treated by gavage with 0.0, 0.3, 1.0, and 3.0 mg/kg of MCT for 28 days.

| | $H_2O_2$ production (Abs 620 nm) | | | |
|---|---|---|---|---|
| | Control (*n*=6)[a] | 0.3 mg/kg (*n*=7) | 1.0 mg/kg (*n*=8) | 3.0 mg/kg (*n*=7) |
| Spontaneous $H_2O_2$ production | 0.054 ± 0.004 | 0.048 ± 0.005 | 0.055 ± 0.004 | 0.099 ± 0.007** |
| PMA-induced $H_2O_2$ production | 0.074 ± 0.005 | 0.068 ± 0.004 | 0.072 ± 0.005 | 0.16 ± 0.005** |

[a] Number of animals per group
** $P < 0.01$ versus control group; Dunnet's test

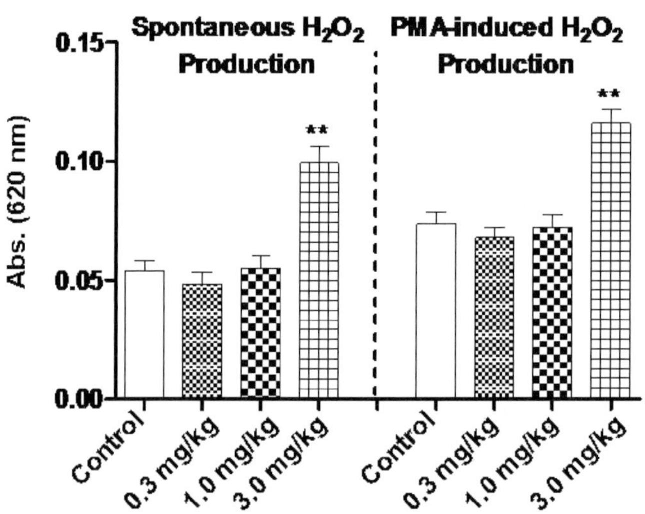


**Figure 1.** Spontaneous and PMA-induced $H_2O_2$ production (Abs 620 nm) by peritoneal macrophages of rats treated with 0.0, 0.3, 1.0, and 3.0 mg/kg of MCT by gavage for 28 days. **$P < 0.01$ versus control group, Dunnet's test.

In a previous study with MCT dosed for 14 days with the same doses used here we found that MO from all MCT-treated animals were not able to produce NO. However, this study showed that NO production was not affected when MCT was dosed for 28 days. On the other hand, we found that $H_2O_2$ production was enhanced by MCT treatment.

It is clear that the duration of MCT administration promotes different responses in MO. It is known that $H_2O_2$ production is induced by a membrane receptor signaling that leads to protein kinase C (PKC) phosphorylation that results in activation of MO to release peroxide. The PMA (phorbol myristate acetate) acts directly on PKC thus forcing peroxide production. MO did not show an enhanced inflammatory phenotype and in the absence of any improvement in phagocytosis activity this suggests that MCT or its metabolite may be interfering in the transduction of peroxidase signaling, probably due to the strong bonding affinity for nucleophilic groups of macromolecules such as DNA (Wang *et al.* 2005).

One may ask why NO production was unaffected in animals treated for 28 days? As suggested by Sasaki *et al.* (2007) endothelial cells may be compromised in producing NO due to the reduction of cytoplasmic L-arginine and it could be that MO were reacting in a similar manner as endothelial cells. Nevertheless, it is difficult to explain why 28 days of MCT treatment did not promote L-arginine deficiency. It could be that the mechanism of NO production and interference in MO is different from that proposed by Sasaki *et al.* (2007).

Due to the discrepancies observed in the present study from previous work it will be necessary to repeat these experiments (14 and 28 days of treatment) to elucidate the actual mechanism by which MCT affects NO and $H_2O_2$ production.

## Acknowledgements

This study was supported by grants from Fundação de Amparo à Pesquisa do Estado de São Paulo – FAPESP, Brazil (Proc. No. 06/60397-8 and 07/51648-0).

## References

Bokhtiar SM, Gafur MA, and Rahman AB (2003). Effects of *Crotalaria* and *Sesbania aculeata* green manures and N-fertilizer on soil fertility and the productivity of sugarcane. *Journal Agricultural Science* 140:305-309.

Figueredo MLA, Rodriguez J, and Alfonso HA (1987). Pathology of experimental acute intoxication of *Crotalaria retusa* and *Crotalaria spectabillis* in chickens. *Revista Cubana Ciências Veterinária* 18:63-71.

Green LC, Wagner DA, Glogowski J, Skipper PL, Wishnok JS, and Tanneubaum SR (1982). Analysis of nitrate, nitrite and [15N] nitrite in biological fluids. *Analytical Biochemistry* 126:131-138.

Hueza IM, Benassi JC, Raspantini PCF, Raspantini LER, Sá LRM, Górniak SL, and Haraguchi M (2009). Low doses of monocrotaline in rats cause diminished bone marrow cellularity and compromised nitric oxide production by peritoneal macrophages. *Journal of Immunotoxicology* 6:11-18.

Molyneux RJ and Roitman JN (1980). Specific detection of pyrrolizidine alkaloids on thin-layer chromatograms. *Journal of Chromatography* 195:412-415.

Pan LC, Wilson DW, Lame MW, Jones AD, and Segall HJ (1993). Cor pulmonale is caused by monocrotaline and dehydromonocrotaline, but not by glutathione or cysteine conjugates of dihydropyrrolizine. *Toxicology Applied Pharmacology* 118:87-97.

Passeti TA (1993). Regulação genética de atividades funcionais de macrófagos inflamatórios. Influência na resistência à infecção experimental pelo *Toxoplasma gondii*, 87 pp. Dissertação de Mestrado, Faculdade de Ciências Farmacêuticas, Universidade de São Paulo, São Paulo.

Pick E and Mizel D (1981). Rapid microassays for the measurement of superoxide and hydrogen peroxide production by MO in culture sing an automatic enzyme immunoassay. *Journal of Immunological Methods* 46:211-226.

Rabinovitch M and Destefano MJ (1973a). Macrophage spreading *in vitro*: I. Inducers of spreading. *Experimental Cell Research* 77:323-334.

Rabinovitch M and Destefano MJ (1973b). Macrophage spreading *in vitro*: II. Manganese and other metals as inducers or as co-factors for induced spreading. *Experimental Cell Research* 79:423-430.

Russo M, Teixeira HC, Marcondes MC, and Barbuto JA (1989). Superoxide independent hydrogen peroxide release by activated macrophages. *Brazilian Journal of Medical Biological Research* 22:1271-1273.

Sasaki A, Doi S, Mizutani S, and Azuma H (2007). Roles of accumulated endogenous nitric oxide synthase inhibitors, enhanced arginase activity, and attenuated nitric oxide synthase activity in endothelial cells for pulmonary hypertension in rats. *American Journal of Physiology - Lung Cellular and Molecular Physiology* 292:L1480-1487.

Wang YP, Yan J, Beger RD, Fu PP, and Chou MW (2005). Metabolic activation of the tumorigenic pyrrolizidine alkaloid, monocrotaline, leading to DNA adduct formation *in vivo*. *Cancer Letters* 226:27-35.

Williams MC and Molyneux RJ (1987). Occurrence, concentration and toxicity of pyrrolizidine alkaloids in *Crotalaria* seeds. *Weed Science* 35: 476-81.

Wilson DW, Segall HJ, Pan LC, Lame MW, Estep JE, and Morin D (1992). Mechanisms and pathology of monocrotaline pulmonary toxicity. *Critical Reviews in Toxicology* 22:307-325.

# Chapter 99

# Effects of Lantadenes on Mitochondrial Bioenergetics

A.F. Garcia[1,2], H.C.D. Medeiros[2], G.A.M. Pasquali[2], M.A. Maioli[2], B.A. Rocha[3], F.B. da Costa[3], C. Curti[4], M. Groppo[5], and F.E. Mingatto[2]

*[1] Programa de Mestrado em Ciência Animal, Universidade Estadual Paulista 'Júlio de Mesquita Filho', Campus de Araçatuba, Araçatuba, SP, Brazil; [2] Laboratório de Bioquímica, Faculdade de Zootecnia, Universidade Estadual Paulista 'Júlio de Mesquita Filho', Campus Experimental de Dracena, Dracena, SP, Brazil; [3] Departamento de Ciências Farmacêuticas; [4] Departamento de Física e Química, Faculdade de Ciências Farmacêuticas de Ribeirão Preto, Universidade de São Paulo, Ribeirão Preto, SP, Brazil; [5] Departamento de Biologia, Faculdade de Filosofia, Ciências e Letras de Ribeirão Preto, Universidade de São Paulo, Ribeirão Preto, SP, Brazil*

## Introduction

Lantana (*Lantana camara* Linn.) is one of the most poisonous weeds in the world. The noxious properties of the plant are well documented: it causes cholestasis, hepatotoxicity, photosensitization, and fatalities in cattle, horses, sheep, dogs, and humans (Wolfson and Solomons 1964; Tokarnia *et al.* 1984; Black and Carter 1985; Fourie *et al.* 1987; Sharma *et al.* 1988; Pass 1991; Brito *et al.* 2004). The most well-known lantana compounds are the lantadenes which belong to the pentacyclic triterpenoid oleanane series. The most abundant triterpene acid is lantadene A (LA); it has been implicated as the main culprit responsible for the toxic effects of the plant (Sharma *et al.* 1991, 2000, 2007) but the mechanism by which it induces toxicity has not yet been clearly established.

Mitochondria carry out a variety of biochemical processes but their main function is to produce a majority (> 90%) of cellular ATP. Mitochondrial dysfunctions can be the main mechanism of induction of hepatic diseases by drugs and/or toxic compounds. These can be divided in two groups: (i) those that affect the function of mitochondria; and (ii) those that primarily target other cellular functions which interact with the mitochondria secondarily. The recognition of the interaction of compounds with the mitochondria as a primary or secondary target can help in the understanding of the mechanisms responsible for the adverse effects and in the development of new drugs that eliminate or minimize these reactions (Szewczyk and Wojtczak 2002).

Since no antidote against the toxic effects of lantana is so far available and treatment of the symptoms has had limited success (Sharma *et al.* 2007), the knowledge of the biochemical mechanism of lantana intoxication at the cellular and molecular levels can help in developing antidotes and more rational therapy in lantana poisoning. In the present work

©CAB International 2011. *Poisoning by Plants, Mycotoxins, and Related Toxins* (eds F. Riet-Correa, J. Pfister, A.L. Schild, and T.L. Wierenga)

we address the action of LA isolated from *L. camara* and its reduced derivative lantadene A (RLA) (Figure 1) on mitochondrial bioenergetics, assessing their effects on respiration, membrane potential, and ATP levels in isolated rat liver mitochondria.

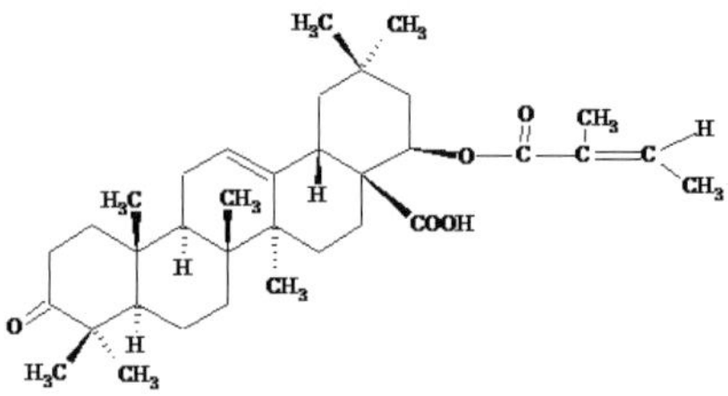

LA

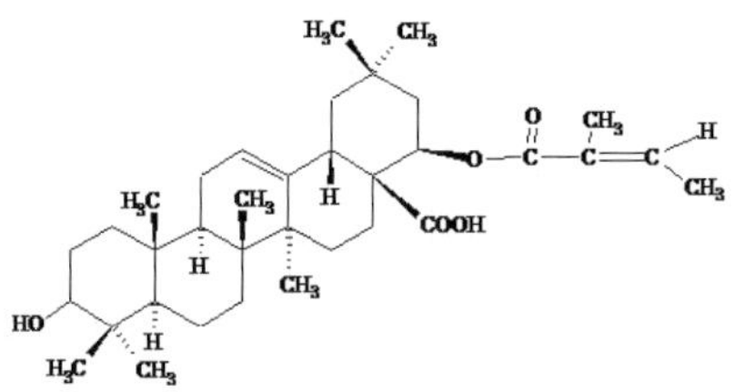

RLA

**Figure 1.** Chemical structures of lantadenes used in this study.

## Effects on Isolated Mitochondria

The action of LA and RLA on mitochondrial bioenergetics was investigated by addressing their effects on respiration, membrane potential ($\Delta\psi$), and ATP levels in succinate-energized isolated rat liver mitochondria. Rat liver mitochondria were isolated by differential centrifugation. Oxygen uptake by the isolated mitochondria was monitored with an oxygraph equipped with a Clark-type oxygen electrode (Strathkelvin Precision Dissolved Oxygen Respirometer). Mitochondrial membrane potential was estimated using the cationic fluorescent probe safranine O (Zanotti and Azzone 1980) and mitochondrial ATP level was determined by means of the firefly luciferin-luciferase assay system (Lemasters and Hackenbrock 1976) where bioluminescence was measured in the supernatant with a Sigma-Aldrich ATP Bioluminescent Assay Kit (Mingatto *et al.* 2007).

At all tested concentrations (5, 10, 15, and 25 μM) RLA significantly stimulated state 4 respiration (7.09±0.59 nmol $O_2$/min/mg protein to control; 12.38±2.71 at 5 μM; 17.76± 0.58 at 10 μM; 15.77±4.40 at 15 μM; and 22.13±2.41 at 25 μM), inhibited state 3 respiration (64.39±1.58 nmol $O_2$/min/mg protein to control; 59.83±2.49 at 5 μM; 43.89± 2.68 at 10 μM; 35.50±7.43 at 15 μM; and 19.51±5.80 at 25 μM), circumvented oligomycin-inhibited state 3 respiration, dissipated membrane potential (89.14±5.64% relative to control at 5 μM; 71.68±5.03 at 10 μM; 28.76±3.21 at 15 μM; and 5.02±2.51 at 25 μM), and depleted ATP (7.28±0.11 nmol/mg protein to control; 4.14±0.62 at 5 μM; 3.76±1.40 at 10 μM; 3.33±1.08 at 15 μM; and 3.05±0.16 at 25 μM) in a dose-dependent manner. LA did not stimulate state 4 respiration but inhibited the state 3 respiration, dissipated $\Delta\psi$, and decreased the mitochondrial ATP levels significantly only at 25 μM (data not shown). These results indicate that RLA is acting as a mitochondrial inhibitory uncoupler while LA acts as an inhibitor of oxidative phosphorilation and RLA is more potent than the parent compound.

## Conclusions

The present study shows that lantadenes in general are potentially very disruptive of mitochondrial bioenergetics. In addition the reduced derivative lantadene A is more potent at decreasing ATP levels via both uncoupling and respiration inhibition, which in turn dissipates the mitochondrial membrane potential. This action of lantadenes may account for the well documented hepatoxicity of lantana to humans and animals.

## References

Black H and Carter RG (1985). Lantana poisoning of cattle and sheep in New Zealand. *New Zealand Veterinary Journal* 33:136-137.

Brito MF, Tokarnia CH, and Döbereiner J (2004). A toxidez de diversas lantanas para bovinos e ovinos no Brasil. *Pesquisa Veterinária Brasileira* 24(3):153-159.

Fourie N, Van der Lugt JJ, Newsholme SJ, and Nel PW (1987). Acute *Lantana camara* toxicity in cattle. *Journal South Africa Veterinary Association* 58:173-178.

Mingatto FE, Dorta DJ, Santos AB, Carvalho I, Silva CHTP, Silva VB, Uyemura SA, Santos AC, and Curti C (2007). Dehydromonocrotaline inhibits mitochondrial complex I. A potential mechanism accounting for hepatotoxicity of monocrotaline. *Toxicon* 50:724-730.

Pass MA (1991). Poisoning of livestock by lantana plants. In *Handbook of Natural Toxins.* In *Toxicology of Plant and Fungal Compounds* (RF Keeler and AT Tu, eds), vol. 6, pp. 297-311. Marcel Dekker, New York.

Sharma OP and Dawra RK (1991). Thin-layer chromatographic separations of lantadenes, the pentacyclic triterpenoids from lantana (*Lantana camara*) plant. *Journal of Chromatography* 587:351-354.

Sharma OP, Makkar HPS, and Dawra RK (1988). A review of the noxious plant *Lantana camara*. *Toxicon* 26:975-987.

Sharma OP, Sharma S, Pattabhi V, Mahato SB, and Sharma PD (2007). A review of the hepatotoxic plant *Lantana camara*. *Critical Review Toxicology* 37:313-352.

Sharma S, Sharma OP, Singh B, and Bhat TK (2000). Biotransformation of lantadenes, the pentacyclic triterpenoid hepatotoxins of lantana plant, in guinea pig. *Toxicon* 38:1191-1202.

Szewczyk A and Wojtczak L (2002). Mitochondria as a pharmacological target. *Pharmacology Review* 54:101-127.

Tokarnia CH, Döbereiner J, Lazzari AA, and Peixoto PV (1984). Intoxicação por *Lantana* spp. (Verbenaceae) em bovinos nos Estados de Mato Grosso e Rio de Janeiro. *Pesquisa Veterinária Brasileira* 4(4):129-141.

Wolfson SL and Solomons TWG (1964). Poisoning by fruit of *Lantana camara*. An acute syndrome observed in children following ingestion of the green fruit. *American Journal Disease Children* 107:173-176.

Zanotti A and Azzone GF (1980). Safranine as membrane potential probe in rat liver mitochondria. *Archives of Biochemistry and Biophysics* 201:255-265.

# Chapter 100

# Determination of the Relative Toxicity of Enantiomers with Cell-Based Assays

B.T. Green[1], S.T. Lee[1], K.D. Welch[1], K.E. Panter[1], and W. Kem[2]

[1]*USDA-ARS Poisonous Plant Research Laboratory, Logan, Utah 84341, USA;*
[2]*Department of Pharmacology and Therapeutics, College of Medicine, University of Florida, Gainesville, Florida 32610, USA*

## Introduction

Many bioactive compounds produced by plants exhibit chirality or 'handedness'. Chirality is a type of molecular asymmetry where two forms of the same molecule exist as non-superimposable mirror images of each other. The alternate forms of the molecule are termed enantiomers that experimentally have the ability to rotate plane polarized light in opposite directions. If the enantiomers are present as an equimolar mixture then it is termed a racemate that experimentally lacks the ability to rotate plane polarized light. This is of biological significance since receptors in animals are stereoselective and are preferentially activated by one enantiomer of a chiral molecule. This was first documented by Pasteur in 1858 and later by Abderhalde and Müller in 1908 whom described the differential effects of (+) and (-)epinepherine on blood pressure (Booth *et al.* 1997). Chirality has also been recognized as an important factor to consider in the design of drugs for the treatment of disease (Agranat *et al.* 2002). However, less attention has been paid to the chiral molecules found in poisonous plants. In poisonous plants chiral molecules of toxins are present as mixtures of enantiomers and the relative concentration of each enantiomer can vary (Lee *et al.* 2008b). Three species of poisonous plants with important chiral molecules include *Conium maculatum* L. (poison hemlock), *Nicotiana glauca* (wild tree tobacco), and *Lupinus* spp. (lupine). The chiral toxins from these plants are well known to cause fetal defects including arthrogyroposis, scoliosis, torticollis, kyposis, and cleft palate (Panter and Keeler 1993). Current estimations of plant toxicities are based on total toxin levels without considering stereochemistry. However, if the predominant enantiomer found in a plant is of a less potent form then the overall toxicity of the plant will be overestimated.

*C. maculatum* L., commonly known as poison hemlock, is found worldwide. There are eight known piperidine alkaloids produced by *C. maculatum*. Clinical signs of intoxication caused by these alkaloids are cholinergic in nature and include salivation, urination, and defecation and can last up to 7 h in intoxicated pregnant animals and effects on the fetus can persist up to 12 h after dosing of the mother (Keeler *et al.* 1980; Panter *et al.* 1990). The most common teratogenic outcome in livestock species exposed to *C. maculatum* is the persistent flexure of a joint known as arthrogryposis and spinal curvature

©CAB International 2011. *Poisoning by Plants, Mycotoxins, and Related Toxins* (eds F. Riet-Correa, J. Pfister, A.L. Schild, and T.L. Wierenga)

(Keeler and Balls 1978; Panter *et al.* 1988). In the mature plant and seed coniine predominates and is both acutely toxic and teratogenic although N-methylconiine is also found in the mature plant at lower levels (Panter *et al.* 1988). The concentration of coniine relative to other piperidine alkaloids in the plant is thought to be dependent on growing conditions and can vary throughout the growing season and by location (Lopez *et al.* 1999). Coniine is a nicotinic acetylcholine receptor (nAChR) agonist and the $IC_{50}$s of coniine for the displacement of ($^{125}$I)-α-bungarotoxin or ($^{3}$H)-cytisine from chick embryonic muscle and brain preparations is in the micromolar range (Forsyth *et al.* 1996). In the plant coniine is found as a mixture of the two enantiomers (Marion 1950). These enantiomers have been separated by preferential crystallization with the enantiomers of mandelic acid and the potencies of the enantiomers assessed with a cell culture-based assay using TE-671 cells which express fetal muscle-type nAChR (Lee *et al.* 2008a).

*N. glauca* commonly known as tree tobacco is indigenous to a region of South America but now has a worldwide distribution that includes parts of southwestern USA (Panter *et al.* 1999; Fluorentine and Westbrooke 2005). *N. glauca* has relatively high concentrations of enantiomers of the piperidine alkaloid anabasine (Keeler and Crowe 1984; DeBoer *et al.* 2009). Clinical signs of *N. glauca* poisoning are also cholinergic in nature and similar to that of *C. maculatum* discussed above and the affinity of anabasine for neuronal nAChR is in the nanomolar range (Panter *et al.* 1999; Daly 2005; Lee *et al.* 2006). Anabasine enantiomers have been separated by reaction with 9-fluorenylmethoxycarbonyl-L-alanine (Fmoc-L-Ala-OH) to give diastereomers which were separated by reversed phase HPLC. The pure R and S-anabasine enantiomers were then obtained by Edman degradation and potencies of the R and S-anabasine enantiomers were assessed with a cell culture-based assay using TE-671 cells (Lee *et al.* 2006).

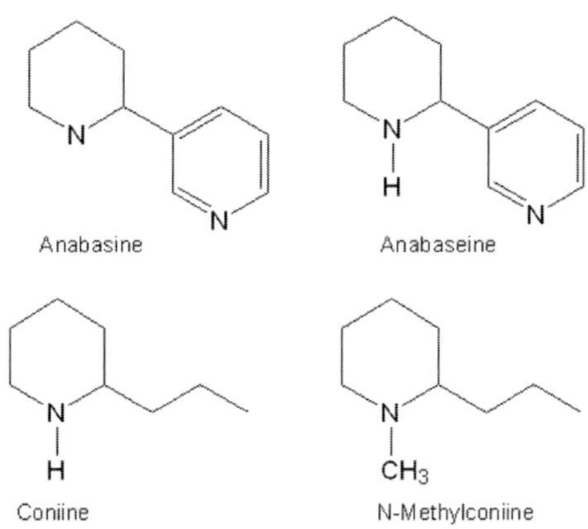


**Figure 1.** Chemical structures of N-methylconiine, coniine, anabasine, and anabaseine.

Many poisonous plants contain teratogenic chiral alkaloids which cause clinical signs consistent with cholinergic over-activation followed by depression in the mother. However, the acute effects of cholinergic overstimulation persist in the developing fetus. There is little information available on actions of these chiral plant toxins on cells with fetal characteristics. In this study two cell lines were used to assess the actions of teratogenic nAChR agonist N-methylconiine: TE-671 cells which express fetal human muscle-type nAChR and SH-SY57 cells which have the characteristics of fetal human sympathetic neurons (Lee *et al.* 2006; Innocent *et al.* 2008). The actions of N-methylconiine were then compared with those of coniine, anabasine, and anabaseine in TE-671 cells.

## Materials and Methods

### Materials

Fetal bovine serum and penicillin/streptomycin were from Media Tech, Inc. (Herndon, VA), Dulbecco's modified Eagle's medium was from the ATCC (American Type Culture Collection (Manassas, VA), and fluorescence dye kits were purchased from Molecular Devices (Sunnyvale, CA). Compounds were obtained as previously described (Lee *et al.* 2006, 2008a, b). Epibatidine was obtained from Sigma Chemical, St Louis, MO USA.

### Nicotinic agonist actions at human nAChR

The rhabdomyosarcoma cell line TE-671 and the neuroblastoma cell line SH-SY5Y were obtained from ATCC (Manassas, VA, USA). The membrane depolarization responses from the addition of nicotinic agonist toxins were measured by changes in fluorescence of a membrane potential-sensitive dye as previously described by Lee *et al.* (2006, 2008a, b). The membrane potential dye solution was prepared by dissolving one vial of the Molecular Devices dye (Catalog number R8042) into 22 ml Hanks' balanced salt solution (HBSS) supplemented with 20 mM Hepes (pH 7.4). Ninety-six-well black-walled cell culture plates were equilibrated to room temperature for 10 min then medium aspirated and replaced with 100 μl of the membrane potential dye solution into each well. The cells were incubated with the dye at room temperature for 30 min before experiments were initiated. Serial dilutions of a compound for concentration-response analysis were prepared in 96-well V-bottom plates by addition of the required volume of a methanolic stock solution. After evaporation of the methanol, the compound in each well was redissolved in membrane potential dye solution. Fluid (agonist or KCl) additions and membrane potential measurements were performed using a Flexstation II (Molecular Devices Corporation, Sunnyvale, CA, USA). Readings were taken every 1.12 s for 255 s, a total of 228 readings per well. The first 17 s were used as a basal reading. At 18 s, 50 μl of a test compound was added to assess agonist activity. At 180 s, 25 μl of KCl in saline was added to attain a final concentration of 40 mM KCl in the dye-HBSS solution bathing the cells. This served as a depolarizing calibrant and to correct for interwell differences in dye loading and cell count. Responses were calculated as equal to $(F_{Max}(\text{Compound})-F_{Basal})/(F_{Max}(\text{Calibrant})-F_{Basal})$. Depolarizing responses to agonists were normalized to the maximum response generated by (±)-epibatidine and fitted to a sigmoidal dose-response equation and graphed with Prism version 4.03 (GraphPad Software, San Diego, CA, USA) to determine $EC_{50}$ using a sigmoidal dose-response equation with variable slope, efficacy (maximal activation), and Hill coefficients with the bottom of the best fit line constrained to baseline.

## Results

The structures of N-methylconiine, coniine, anabasine, and anabaseine are displayed in Figure 1. The concentration-effect relationships of N-methyl-coniine in TE-671 and SH-SY5Y cells are displayed in Figure 2. N-methyl-coniine was more potent and effective in the TE-671 cell line as evidenced by the differences in the $EC_{50}$s (estimated $EC_{50}$s of N-methyl-coniine were 144 and 221 μM for TE-671 and SH-SY5Y cells, respectively) and the amount of maximal activation (50% versus 20% for TE-671 cells and SH-SY5Y cells, respectively).

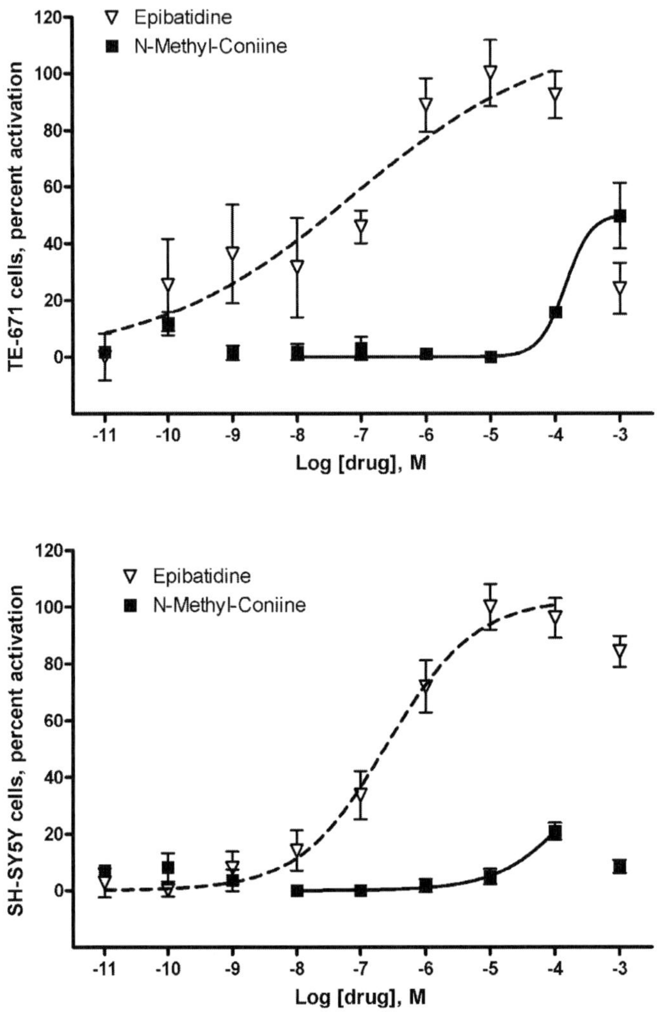


**Figure 2.** The concentration-effect relationships with best-fit lines for the actions of epibatidine and N-methylconiine on membrane potential sensing dye fluorescence in TE-671 and SH-SY5Y cells. Epibatidine is the most potent nAChR agonist known and is used as full agonist control. In each experiment the cells were grown on 96-well black-walled culture plates and the membrane depolarization resulting from addition of either epibatidine or N-methylconiine in $\log_{10}$ molar concentrations indicated was measured and displayed as a percentage of the maximal epibatidine response. Each data point represents six experiments of duplicate wells.

The concentration-effect relationships for the enantiomers of coniine and the racemate in TE-671 cells are displayed in Figure 3 and the $EC_{50}$ are listed in Table 1. The responses of the TE-671 cells to the coniine enantiomers were normalized to the maximal epibatidine response at 10 μM. The maximal activation of the coniine enantiomers at 1mM concentration relative to the maximal epibatidine response were 76 ± 13%, 62 ± 13%, 52 ± 10% for the (-), (±), and (+) forms of coniine, respectively. Therefore the relative order of potency for the enantiomers of coniine in TE-671 cells and the mouse bioassay was (-) coniine > (±) coniine > (+) coniine. The mouse $LD_{50}$ values and the corresponding TE-671 cell $EC_{50}$ values are displayed in Table 1. The relative order of potency for the enantiomers of anabasine in TE-671 cells and the mouse bioassay was (+)-anabasine > (-)-anabasine.

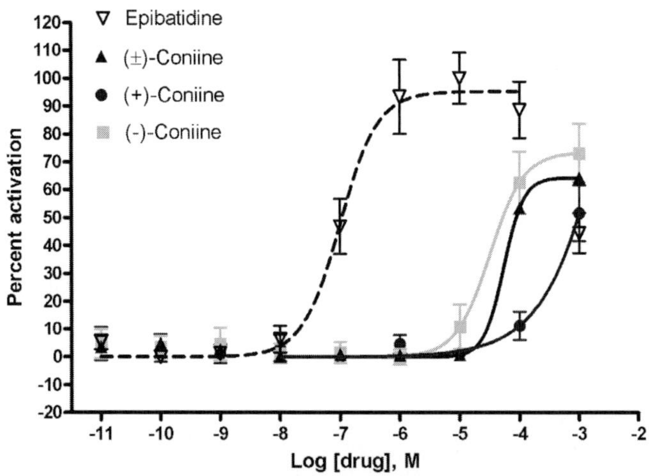


**Figure 3.** The concentration-effect relationships with best-fit lines for the actions of epibatidine and coniine compounds on membrane potential sensing dye fluorescence in TE-671 cells. In each experiment TE-671 cells were grown on 96-well black-walled culture plates and the membrane depolarization resulting from addition of either epibatidine, (±)-coniine, (+)-coniine, or (-)-coniine in $\log_{10}$ molar concentrations indicated was measured and displayed as a percentage of the maximal epibatidine response at 10 μM. Each data point represents three experiments of duplicate wells. Data obtained from Lee *et al.* (2008b).

**Table 1.** $LD_{50}$ values in mice and $EC_{50}$ values in TE-671 cells.

| nAChR Agonist | $LD_{50}$ (mg/kg) | $EC_{50}$ (μM) |
|---|---|---|
| Anabaseine[a] | 0.58 | 0.42 |
| (+)-Anabasine[a] | 16 | 7.1 |
| (-)-Anabasine[a] | 11 | 2.6 |
| (-)-Coniine[b] | 7.1 | 100 |
| (±)-Coniine[b] | 7.7 | 300 |
| (-)-Coniine[b] | 12.1 | 900 |
| N-Methylconiine | | 144[c] |
| (+)-Nicotine[a] | 0.38 | 26 |

[a]Lee *et al.* 2006, [b]Lee *et al.* 2008b, [c]Estimated

## Discussion

Initial screening of N-methylconiine in SH-SY5Y and TE-671 cells has shown there are significant differences in potency and efficacy between the two cell lines. The SH-SY5Y cell line resembles human fetal sympathetic neurons and expresses $\alpha_3$, $\alpha_5$, $\beta_2$, and $\beta_4$ nAChR subunits (Lukas *et al.* 1993; Innocent *et al.* 2008). Due to the lack of potency and efficacy in the SH-SY5Y cell line other nicotinic agonists were screened only in the TE-671 cell line.

The TE-671 cell line expresses a human fetal skeletal muscle nAChR. We have previously suggested (Lee *et al.* 2006) that the teratogenic activities of the enantiomers of coniine, anabasine, and anabaseine were due to their actions at fetal muscle type nAChR. It is interesting to note that for anabasine, the (+)-enantiomer was more potent in both the cell assay and mouse assay than the (-)-enantiomer versus coniine where the (-)-enantiomer was more potent than the (+)-enantiomer. We have proposed (Lee *et al.* 2008b) that the differences in the potency and lethality of the enantiomers between coniine and anabasine is most likely due to stereochemical interaction between the enantiomers and the ligand binding site of the receptor. Chiral recognition of drug enantiomers by receptors is due to asymmetric interactions of weak molecular forces between the receptor and ligand (Booth *et al.* 1997). These enantioselective relationships can vary between ligands as demonstrated in this study by the differences between coniine and anabasine enantiomers. The exact nature of the interactions between the receptor and these ligands has yet to be determined and further investigations are needed.

## Acknowledgements

The authors wish to thank Anita McCollum for her expert technical support.

## References

Agranat I, Caner H, and Caldwell J (2002). Putting chirality to work: The strategy of chiral switches. *Nature Reviews Drug Discovery* 1:753-68.

Booth TD, Wahon D, and Wainer IW (1997). Is chiral recognition a three-point process? *Chirality* 9:96-98.

Daly JW (2005). Nicotinic agonists, antagonists, and modulators from natural sources. *Cellular and Molecular Neurobiology* 25:513-52.

DeBoer KD, Lye JC, Aitken CD, Su AKK, and Hamill JD (2009). The A622 gene in *Nicotiana glauca*: evidence for a functional role in pyridine alkaloid synthesis. *Plant Molecular Biology* 69:299-312.

Florentine SK and Westbrooke ME (2005). Invasion of the noxious weed *Nicotiana glauca* R. Graham after an episodic flooding event in the arid zone of Australia. *Journal of Arid Environments* 60:531-545.

Forsyth CS, Speth RC, Wecker L, Galey FD, and Frank AA (1996). Comparison of nicotinic receptor binding and biotransformation of coniine in the rat and chick. *Toxicological Letters* 89:175-183.

Innocent N, Livingstone PD, Hone A, Kimura A, Young T, Whiteaker P, McIntosh JM, and Wonnacott S (2008). Alpha-conotoxin Arenatus IB[V11L,V16D] [corrected] is a potent and selective antagonist at rat and human native alpha7 nicotinic acetylcholine receptors. *Journal of Pharmacology and Experimental Therapeutics* 327:529-37.

Keeler RF and Balls LD (1978). Teratogenic effects in cattle of Conium maculatum and conium alkaloids and analogs. *Clinical Toxicology* 12:49-64.

Keeler RF and Crowe MW (1984). Teratogenicity and toxicity of wild tree tobacco, *Nicotiana glauca* in sheep. *The Cornell Veterinarian* 74:50-59.

Keeler RF, Balls LD, Shupe JL, and Crowe MW (1980). Teratogenicity and toxicity of coniine in cows, ewes, and mares. *Cornell Veterinarian* 70:19-26.

Lee ST, Panter KE, Gardner DR, Molyneux RJ, Chang C-WT, Kem WR, Wildeboer K, Soti F, and Pfister JA (2006). Relative toxicities and neuromuscular nicotinic receptor agonistic potencies of anabasine enantiomers and anabaseine. *Neurotoxicology and Teratology* 28:220-228.

Lee ST, Gardner DR, Chang CW, Panter KE, and Molyneux RJ (2008a). Separation and measurement of plant alkaloid enantiomers by RP-HPLC analysis of their Fmoc-Alanine analogs. *Phytochemical Analysis* 19:395-402.

Lee ST, Green BT, Welch KD, Pfister JA, and Panter KE (2008b). Stereoselective potencies and relative toxicities of coniine enantiomers. *Chemical Research in Toxicology* 21:2061-2064.

Lopez TA, Cid MS, and Bianchini ML (1999). Biochemistry of hemlock (*Conium maculatum* L.) alkaloids and their acute and chronic toxicity in livestock. A review. *Toxicon* 37:841-865.

Lukas R, Norman S, and Lucero L (1993). Characterization of nicotinic acetylcholine receptors expressed by cells of the SH-SY5Y human neuroblastoma clonal line. *Molecular and Cellular Neuroscience* 4:1-12.

Marion L (1950). IX The alkaloids of hemlock, In: *The Alkaloids* (RHF Manske and HL Holmes, eds) Vol. 1, pp. 211-217. Academic Press New York.

Panter KE and Keeler RF (1993) Quinolizidine and piperidine alkaloid teratogens from poisonous plants and their mechanism of action in animals. *Veterinary Clinics of North America-Food Animal* 9:33-40.

Panter KE, Keeler RF, and Baker DC (1988). Toxicoses in livestock from the hemlocks (*Conium* and *Cicuta* spp.). *Journal of Animal Science* 66:2407-13.

Panter KE, Bunch TD, Keeler RF, Sisson DV, and Callan RJ. (1990). Multiple congenital contractures (MCC) and cleft palate induced in goats by ingestion of piperidine alkaloid-containing plants: reduction in fetal movement as the probable cause. *Journal of Toxicology Clinical Toxicology* 28:69-83.

Panter KE, James LF, and Gardner DR (1999). Lupines, poison-hemlock and Nicotiana spp: Toxicity and teratogenicity in livestock. *Journal of Natural Toxins* 8:117-134.

# Chapter 101

# Rotenoids, Neurotoxic Principles of Seeds from *Aeschynomene indica* (Leguminosae)

G.A. Borghi[1], A.O. Latorre[2], P.L. Lopes[1], K.C. Higa[1], L.M.X. Lopes[3], P.C. Maiorka[2], S.L. Górniak[2], and M. Haraguchi[1]

[1]*Centre for Animal Health, Biological Institute, Av. Conselheiro Rodrigues Alves, 1252, CEP 04014-002, São Paulo, Brazil;* [2]*Dept. of Pathology, Faculty of Veterinary Medicine and Animal Sciences, University of São Paulo, Prof. Dr. Orlando Marques de Paiva 87, 05508-270, São Paulo, Brazil;* [3] *Chemical Institute – UNESP – Araraquara-SP, Brazil*

## Introduction

The invasive plant *Aeschynomene indica* (Leguminosae Fabaceae) occurs in India, Malaysia, Australia, the Pacific Islands, Africa, the southern USA, and southern Brazil. It is prevalent in wetlands in the State of Rio Grande do Sul, often as a weed within rice paddies. The seeds of this plant are harvested with the rice and during rice processing contaminate a byproduct (broken rice) used for animal feeds including swine feed.

Two spontaneous outbreaks in pigs have been reported from ingestion of diets containing 13% and 40% *A. indica* seeds in swine rations in the state of Rio Grande do Sul, Brazil (Riet-Correa *et al.* 2003; Oliveira *et al.* 2004). Clinical signs included uncoordinated gait, sternal recumbence, difficulty in rising, and lateral recumbence followed by death. Histopathological alterations were symmetric focal degeneration in the cerebellar and vestibular nuclei. The poisoning was experimentally reproduced in swine with clinical signs similar to those observed in spontaneous cases (Riet-Correa *et al.* 2003; Oliveira *et al.* 2004). To our knowledge there are no studies in the literature relating toxic compounds in *A. indica* seeds to this disease (intoxication). Bioassay guided fractionation of *A. indica* seeds in swine and mice demonstrated that the ethanol extract and its ethyl acetate fraction were lethal to swine and mice when administered orally (Haraguchi *et al.* 2003). The main objectives of this study were to determine the toxic substance(s) of the *A. indica* seeds by continued bioassay guided studies in mice and to verify their toxicological effects.

## Materials and Methods

### Plant material

*A. indica* seeds were obtained from a rice processing company in Pelotas, Rio Grande do Sul, Brazil. These seeds were used for experimental reproduction of the disease. A

©CAB International 2011. *Poisoning by Plants, Mycotoxins, and Related Toxins* (eds F. Riet-Correa, J. Pfister, A.L. Schild, and T.L. Wierenga)

voucher specimen was identified as Brazil, Rio Grande do Sul, Claudio Timm, N° 17598 (PEL), and deposited in the herbarium at the Federal University of Pelotas, Rio Grande do Sul State.

## Extraction and isolation

The ground seeds were exhaustively extracted first with hexane and then followed by 96% ethanol. After extraction the solvent was removed by rotary evaporation under reduced pressure yielding hexane and ethanol extracts. The ethanol extract, when suspended in 10 ml of 80% ethanol, yielded a white solid residue; this residue was filtered and identified as starch by the Molish test. After concentration under reduced pressure the filtrate yielded an ethanol extract free of white solid residue (EE). The EE was further fractionated by partitioning the EE fraction between water and ethyl acetate to obtain an ethyl acetate residue (EAR) and an aqueous residue after evaporation of solvents. The EAR was applied to a silica gel 60 (70-230 mesh, Merck) chromatographic column and eluted sequentially under reduced pressure with ethyl acetate, mixtures of 5%, 10%, and 50% methanol/ethyl acetate, and finally methanol to obtain five fractions. Each fraction was evaporated under reduced pressure until dryness. The 5% methanol/ethyl acetate fraction was applied to a silica gel 60 (70-230 mesh, Merck) chromatographic column and eluted with mixtures of methanol and ethyl acetate in increasing order of polarity. After evaporation, each fraction was monitored by thin layer chromatography (TLC) employing a plastic plate impregnated with silica gel 60G F254 (Merck) and developed with ethyl acetate:methanol:water (100:13.5:10). The TLC spots were visualized by UV and by treatment of the plates with an alcohol solution of 10% iron chloride and 10% sulphuric acid on a hot plate for 10 min. Fractions that were similar were combined, yielding eight subfractions. The fourth subfraction was further separated by reversed phase HPLC employing a semipreparative C-18 chromatographic column and an acetonitrile:water (1:1) mobile phase. Detection was accomplished using a UV-vis detector at $\lambda$=300 nm and the mobile phase flow rate was 7 ml/min. Three principal substances (<u>**1**</u>-<u>**3**</u>; Figure 1) were obtained. Structural identifications were accomplished using nuclear magnetic resonance (NMR) and infrared (IR) spectroscopic analyses.

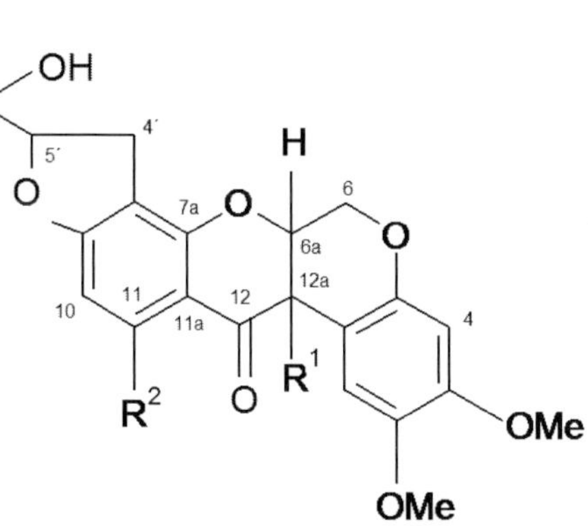


<u>1</u> $R^1$=H $R^2$=H
<u>2</u> $R^1$=OH $R^2$=H
<u>3</u> $R^1$=H $R^2$=OH

**Figure 1.** Structures of the rotenoids from *A. indica.*

### Acute toxicological test

Twenty-one Swiss mice bred at the University of São Paulo Department of Veterinary Medicine and Zootechny were randomly divided into seven groups including a control group. The groups were dosed the following fractions by oral gavage: EE, RAE, ethyl acetate fraction, 5% methanol/ethyl acetate fraction (MEAF), 10% methanol/ethyl acetate fraction, and 4th subfraction from MEAF at the doses of 0.9, 0.45, 0.2, 0.15, and 0.10 g/kg, respectively. All samples were suspended in Tween 80 and water. The control group received a mixture of Tween 80 and water as vehicle. After administration the animals were observed at 1 h, 2 h, 4 h, and 8 h until 30 h and the behavioral changes and lethality in comparison with control animals were recorded. After 30 h the surviving mice were sacrificed by decapitation and the brain collected and fixed in neutral buffered formalin for histopathology. The brains were cut in 2-3 mm transverse sections and all sections were examined microscopically.

**Table 1.** $^1$H NMR data for rotenoids from *A. indica* seeds (ppm, $CDCl_3$, J in Hz, 500 MHz).

| H | 1 | 2 | 3 |
|---|---|---|---|
| 1 | 6.71 s | 6.48 s | 6.78 s |
| 4 | 6.39 s | 6.43 s | 6.38 s |
| 6ax | 4.55 | 4.52 dd J 11.0 e 2.5 | 4.11 d J 12.0 |
| 6eq | 4.11 | 4.41 d J 12.0 | 4.55 dd J 3.1 |
| 6a | 4.86 | 4.5 ddd | 4.8 m |
| 10 | 6.43 | 6.48 d J 8.5 | 5.92 |
| 11 | 7.76 | 7.74 d J 8.5 | - |
| 12a | 3.78 | - | 3.8 |
| OH | | 4.40 s | |
| 2-OMe | 3.72 | 3.68 s | 3.71 s |
| 3-OMe | 3.75 | 3.76 s | 3.74 |
| 4' | 3.06 | | 2.9 d J 9.0 |
| 5' | 4.61 | | |
| 7' | 1.29 | 1.29 s | 1.26 s |
| 8' | 1.16 | 1.16 s | 1.14 s |

## Results and Discussion

### Isolation of toxic principles

The EAR was applied to a silica gel chromatographic column and eluted with ethyl acetate and methanol with increasing polarity to obtain five fractions under low pressure. The fractions obtained with 5% methanol/ethyl acetate (MEAF) showed acute toxicity to mice and were further separated on another silica gel chromatographic column and eluted with ethyl acetate and methanol with increasing polarity to obtain 12 subfractions. The 4th subfraction showed increased concentrations of residue with three dark spots when visualized on a TLC plate sprayed with 10% alcohol iron chloride indicating constituent aromatics at $R_f$ 0.43, 0.36, and 0.33. Spraying with 10% sulfur solutions followed by heating showed additional spots with minor intensity. The purification of the three principle aromatic constituents was accomplished by reversed phase HPLC using a semi-preparative C18 column eluted with acetonitrile and water (1:1). Structural identification of the three principle aromatic compounds was determined by NMR and IR spectroscopy. Proton signals in 6.30 to 7.80 ppm region and $^{13}C$ signals in the100-130 ppm region indicated aromatic groups; signals in the 3.00 to 5.00 ppm region and $^{13}C$ signals in the 73.00-79.00 ppm region indicate carbonyl groups. Proton signals in the 3.60 to 3.80 ppm region and $^{13}C$ signals in the 55.00 to 57.00 region indicate methoxy groups. Proton signals in the 1.20 to 1.40 ppm region and $^{13}C$ signals in the 20-27 ppm region indicate methyl groups among other signals (Tables 1 and 2). The infrared spectra showed an absorption band at 1676 cm indicating the presence of carbonyl groups linked with aromatic group. In comparison with predicted spectral data the compounds were identified as rotenoids dalpanol **1**, 12 α-hydroxydalpanol **2**, and 11-hydroxydalpanol **3** (Figure 1). Compounds **1** and **2** were isolated previously from *Amorpha fructinosa* (Li *et al.* 1993). Compound **3** has not been reported in the literature. Rotenoids **1**, **2**, and **3** described for the first time in *A. indica* in this study are likely the toxic principles that cause neurological signs in mice.

**Table 2.** $^{13}C$ NMR data for rotenoids from *A. indica* seeds (ppm, $CDCl_3$, 75.4 MHz).

| C | **1** | **2** | **3** |
|---|---|---|---|
| 1 | 110.5 | 109.5 | 110.4 |
| 1a | 104.8 | 105.0 | 105.0 |
| 2 | 144.0 | 144.0 | 144.8 |
| 3 | 149.6 | 151.2 | 150.0 |
| 4 | 101.0 | 101.1 | 101.1 |
| 4a | 147.5 | 148.5 | 147.0 |
| 6 | 66.3 | 63.8 | 66.0 |
| 6a | 72.3 | 76.0 | 71.7 |
| 7a | 157.9 | 157.5 | |
| 8 | 113.5 | | 101.1 |
| 9 | 167.2 | | 166.0 |
| 10 | 104.8 | 105.2 | 91.8 |
| 11 | 129.8 | 130.0 | |
| 11a | 113.2 | 112.0 | |
| 12 | | | 190.0 |
| 12a | 44.7 | 67.6 | 43.7 |
| 2-OMe | 56.4 | 56.5 | 56.4 |
| 3-OMe | 55.9 | 55.9 | 55.9 |
| 4' | 27.4 | 114.3 | 26.7 |
| 5' | 91.5 | 129.9 | 91.0 |
| 6' | 71.7 | 79.0 | 71.7 |
| 7' | 26.2 | 21.7 | 26.1 |
| 8' | 24.0 | 22.5 | 24.0 |

## Acute toxicity of extract and fractions in mice

The EE of *A. indica* seeds, free of white solid residue identified as starch by the Molish test, when administered in mice by gavage using 9 g/kg in a single dose caused sternal recumbence, uncoordinated gait, hypothermia, convulsions, and death in 6-8 h. The fraction obtained by partition of EE using water and ethyl acetate yielded EAR. When administered orally, EAR showed similar signs to those caused by EE. Among the fraction, the group that received the ethyl acetate fraction and 5% MEAF showed that these fractions were toxic to mice. Other groups showed no clinical signs. The 4th subfraction obtained from 5% methanol/ethyl acetate as previously described, when administered by gavage in mice, provoked clinical signs such as hypothermia, uncoordinated gait, convulsions, and death at 6 h after administration. After isolation the primary compounds in subfraction 4 mentioned above were identified as derivatives of rotenoids **1**, **2**, and **3**.

Rotenone, a substance isolated from plants, is an insecticide and is also toxic to fish. Although it is safe for most mammals it is toxic to swine, causing neurological signs such as incoordination which progresses from staggering to paralysis of all limbs, respiratory depression, and coma with rapid death and absence of pathological alterations (Oliver and Roe 1957; Manahan 2003). These signs are similar to those observed in Brazil in pigs fed with broken rice contaminated with *A. indica* seeds, further indicating that the toxicity of the seeds is due to rotenoid derivatives.

## References

Haraguchi M, Zambronio F, Górniak SL, Baialardi CEG, and Riet-Correa F (2003). Neurotoxicity to pigs and rodents from different fractions of *Aeschynomene indica* seeds. *Veterinary and Human Toxicology* 45:177-179.

Li L, Wang HK, Chang JJ, McPahail AT, McPhail DR, Terada H, Konoshima T, Kokumai M, Kozuka M, Estes JR, and Lee KH (1993). Antitumor agents, 138. Rotenoids and isoflavones as cytotoxic constituents from *Amorpha fruticosa*. *Journal of Natural Products* 56:690-698.

Manahan SE (2003). *Toxicological Chemistry and Biochemistry*, 389 pp., 3rd edn. Lewis Publishers, London.

Oliveira FN, Rech RR, Rissi DR, Barros RR, and Barros CSL (2005). Intoxicação em suínos pela ingestão de sementes de *Aeschynomene indica* (Leg. Papilionoideae). Poisoning in swine from the ingestion of *Aeschynomene indica* (Leg. Papilionoideae) seeds. *Pesquisa Veterinária Brasileira* 25:135-142.

Oliver WT and Roe CK (1957) Rotenone poisoning of swine. *Journal Association Veterinary Medicine* A 410-411.

Riet-Correa F, Tim CD, Barros SS, and Summers BA (2003). Symmetric focal degeneration in the cerebellar and vestibular nuclei in swine caused by ingestion of *Aeschynomene indica* seeds. *Veterinary Pathology* 40:311-316.

# Chapter 102

# Chemistry of *Dieffenbachia picta*

N.S. Barbi[1], L. Lucchetti[2], N.A. Pereira[3], and A.J.R. da Silva[4]

[1]*Departamento de Análises Clínicas e Toxicológicas, Faculdade de Farmácia, CCS-bloco A, Universidade Federal do Rio de Janeiro, Ilha do Fundão, 21941-590, Rio de Janeiro, RJ, Brazil;* [2]*Laboratório de Química de Produtos Naturais, Instituto de Tecnologia em Fármacos – Farmanguinhos, FIOCRUZ, Rua Sizenando Nabuco 100, 21041-250, Rio de Janeiro, RJ, Brazil;* [3]*Departamento de Farmacologia Básica e Clínica, Instituto de Ciências Biomédicas, CCS-bloco K, Universidade Federal do Rio de Janeiro, Ilha do Fundão, 21941-550, Rio de Janeiro, RJ, Brazil;* [4]*Núcleo de Pesquisas de Produtos Naturais, CCS-bloco H, Universidade Federal do Rio de Janeiro, 21041-250, Rio de Janeiro, RJ, Brazil*

## Introduction

Chronic or acute poisoning caused by plant exposure is a worldwide health problem. As reported by SINITOX (Sistema Nacional de Informações Tóxico-Farmacológicas – National System of Toxic and Pharmacological Information 2009) 1657 single exposures were documented in Brazil in 2007 including three deaths. The actual number of exposures may be much higher considering that most of the poisoning episodes probably do not result in notification to health authorities. Epidemiological studies show that exposures to plants belonging to the Araceae family are among the most common toxic plant exposures reported in Brazil and in the world (Silva and Takemura 2006).

Most of the Araceae display poisonous properties. Some of the ornamental species within this family from the genera *Dieffenbachia*, *Zantedeschia*, *Alocasia*, *Philodendron*, and *Caladium* cause intense pain followed by gastric irritation and swelling of the tongue and glottis when ingested, and produce painful irritation when in contact with the skin, mouth, or eyes. Children and domestic pets are the most frequent victims of toxic events with Araceae species. The chemical nature of the substances involved as well as the toxicokinetic and toxicodynamic mechanisms are not known for most of the species. Some reviews regarding this subject have been written by Hegnauer (1963, 1986), Mitchell and Rook (1979), Mayo *et al.* (1997), Bown (2000), and Froberg *et al.* (2007).

The best known and most toxic member of this family is *Dieffenbachia picta* Schott (= *D. seguine*) commonly referred to as 'comigo-ninguém-pode' in Brazil. As with the other species in the genus *D. picta* is characterized by the presence of many unusual cells (the idioblasts) in its leaves, petioles, and stems (Gardner 1994). The juice of *D. picta* gives rise to an intense inflammatory reaction when in contact with skin, mouth, or eyes.

©CAB International 2011. *Poisoning by Plants, Mycotoxins, and Related Toxins* (eds F. Riet-Correa, J. Pfister, A.L. Schild, and T.L. Wierenga)

## Toxicological Studies

The chemical and toxicological studies on *D. picta* focusing on identification and mechanism of action of the active ingredients are limited. Reported hypotheses about the chemical nature of the toxic principle include proteolytic enzymes, alkaloids, polysaccharides, cyanogenic glycosides, and saponins (Rizzini and Occhioni 1957; Walter 1967; Fochtman *et al.* 1969; Walter and Khanna 1972; Ladeira *et al.* 1975).

In one of the first toxicological reports dealing with *D. picta* by Rizzini and Occhioni (1957) the authors performed *in vivo* experiments (macroscopic and histological examination) with leaves and stems of *D. picta*. They observed that the leaf juice was not toxic and that the petiole juice was less toxic than the stem juice, a conclusion which was previously stated by Barnes and Fox (1955) and later confirmed by Ladeira *et al.* (1975) and Padmanabhan and Shastri (1990). The authors observed that the toxic action was restricted to the insoluble fraction of the juice collected from the stems of the plants. The oral or topic administration of the juice from the stems of *Dieffenbachia* evokes intense allergic reactions leading to inflammation. Death occurs within a few seconds when the administration is done intraperitoneally or intracardiacally. However, when introduced directly to the stomach by intubation no toxic effect is observed. In this case the deactivation of the active principle may occur due to the action of gastric secretions, enzymes, or poor absorption (Rizzini and Occhioni 1957; Fochtman *et al.* 1969; Ladeira *et al.* 1975; Maderosian and Roia Jr 1979). Histological examination of the tongue of animals treated with the juice of *D. picta* showed edema, diffuse vascular congestion, basal membrane degeneration, and inflammatory reaction (Fochtman *et al.* 1969). Traumatic injury of the tongue caused by the calcium oxalate crystal needles was also observed. Lymphocytic and polymorphonuclear cell infiltration are present in the oral cavity, pharynx, and esophagus where hyperemia and subepithelial bleeding were also observed (Rizzini and Occhioni 1957; Gardner 1994). Animals treated with juice kept at 0°C for 30 days showed edema of low intensity (Fochtman *et al.* 1969). The authors verified that the digestion of the juice with trypsin (37°C, 7 min) reduced the irritation and the damage caused in the tongue of the animals. Rizzini and Occhioni (1957) did not obtain the same results; they agree that in some aspects the toxic mechanism could be associated with histamine release – which appears at significant higher levels when animals are treated with the juice of *D. picta*. Animals treated with antihistamine agents presented only a slight edema and a minimal inflammatory response providing some protection against the toxic effects (Rizzini and Occhioni 1957; Fochtman *et al.* 1969). This result is not supported by the works of Ladeira *et al.* (1975) and Carneiro *et al.* (1985). Marderosian and Roia Jr (1979) suggest that this discrepancy can be due to the use of different antihistamine agents or animals of different species.

The suggested role of a protein by Rizzini and Occhioni (1957) was supported by Fochtman *et al.* (1969) and Kuballa *et al.* (1981). Walter and Khanna (1972) proposed that the action of a proteolytic enzyme of the plant (dumbcain) with further release of kynins could be responsible for the toxicity of the species. The kynins display a wide range of physiological properties and act as mediators of inflammation (Kuballa *et al.* 1981); they can act on reproductive ducts, which could explain the sterilizing effect caused by the plant (Dvorjetski 1958; Marderosian and Roia Jr 1979; Pasquale *et al.* 1984). Proteolytic enzymes can cause local necrosis and collapse of little blood vessels thus giving rise to bleeding (Marderosian and Roia Jr 1979). Nevertheless, Ladeira *et al.* (1975) argued against the involvement of a protein as they observed that the active residue of the juice from the stems contained no nitrogen compounds except for amino acids. In order to prove this statement, authors obtained results after incubation of the suspensions of the residue with chymotrypsin (37°C, 1 h) which proved to be unable to inactivate the preparations. Padmanabhan and Shastri (1990) noticed that the proteinase activity was low in several parts of the plants; on the other hand, the inhibitory activity of trypsin was noteworthy. They concluded with some caution that part of the toxic effect of species of *Dieffenbachia* could be due to the presence of amylase inhibitors, responsible for the intense salivation, swelling of the salivary ducts, and temporary speechlessness.

Neves *et al.* (1988) and Carneiro *et al.* (1985, 1989) observed that the juice of *D. picta* after the removal of all crystals when in contact with the oral mucosa of mice caused no edema. However, when injected into their paws an intense reaction was observed. This reaction could be partially inhibited by previous administration of indomethacin or acetylsalicylic acid. These results showed the significance of the raphides (crystals) and suggest the role of prostanoids in inflammatory events.

The possible participation of polysaccharides in the inflammatory event was suggested by Ladeira *et al.* (1979). These compounds could involve the calcium oxalate crystals and be carried by them to the interior of the mucosa cells. Carneiro *et al.* (1989) questioned this theory by presuming the interplay of low polarity substances in the toxic process and showed that the crystals bear an oily material that can be dyed by Sudan III. When washed with petrol the crystals lose the ability to cause inflammation and irritating reactions, suggesting the low polarity nature of the substances. The authors proposed that unsaturated fatty acids ranging from 18 to 22 carbons and associated to the calcium oxalate crystals are responsible for the edematogenic effects and the rise of cutaneous capillary permeability. Padmanabhan and Shastri (1990) pointed out that excess salivation and swelling of salivary ducts caused by ingestion of the plant could be related to the presence of salivary amylase inhibitors. *In vitro* tests with the stem, considered to be the most toxic part of the plant, showed an amylase inhibitory activity higher than that of petiole and leaves.

Recently Dip *et al.* (2004) demonstrated that eugenol was more efficient in inhibition of tongue edema than treatment developed in hospitals (dexametasone + prometazine + adrenaline) for poisoning by *D. picta*. The mechanism by which eugenol inhibits tongue edema remains unclear but it seems to be related to the arachidonic acid metabolism.

## Chemical Studies

The genus *Dieffenbachia* comprises about 135 species, most of them found in South America (Croat 2004). Abundant literature is available for *Dieffenbachia* species toxicity; nevertheless, little information regarding its chemical constituents is available. In one of the few reports for *Dieffenbachia* species, Walter and Khanna (1972) reported the isolation of L-asparagine and a proteolytic enzyme (dumbcain) from an aqueous extract of *D. picta*. The authors suggested that the enzyme was the active compound. The structure of this enzyme still remains unknown. Lanosterol, β-sitosterol, and cycloartenol were also characterized from the petroleum ether extract obtained from *D. picta* (whole plant). The presence of toxic polysaccharides was mentioned by Ladeira *et al.* (1979); however, further studies to determine the identity and the structure of the sugars were not carried out.

### Composition of latex

Although the Araceae family has the largest number of latex-bearing plants of any monocotyledon family, there is scarce information on latex chemical composition and distribution. Aroideae, Philodendroideae, and Colocasioideae are the only subfamilies characterized as lineages with latex in Araceae (Darling 2007). Fox and French (1988) detected free sterols and triterpenes in the latex of many of the abundant New World species examined and suggested that these compounds contribute to the white-opaque appearance of the latex. The same authors detected only esterified sterols in latex of the Old World genera examined, which are consistently clear to cloudy and lack other terpenoid compounds. It is considered that the occurrence and distribution of steroids as latex components can be used as a tool for the taxonomical grouping of plant species (Nielsen *et al.* 1979; Fox and French 1988)

In our work (Barbi 1999), the latex secreted through fresh cut stems (2 kg) was collected (4 ml) and its lipophylic constituents were further extracted with $CH_2Cl_2$ (3 × 4 ml). The oily residue (560 mg) obtained after solvent removal under an $N_2$ stream was divided into two portions of 280 mg each and then fractionated by medium pressure column liquid chromatography (MPLC, silica gel) yielding two main fractions which could be chemically characterized. Fraction 1 (50 mg) was analyzed by GC/MS and found to contain farnesene, δ-cadinene, farnesol and the fatty acids hexadecanoic, tridecanoic, octadecadienoic, octadecenoic, eicosadienoic, and octadecanoic acids. Fraction 2 (70 mg) was further separated by preparative TLC and cycloartenol (4 mg), 24-methylenecyclo-artenol (2 mg), cycloeucalenol (3 mg), and a mixture (1.5 mg) of β-sitosterol, 22-dihydrobrassicasterol, and α- and β-amyrins were identified by comparison of their $^1$H, $^{13}$C-NMR, and mass spectral data with that in the literature (Wehrli and Nishida 1979; McLafferty and Stanffer 1989). This is the first report of the occurrence of triterpenes, sesquiterpenes, and steroids in latex of *Dieffenbachia*. The white-opaque appearance of the latex and the identification of such metabolites are in agreement with Fox and French's (1988) findings about the occurrence of these compounds in Araceae in the special subfamily Aroideae. The methanol extract of the stems was fractionated and found to contain a mixture of steroids: (5α,5β,12α)-ergostan-3-12-diol; [3β,22(Z)]-estigmast-5,22-dien-3-ol, tremulone, β-sitosterol, cycloeucalenol, and cyclolaudenol. An acid triterpene identified as ursolic acid was also isolated.

**Microextraction from the idioblasts**

Microscopic examination of stem plant tissue of *D. picta* shows many specialized cells (idioblasts) morphologically diverse and containing needle-like crystals of calcium oxalate called raphides. The idioblast cells have thin walls, are present in high amounts, and have a yellow oily material inside. This oil embeds the raphides and in contact with Sudan III develops a red color. The oil was collected with microcapillaries directly from the idioblasts. The cell wall of the idioblasts is somewhat rigid and when submitted to pressure releases a great amount of oil-covered raphides. The oily material was removed from the micropipettes with heptane and chloroform and after solvent removal esterified with diazomethane then analyzed by GC/MS. The analysis of this mixture revealed the presence of free fatty acids ($C_8$-$C_{24}$) and linear saturated hydrocarbons ($C_{15}$-$C_{28}$).

GC/MS is considered one of the most important methods for the characterization of fatty acids. However, in unsaturated systems thermally catalyzed migration of double bonds leads to difficulties in the location of double bonds. In order to establish the position of the double bonds of the unsaturated fatty acids, a reaction on the carboxyl group based on the work of Zhang *et al.* (1988) was performed with slight changes on the original procedures. Thus, the 4,4-dimethyloxazoline derivatives (DMOX) were prepared by adding 250 μl of 2-amino-2-methylpropanol (AMP) to 500 μl of the mixture containing the fatty acid methyl esters. This mixture was heated to 180°C for 18 h in a sealed tube. After this period the residue was dissolved with methylene chloride and washed with distilled water. After drying the organic phase with $Na_2SO_4$, the solvent was removed under $N_2$ stream and analyzed by GC/MS. The elution temperature of the fatty acid oxazolines is about 5°C higher that for the corresponding methyl esters. The double bond is located by the change in the sequence of cleavage of methylene units (*m/z* 14). When a difference of 12-mass units appears (instead of 14) between the ions it points to the presence of a double bond between the carbons of that interval. This procedure allowed us to determine the location of the double bonds of the unsaturated *cis*-fatty acids: 9-hexadecenoic, 9-heptadecenoic, 9,12-octadecadienoic, 9-octadecenoic, 11-octadecenoic, and 9,11-octadecadienoic (Barbi 1999).

## Toxicological Test

Female Swiss mice (25-30 g) were used for the assessment of the acute toxicity of the oily fraction of the idioblasts from the stems of *D. picta*. The mice were kept under standard conditions in ventilated animal boxes (water and food *ad libitum*, 12 h dark and light cycle). Different doses of the lipid material extracted by microcapillaries from the idioblasts corresponding to 100, 200, 400, and 800 µg were administered orally to four test groups of five mice per group with an automatic pipette. An additional control group (five animals) received vehicle (commercial vegetable oil) in the same volume as the treated animals (100 µl). The clinical and behavioral signs as well as survival of animals were observed for 24 h. No effects of the vehicle were observed within the control group; in all four test groups treated with the idioblast oil animals were observed with mouth irritation, tongue swelling, intense edema of the lips, and pilo-erection. The edematogenic effects disappeared after 12 h but the animals still showed discomfort and irritation (Barbi 1999).

## Conclusions

The latex collected from the stems of *D. picta* was analyzed and farnesene, δ-cadinene, farnesol, and the fatty acids hexadecanoic, tridecanoic, octadecadienoic, octadecenoic, eicosadienoic, and octadecanoic were characterized by GC/MS. Cycloartenol, 24-methylenecycloartenol, and cycloeucalenol together with a mixture of β-sitosterol, 22-dihydrobrassicasterol, and α- and β-amyrins were also identified. Further fractionation provided the identification of (5α,5β,12α)-ergostan-3-12-diol, [3β,22(Z)]-estigmast-5,22-dien-3-ol, tremulone, β-sitosterol, cycloeucalenol, and cyclolaudenol. An acid triterpene identified as ursolic acid was also isolated. This is the first report of the occurrence of triterpenes, sesquiterpenes, and steroids in the latex of *D. picta*.

The hypothesis that lipophilic toxic substances are present inside the idioblasts was tested. The oily material was collected directly from the idioblasts and subjected to GC/MS analysis providing the identification of free fatty acids ($C_8$-$C_{24}$) and hydrocarbons ($C_{15}$-$C_{28}$). Since these compounds are not described in the literature as being toxic and their concentrations are low, one might assume that there are other uncharacterized and unaccounted for substances present in *D. picta* that may be highly potent but occur at such low concentrations that they were undetectable by the analytical techniques used in this study.

## References


Barbi NB (1999). Estudo Químico de *Dieffenbachia picta* Schott (Araceae), 264 pp. PhD Dissertation, Universidade Federal do Rio de Janeiro, Rio de Janeiro.

Barnes BA and Fox LE (1955). Poisoning with *Dieffenbachia*. *Journal of the History of Medicine* 10:173-181.

Bown D (2000). *Aroids. Plants on the Arum Family,* pp. 275-300. Timber Press, Portland, Oregon, USA.

Carneiro CMTS, Neves LJ, and Pereira NA (1985). Mecanismo Tóxico de Comigo-Ninguém-Pode, *Dieffenbachia picta* Schott (Araceae). *Anais da Academia Brasileira de Ciências* 57:392-393.

Carneiro CMTS, Neves LJ, Pereira EFR, and Pereira NA (1989). Mecanismo Tóxico de Comigo-Ninguém-Pode, *Dieffenbachia picta* Schott, Araceae. *Revista Brasileira de Farmácia* 70:11-13.

Croat TB (2004). Revision of *Dieffenbachia* (Araceae) of Mexico, Central America, and the West Indies. *Annals of the Missouri Botanical Garden* 91(4):668-772.

Darling DC (2007) Holey Aroids: Circular Trenching Behavior by a Leaf Beetle in Vietnam. *Biotropica* 39(4):555-558.

Dip EC, Pereira NA, and Fernandes PD (2004). Ability of eugenol to reduce tongue edema induced by *Dieffenbachia picta* Schott in mice. *Toxicon* 43:729-735.

Dvorjetski M (1958). La Plante stérilisante *Caladium seguinun* et sés Proprietés Pharmacodynamiques. *Revue Française de Gynécologie et Obstetrice* 53:139-150.

Fochtman FW, Manno JE, Winer CL, and Cooper JA (1969). Toxicity of the Genus *Dieffenbachia. Toxicology and Applied Pharmacology* 15:38-45.

Fox MG and French JC (1988). Systematic Occurrence of Sterols in Latex of Araceae: Subfamily Colocasioideae. *American Journal of Botany* 75:132-137.

Froberg B, Ibrahim, and Furbee RB (2007) Plant Poisoning. *Emergency Medicine Clinics of North America* 25:375-433.

Gardner DG (1994). Injury to the mucous membranes caused by the common houseplant, *Dieffenbachia. Oral Surgery, Oral Medicine, Oral Pathology, Oral Radiology and Endodontology* 78(11):631-633.

Hegnauer R (1963). Araceae. In *Chemotaxonomie der Pflanzen,* vol. 2, pp. 11-18, 73-99, 475-476, 497. Birkhäuser Verlag, Basel.

Hegnauer R (1986). Araceae. *Chemotaxonomie der Pflanzen*, vol. 7, pp. 581-591. Birkhäuser Verlag, Basel.

Kuballa B, Lugnier AAJ, and Anton R (1981). Study of *Dieffenbachia*-induced edema in mouse and rat hindpaw: respective role of oxalate needles and trypsin-like protease. *Toxicology and Applied Pharmacology* 58:444-451.

Ladeira AM, Andrade SA, and Sawaya P (1975). Studies on *Dieffenbachia* Schott: toxic effects in guinea pigs. *Toxicology and Applied Pharmacology* 34:363-373.

Ladeira AM, Ornellas SOA, and Sawaya P (1979). *Dieffenbachia picta* Schott. Atividade irritante e tóxica. *Ciência e Cultura* – V Simpósio de Plantas Medicinais do Brasil, pp. 128-129. São Paulo.

Maderosian AD and Roia Jr FR (1979). Literature review and clinical management of household ornamental plants potentially toxic to humans. In *Toxic Plants* (AD Kinghorn, ed.) pp. 101-124. Columbia University Press, USA.

Mayo SJ, Bogner J, and Boyce PC (1997). *The Genera Araceae,* 370 pp. Royal Botanic Gardens, Kew, London.

McLafferty FW and Stanffer DB (1989). *Registry of Mass Spectral Data*, vols. I, II and III. Wiley-Interscience Pub., New York, USA.

Mitchell J and Rook A (1979). Araceae. In *Botanical Dermatology*, pp. 108-121. Greengress, Vancouver, Canada.

Neves LJ, Carneiro CMTS, and Pereira NA (1988). Estudo do mecanismo tóxico em *Dieffenbachia picta. Acta Amazônica* 1/2 (Suppl):171-174.

Nielsen P, Nishimura H, Liang Y, and Calvin M (1979). Steroids from *Euphorbia* and Other Latex Bearing Plants. *Phytochemistry* 18:103-104.

Padmanabhan S and Shastri NV (1990). Studies on Amylase Inhibitors in *Dieffenbachia maculata. Journal of the Science of Food and Agriculture* 52:527-536.

Pasquale RC, Ragusa S, Circosta C, and Forestieri AM (1984). Investigations on *Dieffenbachia amoena* Gentil. I: Endocrine Effects and Contraceptive Activity. *Journal of Ethnopharmacology* 12:293-303.

Rizzini CT and Occhioni P (1957). Ação tóxica das *Dieffenbachia picta* e *D. seguine. Rodriguesia* 20:5-19.

Silva IGR and Takemura OS (2006). Aspectos de intoxicações por *Dieffenbachia* spp (comigo-ninguém-pode) – Araceae. *Revista de Ciências Médicas e Biológicas* 5(2):151-159.

Sistema Nacional de Informações Tóxico-Farmacológicas-Sinitox (2009). Available at http://www.fiocruz.br/sinitox_novo/media/tab06_brasil_2007.pdf. Site visited Sept 13, 2009.

Walter WG (1967). *Dieffenbachia* toxicity. *Journal of The American Medical Association* 201:140-141.

Walter WG and Khanna PN (1972). Chemistry of Aroids. I. *Dieffenbachia seguine, amoena* and *picta*. *Economic Botany* 26:364-372.

Wehrli FW and Nishida T (1979). The use of carbon-13 nuclear magnetic resonance spectroscopy in natural products chemistry. In (Herz W, Grisbach H, and Kirby GW) *Progress in the Chemistry of Organic Natural Products* 36:92-100.

Zhang JY, Yu T, and Huang ZH (1988). Chemical modification in mass spectrometry. IV. 2-alkenyl-4,4-dimethyloxazolines as derivates for the double bond location of long-chain olefinic acids. *Biomedical and Environmental Mass Spectrometry* 15:33-44.

# Chapter 103

# Alkaloid Profiles of *Mimosa tenuiflora* and Associated Methods of Analysis

D.R. Gardner[1], F. Riet-Correa[2], and K.E. Panter[1]

[1]*USDA-ARS Poisonous Plant Research Laboratory, Logan, Utah 84341, USA;* [2]*Centro de Sude e Tecnologia Rural, Universidade Federal de Campina Grande, Patos, Paraíba, Brazil*

## Introduction

*Mimosa tenuiflora* is a common shrub/tree found in many parts of South America and northward into Mexico (Rivera-Arce *et al.* 2007; de Souza *et al.* 2008). In northeastern Brazil it is often eaten by livestock including goats, sheep, and cattle and is believed to be responsible for induced malformations observed in many animals from that region (Pimentel *et al.* 2007; Medeiros *et al.* 2008). In previous work, *M. tenuiflora* fed experimentally to goats was found to produce malformations similar to those observed in field cases and were characterized by cleft lip, unilateral corneal opacity, ocular bilateral dermoids, buphthalmos, and segmental stenosis of the colon. Even though such cases of toxicoses have been associated with the plant *M. tenuiflora* is an accepted forage plant in many regions. There have been a number of reported pharmacological uses of the bark of *M. tenuiflora* and in northeastern Brazil some indigenous uses include making a drink that has psychotropic effects (Rivera-Arce *et al.* 2007; de Souza *et al.* 2008). Most chemical analyses of the plant have focused on the bark. The psychoactive properties of the bark are believed to be caused by the indole alkaloid *N,N*-dimethyltryptamine (DMT) (Nicasio *et al.* 2005). Other indole alkaloids reported from *M. tenuiflora* are 5-hydroxytryptamine (serotonin) (de Souza *et al.* 2008) and the phytoindole yuremamine (Vespsalainen *et al.* 2005) (Figure 1), but generally the alkaloid content of the plant has not been well investigated and especially not in relation to livestock poisonings. The teratogenic principles of *M. tenuiflora* remain unknown.

We report here on the analysis of the alkaloid content from leaves and seeds of *M. tenuiflora* collected from northeastern Brazil. Alkaloids were isolated by classical acid/base extraction procedures and also using cation exchange solid phase extraction. The crude alkaloid fractions were then analyzed by thin layer chromatography (TLC), gas chromatography-mass spectrometry (GC-MS) and by liquid chromatography-mass spectrometry (LC-MS).

©CAB International 2011. *Poisoning by Plants, Mycotoxins, and Related Toxins* (eds F. Riet-Correa, J. Pfister, A.L. Schild, and T.L. Wierenga)

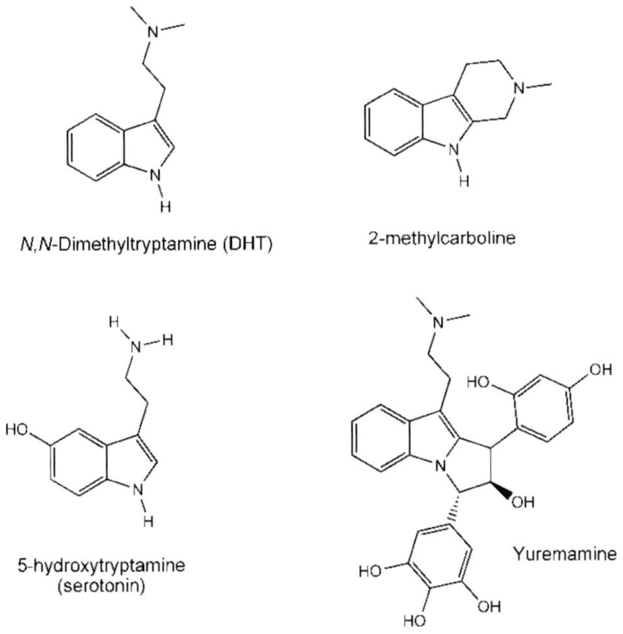


**Figure 1.** Chemical structures of alkaloids from *Mimosa tenuiflora*.

# Material and Methods

## Plant material

Plant leaf material of *M. tenuiflora* was obtained from northeastern Brazil (F. Riet-Correa) near areas of reported cases of poisonings, dried at ambient temperature, and ground for shipping to the Poisonous Plant Research Laboratory for chemical analysis.

## Chemical extraction of plant material for alkaloids

*Method A*

Plant material (100 mg) was placed into a 15 ml glass screw top test tube and 4 ml of 1 N HCl and 4 ml of chloroform were added. The mixture was extracted by mechanical rotation for 1 h and then centrifuged to aid separation of layers. The upper aqueous acid solution was removed to a second test tube and the pH adjusted to ~10 by dropwise addition of concentrated ammonium hydroxide via a Pasteur pipette. The solution was then extracted twice with chloroform (4 ml, 2 ml) each by mechanical rotation (5 min) followed by centrifuging and removal of the chloroform and filtering the chloroform through a small amount of anhydrous sodium sulfate. The chloroform was removed by evaporation under a flow of nitrogen in a heating block (50°C).

*Method B*

Plant material (100 mg) was placed into a 15 ml glass screw top tube and 4 ml of 1 N HCl and 4 ml of chloroform were added. The mixture was extracted by mechanical rotation for 1 h and then centrifuged to aid separation of layers. The upper aqueous layer was removed and added to a solid phase extraction column (Strata-XC, 30 mg, preconditioned by rinsing with 2 ml methanol and 2 ml water). The SPE column was rinsed with an additional 2 ml water and then 2 ml methanol. The alkaloids were then eluted from the column with 4 ml of ammoniated methanol. The solvent was evaporated to dryness under a flow of nitrogen at 60°C in a heat block.

**Methods of analysis (TLC, GC-MS, and LC-MS)**

*Thin Layer Chromatography*

Crude alkaloid extracts were dissolved in chloroform and then 1-5 μl were spotted on pre-coated glass backed plates (5 × 10 cm, Silica gel 60A, 0.25 mm). Plates were developed using chloroform/methanol/ammonium hydroxide (90/10/1). Spots were visualized after spraying with ρ-anisaldehyde reagent and then heating with hot air from a heat gun and monitoring any developing spots and color.

*Gas Chromatography-Mass Spectrometry*

Crude alkaloid extracts were dissolved in chloroform and 2 μl injected for GC-MS using a Polaris Q GC-MS with a splitless injector (250°C) and a DB-5ms capillary column (30 m × 0.25 mm) with helium carrier gas. Column temperature program was 80°C for 1.0 min, increased to 180°C at 40°/min; increased to 260°C at 5°/min; and held at 260°C for 1.5 min. The mass spectrometer scanned a mass range of 50 to 650 amu. The ion source temperature was 200°C and ionization mode was electron impact at 70 eV.

*Liquid Chromatography-Mass Spectrometry*

Crude alkaloid extracts were dissolved in 50% methanol (1.0 ml) and 5 μl injected for LC-MS using LCQ Advantage Max mass spectrometer equipped with a Surveyor photodiode array UV detector, Surveyor autosampler, and Surveyor MS liquid chromatography pump. An atmospheric pressure chemical ionization source (APCI) was used for compound ionization. The mass spectrometer was set to monitor positive ions in the mass range of 70-800 amu. Chromatography conditions included a Aquasil C18 column (100 × 2.1 mm) eluted with a gradient mixture of acetonitrile and 0.1% trifluoroacetic acid (A) and acetonitrile (B) starting with 10% B (0-3 min); linear increase to 70%B (3-10 min); 70% B (10-15 min) at a flow rate of 0.300 ml/min.

## Results

In the initial analysis by GC-MS two major alkaloids were detected and were identified as *N,N*-dimethyltryptamine (DMT) and 2-methylcarboline by comparison to mass spectral database information (Figure 2). A third alkaloid which we believe to be a possible artifact was only detected when using method A extraction procedure and more specifically we believe is created during the solvent removal process (drying of the chloroform). The source of the unknown compound has not been identified. It is clearly not obtained using the solid phase extraction procedure (method B) and yet we do not see a corresponding reduction in one of the other detected components.

TLC analysis of the extracts also showed the presence of two major alkaloids ($R_f$ = 0.42 and 0.67) (Figure 3). It could not be confirmed if these compounds were DMT and 2-methylcarboline that was observed in the GC-MS analysis as standards were not available.

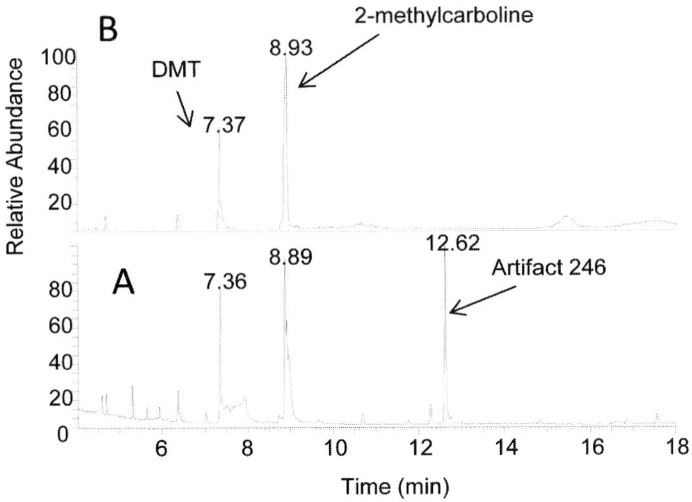


**Figure 2.** GC-MS chromatograms from *M. tenuiflora* leaf material extracted using methods A and B.

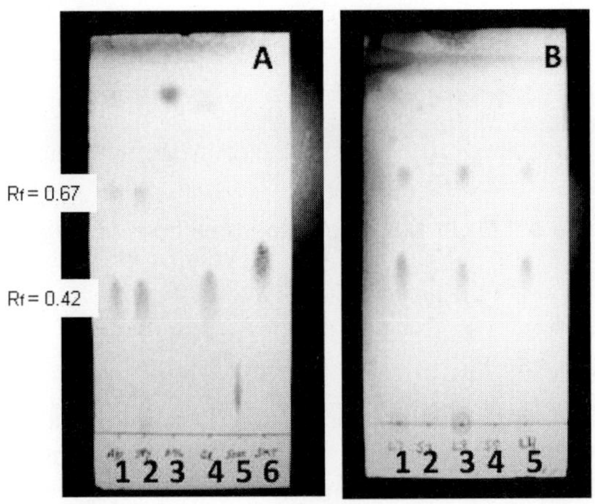


**Figure 3.** Thin layer chromatography (TLC) plates from analysis of: (A) *M. tenuiflora* leaf material lanes 1-2, lane 3 (Artifact 246), lane 4 (gramine), lane 5 (5-hydroxytryptamine), lane 6 (5-methoxytryptamine); (B) lane 1 (leaf 07), lane 2 (seed 07), lane 3 (leaf 08), lane 4 (seed 08), lane 5 (leaf 04).

LC/UV/MS analysis detected five alkaloids with UV and MS data consistent with tryptamine type alkaloids. These included DMT ($MH^+$ = 189) and 2-methylcarboline ($MH^+$ = 187) observed in the GC chromatogram and three unknown alkaloids ($MH^+$ = 175, 201, and 205). In addition at least three possible minor alkaloids were detected ($MH^+$ = 122, 136, and 166) but their UV spectra was not consistent with the tryptamine type alkaloids. The presence of a possible artifact alkaloid ($MH^+$ = 247) was again observed by LC-UV-MS analysis when extracts were prepared using extraction method A.

Seed and leaf samples from several different years were extracted and analyzed by TLC and LC-MS (TLC data Figure 3). There were clear differences in the alkaloid profiles between seed and leaves in that the two major spots in the TLC chromatograms ($R_f = 0.42$ and 0.67) were absent in the seed samples (Figure 4).

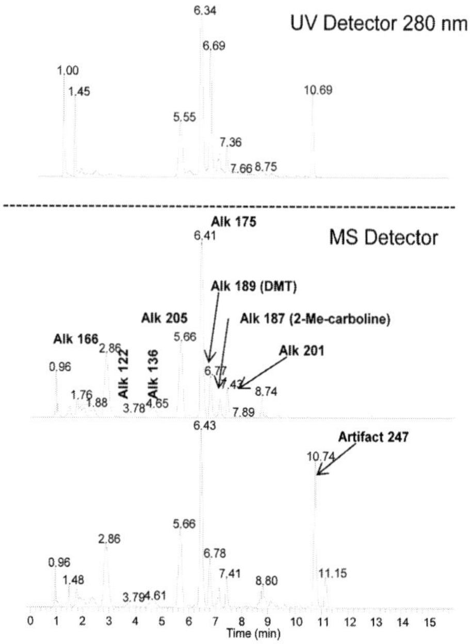


**Figure 4.** LC-UV-MS chromatograms from *M. tenuiflora.*

## Conclusions

The best method of analysis appears to be isolation of alkaloids by solid phase extraction (SPE) and then analysis by LC/UV/MS. The teratogenic agents are still unknown. Tryptamine alkaloids are well known (*Phalaris* spp., reed canary grass, etc.) and no such teratogenic effects are suspected with these plants. Although there are differences in seed versus leaf alkaloids both seed and leaf material are reported to be teratogenic. More work needs to be completed to fully identify the alkaloids in the plant and different plant parts and to correlate data between TLC, GC, and LC methods. If possible a crude alkaloid fraction should be isolated from the plant material and tested in the rat bioassay model (Medeiros *et al.* 2008).

## Acknowledgements

This work has been financially supported by the National Institute for Science and Technology for the Control of Toxic Plants (CNPq), grant number 573534/2008-0.

## References

de Souza RSO, de Albuquerque UP, Monteiro JM, and de Amorim ELC (2008). Jurema-preta (*Mimosa tenuiflora* [Willd.] Poir.): a review of its traditional use, phytochemistry and pharmacology. *Brazilian Archives of Biology and Technology* 51:937-947.

Medeiros RMT, de Figueiredo AMP, Benicio TMA, Dantas FPM, and Riet-Correa F (2008). Teratogenicity of *Mimosa tenuiflora* seeds to pregnant rats. *Toxicon* 51:316-319.

Nicasio MDP, Villarreal ML, Gillet F, Bensaddek L, and Fliniaux MA (2005). Variation in the accumulation levels of *N,N*-dimethyltryptamine in micro-propagated trees and in *in vitro* cultures of *Mimosa tenuiflora*. *Natural Product Research* 19:61-67.

Pimentel LA, Riet-Correa F, Gardner DR, Panter KE, Dantas AFM, Medeiros RMT, Mota RA, and Araujo JAS (2007). *Mimosa tenuiflora* as a cause of malformations in ruminants in the northeastern Brazilian semiarid rangelands. *Veterinary Pathology* 44:928-931.

Rivera-Arce E, Gattuso M, Alvarado R, Zarate E, Aguero J, Feria I, and lozoya X (2007). Pharmacognostical studies of the plant drug *Mimosae tenuiflorae*. *Journal of Ethnopharmacology* 113:400-408.

Vespsalainen JJ, Auriola S, Tukiainen M, Ropponen N, and Callaway JC (2005). Isolation and characterization of yuremamine, a new phytoindole. *Planta Medica* 71:1053-1057.

# Chapter 104

# Distribution of *Delphinium occidentale* Chemotypes and their Potential Toxicity

D. Cook, D.R. Gardner, J.A. Pfister, K.D. Welch, B.T. Green, and S.T. Lee

*USDA-ARS Poisonous Plant Research Laboratory, Logan, Utah 84341, USA*

## Introduction

Larkspurs (*Delphinium* spp.) are poisonous plants on rangelands in the western USA. They are responsible for significant losses to the cattle industry and are the subject of extensive research (Pfister *et al.* 1999, 2002). Total cost to the livestock industry from cattle deaths attributed to larkspur poisoning is estimated to be millions of dollars annually (Nielsen *et al.* 1994). Larkspurs are divided into three groups principally based upon their height: low larkspurs, plains larkspurs, and tall larkspurs. The tall larkspurs are responsible for a greater number of cattle losses than either the plains or low larkspur.

Larkspur-induced poisoning in cattle is attributed to the diterpenoid alkaloids that can represent up to 3% of the plant dry weight. There are two main structural groups of norditerpene alkaloids, the *N*-(methylsuccinimido) anthranoyllycoctonine type (MSAL-type) and the 7,8-methylenedioxylycoconine type (MDL-type) norditerpenoid alkaloids (Figure 1) (Olsen *et al.* 1990). The MSAL-type alkaloids are approximately 20 times more toxic than the MDL-type alkaloids based upon the $LD_{50}$ of the individual compounds in a mouse model (Manners *et al.* 1993, 1995, 1998; Panter *et al.* 2002). Acute larkspur poisoning has been attributed to the MSAL-type alkaloids (Aiyar *et al.* 1979; Pfister *et al.* 1999) and plants high in the MSAL-type alkaloids are thought to be the most toxic to cattle. The concentrations of these alkaloids have been used as a predictor of plant toxicity (Pfister *et al.* 2002; Ralphs *et al.* 2002). The most abundant member of the MSAL-type alkaloids in the tall larkspurs is methyllycaconitine (MLA) (Gardner *et al.* 2002).

An observation of particular interest made by Gardner *et al.* (2002) was the identification of two alkaloid profiles in *D. occidentale*. One alkaloid profile lacked, or displayed very small amounts of the MSAL-type alkaloids whereas the other alkaloid profile displayed large amounts of the MSAL-type alkaloids. The objective of this study was to determine the extent of these two alkaloid profiles throughout the geographical distribution of *D. occidentale*. We report here that *D. occidentale* has two definable chemotypes, with distinct geographical boundaries, that should differ in potential toxicity. These results have important implications in grazing management decisions for *D. occidentale*-infested rangelands and they demonstrate that taxonomic classification alone is not a good indicator to determine the toxic risk of *D. occidentale*. For further details concerning this research one is referred to a recent publication (Cook *et al.* 2009).

©CAB International 2011. *Poisoning by Plants, Mycotoxins, and Related Toxins* (eds F. Riet-Correa, J. Pfister, A.L. Schild, and T.L. Wierenga)

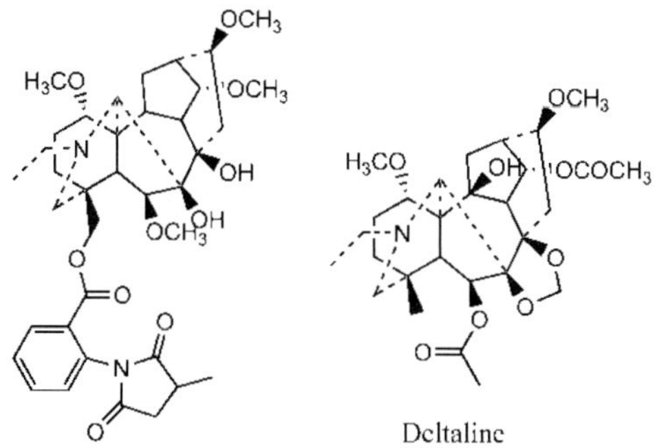

Deltaline

Methyllycaconitine (MLA)

**Figure 1.** Structures of select norditerpene alkaloids in *D. occidentale.*

## Materials and Methods

### Plant material

Analytical samples were prepared from plant material collected from herbarium specimens and resident populations of *D. occidentale*. Herbarium specimens were provided by the Intermountain Herbarium at Utah State University, the Stanley L. Welsh Vascular Plant Herbarium at Brigham Young University, the University of Colorado Museum Herbarium, the Rocky Mountain Herbarium at the University of Wyoming, the University of Washington Herbarium, and the Herbarium at Oregon State University. Specimens of questionable identification were verified to be authentic *D. occidentale* specimens by staff at the Intermountain Herbarium at Utah State University or the Stanley L. Welsh Vascular Plant Herbarium at Brigham Young University. Leaf and flower material were sampled from herbarium specimens and subsequently ground using a Retsch mixer mill MM301 (Haan, Germany).

Field samples of *D. occidentale* populations (1464 plants representing 118 accessions) were collected in the summer of 2007 and 2008. Accessions were collected throughout the geographical distribution of *D. occidentale* including the states of Utah, Idaho, Montana, Wyoming, Colorado, Nevada, and Oregon. Samples were immediately placed on dry ice after collection and stored at -80°C upon return to the laboratory. Samples were frozen for possible use in subsequent research. The samples were freeze dried and ground to pass through a 2 mm screen using a Wiley mill.

### Sample extraction and alkaloid analysis

Individual plant samples were extracted and analyzed by electrospray mass spectrometry using procedures previously described (Gardner *et al.* 1999). In summary, 25 mg plant material from herbarium samples was extracted in 6 ml of methanol for 16 h. Reserpine (125 μg) was added as an internal reference. The sample was mixed then centrifuged. A 200 μl sample was diluted into 800 μl of 1:1 methanol/1% acetic acid. For plant samples collected in the field in the summers of 2007 and 2008, 100 mg plant material was extracted in 6 ml of methanol for 16 h. Reserpine (500 μg) was added as an internal reference. The sample was mixed then centrifuged. A 50 μl sample was diluted into 950 μl of 1:1 methanol/1% acetic acid.

Mass spectra were recorded for each sample over a range of 150-800 *m/z* and averaged across all scans taken at 40% of peak height (total ion current). Data were calculated by recording the abundance of all ions above a relative area of 0.1%. The amount of a compound (as represented by a single mass unit) detected was calculated based on the relative abundance of the internal standard reserpine ($MH^+$=609). The resulting mass spectral data were reduced and tabulated to a final set of quantitative values for 57 different protonated molecules using a method similar to that reported by Gardner *et al.* (2002).

**Data analysis**

Each sample was assigned into group A (samples with MLA concentrations greater than 100 μg/mg) or group B (samples with MLA concentrations less than 100 μg/mg). This cutoff was chosen because it clearly separated the two alkaloid profiles observed previously by Gardner *et al.* (2002). MANOVA and discriminant analysis of the two assigned groups were performed as a pairwise comparison using BioNumerics 4.6 (Applied Maths, Inc.). Two parameters were reported: (i) *L* (Wilk's Lambda likelihood ratio test) is the likelihood of the obtained discrimination with the assumption that the groups are drawn from the same population; a low *L* value infers that the groups are likely to be drawn from different populations; and (ii) *P* is the probability that a random grouping of the groups would yield the same degree of discrimination. All multivariate statistical comparisons were made from plants of the same developmental stage.

## Results and Discussion

Field collections (1464 specimens representing 118 accessions) of *D. occidenatale* were made in the summer of 2007 and 2008. Samples were representative of the reported geographical distribution of *D. occidentale.* In addition 599 herbarium specimens of *D. occidentale* from the cooperating herbaria were sampled. These specimens were collected from the late 1800s to the current year and they were also representative of the reported geographical distribution of *D. occidentale*. All samples were analyzed by electrospray mass spectrometry.

Initially each sample was scored for the presence of MLA (>100 μg/mg) or reduced amounts of MLA (<100 μg/mg). A total of 698 samples were identified that contained greater than 100 μg/mg MLA while 1365 samples were identified that contained less than 100 μg/mg MLA. To confirm that these two groups were unique, multivariate statistical methods (MANOVA and discriminant analysis) were used to test for grouping. The pairwise MANOVA revealed the two groups were different ($P$=0.001%, $L$=0.15). Discriminant analysis was also performed comparing the two groups as a pairwise comparison. Discriminant analysis showed clear separation of the two groups based upon multiple alkaloids. The five most important discriminants were the following masses: *m/z* 683 (MLA), 715, 739, 753, and 699. These two multivariate methods demonstrated that the two groups were clearly different. As a result samples with greater than 100 μg/mg MLA will hereafter be termed chemotype A and those samples that have less than 100 μg/mg MLA will hereafter be termed chemotype B. Representative mass spectra of the two chemotypes are shown in Figure 2. Chemotype A contains the MDL- and MSAL-type alkaloids including MLA while chemotype B contains principally the MDL-type alkaloids and very little, if any, MLA.

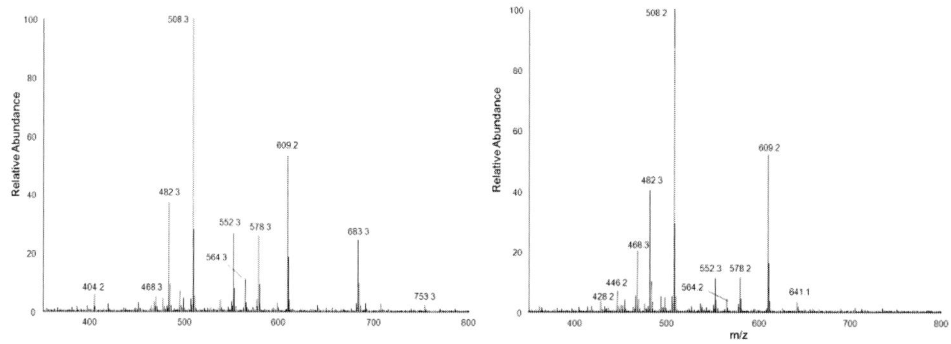


**Figure 2**. Electrospray mass spectra from representative samples of chemotype A (left) and chemotype B (right).

The geographical distribution of each chemotype was investigated by state and counties within each state. A distribution map of the two chemotypes is shown in Figure 3. Most counties were found to only have samples of one chemotype. However, a number of counties such as Caribou (ID), Lincoln (WY), Sublette (WY), and Fremont (WY) counties have samples of each chemotype, indicating that these counties serve as a transition zone from south to north. Although these counties have plants with both chemotypes, individual populations from field collections in each county were either chemotype A or B in most cases. Likewise counties that are in the transition zone from east to west such as Juab and Box Elder counties of Utah also have samples of both chemotypes but are spatially separated; chemotype A to the west and chemotype B to the east.

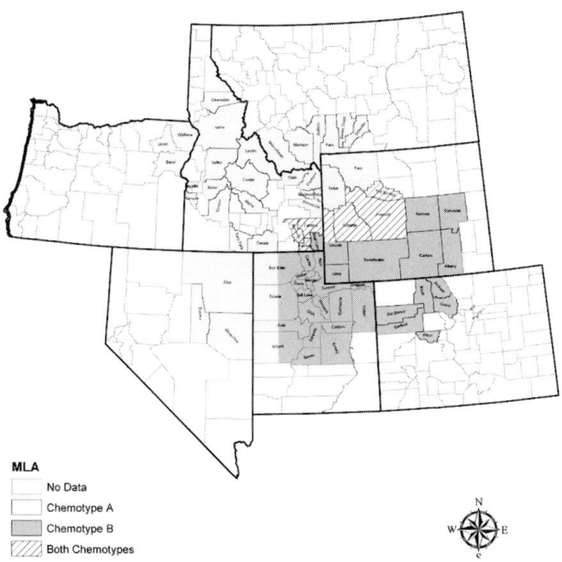


**Figure 3.** Geographical distribution of chemotypes A and B.

Three important conclusions can be drawn from this data:

First, plants representing chemotype B contain very low amounts of or no detectable MSAL-type alkaloids and would likely pose very little risk to grazing cattle based upon current models and recommendations. Current management recommendations state that plants with greater than 3 mg MSAL-type alkaloids/g plant material pose the greatest risk to cattle (Pfister *et al.* 2002). Alternatively, plants representing chemotype A containing the MSAL-type alkaloids may pose considerable risk to cattle based upon current models and recommendations. However, poisoning is always dependent upon the dose and duration. Chemotype B plants contain the less toxic MDL-type alkaloids. Therefore at the proper dose and duration, there is a possible risk of poisoning livestock.

Second, in general each chemotype was found to have a distinct distribution with defined boundaries (Figure 3). In some cases, the chemotypes are separated by notable geographic features. For example, the east to west transition is separated by the desert that runs north to south on the west side of the state of Utah (Figure 3). On the other hand, the north to south transition zone of the two chemotypes is not separated by any notable geographic features. In fact, two populations in Lincoln County, Wyoming representing the two different chemotypes occurred on the same watershed less than 5 miles apart.

Third, the data suggest that the qualitative nature of the alkaloid profiles in *D. occidentale* remains constant at a given location. This conclusion is supported by field collections that have the same qualitative alkaloid profiles as the herbarium specimens from identical locations that were collected up to 100 years earlier. In addition, the data suggest that the alkaloid composition of herbarium specimens is not modified as a result of long term storage at room temperature. Thus, herbarium specimens may serve as useful resources in determining risk of other larkspur species. However, quantitative amounts of these alkaloids may vary between years. Quantitative assessment of the alkaloids over time due to environment and other factors merits further investigation.

In conclusion, the MSAL-type alkaloids such as MLA are the primary factors responsible for the toxicity of larkspur plants. We report here that *D. occidentale* has two defined chemotypes, one (chemotype A) that contains significantly more MSAL-type-alkaloids and one that lacks or contains very small amounts of the MSAL-type alkaloids (chemotype B). In general, the plants with these chemotypes grow in distinct geographical locations; however, there are counties that contain populations of both chemotypes. In addition, this study clearly demonstrates that taxonomic identification of *D. occidentale* is not sufficient to determine risk. Lastly, based upon current toxicity models the results from this study have important implications in making management decisions for *D. occidentale*-infested rangelands. However, more research is needed to determine the exact risk to livestock of each chemotype before these management recommendations can be further refined.

## References

Aiyar VN, Benn MH, Hanna T, Jayco J, Roth SH, and Wilkens JL (1979). The principal toxin of *Delphinium brownii* Rydb., and its mode of action. *Experientia* 35:1367-1368.

Cook D, Gardner DR, Pfister JA, Welch KD, Green BT, and Lee ST (2009). The biogeographical distribution of Duncecap larkspur (*Delphinium occidentale*) chemotypes and their potential toxicity. *Journal of Chemical Ecology* 35:643-652.

Gardner DR, Panter KE, Pfister JA, and Knight AP (1999). Analysis of toxic norditerpenoid alkaloids in *Delphinium* species by electrospray, atmospheric pressure chemical ionization, and sequential tandem mass spectrometry. *Journal of Agriculture and Food Chemistry* 47:5049-5058.

Gardner DR, Ralphs MH, Turner DL, and Welsh SL (2002). Taxonomic implications of diterpene alkaloids in three toxic tall larkspur species (*Delphinium* spp.). *Biochemical Systematics and Ecology* 30:77-90.

Manners GD, Panter KE, Ralphs MH, Pfister JA, Olsen JD, and James LF (1993). Toxicity and chemical phenology of norditerpenoid alkaloids in the tall larkspurs (*Delphinium* species). *Journal of Agriculture and Food Chemistry* 41:96-100.

Manners GD, Panter KE, and Pelletier SW (1995). Structure-activity relationships of norditerpenoid alkaloids occurring in toxic larkspur (*Delphinium*) species. *Journal of Natural Products* 58:863-869.

Manners GD, Panter KE, Pfister JA, Ralphs MH, and James LF (1998). The characterization and structure-activity evaluation of toxic norditerpenoid alkaloids from two *Delphinium* species. *Journal of Natural Products* 61:1086-1089.

Nielsen DB, Ralphs MH, Evans JS, and Call CA (1994) Economic feasibility of controlling tall larkspur on rangelands. *Journal of Range Management* 47:369-372.

Olsen JD, Manners GD, and Pelletier SW (1990). Poisonous properties of Larkspur (*Delphinium spp.*). *Collectanea botanica*, Barcelona, 19:141-151.

Panter KE, Manners GS, Stegelmeier BL, Lee S, Gardner DR, Ralphs MH, Pfister JA, and James LF (2002). Larkspur poisoning: alkaloid structure–activity relationships and toxicity. *Biochemical Systematics and Ecology* 30:113-128.

Pfister JA, Gardner DR, Panter KE, Manners GD, Ralphs MH, Stegelmeier BL, and Schoch TK (1999). Larkspur (*Delphinium* spp.) poisoning in livestock. *Journal of Natural Toxins* 8:81-94.

Pfister JA, Ralphs MH, Gardner DR, Stegelmeier BL, Manners GD, Panter KE, and Lee ST (2002). Management of three toxic *Delphinium* species based on alkaloid concentrations. *Biochemical Systematics and Ecology* 30:129-138.

Ralphs MH, Gardner DR, Turner DL, Pfister JA, and Thacker E (2002). Predicting toxicity of tall larkspur (*Delphinium barbeyi*): measurement of the variation in alkaloid concentration among plants and among years. *Journal of Chemical Ecology* 28:2327-2341.

# CONTROL MEASURES

# Chapter 105

# Conditioned Aversion Induced by *Baccharis coridifolia* in Sheep and Cattle

M.B. Almeida[1], N.D. Assis-Brasil[1], A.L. Schild[1], F. Riet-Correa[2], J.A. Pfister[3], and M.P.S. Soares[1]

*[1]Laboratório Regional de Diagnóstico, Faculdade de Veterinária, UFPel, Campus Universitário s/n, Pelotas, RS, Brazil; [2]Centro de Saúde e Tecnologia Rural, UFCG, Campus de Patos, Patos, PB, Brazil; [3]USDA-ARS Poisonous Plant Research Laboratory, Logan, Utah 84341, USA; [4]Programa de Pós-Graduação em Zootecnia, FAEM, UFPel*

## Introduction

*Baccharis coridifolia* (Compositae) is one of the most important toxic plants in southern Brazil. The natural intoxication occurs mainly in cattle, less frequently in sheep, and rarely in horses and pigs (Barros 1998; Tokarnia *et al.* 2000; Rissi *et al.* 2005; Riet-Correa and Méndez 2007). In Rio Grande do Sul, southern Brazil, outbreaks were described in cattle (Rissi *et al.* 2005), sheep (Rozza *et al.* 2006), and horses (Alda *et al.* 2009) with morbidity varying from 21.73% to 22.51% in cattle and 16.5% in sheep with a 100% case fatality rate in both species. In one outbreak described in horses the morbidity was 100% and the case fatality rate was 66% (Alda *et al.* 2009). The poisoning occurs when animals raised in areas without the plant are transported to and allowed to graze in pastures invaded by *B. coridifolia*. Intoxication risk increases considerably when recently transported animals are stressed, fatigued, hungry, or thirsty (Barros 1998; Riet-Correa and Mendez 2007). In cattle the spontaneous poisoning, causing severe digestive disturbances, occurs between 5 and 30 h after the ingestion of the plant and the animals die between 3 and 23 h after the onset of clinical signs (Rissi *et al.* 2005). In one outbreak described in sheep the clinical signs began 5 days after the introduction into the pastures and the deaths occurred between 5 and 48 hours after the onset of clinical signs (Rozza *et al.* 2006). Farmers prevent *B. coridifolia* poisoning using several unconventional methods to reduce ingestion: (i) burning plant material under an animals' nose and having the animal inhale the resulting smoke; (ii) rubbing the plant on the animals' muzzle and mouth; and (iii) gradually introducing animals into *B. coridifolia*-infested pastures. The aversive effect of *B. coridifolia* was tested in sheep and results demonstrated that *B. coridifolia* is as efficient as LiCl in conditioning an aversion to a previously unknown food by the administration of 50% of a toxic dose (Almeida *et al.* 2009). With the aim to determine the efficiency of unconventional methods used by the owners to prevent *B. coridifolia* poisoning in sheep and cattle, two experiments were performed on two different farms.

©CAB International 2011. *Poisoning by Plants, Mycotoxins, and Related Toxins* (eds F. Riet-Correa, J. Pfister, A.L. Schild, and T.L. Wierenga)

## Experiments in Sheep

To test the efficiency of these three aversive methods to prevent *B. coridifolia* poisoning in sheep, 16 adult ewes were divided into four groups. Two groups were treated by methods 1 and 2 described above. One group was treated by oral gavage of 0.25 g/kg body weight of *B. coridifolia*. The fourth group was used as a control. One day after the treatments all sheep were transferred to a farm where *B. coridifolia* occurs, 400 km from where the treatments were administered. The sheep were placed in a corral overnight and the next day they were transferred to a paddock where flowering *B. coridifolia* composed 50% of the plant composition.

Between 20-24 h after the introduction of sheep into the paddock, two animals of the group that inhaled the smoke from burning plant material died with clinical signs of intoxication. At necropsies characteristic lesions like congestion of rumen and abomasal mucosal and gut hemorrhages were observed. Histologically degeneration and necrosis of epithelial cells with pustule formation and epithelial sloughing were observed in the mucosa of the rumen. Ulceration and hemorrhages of abomasal mucosal were also observed. One sheep from the control group and two from the group treated by rubbing the plant in its nose had anorexia and showed signs of digestive stress. All sheep from the group treated by the administration of *B. coridifolia* did not graze the plant. It was demonstrated that animals which ingested non-toxic doses of *B. coridifolia* were averted to the plant. On the other hand the methods of burning plant material under an animals' nose and having the animal inhale the resulting smoke and rubbing the plant on the animals' muzzle and mouth, both of which are used by livestock producers, had little or no effect on ingestion of the plant by treated animals and did not prevent the intoxication. The administration of non-toxic doses of *B. coridifolia* to sheep before introduction into the invaded areas may not be viable when a great number of animals need to be moved because of the labor and time involved. In these cases the gradual introduction of animals into *B. coridifolia*-infested pastures would be the most efficient method to avoid death losses.

## Experiments in Cattle

Fifty-one naïve heifers from a farm without *B. coridifolia* were orally dosed with 0.5 g/kg body weight of fresh green *B. coridifolia* collected while actively growing. These animals were introduced into a *B. coridifolia*-infested pasture 23-26, 6-10, and 1-3 h after administration of the plant. A control group with ten heifers was introduced into the pasture without treatment.

In this experiment, one heifer introduced 6-10 h after treatment, one introduced 1-3 h after treatment, and five from the control group died after showing signs of poisoning. Necropsy findings characteristic of *B. coridifolia* poisoning included dehydration, large amounts of ruminal fluid, and reddening of the mucosae of the forestomachs. Histologically the main lesions were degeneration and necrosis of the epithelium of the rumen. No cases of poisoning were observed in treated cattle introduced into the paddock infested by *B. coridifolia* 24 h after treatment. Giving cattle 25% of a lethal dose of *B. coridifolia* via oral gavage induces a strong aversion to the plant and prevents poisoning if the animals are introduced into *B. coridifolia*-infested pastures at least 24 h after treatment.

## Discussion and Conclusions

The toxic effect induced by *B. coridifolia* on the gastrointestinal system in sheep and cattle is apparently responsible for the aversive effect of the plant. The mechanism responsible for the development of an aversion is not well established but it is suggested that animals learn which plants or food to eat and which to avoid through interactions between flavor (odor, taste, and texture) and the post-ingestive consequences of nutrients and toxins (Provenza 1996). A conditioned food aversion is probably due to negative gastrointestinal consequences after the ingestion of a plant and the integration of sensory (flavor) and negative post-ingestive consequences (effects of nutrients or toxins on chemo-, osmo-, or mechano-receptors in the gut and brain) (Provenza 1996; Wang and Provenza 1996). The interaction between the timing of plant ingestion and its effect on the digestive system is an important factor for inducing a strong aversion. Sufficient time must elapse between dosing the plant and the exposure of the animals to *B. coridifolia* pastures to allow animals to experience the negative post-dosing consequences. In conclusion, a non-lethal dose of *B. coridifolia* can be used by livestock producers to induce an aversion in naïve cattle and sheep, thus allowing animals to safely graze infested pastures. The practicality of this prevention method will depend on labor costs and other management considerations unique to each ranch.

## Acknowledgements

Financial support by CNPq (Grants Nº 471588/2004-0, Nº 420012/2005-2 and N°501177/2007-8). The assistance of Ana Lucia Schild to the 8th International Symposium on Toxic Plants was financially supported by Conselho Nacional de Desenvolvimento Científico e Tecnológico (CNPq), grant 454084/2008-0, and by Coordenação de Aprefeiçoamento de Pessoal de Nível Superior (CAPES), grant 0017/09-4.

## References

Alda JL, Sallis ESV, Nogueira CEW, Soares MP, Amaral L, Marcolongo-Pereira C, Xavier F, Frey F, and Schild AL (2009). Intoxicação espontânea por *Baccharis coridifolia* (mio-mio) em eqüinos no Rio Grande do Sul. *Pesquisa Veterinária Brasileira* 29:409-414.

Almeida MB, Schild AL, Assis-Brasil ND, Quevedo PS, Fiss L, Pfister JA, and Riet-Correa F (2009). Conditioned aversion in sheep induced by *Baccharis coridifolia* to a previously unknown food. *Applied Animal Behaviour Science* 117:197-200.

Barros CSL (1998). Livestock poisoning by *Baccharis coridifolia.* In *Toxic Plants and Other Natural Toxicants* (T Garland and AC Barr, eds), pp. 569-572. CAB International, Wallingford, England.

Provenza FD (1996). Acquired aversions as the basis for varied diets of ruminants foraging on rangelands. *Journal of Animal Science* 74:2010-2020.

Riet-Correa F and Méndez MC (2007). Intoxicações por Plantas e Micotoxinas. In *Doenças de Ruminantes e Eqüídeos* (F Riet-Correa, AL Schild, RAA Lemos, and JRJ Borges, eds), pp. 99-219. Editora Pallotti, Santa Maria, RS, Brazil.

Rissi DR, Rech RR, Fighera RA, Cagnini DQ, Kommers GD, and Barros CSL (2005). Intoxicação espontânea por *Baccharis coridifolia* em bovinos. *Pesquisa Veterinária Brasileira* 25:111-114.

Rozza DB, Raymundo DL, Corrêa AMR, Leal J, Seitz AL, Driemeier D, and Colodel EM (2006). Intoxicação espontânea por *Baccharis coridifolia* (Compositae) em ovinos. *Pesquisa Veterinária Brasileira* 26:21-25.

Tokarnia CH, Döbereiner J, and Peixoto PV (2000). *Plantas Tóxicas do Brasil,* 310 pp. Editora Helianthus, Rio de Janeiro, Brazil.

Wang J and Provenza FD (1996). Food preference and acceptance of novel foods by lambs depend on composition of the basal diet. *Journal of Animal Science* 74:2349-2354.

# Chapter 106

# A Potential Krimpsiekte Vaccine

C.J. Botha[1], J.E. Crafford[2], V.P. Butler Jr[3], M.N. Stojanovic[3], and L. Labuschagne[4]

*[1]Department of Paraclinical Sciences, Faculty of Veterinary Science, University of Pretoria, Private Bag X04, Onderstepoort, 0110 South Africa; [2]Department of Veterinary Tropical Diseases, Faculty of Veterinary Science, University of Pretoria, Private Bag X04, Onderstepoort, 0110 South Africa; [3]Department of Medicine, College of Physicians and Surgeons, Columbia University, 630 West 168th Street, New York NY 10032, United States of America; [4]Toxicology Division, ARC-Onderstepoort Veterinary Institute, Private Bag X05, Onderstepoort, 0110 South Africa*

## Introduction

Poisoning of livestock by cardiac glycoside-containing plants has the greatest economic impact of all plant-associated poisonings in the Republic of South Africa (Kellerman *et al.* 1996). Collectively they are responsible for 33% of all mortalities from plant poisonings of cattle and 10% of those in small stock. The majority of the cardiac glycoside poisonings in small stock is ascribed to krimpsiekte, arguably the most economically important plant poisoning of small stock in the Little Karoo and southern Great Karoo (Botha 2003; Kellerman *et al.* 2005). It is estimated that more than 26,000 sheep and goats succumb annually (Kellerman *et al.* 1996).

Chemically two major groups of cardiac glycosides, namely the cardenolides and bufadienolides, are recognized. Poisoning by bufadienolide-containing plants surpasses cardenolide-induced poisonings in importance and may be either acute or chronic. Tulp poisoning (induced by various *Moraea* species) and slangkop poisoning (caused by various *Drimia* species) induce only acute intoxication as these species contain non-cumulative bufadienolides (Kellerman *et al.* 1996, 2005).

Members of three genera of the Crassulaceae (*Cotyledon*, *Tylecodon*, and *Kalanchoe*), colloquially referred to as plakkies, may cause either acute or chronic poisoning. Krimpsiekte, the chronic form of the poisoning, is a neuromuscular affliction of small stock following ingestion of plants that contain atypical cardiac glycosides (such as cotyledoside and tyledoside D), generally referred to as cumulative neurotoxic bufadienolides. *T. wallichii* (Harv.) Tölken subsp. *wallichii* and *T. ventricosus* (Burm.f.) Tölken are probably the most important species of the group of plants causing krimpsiekte (Botha *et al.* 1997, 1998). In this syndrome the cardiac, respiratory, and gastrointestinal signs typical of acute poisoning are diminished and the neuromuscular signs increase. Small stock tire easily, lag behind the flock, and frequently lie down. Often they assume a characteristic stance with

©CAB International 2011. *Poisoning by Plants, Mycotoxins, and Related Toxins*
(eds F. Riet-Correa, J. Pfister, A.L. Schild, and T.L. Wierenga)

the back arched, limbs tucked in under the body, and head down, sometimes trembling. The animals may be recumbent for long periods.

Of particular concern is evidence that secondary poisoning of humans can occur if meat or edible tissues obtained from carcasses of animals that have died of krimpsiekte are consumed (Kellerman *et al.* 2005). This is a real danger as rural people with meager means and financial constraints commonly consume animals that die.

Immunoprophylaxis against plant and fungal poisonings is receiving some attention; for instance, Australian researchers are evaluating immunization as a means of preventing the mycotoxicosis lupinosis in sheep (Than *et al.* 1994). Most phytotoxins are not immunogenic due to their small molecular size and in order to induce immunity following parenteral administration to an animal must first be coupled to carrier proteins (Silbart *et al.* 1997). Various attempts at vaccinating stock against plant toxins have been unsuccessful for a number of reasons. One of these is that the immunological response is overwhelmed due to the high toxin concentrations which occur in acute toxic exposures (Edgar 1994). Krimpsiekte, on the other hand, is a chronic manifestation that can be induced by repeated exposures to small doses of the cumulative bufadienolide; a vaccine, therefore, could conceivably prevent this intoxication. If a prophylactic vaccine could be produced the incidence of and mortalities caused by krimpsiekte could be curtailed.

Technically the development of a vaccine to prevent krimpsiekte is quite feasible (Butler and Chen 1966; Schmidt and Butler 1971). Antibodies have been raised against digoxin, a cardenolide cardiac glycoside. Sheep and rabbits immunized by repetitive administration of digoxin-protein conjugates developed high serum titres of antidigoxin antibodies of high specificity and affinity (Schmidt and Butler 1971; Butler *et al.* 1977b). Immunized rabbits were protected from the toxic effects of a lethal dose of digoxin (Butler *et al.* 1977b). However, antibody binding of digoxin interferes with the renal excretion of digoxin, thus greatly prolonging the half-life of this compound (Schmidt *et al.* 1974; Butler *et al.* 1977a, b).

The first objective of this project was to synthesize a cotyledoside-protein conjugate with which to immunize animals. The second aim was to evaluate the efficacy of the conjugate in inducing an immunological response in rabbits and sheep by determining cotyledoside antibody titres in an ELISA. The third objective was to demonstrate the efficacy of the vaccine (conjugate) by challenging sheep with the neurotoxic cumulative bufadienolide cotyledoside.

## Preparation of Cotyledoside-protein Conjugates

Cotyledoside has previously been extracted and purified from *T. wallichii* (Botha *et al.* 1997). Cotyledoside-protein conjugates were prepared by a mixed anhydride procedure (Erlanger *et al.* 1957). Cotyledoside was conjugated to either bovine serum albumin (BSA) or hen ovalbumin (OVA).

## Cotyledoside ELISA

A standard indirect ELISA using 1 h incubation times at room temperature was followed. Maxisorb 96-well plates (NUNC) were coated with antigen (OVA-cotyledoside conjugate) and OVA. Rabbit and sheep sera were loaded in duplicate rows (e.g. row A coated with antigen and row B coated with OVA). When testing rabbit sera a biotinylated goat anti-rabbit serum together with a strepavidin horseradish-peroxidase (HRPO) conjugate was used as an indicator system. For testing sheep sera, rabbit anti-sheep HRPO conjugate was used. Following the addition of substrate the color development was stopped after 10 min. The plates were read in a BioTek EL808 plate reader at 450 nm and results recorded. The net optical density was calculated by subtracting the value obtained for the OVA-coated well from the corresponding antigen-coated well.

## Rabbit Immunization

New Zealand White rabbits (*n*=4) were housed at the University of Pretoria Biomedical Research Centre. They were kept individually in rabbit cages, had free access to water, and were fed a commercial ration. For immunization an emulsion was prepared by adding the antigenic hapten-carrier conjugate (2 mg/ml) or pure BSA (Sigma) dissolved in normal saline (2 mg/ml) to an adjuvant, at first to complete and later to incomplete Freund's (Sigma) (Table 1). Two rabbits (a male and a female) were immunized with BSA-cotyledoside conjugate and the two control rabbits (a male and a female) were immunized with similar volumes of BSA on D (day) 0, D 21, and D 49 (Table 1). A large area on the back of each rabbit was shaved and they were each immunized by intradermal administration of the inoculum at four sites on the back. On D 21 and D 49 the rabbits were weighed and bled from an ear vein (2 ml) to determine antibody titres using the OVA-cotyledoside in an ELISA. On D 61 the rabbits were anesthetized and exsanguinated by inserting a hypodermic needle into the heart. All the blood was collected for serology.

**Table 1.** Immunization regimen for the rabbits and sheep.

| Day | Composition | Dose (ml) | Route |
|---|---|---|---|
| Rabbits: Treated (*n*=2) | | | |
| 0 | 0.5 ml CC+0.5 ml CF | 0.1 | i.d. on back |
| 21 | 0.5 ml CC+ 0.5 ml IF | 0.1 | i.d. on back |
| 49 | 0.25 ml CC+0.25 ml IF | 0.1 | i.d. on back |
| Rabbits: Controls (*n*=2) | | | |
| 0 | 0.5 ml BSA+0.5 ml CF | 0.1 | i.d. on back |
| 21 | 0.5 ml BSA+0.5 ml IF | 0.1 | i.d. on back |
| 49 | 0.25 ml BSA+0.25 ml IF | 0.1 | i.d. on back |
| Sheep: Treated (*n*=2) | | | |
| 0 | 1 ml CC+1 ml CF | 0.4 | s.q. inner thigh |
| 21 | 0.25 m CC+1 ml IF | 0.25 | s.q. brisket |
| Sheep: Controls (*n*=2) | | | |
| 0 | 1 ml BSA+1 ml CF | 0.4 | s.q. inner thigh |
| 21 | 1 ml BSA+1 ml IF | 0.25 | s.q. brisket |

CC=Cotyledoside conjugate; CF=Complete Freund's; IF=Incomplete Freund's; BSA=bovine serum albumin; i.d.=intradermally; s.q.=subcutaneously

High anticotyledoside antibody titres were detected on termination of the experiment (on D 61) in the two rabbits that were immunized with the BSA-cotyledoside conjugate. As antibodies were raised in rabbits following immunization, it was decided to continue with the project in sheep.

## Sheep Immunization

Four mutton Merino ewes weighing 38.5-44.5 kg were housed in a pen with a concrete floor at the Toxicology Biolab, Onderstepoort Veterinary Institute (OVI). They had free access to water and were fed the OVI standard concentrate and *Eragrostis* hay. The sheep were randomly allocated to two groups. Two of the sheep were initially immunized by subcutaneous inoculation of an emulsion containing equal volumes of BSA-cotyledoside plus complete Freund's adjuvant. The other two animals served as controls and were vaccinated with the same volumes of BSA dissolved in saline and complete Freund's (Table 1). A booster vaccination containing cotyledoside conjugate plus incomplete Freund's adjuvant was prepared and inoculated 3 weeks later (Table 1). The serum antibody titres of all four sheep were determined by ELISA before vaccination, 3 weeks after the initial vaccination, and 3 weeks after the booster vaccination (D 42).

All four sheep developed slight to severe injection-site reactions but this did not hamper their habitus. As anticotyledoside antibodies were raised in the two sheep following immunization with the cotyledoside-protein conjugate it was decided to challenge the sheep with purified cotyledoside.

## Cotyledoside Challenge in Sheep

Commencing 45 days after the initial vaccination all four sheep were challenged with daily injections of cotyledoside. Cotyledoside (20 mg), previously extracted and purified from *T. wallichii* (Botha *et al.* 1997), was dissolved in 2 ml ethanol and 78 ml normal saline (0.025% m/v). Each animal received daily intravenous injections of 0.015 mg/kg cotyledoside (Botha *et al.* 1997). During the challenge study clinical examinations were performed daily. In order to avoid any bias the senior author (CJB), who interpreted the clinical presentations, was blinded to the immunization status of the sheep.

The cotyledoside challenge was suspended when clinical signs reminiscent of krimpsiekte were elicited. One of the sheep developed jerky forced abdominal respiration and made grunting noises shortly after the third injection administered in the morning. She became paretic during the afternoon and tended to assume the krimpsiekte posture the next day. Due to ethical considerations no further cotyledoside was administered to this ewe. It transpired that this was a control animal. Another ewe had rumen stasis on the 4th day but did not exhibit signs of paresis and was standing. This animal was not judged to be severely affected and was administered another dose of cotyledoside. However, shortly thereafter her condition rapidly deteriorated and she developed severe respiratory distress and died during the afternoon. It was then revealed that this was also a control sheep.

The daily cotyledoside administrations were continued for a further 2 days (6 daily injections in total) in the other sheep. These sheep remained clinically unaffected and the challenge was stopped as they had received cotyledoside for a longer period (1.5-2 times) than the control animals. These two sheep remained clinically unaffected for 12 months.

## Discussion

This preliminary trial verified that the cotyledoside-protein conjugate used did indeed induce the formation of high anticotyledoside antibody titres in rabbits and sheep. In addition, the prophylactic potential of this 'monovalent vaccine' was demonstrated in the limited challenge study in the sheep. The cotyledoside conjugate induced an immunological response which prevented cotyledoside-induced krimpsiekte in the two sheep immunized with it. The death of one of the control animals was unexpected as none of the sheep in previous experiments died after receiving similar amounts of cotyledoside intravenously and although affected, following cessation of the administration of the toxin they all recovered spontaneously or after administration of strained ruminal contents (Botha *et al.* 1997, 2003).

As the duration of challenge of the two cotyledoside-conjugate immunized sheep was 1.5-2 times longer than that of the other two sheep, the daily cotyledoside administration to them was stopped to prevent saturation of the induced antibodies with cotyledoside. This could eventually have resulted in excess circulating cotyledoside and death as a consequence. It can be concluded that immunization of sheep and goats can most probably be utilized as a means of preventing krimpsiekte.

One possible drawback is that animals with high antibody titres might be able to accumulate toxic amounts of cotyledoside which could be released when the antibody-toxin complex is degraded (Butler *et al.* 1977a, b). However, the cotyledoside-immunized sheep remained asymptomatic for 12 months after the challenge period which provides some evidence that release of cotyledoside from degrading antibodies will not result in recrudescence of toxicity.

Another disadvantage of krimpsiekte immunoprophylaxis is the perceived interference in the elimination of cotyledoside and its subsequent retention which would greatly increase the total body burden of cotyledoside. Carcasses of animals protected by immunization might not be suitable for human consumption (Edgar 1994; Silbart *et al.* 1997; Botha 2003). However, secondary poisoning from consumption of immunized animals that have consumed *T. wallichii* may or may not constitute a problem. Conceivably, such protected animals could dispose of antibody-bound glycoside by metabolic degradation or conjugation e.g. to sulfates and glucuronides which could be excreted in bile.

Another question that needs to be addressed is the degree of cross-reactivity between different krimpsiekte-inducing cumulative bufadienolides. In preliminary trials similar vaccines produced from cardenolides (digoxin, ouabain) and non-cumulative bufadienolides (proscillaridin) were unable to protect small stock from acute poisoning induced by dosing 4-5 g/kg milled, fresh *T. wallichii* leaves, ostensibly through lack of cross-immunity (JPJ Joubert, unpublished data 1985). This lack of cross-immunity can possibly be circumvented by preparing a polyvalent vaccine consisting of the major bufadienolides.

The above drawbacks, however, should not place any constraints on continuing this preliminary investigation. In addition, the antibodies produced could conceivably be used diagnostically in an immunoassay to detect exposure of sheep (live or slaughtered) to cotyledoside and related bufadienolides.

The immunoprophylaxis concept will now be extended to other cardiac glycoside-containing plants such as tulp to explore if naïve cattle can be protected against poisoning. If a prophylactic vaccine can be produced the incidence of and mortalities caused by this economically important plant poisoning could be curtailed.

## References

Botha CJ (2003). Krimpsiekte, a paretic/paralytic syndrome of small stock in South Africa, 101 pp. PhD Thesis, Norwegian School of Veterinary Science, Oslo.

Botha CJ, van der Lugt JJ, Erasmus GL, Kellerman TS, Schultz RA, and Vleggaar R (1997). Krimpsiekte, associated with thalamic lesions, induced by the neurotoxic cardiac glycoside, cotyledoside, isolated from *Tylecodon wallichii* (Harv.) Tölken subsp. *wallichii. Onderstepoort Journal of Veterinary Research* 64:123-128.

Botha CJ, Kellerman TS, Schultz RA, Erasmus GL, Vleggaar R, and Retief E (1998). Krimpsiekte in a sheep following a single dose of *Tylecodon ventricosus* (Burm.f.) Tölken and the isolation of tyledoside D from this plant species. *Onderstepoort Journal of Veterinary Research* 65:17-23.

Botha CJ, Rundberget T, Swan GE, Mülders MSG, and Flåøyen A (2003). Toxicokinetics of cotyledoside following intravenous administered to sheep. *Journal of the South African Veterinary Association* 74:7-10.

Butler VP and Chen JP (1966). Digoxin specific antibodies. *Proceedings of the National Academy of Sciences* 57:71-78.

Butler VP, Schmidt DH, Smith TW, Haber E, Raynor BD, and Demartini P (1977a). The effects of sheep digoxin-specific antibodies and their Fab fragments on digoxin pharmacokinetics in dogs. *The Journal of Clinical Investigation* 59:345-359.

Butler VP, Smith TW, Schmidt DH, and Haber E (1977b). Immunological reversal of the effects of digoxin. *Federation Proceedings* 36:2235-2241.

Edgar JA (1994). Vaccination against poisoning diseases. In *Plant-Associated Toxins, Agricultural, Phytochemical and Ecological Aspects* (SM Colegate and PR Dorling, eds), pp. 421-426. CAB International, Wallingford.

Erlanger BF, Borek F, Beiser SM, and Lieberman S (1957). Steroid-protein conjugates. I. Preparation and characterization of conjugates of bovine serum albumin with testosterone and with cortisone. *Journal of Biological Chemistry* 228:713-727.

Kellerman TS, Naudé TW, and Fourie N (1996). The distribution, diagnosis and estimated economic impact of plant poisonings and mycotoxicoses in South Africa. *Onderstepoort Journal of Veterinary Research* 63:65-90.

Kellerman TS, Coetzer JAW, Naudé TW, and Botha CJ (2005). *Plant poisonings and mycotoxicoses of livestock in Southern Africa,* 2nd edn, 310 pp. Oxford University Press, Cape Town.

Schmidt DH and Butler VP (1971). Immunological protection against digoxin toxicity. *The Journal of Clinical Investigation* 50:866-871.

Schmidt DH, Kaufman BM, and Butler VP (1974). Persistence of hapten-antibody complexes in the circulation of immunized animals after a single intravenous injection of hapten. *The Journal of Experimental Medicine* 139:278-294.

Silbart LK, Rasmussen MV, and Oliver AR (1997). Immunoprophylactic intervention in chemical toxicity and carcinogenicity. *Veterinary and Human Toxicology* 39:37-43.

Than KA, Anderton N, Cockrum PA, Payne AL, Stewart PL, and Edgar JA (1994). Lupinosis vaccine: Positive relationship between anti-phomopsin IgG concentration and protection in Victorian field trials. In *Plant-Associated Toxins, Agricultural, Phytochemical and Ecological Aspects* (SM Colegate and PR Dorling, eds), pp. 433-438. CAB International, Wallingford.

# Chapter 107

# Environmental Effects on Concentrations of Plant Secondary Compounds: Finding a Healthy Balance

A.K. Clemensen[1] and F.D. Provenza[2]

[1]*M.S. Candidate, Wildland Resources, Utah State University, Logan, UT 84322, USA;* [2]*Professor, Animal Behavior and Management, Utah State University, Logan, UT 84322, USA*

## Introduction

The answers to many ecological, economic, and social challenges, many of which arise from a human tendency to simplify complexity, lie in better understanding the ways of nature. Instead of attempting to control nature we may be better off observing and learning from natural systems and using that understanding to continually adapt to ever-changing environments. Nature is diverse and dynamic yet traditional agronomy and livestock production practices have immensely undermined the diversity of plant and animal species. We use only a fraction of available plant species for our consumption and for the consumption by domesticated ruminants (Diamond 1999). The possible beneficial uses of thousands of plant species that contain primary (offering nutrients) and secondary compounds (offering pharmaceuticals) have yet to be fully assessed in husbandry.

Plants have co-evolved with each other and herbivores over millions of years; perhaps they and their cells know what is best for them. Until recently plant secondary compounds (PSCs) were considered metabolic waste products that render foods and forages unpalatable for consumption by humans and other animals. While it is certainly the case that at too high dosages some secondary metabolites can harm animals at appropriate doses others offer numerous potential health benefits to many species of animals including humans (Engel 2002; Crozier *et al.* 2006). For instance, resveratrol, a polyphenolic compound in grapes, wine, soy, and peanuts, helps prevent heart disease and various types of cancer in humans (Crozier *et al.* 2006). Condensed tannins reduce internal parasites and nematodes in ruminants due to the protein-binding characteristics of tannins and also enhance the absorption of amino acids in the small intestine (Barry *et al.* 2000).

These renewable products can also reduce reliance on ever more costly fossil fuels used in vast amounts to produce pesticides, herbicides, and fertilizers, roles once played by secondary compounds in plants. Although beneficial in the short term to provide food for a growing global population these chemicals are having long term pernicious effects on the health of soil, plants, animals, and humans. Chemicals used in conventional agriculture practices inhibit mycorrhizal fungi and soil bacteria which are essential for the

©CAB International 2011. *Poisoning by Plants, Mycotoxins, and Related Toxins*
(eds F. Riet-Correa, J. Pfister, A.L. Schild, and T.L. Wierenga)

sustainability of soil health (Killham 1994; Leake *et al.* 2004; Dunfield 2007). As the health of soil declines so too does the health and nutrient content of plants (Davis *et al.* 2004; Davis 2009) and as animals consume nutritionally deficient foods, their health and ours both decline (Schatzker 2010). When plants grow in mixtures of differing species the diversity of PSCs tend to complement each other, reducing the need for fertilizers, pesticides, and herbicides (Provenza *et al.* 2007). There is enormous potential for plant mixtures that include nitrogen-fixing legumes in organic agriculture.

When offered a variety of forage species, ruminants choose an optimal mix of plants to meet their nutritional needs (Provenza *et al.* 2003). Herbivores can eat more and perform better when offered a variety of plants that vary in concentrations of different secondary compounds (Freeland and Janzen 1974). Arrays of PSCs provide varying nutrients and health benefits to herbivores. When offered a variety of food animals may obtain an overall healthier balance of nutrients (Provenza *et al.* 2002). Nearly everything conceivably edible including water can reach a toxic level or dose when eaten in excessive amounts. When animals consume increasing doses of secondary compounds offered by just one plant their digestive and intercellular networks respond by rendering the taste unpalatable to the animal. This deters herbivores from eating too much of one plant and continually encourages them through transient food aversions to eat a variety of different forages (Provenza 1995, 1996). The concept of eating moderate levels of a variety of foods applies to the health of humans as well as other animals.

## Research

The environment where a plant grows greatly affects the concentrations of PSC it contains. For instance infertile soils typically increase concentrations of carbon-based compounds such as tannins and terpenes (Herms and Mattson 1992) but they may decrease concentrations of nitrogen-based compounds such as alkaloids and cyanogenic glycosides (Herms and Mattson 1992). The reverse is true for fertile soils. For example, increased nutrient uptake in fertile soil increases plant growth and decreases the accumulation of carbohydrates required for production of tannins and terpenes (Herms and Mattson 1992). Likewise, nitrogen fertilizer can increase alkaloid levels in reed canarygrass (Majak *et al.* 1979). Moisture stress, low light, and immature tissues typically elicit higher alkaloid levels in reed canarygrass (Marten 1973; Majak *et al.* 1979). Interestingly, tannins may contribute to drought tolerance by increasing elastic resilience in cell walls (Herms and Mattson 1992). Higher temperatures seem to increase the concentration of ergot alkaloids in tall fescue (Thompson *et al.* 2001). While endophyte frequency and concentration may be higher in months with warmer temperatures it is possible that the endophyte response is related to vernalization and physiological or morphological changes occurring in the plants (Ju *et al.* 2006). However, in southern Missouri ergovaline concentration in tall fescue was highest in mid-December and declined by 85% by the end of the winter (Kallenbach *et al.* 2003). Higher levels of $CO_2$ increase concentrations of phenolic compounds while higher temperatures reduce concentrations of phenolic compounds (Veteli *et al.* 2007). Saponin content in lucerne fluctuates with the seasons, being high in the summer and low in the spring and fall (Cheeke 1998).

The aforementioned findings occur as a result of complex relationships involving allocation of resources within plants, originally proposed in the carbon/nutrient balance (CNB) hypothesis which is based on the premise that nutrient deficiencies limit the rate of plant growth more than they limit the rate of photosynthesis (Bryant *et al.* 1983). Hence, when nutrients are curtailed, creating a high carbon-nutrient ratio, a plant may decrease its growth while still photosynthesizing at an undiminished rate, leading to an accumulation of carbohydrate in excess of what is needed to support immediate growth. This buildup of carbohydrate is believed to provide additional substrate for the production of carbon-based secondary metabolites such as tannins and phenolics. Conversely under low-light conditions when growth is limited by the availability of carbon rather than nutrients, creating a low carbon-nutrient ratio, production of carbohydrates should decline, leading to a reduced formation of carbon-based defenses. The CNB hypothesis further suggests that abundant nutrient availability allows plants to accumulate excess nitrogen in addition to what is needed for primary growth. The nitrogen is then allocated to nitrogen-based secondary compounds. Shade can increase nitrogen-based secondary compounds by decreasing growth (Herms and Mattson 1992).

The CNB hypothesis can account for about 80% of the findings from experimental manipulations where plants have been fertilized or shaded (Reichardt *et al.* 1991). This hypothesis, though consistent with many findings, cannot account for all observations (Hamilton *et al.* 2001). The next iteration of this model, the growth differentiation balance hypothesis (GDB), was originally conceived in 1932 by plant biologist W.E. Loomis. Though the fundamental premise of GDB is similar to that of CNB in that there are physiological trade-offs between growth and differentiation processes including secondary metabolism, GDB is more broadly inclusive of differing intrinsic and extrinsic factors that affect resource allocation in plants (Herms and Mattson 1992). From an evolutionary standpoint both CNB and GDB suggest fast-growing plants differ chemically from slower-growing plants by allocating more energy on growth to out-compete surrounding plants. Slow-growing plants tend to have more chemical defenses and also are more suitable to survive adverse environmental conditions by making more efficient use of limited resources (Coley *et al.* 1985, Herms and Mattson 1992).

We are investigating how various environmental influences affect concentrations of PSCs in tall fescue (*Lolium arundinaceum*) variety endophyte-infected Kentucky 31, reed canarygrass (*Phalaris arundinacea*) variety not specified, lucerne (*Medicago sativa*) variety Vernal, and birdsfoot trefoil (*Lotus corniculatus*) variety Goldie. Tall fescue, a primary forage species in pastures throughout the USA, has two types of alkaloids: those associated with the plant and those due to the fungus *Neotyphodium coenophialum* which lives symbiotically in the intercellular spaces of sheaths, stems, leaves, and seeds of tall fescue (Thompson *et al.* 2001). Reed canarygrass is a cool-season grass with application in irrigated pastures. In its wild form it contains eight alkaloids: four derivatives of tryptamine, gramine, hordine, and two derivatives of β-carboline. Legumes such as lucerne (*Medicago sativa*) and birdsfoot trefoil (*Lotus corniculatus*) also have application in irrigated pastures due to nitrogen fixing capabilities and complementary root profiles. Lucerne contains glycosides such as saponins (Lu and Jorgensen 1987) and birdsfoot trefoil contains tannins (Ramirez-Restrepo *et al.* 2005).

## Results

Livestock producers are increasingly interested in the potential impacts of livestock grazing at high stock densities for improving the biological and chemical characteristics of soil and plants. In an attempt to begin comparative studies of plant responses to nutrient inputs from animal impact, we applied commercial fertilizer, green manure, and fecal manure to each of the aforementioned forage species growing in monoculture once in 2007

(August) and twice in 2008 (May and September). Samples were collected three times in 2007 (August, September, and October) and four times in 2008 (May, June, August, and September). In a related study we evaluated differences when birdsfoot trefoil, lucerne, and tall fescue were grown in mixtures or monocultures planted in the fall of 2005; we collected forage samples three times in 2008 (May, July, and August). Trefoil samples were analyzed for condensed tannins (Terrill *et al.* 1992), fescue samples for ergovaline (Hill *et al.* 1993; Rottinghaus *et al.* 1991), reed canarygrass samples for gramine (Anderton *et al.* 1999), and lucerne for saponins (Patamalai *et al.* 1990).

Preliminary analyses suggest that fertilizer and mixture affected plant chemistry in some forages ($P < 0.05$), and seasonal differences were significant in all forages ($P < 0.05$).

**Birdsfoot trefoil**

In 2007 tannins were highest in August (2.2% by weight) and September (2.2%) and lowest in October (1.5%) while in 2008 tannins were highest in May (5.4%) and dropped throughout the growing season to a low in September (1.0%). In 2007 tannins were highest in unfertilized plots (3.6%), lower with commercial fertilizer (1.15%), and lowest with fecal manure (1.12%). Interestingly, in the mixture vs monoculture study in 2008 tannins were lowest in May (1.4%), rose in July (8.0%), and dropped again in August (4.7%). In this study tannins trended ($P = 0.16$) to be higher in the morning (5.3%) than in the evening (4.1%).

**Lucerne**

In 2007 saponin levels were high in August (3.2% by weight), lower in September (1.1%), and highest in October (3.6%). In 2007 saponin levels were highest with conventional fertilizer (3.3%), dropping slightly with green manure (2.8%), and lowest with fecal manure (1.7%). In 2008 saponins generally declined from May (0.50%) through June (0.35%), August (0.32%), and September (0.22%). In 2008 unfertilized plots (0.3%) and plots fertilized with fecal manure (0.28%) and conventional fertilizer (0.32%) tended ($P = 0.11$) to be lower in saponins than plots fertilized with green manure (0.50%).

**Endophyte-infected tall fescue**

In 2007 ergovaline levels were higher in August (268 ppb) and September (253 ppb) than in October (145 ppb). In 2008 ergovaline levels were low in May (113 ppb), much higher in June (626 ppb), and then lower again in August (204 ppb) and September (133 ppb). In 2008 ergovaline levels trended ($P = 0.17$) to be highest in plots fertilized with conventional fertilizer (347 ppb), lower with fecal manure (265 ppb), and lowest in plots fertilized with green manure (232 ppb) or not fertilized (233 ppb). In the mixture vs monoculture study ergovaline levels were higher when fescue grew adjacent to birdsfoot trefoil (233 ppb) or lucerne (231 ppb) as opposed to a monoculture (125 ppb). Levels were low in May (89 ppb) and higher in July (235 ppb) and August (264 ppb).

**Reed canarygrass**

In 2007, gramine levels declined from August (2239 ppm) though September (1160 ppm) and October (727 ppm). In 2008, gramine levels were much lower in May (507 ppm) and June (582 ppm) than in August (1537 ppm) and then declined once again in September

(1030 ppm). Within fertilizer treatments gramine levels were highest with green manure (1029 ppm), dropping with conventional fertilizer (990 ppm), dropping still in unfertilized plots (880 ppm), and lowest with fecal manure (759 ppm).

In summary our results demonstrate that many factors influence secondary compound concentrations in plants including time of day and season, whether a plant is growing in a mixture or monoculture, and land management practices such as kind of fertilizer. Understanding these fluctuations in secondary compounds in plants will be useful when trying to utilize and balance secondary compound consumption by grazing livestock and in using livestock to enhance soil health, plant chemistry and diversity, and ultimately food for human consumption (Provenza 2008).

# Conclusion

Throughout our research the reoccurring question that inevitably surfaces is why does nature produce such a diverse array of plant secondary compounds? Research covering various ecosystems shows that when animals consume increasing doses of secondary compounds offered by just one plant their digestive and intercellular networks respond by rendering the taste unpalatable to the animal. This deters herbivores from eating too much of one plant. From an ecological standpoint plant secondary compounds offer various benefits to the plant itself. They aid plants in attracting pollinators and seed dispersers, they help plants recover from injury, they help protect plants from ultraviolet radiation, and they defend plants against pathogens, diseases, and herbivores. From an agronomist's standpoint, these compounds are usually considered toxins that render foods and forages unpalatable for consumption by humans and other animals. Nearly everything conceivably edible including water can reach a toxic level or dose when eaten in excessive amounts. The key is a moderate dose of a variety of species. Secondary compounds offer great value to our health and the health of other animals. Natural landscapes are literally nutrition centers and pharmacies with primary compounds that offer nutrition and secondary compounds offering pharmaceuticals, both of which are vital in the nutrition and health of soil, plants, herbivores, and people. The beneficial uses of thousands of plant species that contain primary and secondary compounds have yet to be discovered or rediscovered as the case may be.

In the book *Guns, Germs, and Steel*, Jared Diamond (1999) indicates that of the roughly 200,000 species of wild plants on earth only a few thousand are eaten by humans, a few hundred have been domesticated, and only a dozen account for over 80% of the current annual production of all crops. We thus use only a fraction of plant species for our consumption and for consumption by domesticated ruminants. Shockingly, merely four crops comprise two-thirds of the total agricultural crops grown today and we have selected for high growth rates as opposed to high nutritional quality through both the varieties we developed and the chemicals we use to fertilize the soil upon which we now grow these plants (Davis *et al.* 2004; Davis 2009). We have leaned toward monotypic fields in pasture settings in an attempt to simplify management and keep undesired species out. Inevitably weeds infiltrate and interestingly the livestock don't seem to mind; in fact they eat them. So one must ask, why are we fighting to keep pristine pastures free of undesired species when the animals are using them?

Oddly enough, while we've been learning of the potential benefits of plant secondary compounds we've been breeding these compounds out of plants to enhance growth of monotypic crops and pastures used for agriculture and grazing. Ironically, we are now attempting to genetically engineer compounds with similar benefits back into plants. These compounds are vastly abundant and diverse in natural systems. Instead of attempting to suppress or eliminate them, how might we adapt to them and learn to utilize them in diverse mixtures of species growing on healthy soil?

Through an integrated forage-livestock project underway we are attempting to increase our understanding of the values of secondary compounds for soil, plants, herbivores, and people. We are investigating how various growing factors affect plant secondary compounds and how these compounds interact with soil macro- and microorganisms. We are also investigating how animal impact may affect the levels of compounds in plants as well as what it might do to the soil chemistry. Our research will further evaluate the impact that plant secondary compounds have on ruminants and we will trace the plant secondary compounds through to the meat the cattle provide and determine overall quality and flavor to consumers.

## References

Anderton N, Cockrum PA, Colegate SM, Edgard JA, and Flower K (1999). Assessment of potential for toxicity of *Phalaris* spp. via alkaloid content determination: *P. coerulescens*, a case example. *Phytochemical Analysis* 10:113-118.

Barry TN, Charleston WAG, Hoskin SO, Waghorn GC, and Wilson PR (2000). Effect of forage legumes containing condensed tannins on lungworm and gastrointestinal parasitism in young red deer. *Research in Veterinary Science* 68:223-230.

Bryant JP, Chapin SF, and Klein DR (1983). Carbon / nutrient balance of boreal plants in relation to vertebrate herbivory. *OIKOS*, Copenhagen, 40:357-368.

Cheeke PR (1998). *Natural Toxicants in Feeds, Forages, and Poisonous Plants,* 2nd edn, 196 pp. Interstate Publishers, Danville, Illinois.

Coley PD, Bryant JP, and Chapin III FS (1985). Resource availability and plant antiherbivore defense. *Science* 230:895-899.

Crozier AM, Clifford M, and Ashiharan H (2006). *Plant Secondary Metabolites: Occurrence, Structure and Role in the Human Diet*, pp. 11-16. Blackwell Publishing, UK, USA, and Australia.

Davis DR (2009). Declining fruit and vegetable nutrient composition: What is the evidence. *HortScience* 44:15-19.

Davis DR, Epp MD, and Riordan HD (2004). Changes in USDA food composition data for 43 garden crops, 1950 to 1999. *Journal of the American College of Nutrition* 23:669-682.

Diamond J (1999). *Guns, Germs, and Steel: The Fates of Human Societies*, pp. 130-135. W.W. Norton & Co., New York.

Dunfield PF (2007). The soil methane sink. In *Greenhouse Gas Sinks* (DS Reay, CN Hewitt, KA Smith, and J Grace, eds), pp. 152-170. CAB International, Wallingford, UK.

Engel C (2002). *Wild Health*, pp. 21-37, 108-158. Houghton Mifflin Co., Boston,New York.

Freeland WJ and Janzen DH (1974). Strategies in herbivory by mammals: The role of plant secondary compounds. *The American Naturalist* 108:269-289.

Hamilton JG, Zangeri AR, DeLucia EH, and Berenbaum MR (2001). The carbon-nutrient balance hypothesis: its rise and fall. *Ecology Letters* 4:86-95.

Herms DA and Mattson WJ (1992). The Dilemma of Plants: To Grow or Defend. *The Quarterly Review of Biology* 67(3): 283-335.

Hill NS, Rottinghaus GE, Agee CS, and Schultz LM (1993). Simplified sample preparation for HPLC analysis of ergovaline in tall fescue. *Crop Science* 33:331-333.

Ju HJ, Hill NS, Abbott T, and Ingram KT (2006). Temperature Influences on Endophyte Growth in Tall Fescue. *Crop Science* 46:404-412.

Kallenbach RL, Bishop-Hurley GJ, Massie MD, Rottinghaus GE, and West CP (2003). Herbage Mass, Nutritive Value, and Ergovaline Concentration of Stockpiled Tall Fescue. *Crop Science* 43:1001-1005.

Killham K (1994). *Soil Ecology*, pp. 174-205. Cambridge University Press.

Leake JR, Johnson D, Donnelly DP, Muckle GE, Boddy L, and Read DJ (2004). Networks of power and influence: the role of mycorrhizal mycelium in controlling plant communities and agroecosystem functioning. *Canadian Journal of Botany* 82:1016-1045.

Loomis WE (1932). Growth-differentiation balance vs. carbohydrate-nitrogen ratio. *Proceedings American Society for Horticultural Science (ASHS)* 29:240-245.

Lu CD and Jorgensen NA (1987). Alfalfa saponins affect site and extent of nutrient digestion in ruminants. *Journal of Nurition* 117:919-927.

Majak W, McDiarmid RE, Powell TW, Van Ryswyk AL, Stout DG, Williams RJ, and Tucker RE (1979). Relationships between alkaloids in reed canarygrass (*Phalaris arundinacea*), soil moisture and nitrogen fertility. *Plant, Cell & Environment* 2:335-340.

Marten GC (1973). Alkaloids in Reed Canarygrass. *Anti-quality Components of Forages,* pp.15-31. Crop Science Society of America Inc., Madison, Wisconsin.

Patamalai B, Hejtmancik E, Bridges CH, Hill DW, and Camp BJ (1990). The isolation and identification of steroidal sapogenins in kleingrass. *Veterinary and Human Toxicology* 32(4):314-318.

Provenza FD (1995). Postingestive feedback as an elementary determinant of food preference and intake in ruminants. *Journal Range Management* 48:2-17.

Provenza FD (1996). Acquired aversions as the basis for varied diets of ruminants foraging on rangelands. *Journal Animal Science* 74:2010-2020.

Provenza FD (2008). What does it mean to be locally adapted and who cares anyway? *Journal of Animal Science* 86:E271-E284.

Provenza FD, Villalba JJ, and Bryant JP (2002). Making the match: from biochemical diversity to landscape diversity. In *Landscape Ecology and Resource Management: Making the Match* (JA Bissonette and I Storch, eds), pp. 387-421. Island Press, New York.

Provenza FD, Villalba JJ, and Dziba LE (2003). Linking herbivore experience, varied diets, and plant biochemical diversity. *Small Ruminant Research* 49(3):257-274.

Provenza FD, Villalba JJ, Haskell JH, MacAdam JA, Griggs TC, and Wiedmeier RD (2007). The value to herbivores of plant physical and chemical diversity in time and space. *Crop Science* 47:382-398.

Ramirez-Restrepo CA, Barry TN, Pomroy WE, Lopez-Villalobos N, McNabb WC, and Kemp PD (2005). Use of *Lotus corniculatus* containing condensed tannins to increase summer lamb growth under commercial; dryland farming conditions with minimal anthelmintic drench input. *Animal Feed Science Technology* 122:197-217.

Reichardt PB, Chapin III FS, Bryant JP, Mattes BR, and Clausen TP (1991). Carbon/nutrient balance as a predictor of plant defense in Alaskan balsam poplar: potential importance of metabolite turnover. *Oecologia* 88:401-406.

Rottinghaus GE, Garner GB, Cornell CN, and Ellis JL (1991). HPLC method for quantitating ergovaline in endophyte-infested tall fescue: Seasonal variation of ergovaline levels in stems with leaf sheaths, leaf blades, and seed heads. *Journal of Agriculture Food Chemical* 39:112-115.

Schatzker M (2010). *Steak: One Man's Search for the World's Tastiest Piece of Beef.* Penguin Group, New York.

Terrill TH, Rowan AM, Douglas GB, and Barry TN (1992). Determination of extractable and bound condensed tannin concentrations in forage plants, protein concentrate meals, and cereal grains. *Journal Science Food Agriculture* 58:321-329.

Thompson FN, Stuedemann JA, and Hill NS (2001). Anti-quality factors associated with alkaloids in eastern temperate pasture. *Journal of Range Management* 54:474-489.

Veteli TO, Mattson WJ, Niemela P, Julkunen-Tiitto R, Kellomaki S, Kuokkanen K, and Lavola A (2007). Do elevated temperature and CO2 generally have counteracting effects on phenolic phytochemistry of boreal trees? *Journal of Chemical Ecology* 33(2):287-296.

# Chapter 108

# Maintaining Aversion to *Geigeria ornativa* (Vermeerbos) in Sheep by Means of Continuous Exposure to an Aversive Mixture Presented in a Self-Feeder

L.D. Snyman[1], R.A. Schultz[1], A. Theunissen[2], and K. Mosia[2]

[1]*Toxicology Section, Onderstepoort Veterinary Institute, Private Bag X05, Onderstepoort, 0110 South Africa;* [2]*Vaalharts Research Institute, Northern Cape Department of Agriculture & Land Reform, Private bag X9, Jan Kempdorp, 8550 South Africa*

## Introduction

Conditioned feed aversion is a means of teaching stock to avoid poisonous plants (Provenza *et al.* 1992; Ralphs *et al.* 2001). The induced aversion, however, is easily broken down by the social influence of non-averted animals (Ralphs and Olsen 1990; Ralphs and Provenza 1999). It was also noticed that sheep familiar with *Geigeria ornativa* prior to aversion treatment may lose their aversion to the plant after some time (LD Snyman, unpublished data). Previous studies indicated that continuous exposure to an aversive mixture following the initial aversion treatment might be a means of accomplishing sustained aversion to *G. ornativa* (Snyman *et al.* 2002; Snyman and Joubert 2004). In these studies sheep had to be gathered and penned (kraaled) each day to be exposed to the aversive mixture overnight. Under these circumstances sheep were compelled to eat some of the aversive mixture each night as they had nothing else to eat than the aversive mixture in their immediate vicinity. Gathering the sheep every day, however, was found to be very impractical in the long term. A more feasible alternative would be to present the aversive mixture in a self-feeder placed in the field near the watering point where sheep would have free access to it. However, in contrast to the kraal situation these conditions would not force intake of the aversive mixture and intake would depend on the highly palatable maize meal in the aversive mixture. The objective of this study was to investigate whether continuous exposure to an aversive mixture, presented in a self-feeder in the field, would accomplish persistent intake of the aversive mixture and thereby sustain aversion to *G. ornativa.*

## Materials and Methods

For this investigation, fifteen 2-year-old Dorper wethers averted to *G. ornativa* were continuously offered an aversive mixture presented in a self-feeder compared to 15

©CAB International 2011. *Poisoning by Plants, Mycotoxins, and Related Toxins* (eds F. Riet-Correa, J. Pfister, A.L. Schild, and T.L. Wierenga)

untreated control wethers (Dorpers, 2 years old) grazed in an adjacent camp. The averted sheep were fed hay and maize meal (50 g/sheep/day) for 3 months to acquire a learned safety status to the maize meal. The maize meal was offered in the self-feeder to be used in the trial. For inducing aversion to *G. ornativa* 15 of the sheep, randomly selected, were fasted for 24 h and then confined to a small site in the field heavily infested with *G. ornativa* until most of the *G. ornativa* had been ingested which took them approximately 6 h. The sheep were then drenched with lithium carbonate at 160 mg/kg BW added to an extract of *G. ornativa* (250 g/l extracted in boiling water). This was followed by giving 5 g milled *G. ornativa* in their mouth. The sheep were kept on the site for another 12 h whereafter they were released to the surrounding *G. ornativa*-infested field where they had free access to an aversive mixture presented in a self-feeder placed near the watering point.

The aversive mixture consisted of an aversive concentrate mixed with maize meal. The aversive concentrate contained five parts lithium carbonate and one part milled *G. ornativa* coated with a hexane extract from one part freshly collected *G. ornativa* mixed with one part sodium chloride. Sheep were gradually accustomed to the aversive mixture over a 14 day period by slowly increasing the aversive concentrate until consumption of 50 g aversive mixture/sheep/day (containing 2.5% lithium carbonate) was achieved.

The sheep in both groups were observed on a daily basis for clinical signs of *G. ornativa* poisoning while body weight was determined weekly. Intake of the aversive mixture was measured on a daily basis for the first 14 days and sporadically thereafter. The trial had to be terminated after 62 days as almost all *G. ornativa* had died by that time. The sheep were slaughtered and the width of the flattened esophagus measured where it was most dilated, which in all cases were in the caudal 5 cm before opening in the reticulo-rumen.

Averted and control sheep were compared for the occurrence of clinical and subclinical poisoning as indicated by the following parameters:

1. Clinical signs as manifested by stiffness, paresis, paralysis, and regurgitation of ingesta through the mouth and nose (Kellerman *et al.* 2005)
2. Width of the flattened esophagus as an indication of the extent of esophageal dilatation. Dilatation of the esophagus had been noticed in cases of clinical as well as subclinical *G. ornativa* poisoning (Snyman *et al.* 2008).
3. Body weight change. Negative body weight changes also had been observed in cases of clinical and subclinical *G. ornativa* poisoning (Snyman *et al.* 2008).

## Results and Discussion

### Lithium carbonate intake

Figure 1 shows persistent intake of lithium carbonate (also reflecting intake of the aversive mixture) for the duration of the trial following drenching with lithium carbonate during aversion treatment on Day 0 and subsequent adaptation to the aversive mixture until Day 14. The mean lithium carbonate intake during this period was 20.5±6.3 mg/kg BW/day. The results indicate that sustained aversion to *G. ornativa* should have occurred.

## Clinical signs of *G. ornativa* poisoning

Figure 2 shows that none of the averted sheep showed clinical signs of *G. ornativa* poisoning at any stage during the trial, compared to seven of the control sheep, thus, suggesting avoidance and therefore aversion to *G. ornativa* by averted sheep. The number of sheep showing clinical signs of *G. ornativa* poisoning decreased towards the end of the trial, indicating recovery from *G. ornativa* poisoning. This can be ascribed to the unexpected die-off of *G. ornativa* due to a drought during the latter part of the trial.

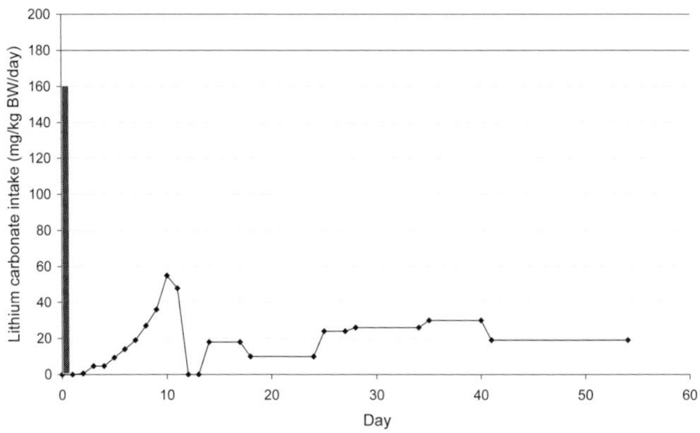


**Figure 1.** Mean lithium carbonate intake as constituent of the aversive mixture.

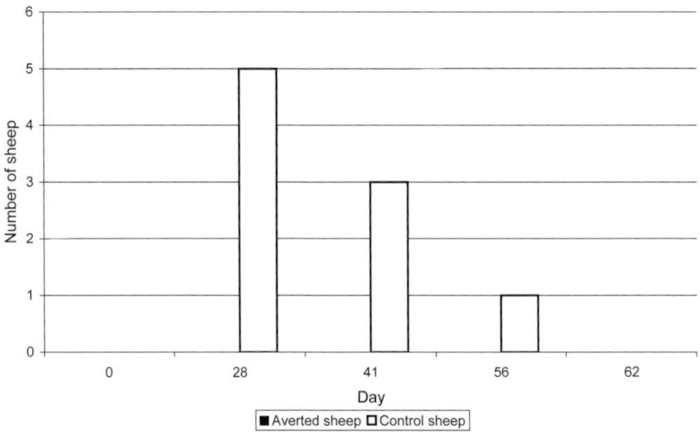


**Figure 2.** Number of control sheep showing clinical signs of *G. ornativa* poisoning at specific stages during the trial. Averted sheep showed no clinical signs.

## Esophageal width

The mean esophageal width of 25.2 mm for averted sheep which can be regarded as normal was significantly lower ($P < 0.05$) than the mean value of 30.9 mm measured for control sheep, again suggesting avoidance and therefore aversion to *G. ornativa* by averted sheep (Figure 3). The individual values of the control sheep show that the larger esophageal width was due to dilated esophagi of more than half of the control sheep.

## Body weight changes

The average daily gain of averted sheep (39.0 g) was significantly ($P < 0.05$) higher than the average daily loss of control sheep (−38.9 g; Figure 4). This once again suggests avoidance and therefore aversion to *G. ornativa* by averted sheep. The initial increase in body weight for the control sheep is in accordance with results regarding *G. ornativa* poisoning (Snyman *et al.* 2008) while the tendency to maintain or increase body weight during the latter part of the trial should be ascribed to diminished availability of *G. ornativa*. Table 1 shows the weight changes of averted and control sheep from Day 0-42 and Day 0-62.

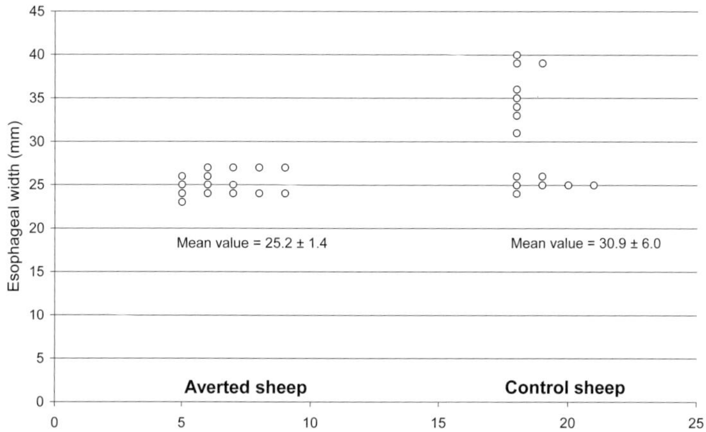


**Figure 3.** Individual values of the esophageal width of averted and control sheep slaughtered on Day 62.

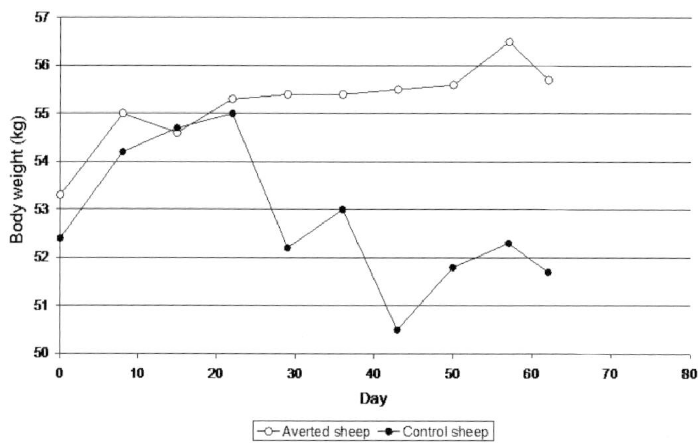


**Figure 4.** Mean body weights of averted and control sheep during the trial.

The seven control sheep that showed clinical signs of *G. ornativa* poisoning during the trial exhibited a loss in body weight of 2.414 kg from Day 0-62 compared to a weight gain of 0.812 kg by the control sheep that did not show clinical signs of *G. ornativa* poisoning. The averted sheep had a body weight gain of 2.420 kg, which was still higher than the 0.812 kg of the apparently unaffected control sheep, suggesting subclinical poisoning of the control sheep not showing clinical signs of *G. ornativa* poisoning. When comparing the body weight changes from Day 0-42, a period when the occurrence of *G. ornativa* was still high, the same effects were obtained except that all differences were highly significant ($P$<0.01). The results indicate that aversion treatment not only prevented a loss in body weight due to clinical poisoning but also prevented decreased growth due to subclinical *G. ornativa* poisoning.

**Table 1.** Weight changes of averted and control sheep over different periods of the trial.

| Treatment | n | Weight change (kg) | | | |
|---|---|---|---|---|---|
| | | Day 0-42 | | Day 0-62 | |
| | | Mean | SEM | Mean | SEM |
| Averted sheep | 15 | 2.353[a] | 0.551 | 2.420[a] | 0.715 |
| Control sheep: without clinical signs | 8 | 0.212[b] | 0.755 | 0.812[a] | 0.980 |
| Control sheep: with clinical signs | 7 | -4.414[c] | 0.807 | -2.414[b] | 1.047 |
| Probability ($P$) | | <0.01 | | <0.05 | |

Superscripts indicate significance between groups.

It should be kept in mind that intake of the aversive mixture might have benefited the averted sheep. Taking a feed conversion ratio of 8:1 for these 2-year-old sheep, 50 g aversive mixture/sheep/day would have contributed 0.25 kg to their weight on Day 42 but does not explain the difference between averted and control sheep.

## Conclusion

The results indicate that continuous exposure to an aversive mixture presented in a self-feeder resulted in persistent intake of the aversive mixture and thereby sustained aversion in sheep averted to *G. ornativa.* Results, however, need to be confirmed over a longer period.

## Acknowledgements

This project was financially supported by the Gauteng Department of Agriculture Conservation and Environment (GDACE) through the Directorate: Technology Development and Support (TDS); the North-West Province Directorate of Veterinary Services (Department of Agriculture, Conservation, Environment and Tourism) and the Northern Cape Department of Agriculture and Land Reform.

Phillemon Fanti and Jacob Veldsman, Koopmansfontein Research station, are thanked for devoted technical assistance.

## References

Kellerman TS, Coetzer JAW, Naudé TW, and Botha CJ (2005). *Plant poisonings and mycotoxicosis of livestock in southern Africa*, 2nd edn, 310 pp. Oxford University Press, Cape Town.

Provenza FD, Pfister JA, and Cheney CD (1992). Mechanisms of learning in diet selection with preference to phytotoxicosis in herbivores. *Journal of Range Management* 45:36-45.

Ralphs MH and Olsen JD (1990). Adverse influence of social facilitation and learning context in training cattle to avoid eating larkspur. *Journal of Animal Science* 68:1944-1952.

Ralphs MH and Provenza FD (1999). Conditioned food aversions: principles and practices, with special reference to social facilitation. *Proceedings of the Nutrition Society* 58:813-820.

Ralphs MH, Provenza FD, Pfister JA, Graham D, Duff G, and Greathouse GC (2001). Conditioned food aversion from theory to practice. *Rangelands* 23:14-17.

Snyman LD and Joubert JPJ (2004). Conditioned feed aversion as a means of preventing sheep from grazing vermeerbos (*Geigeria ornativa*). In *Poisonous Plants and Related Toxins* (T Acamovic, CS Stewart, and TW Pennycott, eds), pp. 421-424. CABI Publishing, Wallingford, UK.

Snyman LD, Schultz RA, Kellerman TS, and Labuschagne L (2002). Continuous exposure to an aversive mixture as a means of maintaining aversion to vermeerbos (*Geigeria ornativa* O. Hoffm.) in the presence of non-averted sheep. *Onderstepoort Journal of Veterinary Research* 69:321-325.

Snyman LD, Carstens A, Schultz RA, Joubert JPJ, and Labuschagne L (2008). Changes in sheep oesophageal diameter and function during *Geigeria ornativa* (vermeerbos) poisoning and subsequent recovery. *Journal of the South African Veterinary Association* 79:178-184.

# Chapter 109

# Conditioned Flavor Aversion and Location Avoidance in Hamsters from Toxic Extract of Tall Larkspur (*Delphinium barbeyi*)

J.A. Pfister[1], C.D. Cheney[2], D.R. Gardner[1], and K.E. Panter[1]

[1]*USDA-ARS Poisonous Plant Research Laboratory, Logan, Utah 84341, USA;* [2]*Dept. of Psychology, Utah State University, Logan, Utah 84322, USA*

## Introduction

The phenomenon of conditioned flavor aversion (CFA) is robust and widespread (Chambers 1990). Garcia (1989) has argued that animals possess dual defense systems: a gut defense system in which compounds causing nausea (negative feedback) produce CFAs through a hedonic shift when the flavor is paired with adverse post-ingestive consequences (toxin defense system) and a skin defense system (for predator avoidance) in which distal cues (e.g. shock, paralysis) produce place avoidance. The conditioned place preference/ avoidance paradigm has gained increased use as a means to test the reinforcing properties of drugs (Hoffman 1989; Bozart 1990). Tests for acquisition of flavor aversions are also being used as indicators of toxicity (Miller and Eckerman 1986).

This experiment addressed CFA and place avoidance learning in hamsters given injections of alkaloid extracts from tall larkspur. Our primary objective was to determine if larkspur had reinforcing or negative properties sufficient to cause place avoidance or preference. Further, we wished to determine if hamsters could acquire CFAs from larkspur alkaloid extract.

Tall larkspur is a widespread toxic plant in the western USA (Ralphs *et al.* 1988). This nutritious and palatable plant (Pfister *et al.* 1990) is responsible for numerous livestock deaths each year (Pfister *et al.* 1988). Cattle often eat sufficient quantities of tall larkspur to become intoxicated yet surviving animals begin eating larkspur anew once the intoxication subsides (Pfister *et al.* 1997). This cyclic consumption is pervasive and raises the question of whether larkspur alkaloids have reinforcing (positive) properties at low doses. We know that at high doses alkaloid extracts from larkspur will condition taste aversions in cattle (Olsen and Ralphs 1986) but the aversion will quickly extinguish at times (Ralphs *et al.* 2001). The primary toxin in tall larkspur is methyllycaconitine (MLA) and related alkaloids (termed MSAL alkaloids). The $LD_{50}$ of MLA in mice is about 4.5 mg/kg BW given i.v. (Manners *et al.* 1995). These MSAL alkaloids block acetylcholine (Ach) receptors at the neuromuscular junction resulting in respiratory paralysis and death (Dobelis *et al.* 1993).

©CAB International 2011. *Poisoning by Plants, Mycotoxins, and Related Toxins* (eds F. Riet-Correa, J. Pfister, A.L. Schild, and T.L. Wierenga)

## Materials and Methods

### Subjects

A total of 24 female Syrian golden hamsters (Charles-River) weighing about 150 g were used for the place avoidance/taste aversion experiment. Thirty other hamsters were used in a preliminary experiment to determine median lethal dose ($LD_{50}$) of the larkspur extract. All animals were housed individually in a temperature-controlled room with a 12 h light-dark cycle. The hamsters were fed rodent chow *ad lib* but were allowed access to flavor solutions for 50 min during training and for 20 min during testing. There was no mortality during the study except during the $LD_{50}$ experiment. All procedures were approved by the Utah State University IACUC committee.

### Apparatus

Our training and testing apparatus followed that given by Reicher and Holman (1977). The rectangular (70 × 30 × 25 cm) shuttlebox was constructed of transparent plexiglass (sides and top) with a stainless steel floor. A plexiglas barrier restricted animals to one side of the shuttlebox during training. The exterior walls of the right side were cross-hatched with 2 cm strips of yellow tape and the walls of the left side were untouched. The floor in the right side was given a texture by gluing 1.5 cm strips of velcro to the floor; the floor in the left side was smooth bare metal. The shuttle boxes were placed at an oblique angle to the ceiling lights. Thus, subjects' cues as to side placement included differential visual and tactile sensations from the lighting pattern and the floor surface.

### Flavors

Two drinking solutions were used: the almond solution contained 0.2% sodium saccharin and 2% Shilling® almond extract in tap water and the lemon solution contained 0.2% sodium saccharin and 2% Shilling® lemon extract in water. Flavors were provided to hamsters in plastic bottles with drinking spouts. Each flavor was presented through a hole in the end of the shuttlebox allowing animal access to the appropriate flavor while in the shuttlebox.

### Larkspur extract and alkaloid analysis

Tall larkspur plants were collected during the flowering stage from the Manti-La Sal National Forest near Manti, Utah at 3400 m elevation. The plant material was air-dried, ground to pass a 2-mm screen, then stored at 20°C in plastic bags until used. Weighed quantities of the plant material were extracted 3× for 3 days each with an 80:20 ethanol:water mixture. The volume was reduced by distilling off the alcohol then blowing the material to dryness under a stream of air while in a water bath at 37°C. The tar-like residue was then reconstituted using a phosphate buffered saline solution and the original pH of 4.6 was adjusted to 6.7 by the addition of 10% NaOH solution. The final extract contained 1 ml of buffered saline with NaOH for each g of plant material extracted. The alkaloid extract was filtered using medium weight filter paper then refrigerated in sterile vials. Quantitative analysis for toxic alkaloids (MLA + other MSAL alkaloids) was done using Fourier transform infrared spectroscopy (Gardner *et al.* 1997).

### Procedures

The acute 24 h $LD_{50}$ dose of the alkaloid extract was determined following single subcutaneous (s.q.) injections of incremental doses (Weil 1952). The alkaloid extract contained 1.51 mg toxic alkaloid/g (dry weight basis); each 1 ml of extract was derived from 1 g of plant material and alkaloid doses were expressed as mg toxic alkaloid/kg body weight (BW). The $LD_{50}$ for the toxic alkaloids was 4.86 mg/kg BW, similar to that noted for mice. In this study we used a s.q. dose of 2.25 mg toxic alkaloid/kg BW. At this dose hamsters exhibited no clinical signs of intoxication.

The training schedule consisted of four injections of larkspur extract and four injections of isotonic saline. On odd-numbered days subjects were removed from the home cage, given a larkspur injection, and placed in one side of the shuttlebox with access to one flavored solution. On even-numbered days, subjects were injected with comparable amounts of saline and placed into the opposite side of the shuttlebox with the opposite flavor. Consumption of flavors was recorded each training day. Half of the hamsters were placed into the left side of the shuttlebox on training days and half on the right; half of the subjects in each group received the almond-flavored solution and half received the lemon-flavored solution. Thus, there were four balanced subgroups of six subjects in each group.

Testing was done on days 9, 10, and 21 except that place avoidance was not retested on day 21. On day 9 saline injections were administered i.p. and the subjects allowed 20 min in the shuttlebox with the interior partition removed, allowing access to both sides. No solutions were available. All subjects were videotaped and the percentage of time spent on each side was determined. Animals were then removed to their home cage. Immediately after the side-preference test subjects were offered both flavored solutions simultaneously for 20 min and consumption of each was recorded. On day 10 the identical procedure was followed except that subjects were injected s.q. with larkspur extract before the side-preference and flavor tests. On day 21 the persistence of the flavor aversion was tested by allowing all subjects access to both flavored solutions for 20 min in their home cages and consumption was recorded. No injections were given on day 21.

### Statistical analysis

The amount of fluid consumed was examined using a mixed linear model that included flavor (almond or lemon), side (right or left), trial day (day 9, 10, 21), and interactions. The same model was used to examine side (place) preference on days 9 and 10.

## Results

### Taste aversion learning

The flavor not associated with larkspur injections during conditioning is referred to as the non-aversive flavor; similarly the flavor associated with larkspur is referred to as the aversive flavor. Testing with saline injection on day 9 showed that subjects had a strong aversion ($P < 0.01$) to the aversive flavor. Mean fluid intake for the non-aversive flavor (either almond or lemon) was 3.5 ml (SEM ± 0.62) on day 9 compared to the 0.6 ml (± 0.45) for the aversive flavor. There were no side effects or side by flavor interactions. Similar results (not shown) were obtained on day 10.

The taste aversion persisted ($P < 0.01$) in all subjects when retested on day 21. Mean fluid intakes were 1.8 m (SEM ± 0.65) for the non-aversive flavor and 0.42 (SEM ± 0.3) for the aversive flavor. There were no side effects or side by flavor interactions. The subjects reduced their day 21 fluid intake of the non-aversive flavor compared to their non-aversive fluid intake in earlier trials (days 9 and 10), suggesting that even though the taste aversion had persisted it had weakened somewhat.

**Place (location) preference/avoidance**

The side of the shuttlebox associated with larkspur injections during conditioning will be referred to as the larkspur side; similarly the side associated with saline injections will be called the saline side. Subjects showed a clear location aversion after being conditioned with larkspur injections. On day 9 (saline injection day) subjects avoided the larkspur side (32.2 ± 3.5% of time spent on that side) compared to the saline side (67.8 ± 3.7% of time spent on saline side) during the 20 min test period. On day 10 (larkspur injection day) subjects again avoided the larkspur side (26.5 ± 5.2% of time spent on that side) compared to the saline side (73.5 ± 5.7% of time spent on saline side). Saline and larkspur injection days did not differ ($P > 0.1$) and there were no flavor effects for almond and saline flavors, respectively, nor were there any interactions ($P > 0.1$).

## Discussion

The alkaloid extract from tall larkspur conditioned both taste aversions and location avoidance. The taste aversion:location preference paradigm used in this study has also detected both rewarding and aversive properties of various drugs, including amphetamine, morphine, and methylscopolamine (Reicher and Holman 1977; Sherman *et al.* 1980, 1983; Hughes *et al.* 1989). Diterpenoid alkaloids in tall larkspur appear to have negative effects on exteroceptive reinforcers associated with location cues while simultaneously negatively affecting interoceptive reinforcers associated with flavor cues (Garcia 1989). The alkaloid extract probably acted both peripherally and centrally to produce the taste aversion and location avoidance. Tall larkspur alkaloids apparently act on both the gut defense system (Garcia 1989) probably by causing nausea (negative feedback) to cause taste aversion. Similarly these extracts probably act on the skin defense system (Garcia 1989) by producing some degree of neuromuscular paralysis to produce location avoidance. In this hamster study, we found no evidence of positive (rewarding) effects from injection of larkspur extracts.

The major toxic alkaloids in tall larkspur, specifically MLA, act at the neuromuscular junction and in the CNS to block acetylcholinergic receptors (Benn and Jacyno 1983; Dobelis *et al.* 1993) thus producing the clinical signs of progressive fatigue and muscular paralysis. In rodents such as rats and mice there appears to be a substantial CNS effect (BL Stegelmeier, personal communication). Rodents but not cattle show grand mal-type seizures at potentially lethal doses. However, at the dose given in this study the hamsters did not show overt symptoms of alkaloid intoxication.

Larkspur alkaloid extracts have been used successfully to condition taste aversions in cattle (Olsen and Ralphs 1986; Ralphs and Olsen 1992). The larkspur extract given to hamsters was less potent than was lithium chloride (LiCl) given to cattle (Ralphs and Olsen 1992) in producing a strong and persistent taste aversion, in part because of variability in susceptibility in hamsters. Larkspur ingestion is at least moderately aversive in grazing cattle as cattle ingestion of larkspur cycles from high to low over several days (cyclic consumption) with mild or severe intoxication followed by avoidance of larkspur for one to several days. This period of detoxification via a transient taste aversion (Pfister *et al.* 1997) is typically followed by renewed consumption of larkspur by grazing cattle as the aversion is extinguished. We have noted that individual grazing cattle may exhibit this transient aversion to larkspur on numerous occasions during a grazing season yet on each occasion they again begin consuming the toxic larkspur. This increase in consumption by grazing cattle may be related to the nutritional content of larkspur (Pfister *et al.* 1997) and the positive post-ingestive consequences from the highly nutritious larkspur (Provenza 1995).

## Acknowledgement

We thank Kermit Price for excellent technical assistance with the study.

## References

Benn MH and Jacyno JM (1983). The toxicology and pharmacology of diterpenoid alkaloids. In *Alkaloids, Chemical and Biological Perspectives* (SW Pelletier, ed.), pp. 155-210. John Wiley & Sons, New York.

Bozart MA (1990). Evidence for the rewarding effects of ethanol using the conditioned place preference method. *Pharmacology Biochemistry and Behavior* 35:485-487.

Chambers KC (1990). A neural model for conditioned taste aversions. *Annual Review of Neuroscience* 13:373-385.

Dobelis P, Madl JE, Manners GD, Pfister JA, and Walrond JP (1993). Effects of *Delphinium* alkaloids on neuromuscular transmission. *Pharmacology and Experimental Therapeutics* 291:538-546.

Garcia J (1989). Food for Tolman: cognition and cathexis in concert. In *Aversion, Avoidance and Anxiety* (T Archer and L Nilsson, eds), pp. 45-85. Lawrence Erlbaum, Hillsdale, New Jersey.

Gardner DR, Manners GD, Ralphs MH, and Pfister JA (1997). Quantitative analysis of diterpenoid alkaloids in larkspur (*Delphinium* spp.) by Fourier transform infrared spectroscopy. *Phytochemical Analysis* 8:55-62.

Hoffman DC (1989). The use of place conditioning in studying the neuropharmacology of drug reinforcement. *Brain Research Bulletin* 23:373-387.

Hughes RN, Blampied NM, Anderson GJ, and Woollett GJ (1989). Methylscopolamine and conditioned location avoidance. *Pharmacology Biochemistry and Behavior* 33:913-914.

Manners GD, Panter KE, and Pelletier SW (1995). Structure-activity relationships of norditerpenoid alkaloids occurring in toxic larkspur (*Delphinium*) species. *Journal of Natural Products* 58:863-866.

Miller DB and Eckerman DA (1986). Learning and memory measures. In *Neurobehavioral Toxicology* (Z Annau, ed.), pp. 94-149. Johns Hopkins Press, Baltimore.

Olsen JD and Ralphs MH (1986). Feed aversion induced by intraruminal infusion with larkspur extract in cattle. *American Journal of Veterinary Research* 47:1829-1832.

Pfister JA, Manners GD, Ralphs MH, Hong ZX, and Lane MH (1988). Effects of phenology, site and rumen fill on tall larkspur consumption by cattle. *Journal of Range Management* 41:509-514.

Pfister JA, Provenza FD, Manners GD, Gardner DR, and Ralphs MH (1997). Tall larkspur ingestion: can cattle regulate intake below toxic levels? *Journal of Chemical Ecology* 23:759-777.

Pfister JA, Provenza FD, and Manners GD (1990). Ingestion of tall larkspur by cattle: separating effects of flavor from postingestive consequences. *Journal of Chemical Ecology* 16:1697-1705.

Provenza FD (1995). Postingestive feedback as an elementary determinant of food preference and intake in ruminants. *Journal of Range Management* 48:2-17.

Ralphs MH and Olsen JD (1992). Comparison of larkspur alkaloid extract and lithium chloride in maintaining cattle aversion to larkspur in the field. *Journal of Animal Science* 70:1116-1120.

Ralphs MH, Olsen JD, Pfister JA, and Manners GD (1988). Plant-animal interactions in larkspur poisoning in cattle. *Journal of Animal Science* 66:2334-2342.

Ralphs MH, Provenza FD, Pfister JA, Graham D, Duff GC, and Greathouse G (2001). Conditioned food aversion:from theory to practice. *Rangelands* 23:14-18.

Reicher MA and Holman EW (1977). Location preference and flavor aversion reinforced by amphetamine in rats. *Animal Learning and Behavior* 5:343-346.

Sherman JE, Roberts T, Roskam SE, and Holman EW (1980). Temporal properties of the rewarding and aversive effects of amphetamine in rats. *Pharmacology Biochemistry and Behavior* 13:597-599.

Sherman JE, Hickis CF, Rice AG, Rusiniak KW, and Garcia J (1983). Preferences and aversions for stimuli paired with ethanol in hungry rats. *Animal Learning and Behavior* 11:101-106.

Weil CS (1952). Tables for convenient calculation of median-effective dose ($LD_{50}$ or $ED_{50}$) and instructions in their use. *Biometrics* 8:249-263.

# Chapter 110

# Conditioning Taste Aversion to *Mascagnia rigida* (Malpighiaceae) in Sheep

I. Pacífico da Silva and B. Soto-Blanco

*Department of Animal Sciences, Universidade Federal Rural do Semi-Árido (UFERSA), BR 110 Km 47, Mossoró, RN, 59625-900, Brazil*

## Introduction

Several species of toxic plants can cause significant losses to livestock. *Mascagnia rigida* from the Malpighiaceae family is the most important poisonous plant in the semiarid region of Brazil, causing significant economic loss to local farmers. Poisoning is characterized by sudden death in sheep, cattle, and goats (Tokarnia *et al.* 1961; Medeiros *et al.* 2002; Pacífico da Silva *et al.* 2008). This plant is reputed to be palatable and poisoning is typically acute, with death occurring up to 48 h after ingestion (Tokarnia *et al.* 1961; Medeiros *et al.* 2002; Pacífico da Silva *et al.* 2008). The toxic compound is not known.

The best way to reduce economic losses is through prevention of animal poisoning and conditioned food aversion is a potential management tool for training livestock to avoid eating poisonous plants (Ralphs and Provenza 1999). Any chemical or physiological state that affects the upper gastrointestinal tract or the emetic center of the brain can cause an aversion (Garcia and Holder 1985). The most used drug for conditioning taste aversions in livestock is lithium chloride (Ralphs and Stegelmeier 1998; Ralphs and Provenza 1999).

The aim of this study was to determine whether lithium chloride-treated sheep could be averted from consuming *M. rigida* in the edaphic and climatic conditions of a semiarid region of northeastern Brazil.

## Methodology

Fresh *M. rigida* leaves were used in this study. They were collected at Mossoró, Rio Grande do Norte State, northeastern Brazil (5°11'15"S and 37°20'39"W) at an altitude of 16 m above sea level. The climate is characterized as semiarid. The mean annual temperature, annual rainfall, and relative humidity are 27.4°C, 674 mm, and 68.9%, respectively.

Twelve 10- to 12-month-old female sheep (41-52 kg) were used. These animals had not previously been familiarized to *M. rigida* as most outbreaks of poisoning by this plant in the region occur in naïve animals. They were fed with 0.2 kg/animal/day of concentrate (0.9 g/kg body weight net energy value; 110 g/kg neutral detergent fiber; 21 g/kg crude

©CAB International 2011. *Poisoning by Plants, Mycotoxins, and Related Toxins* (eds F. Riet-Correa, J. Pfister, A.L. Schild, and T.L. Wierenga)

protein; 73 g/kg PDIA (rumen-undegradable protein truly digestible in the small intestine); 140 g/kg PDIN (true protein truly digestible in the small intestine when N limits microbial-protein synthesis); 120 g/kg PDIE (true protein truly digestible in the small intestine when energy limits microbial-protein synthesis); 15 g/kg Ca; 5 g/kg P) and unlimited quantities of native hay. Water was offered *ad libitum*.

The sheep were randomly allocated to two treatment groups: control group and lithium group. The control group was treated with 15 ml water orally by a drenching gun and the lithium group received 150 mg/kg BW of LiCl in a 50% w/v solution (a mean volume of 16 ml LiCl solution) orally by a drenching gun. Tests were conducted in two pens of 5.0×7.0 m (one pen for each group). Food was removed from the pens the evening before at 18:30 h (13 h before the beginning of the tests) to exclude interference from any other food or flavor. The conditioning trials began at 7:30 h. Sheep were allowed to feed on *M. rigida* leaves (about 1% of BW) for 15 min; at the end of the 15 min period they received a dose of LiCl (lithium group) or water (control group). The amount of time spent on eating *M. rigida* was measured. There was always some residual quantity of the plant after each trial. Duration of time for which the plant was available and the LiCl dose were based on previous studies (Ginane and Dumont 2006; Barbosa *et al.* 2008). The conditioning was repeated daily until the LiCl-treated sheep stopped eating *M. rigida*. Sheep were fed 1 h after trials.

Extinction trials were conducted on day 10, 24, 40, 55, and 70 after conditioning using single-choice intake tests. *M. rigida* leaves were offered at about 1% of BW for 15 min and the time spent eating the available plant was measured. No LiCl was given to sheep in these trials.

Statistical analysis of time spent on eating was carried out using a mixed linear model approach of SAS (SAS Statistical Software V8, 2000, SAS Institute Inc., Cary, NC). Animals were considered as a random factor with each animal nested within treatments and with repeated measurements over time. Various mixed models (e.g. compound symmetry, unstructured, and autoregressive) were compared to determine the covariance structure and the best-fitting model was determined for each dependent variable. Significant day by treatment interactions were examined using the PDIFF procedure of SAS with preplanned comparisons.

## Results

There were no outward signs of poisoning, illness, or distress caused by plant ingestion or LiCl treatment in any sheep for all periods of the experiment.

There was no difference in consumption of *M. rigida* between the two groups on the first day prior to treatment with LiCl (Figure1). On the second day, three out of six sheep from the lithium group abstained from eating the leaves but the other three still consumed a lower amount (15 to 59 s) and therefore received a second LiCl treatment. On the third day, no LiCl-treated sheep ingested *M. rigida*. Groups differed on the third day in time spent eating the leaves following conditioning ($P < 0.001$). There was also a significant day of observation and day by treatment interaction ($P < 0.001$). The decline in consumption by the control group indicates there was a natural aversion caused by the *M. rigida* leaves.

Results of the extinction trials of *M. rigida* aversion are presented in Figure 2. Groups differed on all the evaluated days in time spent eating leaves after conditioning ($P < 0.001$) (Figure 2). One LiCl-treated sheep ate some quantity of *M. rigida* leaves on the 55th day

(20 s) but it did not consume the plant at the next evaluation. Thus, there was no day by treatment interaction. The aversion to *M. rigida* did not extinguish.

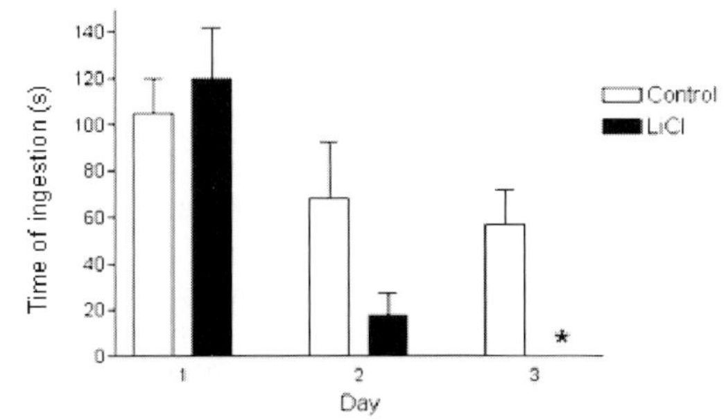


**Figure 1.** Time spent eating (s±SE) *Mascagnia rigida* by the control (open bar) and lithium chloride (LiCl)-treated (solid bar) sheep (*n*=6 for each treatment) during aversion conditioning using LiCl administered at 150 mg/kg BW or 15 ml water (control) orally by a drenching gun. *$P < 0.001$, mixed linear model of SAS.

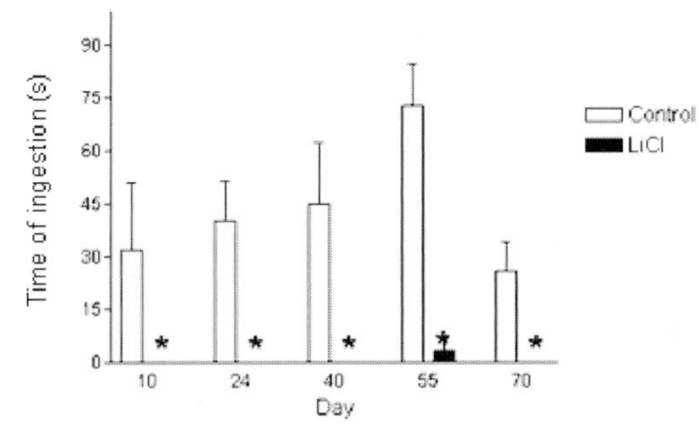


**Figure 2.** Time spent eating (s±SE) *Mascagnia rigida* by the control (open bar) and averted (solid bar) sheep (*n*=6 for each treatment) on the 10th, 24th, 40th, 55th, and 70th day after conditioning. Sheep were averted with 150 mg/kg BW lithium chloride (LiCl); the first day of conditioning was considered day 1. *$P < 0.001$, mixed linear model of SAS.

# Discussion

Aversive taste conditioning can be used for training livestock to avoid the intake of palatable poisonous plants (Ralphs and Provenza 1999; Ralphs 2001) and it may represent an important tool for avoiding plant poisoning. This is particularly important because other ways of control such as digging up the plant and the use of herbicides are limited and not effective. Other traditional management methods to prevent poisoning such as limiting the access of animals to the plant and destroying the plant are not efficient due to the wide distribution of the plant throughout several ranges in the region.

The results of the present study indicate that sheep can be easily conditioned using lithium chloride to avoid eating the poisonous plant *M. rigida.* These results are in agreement to earlier studies conducted with goats that were successfully averted against the same poisonous plant also using lithium chloride (Barbosa *et al.* 2008).

Aversions conditioned by lithium chloride are considered strong and last indefinitely if the animals are not compelled to resample the plant. If averted animals eat the target plant without any adverse feedback, they will continue eating it and eventually the aversion will be extinguished (Ralphs 1997). This can be an important factor interfering with the persistence of the aversion especially because controls in the present study did not stop ingesting *M. rigida* leaves. Social facilitation is the greatest impediment to retaining aversions. Conditioned animals may resume eating a poisonous plant to which they were previously averted if they observe other animals eating it. Therefore, averted animals must graze separately from unconditioned animals to maintain the induced aversion (Ralphs and Provenza 1999).

Another factor that affects the persistence of the conditioned taste aversion is the presence of alternative foods (Kimball *et al.* 2002). *M. rigida* is consumed by animals even when other foods are available (Medeiros *et al.* 2002). Even though *M. rigida* is considered a palatable plant, control sheep reduced the ingestion which was also observed in goats (Barbosa *et al.* 2008). Control sheep reduced ingestion in the conditioning trial after consuming it for 105 s the first day. Consumption by control sheep increased in the persistence trials up to 75 s on day 55 after which consumption declined. This suggests a natural aversion may have been created as a consequence of negative post-ingestive feedback. The toxic compound is not known but the suspected toxin is fluoroacetate (Pacífico da Silva *et al.* 2008). It was found that fluoroacetate injected at a strong but not lethal dose to rats created strong taste aversions (Nachman and Hartley 1975) and so could presumably reduce plant intake by animals if eaten in sublethal amounts. However, it is not possible now to assume whether fluoroacetate or another toxin is the responsible for the negative post-ingestive feedback.

## Conclusions

*Mascagnia rigida* is the most important toxic plant in the semiarid region of northeastern Brazil. Lithium chloride proved effective in creating an aversion to this plant in sheep. However, the efficiency of the treatment on open range and the persistence of the aversion must be evaluated before it is recommended to farmers.

## References

Barbosa RR, Pacífico da Silva I, and Soto-Blanco B (2008). Development of conditioned taste aversion to *Mascagnia rigida* in goats. *Pesquisa Veterinária Brasileira* 28:571-574.

Garcia J and Holder MD (1985). Time, space and value. *Human Neurobiology* 4:81-89.

Ginane C and Dumont B (2006). Generalization of conditioned food aversions in grazing sheep and its implications for food categorization. *Behavioural Processes* 73:178-186.

Kimball BA, Provenza FD, and Burritt EA (2002). Importance of alternative foods on the persistence of flavor aversions: implications for applied flavor avoidance learning. *Applied Animal Behavior Science* 76:249-258.

Medeiros RMT, Geraldo Neto SA, Barbosa RC, Lima EF, and Riet-Correa F (2002). Sudden bovine death from *Mascagnia rigida* in Northeastern Brazil. *Veterinary and Human Toxicology* 44:286-288.

Nachman M and Hartley PL (1975). Role of illness in producing learned taste aversions in rats: A comparison of several rodenticides. *Journal of Comparative and Physiological Psychology* 89:1010-1018.

Pacífico da Silva I, Lira RA, Barbosa RR, Batista JS, and Soto-Blanco B (2008). Intoxicação natural pelas folhas de *Mascagnia rigida* (Malpighiaceae) em ovinos. *Arquivos do Instituto Biológico* 75:229-233.

Ralphs MH (1997). Long term retention of aversions to tall larkspur in naive and native cattle. *Journal of Range Management* 50:367-370.

Ralphs MH (2001). Plant toxicants and livestock: prevention and management. In *Foodborne Disease Handbook* (YH Hui, RA Smith, DG Spoerke Jr, eds), 2nd edn, pp. 441-470. Marcel Dekker, New York.

Ralphs MH and Provenza FD (1999). Conditioned food aversions: principles and practices, with special reference to social facilitation. *Proceedings of the Nutrition Society* 58:813-820.

Ralphs MH and Stegelmeier BL (1998). Ability of apomorphine and lithium chloride to create food aversions in cattle. *Applied Animal Behavior Science* 56:129-137.

Tokarnia CH, Döbereiner J, and Canella CFC (1961). Intoxicação por um 'tingui' (*Mascagnia rigida* Griseb.) em bovinos no Nordeste do Brasil. *Arquivos do Instituto de Biologia Animal* 4:203-215.

# Chapter 111

# Amended Method of Averting Cattle to Yellow Tulp (*Moraea pallida*)

L.D. Snyman, R.A. Schultz, and L. Labuschagne

*Toxicology Section, ARC-Onderstepoort Veterinary Institute, Private bag X05, Onderstepoort, 0110 South Africa*

## Introduction

Conditioned feed aversion, a means of teaching stock to avoid poisonous plants (Ralphs and Provenza 1999; Ralphs *et al.* 2001), has been successfully applied to prevent yellow tulp poisoning in cattle under experimental conditions (Snyman *et al.* 2003, 2007). The standard method of inducing aversion, however, is impractical and costly to apply when large numbers of cattle have to be treated for the following reasons. Firstly, epoxyscillirosidin, the cardioactive bufadienolide and the natural aversive substance of yellow tulp (*Moraea* [=*Homeria*] *pallida*) used in combination with lithium chloride, is expensive to prepare. Pilot trials using sheep (LD Snyman, unpublished data) indicated that an intramuscular injection of one-fifth the epoxyscillirosidin dose might have the same effect. Secondly, the aversive substance lithium chloride, due to its highly reactive nature, has to be administered via a stomach tube to ensure the chemical enters the rumen and mixes with the rumen fluid which is a very cumbersome procedure. Lithium carbonate, another lithium salt which is less soluble in water, can easily be drenched without deleterious effects to the esophagus. Thirdly, the identification factor extracted from yellow tulp with hexane specifically to prevent extracting the toxin as well can also be extracted with propylene glycol (the carrier used to administer the residue of the hexane extract to the animal). We evaluated each of these three alternatives separately then combined them to avert cattle to yellow tulp in a field grazing trial.

## Methods and Results

### Aversion treatment

Nine- to eleven-month-old Nguni steers from a non tulp-infested grazing area were used in this investigation. The steers were basically treated as follows to induce aversion in the different trials: steers were initially accustomized to *Eragrostis curvula* hay in a corral for 2 weeks whereafter they were individually fed in pens with *E. curvula* hay for another week in order to get familiar with the hay and environment prior to testing. The steers were

©CAB International 2011. *Poisoning by Plants, Mycotoxins, and Related Toxins* (eds F. Riet-Correa, J. Pfister, A.L. Schild, and T.L. Wierenga)

fasted overnight (18 h) and then presented with a specific amount of the novel feeds listed below. Directly after total consumption they were treated with the aversive substance listed below and withheld from feed and water for another 6 h whereafter they again received *E. curvula* hay and water *ad lib.* They were evaluated for intake of the feed they were averted to the next morning and for one to several day(s) thereafter. In some trials intake of the hay was also measured.

*Intramuscular injection with epoxyscillirosidin*

Three steers were each presented with 500 g feed pellets (production quality) and injected (i.m.) with 0.00125, 0.0025, and 0.005 mg epoxyscillirosidin/kg BW, respectively, after the pellets had been consumed. Another seven steers were presented on another occasion with 500 g kikuyu grass (*Pennisetum clandestinum*) and each one injected (i.m.) with a different dose of epoxyscillirosidin after the grass had been consumed: 0.01, 0.015, 0.020, 0.025, 0.030, 0.040, and 0.050 mg epoxyscillirosidin/kg BW. The steers were evaluated for intake of feed pellets (500 g) or kikuyu grass (500 g) the next morning and monitored for clinical signs of epoxyscillirosidin poisoning, namely posterior paresis and/or inhibition of rumen movements. Total or partial refusal of the feed pellets or grass presented to the steers was noted as aversion. The aversive and toxic effects induced by the different doses of epoxyscillirosidin are given in Table 1.

**Table 1.** Aversive and toxic effects of different doses of epoxyscillirosidin (one steer per dosage) administered intramuscularly to steers.

| Epoxyscillirosidin (mg/kg BW) | Aversive effect[1] | Toxic effect[2] |
|---|---|---|
| 0.00125 | No aversion | None |
| 0.0025 | No aversion | None |
| 0.005 | Aversion | None |
| 0.010 | Aversion | None |
| 0.015 | Aversion | None |
| 0.020 | Aversion | None |
| 0.025 | Aversion | Poisoning (moderate) |
| 0.030 | Aversion | Poisoning (mild) |
| 0.040 | Aversion | Poisoning (moderate) |
| 0.050 | Aversion | Poisoning (severe) |

[1]Aversive effect was tested utilizing feed pellets (dosages ranged between 0.00125 and 0.005 mg/kg BW) and kikuyu grass (dosages ranged between 0.01 and 0.05 mg/kg BW)
[2]Toxic effect manifested by posterior paresis and/or inhibition of rumen movements

The data in Table 1 show that aversion could not be induced with doses lower than 0.005 mg/kg while doses higher than 0.020 mg/kg were toxic to the animals. The dose of 0.005 mg/kg (a 5× reduction compared to the amount drenched previously) was selected as the optimum dose as a tendency of decreased hay intake was noticed with increasing aversion doses.

*Drenching with lithium carbonate*

Various doses of lithium carbonate were investigated for averting steers to feed pellets (500 g). Six steers were each drenched with a different dose of lithium carbonate: 75, 100, 125, 150, 175, and 200 mg/kg BW. Intake of 500 g feed pellets presented the day after aversion treatment was measured over a 3 day period. Intake (kg/day) of *E. curvula* hay presented *ad libitum* (5 kg/day) was measured over the same period. The results are given in Figure 1.

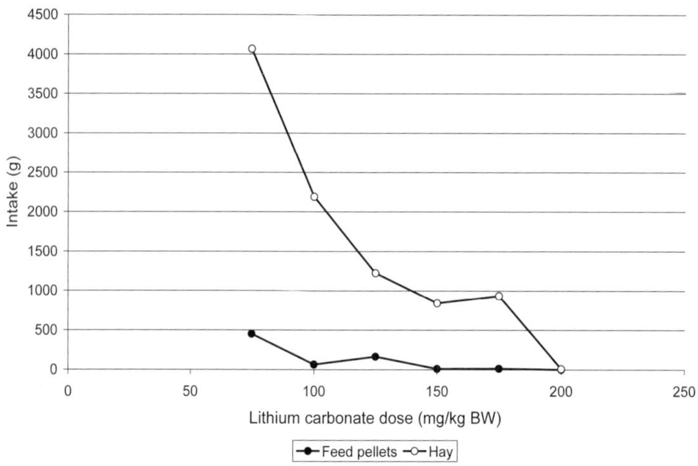


**Figure 1.** Effect of lithium carbonate dose on intake of feed pellets (500 g) and hay (5 kg/day).

Intake of feed pellets varied between zero and 33% for dosages between 100 mg and 200 mg/kg BW indicating total aversion to the feed pellets at most dosages. The steer drenched with 75 mg lithium carbonate, however, consumed 85% of the aversive mixture on Day 1 and the full amount of pellets offered within 3 days. Increasing lithium carbonate doses, however, also reduced the consumption of hay to which the steers were familiar with. As feed pellets are very palatable and partial avoidance of it may still represent strong aversion to an unpalatable plant like yellow tulp, a dosage of 80 mg/kg BW was selected as optimum for averting cattle to yellow tulp.

*Combination of epoxyscillirosidin and lithium carbonate*

A steer was tested for aversion to feed pellets by treatment with a combination of the optimum doses of epoxyscillirosidin (0.005 mg/kg BW) and lithium carbonate (80 mg/kg BW). The effect of this aversion treatment on intake of feed pellets and hay is given in Figure 2. The results indicate that the treatment caused total aversion for 7 days without suppressing hay intake, suggesting that this treatment might be suitable for averting cattle to yellow tulp.

*Propylene glycol extract prepared from low-toxic yellow tulp as identification factor*

The identification factor was prepared by homogenizing 100 g low-toxic yellow tulp (*M. pallida*) collected in the Northern Cape region of the country with 660 ml propylene glycol. A combination of 90 ml of the separated liquid fraction, representing 15 g fresh yellow tulp, together with 10 g dry milled low-toxic yellow tulp placed in the mouth was used as identification factor for each of the steers. Its effectiveness as identification factor in resembling yellow tulp was investigated with four Nguni steers. Prior to aversion treatment the steers were accustomized to freshly cut oats by offering them *ca.* 500 g every morning for 7 days until 2 days prior to aversion treatment on Day 0. During the last offering the steers were tested for intake of 450 g green oats which was totally consumed by all the steers. The steers were fasted the night before aversion treatment by keeping them together in a pen without hay. They were averted to yellow tulp the next morning by injecting (i.m.) 0.01 mg epoxyscillirosidin/kg BW followed by *ca.* 10 g dried low-toxic yellow tulp placed in the mouth and slow drenching with 90 ml propylene glycol extract. After being withheld from food and water for 6 h they were moved back to their pens where they were supplied with hay and water again. Each steer was tested for aversion to yellow tulp the next morning by presenting them with 500 g of a toxic yellow tulp-oats mixture (1:9) followed by testing for willingness to consume pure oats 1 h later (Table 2).

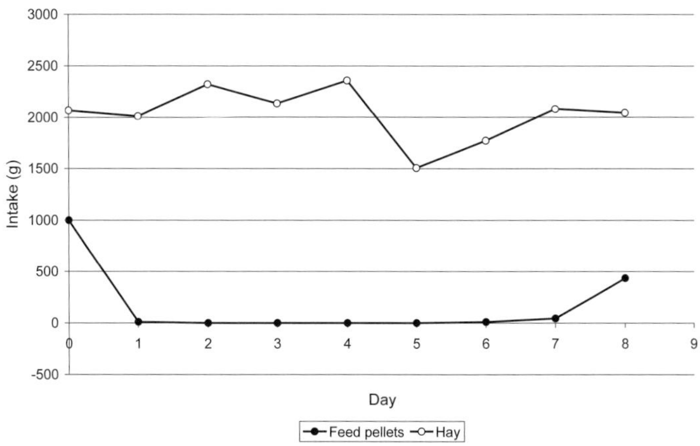


**Figure 2.** Effect of aversion treatment on the intake of feed pellets and hay by one steer.

**Table 2.** Intake of a yellow tulp-oats mixture and oats only by steers averted to yellow-tulp on Day 0 by using a propylene glycol extract prepared from low-toxic yellow tulp as the identification factor.

| | Day -2 | Day 1 | | Day 2 | |
|---|---|---|---|---|---|
| Steer | Oats (500 g) | Tulp-oats mixture (1:9) (500 g) | Oats (100 g) | Tulp-oats mixture (1:9) (500 g) | Oats (100 g) |
| 1 | Consume | Refuse | Consume | Refuse | Consume |
| 2 | Consume | Consume[1] | Consume | | |
| 3 | Consume | Refuse | Consume | Refuse | Consume |
| 4 | Consume | Refuse | Refuse | Refuse | Consume |

[1]The steer refused pure tulp when offered directly thereafter

The data in Table 2 show refusal of the tulp-oats mixture for 2 consecutive days by three of the four steers. The fourth steer consumed the tulp mixture but refused pure tulp thereafter which was consumed by all control animals. All the steers, however, consumed oats, showing that aversion was directed at the tulp specifically. The results indicate that the propylene glycol extract from low-toxic yellow tulp along with the dry milled yellow tulp placed in the mouth was effective as an identification factor in averting the steers to toxic yellow tulp.

**Inducing aversion to yellow tulp by the amended treatment**

Two replications were performed with six averted and six control steers in each replication. The animals, adapted to *E. curvula* hay as previously described, were accustomed to grazing in a non-tulp pasture adjacent to the tulp-infested pasture 1 h a day for 5 days until 2 days prior to aversion treatment on Day 0. After being fasted for 18 h overnight, each steer was injected (i.m.) with epoxyscillirosidin (0.005 mg/kg BW) followed by drenching with lithium carbonate (80 mg/kg BW) shaken up in water. This was followed by slow drenching with a propylene glycol extract (15 g fresh low-toxic yellow tulp/100 ml propylene glycol) and *ca.*10 g milled low-toxic yellow tulp placed in the mouth. Feed and water were withheld for 6 h whereafter the animals were moved into another pen where they had *ad lib* access to *E. curvula* hay and water. The next morning (Day 1) averted and control animals were moved together onto a yellow tulp (flowering stage)-infested grass grazing area at a confined experimental site and monitored for clinical signs of tulp poisoning. Animals remained on the pasture for 2 days but were removed when showing signs of poisoning. The results are presented in Table 3.

**Table 3.** Number of averted and control steers poisoned on a yellow tulp-infested grass pasture.

| | Averted group | | Control group | |
|---|---|---|---|---|
| Replication | Unaffected | Poisoned | Unaffected | Poisoned |
| 1 | 6 | 0 | 3 | 3 |
| 2 | 6 | 0 | 0 | 6 |
| Total | 12 | 0 | 3 | 9 |

The data in Table 3 show that 9 of the 12 control steers were poisoned during the two replications compared to none of the averted steers, indicating effective aversion of the steers to yellow tulp while grazing.

## Discussion

The amended method which was easier and cheaper to apply induced an effective aversion to yellow tulp in cattle on pastures. The combination of epoxyscillirosidin and lithium carbonate is surmised to be effective as the aversive effect of lithium occurs rapidly while that of epoxyscillirosidin is delayed and more prolonged, both properties that favor the induction of aversion (Riley and Tuck 1985). The method needs to be tested over a longer period as natural aversion to yellow tulp while grazing which follows the artificially induced aversion to yellow tulp might temporarily influence consumption of grass with detrimental economical implications to the stock owner.

## Acknowledgements

This project was financially supported by the Gauteng Department of Agriculture Conservation and Environment (GDACE) through the Directorate: Technology

Development and Support (TDS) and the North-West Province Directorate of Veterinary Services (Department of Agriculture, Conservation, Environment and Tourism).

Petrus Senosha, Richard Matshinga, and Johannes Molefe are thanked for devoted technical assistance.

## References

Ralphs MH and Provenza DP (1999). Conditioned food aversions: principles and practices, with special reference to social facilitation. *Proceedings of the Nutrition Society* 58:813-820.

Ralphs MH, Provenza FD, Pfister JA, Graham D, Duff GC, and Greathouse G (2001). Conditioned Food Aversion: From Theory to Practice. *Rangelands* 23:14-18.

Riley A and Tuck D (1985). Conditioned taste aversions: a behavioral index of toxicity. *Annals of the New York Academy of Science* 443:272-292.

Snyman LD, Schultz RA, Joubert JPJ, Basson KM, and Labuschagne L (2003). Conditioned feed aversion as a means to prevent tulp (*Homeria pallida*) poisoning in cattle. *Onderstepoort Journal of Veterinary Research* 70:43-48.

Snyman LD, Schultz RA, Joubert JPJ, Labuschagne L, and Basson KM (2007). Conditioned Feed Aversion as a Means to Prevent Tulp (*Moraea simulans* and *Moraea pallida*) Poisoning of Cattle under Natural Grazing Conditions. In *Poisonous Plants, Global Research and Solutions* (KE Panter, TL Wierenga, and JA Pfister, eds), pp. 420-422. CABI Publishing, Wallingford, UK.

# HERBALS

# Chapter 112

# Reproductive Study of *Chenopodium ambrosioides* Aqueous Extract in Rats

I.U. Medeiros[1], I.M.F. Figueiredo[1], V.F.M. Junior[1], C.N. Oliveira[2], and A. Schwarz[3]

*[1]Pharmacy student from Universidade Federal do Rio Grande do Norte; [2]Pathology Department from Universidade Federal do Rio Grande do Norte; [3]Clinical and Toxicological Analysis Department from Universidade Federal do Rio Grande do Norte, Av, Gal. Gustavo Cordeiro de Farias, s/n, CEP 59010-180, Natal-RN, Brazil*

## Introduction

*Chenopodium ambrosioides* L. (Chenopodiaceae) is an herb known in folk medicine as 'mastruz'. This plant is frequently used for its properties as an anti-helminthic, expectorant, tonic, emmenagogue, abortive agent, and to help heal superficial injuries and fractured bones. It is also used as a pesticide due to its well known repellent and insecticide proprieties (Jorge *et al.* 1986; Mazzonetto and Vendramim 2003).

Studies have demonstrated that *C. ambrosioides* is rich in flavonoids and terpenes, compounds that may promote a range of therapeutic benefits such as antioxidant and chemopreventative effects (Kiuchi *et al.* 2002; DeFeudis *et al.* 2003; Takeoka and Dao 2003; Liu 2004).

A hydroethanol extract from fresh leaves of this plant showed antitumoral activity by inhibition of Erlich tumor cell growth in mice treated with 5 mg (i.p.) of fresh leaves/kg for 2 days (Nascimento *et al.* 2006). Cruz *et al.* (2007) verified that the same extract promoted increased cell proliferation in secondary lymphoid organs such as spleen and lymph nodes. Adesanya *et al.* (2007) showed that an ethanol extract of the plant did not promote mutagenic alterations in lymphocytic chromosomes of mice treated with four different concentrations (2.5, 5, 10, and 20 mg/kg) of the extract. An *in vitro* study developed with two aqueous extracts by infusion and by decoction obtained from fresh leaves demonstrated genotoxic activity with human lymphocytic cell culture (Gadano *et al.* 2002).

Phytochemical analysis of the active compounds of this plant needs to be developed since there are few studies in literature. The known active compounds of *C. ambrosioides* are terpene hydrocarbons (20%), α-terpenes, limonene, p-cimene, and saponins (Vanaclocha and Cañigural 2003). The essential oil (0.6-1% of the plant) consists of basically monoterpenes, mainly ascaridol which is 70% of this fraction (Sagrero-Nieves and Bartley 1995). The oil intake is not recommended due to its toxicity, causing renal parenchyma irritation and death by bulbar respiratory failure after oral administration (Vanaclocha and Cañigueral 2003).

©CAB International 2011. *Poisoning by Plants, Mycotoxins, and Related Toxins* (eds F. Riet-Correa, J. Pfister, A.L. Schild, and T.L. Wierenga)

In traditional folk medicine, an infusion of leaves is used as an abortive agent but no studies have been done to document this effect. Therefore, the aim of this study was to investigate reproductive alterations in rats.

## Material and Methods

Fresh leaves of *C. ambrosioides* were collected in April 2008 at Natal (Rio Grande do Norte, Northeast Brazil; GPS location: lat. -5.900000, long. -35.246333). The plant was identified by Prof Iracema Loiola and a voucher specimen was deposited at Herbarium 'Parque das Dunas' of Federal University of Rio Grande do Norte (registration # 8,168). At the Toxicology Laboratory of the same university the fresh leaves were selected. The aqueous extract was obtained by extracting fresh leaves with distilled water (100ºC). The leaves (200 g) were maintained in boiling water (1 l) for 10 min. The filtered portion was stored in small glass recipients and kept at -20ºC until the moment of use.

Ten female 90-day-old Wistar rats (180-250g) were used. Adult male Wistar rats were employed for copulation. The animals were housed in plastic cages measuring 40×50×20 cm under controlled temperature (22±2°C) with a 12:12 light:dark schedule and free access to food (Purina®) and water. The animals were maintained in accordance with the Ethical Principles in Animal Research adopted by National Ethic Research Committee (CONEP/MS) and approved by the Onofre Lopes University Hospital Research Ethical Committee (protocol no 169/07).

Two female rats were placed together with one male in the afternoon. The next morning females showing evidence of mating (vaginal smear with sperm = gestation day 1) were housed in pairs in plastic cages measuring 40×50×20 cm. The dams were randomly distributed into control and experimental groups ($n$=5/group). The experimental group received the aqueous extract (100 mg/kg) by gavage from gestational day (GD) 1 to GD 20. The control group received tap water by gavage for the same period.

During treatment body weight, food consumption, and water intake were recorded. The body weight gain was also calculated. On day 20 of gestation the dams were anesthetized with ethyl ether. Blood was collected for evaluation of biochemical parameters (ALT, AST, urea, creatinine) by cardiac puncture. After euthanasia by cervical dislocation the ovaries and uterus were removed. The ovaries were weighed and the number of corpora lutea was recorded. The fetuses were counted, removed, weighed, and examined for any malformations. The uterus was weighed. The number of implantation sites, reabsorptions, and dead and live fetuses per dam were recorded. Also, the dam's organ weight was recorded (liver, kidney, spleen, pancreas, uterus, ovaries) and then tissue portions were fixed in 10% formalin for histopathological studies. The reproductive performance was investigated by recording litter weight, total number of fetuses per litter, number of male and female fetuses, fetal body weight for males and females, and number of dead fetuses.

The data were analyzed by the Student $t$ test and by the Dunnett's *posthoc* test when necessary. In all cases results were considered significant when $P < 0.05$.

## Results and Discussion

Statistically significant differences were observed in body weight, body weight gain, and water and food intake of the experimental group when compared to control group. Experimental dams had decreased water intake during on days 5-7 ($P$ < 0.001) and increased water intake on days 7-9 ($P < 0.05$), 9-11 ($P < 0.05$), and 13-15 ($P < 0.05$) (Table 1). Experimental dams showed increased food consumption on days 7-9 ($P < 0.05$) and days 13-15 ($P < 0.05$). The experimental dams also had increased body weight on gestation days 1, 10, 11, 12, and 14 (Figure 1). The statistical analysis revealed increased weight gain of experimental dams on days 7-9 ($P < 0.05$) and days 13-15 ($P < 0.05$) and reduced body weight gain on days 17-19 ($P < 0.05$) compared to the control group (Table 2). However, body weight gain was not different ($P > 0.05$) between the groups when considering the total treatment period (days 1 to 20).

**Table 1.** Food (g) and water (ml) intake of control rats and treated with *C. ambrosioides* aqueous extract (1000 mg/kg /day) from gestation day 1 to day 20 (mean±SEM; *n*=5).

| Interval of days | Food intake | | Water intake | |
|---|---|---|---|---|
| | Control | Experimental | Control | Experimental |
| 01-03 | 25.58±0.01 | 23.87±5.98 | 45.98±0.42 | 48.40±12.15 |
| 03-05 | 28.96±083 | 26.20±2.40 | 56.00±1.23 | 67.18±9.22 |
| 05-07 | 31.00±0.92 | 25.44±5.02 | 58.80±0.73 | 35.18±2.54*** |
| 07-09 | 28.58±0.42 | 33.89±2.02* | 55.96±1.62 | 66.96±5.12* |
| 09-11 | 31.98±1.03 | 41.25±8.95 | 53.18±1.30 | 62.76±3.83* |
| 11-13 | 35.50±0.67 | 37.05±2.55 | 68.38±1.79 | 72.40±3.97 |
| 13-15 | 35.28±1.36 | 40.44±2.38* | 70.80±1.72 | 84.40±5.15* |
| 15-17 | 40.58±1.91 | 41.48±2.37 | 82.02±4.09 | 94.60±6.40 |
| 17-19 | 44.34±0.58 | 43.40±4.42 | 86.00±0.71 | 94.02±8.77 |

*$P < 0.05$, ***$P < 0.001$ – Student *t* test followed by *posthoc* Dunnett's test.

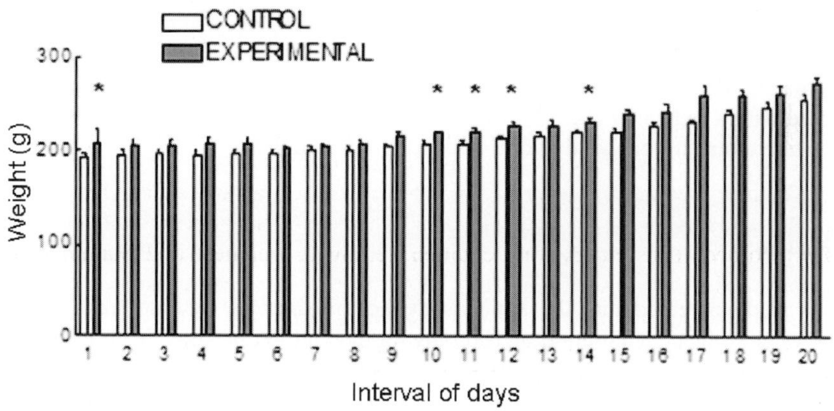


**Figure 1.** Body weight (g) of control rats and rats treated with *C. ambrosioides* aqueous extract (1000 mg/kg/day) from GD01 to GD20 (Student *t* test followed by *posthoc* Dunnett's test: *$P < 0.05$; mean±SEM; *n*=5).

Alterations in organ weight were not observed. Also, no alterations in serum ALT, AST, gamma-GT, cholesterol, urea, or creatinine were observed between experimental and control groups supporting the hypothesis that the aqueous extract did not interfere with hepatic and renal systems of pregnant rats treated with 1000 mg/kg during gestation.

**Table 2.** Body weight gain (g) of rats treated or not (control group) with *C. ambrosioides* aqueous extract (1000mg/kg/day) from GD01 to GD20 (mean ± SEM; *n*=5/group).

| Interval of days | Control | Experimental |
|---|---|---|
| 01-03 | 4.00 ± 1.52 | -5.06 ± 7.14 |
| 03-05 | 0.34 ± 1.85 | 3.96 ± 2.95 |
| 05-07 | 4.20 ± 2.38 | -1.74 ± 2.33 |
| 07-09 | 4.54 ± 1.97 | 9.90 ± 1.58* |
| 09-11 | 2.66 ± 1.21 | 4.36 ± 1.59 |
| 11-13 | 8.20 ± 1.47 | 5.20 ± 4.07 |
| 13-15 | 4.00 ± 1.01 | 10.84 ± 2.48* |
| 15-17 | 8.04 ± 1.64 | 10.96 ± 2.81 |
| 17-19 | 17.82 ± 0.93 | 11.80 ± 1.47** |
| 01-20 | 64.48 ± 4.15 | 62.34 ± 3.09 |

*$P < 0.05$; ** $P < 0.01$ – Student *t* test followed by *posthoc* Dunnett's test.

The statistical analysis revealed no alterations in any reproductive parameter evaluated for the fertility study. The pre- and post-implantation levels along with the number of implantation sites, number of corpora lutea, and number of live fetuses were not different between experimental and control groups.

The reproductive performance study, evaluated by observing the parameters pregnancy duration (days), litter weight, number of male and female pups, and number of dead pups, showed that the aqueous extract at this concentration and administered during gestation did not impair gestation and reproductive performance. Dead fetuses were not observed in the experimental group, suggesting that the aqueous extract (1000 mg/kg) was well tolerated by the dams and was not fetotoxic.

The histopathological study revealed that the *C. ambrosioides* aqueous extract did not cause lesions in liver, kidney, pancreas, adrenal glands, or spleen at this dose (1000 mg/kg) in pregnant rats. Only a very discrete congestion in kidneys and liver was detected in experimental dams. A previous study conducted in our lab revealed that adult female rats treated with a reduced dose (500 mg/kg) of the same aqueous extract for 30 days did not show alterations in renal and liver tissues (unpublished data).

## Conclusions

This study revealed that the aqueous extract obtained by boiling fresh leaves in distilled water at the concentration of 1000 mg/kg/day did not promote maternal and fetal toxicity nor did it impair reproductive performance in rat dams. The fertility study suggests that the aqueous extract (1000 mg/kg/day) administered during gestation to rats does not impair fertility or negatively impact gestation in rats because no alterations were observed in pre and post implantations or any other parameter.

## References

Adesanya SA, Sowemimo AA, Fakoya FA, Awopetu I, and Omobuwajo OR (2007). Toxicity and mutagenic activity of some selected Nigerian plants. *Journal of Ethnopharmacology* 113:427-432.

Cruz GVB, Pereira PVS, Patrício FJ, Costa GC, Souza SM, Frazão JB, Aragão-Filho WC, Maciel MCG, Silva LA, Amaral FMM, Barroqueiro ESB, Guerra RNM, and Nascimento FRF (2007). Increase of cellular recruitment, phagocytosis ability and nitric oxide production induced by hydroalcoholic extract from *Chenopodium ambrosioides* leaves. *Journal of Ethnopharmacology* 111:148-154.

DeFeudis FV, Papadopoulos V, and Drieu K (2003). *Ginkgo biloba* extracts and cancer: a research area in its infancy. *Fundamental Clinical Pharmacology* 17:405-407.

Gadano A, Gurni A, López P, Ferraro G, and Carballo M (2002). *In vitro* genotoxic evaluation of the medicinal plant *Chenopodium ambrosioides* L. *Journal of Ethnopharmacology* 81:11-16.

Jorge LIF, Ferro VO, and Koschtschak MRW (1986). Diagnose comparativa das especies *Chenopodium ambrosioides* L. (Erva-de-santa-maria) e *Coronopus didymus* (l.) Sm (mastruço). Principais caracteristicas morfo-histologicas e quimicas. *Brazilian Journal of Pharmacognosy* 1(2):143-53.

Kiuchi F, Itano Y, Uchiyama N, Honda G, Tsubouchi A, Nakajima-Shimada J, and Aoki T (2002). Monoterpene hydroperoxides with trypanocidal activity from *Chenopodium ambrosioides*. *Journal of Natural Products* 65:509-512.

Liu RH (2004). Potential synergy of phytochemicals in cancer prevention: mechanism of action. *Journal of Nutrition* 134:3479S.

Mazzonetto F and Vendramim JD (2003). Efeito de substâncias de origem vegetal sobre *Acanthoscelides obtectus* (Say) (Col.: Bruchidae) em feijão armazenado. *Neotropical Entomology, Esalq/USP* 32:1.

Nascimento FRF, Cruz GVB, Pereira PVS, Maciel MCG, Silva LA, Azevedo APS, Barroqueiro ESB, and Guerra NM (2006). Ascitic and solid Ehrlich tumor inhibition by *Chenopodium ambrosioides* L. treatment. *Life Sciences* 78: 2650-2653.

Sagrero-Nieves L and Bartley JP (1995). Volatile constituents from the leaves of *Chenopodium ambrosioides* L. *Journal of Essential Oil Research* 7: 221-223.

Takeoka GR and Dao LT (2003). Antioxidant constituents of almond (*Prunus dulcis* (Mill.) D.A.Webb.) hulls. *Journal of Agriculture and Food Chemistry* 51:496-501.

Vanaclocha B and Cañigueral S (2003). *Fitoterapia. Vademécum de Prescripción,* 4th edn, p. 222. Ed. Masson, Spain.

# Chapter 113

# Investigation of *Cereus jamacaru* Ethanol Extract Effects in Rats

F.M. Queiroz[1], I.U. Medeiros[2], D.M.N. Sousa[1], R.N.A. Marinho[1], P.R.S. Pereira[1], C.N. Oliveira[3], and A. Schwarz[4]

[1]*Pharmacy students from Universidade Federal do Rio Grande do Norte;* [2]*Post-graduation student from Post-graduation Program in Pharmaceutical Sciences-Universidade Federal do Rio Grande do Norte;* [3]*Pathology Department from Universidade Federal do Rio Grande do Norte;* [4]*Clinical and Toxicological Analysis Department from Universidade Federal do Rio Grande do Norte, Av. Gal. Gustavo Cordeiro de Farias, S/N, CEP 59010-180, Natal-RN, Brazil*

## Introduction

The cactus *Cereus jamacaru* Mill. (Cactaceae), popularly known as 'mandacaru', is a drought-resistant species commonly found in the Brazilian caatinga in northeastern Brazil and northern Minas Gerais State. This species has irregular stalks, big white flowers, and red juicy fruits (Lima 1996). The *C. mandacaru* cladodes are frequently employed as a food source for livestock (bovine, caprine, and ovine) in drought periods in northeastern Brazil (Braga 1976). In folk medicine *C. jamacaru* roots and cladodes are employed as diuretics. The cladode is also used to reduce arterial blood pressure. A syrup obtained from all parts of the plant is employed to treat ulcers, to avoid scurvy, and to treat respiratory tract affections such as coughs and bronchitis (Scheinvar 1985).

A wide range of compounds–phenylethylamine alkaloids, tyramine, hordenine, mescaline and lophophorine–were found in previous studies in cactus (Starcha *et al.* 1999). Brhun and Lindgren (1976) identify tyramine and 2-hydroxyphenylethylamine in *C. mandacaru* fresh cladodes. Burret *et al.* (1982) observed absence of flavones and presence of kampferol and methyl-3-flavonoids in this species. Steroidal compounds such as ß-sitosterol were identified in cactus (Salt *et al.* 1987). A study detected by carbon nuclear magnetic resonance the compounds ß-sitosterol and tyramine in a crude ethanol extract obtained from *C. jamacaru* cladodes (Davet 2005). Another study detected a methionine-rich protein in *C. jamacaru* seeds (Aragão *et al.* 2000).

Once inside the central nervous system, the phenylethylamine alkaloids mimic dopamine and noradrenaline neurotransmitter effects. This action can promote important behavioral alterations such as the stereotyped behavior observed in rats (Smith 1978; Ortmann *et al.* 1984; Youdim and Tipton 2002; Kitanaka *et al.* 2005). The possible presence of phenylethylamine alkaloids in *C. jamacaru* and the frequent use of this species in folk medicine are points of concern. Considering these facts, the objective of this

©CAB International 2011. *Poisoning by Plants, Mycotoxins, and Related Toxins* (eds F. Riet-Correa, J. Pfister, A.L. Schild, and T.L. Wierenga)

research was to study in adult rats the sub-acute toxicity and behavior alterations of an ethanolic extract obtained from *C. jamacaru* cladodes to contribute to a rational and safe use of this plant in folk medicine or as food source for livestock.

## Material and Methods

The *C. jamacaru* cladodes used in this study were collected in May and June 2008 at Natal (Rio Grande do Norte, northeastern Brazil; GPS location lat −5.900000, long −35.246333). The plant was identified by Prof Iracema Loiola and a voucher specimen was deposited at Herbarium 'Parque das Dunas' of Universidade Federal do Rio Grande do Norte (registration # 4,802). At the Toxicology Laboratory of Universidade Federal do Rio Grande do Norte the fresh cladodes were sliced after thorn elimination, oven dried at 40-50°C, and then milled. The crude ethanol extract was obtained by macerating the dried and milled cladodes (2300 g) with commercial ethyl alcohol 96% (7000 ml) for 7 days. After filtration the solution obtained was concentrated under reduced pressure and reserved. To the vegetal material was again added 96% ethyl alcohol for a maceration period of 24 h. The filtrated portion after concentration at reduced pressure was added to the first. This methodology with the vegetal material was repeated two more times, resulting in four extractions. The combined extractions were concentrated for the experimental extract.

In this study 40 Wistar rats (250-270 g) about 90 days old were used (20 from each gender). The animals were housed in plastic cages (5/cage) measuring 40×50×20 cm under controlled temperature (22±2°C) with a 12:12 light:dark schedule and free access to food and water. The animals used in this study were maintained in accordance with the Ethical Principles in Animal Research adopted by National Ethic Research Committee (CONEP/MS) and approved by the Onofre Lopes University Hospital Research Ethical Committee (protocol no 170/07).

The rats were split into eight groups of five each: four groups of males and four groups of females. The six experimental groups, three male and three female, received by gavage for 21 days 28, 42, or 56 mg/kg/day *C. jamacaru* ethanol extract. The two control groups (males and females) received tap water by gavage during the treatment period. In folk medicine approximately 100 g of the cladode in 1000 ml of water are used to prepare the infusion and this volume is commonly consumed in a single day per person. Considering that 70 kg is the median weight for an adult human the dose corresponds to 1.4 g/kg/day. In this work the experimental animals received by gavage 28, 42, or 56 mg/kg/day of the ethanol extract.

During treatment body weight, food intake, and water ingestion were determined every 2 days. The body weight gain was calculated employing the body weight values. All observations were performed between 8-12 h. At day 22 the animals were observed at the open field and in the elevated plus maze for 5 min in each apparatus, always between 12-17 h. Locomotion, rearing, immobility(s), and number of fecal bolus were evaluated in the open field as described by Broadhurst (1960). The number of entries in open and closed arms and the time(s) spent in each arm were measured in the elevated plus maze.

At the end of the behavioral analysis the animals were anesthetized by ethyl ether inhalation and the blood was collected by cardiac puncture. The serum obtained after blood centrifugation was used to evaluate the biochemical parameters AST, ALT, urea, and creatinine. After blood collection the sedated animals were euthanized by cervical disjoint. The liver, kidney, pancreas, and spleen were exteriorized and the weights were recorded. Tissue portions were fixed in 10% formalin for histopathological studies.

The data were analyzed by the analysis of variance ANOVA and Tukey Kramer *posthoc* test when necessary. In all cases, results were considered significant when $P < 0.05$.

## Results and Discussion

The statistical analysis revealed that the ethanol extract did not promote alterations in body weight, body weight gain, and food and water intake at the three different doses. The body weight gain data is presented in Figure 1. Only liver weight:body weight ratio of female rats treated with the 56 mg/kg/day dose were reduced ($P < 0.05$) when compared to control group as observed in Table 1.

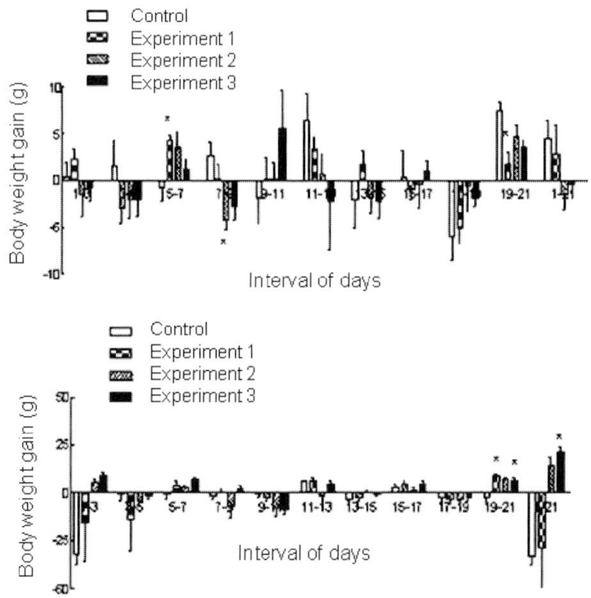

**Figure 1.** Body weight gain of female (A) and male (B) rats treated with 0, 28, 42, or 56 mg/kg/day of the ethanol extract of *C. jamacaru* cladodes for 21 days (mean ± SEM). ANOVA followed by Tukey-Kramer *posthoc* test (*$P < 0.05$).

The ANOVA revealed reduced ALT levels in serum of females treated with 28 and 42 mg/kg/day when compared to control group. Also, reduced AST levels were observed in the serum of the males and females treated with 28 and 56 mg/kg/day when compared to control groups. Only the females treated with 42 mg/kg/day revealed elevated AST levels in serum. Males treated at 28 mg/kg/day presented elevated urea levels when compared to control males and females treated at 56 mg/kg/day dose presented elevated creatinine level when compared to the control group. These data are shown in Table 2.

The histopathologic study revealed that the ethanol extract did not promote alterations in liver, pancreas, and spleen. The lack of alterations in the liver of the experimental females, especially those treated at 56 mg/kg/day dose, was important to confirm that the extract at this dose did not promote hepatotoxicity since these females presented reduced liver weight/body weight ratio. However, the females treated at 56 mg/kg/day presented mild tumefaction of tubular renal cells. This effect was not observed in the experimental males. It is possible that a prolonged exposure to this extract at this dose or an elevated dose may produce renal lesions and that this effect could be gender-dependent. New studies with elevated doses are being developed in our lab to help elucidate this hypothesis.

**Table 1.** Organ weight/body weight ratio of rats treated for 21 days with 0, 28, 42, or 56 mg/kg/day of the ethanol extract obtained from dry and milled cladodes of *C. jamacaru* (mean ($x10^2$) ± SEM ($x10^2$); 5 animals/group).

| | Control | 28 mg/kg/day | 42 mg/kg/day | 56 mg/kg/day |
|---|---|---|---|---|
| Females | | | | |
| Liver | 3.75 ± 0.09 | 3.34 ± 0.15 | 3.44 ± 0.09 | 3.16 ± 0.12* |
| Pancreas | 0.18 ± 0.01 | 0.19 ± 0.02 | 0.21 ± 0.02 | 0.19 ± 0.01 |
| Kidney | 0.38 ± 0.03 | 0.19 ± 0.02 | 0.38 ± 0.01 | 0.37 ± 0.01 |
| Spleen | 0.27 ± 0.02 | 0.22 ± 0.02 | 0.27 ± 0.01 | 0.25 ± 0.02 |
| Males | | | | |
| Liver | 3.32 ± 0.16 | 3.34 ± 0.49 | 3.15 ± 0.05 | 2.98 ± 0.08 |
| Pancreas | 0.175 ± 0.036 | 0.148 ± 0.019 | 0.120 ± 0.010 | 0.130 ± 0.011 |
| Kidney | 0.680 ± 0.305 | 0.338 ± 0.019 | 0.325 ± 0.006 | 0.340 ± 0.017 |
| Spleen | 0.244 ± 0.017 | 0.214 ± 0.014 | 0.188 ± 0.013 | 0.234± 0.009 |

*$P < 0.05$. ANOVA followed by the *posthoc* Tukey-Kramer test.

**Table 2.** Biochemical parameters analyzed in serum of rats treated with 0, 28, 42, or 56 mg/kg/day of the ethanol extract obtained from dry and milled cladodes of *C. jamacaru* (mean ± SEM; 5 animals/group).

| | Control | 28 mg/kg/day | 42 mg/kg/day | 56 mg/kg/day |
|---|---|---|---|---|
| Females | | | | |
| ALT | 75.87 ± 6.66 | 47.80 ± 3.01*** | 51.20 ± 4.96** | 59.70 ± 10.40 |
| AST | 133.13 ± 7.66 | 111.70 ± 5.54* | 203.20 ± 3.36*** | 102.70 ± 2.97*** |
| Urea | 59.50 ± 1.78 | 61.50 ± 4.14 | 60.50 ± 3.95 | 66.90 ± 8.67 |
| Creatinine | 0.75 ± 0.05 | 0.86 ± 0.05 | 0.90 ± 0.12 | 1.10 ± 0.25* |
| Males | | | | |
| ALT | 66.50 ± 46.73 | 76.90 ± 18.80 | 58.00 ± 2.82 | 45.10 ± 6.12 |
| AST | 179.00 ± 9.31 | 116.30 ± 4.56*** | 183.75 ± 15.21 | 144.80 ± 5.25** |
| Urea | 61.60 ± 2.61 | 59.60 ± 3.66 | 72.25 ± 3.38* | 67.60 ± 7.33 |
| Creatinine | 0.76 ± 0.11 | 1.02 ± 0.18 | 0.90 ± 0.22 | 0.88 ± 0.19 |

***$P < 0.001$, **$P < 0.01$, *$P < 0.05$, ANOVA followed by the *posthoc* Tukey-Kramer test.

The statistical analyses revealed that the ethanol extract, at the three different doses, was not able to promote behavioral alterations in the experimental male and female rats observed in the open field (Table 3) and the elevated plus maze (Table 4).

## Conclusions

This study revealed that the ethanol extract obtained from dry and milled cladodes of *C. jamacaru* when administered by gavage to male and female adult rats at the doses of 28, 42, or 56 mg/kg/day did not promote toxicity and behavioral alterations. More studies with elevated doses of the ethanol extract are being developed in our lab to complement these findings.

**Table 3.** Parameters employed for evaluation of general behavior, in an open field, of rats treated with 0, 28, 42, or 56 mg/kg/day of the ethanol extract obtained from *C. jamacaru* (mean ± SEM; *n*=5 rats/group).

| | Control | 28 mg/kg/day | 42 mg/kg/day | 56 mg/kg/day |
|---|---|---|---|---|
| Females | | | | |
| Locomotion | 50.3 ± 12.1 | 46.6 ± 9.9 | 40.2 ± 6.2 | 54.2 ± 10.6 |
| Rearing | 17.0 ± 3.5 | 14.6 ± 3.1 | 11.0 ± 1.6 | 19.8 ± 4.4 |
| Defecation | 1 ± 1 | 0 ± 0 | 2.4 ± 1.4 | 3.4 ± 1.7 |
| Immobility (s) | 67.8 ± 13.0 | 93.0 ± 12.2 | 79.6 ± 29.4 | 57.4 ± 9.1 |
| Males | | | | |
| Locomotion | 37.2 ± 10.0 | 23.0 ± 6.7 | 26.0 ± 8.2 | 20.8 ± 5.6 |
| Rearing | 10.6 ± 3.7 | 6.2 ± 2.1 | 13.3 ± 3.3 | 9.2 ± 1.3 |
| Defecation | 2.8 ± 1.1 | 3.2 ± 1.4 | 2.7 ± 1.1 | 4.2 ± 0.9 |
| Immobility (s) | 89.4 ± 23.5 | 140.2 ± 26.9 | 147.5 ± 15.3 | 125.2± 24.0 |

$P > 0.05$, ANOVA

**Table 4.** Evaluation of the behavior at the elevated plus maze of rats treated with 0, 28, 42, or 56 mg/kg/day of the ethanol extract from *C. jamacaru* cladodes (mean ± SEM; *n*=5 animals/group).

| | Control | 28 mg/kg/day | 42 mg/kg/day | 56 mg/kg/day |
|---|---|---|---|---|
| Females | | | | |
| NE – OA | 0.25 ± 0.25 | 0.20 ± 0.20 | 0.60 ± 0.24 | 0.80 ± 0.20 |
| NE – CA | 1.50 ± 0.28 | 1.40 ± 0.40 | 1.00 ± 0 | 1.00 ± 0 |
| TP – OA (s) | 21.70 ± 10.67 | 6.20 ± 4.24 | 18.6 ± 7.59 | 14 ± 3.49 |
| TP – CA (s) | 78.25 ± 10.49 | 290.20 ± 5.58 | 277.6 ± 8.004 | 285.4 ± 3.7 |
| Males | | | | |
| NE – OA | 1.2 ± 0.2 | 1.0 ± 0 | 1.3 ± 0.3 | 1.4 ± 0.2 |
| NE – CA | 0.2 ± 0.2 | 0.6 ± 0.2 | 0.8 ± 0.5 | 1.2 ± 0.6 |
| TP – OA (s) | 9.8 ± 5.54 | 8.8 ± 2.76 | 24.5 ± 14.14 | 62.8 ± 41.67 |
| TP – CA (s) | 276.6 ± 11.92 | 287 ± 2.34 | 245.25 ± 19.78 | 228 ± 48.3 |

$P > 0.05$, ANOVA (NE = number of entries; TP = time of permanence (s); OA = open arms; CA = closed arms)

## References

Aragão TCFR, Souza PAS, Uchôa AF, Costa IR, Bloch-Jr C, and Campos FAP (2000). Characterization of a methionine-rich protein from the seeds of *Cereus jamacaru* Mill. (Cactaceae). *Brazilian Journal of Medical and Biological Research* 33:897-903.

Braga R (1976). Plantas do Nordeste, Especialmente do Ceará. *Coleção Mossoroense Vol. XLII*. Escola Superior de Agricultura de Mossoró, RN, Brazil, 540 pp.

Broadhurst PL (1960). Experiments in Psychogenetics. In *Experiments in Personality* (HJ Eisenk, ed.), pp. 31-61. Routledge & Kegan Paul, London.

Brhun J and Lindgren NJ (1976). Cactaceae Alkaloids XXIII: alkaloids of Pachycereus pectin-aboriginum and *Cereus jamacaru*. *Lloydia* (*Journal of Natural Products*) 39:175-177.

Burret F, Lebreton P, and Voirin B (1982). Les aglycones flavoniques de Cactees: distribution, signification. *Journal of Natural Products* 45:687-693.

Davet A (2005). Estudo fitoquímico e biológico do cacto – *Cereus jamacaru* de Candolle, Cactaceae, 111 pp. Master's degree Dissertation, Universidade Federal do Paraná, Brazil.

Kitanaka J, Kitanaka N, Tatsuta T, and Takemura M (2005). 2-Phenylethylamine in combination with l-deprenyl lowers the striatal level of dopamine and prolongs the duration of the stereotypy in mice. *Pharmacology Biochemistry and Behavior* 82:488-494.

Lima JLS (1996). *Plantas forrageiras das caatingas - usos e potencialidades,* 44 pp. EMBRAPA-CPATSA.

Ortmann R, Schaub M, Felner A, Lauber J, Christen P, and Waldmeier PC (1984). Phenylethylamine-induced stereotypies in the rat: a behavioral test system for assessment of MAO-B inhibitors. *Psychopharmacology* (Berl) 84:22-27.

Salt TA, Tocker JE, and Adler JH (1987). Dominance of Δ5-sterols in eight species of the Cactaceae. *Phytochemistry* 26:731-733.

Scheinvar L (1985). *Cactáceas*. Flora Ilustrada Catarinense, Itajaí, 25 pp.

Smith DF (1978). The effects of lithium on phenylethylamine behavior in rats are counteracted by monoamine oxidase A and B inhibitors. *Archives Internationales de Pharmacodynamie et de Therapie* 233:221-226.

Starcha R, Chybidziurová A, and Lacný Z (1999). Alkaloids of the genus *Turbinicarpus* (Cactaceae). *Biochemical Systematics and Ecology* 27:839-841.

Youdim MB and Tipton KF (2002). Rat striatal monoamine oxidase-B inhibition by l-deprenyl and rasagiline: its relationship to 2-phenylethylamine-induced stereotypy and Parkinson's disease. *Parkinsonism & Related Disorders* 8:247-253.

# Chapter 114

# Marketing of Boldo (*Plectranthus neochilium* and *Peumus boldus* Molina) by Salesmen of Medical Plants in Campina Grande, Paraíba

R.L. Santos[1], M.S. de C. Nobre[1], G.P. Guimarães[1], K.V.M. Vieira[1], D.C. Felismino[2], I.C. Dantas[2], and L.A. Silva[2]

[1]*Departamento de Farmácia, Universidade Estadual da Paraíba;* [2]*Departamento de Biologia, Universidade Estadual da Paraíba, Campina Grande, Paraíba, Brazil*

## Introduction

In Brazil, the use of medicinal plants is widespread. They are commonly found in backyards, at the herbalist, and in markets being commercialized without a medical prescription. One of the main problems in the use of these products is the belief that plant products are free of adverse reactions and toxicity (Gallo and Koren 2001; Rates 2001) which may cause serious consequences (Navarro Moll 2000).

Some researchers have evaluated commercial herbals according to their current uses. For example, Brandão *et al.* (1998) evaluated the quality of commercial samples of chamomile (*Matricaria recutita* L.) finding that 96% had a lack of standardization and quality. In addition, the quality test with samples of boldo (*Peumus boldus* Molina), chamomile (*Matricaria recutita* L.), cidreira (*Melissa* sp. and *Cymbopogon citratus* (DC) Stapf.), sweet herb (*Pimpinella anisum* L.), and mint (*Mentha* sp.) showed large discrepancies with the standards, confirming that the industry does not follow the principles of quality required for drugs of vegetable origin (Matos 1998). Several species are quite widespread in Brazil for medical use including *P. boldus* Molina (boldo), which is used in industrial phytomedicines in addition to being sold at free markets throughout the country (Melo *et al.* 2004).

There are several species of boldo including *boldo-do-chile* (*Peumus boldus*), *boldo-da-terra* (*Plectranthus barbatus* or *Coleus barbatus*), and *boldo-Bahia* (*Vernonia condensata*) (Brandão *et al.* 2006). *Peumus boldus* is an arboreal species of the Monimiaceae family native to central and southern regions of Chile, where it is very common. Its leaves are used in folk medicine for treatment of digestive and liver problems. In addition to popular use, the base of boldo preparations are described in various official pharmacognostical texts. Boldo is also used in homeopathic medicine (Speisky and Cassels 1994; Brandão *et al.* 2006; Agra *et al.* 2007). *Plectranthus barbatus* is a herb or subshrub with petiolate, elliptical, and velvety leaves popularly known as holy mallow, national boldo, or false-boldo and is often confused with *Peumus boldus* (Mengue *et al.* 2001).

©CAB International 2011. *Poisoning by Plants, Mycotoxins, and Related Toxins* (eds F. Riet-Correa, J. Pfister, A.L. Schild, and T.L. Wierenga)

Some studies have reported that *Plectranthus barbatus* may be toxic in that it causes negative effects on coagulation in addition to increased levels of cholesterol and transaminases and a reduction in total bilirubin and glycemia (Gielen and Goossens 2001; Monzón *et al.* 2004; Izzo *et al.* 2005; Piscaglia *et al.* 2005). Ruiz *et al.* (2008) suggest that consumption of boldo (*P. barbatus*) tea must be done with restraint and care, especially in the first trimester of pregnancy, as this plant may cause teratogenicity and hepatotoxicity.

This study surveyed medical plant salesmen in the city of Campina Grande, Paraíba, Brazil, in order to know the species of boldo commercialized and to evaluate the knowledge of the traders regarding the medicinal uses of the plant including the number of leaves used, preparation methods, and any care that must be taken to avoid the possibility of poisoning during therapy.

## Methodology

The survey was accomplished at free markets of Campina Grande, state of Paraíba, during August-September 2008. The city has about 50 salesmen of medical plants and 12 were selected based on their greater knowledge about medicinal plants. In a cross-sectional study, the data were collected from previously developed questionnaires applied directly to the salesmen of medical plants. Results were treated with descriptive statistics. Samples of boldo were collected from the salesmen and identified in the Botany Laboratory at the State University of Paraíba.

## Results and Discussion

Boldo samples identified in the laboratory confirmed the identifications made by the salesmen. Among the 12 salesmen, 54% sold false boldo (*Plectranthus neochilium*) and 46% Chile boldo (*Peumus boldus*) also known as boldo real. According to Melo *et al.* (2004) boldo, especially the species *P. boldus*, is largely used in Brazil as a medical plant. It is also included in the composition of industrial medical plant products and is traded freely at free markets. In our survey *P. neochilium* was slightly more frequently commercialized than *P. boldus*, suggesting that both species are equally used in the city of Campina Grande.

Most salesmen indicated that boldo could be used to treat liver disorders (35%) and bad digestion (32%). Kringstein and Cederbaum (1995) suggested that the alkaloid boldine prevented the ferric-ATP catalyzed peroxidation of human liver microsomes. As a result it is thought that boldine may be valuable as an antioxidant and hepatoprotective agent because of its strong inhibition of the peroxidation of human liver microsomes. Boldine is thought to relieve intestinal problems due to its actions as cholagogue, choleritic, and cholelithiasis. In addition there are also reports of boldine's diuretic and anti-inflammatory properties, which suggest it may aid in weight loss (Matos 1998).

One salesman said that boldo could be used in the treatment of diabetes. The antioxidant effect of boldine was able to mitigate in mice the development of diabetes induced by streptozotocin by reducing hyperglycemia and weight in the mice, which opens new perspectives of pharmacotherapeutic application of the plant (Jang *et al.* 2000).

The tea was the only form of ingestion recommended by all salesmen for both *P. neochilium* and *P. boldus*. The water and heat of the tea making process results in the extraction of the volatile components and aromatic active ingredients (Castellani 1999).

Although all salesmen recommended the ingestion of both plants as a tea the amount of leaves and water to be used in preparation varied among them: 25% recommended 5 g of leaves in one cup; 25% 5 g of leaves in 1 l; and 50% 10 g of leaves in 1 l. These differences in the proportions of water and plant to make the tea are of considerable concern due to the variability in the dose being ingested and the potential secondary effects.

In relation to the knowledge of medical plant salesmen on boldo toxicity only 8% stated that toxicity may occur due to incorrect use of this plant. This toxicity may be due to high or prolonged dosage, concomitant ingestion of food, or even some reactions of hypersensitivity. According to Alonso (1998) and the Ministry of Health (2005), high doses or prolonged use of boldo may cause visual and auditive disorders, kidney irritation, vomiting, and diarrhea. Boldine may also show narcotics effects or result in convulsions. In consequence, it should not be prescribed to children below 6 years old. A toxicological evaluation of *P. boldus* was performed by Almeida *et al.* (2000); a hydro-alcoholic extract of boldo and boldine dosed to pregnant rats in a single dose of 800 mg/kg caused teratogenic and abortifacient effects. Groups of females and male rats treated orally for 90 days with boldine and the crude extract caused significant increases in serum cholesterol and transaminases and the reduction of total bilirubin, glucose, and urea. Gielen and Goossens (2001) reported an occupational allergic dermatitis caused by boldo in a pharmaceutical among 33 cases of poisoning by plants recorded in the period 1978 to 2001 in the Department of Dermatology, Katholieke Universiteit Leuven, Belgium. Monzón *et al.* (2004) described a case of anaphylactic reaction after the ingestion of boldo tea in a 30 year old man with a history of allergic rhinoconjunctivitis to pollen. Piscaglia *et al.* (2005) reported a case of hepatotoxicity attributed to the consumption of the boldo extract.

Only 8% of the salesmen knew the possible health risks of the use of boldo if the recommendations for use are not followed. The other 92% were not aware of any risk due to boldo consumption. For them boldo is a natural therapy that does not cause health problems regardless of dosage used or use with other medicines. In a human patient treated with warfarin the consumption of boldo increased the anticoagulant effect of the former drug. The interaction between boldo and warfarin was confirmed because the anticoagulant action of warfarin returned to normal levels with the interruption of boldo ingestion and was intensified with readministration of boldo to the patient (Izzo *et al.* 2005).

## Conclusions

The results showed that plant species *Plectranthus neochilium* and *Peumus boldus* are currently commercialized by the salesmen of medical plants in the municipality of Campina Grande for several purposes. However, salesmen have little information about the correct use and dosage of medicinal plants and most do not know the risks of these plants when used incorrectly. It is necessary to provide correct information about these plants to avoid risk of human intoxication.

## References

Agra MF, França PF, and Barbosa-Filho JM (2007). Synopsis of the plants known as medicinal and poisonous in Northeast of Brazil. *Revista Brasileira de Farmacognosia* 17:114-140.

Almeida ER, Melo AM, and Xavier H (2000). Toxicological evaluation of the hydro-alcohol extract of the dry leaves of *Peumus boldus* and boldine in rats. *Phytotherapy Research* 14:99-102.

Alonso JR (1998). *Tratado de fitomedicina: bases clínicas y farmacológicas,* 1039 pp. ISIS Ediciones SRL, Buenos Aires.

Brandão MGL, Freire N, and Soares CDV (1998). Vigilância de fitoterápicos de Minas Gerais. Verficação da qualidade de diferentes amostras comerciais de camomila. *Cadernos de Saúde Pública* 14:613-616.

Brandão MGL, Cosenza GP, Moreira RA, and Monte-Mor RLM (2006). Medicinal plants and other botanical products from the Brazilian Official Pharmacopoeia. *Revista Brasileira de Farmacognosia* 16:408-420.

Castellani DC (1999). *Plantas medicinais.* 20 pp. Agromídia software, Viçosa.

Gallo M and Koren G (2001). Can herbal products be used safely during pregnancy? Focus on *Echinacea. Canadian Family Physician* 47:1727-1728.

Gielen K and Goossens A (2001). Occupational allergic contact dermatitis from drugs in healthcare workers. *Contact Dermatitis* 45:273-279.

Izzo AA, Carlo GD, Borrelli F, and Ernst E (2005). Cardiovascular pharmacotherapy and herbal medicines: the risk of drug interaction. *International Journal of Cardiology* 98:1-14.

Jang YY, Song JH, Shin YK, Han ES, and Lee CS (2000). Protective effect of boldine on oxidative mitochondrial damage in streptozotocin-induced diabetic rats. *Pharmacology Research* 42:361-371.

Kringstein P and Cederbaum AI (1995). Boldine prevents human liver microsomal lipid peroxidation and inactivation of cytochrome P4502E1. *Free Radical Biology and Medicine* 18:559-563.

Matos FJA (1998). *Farmácias Vivas.* 57 pp. Edições Universidade Federal do Ceará.

Melo JG, Nascimento VT, Amorim ELC, Andrade CSL, and Albuquerque UP (2004). Avaliação da qualidade de amostras comerciais de boldo (*Peumus boldus* Molina), pata-de-vaca (*Bauhinia spp.*) e ginco (*Ginkgo biloba* L.). *Revista Brasileira de Farmacognosia* 14(2):111-120.

Mengue SS, Mentz LA, and Schenkel EP (2001). Uso de plantas medicinais na gravidez. In *Manual de Teratogênes* (ATV Anseverino, DI Spreitzer, and L Schüler-Faccini, eds), pp. 42-50. Editora da Universidade, Porto Alegre.

Ministry of Health. Instituto Nacional do Câncer (INCA). *Estimativas da Incidência e Mortalidade por Câncer no Brasil* (monografia online). Brasília; Ministério da Saúde (accessed 14 May 2005). Available at: http:// www.inca.org.br.

Monzón S, Lezaun A, Sáenz D, Marquinez Z, Bernedo N, Uriel O, Colas C, and Duce F (2004). Anaphylaxis to boldo infusion, a herbal remedy. *Allergy* 59:1019-1020.

Navarro Moll MC (2000). Uso racional de las plantas medicinales. *Pharmaceutical Care* Espana 2:9-19.

Piscaglia F, Leoni S, Venturi A, Graziella F, Donati G, and Bolondi L (2005). Caution in the use of boldo in herbal laxatives: a case of hepatotoxicity. *Scandinavian Journal of Gastroenterology* 40:236-239.

Rates SMK (2001). Uso racional de fitoterápicos. *Jornal da Associação dos Farmacêuticos do Rio Grande do Sul* v. Encarte.

Ruiz ALTG, Taffarello D, Sousa VHS, and Carvalho JE (2008). Farmacologia de *Peumus boldus* e *Baccharis genistelloides. Revista Brasileira de Farmacognosia* 18(2):295-300.

Speisky H and Cassels BK (1994). Boldo and boldine: an emerging case of a natural drug development. *Pharmacology Research* 29:1-12.

# Chapter 115

# Evaluation of Hemolytic and Spasmolytic Activities of *Sargassum polyceratium* Montagne (Sargassaceae)

A.C.C. Correia[1], C.L. Macêdo[1], F.S. Monteiro[1], F.H.T. Souza[1], H.L.F. Pessôa[2], G.E.C. de Miranda[4], C.S. Dias[1,3], J.M. Barbosa-Filho[1,3], F.A. Cavalcante[5], and B.A. Silva[1,3]

*[1]Laboratório de Tecnologia Farmacêutica 'Prof. Delby Fernandes de Medeiros', Universidade Federal da Paraíba, PO Box 5009, João Pessoa, Paraíba 58.051-970, Brazil; [2]Departamento de Biologia Molecular/UFPB, Brazil; [3]Departamento de Ciências Farmacêuticas/UFPB, Brazil; [4]Departamento de Sistemática e Ecologia CCEN/UFPB, Brazil; [5]Instituto de Ciências Biológicas e da Saúde/UFAL, Brazil*

## Introduction

Nature in general is responsible for production of the largest number of organic substances. However the plant kingdom is responsible for the greatest chemical diversity reported in the literature. The variety and complexity of molecules which are special metabolites of plants and marine organisms is unattainable by laboratory methods. This is a direct consequence of millions of years of evolution reaching a high degree of refinement in terms of forms of protection and resistance to climate, pollution, and predators (Montanari and Bolzani 2001).

Since more than 70% of the Earth's surface is covered by oceans, many marine plants are used for food and as a source of minerals, dietary fiber, nutrition, and medicine (Haefner 2003). Numerous marine natural products have been found to be useful for pharmacological studies to treat various diseases (Mayer *et al.* 2007).

The Sargassaceae family comprises around 12 genera and approximately 701 species (Sheu *et al.* 1999). Some genera of this family have demonstrated medicinal properties such as antitumor, antiangiogenic (Dias *et al.* 2005), antioxidant, immunostimulant, anticoagulant (Choi *et al.* 2009), and antiviral activities (Zhu *et al.* 2006).

The *Sargassum* genus is a tropical and subtropical brown seaweed (Phaeophyceae class) common to all oceans except Antarctica, comprising 150 species. This genus is well represented on the Brazilian coast (Paula and Eston 1987) and according to Coimbra (2006) is estimated to have about 11 species distributed from the coast of the state of Maranhão to Rio Grande do Sul. The populations of *Sargassum* occur both in protected areas on rocky shores and on shores exposed to wave action (Oliveira-Filho 1977; Széchy and Paula 2000).

©CAB International 2011. *Poisoning by Plants, Mycotoxins, and Related Toxins* (eds F. Riet-Correa, J. Pfister, A.L. Schild, and T.L. Wierenga)

Species of *Sargassum* showed cytotoxic activity in cell cultures as *S. autumnale* (CA-9KB cell) (Arisawa *et al.* 1997) and *S. autumnale* and *S. confusum* (HELA cells) (Hayashi *et al.* 1996). Some species of *Sargassum* (*S. micracanthum* and *S. siliquastrum*) showed vasodilatation effects on the basilar and carotid artery of rabbits (Park *et al.* 2008a, b). Moreover other species such as *S. ilicifolium*, *S. polycystum*, *S. plagiophyllum*, *S. tenerrimum*, and *S. wightii* (Bhakuni *et al.* 1992) also presented significant spasmolytic effects on guinea pig ileum.

Considering that some species of *Sargassum* have shown cytotoxic activity and others have shown spasmolytic activity, there have been no reports on the hemolytic and spasmolytic activities of *S. polyceratium*. We studied the crude ethanolic extract from *S. polyceratium* Montagne (Sargassaceae) on the cytotoxic potential in erythrocytes of rats and investigated the spasmolytic activity on guinea pig ileum.

## Material and Methods

### General

Male Wistar rats (*Rattus norvegicus*) weighing 200-300 g and guinea pigs (*Cavia porcellus*) weighing 300-500 g, from Biotério Prof Thomas George of Laboratório de Tecnologia Farmacêutica (LTF/UFPB) were used in the study. All procedures were approved by the Ethical Committee in Animal Research of LTF/UFPB (protocol/CEPA no 0605/05). Assays with guinea pig ileum used a modified Krebs solution with the following composition (mM): NaCl (117.0), KCl (4.7), $MgSO_4$ (1.3), $NaH_2PO_4$ (1.2), $CaCl_2$ (2.5), glucose (11.0), and $NaHCO_3$ (25.0) (Sun and Benishin 1994) adjusted to pH 7.4, bubbled with a carbogen mixture (95% $O_2$ and 5% $CO_2$), and maintained at 37°C.

### Drugs

Histamine hydrochloride and Triton X-100 were obtained from Sigma-Aldrich (USA). Potassium chloride (KCl) was obtained from Vetec (Brazil). Carbamylcholine hydrochloride (CCh) was obtained from Merck (Brazil). Crude ethanolic extract from *S. polyceratium* Montagne (Sarg-EtOH) was obtained by Dr Celidarque da S. Dias from the Departamento de Ciências Farmacêuticas/UFPB. Botanical material was identified by Dr George Emmanuel Cavalcanti de Miranda from the Departamento de Sistemática e Ecologia CCEN/UFPB and a sample (no 13997) was deposited at the Herbarium Prof. Lauro Pires Xavier (JPB) of Departamento de Sistemática e Ecologia, UFPB, João Pessoa, PB, Brazil. The extract of dried and powdered seaweed was obtained by extraction with 95% EtOH at room temperature for 3 days.

### Toxicological evaluation of the Sarg-EtOH extract

*Effect of the Sarg-EtOH extract on rat erythrocytes*

This procedure followed the methodology described by Rangel *et al.* (1997). A sample of blood was collected from fasting rats (200-300 g). This material was immediately mixed with a solution (pH=7.4) of NaCl (0.9%) and $CaCl_2$ (10 mM) at a ratio of 1:30 in slow and constant agitation to avoid coagulation and centrifuged at 2500 rpm for 5 min to obtain the erythrocytes. Sarg-EtOH extract was added to the suspension of erythrocytes at various concentrations and in different preparations. A negative control was made with a suspension of erythrocytes using more NaCl and $CaCl_2$ (0% hemolysis), a positive control made with a suspension of erythrocytes plus 1% Triton X-100 (100 % hemolysis), and the percentage of hemolysis was calculated relative to this value. The samples were incubated for 1 h at room temperature under slow and constant agitation. After this time samples were centrifuged at 2500 rpm for 5 min and hemolysis was measured by spectrophotometry at 540 nm. The results obtained for the three independent hemolysis experiments are expressed as mean ± SEM. The hemolytic activity was evaluated according to Prokof'eva *et al.* (2004).

### Preliminary pharmacological investigation

*Effect of Sarg-EtOH extract on carbachol- and histamine-induced phasic contractions on the guinea pig ileum*

The guinea pig ileum was prepared according to Daniel *et al.* (2001). Guinea pigs were killed by cervical dislocation followed by exsanguination. The abdomen was opened and a segment of ileum was removed and placed in a modified Krebs solution at pH 7.4. The ileum was cut into small segments (3 cm in length) and mounted vertically in 6 ml isolated organ baths containing a modified Krebs solution bubbled with a carbogen mixture and maintained at 37ºC. After the stabilization period (30 min) two phasic contractions of similar magnitude were obtained with $10^{-6}$ M carbachol or histamine and recorded using isotonic levers coupled to kymographs and smoked drums. The Sarg-EtOH extract was incubated alone for 15 min in different preparations and at various concentrations. Inhibition of response to submaximal carbachol or histamine was assessed by comparing the responses before (control) and after addition to the chamber. The molar concentration of Sarg-EtOH extract that reduced the response to an agonist by 50% ($IC_{50}$) was obtained by non-linear regression.

### Statistical analysis

Data were expressed as mean ± SEM (standard error of mean). Statistical analysis was performed using Graph-Pad Prism 5.00 software (GraphPad Software Inc., San Diego, CA, USA). Differences between means were statistically compared using *t* test or one-way ANOVA followed by Bonferroni's test, as appropriate, and were considered to differ significantly when $P < 0.05$. $IC_{50}$ values were determined by non-linear regression (Jenkinson *et al.* 1995).

## Results and Discussion

### Toxicological evaluation of the Sarg-EtOH extract

*Effect of the Sarg-EtOH extract on rat erythrocytes*

Based on the indication that species of the genus *Sargassum* showed cytotoxic activity in cell cultures (Hayashi *et al.* 1996; Arisawa *et al.* 1997) and that erythrocytes contain high levels of polyunsaturated fats, molecular oxygen, and linked iron ions (Niki *et al.* 1991), its membrane is expected to be highly vulnerable to reactions that involve free radicals and highly susceptible to hemolysis (Brandão *et al.* 2005). Erythrocytes provide a simple model to study protective and toxic effects of a great variety of substances and situations associated with oxidative stress (Eisele *et al.* 2006; Lexis *et al.* 2006).

When evaluating cytotoxicity in rat erythrocytes Sarg-EtOH caused a weak hemolytic activity ($E_{max}$=16.2 ± 0.5%, *n*=3) only at concentration of 500 µg/ml ($P < 0.05$) (Figure 1).

This preliminary study provided information sufficient to choose the concentrations used in experimental protocols designed to investigate a possible spasmolytic effect.

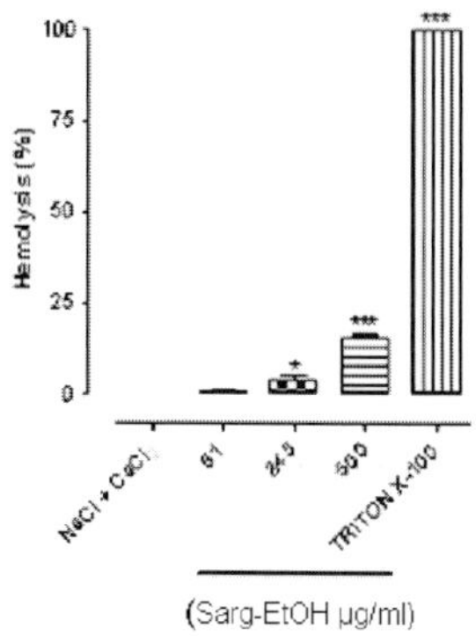


**Figure 1.** Hemolytic effect of Sarg-EtOH extract in rat erythrocytes (n=3). Triton X-100 (positive control) and NaCl + $CaCl_2$ (negative control). Columns and vertical bars represent mean and SEM, respectively. One-way ANOVA followed by Bonferroni's test, *$P < 0.01$ and ***$P < 0.0001$ (negative control vs Sarg-EtOH/positive control).

### Preliminary pharmacological investigation

*Effect of the Sarg-EtOH extract on guinea pig ileum*

Sarg-EtOH inhibited both the carbachol- ($IC_{50}$=319.2±46 µg/ml) and histamine-induced phasic contractions ($IC_{50}$=181.8±34 µg/ml) in a significant and concentration-dependent manner but there were no significant differences in $IC_{50}$ values (Figure 2). This suggests that Sarg-EtOH had non-selective spasmolytic activity on the guinea pig ileum. This activity was also recorded by Bhakuni *et al.* (1992) when screening some marine flora from Indian coasts as some species of the genus *Sargassum* showed spasmolytic activity on the guinea pig ileum.

## Conclusions

Crude ethanolic extract (Sarg-EtOH) does not show significant cytotoxic effects in rat erythrocytes. Therefore, it is reasonable to assume that the Sarg-EtOH extract would have low or no toxicity when tested *in vivo*. In addition, Sarg-EtOH showed significant non-selective spasmodic activity on guinea pig ileum. These results contribute to the body of literature on the species *Sargassum polyceratium* Montagne.

## Acknowledgements

The authors are grateful to José Crispim Duarte for technical assistance and to CNPq and CAPES for financial support.

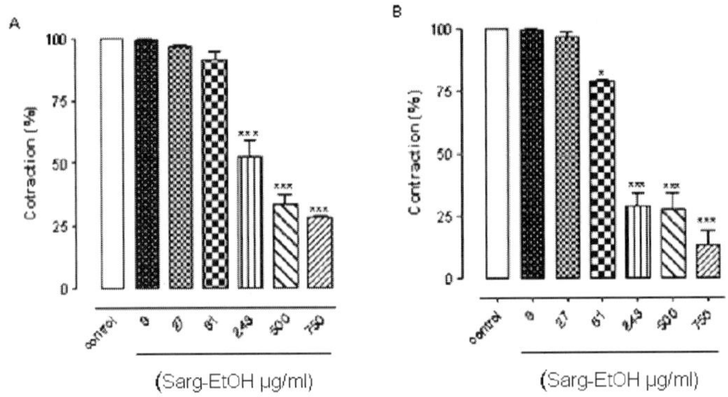


**Figure 2.** Effect of Sarg-EtOH extract on carbachol- (A) and histamine- (B) induced phasic contractions on guinea pig ileum (*n*=5 and 3, respectively). Columns and vertical bars represent mean and SEM, respectively. One-way ANOVA followed by Bonferroni's test, **P* < 0.05 and ****P* < 0.0001 (control vs Sarg-EtOH extract).

## References

Arisawa M, Hayashi K, Nikaido T, Koike K, Fujita D, Nunomura N, Tanaka M, and Sasaki T (1997). Screening of some marine organism extracts for camp phosphodiesterase inhibition, cytotoxicity, and antiviral activity against hsv-1. *International Journal Pharmacognosy* 35(1):6-11.

Bhakuni DS, Dhawan BN, Garg HS, Goel AK, Mehrotra BN, Srimal RC, and Srivastava MN (1992). Bioactivity of marine organisms: part vi–screening of some marine flora from Indian coasts. *Indian Journal Experimental Biology* 30(6):512-517.

Brandão R, Lara FS, Pagliosa LB, Soares FA, Rocha JBT, Nogueira CW, and Farina M (2005). Hemolytic effects of sodium selenite and mercuric chloride in human blood. *Drug Chemical Toxicology* 28:397-407.

Choi EY, Hwang HJ, Kim IH, and Nam TJ (2009). Protective effects of a polysaccharide from *Hizikia fusiformis* against ethanol toxicity in rats. *Food and Chemical Toxicology* 47(4):134-139.

Coimbra CS (2006). Inferências filogenéticas a ordem Fucales (Phaeophyceae), com ênfase no gênero *Sargassum,* 75 pp. Tese de Doutorado, Departamento de Botânica, Instituto de Biociências, Universidade de São Paulo, São Paulo.

Daniel EE, Kwan CY, and Janssen L (2001). Pharmacological techniques for the *in vitro* study of intestinal smooth muscle. *Journal Pharmacological Toxicology* 45:159.

Dias PF, Siqueira JMJR, Vendruscolo LF, Neiva JT, Gagliardi AR, Maraschin M, and Ribeiro-do-Valle RM (2005). Antiangiogenic and antitumoral properties of a polysaccharide isolated from the seaweed *Sargassum stenophyllum. Cancer Chemotherapy and Pharmacology* 56(4):436-46.

Eisele K, Lang PA, Kempe DS, Klarl BA, Niemoller O, Wieder T, Huber SM, Duranton C, and Lang F (2006). Stimulation of erythrocyte phosphatidylserine exposure by mercury ions. *Toxicological Applied Pharmacology* 210:116-122.

Haefner B (2003). Drugs from the deep: marine natural products as drug candidates. *Drug Discovery Today* 8:536-544.

Hayashi K, Hamada J, and Hayashi T (1996). A screening strategy for selection of anti-hsv-1 and anti-hiv extracts from algae. *Phytotherapy Research* 10(3):233-237.

Jenkinson DH, Barnard EA, Hoyer D, Humphrey PPA, Leff P, and Shankley NP (1995). International union of pharmacology committee on receptor nomenclature and drug classification. IX Recommendations on terms and symbols in quantitative pharmacology. *Pharmacology Review* 42:255-266.

Lexis LA, Fassett RG, and Coombes JS (2006). α-tocopherol and α-lipoic acid enhance the erythrocyte antioxidant defence in cyclosporine A-treated rats. *Basic Clinical Pharmacology Toxicology* 98:68-73.

Mayer AMS, Rodriguez AD, Berlinck RGS, and Hamann MT (2007). (Marine pharmacology in 2003-4: Marine compounds with anthelmintic antibacterial, anticoagulant, antifungal, antinflammatory, antimalarial, antiplatelet, antiprotozoal, antituberculosis and antiviral. *Comparative Biochemistry Physiology C: Toxicology & Pharmacology* 145(4):553-581.

Montanari CA and Bolzani VS (2001). Planejamento racional de fármacos baseado em produtos naturais. *Quimica Nova* 24(1):105-111.

Niki E, Yamamoto Y, Komuro E, and Sato K (1991). Membrane damage due to lipid oxidation. *American Journal Clinical Nutrition* 53:201-205.

Oliveira-Filho EC (1977). Algas Marinhas bentônicas do Brasil, 407 pp. Tese de livre-docência em Ficologia, Instituto de Biociências, Universidade de São Paulo, São Paulo.

Park BG, Shin WS, Um Y, Cho S, Park GM, Yeon DS, Kwon SC, Ham J, Choic BW, and Lee S (2008a). Selective vasodilatation effect of sargahydroquinoic acid, an active constituent of *Sargassum micracanthum*, on the basilar arteries of rabbits. *Bioorganic & Medicinal Chemistry Letters* 18:2624-2627.

Park BG, Kwon SC, Park GM, Ham J, Shin WS, and Lee S (2008b). Vasodilatation effect of farnesylacetones, active constituents of *Sargassum siliquastrum*, on the basilar and carotid arteries of rabbits. *Bioorganic & Medicinal Chemistry Letters* 18(24):6324-6326.

Paula EJ and Eston VR (1987). Are there other *Sargassum* species potentially as invasive as *S. muticum*? *Botanica Marina* 30:405-410.

Prokof'eva NG, Utkina NK, Chaikina EL, and Makarchenko AE (2004). Biological activities of marine sesquiterpenoid quinones: structure–activity relationships in cytotoxic and hemolytic assays. *Compendium Biochemical Physiology Part B* 139:169-173.

Rangel M, Malpezzi ELA, Susini SMM, and Freitas JC (1997). Hemolytic activity in extracts of the diatom *Nitzschia. Toxicon* 35(2):305-309.

Sheu J, Wang G, Sung P, and Duh C (1999) New Cytotoxic Oxygenated Fucosterols from the Brown Alga Turbinaria conoides. *Journal of Natural Products* 62(2):224-227.

Sun YD and Benishin CG (1994) $K^+$ channel openers relaxes longitudinal muscle of guinea-pig ileum. *European Journal Pharmacology* 271:453-459.

Széchy MTM and Paula EJ (2000). Padrões estruturais quantitativos de bancos de *Sargassum* (Phaeophyta, Fucales) do litoral dos Estados do Rio de Janeiro e de São Paulo, Brasil. *Revista Brasileira de Botânica* 23(2):121-132.

Zhu W, Chiu LCM, Ooi VEC, Chan PKS, and Ang JRPO (2006). Antiviral property and mechanisms of a sulphated polysaccharide from the brown alga *Sargassum patens* against Herpes simplex virus type 1. *Phytomedicine* 13(9-10):695-701.

# Chapter 116

# Investigation of Hemolytic and Spasmolytic Activities of the Total Alkaloid Fraction from Root Bark of *Solanum paludosum* Moric. (Solanaceae)

A.C.C. Correia[1], F. de S. Monteiro[1], C.L. Macêdo[1], I.J.L.D. Basílio[1], H.L.F. Pessôa[2], M.F. Agra[1,3], J. Bhattacharyya[1], F.A. Cavalcante[4], and B.A. Silva[1,3]

[1]*Laboratório de Tecnologia Farmacêutica 'Prof. Delby Fernandes de Medeiros'/ Universidade Federal da Paraíba, PO Box 5009, João Pessoa, Paraíba 58.051-970, Brazil;* [2]*Departamento de Biologia Molecular/UFPB, Brazil;* [3]*Departamento de Ciências Farmacêuticas/UFPB, Brazil;*[4] *Instituto de Ciências Biológicas e da Saúde/UFAL, Brazil*

## Introduction

The Solanaceae family is distributed worldwide with approximately 106 genera and 2300 species (Olmstead *et al.* 1999). The *Solanum* genus is considered one of the largest among angiosperms with about 1700 species (Hunziker 2001). It is a rich source of secondary metabolites. Many species of *Solanum* are popularly known in Brazil as 'jurubeba.' Plants of this genus are known to produce a great variety of steroidal saponins and glycoalkaloids that play an important role in natural resistance against several pests (Friedman *et al.* 1991).

Many species of *Solanum* (e.g. *S. paniculatum* L., *S. melongena* L., and *S. stipulaceum* Roem. & Schult.) are reported to induce hypotension in rats (Ribeiro *et al.* 1986; Shum and Chiu 1991; Ribeiro 2001). Moreover, other species also had significant spasmolytic effects, for example *S. dulcamara* L. (Boyd 1928), *S. torvum* Sw (Bhakuni *et al.* 1969), *S. paraibanum* Agra (Silva 2007), *S. jabrense* Agra & Nee (Cavalcante 2001; Claudino 2003), *S. agrarium* Sendtn. (Santos *et al.* 2004), *S. megalonyx* Sendtn., *S. asterophorum* Mart. (Oliveira *et al.* 2006a, b), *S. paniculatum* L. (Silva 2006), and *S. asperum* Rich. (Costa 2006; Correia 2007; Garcia 2007).

*Solanum paludosum* Moric. is an herbaceous species popularly known as 'jurubeba-roxa' in northeastern Brazil (Agra and Bhattacharyya 1999). A chemical study with root bark showed the presence of glycoalkaloids (Basílio 2008). As glycoalkaloids are known to be cytotoxic and since many species of *Solanum* have spasmolytic activity, this study was carried out to evaluate if the total alkaloid fraction obtained from the root bark of *S. paludosum* (TAF-SP) had cytotoxic effects on rat erythrocytes as well as to investigate a possible spasmolytic effect on rat uterus and guinea pig trachea.

©CAB International 2011. *Poisoning by Plants, Mycotoxins, and Related Toxins* (eds F. Riet-Correa, J. Pfister, A.L. Schild, and T.L. Wierenga)

## Material and Methods

### General

Virgin female and male Wistar rats (*Rattus norvegicus*) weighing 150-250 g and 200-300 g, respectively, and guinea pigs (*Cavia porcellus*) weighing 300-500 g from Biotério Prof Thomas George of LTF/UFPB were used in the study. All procedures were approved by the Ethical Committee in Animal Research of LTF/UFPB (protocol/CEPA no 0204/08). Nutritive solutions with the following composition (in mM) were used: normal Krebs with NaCl (118.0), $NaHCO_3$ (25.0), KCl (4.6), $MgSO_4$ (5.7), $KH_2PO_4$ (1.1), $CaCl_2$ (2.5) and glucose (11.0); De Jalon: NaCl (154.0), KCl (5.6), $CaCl_2$ (0.27), $NaHCO_3$ (5.9), and glucose (2.8). Solutions were adjusted to pH 7.4 and bubbled with a carbogen mixture (95% $O_2$ and 5% $CO_2$). The guinea pig trachea and the female rat uterus were maintained at 37°C and 31°C, respectively.

### Drugs

Arachidonic acid (AA), diethylstilbestrol, histamine chloride, and Triton X-100 were obtained from Sigma-Aldrich (USA). Potassium chloride (KCl) was obtained from Vetec (Brazil). Acetylcholine chloride (ACh) and carbamylcholine hydrochloride (CCh) were obtained from Merck (Brazil). Oxytocin was obtained from União Química (Brazil). The total alkaloid fraction was assessed from the root bark of *S. paludosum* Moric. A voucher specimen (Agra & Basílio 6734) was deposited in the Herbarium Prof. Lauro Pires Xavier (JPB) of UFPB.

### Toxicological evaluation of the TAF-SP

*Effect of the TAF-SP on rat erythrocytes*

This procedure followed the methodology described by Rangel *et al.* (1997). A sample of blood was collected from fasting rats. Blood was immediately mixed with a solution (pH=7.4) of 0.9% NaCl and 10 mM $CaCl_2$ at a ratio of 1:30 in slow and constant agitation to avoid coagulation and centrifuged at 2500 rpm for 5 min to obtain the erythrocytes. TAF-SP was added to a suspension of erythrocytes at various concentrations and in different preparations. A negative control was obtained with a suspension of erythrocytes using more NaCl and $CaCl_2$ (0% of hemolysis), a positive control was obtained using a suspension of erythrocytes with 1% Triton X-100 (100% hemolysis), and the percentage of hemolysis was calculated relative to this value. The samples were incubated for 1 h at room temperature under slow and constant agitation, then they were centrifuged at 2500 rpm for 5 min and hemolysis was measured by spectrophotometry at 540 nm. The results obtained for the three independent hemolysis experiments are expressed as mean ± SEM. The hemolytic activity was evaluated according to Prokof'eva *et al.* (2004).

### Preliminary pharmacological investigation

*Effect of the TAF-SP on phasic contractions on rat uterus*

The rat uterus preparation was mounted according to De Jalon *et al.* (1945). Rats were killed by cervical dislocation followed by exsanguination. The abdomen was opened and a segment of uterus was removed and placed in a De Jalon solution at pH 7.4. The uterus was mounted vertically in 6 ml isolated organ baths containing De Jalon solution, bubbled with a carbogen mixture, and maintained at 37ºC. After the stabilization period (40 min), two phasic contractions of similar magnitude were obtained with $10^{-5}$ M carbachol or $10^{-2}$ UI/ml oxytocin and recorded using isotonic levers coupled to kymographs and smoked drums. TAF-SP was incubated alone for 15 min in different preparations and at various concentrations. Inhibition of response to submaximal carbachol or oxytocin was assessed by comparing the responses before (control) and after addition to the chamber. The molar concentration of TAF-SP that reduced the response to an agonist by 50% ($IC_{50}$) was obtained by non-linear regression.

*Effect of the TAF-SP on contractions on guinea pig trachea*

The method described by Tschirhart *et al.* (1987) was used to observe the absence or presence of activity on tracheal epithelium. The animals were sacrificed by cerebral concussion followed by exsanguination. The trachea was removed and cleaned of all connective tissue and fat. The trachea segments were kept at a temperature of 37°C and remained at rest for 60 min and the solution was changed every 15 min. After this period of stabilization, we obtained two tonic contractions of similar magnitude induced by $10^{-6}$ M carbachol and considered this as the control. During the third phase of the tonic response to carbachol, the TAF-SP was added in a cumulative manner to the chamber. The relaxation was expressed as the reverse percentage of the carbachol-induced initial contraction. $EC_{50}$ values were assessed by non-linear regression from each individual TAF-SP relaxation curve.

### Statistical analysis

The values were expressed as mean ± SEM. Statistical analysis was performed using Graph-Pad Prism 3.02 software (GraphPad Software Inc., San Diego, CA, USA). Differences between means were statistically compared using *t* test or one-way ANOVA followed by Bonferroni's test, as appropriate, and were considered to differ significantly when $P < 0.05$. The $IC_{50}$ and $EC_{50}$ values were determined by non-linear regression (Jenkinson *et al.* 1995).

## Results and Discussion

### Toxicological evaluation of the TAF-SP

*Effect of the TAF-SP on rat erythrocytes*

When evaluating cytotoxicity to rat erythrocytes, TAF-SP did not induce a significant hemolytic effect at 81, 243, or 500 µg/ml (Figure 1, $n$=3). Based on the indication that glycoalkaloids may be toxic to humans (Morris and Lee 1984) and that erythrocytes contain high levels of polyunsaturated fats, molecular oxygen, and linked iron ions (Niki *et al.* 1991), its membrane is expected to be highly vulnerable to reactions that involve free radicals and highly susceptible to hemolysis (Brandão *et al.* 2005). Erythrocytes provide a simple model to study protective and toxic effects of a great variety of substances and situations associated with oxidative stress (Eisele *et al.* 2006; Lexis *et al.* 2006). This preliminary study provided information on the appropriate concentration of plant extract to be used in experimental protocols designed to investigate a possible spasmolytic effect.

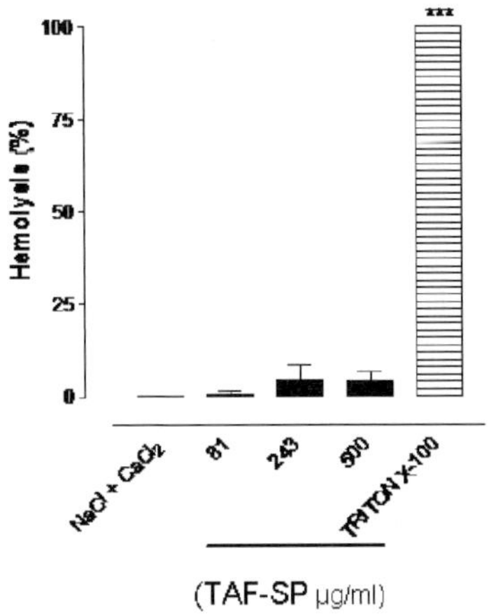


**Figure 1.** Hemolytic effect of TAF-SP on rat erythrocytes (*n*=3). Triton X-100 (positive control) and NaCl + $CaCl_2$ (negative control). Columns and vertical bars represent mean and SEM, respectively. One-way ANOVA followed by Bonferroni's test, ***$P < 0.001$ (negative control vs TAF-SP/positive control).

### Preliminary pharmacological investigation

*Effect of the TAF-SP on phasic contractions on rat uterus*

TAF-SP inhibited phasic contractions induced by $10^{-5}$ M CCh in a significant manner ($IC_{50}$ = 178.8±7.1 µg/ml) but not those by $10^{-2}$ UI/ml oxytocin ($E_{max}$ = 4.0±3.5%) (Figure 2). This result indicates that TAF-SP selectively inhibited CCh-induced contractions when compared to the oxytocin-induced contractions. Similar results were recorded by Silva *et al.* (2002) with ethanol extract from the aerial parts of *S. paludosum.*

*Effect of the TAF-SP on guinea pig trachea*

TAF-SP relaxed CCh-contracted trachea in a concentration-dependent manner, both in the presence and absence of functional epithelium with $EC_{50}$ values of 353.2±15.2 and 159.4±23.0 µg/ml, respectively. These results show an increased potency of 2.2 times in the absence of functional epithelium suggesting that the presence of epithelium is impairing the relaxant effect of TAF-SP. These results are not consistent with those observed by Duarte *et al.* (2003) who showed that the ethyl acetate phase from the aerial parts of *S. paludosum* relaxed guinea pig trachea in an epithelium independent manner.

## Conclusions

The total alkaloid fraction from root bark of *S. paludosum* (TAF-SP) had no significant cytotoxic effects on rat erythrocytes, but showed significant spasmolytic activity on rat uterus and guinea pig trachea. In rat uterus this effect was more selective to carbachol, and in guinea pig trachea the relaxant effect was potentiated in the absence of epithelium.

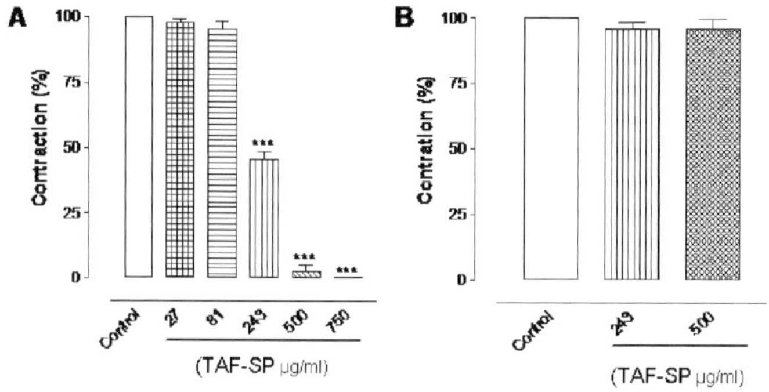


**Figure 2.** Effect of TAF-SP on carbachol- (A) and oxytocin- (B) induced phasic contractions on rat uterus (*n*=5 and 3, respectively). Columns and vertical bars represent mean and SEM, respectively. One-way ANOVA followed by Bonferroni's test, ***$P < 0.001$ (control vs TAF-SP).

## Acknowledgements

The authors are grateful to José Crispim Duarte for technical assistance and to CNPq and CAPES for financial support.

## References

Agra MF and Bhattacharyya J (1999). Ethnomedicinal and phytochemical investigation of the *Solanum* species in the northeast of Brazil. In *Solanaceae IV* (M Nee, DE Symon, RN Lester, and JP Jessop, eds), pp. 341-343. Royal Botanic Gardens, Kew.

Basílio IJLD (2008). Estudo farmacobotânico de orgãos vegetativos e fitoquímico dos alcalóides da casca de raízes de *Solanum paludosum* Moric. (SOLANACEAE), 92 pp. Dissertação de Mestrado, Laboratório de Tecnologia Farmacêutica, Universidade Federal da Paraíba, João Pessoa.

Bhakuni OS, Dhar ML, Dhar MM, Dhawan BN, and Mehrotra BN (1969). Screening of Indian plants for biological activity. Part II. *Indian Journal of Experimental Biology* 7:250-262.

Boyd LJ (1928). Pharmacology of the homeopathic drugs. *Journal of the American Institute of Homeopathy* 21:29.

Brandão R, Lara FS, Pagliosa LB, Soares FA, Rocha JBT, Nogueira CW, and Farina M (2005). Hemolytic effects of sodium selenite and mercuric chloride in human blood. *Drug Chemical Toxicology* 28:397-407.

Cavalcante FA (2001). Mecanismo de ação espasmolítica de solavetivona, sesquiterpeno isolado das partes aéreas de *Solanum jabrense* Agra & Nee (SOLANACEAE), 81 pp. Dissertação de Mestrado, Pós-graduação em Produtos Naturais e Sintéticos e Bioativos (Universidade Federal da Paraíba), João Pessoa, Paraíba.

Claudino FS (2003). Atividade espasmolítica de extratos obtidos de *Solanum jabrense* Agra e Nee (SOLANACEAE), 113 pp. Dissertação de Mestrado, Pós-graduação em Produtos

Naturais e Sintéticos e Bioativos (Universidade Federal da Paraíba), João Pessoa, Paraíba.

Correia ACC (2007). Efeito comparativo entre os extratos obtidos das partes aéreas e dos frutos de *Solanum asperum* rich. (Solanaceae) em músculo liso, 49 pp. Trabalho de Conclusão de Curso, Graduação em Farmácia (Universidade Federal de Alagoas), Maceió, Alagoas.

Costa VCO (2006). Investigação da Atividade Espasmolítica de *Solanum asperum* Rich. e *Solanum paludosum* Moric.: um estudo comparativo, 40 pp. Trabalho de Conclusão de Curso, Graduação em Ciências Biológicas, Universidade Federal da Paraíba, João Pessoa.

De Jalon PG, Bayo JB, and De Jalon MG (1945). *Farmacoterapy Actual* 2:313.

Duarte MC, Silva JLV, Cavalcante FA, Ribeiro LAA, Silva TMS, and Silva BA (2003). Papel dos canais de $Ca^{2+}$ e $K^{+}$ na ação relaxante da fase acetato de etila de *Solanum paludosum* Moric. (SOLANACEAE*)*. In *Encontro de Iniciação Científica da UFPB, 9*, João Pessoa, CD-ROM (Resumos), João Pessoa.

Eisele K, Lang PA, Kempe DS, Klarl BA, Niemoller O, Wieder T, Huber SM, Duranton C, and Lang F (2006). Stimulation of erythrocyte phosphatidylserine exposure by mercury ions. *Toxicological Applied Pharmacology* 210:116-122.

Friedman M, Rayburn JR, and Bantle JA (1991). Developmental toxicology of potato alkaloids in the frog embryo teratogenesis assay – *Xenopus* (FETAX). *Food Chemistry Toxicology* 29:537-547.

Garcia FM (2007). Monitoração da Atividade Espasmolítica de Produtos obtidos de *Solanum asperum* Rich (Solanaceae), 81 pp. Dissertação de Mestrado, Pós-graduação em Produtos Naturais e Sintéticos e Bioativos (Universidade Federal da Paraíba), João Pessoa, Paraíba.

Hunziker AT (2001). *Genera solanacearum: The genera of Solanaceae illustrated, arranged according to a new system,* 500 pp. A.R. Gantner Verlag, Köenigstein, Germany.

Jenkinson DH, Barnard EA, Hoyer D, Humphrey PPA, Leff P, and Shankley NP (1995). International union of pharmacology committee on receptor nomenclature and drug classification. IX Recommendations on terms and symbols in quantitative pharmacology. *Pharmacology Reviews* 42:255-266.

Lexis LA, Fassett RG, and Coombes JS (2006). α-tocopherol and α-lipoic acid enhance the erythrocyte antioxidant defense in cyclosporine A-treated rats. *Basic Clinical Pharmacology Toxicology* 98:68-73.

Morris SC and Lee TH (1984). The toxicity and teratogenicity of Solanaceae glycoalkaloids, particularly those of the potato (*Solanum tuberosum*): a review. *Food Technology in Australia* 36:118-124.

Niki E, Yamamoto Y, Komuro E, and Sato K (1991). Membrane damage due to lipid oxidation. *American Journal Clinical Nutrition* 53:201-205.

Oliveira RCM, Monteiro FS, Silva JLV, Ribeiro LAA, Santos RF, Duarte JC, Agra MF, Silva TMS, Almeida FRC, and Silva BA (2006a). Extratos metanólico e acetato de etila de *Solanum megalonyx* Sedtn. (Solanaceae) apresentam atividade espasmolítica em íleo isolado de cobaia: um estudo comparativo. *Revista Brasileira de Farmacognosia* 16(2):146-151.

Oliveira RCM, Lima JT, Ribeiro LAA, Silva JLV, Monteiro FS, Assis TS, Agra MF, Silva TMS, Almeida FRC, and Silva BA (2006b). Spasmolytic Action of the Methanol Extract and Isojuripidine from *Solanum asterophorum* Mart. (Solanaceae) Leaves in

Guinea-Pig Ileum. *Zeitschrift für Naturforschung CA Journal of Biosciences* 61c:799-805.

Olmstead RGR, Sprangler, E, Bohs L, and Palmer JD (1999). Phylogeny and provisional classification of the Solanaceae based on chloroplast DNA. In *Solanaceae IV* (M Nee and DE Symon, eds), pp. 111-138. Royal Botanic Gardens, Kew, UK.

Prokof'eva NG, Utkina NK, Chaikina EL, and Makarchenko AE (2004). Biological activities of marine sesquiterpenoid quinones: structure–activity relationships in cytotoxic and hemolytic assays. *Compendium Biochemical Physiology, Part B* 139:169-173.

Rangel M, Malpezzi ELA, Susini SMM, and Freitas JC (1997) Hemolytic activity in extracts of the diatom *Nitzschia*. *Toxicon* 35(2):305-309.

Ribeiro R, Fiuza de Melo MMR, Barros F, Gomes C, and Trolin G (1986). Acute antihypertensive effect in conscious rats produced by some medical plants used in the state of São Paulo. *Journal of Ethnopharmacology* 15(3):261-269.

Ribeiro EAN (2001) Estudo das ações cardiovasculares da fração aquosa do extrato etanólico do caule de *Solanum stipulaceum* Roem. & Schult., (SOLANACEAE) em ratos, 166 pp. Dissertação de Mestrado, Pós-graduação em Produtos Naturais e Sintéticos e Bioativos (Universidade Federal da Paraíba), João Pessoa, Paraíba.

Santos RF, Cavalcante FA, Silva JLV, Oliveira RCM, Silva TMS, and Silva BA (2004). Evaluation of spasmolytic action of hexane phase from *Solanum agrarium* Sendt. in rat uterus. In *XXXVI Congresso Brasileiro de Farmacologia e Terapêutica Experimental,* Águas de Lindóia. Programa & Resumos, 290.

Silva JLV, Silva BA, Cavalcante FA, Macêdo LS, Duarte JC, and Silva TMS (2002). Investigação da atividade espamolítica de *Solanum paludosum* Moric. (Solanaceae): estudo comparativo entre os extratos etanólico e metanólico. In *Iniciados*. Editora Universitária/UFPB, João Pessoa, Paraíba.

Silva KN (2006). Estudo farmacobotânico de três espécies de *Solanum L.* (Solanaceae). E triagem farmacológica da atividade espasmolíticade *Solanum paniculatum L.,* 140 pp. Dissertação de Mestrado, Pós-graduação em Produtos Naturais e Sintéticos e Bioativos (Universidade Federal da Paraíba), João Pessoa, Paraíba.

Silva PCB (2007). Investigação da atividade espasmolítica de *Solanum paraibanum* Agra: um estudo comparativo, 65 pp. Trabalho de Conclusão de Curso, Graduação em Farmacia (Universidade Federal de Alagoas), Maceió, Alagoas.

Shum OL and Chiu KW (1991). Hipotensive action of *Solanum melongena* on normotensive rats. *Phytotherapy Research* 5(2):76-81.

Tschirhart E, Frossard N, Bertrand C, and Landry Y (1987). Arachidonic acid metabolites and airway epithelium-dependent relaxant factor. *Journal of Pharmacology and Experimental Therapeutics* 243(1):310-316.

# Chapter 117

# Hemolytic and Spasmolytic Assays of *Solanum asterophorum* Mart. (Solanaceae)

P.C.B. Silva[1], M.A. de Vasconcelos[2], L.C.G.C. Lima[2], L.O. Silva[2], K.M. Silva[2], A.C.C. Correia[3], C.L. Macêdo[3], H.L.F. Pessôa[4], T.M.S. Silva[5], B.A. Silva[3,6], and F.A. Cavalcante[1,2]

[1]*Programa de Pós-Graduação em Nutrição/Faculdade de Nutrição/Universidade Federal de Alagoas (UFAL), Campus A. C. Simões, Maceió, Alagoas, 57072-970, Brazil;* [2]*Instituto de Ciências Biológicas e da Saúde/UFAL;* [3]*Laboratório de Tecnologia Farmacêutica 'Prof. Delby Fernandes de Medeiros'/UFPB;* [4]*Departamento de Biologia Molecular/UFPB;* [5]*Departamento de Química/UFRPE;* [6]*Departamento de Ciências Farmacêuticas/UFPB, Brazil*

## Introduction

The Solanaceae family is comprised of 106 genera and 2300 species (Olmstead *et al.* 1999) with a cosmopolitan distribution and South America is one of the main centers of diversity and endemism (Hunziker 2001). Moreover, this family has substantial economic importance as exemplified by the potato (*Solanum tuberosum* L.) and tomato (*Solanum lycopersicum* L.). Further, this plant family supplies raw material for the production of drugs of pharmacological or toxicological interest, atropine (*Atropa belladonna* L.) and nicotine (*Nicotiana tabacum* L.) (Agra 2000) being prime examples.

The genus *Solanum* L. is the largest of the Solanaceae family with about 1400 species (Bohs 2005) and 5000 epithets (Nee 1999) inhabiting tropical and subtropical regions of the world (Agra 1999). In Brazil, many species of *Solanum* are known popularly as 'jurubeba'; some species are toxic and others are used in folk medicine mainly in the northeast where they are used for the treatment of various diseases including liver and renal disorders (Agra and Bhattacharyya 1999), inflammation (Agra 1999), diarrhea (Abebe 1986; Maikere-Faniyo *et al.* 1989; Chhabra *et al.* 1993), and spasms (Esa *et al.* 2000). Some species present pharmacological activities such as spasmolytic effects. Examples of active species include *S. asperum* (Costa 2006; Correia 2007; Garcia 2007) and *S. paraibanum* Agra (Silva 2007).

*Solanum asterophorum* Mart. is a shrub species, popularly known as 'jurubeba-de-fogo'. In Brazil, it is distributed in the states of Paraíba and Bahia and used in folk medicine for liver problems (Agra and Bhattacharyya 1999). Recent studies with this species observed that the methanol extract obtained from leaves relaxed the guinea pig ileum (Oliveira *et al.* 2006) and the aerial parts and roots did not show molluscicidal activity (Silva *et al.* 2007). Thus, on the basis of chemotaxonomic criteria, we decided to

©CAB International 2011. *Poisoning by Plants, Mycotoxins, and Related Toxins* (eds F. Riet-Correa, J. Pfister, A.L. Schild, and T.L. Wierenga)

investigate a possible hemolytic effect on rat erythrocytes and spasmolytic activity on rat aorta and guinea pig ileum of a methanol extract obtained from roots of *S. asterophorum* (Sast-$MeOH_R$).

## Material and Methods

All experimental procedures were performed in accordance with the guidelines approved by the Research Ethic Committee of the Universidade Federal de Alagoas (Protocol CEP/UFAL No 027241/2008-11). The botanical material was identified by Dr Maria de Fátima Agra, from Setor de Botânica of Laboratório de Tecnologia Farmacêutica of Universidade Federal da Paraíba (LTF/UFPB). A voucher specimen (Agra 6002) is deposited in the Herbarium Prof. Lauro Pires Xavier (JPB) of UFPB.

Initially we evaluated the cytotoxic potential of the Sast-$MeOH_R$ extract on rat erythrocytes. This procedure followed the methodology described by Rangel *et al.* (1997). A sample of blood was collected from fasting Wistar male rats (200-300 g) and immediately mixed with a solution (pH=7.4) of 0.9% NaCl and 10 mM $CaCl_2$ at a ratio of 1:30 in slow and constant agitation to avoid coagulation and centrifuged at 5000 rpm for 3 min to obtain the erythrocytes. Sast-$MeOH_R$ was added to the suspension of erythrocytes at various concentrations and in different preparations. Negative control was prepared using a suspension of erythrocytes with more NaCl and $CaCl_2$ (0% hemolysis); a positive control was prepared with a suspension of erythrocytes with 1% Triton X-100 (100% hemolysis), and the percentage of hemolysis was calculated relative to this value. The samples were incubated for 1 h at room temperature under slow and constant agitation. After this time, they were centrifuged at 2500 g for 5 min and hemolysis was measured by spectrophotometry at 540 nm. The results obtained for the three independent hemolysis experiments are expressed as mean±SEM. The hemolytic activity was evaluated according to Prokof'eva *et al.* (2004).

Monitoring the spasmolytic activity of the Sast-$MeOH_R$ extract was performed on rat aorta and guinea pig ileum. For the aorta experiment the animals were euthanized by cerebral concussion followed by exsanguination. Aorta arteries were dissected free and placed in Krebs solution containing (in mM) NaCl 118.0, $NaHCO_3$ 25.0, KCl 4.6, $MgSO_4$ 5.7, $KH_2PO_4$ 1.1, $CaCl_2$ 2.5, glucose 11.0, pH 7.4. The arteries were cut into small rings (4 mm in length) and mounted on a force transducer vertically in 6 ml isolated organ baths containing Krebs solution bubbled with a carbogen mixture (95% $O_2$ and 5% $CO_2$) and maintained at 37°C. The aortic rings were stabilized for a period of 60 min. During this period the Krebs solution was renewed every 15 min to prevent the interference of metabolites (Altura and Altura 1970). After the stabilization period two contractions of similar magnitude were induced with $3\times10^{-7}$ M phenylephrine and under the tonic component 12 to 15 min. After the second response $10^{-6}$ M acetylcholine (ACh) was added to all preparations to verify the integrity of the endothelium (Furchgott and Zawadzki 1980). The vascular endothelium was considered intact when the aortic rings showed relaxation more than 50% (Ajay *et al.* 2003). The rings were considered to be endothelium-free if more than 90% relaxation was eliminated. After washing, a third response to agonist was induced and the Sast-$MeOH_R$ was added.

For guinea pig ileum experiments the animals were euthanized as described above. The abdomen was opened and a segment of ileum was removed and placed in a modified Krebs solution containing (in mM): NaCl 117.0, $NaHCO_3$ 25.0, KCl 4.7, $MgSO_4$ 1.3, $NaH_2PO_4$ 1.2, $CaCl_2$ 2.5, glucose 11.0, pH 7.4. The ileum was cut into small segments (3 cm in length) and mounted vertically in 6 ml isolated organ baths containing modified Krebs solution bubbled with a carbogen mixture and maintained at 37°C. After the stabilization period (30 min), two phasic contractions of similar magnitude were obtained with $10^{-6}$ M carbachol and recorded using isotonic levers coupled to kymographs and smoked drums. The Sast-MeOH$_R$ was incubated alone for 15 min in different preparations and at various concentrations. Inhibition of response to submaximal carbachol was assessed by comparing the responses before (control) and after addition to chamber. The molar concentration of Sast-MeOH$_R$ that reduces the response to an agonist by 50% ($IC_{50}$) was obtained by non-linear regression.

In other preparations, after stabilization of the guinea pig ileum isometric contraction also was elicited with 40 mM KCl and was measured with a force displacement transducer (World Precision Instruments, USA). After a further 30 min the process was repeated and the Sast-MeOH$_R$ was added cumulatively at the plateau phase in different preparations. The molar concentration of Sast-MeOH$_R$ that produced 50% of its maximal effect ($EC_{50}$) was obtained graphically from concentration-inhibition curves. Relaxation was expressed as reversal percentage of the initial contraction elicited by KCl.

Values were expressed as mean±SEM. Statistical analysis was performed using Graph-Pad Prism 3.03 software (GraphPad Software Inc., San Diego, CA, USA). The $IC_{50}$ and $EC_{50}$ values were determined by non-linear regression (Neubig *et al.* 2003). Differences between means were statistically compared using *t* test and one-way ANOVA followed by Bonferroni's test, as appropriate, and were considered to differ significantly when $P < 0.05$.

## Results and Discussion

Erythrocytes, which are anucleated and without cytoplasmic organelles, have poor repair and biosynthetic mechanisms resulting in oxidative lesions on membrane and hemoglobin which may lead to an oxidative stress condition. Synergistic and cooperative interactions between the antioxidant systems result in the sequential degradation of harmful reactive molecules as well as the protection of cellular proteins and membranes (Martins-Paiva *et al.* 2009). Thus erythrocytes provide a model to study toxic and protective effects of a great variety of substances and situations associated with oxidative stress (Eisele *et al.* 2006; Lexis *et al.* 2006).

In the evaluation of cytotoxicity on rat erythrocytes the Sast-MeOH$_R$ extract did not induce hemolysis at concentrations of 81 and 243 μg/ml but produced a weak hemolytic activity ($E_{max}$=5.2±1.2%) only at a concentration of 500 μg/ml (Figure 1).

The Sast MeOH$_R$ extract showed no damage to the erythrocyte membrane of rats at the concentration used on guinea pig ileum, indicating that this extract probably has low or no toxicity when tested *in vivo.*

In experiments with rat aorta, the Sast-MeOH$_R$ extract did not alter the basal tone of rat aorta both in the presence and absence of functional endothelium. Similarly, this extract (1-27 μg/ml) showed no significant relaxant effects on aorta pre-contracted with phenylephrine. However, at concentrations of 81, 243, and 500 μg/ml the extract Sast-MeOH$_R$ exhibited additional contractile effect on these contractions (data not shown). Similar results were obtained by Oliveira (2006) who observed that up to 750 μg/ml of the methanol extract obtained from the leaves of *Solanum asterophorum* showed no significant effect of relaxing the rat aorta.

Sast-MeOH$_R$ (3-500 μg/ml) antagonized the phasic contractions induced by $10^{-6}$ M carbachol on guinea pig ileum (Figure 2) in a significant and concentration-dependent manner. The $IC_{50}$ values to Sast-MeOH$_R$ were 70.8±20.4 μg/ml. The extract showed an $E_{max}$ = 90.8±3.8 μg/ml.

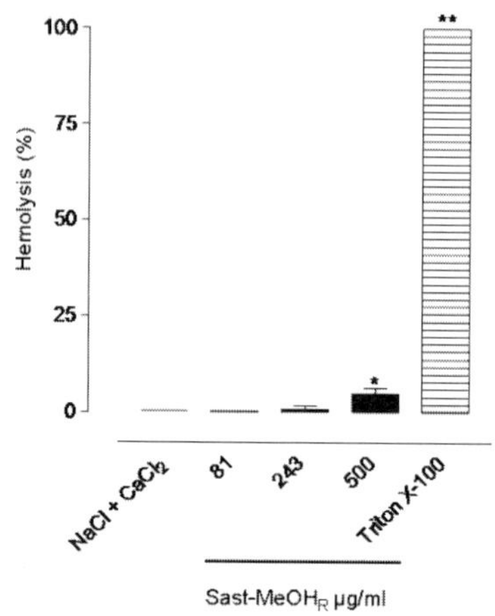


**Figure 1.** Hemolytic effect of Sast-MeOH$_R$ extract on rat erythrocytes (*n*=3). Triton X-100 was a positive control and NaCl + $CaCl_2$ was a negative control. Columns and vertical bars represent the mean and SEM, respectively. One-way ANOVA, followed by Bonferroni's test, **P* < 0.05 and ***P* < 0.001 (negative control vs Sast-MeOH$_R$ or positive control).

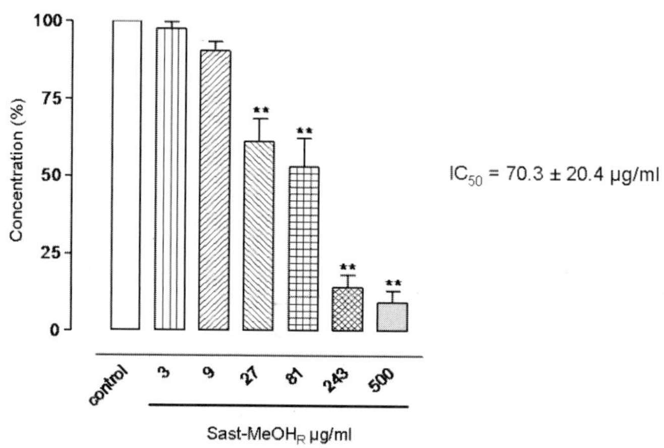


**Figure 2.** Effect of Sast-MeOH$_R$ extract on carbachol-induced phasic contractions on guinea pig ileum (*n*=5). Columns and vertical bars represent the mean and SEM, respectively. One-way ANOVA, followed by Bonferroni's test, ***P* < 0.001 (control vs Sast-MeOH$_R$).

The contraction in smooth muscle in response to various agents is often composed of two phases: a rapid phasic component followed by a slower, more sustained tonic component (Bolton 1979; Van Breemen *et al.* 1979). This biphasic response is due to the dual source of $Ca^{2+}$ in smooth muscle. In guinea pig ileum, muscarinic agonists produce this biphasic response and it is suggested that the phasic contraction is caused by release of $Ca^{2+}$ from intracellular stores mediated by inositol 1,4,5-trisfosfato-$IP_3$ (Abdellatif 1989; Kobayashi *et al.* 1989). On the other hand, the tonic contraction induced by muscarinic agonists on guinea pig ileum is attributed to the influx of $Ca^{2+}$ through voltage-activated calcium channels ($Ca_V$) since the tonic contraction is suppressed by blocking of $Ca_V$ using verapamil (Jim *et al.* 1981). As the mechanisms that trigger the phasic contraction in the ileum are different from those that maintain a sustained tonic contraction (Abdellatif 1989; Kobayashi *et al.* 1989; Honda *et al.* 1996) it may be that the Sast-$MeOH_R$ extract promotes relaxation of ileum pre-contracted by KCl which would be suggestive at a functional level of blocking the influx of $Ca^{2+}$ through the plasma membrane.

As shown in Figure 3, the Sast-$MeOH_R$ extract (1-500 μg/ml) relaxed the guinea pig ileum pre-contracted with 40 mM KCl in a significant and concentration-dependent manner showing an $EC_{50}$ of 45.2±10.2 μg/ml (*n*=3). When the contraction was induced by KCl, the Sast-$MeOH_R$ extract reached an $E_{max}$ of 97.0±2.97.

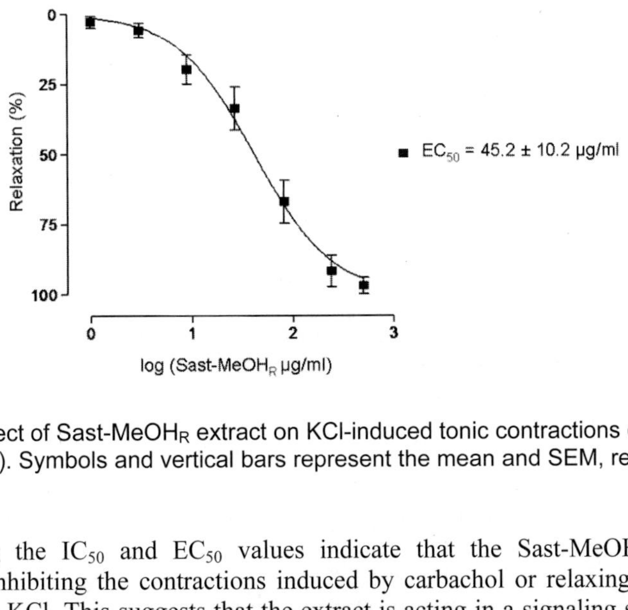


**Figure 3.** Effect of Sast-$MeOH_R$ extract on KCl-induced tonic contractions on guinea pig ileum (*n*=3). Symbols and vertical bars represent the mean and SEM, respectively.

Analyzing the $IC_{50}$ and $EC_{50}$ values indicate that the Sast-$MeOH_R$ extract was equipotent in inhibiting the contractions induced by carbachol or relaxing the ileum pre-contracted with KCl. This suggests that the extract is acting in a signaling step common to these contractile agents. As the extract was able to inhibit the tonic component of contraction induced by KCl this confirmed the earlier hypothesis that Sast-$MeOH_R$ could be blocking the influx of $Ca^{2+}$ through voltage-activated calcium channels ($Ca_V$).

These results when compared with those obtained by Oliveira *et al.* (2006) with the methanol extract of leaves from *S. asterophorum* indicate that the Sast-$MeOH_R$ root extract is 2-4 times more potent.

## Conclusions

The extract inhibited the tonic contractions on guinea pig ileum, suggesting that the blockade of calcium influx is through voltage-activated calcium channels ($Ca_V$), as these channels are responsible for maintaining this contractile response. The Sast-$MeOH_R$ extract caused no damage to the membrane of erythrocytes at concentrations that showed spasmolytic activity on guinea pig ileum, thus it appears that Sast-MeOH$_R$ probably has low or no *in vivo* toxicity.

## Acknowledgements

The authors are grateful to José Crispim Duarte for technical assistance and to PIBIC/UFAL/FAPEAL, CNPq and CAPES for financial support.

## References

Abdellatif AA (1989). Calcium mobilizing receptors, polyphospholinositides, generation of second messengers and contraction in mammalian smooth muscle: historical perspectives and current status. *Life Sciences* 45:757-786.

Abebe W (1986). A survey of prescriptions used in traditional medicine in Gondar region, northwestern Ethiopia: general pharmaceutical practice. *Journal of Ethnopharmacology* 18(2):147-165.

Agra MF (1999). Diversity and distribution of *Solanum* subgenus *Leptostemonum* in north-east Brazil. In *Solanaceae IV* (M Nee, DE Symon, RN Lester, and JP Jessop, eds), pp. 197-203. Royal Botanic Gardens, Kew.

Agra MF (2000). Taxonomic revision of *Solanum* sect. *Erythrotrichum* Child (Solanaceae), 292 pp. Tese de Doutorado, Instituto de Biociências, Universidade de São Paulo, São Paulo.

Agra MF and Bhattacharyya J (1999). Ethnomedicinal and phytochemical investigation on the *Solanum* species in the Northeast of Brazil. In *Solanaceae IV* (M Nee, DE Symon, RN Lester, and JP Jessop, eds), pp. 341-343. Royal Botanic Gardens, Kew.

Ajay M, Gilani AH, and Mustafa MR (2003). Effects of flavonoids on vascular smooth muscle of the isolated rat thoracic aorta. *Life Sciences* 74:603-612.

Altura BM and Altura BT (1970). Differential effects of substrate depletion on drug-induced contractions of rabbit aorta. *American Journal Physiology* 219(6):1698-1705.

Bohs L (2005). Major clades in *Solanum* based on *ndh*F sequence data. In *Solanaceae, William D'Arcy Memorial, Monographs in Systematic Botany from the Missouri Botanical Garden* (RC Keating, VC Hollowell, and TB Croat, eds), pp. 27-49. Missouri Botanical Garden Press, St Louis, Missouri, USA.

Bolton TB (1979). Mechanism of action of transmitters and other substances on smooth muscle. *Physiology Review* 59:606-718.

Chhabra SC, Mahunnah RLA, and Mshiu EN (1993). Plants used in traditional medicine in eastern Tanzania. VI. Angiosperms (Sapotaceae to Zingiberaceae). *Journal of Ethnopharmacology* 39(2):83-103.

Correia ACC (2007). *Efeito comparativo entre os extratos das partes aéreas e dos frutos de* Solanum asperum *Rich. (Solanaceae) em músculo liso*, 49 pp. Monografia, Universidade Federal de Alagoas, Maceió, Alagoas.

Costa VCO (2006). *Research espasmolítica the activity of* Solanum asperum *Rich. and* Solanum paludosum *Moric.: a comparative study*, 40 pp. Monografia, Universidade Federal da Paraíba, João Pessoa, Paraíba.

Eisele K, Lang PA, Kempe DS, Klarl BA, Niemoller O, Wieder T, Huber SM, Duranton C, and Lang F (2006) Stimulation of erythrocyte phosphatidylserine exposure by mercury ions. *Toxicological Applied Pharmacology* 210:116-122.

Esa RD, Vireque AA, Reis JE, and Guerra MD (2000). Evaluation of the toxicity of *Solanum lycocarpum* in the reproductive system of male mice and rats. *Journal of Ethnopharmacology* 73:283-287.

Furchgott RF and Zawadki JV (1980). The obligatory role of endothelium cell in the relaxation of arterial smooth muscle by acetylcholine. *Nature* 286:373-376.

Garcia FM (2007) Monitoração da Atividade Espasmolítica de produtos obtidos de *Solanum asperum* Rich, 81 pp. Dissertação de Mestrado, Pós-Graduação em Produtos Naturais e Sintéticos Bioativos (Universidade Federal da Paraíba), João Pessoa, Paraíba.

Honda K, Takano Y, and Kamiya H (1996). Involvement of protein kinase C in muscarinic agonist-induced contractions of guinea pig ileal longitudinal muscle. *General Pharmacology* 27:957-961.

Hunziker AT (2001) *Genera solanacearum: The genera of Solanaceae illustrated, arranged according to a new system.* A.R. Gantner Verlag, Köenigstein, Germany, 500 pp.

Jim K, Harris A, Rosenberger LB, and Triggle DJ (1981). Stereoselective and non-stereoselective effects of D-600 (methoxyverapamil) in smooth-muscle preparations. *European Journal of Pharmacology* 76:67-72.

Kobayashi S, Kitazawa T, Somlyo AV, and Somlyo AP (1989). Citosolic heparin inhibits muscarinic and adrenergic $Ca^{2+}$-release in smooth muscle: physiological role of inositol 1,4,5-trisphosphate in pharmacomechanical coupling. *Journal of Biological Chemistry* 264:17997-18004.

Lexis LA, Fassett RG, and Coombes JS (2006). α-tocopherol and α-lipoic acid enhance the erythrocyte antioxidant defence in cyclosporine A-treated rats. *Basic Clinical Pharmacology Toxicology* 98:68-73.

Maikere-Faniyo R, Van Puyvelde L, Mutwewingabo A, and Habiyaremye FX (1989). Study of Rwandese medicinal plants used in the treatment of diarrhea. *Journal of Ethnopharmacology* 26:101-109.

Martins-Paiva F, Fernandes J, Rocha S, Nascimento H, Vitorino R, Amado F, Borges F, Belo L, and Santos-Silva A (2009). Effects of olive oil polyphenols on erythrocyte oxidative damage. *Molecular Nutrition & Food Research* 53:609-616.

Nee M (1999). Synopsis of *Solanum* in the New World. In *Solanaceae IV: Advances in Biology and Utilization* (M Nee, DE Symon, RN Lester, and JP Jessop, eds), pp. 285-33. Royal Botanic Gardens, Kew.

Neubig RR, Spedding M, Kenakin T, and Christopoulos A (2003). International union of pharmacology committee on receptor nomenclature and drug classification. XXXVIII. Update on terms and symbols in quantitative pharmacology. *Pharmacological Review* 55:(4)597-606.

Oliveira RCM (2006). Efeito antiinflamatório, antinociceptivo e espasmolítico de *Solanum megalonyx* Sendtn. e *Solanum asterophorum* Mart. (Solanaceae): um estudo comparativo, 123 pp. Tese de Doutorado em Produtos Naturais e Sintéticos Bioativos, Universidade Federal da Paraíba, João Pessoa, Paraíba.

Oliveira RCM, Lima JT, Ribeiro LAA, Silva JLV, Monteiro FS, Assis TS, Agra MF, Silva TMS, Almeida FRC, and Silva BA (2006). Spasmolytic Action of the Methanol Extract and Isojuripidine from *Solanum asterophorum* Mart. (Solanaceae) Leaves in Guinea-Pig Ileum. *Zeitschrift Für Naturforschung C–A Journal of Biosciences* 61:799-805.

Olmstead RGR, Sprangler E, Bohs L, and Palmer JD (1999) Phylogeny and provisional classification of the Solanaceae based on chloroplast DNA. In *Solanaceae IV* (M Nee and DE Symon, eds), pp. 111-138. Royal Botanic Gardens, Kew.

Prokof'eva NG, Utkina NK, Chaikina EL, and Makarchenko AE (2004). Biological activities of marine sesquiterpenoid quinones: structure–activity relationships in cytotoxic and hemolytic assays. *Comparative Biochemistry and Physiology, Part B: Biochemistry & Molecular Biology* 139:169-173.

Rangel M, Malpezzi EL, Susini SM, and Freitas JC (1997). Hemolytic activity in extracts of the diatom *Nitzschia*. *Toxicon* 35:305-309.

Silva PC (2007). *Efeito comparativo entre os extratos obtidos das partes aéreas e dos frutos de* Solanum paraibanum *Rich. (Solanaceae) em músculo liso,* 65 pp. Monografia, Universidade Federal de Alagoas, Maceió, Alagoas.

Silva TM, Nascimento RJB, Batista MM, Agra MF, and Camara CA (2007). Brine shrimp bioassay of some species of *Solanum* from Northestern Brazil. *Revista Brasileira de Farmacognosia* 17:35-38.

Van Breemen C, Aaronson P, and Loutzenhiser R (1979). Sodium-calcium interaction in mammalian smooth muscle. *Pharmacological Reviews* 30:167-208.

Chapter 118

# Evaluation of the Cytotoxic and Spasmolytic Activities of *Solanum asperum* Rich. (Solanaceae)

L.O. Lima[1], A.D.S. Silva[1], P.C.B. Silva[2], A.C.C. Correia[3], C.L. Macêdo[3], H.L.F. Pessôa[4], T.M.S. Silva[5], B.A. Silva[3,6], and F.A. Cavalcante[1,2]

[1]*Instituto de Ciências Biológicas e da Saúde/Universidade Federal de Alagoas, Maceió, Alagoas, 57010-020, Brazil;* [2]*Programa de Pós-Graduação em Nutrição/FANUT/UFAL;* [3]*Laboratório de Tecnologia Farmacêutica 'Prof. Delby Fernandes de Medeiros'/UFPB;* [4]*Departamento de Biologia Molecular/UFPB;* [5]*Departamento de Química/UFRPE;* [6]*Departamento de Ciências Farmacêuticas/UFPB*

## Introduction

The genus *Solanum* belongs to the Solanaceae family. This family is comprised of 106 genera and 2300 species (Olmstead *et al.* 1999). Economically it is one of the most important families including numerous ornamental, edible, spicy, medicinal, narcotic, and poisonous species (Agra 2000). *Solanum* is well represented in Brazil and is widely distributed from north to south in diverse phytogeographic regions (Roe 1972).

Many of the species of *Solanum* are endemic in South America (Hunziker 2001) and in northeastern Brazil many are used in folk medicine for the treatment of various diseases including diarrhea (Coune and Denoel 1975; Hedberg *et al.* 1983; Abebe 1986; Maikere-Faniyo *et al.* 1989; Chhabra *et al.* 1993), asthma (Hope *et al.* 1993), liver and renal disorders (Agra 1996; Agra and Bhattacharyya 1999), and inflammation (Agra 1999). Many species of *Solanum* have shown spasmolytic activity and among them some have also shown toxic activities and are commonly known as 'jurubeba.' *S. asperum* Rich. is an erect shrub popularly known as 'jussara' or 'coça-coça' (Agra and Bhattacharyya 1999).

Several of these activities have been attributed to the presence of the steroidal alkaloid solasodine. This alkaloid is characteristic of the genus *Solanum*, in particular of the species *S. asperum* and it can be an important starting material for the synthesis of steroidal hormones. Recent studies carried out with the methanol extract of aerial parts and fruits from *S. asperum* have shown spasmolytic activity (Costa 2006; Correia 2007; Garcia 2007), therefore it is interesting to further investigate this species.

Based on chemotaxonomic criteria we decided to investigate if methanol extract obtained from roots of *S. asperum* (SAr-MeOH) shows hemolytic activity on rat erythrocytes and spasmolytic activity on guinea pig trachea and ileum.

©CAB International 2011. *Poisoning by Plants, Mycotoxins, and Related Toxins*
(eds F. Riet-Correa, J. Pfister, A.L. Schild, and T.L. Wierenga)

## Material and Methods

All experimental procedures were performed in accordance with the guidelines approved by the Research Ethic Committee of the Universidade Federal de Alagoas (Protocol CEP/UFAL No 007140/2004-91). Thc botanical material was identified by Dr Maria de Fátima Agra from Setor de Botânica of LTF. A voucher specimen (Agra *et al.* 6487) is deposited in the Herbarium Prof. Lauro Pires Xavier (JPB) of UFPB.

Evaluation of cytotoxic potential of the methanol extract obtained from roots of *S. asperum* Rich. (SAr-MeOH) on rat erythrocytes followed the methodology described by Rangel *et al.* (1997) in which the rats (*Rattus norvegicus*) weighing between 200 and 300 g were fasted for a period of 12 h. After this period a blood sample was collected through cardiac puncture and immediately mixed with a solution (pH=7.4) of 0.9% NaCl and 10 mM $CaCl_2$ at a ratio of 1:30 by slow and constant shaking to avoid coagulation and centrifuged at 2500 rpm for 5 min to obtain the erythrocytes. This procedure was repeated and the sediment of the last centrifugation was resuspended in a 0.5% solution of NaCl and $CaCl_2$. The SAr-MeOH extract was added to suspension of erythrocytes at various concentrations (81, 243, and 500 μg/ml). The negative control was prepared using a suspension of erythrocytes with more NaCl and $CaCl_2$ (0% hemolysis) and positive control prepared with a suspension of erythrocytes to which was added 1% Triton X-100 (100% hemolysis) and the percentage of hemolysis was calculated relative to this value. The samples were incubated for 1 h at room temperature under slow and constant agitation. After this time they were centrifuged at 2500 rpm for 5 min and hemolysis was measured by spectrophotometry at 540 nm. The results obtained for the three independent hemolysis experiments are expressed as mean ± SEM. When hemolysis was quantified 20% was considered low hemolytic activity, between 20 and 50% was considered moderate hemolytic activity, and greater than 50% was considered high (Prokof'eva *et al.* 2004).

For guinea pig trachea experiments the animals were killed by cerebral concussion followed by exsanguination. Trachea was dissected free and placed in Krebs solution containing (in mM) NaCl 118.0, $NaHCO_3$ 25.0, KCl 4.6, Mg $SO_4$ 5.7, $KH_2PO_4$ 1.1, $CaCl_2$ 2.5, glucose 11.0, pH 7.4. The trachea was removed and cleaned of all connective tissue and fat. The trachea were mounted in isolated organ baths containing modified Krebs solution bubbled with a carbogen mixture and maintained at 37°C. The method described by Tschirhart *et al.* (1987) was used to observe the absence or presence of tracheal epithelium. The trachea segments were kept at a temperature of 37°C and remained at rest for 60 minutes, the Krebs solution being changed every 15 min. After this period of stabilization two tonic contractions were obtained of similar magnitude induced by $10^{-6}$ M carbachol and considered as controls. During the third phase of the tonic response to carbachol the extract was added in a cumulative manner to the chamber. The relaxation was expressed as the percentage of reverse carbachol-induced initial contraction.

For guinea pig ileum experiments the animals were euthanized as described above. The abdomen was opened and a segment of ileum was removed and placed in a modified Krebs solution containing (in mM) NaCl 117.0, $NaHCO_3$ 25.0, KCl 4.7, $MgSO_4$ 1.3, $NaH_2PO_4$ 1.2, $CaCl_2$ 2.5, glucose 11.0, pH 7.4. The ileum was cut into small segments (3 cm in length) and mounted vertically in 6 ml isolated organ baths containing modified Krebs solution bubbled with a carbogen mixture and maintained at 37°C. After the stabilization period (30 min) two phasic contractions of similar magnitude were obtained with $10^{-6}$ M carbachol and recorded using isotonic levers coupled to kymographs and smoked drums. The SAr-MeOH was incubated alone for 15 min in different preparations and at various concentrations. Inhibition of response to submaximal carbachol was assessed by comparing the responses before (control) and after addition to the chamber. The molar concentration of SAr-MeOH that reduces the response to an agonist by 50% ($IC_{50}$) was obtained by non-linear regression.

Values were expressed as mean ± SEM. Statistical analysis was performed using Graph-Pad Prism 3.03 software (GraphPad Software Inc., San Diego, CA, USA). The $IC_{50}$ values were determined by non-linear regression (Neubig *et al.* 2003). Differences between means were statistically compared using *t*-test and one-way ANOVA followed by Bonferroni's test, as appropriate, and were considered to differ significantly when $P < 0.05$.

## Results and Discussion

Erythrocytes, anucleated and without cytoplasmatic organelles, have poor repair and biosynthetic mechanisms resulting in oxidative lesions on membrane and hemoglobin which may lead to an oxidative stress condition (Martins-Paiva *et al.* 2009). The oxidative stress can be associated with formation of reactive oxygen species (ROS) (Anderson 1996), oxidation of hemoglobin to methemoglobin (Mansouri and Lurie 1993), lipid peroxidation (Hochstein 1988; Comporti 1993), and damage to cytoskeletal proteins (Grossman *et al.* 1992, McMillan *et al.* 2001) leading ultimately to hemolysis. Nevertheless, erythrocytes are equipped with several antioxidant defense mechanisms e.g. antioxidant enzymes, glutathione, tocopherol, and ascorbate (Martins-Paiva *et al.* 2009). Synergistic and cooperative interactions between these antioxidant systems results in the sequential degradation of harmful reactive molecules as well as the protection of cellular proteins and membranes (Arbos *et al.* 2008; Martins-Paiva *et al.* 2009). Thus erythrocytes provide a model to study toxic and protective effects of a great variety of substances and situations associated with oxidative stress (Eisele *et al.* 2006; Lexis *et al.* 2006).

In the evaluation of cytotoxicity on rat erythrocytes the SAr-MeOH extract did not induce hemolysis at concentrations of 81 and 243 μg/ml. However, SAr-MeOH induced moderately significant lysis activity ($E_{max} = 33.8 \pm 11.8\%$) only at the highest concentration tested of 500 μg/ml (Figure 1). This preliminary study demonstrated a reasonable choice of concentrations for use in experimental protocols designed to possible spasmolytic effects.

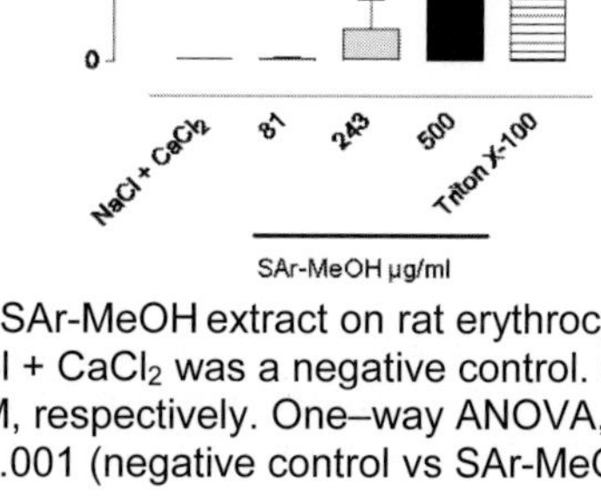


**Figure 1.** Hemolytic effect of SAr-MeOH extract on rat erythrocytes ($n$=3). Triton X-100 was a positive control, and NaCl + $CaCl_2$ was a negative control. Columns and vertical bars represent the mean and SEM, respectively. One–way ANOVA, followed by Bonferroni's test, $^*P < 0.05$ and $^{**}P < 0.001$ (negative control vs SAr-MeOH or positive control).

The SAr-MeOH was tested on basal tone of guinea pig trachea, however, no spasmogenic or spasmolytic effect was observed (data not shown). Conversely, at concentrations of 243 and 500 μg/ml the SAr-MeOH extract showed significant spasmolytic effect on the tonic contractions induced by $10^{-6}$ M carbachol on guinea pig trachea with and without functional epithelium (*n*=3). However, the extract did not produce a relaxation of more than 50% of maximum effect (Figure 2). These results are consistent with those found by Costa (2006) using the aerial parts of *S. asperum* who reported that the methanol extract did not present any effect on guinea pig trachea with and without functional epithelium and that the ethyl acetate extract presented a relaxation of more than 60%.

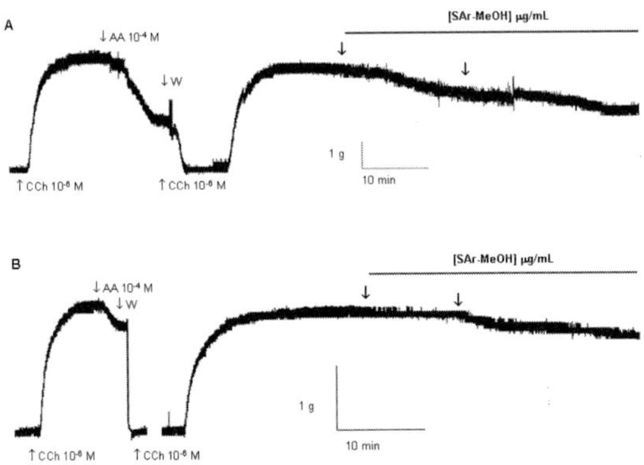


**Figure 2.** Original record representative of relaxant effect of SAr-MeOH (243 and 500 μg/ml) on guinea pig trachea pre-contracted with $10^{-6}$ M carbachol (CCh) in the presence (A) and absence (B) of functional epithelium. AA = arachidonic acid, W = washing.

The SAr-MeOH extract (3 to 81 μg/ml) also antagonized the phasic contractions induced by $10^{6}$ M carbachol on guinea pig ileum (Figure 3) in a significant and concentration-dependent manner ($IC_{50}$ = 43.7±5.6 μg/ml). The SAr-MeOH extract showed an $E_{max}$ of 91.9±6.1 at the concentration 81 μg/ml.

Interestingly, this extract obtained from roots was more potent than methanol extract obtained from fruits (Correia 2007) and aerial parts (Costa 2006) of *S. asperum*, suggesting that secondary metabolites may be more concentrated in roots than in other parts tested.

## Conclusions

The results demonstrate the SAr-MeOH showed low damage to the rat erythrocytes and the concentration tested on guinea pig ileum had no toxic effect, suggesting that SAr-MeOH would probably provide low or no toxicity when tested *in vivo*. In conclusion, these results indicate that the SAr-MeOH extract has secondary metabolites with low toxicity and a potential spasmolytic activity on guinea pig ileum. On the other hand the extract did not show any effect on guinea pig trachea.

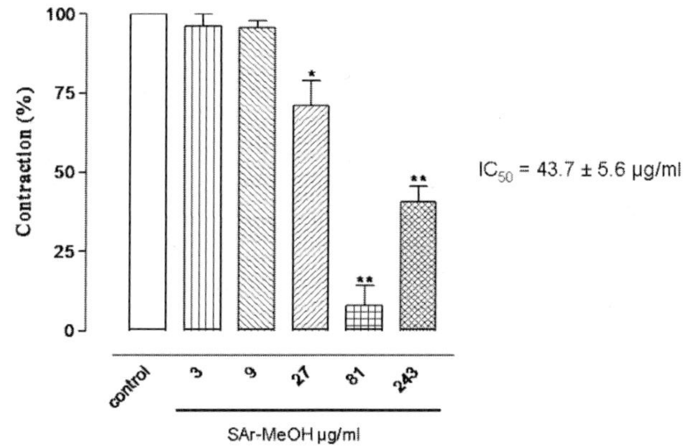


**Figure 3.** Effect of SAr-MeOH extract on phasic contractions induced by $10^{6}$ M carbachol on guinea pig ileum (*n*=3). Columns and vertical bars represent the mean and SEM, respectively. One-way ANOVA, followed by Bonferroni's test, *$P < 0.05$ and **$P < 0.001$ (control vs SAr-MeOH).


## Acknowledgements

The authors are grateful to José Crispim Duarte for technical assistance and to PIBIC/UFAL/FAPEAL, CNPq and CAPES for financial support.


## References


Abebe W (1986). A survey of prescriptions used in traditional medicine in Gondar region, northwestern Ethiopia: general pharmaceutical practice. *Journal of Ethnopharmacology* 18(2):147-165.

Agra MF (1996). *Plants of the folk medicine of Cariris Velhos, Paraíba, Brazil: Species most common,* 125 pp. Editora União, João Pessoa.

Agra MF (1999) Diversity and distribution of *Solanum* subgenus *Leptostemonum* in north-east Brazil. In *Solanaceae IV* (M Nee, DE Symon, RN Lester, and JP Jessop, eds), pp. 197-203. Royal Botanic Gardens, Kew

Agra MF (2000). Revisão taxonômica de *Solanum* sect. *Erythrotrichum* Child (Solanaceae), 292 pp. Tese de Doutorado, Instituto de Biociências da Universidade de São Paulo, São Paulo.

Agra MF and Bhattacharyya J (1999). In: M Nee, DE Symon, RN Lester, and JP Jessop, editors. *Solanaceae IV*, pp. 341-343. Royal Botanic Gardens, Kew.

Anderson D (1996). Antioxidant defenses against reactive oxygen species causing genetic and other damage. *Mutation Research* 350:103-108.

Arbos KA, Claro LM, Borges L, Santos CAM, and Weffort-Santos AM (2008). Human erythrocytes as a system for evaluating the antioxidant capacity of vegetables extracts. *Nutrition Research* 28:457-463.

Chhabra SC, Mahunnah RLA, and Mshiu EN (1993) Plants used in traditional medicine in eastern Tanzania. VI. Angiosperms (Sapotaceae to Zingiberaceae). *Journal of Ethnopharmacology* 39(2):83-103.

Comporti M (1993). Lipid peroxidation. Biopathological significance. *Molecular Aspects of Medicine* 14:199-207.

Correia ACC (2007). *Efeito comparativo entre os extratos obtidos das partes aéreas e dos frutos de* Solanum asperum *Rich. (Solanaceae) em músculo liso*, 49 pp. Monografia, Universidade Federal de Alagoas, Maceió, Alagoas.

Costa VCO (2006). *Investigação da atividade espasmolítica de* Solanum asperum *Rich. e* Solanum paludosum *Moric.: um estudo comparativo,* 40 pp. Monografia, Universidade Federal da Paraíba, João Pessoa, Paraíba.

Coune C and Denoel A (1975). Pytochemical Study of Central African Solanaceae. 1. Alkaloids of *Solanum dasyphyllum*. *Planta Medica* 28(5):168.

Eisele K, Lang PA, Kempe DS, Klarl BA, Niemoller O, Wieder T, Huber SM, Duranton C, and Lang F (2006). Stimulation of erythrocyte phosphatidylserine exposure by mercury ions. *Toxicology and Applied Pharmacology* 210:116-122.

Garcia FM (2007). Monitoração da Atividade Espasmolítica de produtos obtidos de *Solanum asperum* Rich, 81 pp. Dissertação de Mestrado, Pós-Graduação em Produtos Naturais e Sintéticos Bioativos (Universidade Federal da Paraíba), João Pessoa, Paraíba.

Grossman SJ, Simson J, and Jollow DJ (1992). Dapsone-induced hemolytic anemia: Effect of N-hydroxy dapsone on the sulfhydryl status and membrane proteins of rat erythrocytes. *Toxicology and Applied Pharmacology* 117:208-217.

Hedberg I, Hedberg O, Madati PJ, Mshigeni KE, Mshiu EN, and Samuelsson G (1983). Inventory of plants used in traditional medicine in Tanzania. Part iii. Plants of the families Papilionaceae-Vitaceae. *Journal of Ethnopharmacology* 9:237-260.

Hochstein P (1988). Perspectives on hydrogen peroxide and drug-induced hemolytic anemia in glucose-6-phosphate dehydrogenase deficiency. *Free Radical Biology and Medicine* 5:387-392.

Hope BE, Massey DG, and Fournier-Massey G (1993). Hawaiian Materia Medica for Asthma. *Hawaii Medical Journal* 52(6):160-166.

Hunziker AT (2001). *Genera solanacearum: The genera of Solanaceae illustrated, arranged according to a new system,* 500 p. A. R. Gantner Verlag, Köenigstein, Germany.

Lexis LA, Fassett RG, and Coombes JS (2006). α-tocopherol and α-lipoic acid enhance the erythrocyte antioxidant defence in cyclosporine A-treated rats. *Basic Clinical Pharmacology Toxicology* 98:68-73.

Maikere-Faniyo R, Van Puyvelde L, Mutwewingabo A, and Habiyaremye FX (1989). Study of Rwandese medicinal plants used in the treatment of diarrhea. *Journal of Ethnopharmacology* 26:101-109.

Mansouri A and Lurie AA (1993). Concise review: methemoglobinemia. *American Journal Hematology* 42:7-12.

Martins-Paiva F, Fernandes J, Rocha S, Nascimento H, Vitorino R, Amado F, Borges F, Belo L, and Santos-Silva A (2009). Effects of olive oil polyphenols on erythrocyte oxidative damage. *Molecular Nutrition and Food Research* 53:609-616.

McMillan DC, Bolchoz LJ, and Jollow DJ (2001). Favism: Effect of divicine on rat erythrocyte sulfhydryl status, hexose monophosphate shunt activity, morphology, and membrane skeletal proteins. *Toxicological Science* 62:353-359.

Neubig RR, Spedding M, Kenakin T, and Christopoulos A (2003). International union of pharmacology committee on receptor nomenclature and drug classification. XXXVIII. Update on terms and symbols in quantitative pharmacology. *Pharmacological Reviews* 55:597-606.

Olmstead RGR, Sprangler E, Bohs L, and Palmer JD (1999). Phylogeny and provisional classification of the Solanaceae based on chloroplast DNA. In *Solanaceae IV* (M Nee and DE Symon, eds), pp. 111–138. Royal Botanic Gardens, Kew.

Prokof'eva NG, Utkina NK, Chaikina EL, and Makarchenko AE (2004). Biological activities of marine sesquiterpenoid quinones: structure–activity relationships in cytotoxic and hemolytic assays., *Comparative Biochemistry and Physiology, Part B: Biochemistry and Molecular Biology* 139:169-173.

Rangel M, Malpezzi ELA, Susini SMM, and Freitas JC (1997). Hemolytic activity in extracts of the diatom *Nitzschia*. *Toxicon* 35:305-309.

Roe KEA (1972). Revision of *Solanum* section *Brevantherum* (Solanaceae). *Brittonia* 29:239-278.

Tschirhart E, Frossard N, Bertrand C, and landry Y (1987). Arachidonic acid metabolites and airway epithelium-dependent relaxant factor. *Journal of Pharmacology and Experimental Therapeutics* 243(1):310-316.

# Chapter 119

# Chemical Analysis of Toxic Principles in Preparations of *Ruta graveolens* and *Petiveria alliacea*

A.C.F. Amaral[1], J.R. de A. Silva[2], D.Q. Falcão[1], L.G. Ferreirinha[1], A.R. dos Santos[1], R.B. Araújo[1], and J.L.P. Ferreira[1,3]

[1]*Laboratory of Medicinal Plants and Derivatives, Dept. of Natural Products, Farmanguinhos, Fiocruz, Rio de Janeiro, RJ, 21041-250;* [2]*Laboratory of Chromatography, Dept. Chemistry, UFAM, Manaus, AM, 69077-000;* [3]*Laboratory of Pharmacognosy, Fac. of Pharmacy, UFF, Niterói, RJ, 24241-000, Brazil*

## Introduction

Since time immemorial, ancient civilizations have used plants for various purposes including the cure of diseases, as foods, and in religious rituals. Two species, *Petiveria alliacea* L. (Phytolaccaceae), a plant native to Africa and tropical America, and *Ruta graveolens* L. (Rutaceae), originally from meridional Europe, continue to have broad popular use in Brazil with applications ranging from religious ceremonies to various medicinal purposes. Preparations made from these species are used to treat several disorders. However, these preparations have shown signs of toxicity. For example, sheep developed muscular dystrophy and subsequently died after consuming *P. alliacea* daily (Hoyos *et al.* 1992). In addition, extracts of both species have shown mutagenic and carcinogenic effects after long periods of use and have caused abortion because of their ability to induce contractions of the uterus muscle (Paulini *et al.* 1987; Hoyos *et al*, 1992; El Agraa *et al.* 2002; Freitas *et al.* 2005; Gomes *et al.* 2005; Ivanova *et al.* 2005).

*Ruta graveolens*, popularly known as rue, is a hardy evergreen shrub of up to 1 m tall with a characteristic grayish green color and sharp unpleasant smell. The stems are ramified and flowers are small, yellow, and clustered during spring and summer. The taste is slightly stinging but is masked by a strong bitter odor (Harat *et al.* 2008). Chemically, this species is rich in flavonoids, quinolines, coumarins, and acridone alkaloids and is the main source of four commercially important linear furanocoumarins: psoralen, xanthotoxin (8-methoxypsoralen), bergapten (5-methoxypsoralen), and isopimpinelin (5,8-dimethoxypsoralen) which have gained wide applications in the pharmaceutical industry because of their photoreactive and potassium channel blocker properties. The yield of these metabolites is probably related to *R. graveolens* biological activities (Milesi *et al.* 2001; Diwan and Malpathak 2008; Harat *et al.* 2008). However, *R. graveolens* has shown high toxic potential for both animals and humans. Several studies related the consumption of this

©CAB International 2011. *Poisoning by Plants, Mycotoxins, and Related Toxins* (eds F. Riet-Correa, J. Pfister, A.L. Schild, and T.L. Wierenga)

plant to cases of intoxication which led to multi-organ damages and death (El Agraa *et al.* 2002; Seak and Lin 2007). Despite the high toxicity this species is considered to be a medicinal plant and has important uses in traditional medicines around the world including those formulated for heart protection (Seak and Lin 2007); preparations with antimicrobial, fungicide, anti-inflammatory, and hypotensive activities; male contraceptive formulations; female antifertility preparations; abortifacient formulations; stimulant agents; and preparations for treatment of rheumatism, gastric disorders, amenorrhea, menorrhagia, and headache (El Agraa *at al.* 2002; Ivanova *et al.* 2005; Raghav *et al.* 2006; Harat *et al.* 2008).

*Petiveria alliacea* is a perennial shrub that can reach 1 m in height and tastes and smells similar to garlic. This shrub is known by various popular names, such as 'erva-guiné,' 'guiné,' 'pipi,' and 'tipi' (Kubec and Musah 2001). The stems are thin and angled; the leaves are elliptic and pointed. The inflorescences are in thin clusters and the fruits are linear. The roots of *P. alliacea* are the most studied part of the plant but both roots and leaves are used in folk medicines. Crude extracts or infusions administered topically and/or orally are used to treat influenza, tumors, gastrointestinal disorders, respiratory diseases, and other disorders. Coumarins, triterpenes, flavonoids, alkaloids, amino acids, steroids, thiosulfinates, and sulfines have been isolated and characterized from this species (Lopes-Martins *et al.* 2002; Garcia-González *et al.* 2006; Kim *et al.* 2006). Dibenzyl trisulfide, found in the roots, has anti-proliferative, cellular differentiation, and anti-tumor activities (Bao *et al.* 2008; Gu *et al.* 2008). The lachrymatory sulfine, (*Z*)-thiobenzaldehyde *S*-oxide, is the main sulfur-containing component identified in fresh roots of *P. alliacea* (Kubec *et al.* 2003). This and other thiosulfinates and sulfines are labile and can decompose into other compounds such as sulfides, benzylsulfinic and sulfonyl acids, benzaldehyde, and other benzenoids. These decomposition products have been associated with variations in the procedures used to prepare plant samples, leading to differences in the results of various pharmacological assays (Kice *et al.* 1960; Kim *et al.* 2006).

The aim of this work was to evaluate the chemical composition of preparations from the leaves of *P. alliacea* and *R. graveolens* and to correlate chemical compositions with their toxicological effects already described.

## Material and Methods

Plants of both species were grown and collected in Fiocruz experimental plots (Rio de Janeiro, Brazil). For each procedure, 200 g of fresh leaves were minced and extracted with water (by steam distillation), $CH_2Cl_2$/shaker (80 ml, 40 min of extraction), and 70% ethanol tincture (80 ml with maceration during 2 h of extraction). A 5 ml aliquot of the shaker preparation was subjected to direct chromatographic analysis and the rest of the solvent was evaporated at 27°C. After the addition of 20 ml water the ethanol extract was partitioned with $CH_2Cl_2$ (2 × 40 ml) which was then evaporated at 35°C. All samples were analyzed by gas chromatography/mass spectroscopy (GC/MS) using a gas chromatograph model 6890N (Agilent Technologies) equipped with a mass detector (model 5973 Network; Agilent Technologies), an injector of the 7683B Series (Agilent Technologies), and a DB-5MS column (30 m × 0.32 mm, 0.25 μm film thickness). Mass range was from *m/z* 40 to 600. The components were identified via peak matching with a Wiley mass spectra library, compared with the literature and by utilizing their retention times (RT) on a DB-5MS column.

## Results and Discussion

The compounds extracted depended on extraction methods for both *P. alliacea* (Table 1) and *R. graveolens* (Table 2).

**Table 1.** Composition of different extracts from fresh leaves of *P. alliacea* cultivated in southeastern Brazil.

| Compounds | RT (min) | Area (%) | | |
|---|---|---|---|---|
| | | Steam distillation | Evaporated $CH_2Cl_2$ from 70% EtOH | Evaporated $CH_2Cl_2$/ shaker |
| 1,2-Ethanedithiol | 4.5 | 0.37 | - | - |
| 2-Hexenal | 5.4 | 0.32 | - | - |
| 3-Hexen-1-ol | 5.5 | 0.90 | - | - |
| 1-Hexanol | 6.1 | 0.29 | - | - |
| Benzaldehyde | 10.3 | 66.60 | 6.32 | 5.58 |
| 2-Hydroxy-benzaldehyde | 13.2 | 0.16 | - | - |
| Benzenemethanethiol | 14.8 | 2.19 | - | - |
| Isobutyrophenone | 18.3 | 0.67 | - | - |
| 3-Methyl-1,2,4-trithiane | 21.0 | 2.82 | - | - |
| **(Z)-thiobenzaldehyde S-oxide** | **22.1** | **0.50** | **5.42** | **26.94** |
| 2-Methoxy-4-vinylphenol | 23.5 | 0.69 | - | - |
| 1,2,5-Trithiepane | 25.3 | 0.59 | - | - |
| Benzyl methyl disulfide | 25.9 | 0.57 | - | - |
| Dihydroactinidiolide | 27.2 | - | - | - |
| α-Ionone | 27.3 | 0.39 | - | - |
| Geranylacetone | 28.2 | 0.20 | - | - |
| β-Ionone | 29.1 | 0.83 | - | - |
| (-) Loliolide | 29.9 | - | - | - |
| Hexahydropseudoionone | 30.1 | - | - | - |
| Hexadecane | 33.0 | - | 0.96 | - |
| Benzyl sulfide | 34.9 | - | - | 1.80 |
| (E)-Stilbene | 35.9 | 1.15 | 5.58 | 9.25 |
| Octadecane | 38.6 | - | 3.28 | 2.90 |
| Phytane | 38.8 | - | 1.87 | - |
| Neophytadiene | 39.5 | - | 0.79 | - |
| Dibenzyl disulfide | 44.8 | 3.53 | - | - |
| Dibenzyl trisulfide | 50.7 | 0.48 | - | 2.20 |
| Nonadecane | 41.2 | - | 2.88 | 2.81 |
| Eicosane | 43.7 | - | 1.72 | 1.71 |
| Heneicosane | 46.1 | - | 1.01 | - |
| Phytol | 46.2 | 1.82 | 4.01 | - |
| 14-β-pregnane | 47.5 | - | 0.83 | - |
| Docosane | 48.3 | - | 0.80 | - |
| Tricosane | 50.5 | - | 0.78 | - |
| Squalene | 60.3 | - | 9.83 | 7.43 |
| γ-Tocopherol | 64.2 | - | 1.91 | - |
| Vitamin E | 65.5 | - | 31.73 | 35.78 |
| Identified compounds (%) | | 85.1 | 80.5 | 96.4 |

RT = retention time

**Table 2.** Different extracts composition from the leaves of *Ruta graveolens*.

| Compounds | RT (min) | Area (%) | | |
|---|---|---|---|---|
| | | Steam distillation | Evaporated $CH_2Cl_2$ from 70% EtOH | Evaporated $CH_2Cl_2$/ shaker |
| 2-Octanone | 5.5 | 0.85 | - | - |
| **2-Nonanone** | **8.0** | **53.03** | - | **4.55** |
| 2-Nonanol | 8.3 | 0.30 | - | - |
| Geijerene | 9.3 | 0.81 | - | - |
| 2-Decanone | 10.7 | 2.90 | - | - |
| **2-Undecanone** | **13.5** | **37.99** | - | **2.84** |
| 2-Dodecanone | 16.30 | 0.44 | - | - |
| 2-Tridecanone | 18.8 | 0.24 | - | - |
| Angelicin | 26.3 | - | 8.12 | 9.83 |
| 6,10,14-Trimethyl-2-pentadecanone | 26.7 | - | 1.14 | - |
| Ethyl palmitate | 29.9 | - | 1.81 | - |
| 8-(3',5'-Benzodioxyl)-2-octanone | 30.60 | - | 5.79 | 2.15 |
| Xanthotoxin | 30.8 | - | 4.79 | 3.45 |
| Phytol | 32.0 | - | 3.14 | - |
| Chalepensin | 33.0 | - | 19.34 | 3.10 |
| Pteleine | 33.2 | - | 1.17 | 0.67 |
| γ-Fagarine | 33.6 | - | 2.48 | 4.31 |
| Kokusaginine | 37.6 | - | 3.25 | 8.38 |
| (*E*)-5-Eicosene | 38.3 | - | - | 0.93 |
| **Chalepin** | **39.9** | - | **39.77** | **11.03** |
| Heptacosane | 40.5 | - | - | 0.76 |
| 1,2-Epoxynonadecane | 41.7 | - | - | 1.22 |
| 1,2-Epoxyoctadecane | 42.5 | - | - | 0.35 |
| Arborinine | 42.6 | - | 2.70 | 3.64 |
| 17-Octadecenal | 43.3 | - | - | 5.37 |
| Rutamine | 43.6 | - | - | 2.94 |
| Octacosane | 43.7 | - | - | 6.38 |
| Eicosane | 44.4 | - | - | 0.44 |
| Campesterol | 44.6 | - | 2.21 | - |
| Heneicosane | 45.2 | - | - | 16.06 |
| β-Sitosterol | 45.3 | - | 2.42 | - |
| Identified compounds (%) | | 96.5 | 98.1 | 88.4 |

RT = retention time

In the *P. alliacea* extractions the most differences were observed when sulfur compounds were examined. The lability of (Z)-thiobenzaldehyde S-oxide (TBSO), the toxic lachrymatory principle of the plant, arose principally from effects of solvent and temperature. TBSO decomposes to benzaldehyde, stilbene, and other less toxic compounds. The $CH_2Cl_2$ from tincture contained only 5.42% of TBSO whereas following $CH_2Cl_2$/ shaker extraction and solvent removal it was 26.94% TBSO. An aliquot of this last material prior to $CH_2Cl_2$ evaporation showed the major content of TBSO (49.6%), confirming the effect of temperature on the loss and transformation of this compound as described by Kim and coworkers (2006). The higher temperature used in steam distillation resulted in only

0.5% of TBSO with benzaldehyde (66.6%) the major component extracted from the hydrolate with $CH_2Cl_2$.

Other sulfur compounds were identified following steam distillation (10.5%) and $CH_2Cl_2$/shaker extraction (4.0%) probably arising from TBSO decomposition. Rapid solvent extractions resulted in preparations with a high content of vitamin E which can be degraded during long extractions at higher temperatures. To our knowledge this is the first report of TBSO in the fresh leaves of *P. alliacea* cultivated in southeastern Brazil.

By the steam distillation method *R. graveolens* produced an essential oil composed almost entirely of two compounds, 2-nonanone (53.03%) and 2-undecanone (37.99%), in agreement with previous findings (De Feo *et al.* 2002).

An earlier study found that the 2-nonanone content was lower than that of 2-undecanone; this may have been related to the use of dry plant material. The acute toxicity and cytotoxicity of these and other aliphatic ketones were studied in different assays and all results indicated toxicological potential which increased as the carbon chain length rose (Martin and Young 2001; Mohajeri and Dinpajooh 2008). The aliphatic ketones 2-nonanone and 2-undecanone were also observed following the $CH_2Cl_2$/shaker extraction but in smaller percentages of contents: 4.55% and 2.84%, respectively. These and other volatile compounds present in the essential oil were probably lost during solvent evaporation. The last extract showed a high content of toxic furanocoumarins (27.41%) such as angelicin, xanthotoxin, chalepensin, and chalepin which were also found in the $CH_2Cl_2$ extraction at high concentration (72.02%). Chalepin, the major constituent of both extractions (11.03% and 39.77%), showed hepatotoxicity in rats after intraperitoneal administration (100 mg/kg) for 2 days with a 4% death rate (Emerole *et al.* 1981). Two different alkaloids, furoquinoline and acridone, were also identified in both extracts at concentrations of 20.02% and 9.60%, respectively. The furoquinoline alkaloids identified in both extracts, especially γ-fagarine, have been described as mutagenic compounds (Paulini *et al.* 1987). Neither furanocoumarins nor alkaloids were identified in the essential oil.

## Conclusion

All extraction methods used for *R. graveolens* resulted in compounds with previously described toxicological potential. The extraction methods employed for *P. alliacea* showed that temperature and solvent influenced the percentage content of the toxic principle TBSO. The variation in compounds content showed the importance of plant manipulation especially when considering medicinal uses. Further chemical and toxicological studies are necessary to evaluate the safe use of preparations from these species.

## References

Bao Y, Mo X, Xu X, He Y, Xu X, and An H (2008). Stability studies of anticancer agent bis (4-fluorobenzyl) trisulfide and synthesis of related substances. *Journal of Pharmaceutical and Biomedical Analysis* 46:206-210.

De Feo V, De Simone F, and Senatore F (2002). Potential allelochemicals from the essential oil of *Ruta graveolens*. *Phytochemistry* 61:573-578.

Diwan R and Malpathak N (2008). Novel technique for scaling up of micropropagated *Ruta graveolens* shoots using liquid culture systems: A step towards commercialization. *New Biotechnology* 25:85-91.

El Agraa SEI, El Badwi SMA, and Adam SEI (2002). Preliminary observations on experimental *Ruta graveolens* toxicosis in nubian goats. *Tropical Animal Health and Production* 34:271-281.

Emerole G, Thabrew MI, Anosa V, and Okorie DA (1981). Structure-activity relationship in the toxicity of some naturally occurring coumarins–chalepin, imperatorin and oxypeucedanine. *Toxicology* 20:71-80.

Freitas TG, Augusto PM, and Montanari T (2005). Effect of *Ruta graveolens* L. on pregnant mice. *Contraception* 71:74-77.

Garcia-González M, Morales TC, Ocampo R, and Pazos L (2006). Subchronic and acute preclinic toxicity and some pharmacological effects of the water extract from leaves of *Petiveria alliacea* (Phytolaccaceae). *Revista de Biologia Tropical* 54(4):1323-1326.

Gomes PB, Oliveira MMS, Nogueira CRA, Noronha, EC, Carneiro, LMV, Bezerra JNS, Neto MA, Vasconcelos SMM, Fonteles MMF, Viana GSB, and Sousa FCF (2005). Study of antinociceptive effect of isolated fractions from *Petiveria alliacea* L. (tipi) in mice. *Biological Pharmaceutical Bulletin* 28(1):42-46.

Gu L, Li L, Chen Z, Pan H, Jiang H, Zeng S, Xu X, and An H (2008). Determination of anti-tumor agent bis (*p*-fluorobenzyl) trisulfide and its degraded compound in rat blood using reversed phase high-performance liquid chromatography. *Journal of Chromatography B* 868:77-82.

Harat ZN, Sadeghi MR, Sadeghipout HR, Kamalinejad M, and Eshraghian MR (2008). Immobilization effect of *Ruta graveolens* L. on human sperm: A new hope for male contraception. *Journal of Ethnopharmacology* 115:36-41.

Hoyos LS, Au WW, Heo MY, Morris DL, and Legator MS (1992). Evaluation of the genotoxic effects of a folk medicine, *Petiveria alliacea* (Anamu). *Mutation Research* 280:29-34.

Ivanova A, Mikhova B, Najdenski H, Tsvetkova I, and Kostova I (2005). Antimicrobial and cytotoxic activity of *Ruta graveolens*. *Fitoterapia* 76:344-347.

Kice JL, Parham FM, and Simons RM (1960). The thermal decomposition of thiolsulfonates. *Journal of the American Chemical Society* 82:834-842.

Kim S, Kubec R, and Musah RA (2006). Antibacterial and antifungal activity of sulfur-containing compounds from *Petiveria alliacea* L. *Journal of Ethnopharmacology* 104:188-192.

Kubec R and Musah RA (2001). Cysteine sulphoxide derivatives in *Petiveria alliacea. Phytochemistry* 58:981-985.

Kubec R, Kim S, and Musah RA (2003). The lachrymatory principle of *Petiveria alliacea. Phytochemistry* 63:37-40.

Lopes-Martins RAB, Pegoraro DH, Woisky R, Penna SC, and Sertié JAA (2002). The anti-inflammatory and analgesic effects of a crude extract of *Petiveria alliacea* L. (Phytolaccaceae). *Phytomedicine* 9:245-248.

Martin TM and Young DM (2001). Prediction of the acute toxicity (96h LC50) of organic compounds to the feathead minnow (*Pimephales promelas*) using a group contribution method. *Chemical Research in Toxicology* 14:1378-1385.

Milesi S, Massot B, Gontier E, Bourgaud F, and Guckert A (2001). *Ruta graveolens* L.: a promising species for the production of furanocoumarins. *Plant Science* 161:189-199.

Mohajeri A and Dinpajooh MH (2008). Structure-toxicity relationship for aliphatic compounds using quantum topological descriptors. *Journal of Molecular Structure:THEOCHEM* 855:1-5.

Paulini H, Eilert U, and Schimmer O (1987). Mutagenic compounds in an extract from Rutae Herba (*Ruta graveolens* L.). I. Mutagenicity is partially caused by furoquinoline alkaloids. *Mutagenesis* 2(4):271-273.

Raghav SK, Gupta B, Agrawal C, Goswami K, and Das HR (2006). Anti-inflammatory effect of *Ruta graveolens* L. in murine macrophage cells. *Journal of Ethnopharmacology* 104:234-239.

Seak CJ and Lin CC (2007). *Ruta graveolens* intoxication. *Clinical Toxicology* 45(2):173-175.

# Chapter 120

# Antimicrobial Effect of an Extract of *Anacardium occidentale* Linn against Clinical Isolates of Multidrug-Resistant *Staphylococcus aureus*

J.T.J.G. de Lacerda, J.G. da Silva, J.S. Higino, J.V.Pereira, and M.S.V. Pereira

*Department of Molecular Biology, Laboratory of Genetics of Microorganisms, Federal University of Paraíba, João Pessoa, Paraíba, Brazil. University city, João Pessoa, Paraíba - CEP - 58059-900*

## Introduction

The plant *Anacardium occidentale* Linn, known popularly as the cashew tree, has wide distribution throughout Brazil, mainly in northeastern Brazil. Various pharmacological properties are attributed to the extract of the cashew such as the treatment of cough, syphilis, bronchitis, arthritis, intestinal colic, jaundice, excess fluids (as diuretic), and wounds (Olajide *et al.* 2004; Morais *et al.* 2005; Agra *et al.* 2007). Proven pharmacological activities for the cashew plant can be found in the literature: anti-inflammatory (Olajide *et al.* 2004; Falcão *et al.* 2005); antidiabetic (Oliveira and Salto 1989; Kamtchouing *et al.* 1998; Barbosa-Filho *et al.* 2005); and acetylcholinesterase inhibitor (Barbosa-Filho *et al.* 2006). In addition, substances isolated from the fruit have been demonstrated to inhibit tyrosinase (Kubo *et al.* 1994). The screening of various medicinal plants for antibacterial activity was carried out in Nigeria and the extract of leaves and bark of *A. occidentale* showed good activity against *Escherichia coli* and *Pseudomonas aeruginosa*, gram-negative bacteria (Kudi *et al.* 1999).

The interest in the use of natural products for health has been around for ages and humanity has always made use of the therapeutic potential of plant extracts in the form of teas, ointments, poultices, vapors, tinctures, and even incense where they are utilized still today mainly in developing countries or by those seeking non-conventional therapies. This is the aim of the study of ethnopharmacology, a science that uses the scientific method to determine the therapeutic activity of plant products that are popularly used.

The indiscriminate use of antibiotics has resulted in the selection of pathogens that have acquired antibiotic resistance. In *Staphylococcus aureus* multiple resistance to antimicrobials results mainly due to the presence of plasmids; another strategy of resistance in *S. aureus* is the acquisition of resistance genes in the chromosome thereby producing a

©CAB International 2011. *Poisoning by Plants, Mycotoxins, and Related Toxins*
(eds F. Riet-Correa, J. Pfister, A.L. Schild, and T.L. Wierenga)

multidrug resistance that is more stable, for example in the *S. aureus* strain resistant to methicillin (MRSA) which is usually resistant not only to penicillins and cephalosporins but also to aminoglycosides, lincomycins, tetracyclines, rifampicin, and more recently vancomycin.

Therefore, plants with therapeutic properties are of great importance in medicine throughout world, mainly in developing countries where drugs are imported, costly, and in most cases inaccessible to a large part of the population. The objective of the present study was to determine the *in vitro* antimicrobial activity of an extract of the cashew tree stem (*Anacardium occidentale* Linn.) against hospital isolates of *S. aureus*.

## Material and Methods

### Botanical material

The extract was obtained from the stem of the cashew tree (*A. occidentale*) and identified botanically in the Toxicology Laboratory of the Federal University of Pernambuco. The extract was obtained from the stem bark of the plant. The bark was dried in an oven at 33°C for 1 week, triturated into a powder in an electric mill, and then extracted. The extraction method used was leaching or percolation with continuous flow at room temperature. Leaching with continuous flow involved a process where there is constant renewal of the solvent (hydroalcoholic solution at 70% v/v). Afterward, the extract was concentrated in a rotary evaporator (Model Ika-Werk) at a constant temperature of 45°C. The concentration obtained at the level of fluid extract was 1:1 (w/v).

### Bacterial strains

This study used 20 strains of *S. aureus* obtained from patients admitted to Hospital Universitario Lauro Wanderley¨/UFPB and antibacterial activity was determined and phenotypically characterized as sensitive and resistant to methicillin (Freitas 1992). The strain ATCC was utilized as the reference control in all experiments performed (NCCLS 1988).

## Results and Discussion

The results on antimicrobial activity of the extract of the cashew plant on the specimens of *S. aureus* studied are presented in Table 1.

The surge in multidrug-resistant staphylococci in the last years is an important problem in the control of infections and treatment and therefore it is extremely important to study ways to combat a staphylococcal epidemic since the transmission of *S. aureus* is intense and easy, principally in large medical centers. Generally, these infections are associated with a lengthy hospitalization and prolonged antibiotic treatment. *S. aureus* resistant to methicillin (MRSA) has been described as the most common nosocomial pathogen in the world. Some MRSA strains, called epidemic strains (EMRSA), are capable of rapidly spreading among patients. Once introduced in an institution these EMRSA strains are difficult to control and eradicate.

To date there have been many studies in relation to the various activities of natural products, mainly in developing countries which need more economical alternatives for the control of disease. Drugs have failed in the treatment of long term infections in degenerative diseases associated with the development of resistance to antimicrobial agents. Thus, a combination of preventive and curative measure are needed, particularly the search for new antimicrobial agents. This is the most important role of natural products potentially capable of resolving some chronic conditions and constituting an excellent alternative for the solution to the problem of resistance to antibiotics available for use in clinical practice.

**Table 1.** Minimum inhibitory concentration of hydroalcoholic extract of cashew plant (*Anacardium occidentale* Linn) against hospital isolates of *Staphylococcus aureus*.

| Bacterial strain | Diameter of inhibition halo (mm) Dilution of extract | | | | | | | | | |
|---|---|---|---|---|---|---|---|---|---|---|
| | PE | 1:2 | 1:4 | 1:8 | 1:16 | 1:32 | 1:64 | 1:128 | 1:256 | NOR |
| 01H | 13 | 10 | 10 | 0 | 0 | 0 | 0 | 0 | 0 | - |
| 02H* | 14 | 13 | 11 | 10 | 0 | 0 | 0 | 0 | 0 | 23 |
| 03H | 18 | 14 | 11 | 10 | 0 | 0 | 0 | 0 | 0 | - |
| 05H* | 13 | 10 | 10 | 0 | 0 | 0 | 0 | 0 | 0 | - |
| 09H | 13 | 12 | 10 | 0 | 0 | 0 | 0 | 0 | 0 | 36 |
| 104H | 15 | 14 | 13 | 0 | 0 | 0 | 0 | 0 | 0 | 31 |
| 06cc* | 15 | 10 | 10 | 10 | 0 | 0 | 0 | 0 | 0 | 25 |
| 07cc | 14 | 13 | 10 | 10 | 0 | 0 | 0 | 0 | 0 | - |
| 08cc | 14 | 13 | 11 | 0 | 0 | 0 | 0 | 0 | 0 | - |
| 09cc* | 13 | 10 | 10 | 0 | 0 | 0 | 0 | 0 | 0 | 23 |
| 10cc* | 15 | 14 | 12 | 11 | 10 | 0 | 0 | 0 | 0 | 24 |
| 102cc | 18 | 15 | 14 | 13 | 10 | 10 | 0 | 0 | 0 | - |
| 117H | 14 | 12 | 12 | 11 | 10 | 10 | 0 | 0 | 0 | 22 |
| 08Sn | 14 | 13 | 11 | 0 | 0 | 0 | 0 | 0 | 0 | - |
| 11Sn | 16 | 14 | 13 | 10 | 0 | 0 | 0 | 0 | 0 | 25 |
| 13Sn | 14 | 13 | 11 | 10 | 10 | 0 | 0 | 0 | 0 | 25 |
| 19Lab* | 17 | 15 | 13 | 10 | 10 | 0 | 0 | 0 | 0 | 11 |
| 149Lab* | 16 | 14 | 13 | 10 | 10 | 0 | 0 | 0 | 0 | 26 |
| 171cc* | 17 | 16 | 14 | 14 | 12 | 0 | 0 | 0 | 0 | 24 |
| 296I* | 17 | 16 | 14 | 13 | 13 | 0 | 0 | 0 | 0 | 25 |
| ATCC29213 | 17 | 16 | 14 | 13 | 12 | 0 | 0 | 0 | 0 | - |

Origin of strains: H = hospital; cc = surgical center; Sn = outpatient clinic; Lab = laboratory; * = *S. aureus* resistant to methicillin; NOR = norfloxacin; PE = pure extract

The search for alternative therapeutic substances represents the main goal in the study of the properties of plant extracts. A wide variety of products of plant origin have been shown to be potentially efficacious with regard to antimicrobial effects on various species of microorganisms. Considering the richness of plant constituents, the positive antimicrobial activity of the cashew plant extract against *S. aureus* isolates could be due to the presence of compounds such as tannins (polyphenolic compounds) and alkaloids previously found in the plant. These compounds have demonstrated antimicrobial action and phenolic compounds possess a nonspecific action on microorganisms by disrupting the bacterial cell wall and inhibiting enzyme systems for its formation (Haslam 1995; Jorge *et al.* 1996; Akinpelu 2001).

In the present study, the hospital strains of *S. aureus* were susceptible to the action of the cashew stem extract (*Anacardium occidentale* Linn.). It is important to note that based on these results all specimens were sensitive to the extract of *A. occidentale*.

## Conclusion

The results showed the potent antimicrobial activity of the extract of cashew stem bark against *Staphylococcus aureus* strains resistant and sensitive to methicillin as well as the importance of evaluating alternative and economically viable measures for the control of staphylococcal infections involving multidrug-resistant strains.

## References

Agra MF, França PF, and Barbosa-Filho JM (2007). Synopsis of the plants known as medicinal and poisonous in Northeast of Brazil. *Revista Brasileira de Farmacognósia* 17:114-140.

Akinpelu DA (2001). Antimicrobial activity of *Anacardium occidentale* bark. *Fitoterapia* 72:286-287.

Barbosa-Filho JM, Vasconcelos THC, Alencar AA, Batista LM, Oliveira RAG, Guedes DN, Falcão HS, Moura MD, Diniz MFFM, and Modesto-Filho J (2005). Plants and their active constituents from South, Central, and North America with hypoglycemic activity. *Revista Brasileira de Farmacognósia* 15:392-413.

Barbosa-Filho JM, Medeiros KCP, Diniz MFFM, Batista LM, Athayde-Filho PF, Silva MS, Cunha EVL, Almeida JRGS, and Quintans-Júnior LJ (2006). Natural products inhibitors of the enzyme acetylcholinesterase. *Revista Brasileira de Farmacognósia* 16:258-285.

Falcão HS, Lima IO, Santos VL, Dantas HF, Diniz MFFM, Barbosa-Filho JM, and Batista LM (2005). Review of the plants with anti-inflammatory activity studied in Brazil. *Revista Brasileira de Farmacognósia 15*:381-391

Freitas FIS (1992). Caracterização fenotípica de amostras hospitalares de *Staphylococcus aureus* isoladas no Estado da Paraíba, 56 pp. Dissertação de Mestrado, Universidade Federal da Paraíba, João Pessoa.

Haslam E (1995). Natural polyphenols (vegetable tannins) as drugs: Possible modes of action. *Journal of Natural Products* 59:205-215

Jorge LIF, Silva GA, and Ferro VO (1996). Diagnose laboratorial dos frutos de *Anacardium occidentale* L. (caju). *Revista Brasileira de Farmacognósia* 5:55-69.

Kamtchouing P, Sokeng DS, Moundipa FP, Watcho P, Jatsa BH, and Lontsi D (1998). Protective role of *Anacardium occidentale* extract against streptozotocin-induced in rats. *Journal of Ethnopharmacology* 62:95-99

Kubo I, Kinst-Hori I, and Yokokawa Y (1994). Tyrosinase inhibitors from *Anacardium occidentale* fruits. *Journal Natural Products* 57:545-551

Kudi AC, Umoh JU, Eduvie LO, and Gefu J (1999). Screening of some Nigerian medicinal plants for antibacterial activity. *Journal of Ethnopharmacology* 67:225-228.

Morais SM, Dantas JDP, Silva ARA, and Magalhães EF (2005). Plantas medicinais usadas pelos índios Tapebas do Ceará. *Revista Brasileira de Farmacognósia* 15:169-177.

NCCLS (1988). National Committee for Clinical Laboratory Standards. *Methods for dilution antimicrobial susceptibility test for bacteria that grow aerobically*, 2nd edn. Tentative Standard. Document M7 - T2.

Olajide OA, Aderogba MA, Adedapo AD, and Makinde JM (2004). Effects of *Anacardium occidentale* stem bark extract on *in vivo* inflammatory models. *Journal of Ethnopharmacology* 95:139-142.

Oliveira F and Salto ML (1989). Alguns vegetais brasileiros empregados no tratamento de diabetes. *Revista Brasileira de Farmacognósia* 2/4:170-196.

# Chapter 121

# Evaluation of Hepatotoxicity Induced by *Piper methysticum*

M.F. Amorim[1], M.S. Trigueiro[1], M.F.F.M. Diniz[1], R.N. de Almeida[1], H.B. Santos[1], J.R. do Amaral[1], L.M. Batista[1], M.S.C. Branco[2], and J.M. Diniz[2]

[1]*Laboratório de Ensaios Toxicológicos/LTF, Programa de Pós-Graduação em Produtos Naturais e Sintéticos Bioativos, Universidade Federal da Paraíba;* [2]*Programa Institucional de Bolsa de Iniciação Científica*

## Introduction

*Piper methysticum* G. Foster (kava or kava-kava) is a medicinal herb used to treat insomnia and anxiety that contains kavalactones and is freely available in many countries (Russmann *et al.* 2001). Cases of toxic hepatitis, some progressing to fulminant hepatitis, are associated with its use (Russmann *et al.* 2001; Humbertston *et al.* 2003). Histological findings of liver injury related to kava ingestion are mainly cholestatic hepatitis and hepatocellular necrosis (Stickel *et al.* 2003). There are reports of hepatotoxicity with portal infiltration, bridging necrosis, destruction of interlobular bile ducts, and ductular cholestasis in addition to mixed cell infiltrate with lymphocytes, eosinophils, and activated macrophages involving interlobular bile ducts (Russman *et al.* 2001). Gastrointestinal disturbances may occur. Allergic skin reactions, headaches, dizziness, tiredness, restlessness, tremors, and neurological complications are rare events and are in general moderate and reversible, only occasionally becoming more severe (Stevenson *et al.* 2002).

Recently Lüde *et al.* (2008) studied the cytotoxicity of kava, treating HepG2 cells and liver mitochondria isolated from rats with methanol and acetone extracts of kava showing that they were toxic to the mitochondria, causing respiratory chain inhibition, increasing ROS (reactivate oxygen species) production, decreasing the mitochondrial membrane potential, and eventually apoptosis. Since the pharmacokinetic properties of kava are not fully known they suggest that because of the toxic effects found *in vitro*, under certain conditions individuals may develop hepatotoxicity after ingestion of kava extracts (Lüde *et al.* 2008). In this study we observe experimentally the occurrence of liver damage induced by an extract of *P. methysticum* using chronic pre-clinical toxicological tests in rats.

©CAB International 2011. *Poisoning by Plants, Mycotoxins, and Related Toxins* (eds F. Riet-Correa, J. Pfister, A.L. Schild, and T.L. Wierenga)

## Material and Methods

The trial was conducted in the Laboratory of Toxicological Assays in the Federal University of Paraíba (UFPB). Biochemical analysis including AST (aspartate aminotransferase), ALT (alanine aminotransferase), and LDH (lactate dehydrogenase) were performed in the Vivarium Laboratory of Pharmaceutical Technology (LTF). Histological analysis was performed in the Virchow Medical Laboratory of Cellular Pathology. This research was approved by the Ethics Committee on Animal Research of the LTF, UFPB.

The animals included adult male and female Wistar rats (*Rattus norvegicus*). Groups of ten animals each were subjected to daily doses of kava extract WS 1490 orally by gavage for 13 weeks, corresponding to the commonly used dose (4.3 mg/kg) and three and nine times the common dose (12.9 mg/kg and 38.7 mg/kg, respectively). A satellite group for each dose was preserved and observed for 45 days after kava withdrawal. Two groups were used as controls: one with the animals euthanized immediately after dosing and another as control for the satellite groups.

At the end of 13 weeks the rats were euthanized by cervical dislocation and blood was collected subsequently from the brachial plexus for laboratory tests. After performing an examination of the thoracic and abdominal cavities, liver resection was performed for macroscopic and histologic examination. Liver samples were fixed in buffered formalin, embedded in paraffin, cut into 3-4 μm sections, and stained by hematoxylin-eosin (HE), Masson's trichrome, and picrosirius red (Michalany 1998). For liver histological analysis the protocol established by the Brazilian Society of Pathology (BSP) and the Brazilian Society of Hepatology (BSH) was adapted (Gayotto 2000).

For statistical analysis data were entered into Excel (Microsoft Office) and exported to statistical packages Prism version 3.0 and SPSS for Windows.

## Results

Results of blood biochemistry from the treated group were similar to controls except for ALT which was significantly lower in males of the group that received a dose of 4.3 mg/kg. No macroscopic abnormalities were detected in the examined organs.

The liver histological study of treated and control groups showed parenchyma with a normal lobular architecture with thin walled regularly distributed terminal hepatic veins. Such aspects in treated animals were associated with microvascular changes and Kupffer cell reactivity. Terminal hepatic vein congestion and sinusoidal dilatation were the only histological changes noted in males and females treated with 4.3 mg/kg. Diffuse microvascular changes were observed in rats treated with 12.9 mg/kg including venule-sinusoidal congestion and dilatation in addition to sinusoidal inflammation. The vascular congestion and the sinusoidal inflammation achieved zones 2 and 3 with the maximum dose used (38.7 mg/kg). Hypertrophy and sometimes hyperplasia of Kupffer cells was observed in rats from all groups.

Focal hepatocellular single necrosis was observed at the dose of 4.3 mg/kg. The liver of female and male rats treated with 12.9 mg/kg and 38.7 mg/kg of *P. methysticum* showed increased number of portal lymphocytes (scores 1 and 2). The entire population of this group showed signs of hydropic degeneration, individual acidophilic retraction of hepatocytes, and necrosis of groups of hepatocytes with lymphohistiocytic infiltration (score 2).

The histological study of the liver of all animals in the satellite group showed parenchyma with normal lobular architecture. Vascular-congestive and venular-sinusoidal phenomena, with or without sinusoidal dilatation, involving zones 2 and 3 were found in males and females treated with doses of 12.9 mg/kg and 38.7 mg/kg. Such changes were associated with sinusoidal inflammatory cell infiltration in the same locations, especially in animals subjected to higher doses of kava. Kupffer cell reactivity was also observed. In all animals regardless of the dose used and sex, there was persistence of portal and lobular lymphocytic infiltrate, the latter as a consequence of necrosis. For the typical daily dose of *P. methysticum*, in females and males there was discrete periportal infiltration of lymphocytes and some lobular foci of necrosis associated with mild lymphohistiocytic infiltration. In the satellite group treated with 38.7 mg/kg of *P. methysticum* the score that represents the portal inflammation was kept 1 in females while the score in the males decreased from 2 to 1. In the entire population of this group hydrotropic degeneration and individual acidophilic retraction was decreased. The hepatic necrosis with lymphohistiocytic infiltration decreased from score 2 to score 1.

## Discussion

Despite the numerous reports on the potentially deleterious effects of *P. methysticum* on the liver, mechanisms involved in the hepatotoxicity of kava or its constituents are in need of elucidation (Amorim *et al.* 2007). In this study biochemical parameters showed no differences compared to controls except for ALT which was significantly lower in males of the group that received a dose of 4.3 mg/kg, suggesting no toxicity. However, histological evaluation showed the occurrence of hepatic lesions including vascular-venular and sinusoidal congestion, with or without sinusoidal dilatation, sinusoidal inflammation, and Kupffer cell reactivity, and hepatocellular single necrosis at the dose of 4.3 mg/kg and multifocal necrosis at doses of 12.9 mg/kg and 38.7 mg/kg. These findings in the liver suggest a drug-induced liver disease. Hepatic microvascular changes, notably the sinusoidal dilatation accompanied by inflammation, similar to those observed in this trial have been associated with drug-induced injury upon histological and ultrastructural analysis (Araújo *et al.* 1993).

Some issues are relevant concerning the histological results. Due to the severe development of toxic hepatitis in cases related to kava, culminating sometimes with liver transplants, histologic examination during different stages of the disease is generally not performed in clinical practice (Stickel *et al.* 2003). Furthermore, in many experiments with animals histologic lesions were examined only immediately after treatment with kava without a delayed evaluation after treatments ended (Singh and Devkota 2003; Carvalho 2005; Sorrentino *et al.* 2006; DiSilvestro *et al.* 2007). In this study lesions were also present in the satellite group, demonstrating that the lesions persist for at least 45 days. Because of these variations it is difficult to make comparisons between experiments. Also, comparing studies on hepatotoxicity between animal species and humans is complex since the latter are exposed to numerous factors not reproducible in animal studies in addition to varying greatly among people (DiSilvestro *et al.* 2007).

## Conclusions

It is concluded that, despite a lack of effect on serum biochemistry, *Piper methysticum* causes liver lesions in rats, therefore it may elicit a toxic effect in humans depending on the dose and period of administration.

## References

Amorim MFD, Araújo MST, Diniz MFFM, Carvalho AC, and Araújo MST (2007). The controvertible role of kava (*Piper methysticum* G. Foster), an anxiolitic herb, on toxic hepatitis. *Revista Brasileira de Farmacognosia* 17(3):451-457.

Araújo MST, Gerard F, Chossegros P, Guerret S, Barlet P, Adeleine P, and Grimaud JA (1993). Cellular and matrix changes in drug abuser liver sinusoids: a semiquantitative and morphometric ultrastructural study. *Virchows Archivs* 422:145-152.

Carvalho ACB (2005). Avaliação legal da publicidade de produtos naturais e investigação toxicológica dos produtos anunciados. 246 pp. Dissertação de mestrado, Programa de Pós-graduação em Produtos Naturais e Sintéticos Bioativos do Laboratório de Tecnologia Farmacêutica, Universidade Federal da Paraíba, João Pessoa.

DiSilvestro RA, Zhang W, and DiSilvestro DJ (2007). Kava feeding in rats does not cause liver injury nor enhance galactosamine-induced hepatitis. *Food and Chemical Toxicology* 45:1293-1300.

Gayotto LCC (2000). Comitê SBP/SBH. Visão histórica e consenso nacional sobre a classificação das hepatites crônicas. *Gastroenterol and Endoscopy Diagnosis* 19(3):137-140.

Humberston CL, Akhtar J, and Krenzelok EP (2003). Acute hepatitis induced by kava. *Journal of Clinical Toxicology* 41:109-113.

Lüde S, Torok M, Dieterle S, Jaggi R, Büter KB, and Krahenbühl S (2008). Hepatocellular toxicity of kava leaf and root extracts. *Phytomedicine* 15:120-131.

Michalany J (1998). *Técnica histológica em anatomia patológica: com instruções para o cirurgião, enfermeira e citotécnico*, 3rd edn, 295 pp. Michalany, São Paulo.

Russmann S, Lauterburg BH, and Helbling A (2001). Kava hepatotoxicity. *Internal Medicine* 135:68-69.

Singh YN and Devkota AK (2003). Aqueous kava extracts do not affect liver function tests in rats. *Planta Medica* 69:496-499.

Sorrentino L, Capasso A, and Schmidt M (2006). Safety of ethanolic kava extract: Results of a study of chronic toxicity in rats. *Phytomedicine* 13:542-549.

Stevenson C, Huntly A, and Ernst E (2002). A systematic review of the safety of kava extract in the treatment of anxiety. *Drug Safety* 25:251-261.

Stickel F, Baumüller HM, Seitz K, Vasilakis D, Seitz G, Seitz HK, and Schuppan D (2003). Hepatitis induced by Kava (*Piper methysticum* rhizoma). *Journal of Hepatology* 39:62-67.

# Chapter 122

# Toxic Effects of *Baccharis trimera* on Pregnant Rats and Their Conceptuses

S.R.M. Grance[1], M.A. Teixeira[2], and R.J. Oliveira[3]

*[1]Veterinary Physician Master's Degree in Program in Animal Science at the Universidade Federal de Mato Grosso do Sul (UFMS), Av. Senador Filinto Muller, 2443, Campo Grande, MS 79074-460, Brazil; [2]Central Animal Facility, UFMS, Av. Senador Filinto Muller, 1555, Campo Grande, MS, 79074-460, Brazil; [3]Department of Biomedicine, Centro Universitário Filadélfia, Rua Alagoas, 2050, Londrina, PR, 86020-430, Brazil*

## Introduction

*Baccharis trimera* (Less) DC, known as 'carqueja,' is widely used in folk medicine in the treatment of indigestion, gastritis, inflammation, diabetes, and rheumatism (Simões *et al.* 1998). In Campo Grande, Mato Grosso do Sul, Brazil, *B. trimera* is the second most commonly used plant by local people (Nunes *et al.* 2003) even during pregnancy.

Maternal toxicity is associated with the incidence of malformations. Therefore, in order to evaluate the incidence of toxic effects of chemicals during pregnancy, we first analyzed the effects of these substances on the maternal organism (Khera 1985).

Exposure to chemicals during pregnancy may have different effects on prenatal development according to the administered dose, exposure time, and gestational period with the period of organogenesis being the most sensitive to the action of external agents since it is the stage of tissue and organ formation (Sadler 2005).

This study evaluated the effects of the hydroethanolic extract of *B. trimera* (HEBT) in pregnant rats when administered during the period of organogenesis and the entire pregnancy and the occurrence of malformations and/or variations in their litters.

## Material and Methods

Aerial parts of *B. trimera* were dried, steeped in ethanol 70% v/v, macerated, percolated, and the solvent was eliminated, yielding 6.57 g (11.7%, w/w yield) of crude extract. The resulting mass was divided into 8.4 mg/kg aliquots, each of which was diluted in 0.3 ml of distilled water and stored at –5°C until processing.

The 35 nulliparous outbred female Wistar-strain rats (*Rattus norvegicus*) at an average age of 4 months and weighing an average of 250 g were housed in conformity with the US National Research Council (1996) guidelines. The experimental protocol No 112/2006 was approved by UFMS's Ethics Committee on the Use of Animals.

©CAB International 2011. *Poisoning by Plants, Mycotoxins, and Related Toxins* (eds F. Riet-Correa, J. Pfister, A.L. Schild, and T.L. Wierenga)

Females were mated and pregnancy was confirmed according to Taylor (1986). The animals were randomly assigned to three groups: treatment 1 (*n*=12) received 0.3 ml/day of the HEBT orally from gestational day (GD) 1 to 19; those in treatment 2 (*n*=11) received 0.3 ml/day of the HEBT orally from GD 6 to 15; and those in treatment 3 (control; *n*=12) received 0.3 ml/day of distilled water orally from GD 1 to 19.

The female rats were inspected twice a day for clinical signs of toxicity (Manson and Kang 1989) and weighed on GD 1, 6, 15, and 20. At GD 20 the animals were anesthetized and ovariohysterectomies were performed. Ovaries were weighed and the number of corpora lutea was counted. The uterus was weighed with the conceptuses and after the conceptuses were removed the uterus was weighed again to determine the corrected weight gain (weight gain from GD 1-20 less weight of conceptuses). The number of live and dead fetuses, implantation sites, and reabsorptions were counted. All fetuses and respective placentas were individually weighed, measured, and systematically inspected with a stereomicroscope for external structural anomalies. The following determinations were made: implantation rate, reabsorption rate, preimplantation loss, postimplantation loss, and placental index.

The rats were euthanized and their kidneys and liver were removed and weighed.

The litters were randomly assigned to three groups. Group 1 (40% of the litters) was fixed in Bodian's solution for visceral examination using the technique described by Wilson (1965), Barrow and Taylor (1969), and Taylor (1986). Group 2 (35% of the litters) underwent skeletal examinations (Staples and Schnell 1964). Group 3 (25%) was fixed in Bouin's solution for histological analysis.

For the statistical analysis the results were compared by the use of parametric (ANOVA) and nonparametric (Kruskal-Wallis test) tests followed by the Tukey's test and Dunn's tests, respectively. Based on the literature (Manson and Kang 1989) the litter was used as the experimental unit for the fetal analyses. Proportions were analyzed by the chi-square test or Fisher's exact test. The difference was considered significant at $P < 0.05$.

## Results and Discussion

During the period of treatment, no maternal deaths and no clinical signs of toxicity were recorded. Significant changes in maternal body weight were observed in treatment 1 only in the period from GD 15 to 20 (Figure 1).

The corrected maternal weight gain was similar in all three study groups. The significant reduction in the maternal body weight gain seen in treatment 1 is indicative of the toxic effects of HEBT on the fetal weight. This period (GD 15 to 20) is characterized by a rapid fetal weight gain due to a large accumulation of subcutaneous adipose tissue, according to Moore and Persaud (2004).

The weight of the kidneys of the animals in treatment 1 and the weight of the uterus with and without conceptuses of the animals from both treatments 1 and 2 showed significant differences when compared with the control group (Table 1).

Microscopic examination of maternal organs revealed histopathological changes in the kidneys in 91.7%, 81.8%, and 25% of the animals in treatments 1, 2, and controls, respectively, and liver in 58.3%, 63.6%, and 16.7% of the animals in treatments 1, 2, and controls, respectively. The percentage of histopathological changes observed in treatments 1 and 2 were significantly different from those observed in the control group (chi-square test, $P < 0.05$). These alterations indicate that HEBT probably exceeded the maternal capacity for detoxification and elimination of the chemical thus changing significantly maternal homeostasis (Cotran *et al.* 1996).

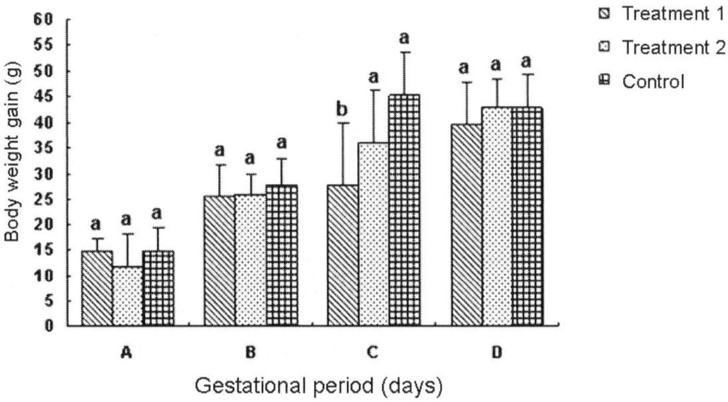


**Figure 1.** Maternal body weight gain during gestation. Gestational days (GD): A: 1 to 6; B: 6 to 15; C: 15 to 20; D: 1 to 20 (conceptus weight excluded). Values with different letters in the same gestational period differ significantly (Tukey, $P < 0.05$).

**Table 1.** Organ weights (mean ± standard deviation) in pregnant rats treated with a hydroethanolic extract of *Baccharis trimera*.

| Weight (g) | Treatment 1 (*n*=12) | Treatment 2 (*n*=11) | Controls (*n*=12) |
|---|---|---|---|
| Right kidney | 0.82 ± 0.08 [b] | 0.76 ± 0.09 [a] | 0.73 ± 0.05 [a] |
| Left kidney | 0.76 ± 0.08 [b] | 0.74 ± 0.08 [a] | 0.72 ± 0.06 [a] |
| Liver | 13.43 ± 1.86 [a] | 12.32 ± 1.37 [a] | 12.92 ± 1.42 [a] |
| Ovaries | 0.13 ± 0.02 [a] | 0.12 ± 0.02 [a] | 0.11 ± 0.01 [a] |
| Uterus with conceptuses | 33.12 ± 15.48 [b] | 34.82 ± 12.29 [b] | 51.27 ± 10.20 [a] |
| Uterus without conceptuses | 3.80 ± 1.07 [b] | 4.05 ± 0.74 [b] | 5.04 ± 0.56 [a] |

Values on the same row followed by different letters differ significantly (Tukey, $P < 0.05$).

Maternal and developmental parameters are shown in Table 2. The number of live fetuses, rate of implantation and preimplantation loss were significantly different between the animals in treatment 1 and those in controls. The fetal weight and length in treatment 1 were significantly smaller than those in controls. Placentomegaly was observed in both treatments 1 and 2.

The decrease in implantation rate and increase in preimplantation loss indicate that HEBT affected the implantation process. Fertilized oocytes did not result in implanted blastocysts, possibly due to the relaxing effect of *B. trimera* on the smooth muscle of the uterine tube. The decreases in fetal weight and length in treatment 1 associated with reduced ossification of the skull and sternebrae are strong indicators of intrauterine growth retardation (IUGR) (Aliverti *et al.* 1979).

The increases in placental weight and the placental index observed in animals from treatments 1 and 2 indicate that HEBT alters placental development thus inducing fetal hypoxia and nutritional insufficiency. To compensate for the IUGR, the expansion of the spongy trophoblast layer, which is essential for the production of hormones and growth factors (Ishikawa *et al.* 2006), resulted in placentomegaly.

**Table 2.** Gestational and developmental parameters (mean ± standard deviation).

| Parameters | Treatment 1 (*n*=12) | Treatment 2 (*n*=11) | Controls (*n*=12) |
|---|---|---|---|
| Number of corpora lutea | 13.00 ± 1.71 [a] | 11.73 ± 1.10 [a] | 12.42 ± 0.90 [a] |
| Number of live fetuses | 6.67 ± 4.31 [b] | 6.82 ± 2.71 [a] | 9.92 ± 1.51 [a] |
| Implantation (%) | 78.29 ± 21.17 [b] | 84.16 ± 19.93 [a] | 96.05 ± 4.13 [a] |
| Reabsorption (%) | 33.37 ± 24.87 [a] | 31.07 ± 14.84 [a] | 16.06 ± 12.29 [a] |
| Preimplantation loss (%) | 21.71 ± 21.17 [b] | 15.84 ± 19.93 [a] | 3.95 ± 4.13 [a] |
| Postimplantation loss (%) | 34.21 ± 25.64 [a] | 32.21 ± 15.83 [a] | 16.76 ± 11.51 [a] |
| Fetal weight (g) | 2.68 ± 0.51 [b] | 2.90 ± 0.56 [a] | 3.24 ± 0.25 [a] |
| Fetal length (cm) | 3.37 ± 0.29 [b] | 3.48 ± 0.30 [a] | 3.65 ± 0.12 [a] |
| Placental weight (g) | 0.55 ± 0.11 [b] | 0.51 ± 0.07 [b] | 0.46 ± 0.04 [a] |
| Placental index | 0.21 ± 0.06 [b] | 0.18 ± 0.02 [b] | 0.14 ± 0.01 [a] |

Values on the same row followed by different letters differ significantly (Tukey, $P < 0.05$).

Visceral examination revealed that the occurrence of moderate hydrocephalus was significantly higher in treatment 2 while the occurrence of severe hydrocephalus in both treatments 1 and 2 was significantly different from that in control group (Table 3).

**Table 3.** Number of fetuses analyzed and occurrence (%) of visceral abnormalities in the litter of rats treated with a hydroethanolic extract of *Baccharis trimera*.

| | Treatment 1 | Treatment 2 | Controls |
|---|---|---|---|
| Number of fetuses analyzed | 39 | 31 | 49 |
| Slight hydrocephalus | 15.4 | 9.7 | 10.2 |
| Moderate hydrocephalus | 17.9 | 29.0 * | 10.2 |
| Severe hydrocephalus | 23.1 * | 29.0 * | 6.12 |
| Cleft palate | 0 | 3.2 | 2.0 |
| Hydronephrosis | 17.9 | 32.3 | 14.3 |
| Globular shaped heart | 5.1 | 9.7 | 0 |

*$P < 0.05$, as compared with controls (chi-square test or, alternatively, Fisher's exact test).

These results indicate embryo-fetal toxicity associated with gestational period and exposure time to the HEBT. The critical period in brain development occurs from the beginning of organogenesis to approximately the second third of the fetal period (Moore and Persaud 2004). Exposure to HEBT from GD 6 to 15 affected a larger number of animals causing moderate and severe lesions since this time interval is the most susceptible, while exposure to HEBT from GD 1 to 19, although affecting the cerebral development of a smaller number of animals, caused more severe and irreversible lesions due to gradual and continued accumulation of fluid in the third and lateral ventricles (Solecki *et al.* 2003).

Skeletal examinations revealed significant differences between the treatment groups and controls (Table 4). In treatment 1 ossification of the tympanic bulla was absent; reduced ossification was seen in the pterygoid, basisphenoid, parietal, interparietal, and sternebrae bones. The presence of asymmetrical parietal, interparietal, and supraoccpital bones in treatment 2 was significantly different from that in control group.

**Table 4.** Number of fetuses analyzed and occurrence (%) of skeletal abnormalities in the litter of rats treated with a hydroethanolic extract of *Baccharis trimera*.

| | Treatment 1 (%) | Treatment 2 (%) | Controls (%) |
|---|---|---|---|
| Number of fetuses analyzed | 28 | 25 | 40 |
| Skull | | | |
| Tympanic bulla (absence of ossification) | 21.4* | 4.3 | 2.5 |
| pterygoid (reduced ossification) | 21.4* | 4.3 | 2.5 |
| Basisphenoid (reduced ossification) | 21.4* | 4.3 | 2.5 |
| Hyoid (absence of the corpus) | 0 | 1.1 | 0 |
| Frontal (reduced ossification) | 10.7 | 1.1 | 0 |
| Parietal | | | |
| (reduced ossification) | 25.0* | 3.2 | 5.0 |
| (asymmetrical) | 0 | 4.3* | 0 |
| (separate) | 3.6 | 1.1 | 2.5 |
| Interparietal | | | |
| (absence of ossification) | 0 | 1.1 | 0 |
| (reduced ossification) | 21.4* | 3.2 | 2.5 |
| (asymmetrical) | 0 | 4.3* | 0 |
| (separate) | 3.6 | 1.1 | 2.5 |
| Supraoccipital | | | |
| (absent) | 0 | 2.2 | 5.0 |
| (absence of ossification) | 0 | 1.1 | 0 |
| (reduced ossification) | 0 | 3.2 | 0 |
| (asymmetrical) | 7.1 | 4.3* | 0 |
| (separate) | 10.7 | 0 | 0 |
| Sternum | | | |
| (number of malformed sternebrae) | 17.9 | 9.7 | 20.0 |
| (reduced ossification) | 57.1* | 3.2 | 2.5 |
| (asymmetrical sternebrae) | 10.7 | 4.3 | 15.0 |
| (fused sternebrae) | 3.6 | 1.1 | 2.5 |
| (reduced sternebrae) | 28.6 | 4.3 | 17.5 |
| (separate sternebrae) | 7.1 | 1.1 | 0 |
| (dumbbell-shaped sternebrae) | 3.6 | 0 | 2.5 |
| (ladder-shaped sternebrae) | 0 | 0 | 7.5 |
| (absent) | 0 | 1.1 | 2.5 |
| Vertebrae (number of malformed vertebrae) | 0 | 3.2 | 0 |
| Ribs (rudimentary) | 3.6 | 1.1 | 0 |

*$P < 0.05$, as compared with controls (chi-square test or, alternatively, Fisher's exact test)

## Conclusions

The treatment-related skeletal abnormalities may be classified as variations instead of malformations since they are temporary alterations that do not adversely affect survival (Chahoud *et al.* 1999). Reduced ossification of the parietal and interparietal bones and of sternebrae (treatment 1) is reversible and indicative of a slight delay in fetal development and is usually associated with decreased fetal weight (Manson and Kang 1989). The fetuses in treatment 2 showed irregular ossification of the parietal, interparietal, and supraoccipital bones, suggesting interference in the normal process of ossification. The absence of ossification of the tympanic bulla and reduced ossification of the pterygoid and basisphenoid bones, enhanced in fetuses from treatment 1, may also be classified as variations due to their transitory nature (Solecki *et al.* 2001). According to Manson and Kang (1989) these variations are usually secondary to alterations in maternal homeostasis.

The results presented above suggest that HEBT when administered at daily oral doses of 8.4 mg/kg is toxic to the kidney and liver in the rat and the severity of changes is directly associated with exposure time to the toxic agent. The extract also interferes with blastocyst implantation. The administration of the extract throughout pregnancy increases the incidence of visceral malformations and skeletal variations and causes a decrease in fetal weight, size, and ossification. However, the fetal abnormalities may not have been due to the direct effect of the extract but rather to the maternal toxicity induced by it. The occurrence of maternal and developmental toxicity should not, in principle, serve to diminish the significance of the toxic effects on embryos and fetuses.

## References

Aliverti V, Bonanomi L, Giavini E, Leone VG, and Mariani L (1979). The extent of fetal ossification as an index of delayed development in teratogenic studies on the rat. *Teratology* 20:237-242.

Barrow MV and Taylor WI (1969). A rapid method for detecting malformation in rat fetuses. *Journal of Morphology* 127:291-306.

Chahoud I, Buschmann J, Clark R, Druga A, Falke H, Faqi A, Hansen E, Heinrich-Hirsch B, Hellwig J, Lingk W, Parkinson M, Paumgartten FJR, Pfeil R, Platzek T, Scialli AR, Seed J, Staiilmann R, Ulbrich B, Wu X, Yasuda M, Younes M, and Solecki R (1999). Classification terms in developmental toxicology: need for harmonisation. *Reproductive Toxicology* 13(1):77-82.

Cotran RS, Kumar V, and Robbins SL (1996). *Robbins Patologia Estrutural e Funcional*, pp. 834-891. Guanabara Koogan, Rio de Janeiro.

Ishikawa H, Seki R, Yokonishi S, Yamauchi T, and Yokoyama K (2006). Relationship between fetal weight, placental growth and litter size in mice from mid- to late-gestation. *Reproductive Toxicology* 21:267-270.

Khera KS (1985). Maternal toxicity: a possible etiological factor in embryo-fetal deaths and fetal malformations of rodent-rabbit species. *Teratology* 31:129-153.

Manson JM and Kang YJ (1989). Test methods for assessing female reproductive and developmental toxicology. In *Principles and Methods of Toxicology* (AW Hayes, ed.), pp. 311-359. Raven Press, New York.

Moore KL and Persaud TVN (2004). *Embriologia Clínica*, 609 pp. Elsevier, Rio de Janeiro.

Nunes GP, Silva MF, Resende UM, and Siqueira JM (2003). Plantas medicinais comercializadas por raizeiros no Centro de Campo Grande, Mato Grosso do Sul. *Revista Brasileira de Farmacognosia* 13(2):83-92.

Sadler TW (2005). *Langman Embriologia Médica*, 347 pp. Guanabara Koogan, Rio de Janeiro.

Simões CMO, Mentz LA, Schenkel EP, Irgang BE, and Stehmann JR (1998). *Plantas da Medicina Popular no Rio Grande do Sul*, 173 pp. Ed. UFRGS, Porto Alegre.

Solecki R, Bürgin H, Buschmann J, Clark R, Duverger M, Fialkowski O, Guittin P, Hazelden KP, Hellwig J, Hoffmann E, Hofmann T, Hubel U, Khalil S, Lingk W, Mantovani A, Moxon M, Müller S, Parkinson M, Paul M, Paumgartten F, Pfeil R, Platzek T, Rauch-Ernst M, Scheevelenbos A, Seed J, Talsness CE, Yasuda M, Younes M, and Chahoud I (2001). Harmonisation of rat fetal skeletal terminology and classification. Report of the third Workshop on the terminology in developmental toxicology, Berlin, 14-16 September 2000. *Reproductive Toxicology* 15:713-721.

Solecki R, Bergmann B, Bügin H, Buschmann J, Edwards J, Freudenberger H, Guittin P, Hakaite P, Khali S, Klaus A, Kudicke S, Lingk W, Meredith T, Moxon M, Müller S, Paul M, Paumgartten F, Röhrdanz E, Pfeil R, Rauch-Ernst M, Seed J, Spezia F, Vickers C, Woelffel B, and Chahoud I (2003). Harmonisation of rat fetal external and visceral terminology and classification. Report of the fourth Workshop on the terminology in developmental toxicology, Berlin, 18-20 April 2002. *Reproductive Toxicology* 17:625-637.

Staples RE and Schnell VL (1964). Refinements in rapid clearing technic in the KOH-alizarin red S method for fetal bone. *Stain Technology* 39:61-63.

Taylor P (1986). *Practical Teratology*, 171 pp. Academic Press, New York.

US National Research Council (1996). *Guide for the Care and Use of Laboratory Animals*, 140 pp. National Academy Press, Washington.

Wilson JG (1965). Methods for administering agents and detecting malformations in experimental animals. In *Teratology: Principles and Techniques* (JG Wilson and J Warkany, eds), pp. 262-277. University of Chicago Press, Chicago.

# Chapter 123

# Toxicity in Mice of the Total Alkaloid Fraction of *Chondrodendron platyphyllum*

F.C. Montenegro, M.C.P. Sena, J.M.B. Filho, D.A.M. Araújo, and R.N. Almeida

*Laboratory of Pharmaceutical Technology, Federal University of Paraíba, João Pessoa-Paraíba, PO Box 58051-900, Brazil*

## Introduction

The Menispermaceae family belongs to the order Ranunculales and has 72 genera with over 400 species. These species can be found all over America, especially in tropical and subtropical regions. In Brazil, the Menispermaceae family has 106 species and most of them are found in the Amazonian region (Barroso 1978).

Plants of the Menispermaceae family contain curare alkaloids which are used by South American Indians for hunting. These plant-derived compounds are used to coat the tips of hunting arrows or blow-pipe darts. Very little poison is absorbed after oral ingestion and the meat from animals killed with curare is harmless. The mechanism of action of curare alkaloids is to block the nicotinic acetylcholine receptor (nAChR) at motor endplates (Bisset 1992; Bowman 2006). D-tubocurarine is the most important curare alkaloid. In the past, d-tubocurarine was used to determine the role of acetylcholine in neuromuscular transmission and it is commonly used as a muscle relaxant during surgical procedures (Aguayo *et al.* 2006; Bowman 2006).

It has been reported that some species from the Menispermaceae family have a toxic effect on the central nervous system (CNS) (Almeida *et al.* 1998, 2001; Nsour *et al.* 2000; Akah *et al.* 2002) while other species from this family have both a neurotoxic and cytotoxic effect (Hwi and Lay 1998; Melo *et al.* 2003; Wattanathorn *et al.* 2006). *Chondrodendron platyphyllum* is a South American species used in folk medicine for the treatment of fever caused by malaria and as an antispasmodic (Tang *et al.* 1980). Bisbenzylisoquinoline alkaloids (BBA) such as curine and 12-O-methylcurine have been isolated from the root bark of this plant. Both alkaloids act as vasodilators in rat aorta (Dias *et al.* 2002; Guedes *et al.* 2002). Due to the lack of records in the literature about the toxic effects of *C. platyphyllum* and its possible activity on the central nervous system, the present work was conducted with the objective to evaluate the toxicity of *C. platyphyllum* and its possible effects on the CNS.

©CAB International 2011. *Poisoning by Plants, Mycotoxins, and Related Toxins* (eds F. Riet-Correa, J. Pfister, A.L. Schild, and T.L. Wierenga)

## Material and Methods

Adult male albino Swiss mice weighing 25-40 g were used in all the studies conducted. Animals were housed in appropriate cages and the temperature was 21±1°C. The mice were submitted to a 12/12-h light/dark cycle (light from 06:00h to 18:00h) and had free access to standard rodent chow (Purina, Brazil) and water. All animals were acclimatized before the experiments and all experimental observations were conducted between 12:00 h and 17:00 h. All substances tested were injected via intraperitoneal procedures. The mice were euthanized by cervical dislocation and all procedures were carried out in accordance with Ethical Committee guidelines (CEPA/LTF-UFPB, process number 0206/08).

To evaluate the toxicity of this plant and its effects on the CNS we used pharmacological screening based on changes in animal behavior (Almeida *et al.* 1999). Behaviors that indicate activity in the CNS, the autonomic nervous system, and the number of deaths were recorded using a previously reported procedure (Almeida *et al.* 1999). Animals were divided into four groups, one control group and three experimental groups ($n$=8). The control group received a solution of distilled water plus 5% Tween 80. The three experimental groups received doses of the total alkaloid fraction (TAF) given at 100, 200, and 400 mg/kg. The alkaloid fraction was extracted from the root bark of *C. platyphyllum* following the process described by Filho *et al.* (1997). The animals were observed at various time points up to 4 h and after administration of the TAF. Changes in animal behavior (such as writhing, rearing, grooming, anesthesia, analgesia, convulsions, cyanosis, salivation, and hypnosis) that indicate CNS activity or toxicity and number of deaths were recorded later at 24, 48, and 72 h.

Motor activity was evaluated in a separate experiment using rotarod methodology (Gonçalves *et al.* 2008). Initially, animals were pre-selected based on their ability to perform on the rotarod apparatus (Insight, Model EF 412, Brazil). The animals that were able to remain on the gyratory bar (7 rpm) for 180 s were selected. The animals were tested and selected 24 h before the experiment. The experimental ($n$=8) group received a 200 mg/kg dose of TAF; controls were given distilled water. Each animal was tested on the rotarod and the length of time they remained on the gyratory bar up to 180 s was determined at 30, 60, and 120 min after drug administration. Experimental data obtained were evaluated using the Mann-Whitney test. Differences were considered to be statistically significant when $P < 0.05$.

## Results and Discussion

Changes observed during the pharmacological screening procedure of the treated and control animals are reported in Table 1. Animals treated with TAF at 100 mg/kg had many changes in behavior by 4 h post-dosing including catatonia, reduced spontaneous locomotion, loss of corneal and ear reflexes, reduced response to touch, and writhing. Defecation and micturition were also altered. Mice treated with TAF at 200 mg/kg had reduced spontaneous locomotion and corneal and ear reflexes were lost. Response to noise was reduced for up to 2 h post-dosing and analgesia was present for up to 0.5 h post-dosing. Writhing was observed for up to 1 h post-dosing. Defecation was reduced for up to 4 h post-dosing. Treatment with TAF at 400 mg/kg showed analgesia from 2 to 4 h post-dosing, reduction of spontaneous locomotion, loss of corneal and ear reflexes, and reduced response to touch up to 2 h post-dosing. Writhing was observed for up to 1 h post-dosing.

Defecation and micturition were reduced for up to 4 h post-dosing. Changes in the observed behaviors indicate that treatment with TAF exerts depressor activity on central nervous system. Additionally, seven mice treated with TAF at 400 mg/kg died 4 h post-dosing.

**Table 1.** Changes in behavior observed during pharmacological screening in mice treated with the total alkaloid fraction (TAF) from *Chondrodendron platyphyllum*.

| Treatment (mg/kg) | Observation time (h) | Changes in behavior |
|---|---|---|
| TAF 100* | 0.5 h | Catatonia, spontaneous ambulation reduced, loss of corneal and ear reflexes, response to touch reduced, writhing, defecation and micturition altered. |
| | 1 h | Catatonia, defecation and micturition altered. |
| | 2 h | Catatonia, defecation and micturition altered. |
| | 3 h | Catatonia, defecation and micturition altered. |
| | 4 h | Catatonia, defecation and micturition altered. |
| TAF 200 | 0.5 h | Loss of corneal and ear reflexes, analgesia, writhing, spontaneous ambulation, response to touch and defecation reduced. |
| | 1 h | Loss of corneal and ear reflexes, analgesia, writhing, spontaneous ambulation, response to touch and defecation reduced. |
| | 2 h | Loss of corneal and ear reflexes, analgesia, writhing, spontaneous ambulation, response to noise and defecation reduced. |
| | 3 h | Defecation reduced. |
| | 4 h | Defecation reduced. |
| TAF 400† | 0.5 h | Writhing, loss of corneal and ear reflexes, spontaneous ambulation, response to touch, defecation and micturition reduced. |
| | 1 h | Writhing, loss of corneal and ear reflexes, spontaneous ambulation, response to touch, defecation and micturition reduced. |
| | 2 h | Writhing, analgesia, loss of corneal and ear reflexes, spontaneous ambulation, response to touch, defecation and micturition reduced. |
| | 3 h | Analgesia, defecation and micturition reduced. |
| | 4 h | Analgesia, defecation and micturition reduced. |

*5/8 mice tested displayed catatonia. The other signs were observed in all mice; †7/8 mice died at 4 h post-dosing but all of them showed all of the mentioned clinical signs.

According to the literature some species of the Menispermaceae family have a depressive effect on activity in the central nervous system. *Cissampelos mucronata* has sedative and anticonvulsive activities and *Cissampelos sympodialis* has antidepressant effects in animal models (Almeida *et al.* 1998; Akah *et al.* 2002). Other species are toxic: *Limacia scanden* was toxic in animal models and *Coscinium fenestratum* is neurotoxic and is able to induce neurobehavioral changes in rats. Two bisbenzylisoquinoline alkaloids from *C. sympodialis*, a compound named warifteine and another named milonine, induced toxic effects in hepatic cell models (Hwi and Lay 1998; Melo *et al.* 2003; Wattanathorn *et al.* 2006; Wongcome *et al.* 2007). Another member of the Menispermaceae family, *Chondrodendron platyphyllum*, exerts central nervous system activity and toxic effects.

According to our observations the alkaloid fraction from *C. platyphyllum* exerted a depressor activity on the CNS at all doses tested and was lethal to 60% of the mice when the TAF was dosed at 400 mg/kg.

In the second study, motor activity was evaluated using a rotarod test. Results showed that animals treated with TAF at 200 mg/kg were able to remain on the rotarod apparatus for 180 s and treated animals did not differ from control animals. Curare alkaloids like d-tubocurarine generally act as muscle relaxants (McManus 2001; Zlotos 2005). In this experiment motor activity was evaluated by the gyratory bar of a rotarod test (Dunham and Miya 1957). Since the fraction used should contain the curare alkaloid the absence of a muscle relaxant action of the *C. platyphyllum* treatment, which did not alter performance on the rotarod apparatus, suggests that the plant may not contain this substance or that it does not have the same effect in this model at the dose used.

## Conclusions

The results of this study indicate that the total alkaloid fraction from *C. platyphyllum* has a toxic effect in mice at doses from 100 to 400 mg/kg. The mice showed clinical signs that are representative of central nervous system depressor activity. A dose of TAF of 400 mg/kg was often lethal. These results are similar to those observed in other studies using plants from the Menispermaceae family. No effect was found, however, on motor activity.

## References

Aguayo LG, Guzman L, Perez C, Aguayo LJ, Silva M, Becerra J, and Fuentealba J (2006). Historical and current perspectives of neuroactive compounds derived from Latin America. *Mini Reviews in Medical Chemistry* 6(9):997-1008.

Akah PA, Nwafor SV, Okoli CO, and Egbogha CU (2002). Evaluation of the sedative properties of the ethanolic root extract of *Cissampelos mucronata*. *Bollettino Chimico Farmaceutico* 141(3):243-246.

Almeida RN, Navarro DS, Assis TS, Medeiros IA, and Thomas G (1998). Antidepressant effect of an ethanolic extract of the leaves of *Cissampelos sympodialis* in rats and mice. *Journal of Ethnopharmacology* 63(3):247-252.

Almeida RN, Falcão ACGM, Diniz RT, Júnior LJQ, Polari RMP, Filho JMB, Agra MF, Duarte JCD, Ferreira DF, Antaniole A, and Araújo CA (1999). Metodologia para avaliação de plantas com atividade no sistema nervoso central e alguns dados experimentais. *Revista Brasileira de Ciências Farmacêuticas* 80(3/4):72-76.

Almeida RN, Navarro DS, and Filho JMB (2001). Plants with central analgesic activity. *Phytomedicine* 8(4):310-32.

Barroso GM (1978). *Sistematica de Angiospermas do Brasil*, 255 pp. EPUSP, São Paulo.

Bisset NG (1992). War and hunting poisons of the New World. Part 1. Notes on the early history of curare. *Journal of Ethnopharmacology* 36(1):1-26.

Bowman WC (2006). Neuromuscular block. *British Journal of Pharmacology* 147:S277-S286.

Dias CS, Filho JMB, Lemos VS, and Côrtes SF (2002). Mechanisms involved in the vasodilator effect of curine in rat resistance arteries. *Planta Medica* 68(11):1049-1051.

Dunham NW and Miya TSA (1957). A note on a simple apparatus for detecting neurological deficit in rats and mice. *Journal of the American Pharmaceutical Association* 16(3):208-209.

Filho JMB, Cunha EVL, Cornélio ML, Dias CS, and Gray AI (1997). Cissaglaberrimine, an aporphine alkaloid from *Cissampelos glaberrima. Phytochemistry* 44(5):959-961.

Gonçalves JCR, Oliveira FS, Benedito RB, Souza DP, Almeida RN, and Araújo DAM (2008). Antinociceptive Activity of (-)-Carvone: Evidence of Association with Decreased Peripheral Nerve Excitability. *Biological & Pharmaceutical Bulletin* 31(5):1017-1020.

Guedes DN, Filho JMB, Lemos VS, and Côrtes SF (2002). Mechanism of the vasodilator effect of 12-O-methylcurine in rat aortic rings. *Journal Pharmacy and Pharmacology* 54(6):853-858.

Hwi KK and Lay WB (1998). Pharmacological, electrophysiological and toxicity studies of *Limacia scanden* Lour (Menispermaceae). *Journal of Ethnopharmacology* 62:137–148.

McManus MC (2001). Neuromuscular blockers in surgery and intensive care, Part 2. *American Journal of Health-System Pharmacy* 58(24):2381-2395.

Melo PS, Cavalcante HMM, Filho JMB, Diniz MFFM, Medeiros IA, and Haun M (2003). Warifteine and milonine, alkaloids isolated from *Cissampelos sympodialis* Eichl: cytotoxicity on rat hepatocyte culture and in V79 cells. *Toxicology Letters* 142(1-2):143-151.

Nsour WM, Lau CBS, and Wong ICK (2000). Review on phytotherapy in epilepsy. *Seizure* 9:96-107.

Tang XC, Feng J, Wang YE, and Liu MZ (1980). Neuromuscular blocking activity of alkaloids of *Cyclea hainanensis. Acta Pharmacologica* 1:17-22.

Wattanathorn J, Uabundit N, Itarat W, Mucimapura S, Laopatarakasem P, and Sripanidkulchai B (2006). Neurotoxicity of *Coscinium fenestratum* stem, a medicinal plant used in traditional medicine. *Food and Chemical Toxicology* 44(8):1327-1333.

Zlotos DP (2005). Recent advances in neuromuscular blocking agents. *Mini Reviews in Medical Chemistry* 5(6):595-606.

# Chapter 124

# Evaluation of Anticholinesterasic Activity of Strain SPC 920–*Geitlerinema unigranulatum* (Oscillatoriales, Cyanobacteria)

C.R. Dogo[1,2], M. Rangel[3], E.M. Cardoso-Lopes[4], C.L. Sant'Anna[2], F.M. Bruni[5], M. Lopes-Ferreira[5], and L.R. de Carvalho[2]

*[1]Post-Graduate Program in Plant Biodiversity and Environment, Botanic Institute of São Paulo; [2]Phycology Section, Botanic Institute of São Paulo; [3]Laboratory of Immunopathology, Butantan Institute; [4]Plants Biochemistry and Physiology, Botanic Institute of São Paulo, Av. Miguel Estéfano, 3687 Água Funda 04301-902 - São Paulo, SP, Brazil; [5] Center for Applied Toxinology, Butantan Institute, Av. Vital Brasil, 1500 05503-900 - São Paulo, SP, Brazil*

## Introduction

Cyanobacteria are gram-negative and photosynthetic microorganisms present in terrestrial and aquatic environments, even in those with extreme conditions. However, most of them inhabit continental water bodies which include drinking water supplies where they are present as phytoplankton components. Frequently, under favorable environmental conditions, these organisms form blooms which turn the water green and with unpleasant odor and taste. This ability combined with that of producing highly toxic secondary metabolites makes the cyanobacteria responsible for acute poisoning in animals and humans and also for promoting cancer and neuro-degenerative disease (Nishiwaki-Matsushima *et al.* 1992; Azevedo *et al.* 2002; Murch *et al.* 2004).

The cyanobacterial toxins form a chemically heterogeneous group that belongs to three different chemical classes: cyclic peptides, alkaloids, and lipopolysaccharides. According to their pharmacological action, they are characterized as hepatotoxins, neurotoxins, cytotoxins, and dermatotoxins.

The hepatotoxins include the cyclic peptide microcystins and nodularins. The neurotoxins are alkaloids that are grouped into three classes according to their pharmacological action: the saxitoxins are blockers of nerve cells sodium channels, the anatoxin-a (S) blocks the activity of acetylcholinesterase, and anatoxin-a has a similar effect on acetylcholine. The cytotoxins belong to an alkaloid class formed by cylindrospermop-sins that cause lesions in the kidneys, heart, lungs, and gastric mucosa. Dermatotoxins are lipopolysaccharides that are cell wall components of all gram-negative bacteria and cyanobacteria which induce allergies and irritation to the skin, eyes, and throat (Codd 2000; Carvalho *et al.* 2008).

©CAB International 2011. *Poisoning by Plants, Mycotoxins, and Related Toxins* (eds F. Riet-Correa, J. Pfister, A.L. Schild, and T.L. Wierenga)

In addition to these toxins, the neurotoxic amino acid β-methyl-L-amino-alanine (BMAA) which has cumulative effects and causes poisoning and death in humans was recently isolated from the cyanobacterial genus *Nostoc* (Cox *et al.* 2003).

In the past decades, the search for other cyanobacterial properties was almost nonexistent because all attention was focused on toxicity. This scenery has changed after some varieties of the genus *Spirulina* were recognized as 'the most nutritious food in the world' (Becker 2004).

Currently, it is known that cyanobacteria which can also be used as fertilizer (Valiente *et al.* 2004) and for bioremediation of aquatic and terrestrial environments produce not only toxins but also countless compounds with potential application in the treatment of diseases such as cancer, acquired immunodeficiency syndrome, asthma, and diabetes (Luesch *et al.* 2001; Gademann and Portmann 2008; Svircev *et al.* 2008). These compounds can be used also in cosmiatry and as biofuel (Chisti 2007; Hellingwerf and Teixeira de Mattos 2009).

During the search for bioactive substances from strains maintained in the Cyanobacteria Culture Collection of the Institute of Botany, São Paulo, Brazil, the strain SPC 920–*Geitlerinema unigranulatum* showed toxicity when tested by mouse bioassay (Dogo and Carvalho 2005). The symptoms observed in this bioassay were very distinct from those presented by animals poisoned with the known cyanotoxins (Harada *et al.* 1999).

The mouse bioassay, previously used to detect saxitoxins (neurotoxin) in samples of seafood (AOAC 1995), is the toxicological test recommended by the World Health Organization for the detection of cyanotoxins and is also a powerful aid in the indication of the chemical class of the toxin under study. The signs of poisoning, the time to death, and the autopsy findings clearly show if the toxin in question is a hepatotoxin (nodularinas and microcystins) or a neurotoxin (saxitoxins and anatoxins) (Carvalho 2006). Thus, our aim is to search for toxic compounds and potential pharmaceutical drugs in extracts of SPC 920–*Geitlerinema unigranulatum*.

## Material and Methods

SPC 920–*G. unigranulatum* strain is kept in the Cyanobacteria Culture Collection, Institute of Botany, São Paulo, Brazil. This filamentous species belongs to the family Anabaenaceae, order Oscillatorialles.

### Culture of *G. unigranulatum*

Lyophilized cells of *G. unigranulatum* were obtained from 5 l batch cultures. These cultures were grown in ASM-1 medium at 22±1°C with continuous illumination (15-20 μmo/m/s) and aeration (Azevedo and Sant'Anna 2003). Cells were harvested at the end of the late-exponential growth phase (about 3 weeks).

### Extract preparation and fractionation

Freeze-dried cells were extracted with MeOH/$H_2O$ 75:25 v/v (5×) by sonication (40 × 30 s, 50 W). The combined extracts were centrifuged (1830 g, 50 min), collected, and concentrated at low pressure and the remaining aqueous residue was lyophilized (Fastner *et al.* 1998).

The dried methanolic extract was submitted to Preparative Planar Chromatography (PPC). The plates (silica gel, Analtech $F_{254}$, 20×20 cm; 1000 microns) were developed in $CHCl_3/MeOH/H_2O$ 64:36:08 (v/v/v). The bands were detected under UV light (254 and 365 nm), removed and extracted with $CHCl_3/MeOH/H_2O$ 64:36:08 (v/v/v) and acetic acid 0.1 M according to the band polarity. The extract from the band with anti-cholinesterase activity was submitted to High Performance Liquid Chromatography (HPLC) analysis

**HPLC analysis of band with anticholinesterasic action**

HPLC analyses were performed in a Dionex (P680–Chromeleon 6.80) apparatus equipped with UVD340UV and Ultimate 3000 Autosampler. The separation was carried out in a multi-step gradient in an analytical Zorbax ODS column (250×4.6 mm with particle size of 5 μm; Agilent). The eluents used were distilled and deionized water (solvent A) and ACN (solvent B), each containing 0.1% (v/v) TFA. Elution was monitored at 220 nm and the flow rate was 1.0 ml/min.

**Toxicity assay**

Toxicity of lyophilized cells and chromatographic bands were checked by mouse assay. These tests ($n$=3) were performed using line 'Swiss' male mice weighing between 18 and 22 g (Harada *et al.* 1999). Doses of 1000, 825, 500, and 250 mg/kg of body weight (BW) of *G. unigranulatum* methanolic extract were administered intraperitoneally (ip) after being lyophilized for total removal of organic solvent and suspended in 500 μl of sterile saline solution (NaCl 0.9%).

Band extracts obtained from 20 mg of lyophilized cells free of solvents and diluted in 500 μl of sterile saline solution (NaCl 0.9%) were also assayed. The signs of intoxication and the survival time were recorded; necropsies were performed on mice that died from the administration of the extract.

**Anti-cholinesterase assay**

The procedure by Marston *et al.* (2002) was used for this bioassay. Briefly, acetylcholinesterase type V (Sigma product no. C 2888, 1000 U) was dissolved in Tris-HCl buffer (pH 7.8) and stabilized by the addition of bovine serum albumin fraction V (0.1%, Sigma, product no. A-4503). TLC layers were spotted with 200 μg/spot of SPC 920–*G. unigranulatum* crude extract and galanthamine (Sigma, 1 μg/spot) and eserine (Sigma, 0.3 μg/spot) were used as positive controls. TLC layers (Merck cromatoplate, silica gel 60 $F_{254}$, 20×20cm) were developed with $CHCl_3/MeOH/H_2O$ 64:36:08 (v/v/v) and subsequently dried. The plates were then sprayed with the enzyme solution (6.66 U/ml), thoroughly dried, and incubated at 37°C for 20 min (moist atmosphere). Enzyme activity was detected by spraying with a solution consisting of 0.25% of 1-naphtyl acetate in EtOH (5 ml) plus 0.25% aqueous solution of Fast Blue B salt (20 ml). Potential acetylcholinesterase inhibitors appeared as clear zones on a purple colored background.

## Results and Discussion

The *G. unigranulatum* crude extract showed, by bioautography test (anti-AChE assay), a band at Rf 0.89 capable of inhibiting the enzyme AChE. In the mouse bioassay, the methanolic extract of lyophilized cells had a $MLD_{100}$ of 500 mg/kg BW. Mouse deaths occurred from 50 min to 19 h; the lungs were consolidated but the livers had a normal appearance without typical signs of hepatotoxic intoxication (Table 1).

**Table 1.** Results of mouse bioassays with SPC 920 *G. unigranulatum* crude extract (*n*=3).

| mg Crude extract | Survival time | Postmortem observations |
|---|---|---|
| 20 | 2 to 19 h | Hepatized lungs; liver with normal size and absence of blood in the abdominal cavity |
| 15 | 2 h | Hepatized lungs; liver with normal size and absence of blood in the abdominal cavity |
| 10 | Between 50 min to 2 h | Hepatized lungs; liver with normal size and absence of blood in the abdominal cavity |
| 5 | Death did not occur | - |

All band extracts from PPC were examined using a mouse bioassay. The band at Rf 0.2 caused the death of animals with the same symptoms and postmortem observations as the methanolic extract but the band at Rf 0.89 (with anti-AChE activity) was not toxic. These results indicated that the anti-AChE compound was not the toxin produced by *G. unigranulatum.* The active band (Rf 0.89) chromatogram obtained by HPLC-PDA showed the presence of five major peaks detected at 220 nm. The UV spectra of these peaks were recorded (Figure 1).

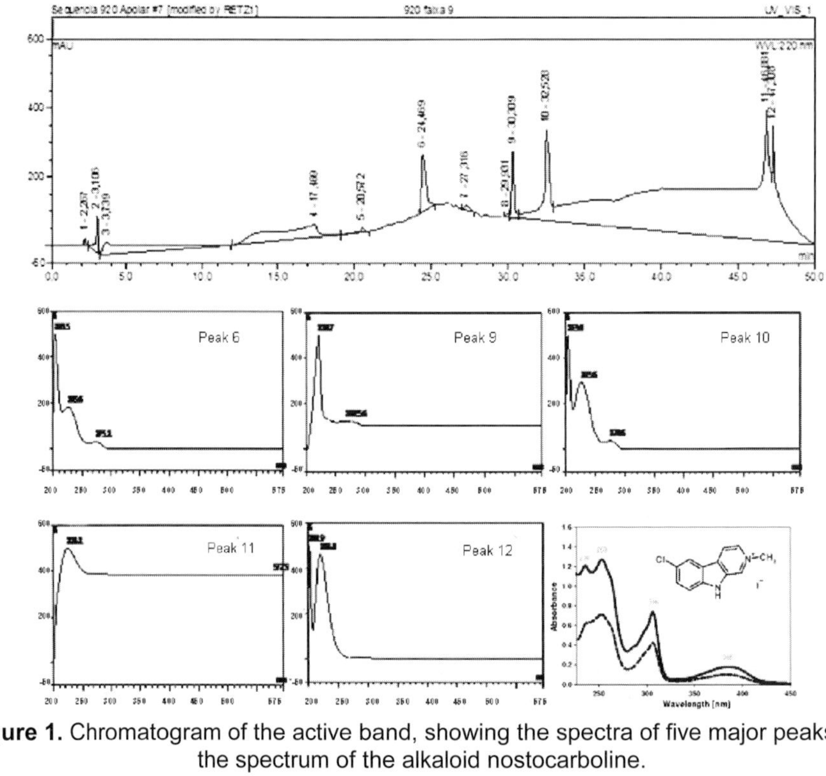


**Figure 1.** Chromatogram of the active band, showing the spectra of five major peaks and the spectrum of the alkaloid nostocarboline.

Until now only two substances from cyanobacteria have been shown to display anti-cholinesterase activity: anatoxin-a(S), a natural organophosphate (methyl ester of N-hydroxyguanidine phosphate) (Rodriguez *et al.* 2006), and nostocarboline, an alkaloid isolated from species of the genus *Nostoc* (Becher *et al.* 2009). The anatoxin-a (S) is a very potent neurotoxic alkaloid with different symptoms and a survival time shorter than the lethal dose of *G. unigranulatum* toxin.

As the UV spectra obtained from *G. unigranulatum* active band are different from that of nostocarboline we conclude that we are dealing with new bioactive compounds.

These results may have important biomedical implications. For example, substances with anti-cholinesterase action raise the level of acetylcholine in the brain and are important in the symptomatic treatment of Alzheimer's disease (AD), a neurodegenerative disorder characterized by progressive loss of memory and cognitive functions.

## References


AOAC Official methods of analysis (1995). *Paralytic shellfish poison*. 35:21-22.

Azevedo MTP and Sant'Anna CL (2003). *Sphaerocavum brasiliense*, a new planktic genus and species of Cyanobacteria from reservoirs of São Paulo State, Brazil. *Algological Studies* 109:79-92.

Azevedo SMFO, Carmichael WW, Joghimsen EM, Rinehart KL, Lau S, Shaw GR, and Eaglesham GK (2002). Human intoxication by microcystins during renal dialysis treatment in Caruaru – Brazil. *Toxicology* 181-182:441-446.

Becher PG, Baumann HI, Gademann K, and Jüttner F (2009). The cyanobacterial alkaloid nostocarboline: an inhibitor of acetylcholinesterase and trypsin. *Journal of Applied Phycology* 20:1-8.

Becker W (2004). Microalgae in human and animal nutrition. In *Handbook of Microalgae Culture: Biotechnology and Applied Phycology* (A Richmond, ed.), pp. 312-351. Blackwell Publishing Company.

Carvalho LR (2006). Cianotoxinas. In *Manual ilustrado para identificação e contagem de cianobactérias planctônicas de águas continentais brasileiras* (Sant'Anna CL, Azevedo MTP, Agujaro LF, Carvalho MC, Carvalho LR, and Souza RCR, eds), pp. 9-19. Interciência, Rio de Janeiro.

Carvalho LR, Haraguchi M, and Górniak SL (2008). Intoxicação produzida por algas de água doce. In *Toxocologia Aplicada à Medicina Veterinária* (HS Spinosa, SL Górniak, and J Palermo Neto, eds), pp. 621-640. Editora Manole, Barueri.

Chisti Y (2007). Biodiesel from microalgae. *Biotechnology Advances* 25(3):294-306.

Codd GA (2000). Cyanobacterial toxins, the perception of water quality, and the prioritisation of eutrophication control. *Ecological Engineering* 16:51-60.

Cox PA, Banack SA, and Murch SJ (2003). Biomagnification of cyanobacterial neurotoxins and neurodegenerative disease among the Chamorro people of Guam. *Proceedings of the National Academy of Sciences of the United States of America* 110(23):13380-13383.

Dogo CR and Carvalho LR (2005). *Estudo químico de cepas do Banco de Cultura de Cianobactérias da Seção de Ficologia*. Relatório PIBIC, São Paulo.

Fastner J, Flieger I, and Neumann V (1998). Optimized extraction of microcystins from field samples: a comparison of different solvents and procedures. *Water Research* 32:3177-3181.

Gademann K and Portmann C (2008). Secondary metabolites from Cyanobacteria : Complex structures and Powerful bioactivities. *Current Organic Chemistry* 12:326-341.

Harada K, Kondo F, and Lawton L (1999). Laboratory analysis of cyanotoxins. In *Toxic Cyanobacteria in Water. A guide to their public health consequences, monitoring and management* (I Chorus and J Bartram, eds), pp. 369-405. E & FN SPON, New York.

Hellingwerf KJ and Teixeira de Mattos MJ (2009). Alternative routes to biofuels: Light-driven biofuel formation from $CO_2$ and water based on the 'photanol' approach. *Journal of Biotechnology* doi:10.1016/j.jbiotec.2009.02.002

Luesch H, Yoshida WY, Moore RE, Paul VJ, and Corbett TH (2001). Total Structure Determination of Apratoxin A, a Potent Novel Cytotoxin from the Marine Cyanobacterium *Lyngbya majuscula. Journal of the American Chemical Society* 123(23):5418-5423.

Marston A, Kissling J, and Hostettmann K (2002). A rapid TLC bioautographic method for the detection of acetylcholinesterase and butyrylcholinesterase inhibitors in plants. *Phytochemical Analysis* 13:51-54.

Murch SJ, Cox PA, Banack SA, Steele JC, and Sacks OW (2004). Occurrence of β-methylamino-L-alanine (BMAA) in ALS/PDC patients from Guam. *Acta Neurologica Scandinavica* 110:267-269.

Nishiwaki-Matsushima R, Ohta T, Nishiwaki S, Suganuma M, Kohyama K, Ishikawa T, Carmichael WW, and Fujiki H (1992). Liver tumor promotion by the cyanobacterial cyclic peptide toxin microcystin LR. *Journal of Cancer Research and Clinical Oncology* 118(6):420-424.

Rodriguez V, Moura S, Pinto E, Pereira CMP, and Braga MC (2006). Aspectos tóxicos e químicos da anatoxina-a e seus análogos. *Química Nova* 29(6):1365-1371.

Svircev Z, Cetojevic-Simin D, Simeunovic J, Karaman M, and Stojanovic D (2008). Antibacterial, antifungal and cytotoxic activity of terrestrial cyanobacterial strains from Serbia. *Science in China Series C: Life Sciences* 51(10):941-947.

Valiente OLD, Andueza A, de Vega G, and Muñoz F (2004). The use of NIRS for prediction of intake, digestibility and diet composition in sheep fed mixed grain:roughage diets. *Journal of Animal and Feed Sciences* 13:227-230.

# Index

# Index of Authors